Introduction to Effective Field Theory

Using examples from across many disciplines of physics, this introduction shows why effective field theories are the language in which physical laws are written.

The tools of effective field theory are presented and illustrated using worked examples from areas including particle, nuclear, atomic, condensed matter and gravitational physics.

To bring the subject within reach of scientists with a wide variety of backgrounds and interests, there are clear physical explanations, rigorous derivations and extensive appendices on background material like scattering and quantum field theory. Starting from undergraduate-level quantum mechanics, the book gets to state-of-the-art calculations using both relativistic and nonrelativistic few-body and many-body examples, and numerous end-of-chapter problems derive classic results not covered in the main text.

Graduate students and researchers in particle physics, condensed matter physics, nuclear physics, string theory, and mathematical physics more generally will find this book ideal for both self-study and organized courses on effective field theory.

Cliff Burgess is a professor at both McMaster University and Perimeter Institute for Theoretical Physics, and co-author of the book *The Standard Model: A Modern Primer*. He is a fellow of the Royal Society of Canada and has been awarded the CAP/CRM medal for Theoretical Physics.

After initially learning about effective field theories from his PhD supervisor, Nobel Laureate Steven Weinberg, he is now a world expert on their applications throughout physics.

Introduction to Effective Field Theory

Thinking Effectively about Hierarchies of Scale

C. P. BURGESS
McMaster University and Perimeter Institute for Theoretical Physics

CAMBRIDGE
UNIVERSITY PRESS

CAMBRIDGE
UNIVERSITY PRESS

University Printing House, Cambridge CB2 8BS, United Kingdom

One Liberty Plaza, 20th Floor, New York, NY 10006, USA

477 Williamstown Road, Port Melbourne, VIC 3207, Australia

314–321, 3rd Floor, Plot 3, Splendor Forum, Jasola District Centre, New Delhi – 110025, India

79 Anson Road, #06–04/06, Singapore 079906

Cambridge University Press is part of the University of Cambridge.

It furthers the University's mission by disseminating knowledge in the pursuit of education, learning, and research at the highest international levels of excellence.

www.cambridge.org
Information on this title: www.cambridge.org/9780521195478
DOI: 10.1017/9781139048040

First published 2021

A catalogue record for this publication is available from the British Library.

ISBN 978-0-521-19547-8 Hardback

Contents

Illustrations

Tables

Preface

It is an everyday fact of life that Nature comes to us with a variety of scales: from quarks, nuclei and atoms through planets, stars and galaxies up to the overall Universal large-scale structure. Science progresses because we can understand each of these on its own terms, and need not understand all scales at once. This is possible because of a basic fact of Nature: most of the details of small distance physics are irrelevant for the description of longer-distance phenomena.

Our description of Nature's laws use quantum field theories, which share this property that short distances mostly decouple from larger ones. Effective Field Theories (EFTs) are the tools developed over the years to show why they do. These tools have immense practical value: knowing which scales are important and why the rest decouple allows hierarchies of scale to be used to simplify the description of many systems. This book provides an introduction to these tools, and to emphasize their great generality illustrates them using applications from many parts of physics: relativistic and nonrelativistic; few-body and many-body.

The book is broadly appropriate for an introductory graduate course, though some topics could be done in an upper-level course for advanced undergraduates. It should appeal to physicists interested in learning these techniques for practical purposes as well as those who enjoy the beauty of the unified picture of physics that emerges.

It is to emphasize this unity that a broad selection of applications is examined, although this also means no one topic is explored in as much depth as it deserves. The book's goal is to engage the reader's interest, but then to redirect to the appropriate literature for more details. To this end references in the main text are provided mostly just for the earliest papers (that I could find) on a given topic, with a broader – probably more useful – list of textbooks, reviews and other sources provided in the bibliography. There will be inevitable gems about which I am unaware or have forgotten to mention, and I apologize in advance to both their authors and to you the reader for their omission.

An introductory understanding of quantum and classical field theory is assumed, for which an appendix provides a basic summary of the main features. To reconcile the needs of readers with differing backgrounds – from complete newbies through to experts seeking applications outside their own areas – sections are included requiring differing amounts of sophistication. The background material in the appendices is also meant to help smooth out the transitions between these different levels of difficulty.

The various gradations of sophistication are flagged using the suits of playing cards: ◇, ♡, ♠ and ♣ in the titles of the chapter sections. The flag ◇ indicates good value and labels sections that carry key ideas that should not be missed by any student of effective theories. ♡ flags sections containing material common to most quantum field theory classes, whose familiarity may warm a reader's heart but can be skipped by aficianados in a hurry. The symbol ♠ indicates a section which may require a bit more digging for new students to digest, but which is reasonably self-contained and worth a bit of spadework. Finally, readers wishing to beat their heads against sections containing more challenging topics should seek out those marked with ♣.

The lion's share of the book is aimed at applications, since this most effectively brings out both the utility and the unity of the approach. The examples also provide a pedagogical framework for introducing some specific techniques. Since many of these applications are independent of one another, a course can be built by starting with Part I's introductory material and picking and choosing amongst the later sections that are of most interest.

Acknowledgements

This book draws heavily on the insight and goodwill of many people: in particular my teachers of quantum and classical field theory – Bryce De Witt, Willy Fischler, Joe Polchinski and especially Steven Weinberg – who shaped the way I think about this subject.

Special thanks go to Fernando Quevedo for a life-long collaboration on these subjects and his comments over the years on many of the topics discussed herein.

I owe a debt to Patrick Labelle, Sung-Sik Lee, Alexander Penin and Ira Rothstein for clarifying issues to do with nonrelativistic EFTs; to John Donoghue for many insights on gravitational physics; to Thomas Becher for catching errors in early versions of the text; to Jim Cline for a better understanding of the practical implications of Goldstone boson infrared effects; to Claudia de Rham, Luis Lehner, Adam Solomon, Andrew Tolley and Mark Trodden for helping better understand applications to time-dependent systems; to Subodh Patil and Michael Horbatsch for helping unravel multiple scales in scalar cosmology; to Michele Cicoli, Shanta de Alwis, Sven Krippendorf and Anshuman Maharana for shepherding me through the perils of string theory; to Mike Trott for help understanding the subtleties of power-counting and SMEFT; to Peter Adshead, Richard Holman, Greg Kaplanek, Louis Leblond, Jerome Martin, Sarah Shandera, Gianmassimo Tasinato, Vincent Vennin and Richard Woodard for understanding EFTs in de Sitter space and their relation to open systems, and to Ross Diener, Peter Hayman, Doug Hoover, Leo van Nierop, Duncan Odell, Ryan Plestid, Markus Rummel, Matt Williams and Laszlo Zalavari for helping clarify how EFTs work for massive first-quantized sources.

Collaborators and students too numerous to name have continued to help deepen my understanding in the course of many conversations about physics.

CERN, ICTP, KITP Santa Barbara and the Institute Henri Poincaré have at various times provided me with pleasant environs in which to focus undivided time on writing, and with stimulating discussions when taking a break from it. The book would not have been finished without them. The same is true of McMaster University and Perimeter Institute, whose flexible work environments allowed me to take on this project in the first place.

Heaven holds a special place for Simon Capelin and his fellow editors, both for encouraging the development of this book and for their enormous patience in awaiting it.

Most importantly, I am grateful to my late parents for their gift of an early interest in science, and to my immediate family (Caroline, Andrew, Ian, Matthew and Michael) for their continuing support and tolerance of time taken from them for physics.

Part I

Theoretical Framework

About Part I

This first part of the book sets up the basic framework of effective field theories (EFTs), developing along the way the main tools and formalism that is used throughout the remainder of the text. An effort is made to discuss topics that are sometimes left out in reviews of EFT methods, such as how to work with time-dependent backgrounds or in the presence of boundaries. This part of the book is meant to be relatively self-contained, and so can be studied on its own given limited time.

Discussions of formalism can easily descend into obscurity if not done with concrete questions in mind. To keep things focussed, the first chapter here introduces a toy model in which most of the conceptual issues arise in a simple way. As each subsequent chapter in Part I introduces a new concept, its formal treatment is accompanied by a short illustrative discussion about how that particular issue arises within the toy model. Hopefully, by the end of Part I the reader will be familiar with the main EFT tools, and will know the toy model inside and out.

The book's remaining major parts then work through practical examples of EFT reasoning throughout physics. Because the formalism is largely handled in Part I, the focus of the rest of the book is both on illustrating some of the techniques introduced in Part I, and on physical insights that emerge when these tools are used to study specific problems.

The later parts are grouped into three categories that share similar features:

- Part II studies relativistic applications, studying first examples where both the low- and high-energy parts of the theory are well-understood and then switching to problems for which the high-energy sector is either unknown or is known but difficult to use precisely (such as if it involves strong interactions).
- Part III switches to nonrelativistic applications, such as to slowly moving systems of a small number of particles, like atoms, or systems involving lots of particles for which only gross features like the centre-of-mass motion are of interest, like for planetary orbits in the solar system.
- Part IV then examines many-body and open systems, for which many particles are involved and more degrees of freedom appear in the coarse-grained theory. Part IV starts with calculations for which dissipative effects are chosen to be negligible, but closes with a discussion of open systems for which dissipation and decoherence can be important.

1 Decoupling and Hierarchies of Scale

The world around us contains a cornucopia of length scales, ranging (at the time of writing) down to quarks and leptons at the smallest and up to the universe as a whole at the largest, with qualitatively new kinds of structures – nuclei, atoms, molecules, cells, organisms, mountains, asteroids, planets, stars, galaxies, voids, and so on – seemingly arising at every few decades of scales in between. So it is remarkable that all of this diversity seems to be described in all of its complexity by a few simple laws.

How can this be possible? Even given that the simple laws exist, why should it be possible to winkle out an understanding of what goes on at one scale without having to understand everything all at once? The answer seems to be a very deep property of nature called *decoupling*, which states that most (but not all) of the details of very small-distance phenomena tend to be largely irrelevant for the description of much larger systems. For example, not much needs to be known about the detailed properties of nuclei (apart from their mass and electrical charge, and perhaps a few of their multipole moments) in order to understand in detail the properties of electronic energy levels in atoms.

Decoupling is a very good thing, since it means that the onion of knowledge can be peeled one layer at a time: our initial ignorance of nuclei need not impede the unravelling of atomic physics, just as ignorance about atoms does not stop working out the laws describing the motion of larger things, like the behaviour of fluids or motion of the moon.

It so happens that this property of decoupling is also shared by the mathematics used to describe the laws of nature [1]. Since nowadays this description involves quantum field theories, it is gratifying that these theories as a group tend to predict that short distances generically decouple from long distances, in much the same way as happens in nature.

This book describes the way this happens in detail, with two main purposes in mind. One purpose is to display decoupling for its own sake since this is satisfying in its own right, and leads to deep insights into what precisely is being accomplished when writing down physical laws. But the second purpose is very practical; the simplicity offered by a timely exploitation of decoupling can often be the difference between being able to solve a problem or not. When exploring the consequences of a particular theory for short distance physics it is obviously useful to be able to identify efficiently those observables that are most sensitive to the theory's details and those from which they decouple. As a consequence the mathematical tools – *effective field theories* – for exploiting decoupling have become ubiquitous in some areas of theoretical physics, and are likely to become more common in many more.

The purpose of the rest of Chapter 1 is twofold. One goal is to sketch the broad outlines of decoupling, effective lagrangians and the physical reason why they work, all in one place. The second aim is to provide a toy model that can be used as a concrete example as the formalism built on decoupling is fleshed out in more detail in subsequent chapters.

1.1 An Illustrative Toy Model ◇

The first step is to set up a simple concrete model to illustrate the main ideas. To be of interest this model must possess two kinds of particles, one of which is much heavier than the other, and these particles must interact in a simple yet nontrivial way. Our focus is on the interactions of the two particles, with a view towards showing precisely how the heavy particle decouples from the interactions of the light particle at low energies.

To this end consider a complex scalar field, ϕ, with action[1]

$$S := -\int d^4x \left[\partial_\mu \phi^* \partial^\mu \phi + V(\phi^*\phi)\right], \tag{1.1}$$

whose self-interactions are described by a simple quartic potential,

$$V(\phi^*\phi) = \frac{\lambda}{4}\left(\phi^*\phi - v^2\right)^2, \tag{1.2}$$

where λ and v^2 are positive real constants. The shape of this potential is shown in Fig. 1.1.

1.1.1 Semiclassical Spectrum

The simplest regime in which to explore the model's predictions is when $\lambda \ll 1$ and both v and $|\phi|$ are $\mathcal{O}\left(\lambda^{-1/2}\right)$. This regime is simple because it is one for which the semiclassical approximation provides an accurate description. (The relevance of the semiclassical limit in this regime can be seen by writing $\phi := \varphi/\lambda^{1/2}$ and $v := \mu/\lambda^{1/2}$ with φ and μ held fixed as $\lambda \to 0$. In this case the action depends on λ only through an overall factor: $S[\phi, v, \lambda] = (1/\lambda)S[\varphi, \mu]$. This is significant because the action appears in observables only in the combination $S/\hbar$, and so the small-λ limit is equivalent to the small-$\hbar$ (classical) limit.)[2]

In the classical limit the ground state of this system is the field configuration that minimizes the classical energy,

$$E = \int d^3x \left[\partial_t \phi^* \partial_t \phi + \nabla\phi^* \cdot \nabla\phi + V(\phi^*\phi)\right]. \tag{1.3}$$

Since this is the sum of positive terms it is minimized by setting each to zero; the classical ground state is any constant configuration (so $\partial_t\phi = \nabla\phi = 0$), with $|\phi| = v$ (so $V = 0$).

[1] Although this book presupposes some familiarity with quantum field theory, see Appendix C for a compressed summary of some of the relevant ideas and notation used throughout. Unless specifically stated otherwise, units are adopted for which $\hbar = c = 1$, so that time ~ length and energy ~ mass ~ 1/length, as described in more detail in Appendix A.

[2] The connection between small coupling and the semi-classical limit is explored more fully once power-counting techniques are discussed in §3.

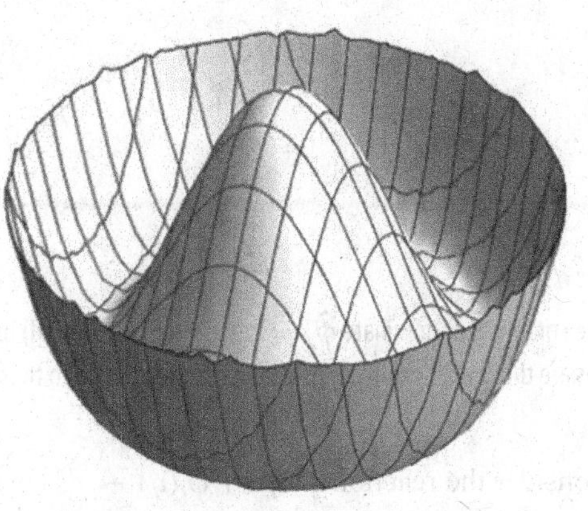

Fig. 1.1 The potential $V(\phi_R, \phi_I)$, showing its sombrero shape and the circular line of minima at $|\phi| = v$.

In the semi-classical regime, particle states are obtained by expanding the action about the classical vacuum, $\phi = v + \tilde{\phi}$,

$$S = -\int d^4x \left\{ \partial_\mu \tilde{\phi}^* \partial^\mu \tilde{\phi} + \frac{\lambda}{4} \left[v(\tilde{\phi} + \tilde{\phi}^*) + \tilde{\phi}^* \tilde{\phi} \right]^2 \right\}, \tag{1.4}$$

and keeping the leading (quadratic) order in the quantum fluctuation $\tilde{\phi}$. In terms of the field's real and imaginary parts, $\tilde{\phi} = \frac{1}{\sqrt{2}}(\tilde{\phi}_R + i\tilde{\phi}_I)$, the leading term in the expansion of S is

$$S_0 = -\frac{1}{2} \int d^4x \left[\partial_\mu \tilde{\phi}_R \partial^\mu \tilde{\phi}_R + \partial_\mu \tilde{\phi}_I \partial^\mu \tilde{\phi}_I + \lambda v^2 \tilde{\phi}_R^2 \right]. \tag{1.5}$$

The standard form (see §C.3.1) for the action of a free, real scalar field of mass m is proportional to $\partial_\mu \psi \, \partial^\mu \psi + m^2 \psi^2$, and so comparing with Eq. (1.5) shows $\tilde{\phi}_R$ represents a particle with mass $m_R^2 = \lambda v^2$ while $\tilde{\phi}_I$ represents a particle with mass $m_I^2 = 0$. These are the heavy and light particles whose masses provide a hierarchy of scales.

1.1.2 Scattering

For small λ the interactions amongst these particles are well-described in perturbation theory, by writing $S = S_0 + S_{\rm int}$ and perturbing in the interactions

$$S_{\rm int} = -\int d^4x \left[\frac{\lambda v}{2\sqrt{2}} \tilde{\phi}_R \left(\tilde{\phi}_R^2 + \tilde{\phi}_I^2 \right) + \frac{\lambda}{16} \left(\tilde{\phi}_R^2 + \tilde{\phi}_I^2 \right)^2 \right]. \tag{1.6}$$

Using this interaction, a straightforward calculation – for a summary of the steps involved see Appendix B – gives any desired scattering amplitude order-by-order in λ. Since small λ describes a semiclassical limit (because it appears systematically together with $\hbar$ in $S/\hbar$, as argued above), the leading contribution turns out to come from evaluating Feynman graphs with no loops[3] (*i.e.* tree graphs).

[3] A connected graph with no loops (or a 'tree' graph) is one which can be broken into two disconnected parts by cutting any internal line. Precisely how to count the number of loops and why this is related to powers of the small coupling λ is the topic of §3.

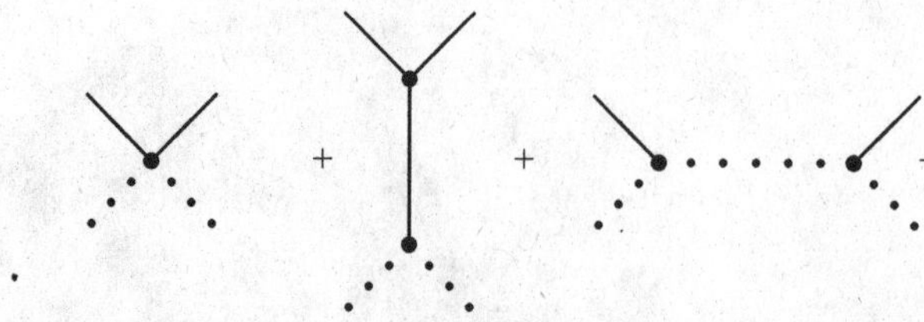

Fig. 1.2 The tree graphs that dominate $\tilde{\phi}_R\,\tilde{\phi}_I$ scattering. Solid (dotted) lines represent $\tilde{\phi}_R$ ($\tilde{\phi}_I$), and 'crossed' graphs are those with external lines interchanged relative to those displayed.

Consider the reaction $\tilde{\phi}_R(p) + \tilde{\phi}_I(q) \to \tilde{\phi}_R(p') + \tilde{\phi}_I(q')$, where $p^\mu = \{p^0, \mathbf{p}\}$ and $q^\mu = \{q^0, \mathbf{q}\}$ respectively denote the 4-momenta of the initial $\tilde{\phi}_R$ and $\tilde{\phi}_I$ particle, while p'^μ and q'^μ are 4-momenta of the final $\tilde{\phi}_R$ and $\tilde{\phi}_I$ states. The Feynman graphs of Fig. 1.2 give a scattering amplitude proportional to[4] $\mathcal{A}_{RI\to RI}\,\delta^4(p+q-p'-q')$, where the Dirac delta function, $\delta^4(p+q-p'-q')$, expresses energy–momentum conservation, and

$$\begin{aligned}\mathcal{A}_{RI\to RI} &= 4\mathrm{i}\left(-\frac{\lambda}{8}\right) + \left(\frac{\mathrm{i}^2}{2}\right)\left(-\frac{\lambda v}{2\sqrt{2}}\right)^2 \left[\frac{24(-\mathrm{i})}{(p-p')^2 + m_R^2} + \frac{8(-\mathrm{i})}{(p+q)^2} + \frac{8(-\mathrm{i})}{(p-q')^2}\right] \\ &= -\frac{\mathrm{i}\lambda}{2} + \frac{\mathrm{i}(\lambda v)^2}{2m_R^2}\left[\frac{3}{1-2q\cdot q'/m_R^2} - \frac{1}{1-2p\cdot q/m_R^2} - \frac{1}{1+2p\cdot q'/m_R^2}\right].\end{aligned} \tag{1.7}$$

Here the factors like 4, 24 and 8 in front of various terms count the combinatorics of how many ways each particular graph can contribute to the amplitude. The second line uses energy–momentum conservation, $(p-p')^\mu = (q'-q)^\mu$, as well as the kinematic conditions $p^2 = -(p^0)^2 + \mathbf{p}^2 = -m_R^2$ and $(q')^2 = q^2 = -(q^0)^2 + \mathbf{q}^2 = 0$, as appropriate for relativistic particles whose energy and momenta are related by $E = p^0 = \sqrt{\mathbf{p}^2 + m^2}$.

Notice that the terms involving the square bracket arise at the same order in λ as the first term, despite nominally involving two powers of S_{int} rather than one (provided that the square bracket itself is order unity). To see this, keep in mind $m_R^2 = \lambda v^2$ so that $(\lambda v/m_R)^2 = \lambda$.

For future purposes it is useful also to have the corresponding result for the reaction $\tilde{\phi}_I(p) + \tilde{\phi}_I(q) \to \tilde{\phi}_I(p') + \tilde{\phi}_I(q')$. A similar calculation, using instead the Feynman graphs of Fig. 1.3, gives the scattering amplitude

$$\begin{aligned}\mathcal{A}_{II\to II} &= 24\mathrm{i}\left(-\frac{\lambda}{16}\right) + 8\left(\frac{\mathrm{i}^2}{2}\right)\left(-\frac{\lambda v}{2\sqrt{2}}\right)^2 \\ &\quad\times\left[\frac{-\mathrm{i}}{(p+q)^2 + m_R^2} + \frac{-\mathrm{i}}{(p-p')^2 + m_R^2} + \frac{-\mathrm{i}}{(p-q')^2 + m_R^2}\right] \\ &= -\frac{3\mathrm{i}\lambda}{2} + \frac{\mathrm{i}(\lambda v)^2}{2m_R^2}\left[\frac{1}{1+2p\cdot q/m_R^2} + \frac{1}{1-2q\cdot q'/m_R^2} + \frac{1}{1-2p\cdot q'/m_R^2}\right].\end{aligned} \tag{1.8}$$

[4] See Exercise 1.1 and Appendix B for the proportionality factors.

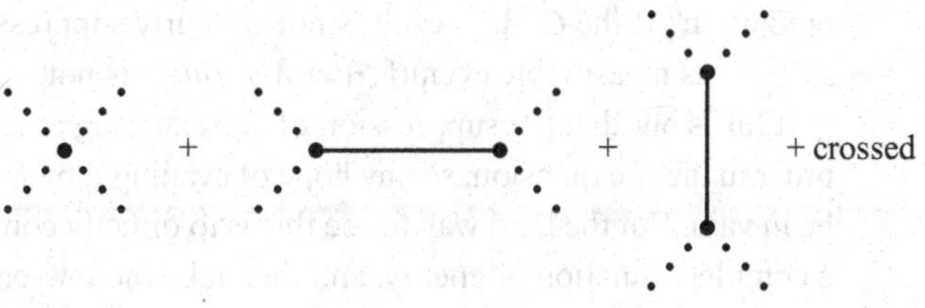

Fig. 1.3 The tree graphs that dominate the $\tilde{\phi}_I \tilde{\phi}_I$ scattering amplitude. Solid (dotted) lines represent $\tilde{\phi}_R$ and $\tilde{\phi}_I$ particles.

1.1.3 The Low-Energy Limit

For the present purposes it is the low-energy regime that is of most interest: when the centre-of-mass kinetic energy and momentum transfers during scattering are very small compared with the mass of the heavy particle. This limit is obtained from the above expressions by taking $|p\cdot q|$, $|p\cdot q'|$ and $|q\cdot q'|$ all to be small compared with m_R^2.

Taylor expanding the above expressions shows that both $\mathcal{A}_{RI\to RI}$ and $\mathcal{A}_{II\to II}$ are suppressed in this limit by powers of $(q$ or $q')/m_R$, in addition to the generic small perturbative factor λ:

$$\mathcal{A}_{RI\to RI} \simeq 2i\lambda \left(\frac{q\cdot q'}{m_R^2}\right) + \mathcal{O}\left(m_R^{-4}\right), \tag{1.9}$$

while

$$\mathcal{A}_{II\to II} \simeq 2i\lambda \left[\frac{(p\cdot q)^2 + (p\cdot q')^2 + (q\cdot q')^2}{m_R^4}\right] + \mathcal{O}\left(m_R^{-6}\right). \tag{1.10}$$

Both of these expressions use 4-momentum conservation, and kinematic conditions like $q^2 = 0$ *etc.* to simplify the result, and both expressions end up being suppressed by powers of q/m_R and/or q'/m_R once this is done.

The basic simplicity of physics at low energies arises because physical quantities typically simplify when Taylor expanded in powers of any small energy ratios (like scattering energy/m_R in the example above). It is this simplicity that ultimately underlies the phenomenon of decoupling: in the toy model the low-energy implications of the very energetic $\tilde{\phi}_R$ states ultimately can be organized into a sequence in powers of m_R^{-2}, with only the first few terms relevant at very low energies.

1.2 The Simplicity of the Low-Energy Limit ◇

Now imagine that your task is to build an experiment to test the above theory by measuring the cross section for scattering $\tilde{\phi}_I$ particles from various targets, using only accelerators whose energies, E, do not reach anywhere near as high as the mass m_R. Since the experiment is more difficult if the scattering is rare, the suppression of the order-λ cross sections by powers of q/m_R and/or q'/m_R at low energies presents a potential problem. But maybe this suppression is an accident of the leading, $\mathcal{O}(\lambda)$, prediction? If the $O(\lambda^2)$ result is not similarly suppressed, then it might happen that $\mathcal{A} \simeq \lambda^2$ is measurable even if $\mathcal{A} \simeq \lambda(E/m_R)^2$ is not.

It turns out that the suppression of $\tilde{\phi}_I$ scattering at low energies persists order-by-order in the λ expansion, so any hope of evading it by working to higher orders would be in vain. But the hard way to see this is to directly compute the $O(\lambda^n)$ amplitude as a complete function of energy, and then take the low-energy limit. It would be much more efficient if it were possible to zero in directly on the low-energy part of the result *before* investing great effort into calculating the complete answer. Any simplicity that might emerge in the low-energy limit then would be much easier to see.

Indeed, a formalism exists precisely for efficiently identifying the nature of physical quantities in the low-energy limit – *effective field theories* – and it is this formalism that is the topic of this book. This formalism exists and is so useful because one is often in the situation of being faced with a comparatively simple low-energy limit of some, often poorly understood, more complicated system.

The main idea behind this formalism is to take advantage of the low-energy approximation as early as possible in a calculation, and the best way to do so is directly, once and for all, in the action (or Hamiltonian or Lagrangian), rather than doing it separately for each independent observable. But how can the low-energy expansion be performed directly in the action?

1.2.1 Low-Energy Effective Actions

To make this concrete for the toy model discussed above, a starting point is the recognition that the low-energy limit, Eq. (1.10), of $\mathcal{A}_{II\to II}$ has precisely the form that would be expected (at leading order of perturbation theory) if the $\tilde{\phi}_I$ particles scattered only through an *effective* interaction of the form $S_{\rm eff} = S_{\rm eff\,0} + S_{\rm eff\,int}$, with

$$S_{\rm eff\,0} = -\frac{1}{2}\int d^4x\; \partial_\mu\tilde{\phi}_I\,\partial^\mu\tilde{\phi}_I, \tag{1.11}$$

and

$$S_{\rm eff\,int} = \frac{\lambda}{4m_R^4}\int d^4x\; (\partial_\mu\tilde{\phi}_I\,\partial^\mu\tilde{\phi}_I)(\partial_\nu\tilde{\phi}_I\,\partial^\nu\tilde{\phi}_I), \tag{1.12}$$

up to terms of order λ^2 and/or m_R^{-6}.

What is less obvious at this point, but nonetheless true (and argued in detail in the chapters that follow), is that this same effective interaction, Eqs. (1.11) and (1.12), also correctly captures the leading low-energy limit of other scattering processes, such as for $\tilde{\phi}_I\tilde{\phi}_I \to \tilde{\phi}_I\tilde{\phi}_I\tilde{\phi}_I\tilde{\phi}_I$ and reactions involving still more $\tilde{\phi}_I$ particles. That is, *all* amplitudes obtained from the full action, Eqs. (1.5) and (1.6), precisely agree with those obtained from the effective action, Eqs. (1.11) and (1.12), provided that the predictions of both theories are expanded only to leading order in λ and m_R^{-2} [2].

Given that a low-energy action like $S_{\rm eff}$ exists, it is clear that it is much easier to study the system's low-energy limit by first computing $S_{\rm eff}$ and then using $S_{\rm eff}$ to work out any observable of interest, than it is to calculate all observables using $S_0 + S_{\rm int}$ of Eqs. (1.5) and (1.6), and only then expanding them to find their low-energy form.

As an example of this relative simplicity, because each factor of $\tilde{\phi}_I$ appears differentiated in Eq. (1.12), it is obvious that the amplitudes for more complicated scattering processes computed with it are also suppressed by high powers of the low-energy scattering scale. For instance, the amplitude for $\tilde{\phi}_I\tilde{\phi}_I \to N\tilde{\phi}_I$ (into N final particles) computed using tree graphs built using just the quartic interaction $S_{\text{eff int}}$ would be expected to give an amplitude proportional to at least

$$\mathcal{A}_{II\to I\cdots I} \propto \lambda^{N/2}\left(\frac{\text{scattering energy}}{m_R}\right)^{N+2} \tag{1.13}$$

in the low-energy limit. Needless to say, this type of low-energy suppression is much harder to see when using the full action, Eqs. (1.5) and (1.6).

It may seem remarkable that an interaction like S_{eff} exists that completely captures the leading low-energy limit of the full theory in this way. But what is even more remarkable is that a similar effective action also exists that reproduces the predictions of the full theory to *any* fixed higher order in λ and m_R^{-2}. This more general effective action replaces Eq. (1.12) by

$$S_{\text{eff int}} = \int d^4x\, \mathfrak{L}_{\text{eff int}}, \tag{1.14}$$

where

$$\mathfrak{L}_{\text{eff int}} = a\,(\partial_\mu\tilde{\phi}_I\partial^\mu\tilde{\phi}_I)(\partial_\nu\tilde{\phi}_I\partial^\nu\tilde{\phi}_I) + b\,(\partial_\mu\tilde{\phi}_I\partial^\mu\tilde{\phi}_I)(\partial_\nu\tilde{\phi}_I\partial^\nu\tilde{\phi}_I)(\partial_\rho\tilde{\phi}_I\partial^\rho\tilde{\phi}_I) + \cdots, \tag{1.15}$$

where the ellipses represent terms involving additional powers of $\partial_\mu\hat{\phi}_I$ and/or its derivatives, though only a finite number of such terms is required in order to reproduce the full theory to a fixed order in λ and m_R^{-2}.

In principle, the coefficients a and b in Eq. (1.15) are given as a series in λ once the appropriate power of m_R is extracted on dimensional grounds,

$$a = \frac{1}{m_R^4}\left[\frac{\lambda}{4} + a_2\lambda^2 + O(\lambda^3)\right] \quad\text{and}\quad b = \frac{1}{m_R^8}\left[b_1\lambda + b_2\lambda^2 + b_3\lambda^3 + O(\lambda^4)\right], \tag{1.16}$$

which displays explicitly the order-λ value for a found above that reproduces low-energy scattering in the full theory. Explicit calculations in later sections also show $b_1 = 0$. More generally, to the extent that the leading (classical, or tree-level) part of the action should be proportional to $1/\lambda$ once m_R is eliminated for v using $m_R^2 = \lambda v^2$ (as is argued above, and in more detail in Eq. (2.24) and §3), it must also be true that b_2 vanishes.

1.2.2 Why It Works

Why is it possible to find an effective action capturing the low-energy limit of a theory, along the lines described above? The basic idea goes as follows.

It is not in itself surprising that there is some sort of Hamiltonian describing the time evolution of low-energy states. After all, in the full theory time evolution is given by a unitary operation

$$|\psi_f(t)\rangle = U(t,t')\,|\psi_i(t')\rangle, \tag{1.17}$$

where $U(t,t') = \exp[-iH(t-t')]$ with a Hamiltonian[5] $H = H(\hat{\phi}_R, \hat{\phi}_l)$ depending on both the heavy and light fields. But if the initial state has an energy $E_i < m_R$ it cannot contain any $\hat{\phi}_R$ particles, and energy conservation then precludes $\hat{\phi}_R$ particles from ever being produced by subsequent time evolution.

This means that time evolution remains a linear and unitary transformation even when it is restricted to low-energy states. That is, suppose we define

$$U_{\rm eff}(t,t') := P_\Lambda\, U(t,t')\, P_\Lambda := \exp\left[-iH_{\rm eff}(t-t')\right], \tag{1.18}$$

with $P_\Lambda^2 = P_\Lambda$ being the projection operator onto states with low energy $E < \Lambda \ll m_R$. P_Λ commutes with H and so also with time evolution. Because $H_{\rm eff} = P_\Lambda H P_\Lambda$ if H is hermitian then so must be $H_{\rm eff}$ and so if $U(t,t')$ is unitary then so must be $U_{\rm eff}(t,t')$ when acting on low-energy states.

Furthermore, because the action of $H_{\rm eff}$ is well-defined for states having energy $E < \Lambda$, it can be written as a linear combination of products of creation and annihilation operators for the $\hat{\phi}_l$ field only (since these form a basis for operators that transform among only low-energy states).[6] As a consequence, it must be possible to write $H_{\rm eff} = H_{\rm eff}[\hat{\phi}_l]$, without making any reference to the heavy field $\hat{\phi}_R$ at all.

But there is no guarantee that the expression for $H_{\rm eff}[\hat{\phi}_l]$ obtained in this way is anywhere as simple as is $H[\hat{\phi}_R, \hat{\phi}_l]$. So the real puzzle is why the effective interaction found above is so simple. In particular, why is it local,

$$H_{\rm eff}[\hat{\phi}_l] = \int {\rm d}^3x\, \mathcal{H}_{\rm eff}(x), \tag{1.19}$$

with $\mathcal{H}_{\rm eff}(x)$ a simple polynomial in $\hat{\phi}_l(x)$ and its derivatives, all evaluated at the same spacetime point?

Ultimately, the simplicity of this local form can be traced to the uncertainty principle. Interactions, like Eq. (1.12), in $H_{\rm eff}$ not already present in H describe the influence on low-energy $\hat{\phi}_l$ particles of virtual processes involving heavy $\hat{\phi}_R$ particles. These virtual processes are not ruled out by energy conservation even though the production of real $\hat{\phi}_R$ particles is forbidden. One way to understand why they are possible is because the uncertainty principle effectively allows energy conservation to be violated,[7] $E_f = E_i + \Delta E$, but only over time intervals that are sufficiently short, $\Delta t \lesssim \hbar/\Delta E$. The effects of virtual $\hat{\phi}_R$ particles are necessarily localized in time over intervals that are of order $1/m_R$, which are unobservably short for observers restricted to energies $E \ll m_R$. Consequently, they are described at these energies by operators all evaluated at effectively the same time.

In relativistic theories, large momenta necessarily involve large energies and since the uncertainty principle relates large momenta to short spatial distances, a similar argument can be made that the effect of large virtual momentum transfers on the

[5] The convention here is to use $\tilde{\phi}$ to denote the fluctuation when this is a non-operator field (appearing within a path integral, say) and instead use $\hat{\phi}$ for the quantum operator fluctuation field.
[6] See the discussion around Eq. (C.9) of Appendix C for details.
[7] More precisely, energy need not be conserved at each vertex when organized in old-fashioned Rayleigh–Schrödinger perturbation theory from undergraduate quantum mechanics classes. Once reorganized into manifestly relativistic Feynman–Schwinger–Dyson perturbation theory energy actually *is* preserved at each vertex, but internal particles are not on-shell: $E \neq \sqrt{\mathbf{p}^2 + m^2}$. Either way the locality consequences are the same.

low-energy theory can also be captured by effective interactions localized at a single spatial point. Together with the localization in time just described, this shows that the effects of very massive particles are local in both space and time, as found in the toy model above.

Locality arises explicitly in relativistic calculations when expanding the propagators of massive particles in inverse powers of m_R, after which they become local in spacetime since

$$G(x,y) := \langle 0|T\hat{\phi}_R(x)\hat{\phi}_R(y)|0\rangle = -\mathrm{i}\int \frac{\mathrm{d}^4p}{(2\pi)^4}\,\frac{e^{ip(x-y)}}{p^2+m_R^2} \tag{1.20}$$

$$\simeq -\frac{\mathrm{i}}{m_R^2}\sum_{k=0}^{\infty}\int \frac{\mathrm{d}^4p}{(2\pi)^4}\left(-\frac{p^2}{m_R^2}\right)^k e^{ip(x-y)} = -\frac{\mathrm{i}}{m_R^2}\sum_{k=0}^{\infty}\left(\frac{\Box}{m_R^2}\right)^k \delta^4(x-y),$$

where the 'T' denotes time ordering, $p(x-y) := p\cdot(x-y) = p_\mu(x-y)^\mu$ and $\Box = \partial_\mu\partial^\mu = -\partial_t^2 + \nabla^2$ is the covariant d'Alembertian operator.

The upshot is this: to any fixed order in $1/m_R$ the full theory usually can be described by a local effective lagrangian.[8] The next sections develop tools for its efficient calculation and use.

1.2.3 Symmetries: Linear vs Nonlinear Realization

Before turning to the nitty gritty of how the effective action is calculated and used, it is worth first pausing to extract one more useful lesson from the toy model considered above. The lesson is about symmetries and their low-energy realization, and starts by asking why it is that the self-interactions among the light $\hat{\phi}_I$ particles – such as the amplitudes of Eqs. (1.9) and (1.10) – are so strongly suppressed at low energies by powers of $1/m_R^2$.

That is, although it is natural to expect some generic suppression of low-energy interactions by powers of $1/m_R^2$, as argued above, why does nothing at all arise at zeroeth order in $1/m_R$ despite the appearance of terms like $\lambda\hat{\phi}_I^4$ in the full toy-model potential? And why are there so very many powers of $1/m_R$ in the case of $2\hat{\phi}_I \to N\hat{\phi}_I$ scattering in the toy model? (Specifically, why is the amplitude for two $\hat{\phi}_I$ particles scattering to N $\hat{\phi}_I$ particles suppressed by $(1/m_R)^{N+2}$?)

This suppression has a very general origin, and can be traced to a symmetry of the underlying theory [3–5]. The symmetry in question is invariance under the $U(1)$ phase rotation, $\phi \to e^{\mathrm{i}\omega}\phi$, of Eqs. (1.1) and (1.2). In terms of the real and imaginary parts this acts as

$$\begin{pmatrix}\phi_R\\ \phi_I\end{pmatrix} \to \begin{pmatrix}\cos\omega & -\sin\omega\\ \sin\omega & \cos\omega\end{pmatrix}\begin{pmatrix}\phi_R\\ \phi_I\end{pmatrix}. \tag{1.21}$$

A symmetry such as this that acts linearly on the fields is said to be *linearly realized*. As summarized in Appendix C.4, if the symmetry is also linearly realized on particle states then these states come in multiplets of the symmetry, all elements of which share the same couplings and masses. However (as is also argued in

[8] For nonrelativistic systems locality sometimes breaks down in space (*e.g.* when large momenta coexist with low energy). It can also happen that the very existence of a Hamiltonian (without expanding the number of degrees of freedom) breaks down for open systems – the topic of §16.

Appendix C.4) linear transformations of the fields – such as (1.21) – are insufficient to infer that the symmetry also acts linearly for particle states, $|\mathbf{p}\rangle = \mathfrak{a}^*_{\mathbf{p}}|0\rangle$, unless the ground-state, $|0\rangle$, is also invariant. If a symmetry of the action does not leave the ground state invariant it is said to be *spontaneously broken*.

For instance, in the toy model the ground state satisfies $\langle 0|\phi(x)|0\rangle = v$, and so the ground state is only invariant under $\phi \to e^{i\omega}\phi$ when $v = 0$. Indeed, for the toy model if $v = 0$ both particle masses are indeed equal: $m_R = m_I = 0$, as are all of their self-couplings. By contrast, when $v \neq 0$ the masses of the two types of particles differ, as does the strength of their cubic self-couplings. Although $\phi \to e^{i\omega}\phi$ always transforms linearly, the symmetry acts inhomogeneously on the deviation $\hat{\phi} = \phi - v = \frac{1}{\sqrt{2}}(\hat{\phi}_R + i\hat{\phi}_I)$ that creates and destroys the particle states. It is because the deviation does not transform linearly (and homogeneously) that the arguments in Appendix C.4 no longer imply that particle states need have the same couplings and masses when $v \neq 0$.

To see why this symmetry should suppress low-energy $\hat{\phi}_I$ interactions, consider how it acts within the low-energy theory. Even though ϕ transforms linearly in the full theory, because the low-energy theory involves only the single real field $\hat{\phi}_I$, the symmetry cannot act on it in a linear and homogeneous way. To see what the action of the symmetry becomes purely within the low-energy theory, it is useful to change variables to a more convenient set of fields than $\hat{\phi}_R$ and $\hat{\phi}_I$.

To this end, define the two real fields χ and ξ by[9]

$$\phi = \left(v + \frac{\chi}{\sqrt{2}}\right) e^{i\xi/\sqrt{2}v}. \tag{1.22}$$

These have the advantage that the action of the $U(1)$ symmetry, $\phi \to e^{i\omega}\phi$ takes a particularly simple form,

$$\xi \to \xi + \sqrt{2}\, v\, \omega, \tag{1.23}$$

with χ unchanged, so ξ carries the complete burden of symmetry transformation.

In terms of these fields the action, Eq. (1.1), becomes

$$S = -\int d^4x \left[\frac{1}{2}\,\partial_\mu\chi\partial^\mu\chi + \frac{1}{2}\left(1 + \frac{\chi}{\sqrt{2}\,v}\right)^2 \partial_\mu\xi\partial^\mu\xi + V(\chi)\right], \tag{1.24}$$

with

$$V(\chi) = \frac{\lambda}{4}\left(\sqrt{2}\,v\,\chi + \frac{\chi^2}{2}\right)^2. \tag{1.25}$$

Expanding this action in powers of χ and ξ gives the perturbative action $S = S_0 + S_{\rm int}$, with unperturbed contribution

$$S_0 = -\frac{1}{2}\int d^4x \left[\partial_\mu\chi\partial^\mu\chi + \partial_\mu\xi\partial^\mu\xi + \lambda v^2\,\chi^2\right]. \tag{1.26}$$

This shows that χ is an alternative field representation for the heavy particle, with $m_\chi^2 = m_R^2 = \lambda v^2$. ξ similarly represents the massless field.

It also shows the symmetry is purely realized on the massless state, as an inhomogeneous shift (1.23) rather than a linear, homogeneous transformation.

[9] Numerical factors are chosen here to ensure fields are canonically normalized.

Such a transformation – often called a *nonlinear realization* of the symmetry (both to distinguish it from the linear realization discussed above, and because the transformations turn out in general to be nonlinear when applied to non-abelian symmetries) – is a characteristic symmetry realization in the low-energy limit of a system which spontaneously breaks a symmetry.

The interactions in this representation are given by

$$S_{\rm int} = - \int d^4x \left[\left(\frac{\chi}{\sqrt{2}\, v} + \frac{\chi^2}{4\, v^2} \right) \partial_\mu \xi \partial^\mu \xi + \frac{\lambda v}{2\sqrt{2}} \chi^3 + \frac{\lambda}{16} \chi^4 \right]. \tag{1.27}$$

For the present purposes, what is important about these expressions is that ξ always appears differentiated. This is a direct consequence of the symmetry transformation, Eq. (1.23), which requires invariance under constant shifts: $\xi \to \xi + \text{constant}$. Since this symmetry forbids a ξ mass term, which would be $\propto m_I^2\, \xi^2$, it ensures ξ remains exactly massless to all orders in the small expansion parameters. ξ is what is called a *Goldstone boson* for the spontaneously broken $U(1)$ symmetry: it is the massless scalar that is guaranteed to exist for spontaneously broken (global) symmetries. Because ξ appears always differentiated it is immediately obvious that an amplitude describing N_i ξ particles scattering into N_f ξ particles must be proportional to at least $N_i + N_f$ powers of their energy, explaining the low-energy suppression of light-particle scattering amplitudes in this toy model.

For instance, explicitly re-evaluating the Feynman graphs of Fig. 1.3, using the interactions of Eq. (1.27) instead of (1.6), gives the case $N_i = N_f = 2$ as

$$\begin{aligned} \mathcal{A}_{\xi\xi\to\xi\xi} &= 0 + 8 \left(\frac{i^2}{2}\right)\left(-\frac{1}{\sqrt{2}\, v}\right)^2 \left[\frac{-i(p\cdot q)(p'\cdot q')}{(p+q)^2 + m_R^2} + \frac{-i(p\cdot p')(q\cdot q')}{(p-p')^2 + m_R^2} + \frac{-i(p\cdot q')(q\cdot p')}{(p-q')^2 + m_R^2} \right] \\ &= \frac{2i\lambda}{m_R^4} \left[\frac{(p\cdot q)^2}{1 + 2p\cdot q/m_R^2} + \frac{(q\cdot q')^2}{1 - 2q\cdot q'/m_R^2} + \frac{(p\cdot q')^2}{1 - 2p\cdot q'/m_R^2} \right], \end{aligned} \tag{1.28}$$

in precise agreement with Eq. (1.8) – as may be seen explicitly using the identity $(1+x)^{-1} = 1 - x + x^2/(1+x)$ – but with the leading low-energy limit much more explicit.

This representation of the toy model teaches several things. First, it shows that scattering amplitudes (and, more generally, arbitrary physical observables) do not depend on which choice of field variables are used to describe a calculation [8–10]. Some kinds of calculations (like loops and renormalization) are more convenient using the variables $\hat{\phi}_R$ and $\hat{\phi}_I$, while others (like extracting consequences of symmetries) are easier using χ and ξ.

Second, this example shows that it is worthwhile to use the freedom to perform field redefinitions to choose those fields that make life as simple as possible. In particular, it is often very useful to make symmetries of the high-energy theory as explicit as possible in the low-energy theory as well.

Third, this example shows that once restricted to the low-energy theory it need not be true that a symmetry remains linearly realized by the fields [11–13], even if this were true for the full underlying theory including the heavy particles. The necessity of realizing symmetries nonlinearly arises once the scales defining the low-energy theory (*e.g.* $E \ll m_R$) are smaller than the mass difference (*e.g.* m_R) between particles that are related by the symmetry in the full theory, since in this case some of the states required to fill out a linear multiplet are removed as part of the high-energy theory.

1.3 Summary

This first chapter defines a toy model, in which a complex scalar field, ϕ, self-interacts *via* a potential $V = \frac{\lambda}{4}(\phi^*\phi - v^2)^2$ that preserves a $U(1)$ symmetry: $\phi \to e^{i\omega}\phi$. Predictions for particle masses and scattering amplitudes are made as a function of the model's two parameters, λ and v, in the semiclassical regime $\lambda \ll 1$. This model is used throughout the remaining chapters of Part I as a vehicle for illustrating how the formalism of effective field theories works in a concrete particular case.

The semiclassical spectrum of the model has two phases. If $v = 0$ the $U(1)$ symmetry is preserved by the semiclassical ground state and there are two particles whose couplings and masses are the same because of the symmetry. When $v \neq 0$ the symmetry is spontaneously broken, and one particle is massless while the other gets a nonzero mass $m = \sqrt{\lambda}\, v$.

The model's symmetry-breaking phase has a low-energy regime, $E \ll m$, that provides a useful illustration of low-energy methods. In particular, the massive particle decouples at low energies in the precise sense that its virtual effects only play a limited role for the low-energy interactions of the massless particles. In particular, explicit calculation shows the scattering of massless particles at low energies in the full theory to be well-described to leading order in λ and E/m in terms of a simple local 'effective' interaction with lagrangian density $\mathfrak{L}_{\rm eff} = a_{\rm eff}(\partial_\mu \xi\, \partial^\mu \xi)^2$, with effective coupling: $a_{\rm eff} = \lambda/(4m^4)$. The $U(1)$ symmetry of the full theory appears in the low-energy theory as a shift symmetry $\xi \to \xi + \text{constant}$.

Exercises

Exercise 1.1 Use the Feynman rules coming from the action $S = S_0 + S_{\rm int}$ given in Eqs. (1.5) and (1.6) to evaluate the graphs of Fig. 1.2. Show from your result that the corresponding S-matrix element is given by

$$\langle \hat\phi_R(p'), \hat\phi_I(q')|\mathcal{S}\,|\,\hat\phi_R(p), \hat\phi_I(q)\rangle = -i(2\pi)^4 \mathcal{A}_{RI\to RI}\, \delta^4(p+q-p'-q'),$$

with $\mathcal{A}_{RI\to RI}$ given by Eq. (1.7). Taylor expand your result for small q, q' to verify the low-energy limit given in Eq. (1.9). [Besides showing the low-energy decoupling of Goldstone particles, getting right the cancellation that provides this suppression in these variables is a good test of – and a way to develop faith in – your understanding of Feynman rules.]

Exercise 1.2 Using the Feynman rules coming from the action $S = S_0 + S_{\rm int}$ given in Eqs. (1.5) and (1.6) evaluate the graphs of Fig. 1.3 to show

$$\langle \hat\phi_I(p'), \hat\phi_I(q')|\mathcal{S}\,|\,\hat\phi_I(p), \hat\phi_I(q)\rangle = -i(2\pi)^4 \mathcal{A}_{II\to II}\, \delta^4(p+q-p'-q'),$$

with $\mathcal{A}_{II\to II}$ given by Eq. (1.8). Taylor expand your result for small q, q' to verify the low-energy limit given in Eq. (1.10).

Exercise 1.3 Using the toy model's leading effective interaction $S = S_{\text{eff}\,0} + S_{\text{eff int}}$, with Feynman rules drawn from (1.11) (1.12), draw the graphs that produce the dominant contributions – *i.e.* carry the fewest factors of λ and (external energy)/m_R – to the scattering process $\hat{\phi}_I + \hat{\phi}_I \to 4\hat{\phi}_I$. Show that these agree with the estimate (1.13) in their prediction for the leading power of λ and of external energy.

Exercise 1.4 Using the Feynman rules coming from the action $S = S_0 + S_{\text{int}}$ given in Eqs. (1.26) and (1.27) evaluate the graphs of Fig. 1.3 to show

$$\langle \xi(p'), \xi(q') | S \,|\, \xi(p), \xi(q) \rangle = -\mathrm{i}(2\pi)^4 \mathcal{A}_{\xi\xi\to\xi\xi}\, \delta^4(p + q - p' - q'),$$

with $\mathcal{A}_{\xi\xi\to\xi\xi}$ given by Eq. (1.28). [Comparing this result to the result in Exercise 1.2 provides an illustration of Borcher's theorem [8–10], which states that scattering amplitudes remain unchanged by a broad class of local field redefinitions.]

2 Effective Actions

Having seen in the previous chapter how the low-energy limit works in a specific example, this chapter gets down to the business of defining the low-energy effective theory more explicitly. The next chapter is then devoted to calculational issues of how to use and compute with an effective theory. The first two sections start with a brief review of the formalism of generating functionals, introducing in particular the generator of one-particle irreducible correlation functions. These can be skipped by field theory aficionados interested in cutting immediately to the low-energy chase.

2.1 Generating Functionals – A Review ♡

The starting point is a formalism convenient for describing the properties of generic observables in a general quantum field theory. Consider a field theory involving N quantum fields $\hat{\phi}^a(x)$, $a = 1, \ldots, N$, governed by a classical action $S[\phi]$, and imagine computing the theory's 'in-out' correlation functions,[1]

$$G^{a_1 \cdots a_n}(x_1, \cdots, x_n) := {}_o\langle\Omega|T[\hat{\phi}^{a_1}(x_1) \cdots \hat{\phi}^{a_n}(x_n)]|\Omega\rangle_i, \tag{2.1}$$

where $|\Omega\rangle_i$ (or $|\Omega\rangle_o$) denotes the system's ground state in the remote past (or future) and T denotes time ordering.[2] These correlation functions are useful to consider because general observables including (but not limited to) scattering amplitudes can be extracted from them using standard procedures.

All such correlation functions can be dealt with at once by working with the generating functional, $Z[J]$, defined by

$$Z[J] := \sum_{n=0}^{\infty} \frac{\mathrm{i}^n}{n!} \int \mathrm{d}^4x_1 \cdots \mathrm{d}^4x_n \, G^{a_1 \cdots a_n}(x_1, \cdots, x_n) J_{a_1}(x_1) \cdots J_{a_n}(x_n), \tag{2.2}$$

from which each correlation function can be obtained by functional differentiation

$$G^{a_1 \cdots a_n}(x_1, \cdots, x_n) = (-\mathrm{i})^n \left(\frac{\delta^n Z[J]}{\delta J_{a_1}(x_1) \cdots \delta J_{a_n}(x_n)} \right)_{J=0}. \tag{2.3}$$

A useful property of $Z[J]$ is that it has a straightforward expression in terms of path integrals, which in principle could be imagined to be computed numerically, but in practice usually means that it is relatively simple to calculate perturbatively in terms of Feynman graphs.

[1] In what follows these are assumed to be bosonic fields, though a similar treatment goes through for fermions.

[2] Strictly speaking, for relativistic theories T denotes the T^* ordering, which includes certain additional equal-time 'seagull' contributions required to maintain Lorentz covariance [14].

The basic connection between operator correlation functions and path integrals is the expression

$$G^{a_1 \cdots a_n}(x_1, \cdots, x_n) = \int \mathcal{D}\phi \left[\phi^{a_1}(x_1) \cdots \phi^{a_n}(x_n)\right] \exp\left\{\mathrm{i}S[\phi\,]\right\}, \tag{2.4}$$

where $\mathcal{D}\phi = \mathcal{D}\phi^{a_1} \cdots \mathcal{D}\phi^{a_n}$ denotes the functional measure for the sum over all field configurations, $\phi^a(x)$, with initial and final times weighted by the wave functional, $\Psi_i[\phi]$ and $\Psi^*_o[\phi]$, appropriate for the initial and final states, ${}_o\langle\Omega|$ and $|\Omega\rangle_i$. The special case $n = 0$ is the example most frequently encountered in elementary treatments, for which

$${}_o\langle\Omega|\Omega\rangle_i = \int \mathcal{D}\phi \; \exp\left\{\mathrm{i}S[\phi]\right\}. \tag{2.5}$$

Direct use of the definitions then leads to the following expression for $Z[J]$:

$$Z[J] = \int \mathcal{D}\phi \; \exp\left\{\mathrm{i}S[\phi\,] + \mathrm{i}\int \mathrm{d}^4x \, \phi^a(x) J_a(x)\right\}, \tag{2.6}$$

and so $Z[J = 0] = {}_o\langle\Omega|\Omega\rangle_i$.

Semiclassical Evaluation

Semiclassical perturbation theory can be formulated by expanding the action within the path integral about a classical background:[3] $\phi^a(x) = \varphi^a_{\rm cl}(x) + \tilde\phi^a(x)$, where $\varphi^a_{\rm cl}$ satisfies

$$\left(\frac{\delta S}{\delta\phi^a}\right)_{\phi=\varphi_{\rm cl}} + J_a = 0. \tag{2.7}$$

The idea is to write the action, $S_J[\phi\,] := S[\phi\,] + \int \mathrm{d}^4x\,(\phi^a J_a)$, as

$$S_J[\varphi_{\rm cl} + \tilde\phi\,] = S_J[\varphi_{\rm cl}] + S_2[\varphi_{\rm cl}, \tilde\phi\,] + S_{\rm int}[\varphi_{\rm cl}, \tilde\phi\,], \tag{2.8}$$

with

$$S_2 = -\int \mathrm{d}^4x \, \tilde\phi^a \, \Delta_{ab}(\varphi_{\rm cl}) \tilde\phi^b, \tag{2.9}$$

being the quadratic part in the expansion (for some differential operator Δ_{ab}). The 'interaction' term, $S_{\rm int}$, contains all terms cubic and higher order in $\tilde\phi^a$; no linear terms appear because the background field satisfies Eq. (2.7).

The relevant path integrals can then be evaluated by expanding

$$\exp\left\{\mathrm{i}S[\phi\,] + \mathrm{i}\int \mathrm{d}^4x\,\phi^a J_a\right\} = \exp\left\{\mathrm{i}S_J[\varphi_{\rm cl}] + \mathrm{i}S_2[\varphi_{\rm cl}, \tilde\phi\,]\right\} \sum_{r=0}^{\infty} \frac{1}{r!}\left(\mathrm{i}S_{\rm int}[\varphi_{\rm cl}, \tilde\phi\,]\right)^r, \tag{2.10}$$

in the path integral (2.6) and explicitly computing the resulting gaussian functional integrals.

[3] It is sometimes useful to make a more complicated, nonlinear, split $\phi = \phi(\varphi_{\rm cl}, \tilde\phi)$ in order to make explicit convenient properties (such as symmetries) of the action.

$$Z[J] = N\,(\det{}^{-1/2}\Delta)\Big[\,1 + \cdots\Big]$$

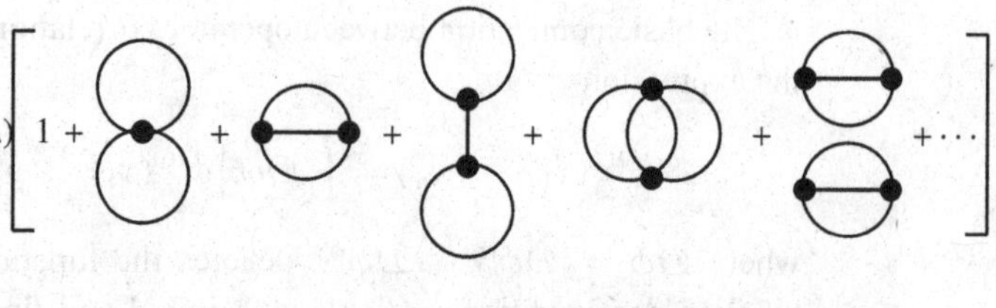

Fig. 2.1 A sampling of some leading perturbative contributions to the generating functional $Z[J]$ expressed using Eq. (2.11) as Feynman graphs. Solid lines are propagators (Δ^{-1}) and solid circles represent interactions that appear in $S_{\rm int}$. 1-Particle reducible and 1PI graphs are both shown as examples at two loops and a disconnected graph is shown at four loops. The graphs shown use only quartic and cubic interactions in $S_{\rm int}$.

This process leads in the usual way to the graphical representation of any correlation function. Gaussian integrals ultimately involve integrands that are powers of fields, leading to integrals of the schematic form[4]

$$\int \mathcal{D}\tilde\phi\; e^{i\tilde\phi^a\Delta_{ab}\tilde\phi^b}\,\tilde\phi^{c_1}(x_1)\cdots\tilde\phi^{c_n}(x_n) \propto \left(\det{}^{-1/2}\Delta\right) \times \Big[(\Delta^{-1})^{c_1c_2}\cdots(\Delta^{-1})^{c_{n-1}c_n} + (\text{perms})\Big], \tag{2.11}$$

if n is even, while the integral vanishes if n is odd. Here, the evaluation ignores a proportionality constant that is background-field independent (and so isn't important in what follows). The interpretation in terms of Feynman graphs comes because the combinatorics of such an integral correspond to the combinatorics of all possible ways of drawing graphs whose internal lines represent factors of Δ^{-1} and whose vertices correspond to interactions within $S_{\rm int}$.

Within this type of graphical expression $Z[J]$ is given as the sum over all vacuum graphs, with no external lines. All of the dependence on J appears through the dependence of the result on $\varphi_{\rm cl}$, which depends on J because of (2.7). The graphs involving the fewest interactions (vertices) first arise with two loops, a sampling of which are shown in Fig. 2.1 that can be built using cubic and quartic interactions within $S_{\rm int}$.

As mentioned earlier, this includes *all* graphs, including those that are disconnected, like the right-most four-loop graph involving four cubic vertices in Fig. 2.1. Graphs like this are disconnected in the sense that it is not possible to get between any pair of vertices along some sequence of contiguous internal lines.

Although simple to state, the perturbation expansion outlined above in terms of vacuum graphs is not yet completely practical for explicit calculations. The problem is the appearance of the background field φ in the propagator $(\Delta^{-1})_{ab}$. Although $\Delta_{ab}(x,y) = -\delta^2 S/\delta\phi^a(x)\delta\phi^b(y)$ itself is easy to compute, it is often difficult to invert explicitly for generic background fields. For instance, for a single scalar field interacting through a scalar potential $U(\phi)$ one has $\Delta(x,y) = [-\Box + U''(\varphi)]\,\delta^4(x-y)$ and although this is easily inverted in momentum space when φ is constant, it is more difficult to invert for arbitrary $\varphi(x)$.

This problem is usually addressed by expanding in powers of $J_a(x)$, so that the path integral is evaluated as a semiclassical expansion about a simple background

[4] This expression assumes a bosonic field, but a similar expression holds for fermions.

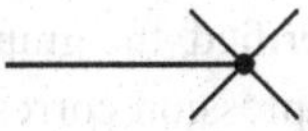

Fig. 2.2 The Feynman rule for the vertex coming from the linear term, $S_{\rm lin}$, in the expansion of the action. The cross represents the sum $\delta S/\delta\varphi^a + J_a$.

configuration, $\varphi^a_{\rm cl}$, that satisfies $(\delta S/\delta\phi^a)(\phi = \varphi_{\rm cl}) = 0$ instead of Eq. (2.7). The Feynman graphs for this modified expansion differ in two ways from the expansion described above: *(i)* the propagators Δ^{-1} now are evaluated at a J-independent configuration, $\varphi^a_{\rm cl}$, which can be explicitly evaluated if this configuration is simple enough (such as, for instance, if $\varphi^a_{\rm cl} = 0$); and *(ii)* the term $\phi^a J_a$ in the exponent of the integrand in (2.6) is now treated as an interaction. Since this interaction is linear in ϕ^a it corresponds graphically to a 'tadpole' contribution (as in Fig. 2.2), with the line ending in a cross whose Feynman rule is $J_a(x)$.

Within this framework, the Feynman graphs giving $Z[J]$ are obtained from those given in Fig. 2.1 by inserting external lines in all possible ways (both to propagators and vertices), with the understanding that the end of each external line represents a factor of $J_a(x)$. This kind of modified expansion gives $Z[J]$ explicitly as a Taylor expansion in powers of J.

2.1.1 Connected Correlations

As Fig. 2.1 shows, the graphical expansion for $Z[J]$ in perturbation theory includes both connected and disconnected Feynman graphs. It is often useful to work instead with a generating functional, $W[J]$, whose graphical expansion contains only connected graphs. As shown in Exercise 2.4, this is accomplished simply by defining $Z[J] := \exp\{iW[J]\}$ [5, 15], since taking the logarithm has the effect of subtracting out the disconnected graphs. This implies the path integral representation

$$\exp\{iW[J]\} = \int \mathcal{D}\phi \, \exp\left\{iS[\phi] + i\int d^4x\, \phi^a J_a\right\}. \tag{2.12}$$

The connected, time-ordered correlation functions are then given by functional differentiation:

$$\langle T[\phi^{a_1}(x_1)\cdots\phi^{a_n}(x_n)]\rangle_c := (-i)^{n-1}\left(\frac{\delta^n W[J]}{\delta J_{a_1}(x_1)\cdots\delta J_{a_n}(x_n)}\right)_{J=0}. \tag{2.13}$$

For example,

$$\langle\phi^a(x)\rangle_c = \left(\frac{\delta W[J]}{\delta J_a(x)}\right)_{J=0} = -i\left(\frac{1}{Z[J]}\frac{\delta Z[J]}{\delta J_b(y)}\right)_{J=0} = \frac{{}_o\langle\Omega|\phi^a(x)|\Omega\rangle_i}{{}_o\langle\Omega|\Omega\rangle_i}, \tag{2.14}$$

while,

$$\begin{aligned}\langle T[\phi^a(x)\,\phi^b(y)]\rangle_c &= -i\left(\frac{\delta^2 W[J]}{\delta J_a(x)\delta J_b(y)}\right)_{J=0}\\ &= \frac{{}_o\langle\Omega|T[\phi^a(x)\,\phi^b(y)]|\Omega\rangle_i}{{}_o\langle\Omega|\Omega\rangle_i}\\ &\quad -\left(\frac{{}_o\langle\Omega|\phi^a(x)|\Omega\rangle_i}{{}_o\langle\Omega|\Omega\rangle_i}\right)\left(\frac{{}_o\langle\Omega|\phi^b(y)|\Omega\rangle_i}{{}_o\langle\Omega|\Omega\rangle_i}\right),\end{aligned} \tag{2.15}$$

and so on. As is easily verified, the graphical expansion of the factor ${}_o\langle\Omega|T[\phi^a(x)\,\phi^b(y)]|\Omega\rangle_i$ in this last expression corresponds to the sum over all Feynman graphs with precisely two external lines, corresponding to the fields $\phi^a(x)$ and $\phi^b(y)$. The graphical representation of a term like ${}_o\langle\Omega|\phi^a(x)|\Omega\rangle_i$ is similarly given by the sum over all Feynman graphs (called tadpole graphs) with precisely one external line, corresponding to $\phi^a(x)$.

Dividing all terms by the factors of ${}_o\langle\Omega|\Omega\rangle_i$ in the denominator is precisely what is needed to cancel disconnected vacuum graphs (*i.e.* those disconnected subgraphs having no external lines). But this does not remove graphs in ${}_o\langle\Omega|T[\phi^a(x)\,\phi^b(y)]|\Omega\rangle_i$ corresponding to a pair of disconnected 'tadpole' graphs, each of which has a single external line. These disconnected graphs precisely correspond to the product ${}_o\langle\Omega|\phi^a(x)|\Omega\rangle_i\,{}_o\langle\Omega|\phi^b(y)|\Omega\rangle_i$ in (2.15), whose subtraction therefore cancels the remaining disconnected component from $\langle T[\phi^a(x)\,\phi^b(y)]\rangle_c$.

A similar story goes through for the higher functional derivatives, and shows how correlations obtained by differentiating W have their disconnected parts systematically subtracted off. Indeed Eqs. (2.13) and (2.12) can be used as non-perturbative definitions of what is meant by connected correlations functions and their generators [15].

2.1.2 The 1PI (or Quantum) Action ♠

As Eqs. (2.12) and (2.4) show, the functional $Z[J] = \exp\{iW[J]\}$ can be physically interpreted as the 'in-out' vacuum amplitude in the presence of an applied current $J_a(x)$. Furthermore, the applied current can be regarded as being responsible for changing the expectation value of the field, since not evaluating Eq. (2.14) at $J_a = 0$ gives

$$\varphi^a(x) := \langle\phi^a(x)\rangle_J = \frac{\delta W}{\delta J_a(x)}, \tag{2.16}$$

as a functional of the current $J_a(x)$. However, it is often more useful to have the vacuum-to-vacuum amplitude expressed directly as a functional of the expectation value, $\varphi^a(x)$, itself, rather than $J_a(x)$. This is accomplished by performing a Legendre transform, as follows.

Legendre Transform

To perform a Legendre transform, define [15]

$$\Gamma[\varphi\,] := W[J] - \int d^4x\, \varphi^a J_a, \tag{2.17}$$

with $J_a(x)$ regarded as a functional of $\varphi^a(x)$, implicitly obtained by solving Eq. (2.16). Once $\Gamma[\varphi\,]$ is known, the current required to obtain the given $\varphi^a(x)$ is found by directly differentiating the definition, Eq. (2.17), using the chain rule together with Eq. (2.16) to evaluate the functional derivative of $W[J]$:

$$\frac{\delta\Gamma}{\delta\varphi^a(x)} = \int d^4y\, \frac{\delta J_b(y)}{\delta\varphi^a(x)}\,\frac{\delta W}{\delta J_b(y)} - J_a(x) - \int d^4y\, \varphi^b(y)\,\frac{\delta J_a(y)}{\delta\varphi^a(x)} = -J_a(x). \tag{2.18}$$

In particular, this last equation shows that the expectation value for the 'real' system with $J_a = 0$ is a stationary point of $\Gamma[\varphi]$. In this sense $\Gamma[\varphi]$ is related to $\langle\phi^a\rangle$ in the same way that the classical action, $S[\phi]$, is related to a classical background configuration, $\varphi^a_{\rm cl}$. For this reason $\Gamma[\varphi]$ is sometimes thought of as the quantum generalization of the classical action, and known as the theory's *quantum action*.

This similarity between $\Gamma[\varphi]$ and the classical action is also reinforced by other considerations. For instance, because the classical action is usually the difference, $S = K - V$, between kinetic and potential energies, for time-independent configurations (for which the kinetic energy is $K = 0$) the classical ground state actually minimizes $V = -S$. It can be shown that for time-independent systems – *i.e.* those where the ground state $|\Omega\rangle$ is well-described in the adiabatic approximation – the configuration $\varphi^a = \langle\phi^a\rangle$ similarly minimizes the quantity $-\Gamma$. In particular, for configurations φ^a independent of spacetime position the configuration minimizes the quantum 'effective potential' $V_{\rm q}(\varphi) = -\Gamma[\varphi]/({\rm Vol})$, where 'Vol' is the overall volume of spacetime.

One way to prove this [16, 17] is to show that, for any static configuration, φ^a, the quantity $-\Gamma[\varphi]$ can be interpreted as the minimum value of the energy, $\langle\Psi|H|\Psi\rangle$, extremized over all normalized states, $|\Psi\rangle$, that satisfy the condition $\langle\Psi|\phi^a(x)|\Psi\rangle = \varphi^a(x)$. The global minimum to $-\Gamma[\varphi]$ then comes once φ^a is itself varied over all possible values.

Semiclassical Expansion

How is $\Gamma[\varphi]$ computed within perturbation theory? To find out, multiply the path integral representation for $W[J]$, Eq. (2.12), on both sides by $\exp\left\{-{\rm i}\int {\rm d}^4x\,(\varphi^a J_a)\right\}$. Since neither φ^a nor J_a are integration variables, this factor may be taken inside the path integral, giving

$$\begin{aligned}\exp\Big\{{\rm i}\Gamma[\varphi]\Big\} &= \exp\left\{{\rm i}W[J] - {\rm i}\int {\rm d}^4x\,\varphi^a J_a\right\}\\ &= \int \mathcal{D}\phi\,\exp\left\{{\rm i}S[\phi] + {\rm i}\int {\rm d}^4x\,(\phi^a - \varphi^a)J_a\right\}\\ &= \int \mathcal{D}\tilde\phi\,\exp\left\{{\rm i}S[\varphi + \tilde\phi] + {\rm i}\int {\rm d}^4x\,\tilde\phi^a J_a\right\}.\end{aligned} \tag{2.19}$$

The last line uses the change of integration variable $\phi^a \to \tilde\phi^a := \phi^a - \varphi^a$.

At face value, Eq. (2.19) doesn't seem so useful in practice, since the dependence on J_a inside the integral is to be regarded as a functional of φ^a, using Eq. (2.18). This means that $\Gamma[\varphi]$ is only given implicitly, since it appears on both sides. But on closer inspection, the situation is much better than this, because the implicit appearance of Γ through J_a on the right-hand side is actually very easy to implement in perturbation theory.

To see how this works, imagine evaluating Eq. (2.19) perturbatively by expanding the action inside the path integral about the configuration $\phi^a = \varphi^a$, using

$$S[\varphi + \tilde\phi] = S[\varphi] + S_2[\varphi, \tilde\phi] + S_{\rm lin}[\varphi, \tilde\phi] + S_{\rm int}[\varphi, \tilde\phi]. \tag{2.20}$$

This is very similar to the expansion in Eq. (2.10), apart from the term linear in $\tilde\phi^a$,

$$S_{\rm lin}[\varphi, \tilde{\phi}\,] = \int {\rm d}^4x \left[\left(\frac{\delta S}{\delta \phi^a(x)} \right)_{\phi=\varphi} + J_a(x) \right] \tilde{\phi}^a(x), \tag{2.21}$$

which does not vanish as it did before because $\delta\varphi^a := \varphi^a - \varphi^a_{\rm cl} \neq 0$. However, because the difference $\delta\varphi^a$ is perturbatively small, it may be grouped with the terms in $S_{\rm int}$ and expanded within the integrand, and not kept in the exponential.

The resulting expansion for $\Gamma[\varphi\,]$ then becomes

$$e^{{\rm i}\Gamma[\varphi\,]} = e^{{\rm i}S[\varphi\,]} \int \mathcal{D}\tilde{\phi}\; e^{{\rm i}S_2[\varphi,\tilde{\phi}\,]} \sum_{r=0}^{\infty} \frac{1}{r!} \Big({\rm i}S_{\rm int} + {\rm i}S_{\rm lin} \Big)^r, \tag{2.22}$$

and so

$$\Gamma[\varphi\,] = S[\varphi\,] + \frac{\rm i}{2} \ln\det\Delta + (\text{2-loops and higher}), \tag{2.23}$$

where $\Delta(\varphi)$ is the operator appearing in $S_2 = -\int {\rm d}^4x\, \tilde{\phi}^a \Delta_{ab} \tilde{\phi}^b$, and the contribution called '2-loops and higher' denotes the sum of all Feynman graphs involving two or more loops (and no external lines) built with internal lines representing the propagator Δ^{-1}, and vertices built using $S_{\rm int} + S_{\rm lin}$.

Expression (2.23) shows why the perturbative expansion of $\Gamma[\phi]$ is often related to the semiclassical approximation and loop expansion. As argued above Eq. (1.3), whenever a large dimensionless parameter pre-multiplies the classical action – such as if $S = \tilde{S}/\lambda$ for $\lambda \ll 1$ – then each additional loop costs a factor of λ. In detail this is true because each vertex in a Feynman graph carries a factor $1/\lambda$, while each propagator is proportional to $\Delta^{-1} \propto \lambda$. Consequently, a graph with $\mathcal{I}$ internal lines and $\mathcal{V}$ vertices is proportional to λ^x with

$$x = \mathcal{I} - \mathcal{V} = \mathcal{L} - 1, \tag{2.24}$$

where $\mathcal{L}$ is the total number of loops[5] in the graph (more about which below).

The first two contributions to (2.23) are the classical and one-loop results, while the last term turns out to consist of the sum over all Feynman graphs with two or more loops. This can be seen because the first two terms of (2.23) are proportional to λ^{-1} and λ^0, respectively, while all terms built using $S_{\rm int}$ start at order λ, with two loop graphs being order λ; 3-loop graphs are order λ^2 and so on. The connection between the loop and semiclassical expansions comes because the semiclassical expansion is normally regarded as a series in powers of $\hbar$. In ordinary units it is $S/\hbar$ that appears in the path integral, so the argument just given shows that powers of $\hbar$ also count loops. Counting powers of small dimensionless quantities is more informative than counting powers of $\hbar$ because the semiclassical expansion is really an expansion in powers of the dimensionless ratio $\hbar/S$, and this ultimately is small *because* of its proportionality to small dimensionless parameters.

[5] When restricted to graphs that can be drawn on a plane, this identity agrees with an intuitive definition of what the number of loops in a graph should be (because the combination $\mathcal{L} - \mathcal{I} + \mathcal{V} = 1$ is a topological invariant for all such graphs). For graphs that cannot be drawn on a plane this expression defines the number of loops.

Now comes the main point. Because $S_{\rm lin}$ is linear in $\tilde{\phi}^a$, its Feynman rule is as given in Fig. 2.2, which inserts a 'tadpole' contribution proportional to $(\delta S/\delta\varphi^a) + J_a$. But something wonderful happens once this is evaluated at $J_a = -\delta\Gamma/\delta\varphi^a$, since Eq. (2.23) implies that

$$\begin{aligned}\frac{\delta S}{\delta\varphi^a(x)} + J_a(x) &= \frac{\delta}{\delta\varphi^a(x)}\Big(S[\varphi\,] - \Gamma[\varphi\,]\Big)\\ &= -\frac{\delta}{\delta\varphi^a(x)}\left[\frac{\mathrm{i}}{2}\ln\det\Delta + (\text{2-loops and higher})\right]\\ &= -(\text{sum of tadpole graphs}),\end{aligned} \tag{2.25}$$

where 'tadpole graphs' mean all those graphs involving one or more loops having one dangling unconnected internal propagator that ends at the point x.

What is wonderful about this condition is it is easy to implement without having a detailed expression for $\Gamma[\varphi\,]$. The condition $J_a = -\delta\Gamma/\delta\varphi^a$ simply ensures that all graphs involving explicit dependence on J_a precisely cancel all graphs without J_a that are *1-particle reducible*: that is, all graphs that can be cut into two disconnected pieces by breaking a single internal line. (A graph that cannot be broken into two in this way is called *1-particle irreducible*, or 1PI.)

The upshot is very simple: $\Gamma[\varphi\,]$ is computed by calculating graphs that do not involve the vertex $S_{\rm lin}$ at all, but evaluating only those graphs that are 1-particle irreducible. For this reason $\Gamma[\varphi\,]$ is often called the generator of 1-particle irreducible correlations, or the *1PI action* for short. In the semiclassical expansion about an arbitrary configuration φ^a_J Eq. (2.23) is evaluated as a sum of 1PI connected vacuum graphs (*i.e.* connected graphs having no external lines), obtained by dropping all disconnected and one-particle reducible graphs from the sum sketched in Fig. 2.1.

Recognizing that $\Gamma[\varphi\,]$ involves only 1PI graphs gives another way in which the quantity $\Gamma[\varphi\,]$ generalizes the classical action [17]. In a perturbative expansion the leading, classical, approximation corresponds to using the leading term of Eq. (2.23): $\Gamma[\varphi\,] \simeq S[\varphi\,]$. As discussed in Appendix C, when applied to scattering amplitudes this amounts to just summing the tree graphs built from vertices coming from the classical interactions in $S_{\rm int}$. By contrast, imagine instead computing Feynman graphs using the expansion of $\Gamma[\varphi + \tilde{\phi}\,]$, rather than $S[\varphi + \tilde{\phi}\,]$, to generate the propagators and vertices. In this case, the all-loops result for any correlation function constructed with Feynman rules built from $S[\varphi\,]$ is precisely the same as the quantity obtained by summing just tree graphs constructed from the Feynman rules built from $\Gamma[\varphi\,]$. In this sense, the full quantum amplitude is obtained by calculating using $\Gamma[\varphi\,]$ but working within the classical approximation. (It is this property that makes $\Gamma[\varphi\,]$ useful for discussing ultraviolet (UV) divergences, since the absence of new UV divergences in tree graphs makes it sufficient to renormalize divergences in $\Gamma[\varphi\,]$ in order to ensure all amplitudes are finite.)

Similar to the discussion for $Z[J]$, an important practical issue arises when φ is too complicated to allow an explicit calculation of the propagators $\Delta^{-1}_{ab} = -(\delta^2 S/\delta\phi^a\delta\phi^b)_{\phi=\varphi}$ used in the Feynman rules. This is often dealt with by expanding the result in powers of $\varphi^a(x)$ (or, more generally, in powers of the displacement of φ^a away from a sufficiently simple background for which Δ^{-1}_{ab} can be evaluated).

In this case, using

$$\left(\frac{\delta^2 S}{\delta\phi^a(x)\,\delta\phi^b(y)}\right)_\varphi = \left(\frac{\delta^2 S}{\delta\phi^a(x)\,\delta\phi^b(y)}\right)_0 + \int d^4z \left(\frac{\delta^3 S}{\delta\phi^a(x)\,\delta\phi^b(y)\,\delta\phi^c(z)}\right)_0 \varphi^c(z) + \cdots, \tag{2.26}$$

and $(\Delta_0 - \delta\Delta)^{-1} = \Delta_0^{-1}\sum_{n=0}^{\infty}(\delta\Delta\,\Delta_0^{-1})^n$ shows that the required Feynman graphs are obtained by inserting external lines in all possible ways (both to the internal lines and the vertices) in the 1PI vacuum graphs of Fig. 2.1, with the external lines representing the Feynman rule $\varphi^a(x)$.

2.2 The High-Energy/Low-Energy Split $^\diamond$

So far, so good, but how can the above formalism be used to compute and use low-energy effective actions? The rest of this chapter specializes to theories having two very different intrinsic mass scales – like $m_l \ll m_R$ of the toy model in Chapter 1 – in order to address this question. After formalizing the split into low- and high-energy theory in this section, the following two sections identify two useful ways of defining a low-energy effective action.

2.2.1 Projecting onto Low-Energy States

The starting point, in this section, is to define more explicitly the split between low- and high-energy degrees of freedom. There are a variety of ways to achieve this split. Most directly, imagine dividing the quantum field ϕ^a into a low-energy and high-energy part: $\phi^a(x) = l^a(x) + \mathfrak{h}^a(x)$, with

$$l^a(x) := P_\Lambda \phi^a(x) P_\Lambda, \tag{2.27}$$

where $P_\Lambda = P_\Lambda^2$ denotes the projector onto states having energy $E < \Lambda$. To be of practical use, the scale Λ should lie somewhere between the two scales (such as m_l and m_R) that define the underlying hierarchy ($m_l \ll m_R$) in terms of which the low-energy limit is defined for the theory of interest.

This can be made more explicit in semiclassical perturbation theory, where $\phi^a = \varphi^a_{\rm cl} + \hat\phi^a$. Since in the interaction representation the quantum field satisfies the linearized field equation, $\Delta_{ab}\hat\phi^b = 0$, one can decompose $\hat\phi^a(x)$ in terms of a basis of eigenmodes, $u_p(x)$,

$$\hat\phi^a(x) = \sum_p \left[\mathfrak{a}_p u_p^a(x) + \mathfrak{a}_p^* u_p^{a*}(x)\right]. \tag{2.28}$$

For time-independent backgrounds these eigenmodes can be chosen to simultaneously diagonalize the energy, $i\partial_t u_p = \varepsilon_p\, u_p$, and so the low-energy part of the field is that part of the sum for which the mode energies are smaller than the reference scale Λ. That is,

$$\hat{l}^a(x) := \sum_{\varepsilon_p<\Lambda} \left[\mathfrak{a}_p u_p^a(x) + \mathfrak{a}_p^* u_p^{a*}(x)\right], \tag{2.29}$$

and so

$$\hat{\mathfrak{h}}^a := \hat{\phi}^a - \hat{l}^a = \sum_{\varepsilon_p>\Lambda} \left[\mathfrak{a}_p u_p^a(x) + \mathfrak{a}_p^* u_p^{a*}(x)\right]. \tag{2.30}$$

Of course, one might also implement a cutoff more smoothly, by weighting high-energy states in amplitudes by some suitably decreasing function of energy rather than completely cutting them off above Λ.

It is natural at this point to worry that a division into high- and low-energy modes introduces an explicit frame-dependence into the problem. After all, a collision that appears to involve only low energies to one observer would appear to involve very high energies to another observer who moves very rapidly relative to the first one. Although this is true in principle, in practice frame-independent physical quantities (like the scattering amplitudes examined for the toy model in Chapter 1) only depend on invariant quantities like centre-of-mass energies, and all observers agree when these are large or small. For scattering calculations the natural split between low- and high-energies is therefore made in the centre-of-mass frame. The point is that in order to profit from the simplification of physics at low energies, it suffices that there exist *some* observers who see a process to be at low energies; it is not required that *all* observers do so.

Notice that correlation functions of low-energy states necessarily do not vary very quickly with time. This may be seen by inserting a complete set of intermediate energy eigenstates between any two pairs of low-energy fields, such as

$$\begin{aligned}
{}_o\langle\Omega|\,\hat{l}^a(x)\,\hat{l}^b(y)\,|\Omega\rangle_i &= {}_o\langle\Omega|P_\Lambda\hat{\phi}^a(x)P_\Lambda^2\hat{\phi}^b(y)P_\Lambda|\Omega\rangle_i \\
&= \sum_{\varepsilon_p<\Lambda} {}_o\langle\Omega|\hat{\phi}^a(x)|p\rangle\langle p|\hat{\phi}^b(y)|\Omega\rangle_i,
\end{aligned} \tag{2.31}$$

which uses $P_\Lambda|\Omega\rangle = |\Omega\rangle$ and $P_\Lambda|p\rangle = |p\rangle$ for low-energy states, while $P_\Lambda|p\rangle = 0$ for high-energy states. This result clearly has support only for frequencies $\omega = \varepsilon_p < \Lambda$. In relativistic and translation-invariant theories, for which low energy also means low momentum, the same argument shows that correlations also have slow spatial variation.

Example: The Toy Model

To make this concrete, consider the toy model of Chapter 1. In this case, there are two quantum fields, $\hat{\phi}_I$ and $\hat{\phi}_R$ (or equivalently, $\hat{\xi}$ and $\hat{\chi}$), and the two intrinsic mass scales are $m_I = 0 \ll m_R$. The energy eigenmodes for these are labeled by 4-momentum, $u_p(x) \propto e^{ipx}$, and the linearized field equation $(-\Box + m^2)\,\hat{\phi} = 0$, relates the energy to the momentum by $q^0 = \varepsilon_q = q$ for $\hat{\phi}_I$ and $p^0 = \varepsilon_p = \sqrt{p^2 + m_R^2}$ for $\hat{\phi}_R$.

In this case, the useful choice is $m_I \ll \Lambda \ll m_R$, which is possible because of the hierarchy $m_I \ll m_R$. With this choice, the light fields consist only of the long-wavelength modes of $\hat{\phi}_I$,

$$\hat{\mathfrak{l}}(x) = \sum_{\epsilon_q < \Lambda} \left[\mathfrak{c}_q \, e^{iqx} + \mathfrak{c}_q^* \, e^{-iqx} \right], \tag{2.32}$$

while the heavy fields contain all of the $\hat{\phi}_R$ modes together with the short-wavelength modes of $\hat{\phi}_l$:

$$\hat{\mathfrak{h}}_l(x) = \sum_{\epsilon_q > \Lambda} \left[\mathfrak{c}_q \, e^{iqx} + \mathfrak{c}_q^* \, e^{-iqx} \right] \quad \text{and} \quad \hat{\mathfrak{h}}_R(x) = \sum_p \left[\mathfrak{b}_p \, e^{ipx} + \mathfrak{b}_p^* \, e^{-ipx} \right]. \tag{2.33}$$

2.2.2 Generators of Low-Energy Correlations ♠

The next step is to seek generating functionals specifically for low-energy correlation functions, and to investigate their properties. The key tool for this purpose is the observation made above that the correlation functions themselves can vary only over time and length scales larger than Λ^{-1}.

Imagine now defining the generating functional, $Z_{\rm le}[J]$, for the time-ordered in-out correlations of only the low-energy fields, $\hat{\mathfrak{l}}^a(x)$. This can be done simply by restricting the definition, Eq. (2.2), of $Z[J]$, to include only correlation functions that vary slowly in space and time (*i.e.* only over scales larger than Λ^{-1}), leading to the result

$$Z_{\rm le}[J] := \sum_{n=0}^{\infty} \frac{i^n}{n!} \int {\rm d}^4 x_1 \cdots {\rm d}^4 x_n G_{\rm le}^{a_1 \cdots a_n}(x_1, \cdots, x_n) J_{a_1}(x_1) \cdots J_{a_n}(x_n). \tag{2.34}$$

Because the low-energy correlation functions only vary slowly in space and time, the same is true of any currents, $J_a(x)$, appearing in $Z_{\rm le}[J]$. That is, if the current is split into long- and short-wavelength Fourier modes, $J_a(x) = j_a(x) + \mathcal{J}_a(x)$, with

$$j_a(x) = \sum_{\text{slowly varying}} j_a(p) \, e^{ipx}, \tag{2.35}$$

then the generating functional for low-energy correlations, $Z_{\rm le}[J]$, is simply the restriction of the full generating functional to slowly varying currents:

$$Z_{\rm le}[j] = Z[j, \mathcal{J} = 0]. \tag{2.36}$$

Here, the precise definition of 'slowly varying' in Eq. (2.35) depends on the details of the particle masses and the way the cutoff Λ is implemented – *c.f.* Eq. (2.29) for example – for the quantum states.

It might seem bothersome that the generating functionals for low-energy correlation functions depend explicitly on the value of Λ, as well as on all of the details of precisely how the high-energy modes are cut off. One of the tasks of later chapters is to show how this dependence can be absorbed into appropriate renormalizations of effective couplings, so that predictions for physical processes (like scattering amplitudes) only depend on physical mass scales like m_R (rather than Λ or other definitional details).

For relativistic, translationally invariant theories a slightly more convenient way to break Fourier modes into slowly and quickly varying parts is to Wick rotate [18] to euclidean signature, $\{x^0, \mathbf{x}\} \to \{ix^4, \mathbf{x}\}$. In this case, the time-components of any 4-vector must be similarly rotated, so the invariant inner product of two 4-vectors becomes

$$p \cdot q = p_\mu q^\mu = -p^0 q^0 + \mathbf{p} \cdot \mathbf{q} \to +p^4 q^4 + \mathbf{p} \cdot \mathbf{q} = (p \cdot q)_E. \tag{2.37}$$

This ensures that the invariant condition $p_\mu p^\mu = (p^4)^2 + \mathbf{p}^2 < \Lambda^2$ excludes large values of both $|\mathbf{p}|$ and p^4 (unlike for Minkowski signature, where $p_\mu p^\mu = -(p^0)^2 + \mathbf{p}^2 < \Lambda^2$ allows both $|\mathbf{p}|$ and p^0 to be arbitrarily large but close to the light cone).

The generator, $W_{\rm le}[j]$, of low-energy connected correlations can be defined as before, by taking the logarithm of $Z_{\rm le}[j]$, leading to the path integral representation

$$\exp\left\{{\rm i}W_{\rm le}[j]\right\} = \int \mathcal{D}\phi \, \exp\left\{{\rm i}S[\phi] + {\rm i}\int {\rm d}^4x \, \phi^a j_a\right\}. \tag{2.38}$$

The main difference between this and the expression for $W[J]$ is the absence of currents coupled to the high-frequency components of ϕ^a. That is, if $\phi^a = l^a + \mathfrak{h}^a$ is split into slowly varying ('light', l^a) and rapidly varying ('heavy', $\mathfrak{h}^a$) parts, along the same lines as Eq. (2.35) for J_a, then Eq. (2.38) becomes

$$\exp\left\{{\rm i}W_{\rm le}[j]\right\} = \int \mathcal{D}l\mathcal{D}\mathfrak{h} \, \exp\left\{{\rm i}S[l+\mathfrak{h}] + {\rm i}\int {\rm d}^4x \, l^a j_a\right\}. \tag{2.39}$$

Physically, this states that a restriction to low-energy correlations can be obtained simply by restricting oneself only to probing the system with slowly varying currents.

2.2.3 The 1LPI Action

At this point, it is hard to stop from performing a Legendre transformation to obtain the generating functional, $\Gamma_{\rm le}[\ell]$, directly in terms of the low-energy field configurations, ℓ^a, rather than j_a. To this end, define

$$\Gamma_{\rm le}[\ell] := W_{\rm le}[j] - \int {\rm d}^4x \, \ell^a j_a, \tag{2.40}$$

with $j_a = j_a[\ell]$ regarded as a functional of ℓ^a found by inverting the relation $\ell^a = \ell^a[j]$ obtained from

$$\ell^a := \frac{\delta W_{\rm le}}{\delta j_a}, \tag{2.41}$$

with the result — *c.f.* Eq. (2.18)

$$j_a = -\frac{\delta \Gamma_{\rm le}}{\delta \ell^a}. \tag{2.42}$$

It is important to realize that although $\Gamma_{\rm le}[\ell]$ obtained in this way only has support on slowly varying field configurations, $\ell^a(x)$, it is *not* simply the restriction of $\Gamma[\varphi] = \Gamma[\ell, h]$ to long-wavelength configurations: $h^a := \delta W/\delta \mathcal{J}_a = 0$. To see why not, consider its path integral representation:

$$\begin{aligned}\exp\left\{{\rm i}\Gamma_{\rm le}[\ell]\right\} &= \int \mathcal{D}l\mathcal{D}\mathfrak{h} \, \exp\left\{{\rm i}S[l,\mathfrak{h}] + {\rm i}\int {\rm d}^4x \, (l^a - \ell^a) j_a\right\} \\ &= \int \mathcal{D}\tilde{l}\mathcal{D}\mathfrak{h} \, \exp\left\{{\rm i}S[\ell + \tilde{l}, \mathfrak{h}] + {\rm i}\int {\rm d}^4x \, \tilde{l}^a j_a\right\}.\end{aligned} \tag{2.43}$$

For comparison, the earlier result, Eq. (2.4), for $\Gamma[\varphi] = \Gamma[\ell, h]$ is

$$\exp\left\{{\rm i}\Gamma_{\rm le}[\ell, h]\right\} = \int \mathcal{D}\tilde{l}\mathcal{D}\tilde{\mathfrak{h}} \, \exp\left\{{\rm i}S[\ell + \tilde{l}, h + \tilde{\mathfrak{h}}] + {\rm i}\int {\rm d}^4x \, (\tilde{l}^a j_a + \tilde{\mathfrak{h}}^a \mathcal{J}_a)\right\}. \tag{2.44}$$

The key point is that the condition $h = 0$ is *not* generically equivalent to the condition $\mathcal{J}_a(x) = 0$ that relates $\Gamma_{\rm le}[\ell]$ to $\Gamma[\ell, h]$. Instead, the condition $\mathcal{J}_a = 0$ states that the short-wavelength part of the field should be chosen as that configuration, $h^a = h^a_{\rm le}(\ell)$, that satisfies

$$\left(\frac{\delta\Gamma}{\delta h^a}\right)_{h=h_{\rm le}(\ell)} = 0. \tag{2.45}$$

In particular, the vanishing of $\mathcal{J}_a$ means that the short-wavelength components of the current are not available to take the values $\mathcal{J}_a = -\delta\Gamma/\delta h^a$ they would have taken in the Legendre transform of $W[J] = W[j, \mathcal{J}]$. They are therefore not able to cancel the one-particle-reducible graphs that can be broken in two by cutting a single $\hat{\mathfrak{h}}^a$ line. The quantity $\Gamma_{\rm le}[\varphi]$ is therefore given as the sum of *one-light-particle irreducible* (or 1LPI) graphs, which are only irreducible in the sense that they cannot be broken into two disconnected pieces by cutting a *light*-particle, $\hat{l}^a$, line.

Example: The Toy Model

How does all this look in the toy model of Chapter 1? In this case, with Λ chosen to satisfy $m_I \ll \Lambda \ll m_R$, the 'light' fields consist only of the low-energy modes of the massless field, ξ (or $\hat{\phi}_I$), and the 'heavy' fields consist of both the high-energy modes of ξ together with all of the modes of the massive field χ (or $\hat{\phi}_R$). The 1LPI generator of low-energy connected correlation functions then is

$$\exp\left\{i\Gamma_{\rm le}[\xi]\right\} = \int \mathcal{D}\tilde{\xi}\mathcal{D}\tilde{\chi}\, \exp\left\{iS[\xi + \tilde{\xi}, \tilde{\chi}] + i\int d^4x\, \tilde{\xi}^a j_a\right\}, \tag{2.46}$$

with ξ^a and $j_a = -\delta\Gamma_{\rm le}/\delta\xi^a$ only varying over times and distances longer than Λ^{-1}.

Recall that small λ controls a semiclassical expansion, and imagine computing $\Gamma_{\rm le}[\xi]$ in the leading, classical approximation. As argued earlier (and elaborated in §3 below), in this limit the full 1PI generator reduces to the classical action: $\Gamma[\xi, \chi] \simeq S[\xi, \chi]$, explicitly given in Eq. (1.24),

$$S[\xi, \chi] = -\int d^4x \left[\frac{1}{2}\,\partial_\mu\chi\partial^\mu\chi + \frac{1}{2}\left(1 + \frac{\chi}{\sqrt{2}\,v}\right)^2 \partial_\mu\xi\partial^\mu\xi + V(\chi)\right], \tag{2.47}$$

with

$$V(\chi) = \frac{m_R^2}{2}\,\chi^2 + \frac{\lambda v}{2\sqrt{2}}\,\chi^3 + \frac{\lambda}{16}\,\chi^4. \tag{2.48}$$

In general, the above arguments say that $\Gamma_{\rm le}[\xi] = \Gamma_{\rm le}[\xi, \chi_{\rm le}(\xi)]$, where $\chi_{\rm le}(\xi)$ is obtained by solving $\delta\Gamma[\xi, \chi]/\delta\chi = 0$ (*c.f.* Eq. (2.45)). So in the classical approximation $\Gamma_{\rm le}[\xi] \simeq S[\xi, \chi_{\rm le}(\xi)]$, where Eq. (2.45) in the classical approximation says $\chi_{\rm le}(\xi)$ is found by solving the classical field equation

$$\left(-\Box + m_R^2 + \frac{1}{2v^2}\,\partial_\mu\xi\partial^\mu\xi\right)\chi_{\rm le} = -\frac{1}{\sqrt{2}\,v}\,\partial_\mu\xi\partial^\mu\xi - \frac{3\lambda v}{2\sqrt{2}}\,\chi_{\rm le}^2 - \frac{\lambda}{4}\,\chi_{\rm le}^3. \tag{2.49}$$

Using this in the classical action leads (after an integration by parts) to

$$S[\xi, \chi_{\rm le}(\xi)] = \int d^4x \left[-\frac{1}{2}\left(1 + \frac{\chi_{\rm le}}{\sqrt{2}\,v}\right)\partial_\mu\xi\partial^\mu\xi + \frac{\lambda v}{4\sqrt{2}}\,\chi_{\rm le}^3 + \frac{\lambda}{16}\,\chi_{\rm le}^4\right], \tag{2.50}$$

where $\chi_{\rm le}(\xi)$ is to be interpreted as the function of ξ obtained by solving (2.49).

To proceed further, expand the solution $\chi_{\rm le}$ in powers of $\partial_\mu \xi \partial^\mu \xi$ and $\Box$, using

$$(-\Box + m_R^2 + X)^{-1} \simeq \frac{1}{m_R^2} - \frac{(-\Box + X)}{m_R^4} + \cdots , \tag{2.51}$$

and so on. This gives

$$\chi_{\rm le} \simeq -\frac{1}{\sqrt{2}\, v m_R^2}\left(1 + \frac{\Box}{m_R^2} + \frac{\Box^2}{m_R^4} + \cdots\right)(\partial_\mu \xi \partial^\mu \xi) - \frac{1}{4\sqrt{2}\, v^3 m_R^4}\left(1 + \frac{\Box}{m_R^2} + \cdots\right)(\partial_\mu \xi \partial^\mu \xi)^2 + \cdots , \tag{2.52}$$

where the ellipses involve terms with more powers of $\Box$ and/or more powers of $(\partial_\mu \xi \partial^\mu \xi)$.

Combining results then gives

$$\Gamma_{\rm le}[\xi] \simeq \int {\rm d}^4 x \left[-\frac{1}{2}\,\partial_\mu \xi \partial^\mu \xi + a\left(\partial_\mu \xi \partial^\mu \xi\right)\left(1 + \frac{\Box}{m_R^2}\right)\left(\partial_\mu \xi \partial^\mu \xi\right) + b\,(\partial_\mu \xi \partial^\mu \xi)^3 + \cdots\right], \tag{2.53}$$

where again ellipses include terms with more powers of $\Box$ and/or $\partial_\mu \xi \partial^\mu \xi$ than those shown. In the classical approximation used here, the coefficients a and b evaluate to

$$a = \frac{\lambda}{4m_R^4} = \frac{1}{4\lambda v^4} \quad \text{and} \quad b = \left(\frac{1}{16} - \frac{1}{16}\right)\frac{\lambda^2}{m_R^8} = 0. \tag{2.54}$$

In particular, b vanishes due to a cancelation between the $\chi_{\rm le}(\partial_\mu \xi \partial^\mu \xi)$ and $\chi_{\rm le}^3$ terms, while $a = \lambda/4m_R^4$ precisely agrees with Eq. (1.12) (and is proportional to λ^{-1} when expressed in terms of v, as expected for a classical contribution). As shown earlier, when used in the classical – no loop – approximation, this value ensures that the interactions quadratic in $\partial_\mu \xi \partial^\mu \xi$ accurately reproduce the first two terms of the low-energy limit for $\xi\xi \to \xi\xi$ scattering found earlier.

Of course, Feynman graphs greatly simplify calculations such as these, particularly once one progresses beyond the leading classical approximation, and it is useful to see how these reproduce the above calculation. To this end, imagine working with the Feynman rules described around Eq. (2.26), in which one perturbs in powers of the background field (in this case ξ). As shown in the next chapter, at the classical level one seeks tree (*i.e.* no loops) graphs whose external lines correspond to the fields appearing in the appropriate effective interaction. Since $\Gamma_{\rm le}[\xi]$ is 1LPI at tree level, only heavy states can appear as internal lines.

For instance, for the calculation performed above the term quadratic in $\partial_\mu \xi\, \partial^\mu \xi$ arises from the Feynman graph shown in panel (a) of Fig. 2.3,[6] once the heavy-particle propagator is expanded in powers of $\Box/m_R^2$. The result found in this way for the parameter a is

$$i a_{\text{graph (a)}} = \left[\left(\frac{{\rm i}^2}{2!}\right)\right]\left(-\frac{1}{\sqrt{2}\, v}\right)^2\left(\frac{-{\rm i}}{m_R^2}\right) = \frac{\rm i}{4v^2 m_R^2}, \tag{2.55}$$

where the square bracket contains the numerical factors from the expansion of $\exp[{\rm i}S_{\rm int}]$, the next factor is the coupling constant Feynman rule for the vertices and

[6] Note that no combinatorial factors are in this case necessary for external lines – in contrast to the scattering evaluation of Fig. 1.3 – because external lines here represent factors of $\partial_\mu \xi$ all evaluated at the same position, rather than scattering states with distinguishable momenta.

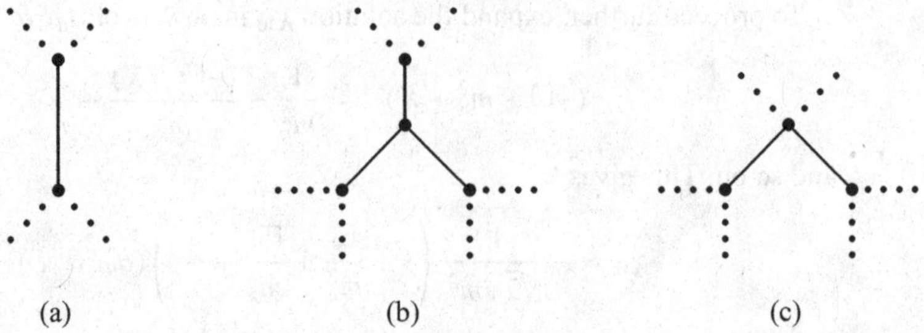

Fig. 2.3 The tree graphs that dominate the $(\partial_\mu \xi\, \partial^\mu \xi)^2$ (panel a) and the $(\partial_\mu \xi\, \partial^\mu \xi)^3$ (panels b and c) effective interactions. Solid lines represent χ propagators, while dotted lines denote external ξ fields.

the final factor is the leading contribution from the propagator. The left-hand side is obtained by expanding $\exp\{i\Gamma[\xi]\}$ using (2.53) and identifying the coefficient of the $(\partial_\mu \xi\, \partial^\mu \xi)^2$ term. This result for a agrees with the one found in (2.54) by eliminating $\chi_{\rm le}(\xi)$ from the action, and in (1.16) by demanding the correct result for low-energy ξ-particle scattering.

The contribution cubic in $\partial_\mu \xi\, \partial^\mu \xi$ is similarly obtained from the graphs in panels (b) and (c) of Fig. 2.3, which contribute the following to the effective coupling b:

$$
\begin{aligned}
i b_{\text{graph (b)}} &= 6\left[4\left(\frac{i^4}{4!}\right)\right]\left(-\frac{1}{\sqrt{2}\,v}\right)^3\left(-\frac{\lambda v}{2\sqrt{2}}\right)\left(\frac{-i}{m_R^2}\right)^3 = \frac{i\lambda}{8v^2 m_R^6}\\
i b_{\text{graph (c)}} &= 2\left[3\left(\frac{i^3}{3!}\right)\right]\left(-\frac{1}{\sqrt{2}\,v}\right)^2\left(-\frac{1}{4v^2}\right)\left(\frac{-i}{m_R^2}\right)^2 = -\frac{i}{8v^4 m_R^4}.
\end{aligned}
\tag{2.56}
$$

Starting with the square bracket, each factor here has the same origin as its counterpart in Eq. (2.55) for panel (a), and the first number arises as the number of ways of connecting the relevant external lines and vertices into the given graph. These again cancel once the relation $m_R^2 = \lambda v^2$ is used, giving the tree-level prediction $b = 0$.

The above arguments make clearer why – if low-energy observables are the only things of interest – the physics of the two fields in the toy model can be traded for a collection of effective interactions involving only the light particle: it suffices to know $\Gamma_{\rm le}[\ell]$ to capture all low-energy physical applications. What remains is to find a way to do so more efficiently than by first computing the full generating functional $\Gamma[\ell, h]$, and then eliminating $h = h_{\rm le}(\ell)$ to obtain $\Gamma_{\rm le}[\ell] = \Gamma[\ell, h_{\rm le}(\ell)]$.

2.3 The Wilson action $^\diamond$

Although the low-energy 1LPI generator captures all of a theory's low-energy observables, what remains elusive is an efficient means for capturing, *as early in a calculation as possible*, how high-energy physics appears in low-energy processes. The main tool for doing this – the *Wilson action*[7] – is topic of this section.

[7] Named for Ken Wilson, a pioneer in the development of renormalization techniques (see the brief historical notes in the Bibliography).

2.3.1 Definitions

A good starting point for describing the Wilson action is the path integral expression for the 1LPI generator, Eq. (2.43):

$$\exp\left\{i\Gamma_{le}[\ell]\right\} = \int \mathcal{D}\tilde{l}\,\mathcal{D}\mathfrak{h}\,\exp\left\{iS[\ell+\tilde{l}+\mathfrak{h}] + i\int d^4x\,\tilde{l}^a j_a\right\}. \tag{2.57}$$

What is noteworthy about this expression is that – because the currents are chosen to explore only low-energy quantities – the heavy field, $\mathfrak{h}^a$, appears only in the classical action and not in the current term. As a consequence, all possible low-energy influences of the heavy field must be captured in the quantity

$$\exp\left\{iS_W[l]\right\} := \int \mathcal{D}\mathfrak{h}\,\exp\left\{iS[l+\mathfrak{h}]\right\}, \tag{2.58}$$

in terms of which the full 1LPI action is given by

$$\exp\left\{i\Gamma_{le}[\ell]\right\} = \int \mathcal{D}\tilde{l}\,\exp\left\{iS_W[\ell+\tilde{l}] + i\int d^4x\,\tilde{l}^a j_a\right\}. \tag{2.59}$$

Eq. (2.58) defines the Wilson action, obtained by *integrating out* all heavy degrees of freedom having energies above the scale Λ. It has several noteworthy features, which are explored in detail throughout the rest of the book.

- As the definition shows, the Wilson action provides the earliest place in a calculation to systematically identify, once and for all, the low-energy influence of the heavy degrees of freedom $\mathfrak{h}$. Best of all, this can be done in one fell swoop, before choosing precisely which observable or correlation function is of interest in a particular application.
- For practical applications, most real interest is in obtaining the Wilson action as a series expansion in inverse powers of the heavy mass scales in the problem of interest. As shall be seen in some detail, at any fixed order in this expansion the Wilson action is a local functional, $S_W = \int d^4x\,\mathfrak{L}_W(x)$, with $\mathfrak{L}_W(x)$ being a function of the light fields and their derivatives all evaluated at the same spacetime point.
- What is striking about Eq. (2.59) is that the Wilson action, S_W, appears in the expression for the generator, Γ_{le}, of low-energy correlators, in precisely the way that the classical action, S, appears in the expression, Eq. (2.19), for the generator, Γ, of generic correlators. This suggests that the classical action of the full theory might itself be better regarded as the Wilson action from some even higher-energy theory.
- Eq. (2.58) shows that S_W depends in detail on things like Λ and precisely how the split is made between the high- and low-energy sectors, since these are buried in the definitions of the split between $\mathfrak{h}^a$ and l^a. So it is misleading to speak about 'the' Wilson action, rather than 'a' Wilson action. Yet we know that Λ cannot appear in any physical observables, because it is just an arbitrary artificial scale that is introduced for calculational convenience. Part of the story to follow must therefore be why all these calculational details in S_W drop out of observables. The outlines of this argument are already clear in Eq. (2.59), which shows that the Λ dependence introduced by performing the integration over $\mathcal{D}\mathfrak{h}$ is later canceled when integrating over the rest of the fields, $\mathcal{D}l$, since the total measure $\mathcal{D}\phi = \mathcal{D}l\,\mathcal{D}\mathfrak{h}$ is Λ-independent.

In semiclassical perturbation theory, the arguments of earlier sections show that Eq. (2.58) gives S_W as the sum over all connected vacuum graphs – not just 1PI graphs, say – using Feynman rules computed for the 'high-energy' fields with the 'low-energy' fields regarded as fixed background values. (Recall in this split that high-energy fields can include the high-energy modes of particles with small masses.) Eq. (2.59) then says to construct Feynman graphs using propagators and vertices for the light fields defined from S_W, with $\Gamma_{\rm le}$ then obtained by computing all 1PI graphs in this low-energy theory. This combination reproduces the set of 1LPI graphs using the Feynman rules of the full theory.

In particular, since any tree graph with an internal line is one-particle reducible, this means that $\Gamma_{\rm le}[\ell] \simeq S_W[\ell]$ within the classical approximation (no loops). Furthermore, in the same approximation both are related to the classical action of the full theory by

$$\Gamma_{\rm le}[\ell] \simeq S_W[\ell] \simeq S[\ell, h_{\rm le}(\ell)] \qquad \text{(classical approximation)}, \tag{2.60}$$

where $h^a = h^a_{\rm le}(\ell)$ is obtained (in the classical approximation) by solving $(\delta S/\delta h^a)_{h=h_{\rm le}(\ell)} = 0$. But – as is seen more explicitly below – $\Gamma_{\rm le}[\ell]$ and $S_W[\ell]$ need no longer agree once loops are included.

It is the Wilson action that is the main tool used in the rest of this book. But why bother with S_W, given that $\Gamma_{\rm le}$ also captures all of the information relevant for low-energy predictions? As later examples show in more detail, in real applications it is often the Wilson action that is the easier to use, since it exploits the simplicity of the low-energy limit as early as possible. It plays such a central role because it has two very attractive properties.

First, it contains enough information to be useful. That is, any low-energy observable can be constructed from low-energy correlation functions (and so also from $\Gamma_{\rm le}$), and because $\Gamma_{\rm le}$ can be computed from S_W using only low-energy degrees of freedom, it follows that S_W carries all of the information necessary to extract the predictions for any low-energy observable.

But it is the second property that makes it such a practical tool: it doesn't contain too much information. That is, the Wilson action is the bare-bones quantity that contains all of the information about the system's high-energy degrees of freedom without polluting it with any low-energy details. Unlike $\Gamma_{\rm le}$, the Wilson action is constructed by integrating only over the high-energy degrees of freedom. This means that there is a maximal labour saving in exploiting any simplicity in S_W, since this simplicity is present *before* performing the rest of the low-energy part of the calculation.

Example: The Toy Model

To better understand how the Wilson action is defined, and how it is related to the low-energy 1LPI generator, it is useful to have a concrete example to examine in detail. Once again the toy model of Chapter 1 provides a useful place to start.

Since $\Gamma_{\rm le}$ and S_W only begin to differ beyond the classical approximation, imagine computing both $\Gamma_{\rm le}$ and S_W at one loop. According to its definition, the Feynman graphs contributing to S_W can involve only the high-energy degrees of freedom in the internal lines, while those contributing to $\Gamma_{\rm le}$ also involve virtual low-energy states.

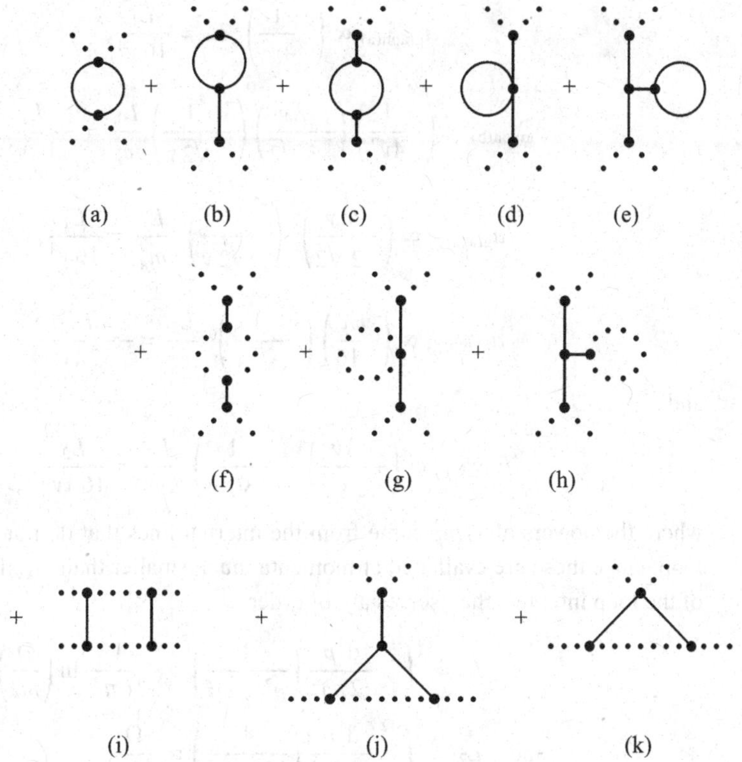

Fig. 2.4 One-loop graphs that contribute to the $(\partial_\mu \xi \partial^\mu \xi)^2$ interaction in the Wilson and 1LPI actions using the interactions of Eqs. (1.24) and (1.25). Solid (dotted) lines represent χ (and ξ) fields. Graphs involving wave-function renormalizations of ξ are not included in this list.

For both S_W and $\Gamma_{\rm le}$ the graphs can be one-particle reducible when cutting heavy-particle lines, but for $\Gamma_{\rm le}$ the graphs must be one-particle irreducible when light-particle lines are cut.

For concreteness' sake, for the toy model consider the one-loop contributions to the effective interaction

$$a \int d^4x \, (\partial_\mu \xi \partial^\mu \xi)^2, \tag{2.61}$$

in both $\Gamma_{\rm le}$ and S_W. The relevant Feynman graphs are shown in Fig. 2.4, using Feynman rules appropriate for the χ and ξ fields using the interactions given in Eqs. (1.24) and (1.25). (An equivalent set of graphs could also be written for the variables $\hat{\phi}_R$ and $\hat{\phi}_I$. Although both ultimately give the same physical results, they can differ in intermediate steps, and which is more useful depends on the application one has in mind.)

Since all of the internal lines for Feynman graphs (a) through (e) involve only the field χ, and since all modes of this field are classified as 'high-energy' – *c.f.* the discussion in §2.2.1 – all five of these graphs contribute to both S_W and $\Gamma_{\rm le}$. In order of magnitude, each contributes to the effective interaction an amount, as follows

$$a_{\text{graph(a)}} \propto \left(-\frac{1}{4v^2}\right)^2 L_1 = \frac{L_1}{16\,v^4}, \tag{2.62}$$

$$a_{\text{graph(b)}} \propto \left(-\frac{1}{4v^2}\right)\left(-\frac{\lambda v}{2\sqrt{2}}\right)\left(-\frac{1}{\sqrt{2}\,v}\right)\frac{L_1}{m_R^2} = -\frac{L_1}{16\,v^4}, \tag{2.63}$$

$$a_{\text{graph(c)}} \propto \left(-\frac{\lambda v}{2\sqrt{2}}\right)^2\left(-\frac{1}{\sqrt{2}\,v}\right)^2\frac{L_1}{m_R^4} = \frac{L_1}{16\,v^4}, \tag{2.64}$$

$$a_{\text{graph(d)}} \propto \left(-\frac{\lambda}{16}\right)\left(-\frac{1}{\sqrt{2}\,v}\right)^2\frac{L_2}{m_R^4} = -\frac{L_2}{32\lambda v^6}, \tag{2.65}$$

and

$$a_{\text{graph(e)}} \propto \left(-\frac{\lambda v}{2\sqrt{2}}\right)^2\left(-\frac{1}{\sqrt{2}\,v}\right)^2\frac{L_2}{m_R^6} = \frac{L_2}{16\lambda v^6}, \tag{2.66}$$

where the powers of $1/m_R^2$ come from the internal lines that do not appear within a loop, since these are evaluated at momenta much smaller than m_R. The contribution of the loop integrals themselves are of order

$$L_1 = \int^{\Omega} \frac{\mathrm{d}^4 p}{(2\pi)^4}\left(\frac{1}{p^2+m_R^2}\right)^2 \propto \frac{1}{16\pi^2}\ln\left(\frac{\Omega}{m_R}\right)$$

$$\text{and}\quad L_2 = \int^{\Omega} \frac{\mathrm{d}^4 p}{(2\pi)^4}\left(\frac{1}{p^2+m_R^2}\right) \propto \frac{\Omega^2}{16\pi^2}. \tag{2.67}$$

Here $\Omega \gg m_R$ is a cutoff that is introduced because the loop in the full theory is UV divergent. This divergence is ultimately dealt with by renormalizing the couplings of the microscopic theory; a point to be returned to in more detail shortly.

For the present purposes – keeping in mind that $m_R^2 = \lambda v^2$ – what is important is that graphs (a) through (c) clearly contribute to the coupling a (in both S_W and Γ_{le}) an amount of order $a_{1\text{-loop}} \propto L_1/v^4 \propto (1/4\pi v^2)^2 \ln\left(\Omega^2/m_R^2\right)$. Graphs (d) and (e) instead contribute an amount of order $a_{1\text{-loop}} \propto (1/4\pi v^2)^2 \left(\Omega^2/m_R^2\right)$. Once the UV divergent function of Ω/m_R is renormalized into an appropriate coupling, the remaining coefficient for each of these loop contributions is suppressed by a factor of $\lambda/16\pi^2$ relative to the tree-level result, $a_{\text{tree}} = 1/(4\lambda v^4)$, in agreement with the discussion surrounding Eq. (2.24). As such, they all contribute to the coefficient a_2 of Eq. (1.16).

The difference between S_W and Γ_{le} arises in graphs (f) through (k), with S_W only receiving contributions where the momentum in the internal ξ propagators is larger than Λ, whereas there is no such a restriction for Γ_{le}. The contribution from graphs (i) through (k) when all loop momenta are large is of order

$$a_{\text{graph(i)}} \propto \left(-\frac{1}{\sqrt{2}\,v}\right)^4 L_1 = \frac{L_1}{4v^4}$$

$$a_{\text{graph(j)}} \propto \left(-\frac{1}{\sqrt{2}\,v}\right)^3\left(-\frac{\lambda v}{2\sqrt{2}}\right)\frac{L_1}{m_R^2} = \frac{L_1}{8v^4} \tag{2.68}$$

$$a_{\text{graph(k)}} \propto \left(-\frac{1}{\sqrt{2}\,v}\right)^2\left(-\frac{1}{4v^2}\right) L_1 = -\frac{L_1}{8v^4},$$

where the new loop integrals are also logarithmically divergent in the UV, and so up to numerical factors are again or order L_1 in size. These contribute to a an amount comparable to the size of graphs (a) through (c). By contrast, graphs (f) through (h) give the results,

$$\begin{aligned} a_{\text{graph(f)}} &\propto \left(-\frac{1}{\sqrt{2}\,v}\right)^4 \frac{L_3}{m_R^4} = \frac{L_3}{4v^4 m_R^4} \\ a_{\text{graph(g)}} &\propto \left(-\frac{1}{\sqrt{2}\,v}\right)^2 \left(-\frac{1}{4v^2}\right) \frac{L_3}{m_R^4} = -\frac{L_3}{8v^4 m_R^4} \\ a_{\text{graph(h)}} &\propto \left(-\frac{1}{\sqrt{2}\,v}\right)^3 \left(-\frac{\lambda v}{2\sqrt{2}}\right) \frac{L_3}{m_R^6} = \frac{L_3}{8v^4 m_R^4}, \end{aligned} \tag{2.69}$$

and so are of order $(1/v^4)(L_3/m_R^4)$, where the ξ loop has the ultraviolet behaviour

$$L_3 = \int^{\Omega} \frac{d^4p}{(2\pi)^4} \left(\frac{p^2}{p^2}\right)^k \propto \frac{\Omega^4}{16\pi^2}, \tag{2.70}$$

where $k = 2$ for graph (f) and $k = 1$ for graphs (g) and (h).

All of these graphs are dominated by large momenta (small wavelengths), which is why they diverge for large Ω. Although a more systematic treatment of these UV divergences (in particular how to treat them using dimensional regularization) is given in Chapter 3, there is a conceptual point to be made concerning their *lower* limit of integration. The point is that for graphs (f) through (k) this lower limit differs when computing Γ_{le} and S_W. For Γ_{le} the contributions to the effective interaction

$$a_{le} \int d^4x\, (\partial_\mu \xi \partial^\mu \xi)^2 \subset \Gamma_{le} \tag{2.71}$$

integrate over all momenta. But for S_W, in the contribution to

$$a_W \int d^4x\, (\partial_\mu \xi \partial^\mu \xi)^2 \subset S_W, \tag{2.72}$$

the integrations exclude momenta smaller than Λ (for which the internal ξ propagators are then 'light' degrees of freedom) that are not integrated out in the path integral representation of S_W.

Take, for instance, graphs (i) through (k) of Fig. 2.4. Since $\Lambda \ll m_R$, the predicted coefficient for Γ_{le} differs from the coefficient in S_W by an amount of order

$$a_{le} - a_W(\Lambda) \simeq \frac{1}{v^4} \int_0^{\Lambda} \frac{d^4p}{(2\pi)^4} \left(\frac{1}{p^2 + m_R^2}\right)^2 \left(\frac{p^2}{p^2}\right)^k \propto \frac{1}{16\pi^2 v^4} \left(\frac{\Lambda^4}{m_R^4}\right). \tag{2.73}$$

The suppression by powers of Λ/m_R ensures this difference is numerically small, although that turns out to be an artefact of this particular example. It is nonetheless conceptually important. In particular, the Λ-dependence of the right-hand side is associated with the Wilsonian coupling $a_W(\Lambda)$ because the scale Λ does not appear at all in the definition of a_{le} (which, after all, is defined in terms of integrations over modes at *all* scales).

How can these different values, $a_{le} \neq a_W$, lead to the same physical predictions for observables? The answer lies in Eq. (2.59), which states that Γ_{le} is obtained from S_W by integrating over the light degrees of freedom, using S_W rather than S as the action. It is this that fills in those parts of Γ_{le} not produced through loops

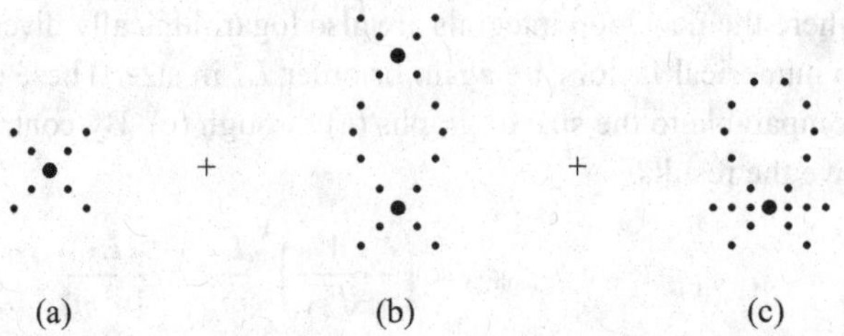

Fig. 2.5 The tree and one-loop graphs that contribute to the $(\partial_\mu \xi \, \partial^\mu \xi)^2$ interaction in the 1LPI action, using Feynman rules built from the Wilson action. All dotted lines represent ξ particles, and the 'crossed' versions of (b) are not drawn explicitly.

involving heavy degrees of freedom. The relevant one-particle-irreducible one-loop graphs for generating the 4-point interaction in $\Gamma_{\rm le}$ using the interactions of S_W are shown in Fig. 2.5.

To see how things work in detail, consider each of the graphs in Fig. 2.5 in turn. Graph (a) just contributes an amount

$$a_{\rm graph(a)} = a_W(\Lambda), \tag{2.74}$$

where a_W is computed up to one loop order (using only short-distance scales in the loop). This is the way that the high-energy parts of graphs (a) through (k) of Fig. 2.4 contribute when using the Wilson action. The dependence on the scale Λ is emphasized, since (as described above) this appears once a_W is computed at the loop level.

Next, consider the contribution of the one-loop graph (b) of Fig. 2.5. This corresponds to the low-energy contribution to $a_{\rm le}$ from graphs (f) and (i) of Fig. 2.4, as can be seen if the χ internal lines in these graphs are shrunk down to a point, since the position-space version of the χ propagator is $G(x,y) \propto m_R^{-2}\delta^4(x-y)$. The vertex appearing at both ends in graph (b) is again the effective 4-point interaction, $\mathfrak{L}_{\rm eff} = a_W\,(\partial_\mu \xi \partial^\mu \xi)^2$, but because it is used in a loop, its effective coupling should only be kept to tree-level accuracy: $a_{\rm tree} = \lambda/(4m_R^4)$. Evaluating graph (b) of Fig. 2.5 gives a contribution to $a_{\rm le}$ of order

$$a_{\rm graph(b)} \propto a_{\rm tree}^2 \int_0^\Lambda \frac{{\rm d}^4 p}{(2\pi)^4}\,\frac{p^4}{p^4} \propto \left(\frac{\lambda}{m_R^4}\right)^2 \frac{\Lambda^4}{16\pi^2} = \frac{1}{16\pi^2 v^4}\left(\frac{\Lambda^4}{m_R^4}\right). \tag{2.75}$$

Schematically, graph (b) of Fig. 2.5 is obtained from the low-frequency part of graphs (f) and (i) of Fig. 2.4 by contracting the heavy internal χ lines down into the effective point-like quartic self-interaction in the Wilson action. Notice that Eq. (2.75) depends on the parameters Λ, m_R and v in precisely the way that is required to capture the low-energy part of graph (i), as estimated in Eq. (2.73). It also captures the low-energy part of graph (f), since in this case the contribution of the lower limit of integration in L_3, Eq. (2.70), is also quartic in Λ.

Similarly, graph (c) of Fig. 2.5, using the tree-level 6-point ξ interaction of S_W as the vertex, captures the low-energy contributions to $a_{\rm le}$ of graphs (g), (h), (j) and (k) of Fig. 2.4. The relevant 6-point coupling, b, arises (in principle) at tree level due to the graphs of Fig. 2.3, and so the result for graph (c) scales with the parameters v, m_R and Λ in precisely the way required,

$$a_{\text{graph(c)}} \propto b_{\text{tree}} \int_0^{\Lambda} \frac{d^4p}{(2\pi)^4} \frac{p^2}{p^2} \propto \left(\frac{1}{v^4 m_R^4}\right) \frac{\Lambda^4}{16\pi^2} = \frac{1}{16\pi^2 v^4} \left(\frac{\Lambda^4}{m_R^4}\right), \tag{2.76}$$

to reproduce the missing low-energy contributions of graphs (g), (h), (j) and (k). When calculated earlier – *c.f.* §2.2.3 – it transpired that the two tree graphs contributing to b cancel, so $b_{\text{tree}} = 0$. This corresponds to a similar cancelation in the low-energy limits of graphs (g), (h), (j) and (k) of Fig. 2.4.

When the dust settles, graphs (b) and (c) of Fig. 2.5 fill in those low-frequency parts of a_{le} that are missing from the contribution, a_W, of graph (a). The final result obtained for a_{le} using the Wilson action in Fig. 2.5 agrees with the one obtained directly from the full theory using Fig. 2.4.

In this particular example the difference $a_{\text{le}} - a_W$ vanishes as $\Lambda \to 0$, because the quantity being computed is largely insensitive to long-wavelength modes. As a result, the explicit Λ-dependence appearing in a_W cancels with the Λ-dependence implicit as cutoffs to the loop momenta in graphs (b) and (c) of Fig. 2.5. As shall be seen, in less simple examples loop contributions from long-wavelength modes can be important, and this cancellation of Λ in physical quantities is more subtle (involving renormalizations of low-energy effective couplings).

2.4 Dimensional Analysis and Scaling ◇

Although the discussion and examples described to this point contain the basic definitions of the Wilson action, many details remain to be filled in about its systematic use. Before developing the tools required for more systematic heavy lifting in the next chapter, this section first highlights some useful general properties of effective interactions, illustrated along the way using the toy model of Chapter 1.

2.4.1 Dimensional Analysis

Dimensional analysis plays an important role in what follows, since it can be used to track the appearance of large mass scales in physical predictions. This section reviews the dimensions of the various ingredients from which the Wilson action is built.

To this end, imagine writing down the Wilson action, $S_W = \int d^4x \, \mathfrak{L}_W$, for some theory, describing physics below some high-energy mass scale: $E < M$. Suppose also this low-energy action comes as a functional of some field, ϕ, and its derivatives. As discussed in previous chapters, once organized into powers of the inverse of the heavy scale, $1/M$, the effective interactions in $\mathfrak{L}_W$ must be local. These conditions require the action to have the general form

$$S_W[\phi] = \int d^4x \, \mathfrak{L}_W, \tag{2.77}$$

with

$$\mathfrak{L}_W = \sum_n c_n \, O_n(\phi, \partial\phi, \cdots), \tag{2.78}$$

a sum of powers of ϕ and its derivatives all evaluated at the same point, and (for relativistic systems) built to transform like a Lorentz scalar. If the low-energy theory is unitary then $\mathfrak{L}_W$ should also be real. The goal is to use dimensional analysis to identify the power of M appearing in each effective coupling, c_n.

This book uses 'fundamental' units for which[8] $\hbar = c = 1$, and so the (engineering) dimension of any quantity can be regarded as a power of energy or mass (see Appendix A for conversions between these and more conventional units). In these units the action, S_W, itself is dimensionless – *i.e.* has dimension (energy)^0 – or, more precisely, $S_W/\hbar$ is dimensionless. Similarly, time and space coordinates, t and $\mathbf{x}$, have dimension (energy)$^{-1}$, while derivatives like ∂_μ have dimension of energy. It is common to use the notation $[A] = p$ as a short form for the statement 'quantity A has dimension (energy)^p in units where $\hbar = c = 1$', and in this notation $[S_W] = 0$, $[x^\mu] = -1$ and $[\partial_\mu] = 1$.

Because the action is related to the lagrangian density by Eq. (2.77), in four spacetime dimensions it follows that $\mathfrak{L}_W$ has dimension (energy)^4 – *i.e.* $[\mathfrak{L}_W] = 4$ – because the measure, d^4x, satisfies $\left[\mathrm{d}^4x\right] = -4$. If M is the only relevant mass scale in the problem and if a particular interaction, O_n, has dimension $[O_n] = \Delta_n$, then because $[c_n O_n] = 4$ it follows that $[c_n] = 4 - \Delta_n$, and so one expects

$$[O_n] = \Delta_n \quad \Rightarrow \quad c_n = \frac{a_n}{M^{p_n}} \quad \text{with} \quad p_n = \Delta_n - 4, \tag{2.79}$$

where a_n is dimensionless. To the extent that it is M that sets the scale of c_n in this way (and much of the next chapter is devoted to showing that the low-energy theory often can be set up so that it is), higher-dimensional interactions in S_W should be expected to be more suppressed at low energies by higher powers of M.

Further progress requires a way to compute the dimension, Δ_n, of a given operator, O_n. For weakly interacting systems dimensions can be computed in perturbation theory. To see how this works, consider a real scalar field, ϕ, and suppose the regime of interest is one where it is relativistic and very weakly interacting. This means the action $S_W = S_0 + S_{\rm int}$ is dominated by its kinetic term

$$S_0 = -\frac{1}{2}\int \mathrm{d}^4x \; \partial_\mu\phi \, \partial^\mu\phi, \tag{2.80}$$

while all remaining terms, lumped together into $S_{\rm int} = \int \mathrm{d}^4x \; \mathfrak{L}_{\rm int}$ with

$$\mathfrak{L}_{\rm int} = -\frac{m^2}{2}\phi^2 + c_{4,0}\,\phi^4 + \frac{c_{6,0}}{\Lambda^2}\phi^6 + \frac{c_{4,2}}{\Lambda^2}\phi^2\,\partial_\mu\phi\,\partial^\mu\phi + \frac{c_{4,4}}{\Lambda^4}(\partial_\mu\phi\,\partial^\mu\phi)^2 + \cdots, \tag{2.81}$$

are assumed to be small. In this expression a symmetry of the form $\phi \to -\phi$ is imposed (for simplicity) so that only terms involving an even power of ϕ need to be considered. Furthermore, appropriate powers of the cutoff, Λ, for the Wilsonian EFT

[8] When temperatures are considered, units are also chosen with Boltzmann's constant satisfying $k_B = 1$, so temperature can also be measured in units of energy.

are included explicitly in the coefficients for each effective interaction for reasons now to be explained.

Any effective coupling premultiplying S_0 is imagined to be removed by appropriately rescaling ϕ, with the choice of $\frac{1}{2}$ in Eq. (2.80) called canonical normalization. (The reasons for using this normalization are elaborated below, and in Appendix C.3.) Given this normalization, the dimension, $[\phi]$, of the (scalar) field ϕ is then determined by demanding that $[\mathfrak{L}_0] = 4$, and so

$$4 = [\partial_\mu \phi \, \partial^\mu \phi] = 2 + 2[\phi], \tag{2.82}$$

which implies that $[\phi] = 1$ (or ϕ has dimensions of energy). With this choice, an identical argument shows $\left[m^2\right] = 2$, and so m also has dimensions of energy (as expected, since m is interpreted as the ϕ-particle's mass).

A similar story applies to all of the other terms in $S_{\rm int}$, and shows that the factors of Λ in (2.81) are extracted so that the remaining effective couplings are dimensionless: $[c_{n,d}] = 0$. For later purposes notice that a term in $S_{\rm int}$ involving n powers of ϕ and d derivatives comes premultiplied by Λ^p with $p = 4 - n - d$, and so the infinite number of local interactions that are not written explicitly in (2.81) have effective couplings with only negative powers of Λ.

The goal is to identify the domain of validity of the assumed perturbative hierarchy between S_0 and $S_{\rm int}$. The next few paragraphs argue that perturbative arguments are appropriate when the dimensionless couplings are assumed to be small: $|c_{n,d}| \ll 1$, following arguments made in [19]. To this end, consider evaluating $S_w[\phi_k]$ at a wave-packet configuration $\phi_k(x) = f_k(x)\, e^{ikx}$, where $f_k(x)$ is a smooth envelope that is order unity for a spacetime region of linear size $2\pi/k$ in all four rectangular spacetime directions. For such a configuration spacetime derivatives are of order $\partial_\mu \phi_k \sim k_\mu \phi_k$ and the spacetime volume integral is of order $\int \mathrm{d}^4x \sim (2\pi/k)^4$, and so

$$\begin{aligned} S_w[\phi_k] &\sim \left(\frac{2\pi}{k}\right)^4 \left[k^2\,\phi_k^2 + m^2\phi_k^2 + c_{4,0}\,\phi_k^4 + \frac{c_{6,0}}{\Lambda^2}\,\phi_k^6 \right. \\ &\qquad \left. + \, c_{4,2}\left(\frac{k^2}{\Lambda^2}\right)\phi_k^4 + c_{4,4}\left(\frac{k^4}{\Lambda^4}\right)\phi_k^4 + \cdots\right] \\ &\sim (2\pi)^4\left[\varphi_k^2 + \frac{m^2}{k^2}\,\varphi_k^2 + c_{4,0}\,\varphi_k^4 + c_{6,0}\left(\frac{k^2}{\Lambda^2}\right)\varphi_k^6 \right. \\ &\qquad \left. + c_{4,2}\left(\frac{k^2}{\Lambda^2}\right)\varphi_k^4 + c_{4,4}\left(\frac{k^4}{\Lambda^4}\right)\varphi_k^4 + \cdots\right], \end{aligned} \tag{2.83}$$

where $\varphi_k := \phi_k/k$ is a new dimensionless variable. What is key about φ_k is that the path integral over φ_k would be dominated by $\varphi_k \lesssim O(1)$ in the absence of interactions[9] (*i.e.* when $S_w = S_0$). This conclusion that dominant configurations for φ_k are order unity is contingent on the coefficient of $(\partial\phi)^2$ in S_0 being order unity due to the choice of canonical normalization.

Perturbation theory in $S_{\rm int}$ requires $|S_{\rm int}| \ll |S_0|$ throughout the regime from which the path integral receives significant contributions, which from the above

[9] This is clearest if the problem is Wick-rotated to euclidean signature by going to imaginary time, so that $e^{iS} \to e^{-S}$. This estimate also ignores factors of 2π, though their inclusion somewhat broadens the domain of validity of perturbative methods.

considerations is the regime $\varphi_k \lesssim O(1)$. Consider first choosing k as large as possible: near the UV cutoff $|k^2| \sim \Lambda^2$. Since $\Lambda \gg m$, it follows that $|k^2| \gg m^2$, and so perturbation theory in this regime requires $|c_{n,d}| \ll 1$.

How does this conclusion change if k is now dialled down to smaller values? Since all interactions except the ϕ^2 and ϕ^4 interactions come pre-multiplied by positive powers of k^2/Λ^2, they become less and less important for smaller k. Interactions like this, which are less important at lower energies, are called *irrelevant*. Irrelevant interactions are also often called 'non-renormalizable'.[10]

By contrast, the ϕ^4 interaction is k-independent and so has strength controlled by $c_{4,0}$ for all k. Interactions like this, whose strength does not vary with k, are called *marginal*.

Finally, the mass (or ϕ^2) term is the only interaction that grows in importance for smaller k, the defining property of a *relevant* interaction. Once $|k^2| \lesssim m^2$, the mass term competes with S_0 and so changes the nature of the dominant path integral configurations. This nonrelativistic regime is, of course, important to many applications, and so is returned to in some detail as the topic of Part III. Relevant interactions are sometimes also called 'super-renormalizable', while marginal and relevant interactions taken together are called 'renormalizable'.

A similar story goes through for fields representing other spins at weak coupling. For instance, a field, ψ, describing a free relativistic spin-half particle with lagrangian density

$$S_{1/2} = -\int \mathrm{d}^4x\, \overline{\psi}\, \partial\!\!\!/\, \psi, \tag{2.84}$$

with $\partial\!\!\!/ = \gamma^\mu \partial_\mu$ for dimensionless Dirac matrices, γ^μ (see Appendices A.2.3 and C.3.2), must have dimension $[\psi] = 3/2$. The kinetic term for an electromagnetic potential, A_μ, is

$$S_1 = -\frac{1}{4}\int \mathrm{d}^4x\, F_{\mu\nu}F^{\mu\nu} = -\frac{1}{2}\int \mathrm{d}^4x\, \partial_\mu A_\nu(\partial^\mu A^\nu - \partial^\nu A^\mu), \tag{2.85}$$

and so the potential has dimension $[A_\mu] = 1$, while the field-strength satisfies $[F_{\mu\nu}] = 2$.

It is an important fact that all of the weakly interacting fields most commonly dealt with – such as ϕ, ψ and A_μ, as well as the derivative ∂_μ – have positive dimension, so more complicated interactions involving more powers of fields and/or derivatives always have higher and higher dimension. The corresponding effective couplings must therefore be proportional to more and more powers of $1/\Lambda$ (and so be less and less important at low energies). This is what ensures that all but a handful of effective interactions are irrelevant at low energies, in the sense defined above. Precisely how irrelevant they are for any given k depends on the power of k^2/Λ^2 involved, so it makes sense to organize any list of potential interactions in order of increasing operator dimension, since the leading terms on the list are likely to be the most important at low energies.

10 The recognition that it is useful to classify interactions according to their dimension came early [6], as did the connection to renormalizable and non-renormalizable interactions [7].

From this point of view it is clear that the limited number of renormalizable (marginal and relevant) interactions with dimension $[O_n] \leq 4$ are special, since their importance is not diminished (and can be enhanced) at low energies.

For concreteness' sake the introductory discussion given above is phrased in perturbation theory, so it is worth mentioning in passing that this is not, in principle, necessary. That is, although EFT methods always exploit expansions in ratios of energy scales (like E/m_R for the toy model), it is *not* a requirement of principle that the dimensionless couplings of the underlying UV theory (like λ for the toy model) are perturbatively small.[11] Although it goes beyond the scope of this chapter to show in detail, strong underlying couplings can change some of the detailed statements used above, such as by changing the dimension of the field to be $[\phi] = 1 + \delta$, with $|\delta| \to 0$ as these couplings are taken to zero. (Examples along these lines where δ is perturbatively small are considered in later sections.) Differences like δ are called 'anomalous dimensions' for the quantities involved. What counts in the dimensional arguments to follow is that the full scaling dimensions (including these anomalous contributions) are used, rather than the lowest-order 'naive' scaling dimension.

Example: The Toy Model

As applied to the toy model, because the kinetic terms for the two fields ξ and χ have the form of Eq. (2.80), the dimensions of both are $[\chi] = [\xi] = 1$. Using this with the full classical action, Eqs. (1.24) and (1.25), shows that $[\lambda] = 0$ and $[v] = [m_R] = 1$, as expected.

Applied to the Wilson action, the effective coupling a appearing in the interaction $\mathfrak{L}_W \supset a\,(\partial_\mu \xi \partial^\mu \xi)^2$ must have dimension (energy)$^{-4}$, consistent with its computed tree-level value $a_{\rm tree} = \lambda/4m_R^4$. This shows that at leading order it is explicitly the heavy scale m_R that plays the role of the dimensional parameter of the general discussion. Powers of $\Lambda \ll m_R$ also arise once loops are included, and subsequent sections are devoted to identifying which scale is important in any particular application.

2.4.2 Scaling

It is worth rephrasing the above discussion more formally in terms of a scaling transformation. This is useful for several reasons: because it sets up the use of renormalization-group methods; and because it provides a framework that is more easy to generalize in more complicated settings, such as in the nonrelativistic limit considered in Part III.

To this end, consider again the scalar-field Wilson action of Eqs. (2.80) and (2.81):

$$S_W[\phi(x); m, c_{4,0}, c_{6,0}, \cdots] = S_0[\phi(x)] + S_{\rm int}[\phi(x); m, c_{4,0}, \cdots], \qquad (2.86)$$

[11] In an unfortunate use of language the breakdown of the low-energy (*e.g.* E/m_R) expansion has in some quarters come to be called 'strong coupling'. This is misleading because it can happen that the physics appropriate to these energies is weakly coupled, in the sense that it involves dimensionless couplings that are small. For this reason in this book 'strong coupling' never means 'breakdown of the low-energy limit', and it is reserved for situations where underlying dimensionless couplings (like λ in the toy model) are not small.

where the notation explicitly highlights the dependence on the effective couplings as well as on the field ϕ. Now perform the scale transformation, $x^\mu \to sx^\mu$ and $\phi(x) \to \phi(sx)$, where s is a real parameter. For a configuration like $\phi_k(x) \propto e^{ikx}$ this becomes $\phi_k(sx) \propto e^{iskx} \propto \phi_{sk}(x)$ and so taking $s \to 0$ corresponds to taking the infrared limit where $k \to 0$.

Inserting these definitions into S_0 gives

$$S_0[\phi(sx)] = -\frac{1}{2}\int \mathrm{d}^4x\; \partial_\mu\phi(sx)\,\partial^\mu\phi(sx)$$
$$= -\frac{1}{2}\int \frac{\mathrm{d}^4x'}{s^4}\,\left[s^2\partial_{\mu'}\phi(x')\partial^{\mu'}\phi(x')\right], \tag{2.87}$$

in which the spacetime integration variable is changed from x^μ to $x'^\mu = sx^\mu$. This shows that S_0 remains unchanged if the scalar field variable is also rescaled according to $\phi(x) \to \phi_s(x) := \phi(x)/s$. Requiring $S_0[\phi(sx)] = S_0[\phi_s(x)]$ is natural in the weak-coupling limit, since this keeps fixed the configurations that dominate in the path integral over ϕ and ϕ_s.

With these choices, the effects of rescaling on the interaction terms can be read off, giving

$$S_{\rm int}[\phi(sx); m, c_{4,0}, c_{6,0}, c_{4,2}, \cdots] = \int \mathrm{d}^4x'\left[-\frac{m^2}{2s^2}\,\phi_s^2 + c_{4,0}\phi_s^4 + \frac{s^2 c_{6,0}}{\Lambda^2}\,\phi_s^6 + \frac{s^2 c_{4,2}}{\Lambda^2}\,\phi_s^2(\partial_{\mu'}\phi_s\partial^{\mu'}\phi_s) + \cdots\right]$$
$$= S_{\rm int}\left[\phi_s(x); \frac{m}{s}, c_{4,0}, s^2c_{6,0}, s^2c_{4,2}\,\cdots\right]. \tag{2.88}$$

This shows that changes of scale can be compensated by appropriately rescaling all effective couplings. For instance, for an interaction $S_n[\phi; a_n] = \int \mathrm{d}^4x\; a_n O_n[\phi] \in S_{\rm int}$, where O_n has engineering dimension $[O_n] = \Delta_n$, the required scaling is

$$S_n[\phi(sx); a_n] = S_n[\phi_s(x); s^{p_n}a_n], \tag{2.89}$$

where $p_n = -[a_n] = \Delta_n - 4 =$ (so that $a_n = c_n/\Lambda^{p_n}$ for dimensionless c_n).

Rescalings can be regarded as motions within the space of coupling constants. This provides an alternative way to define relevant, marginal and irrelevant interactions. Since low energies correspond to $s \to 0$, an effective interaction is irrelevant if $p_n > 0$, it is marginal if $p_n = 0$ and it is relevant if $p_n < 0$. This definition clearly agrees with the one presented earlier.

2.5 Redundant Interactions ◇

It is generally useful to have in mind what are the most general possible kinds of effective interactions that can arise in a Wilson action at any given dimension, and it is tempting to think that this means simply listing all possible combinations of local interactions involving the given fields and their derivatives. In practice, such a list over-counts the number of possible interactions because there is considerable freedom to rewrite the effective action in superficially different, but equivalent, ways.

Because of this freedom, some combinations of interactions can turn out not to influence observables at all (or only do so in special situations), and it is important to identify these to avoid over-counting the couplings that are possible. Ignorable interactions like this that have no physical effects are called *redundant* interactions, and this section describes two generic kinds of redundancy that commonly arise.

Total Derivatives

The first category of often-ignorable interactions are total derivatives – such as, for the generic low-energy field ϕ of the last section

$$S_{\rm int} = -g \int d^4x\, \partial_\mu \left(\phi^2 \partial^\mu \phi\right). \tag{2.90}$$

Stokes' theorem allows this kind of interaction to be written as a function only of boundary data

$$\int_M d^4x\, \partial_\mu \left(\phi^2 \partial^\mu \phi\right) = \int_{\partial M} d^3x\, n_\mu \left(\phi^2 \partial^\mu \phi\right), \tag{2.91}$$

where n_μ denotes the outward-directed unit normal to the boundary ∂M of the initial integration region M. To the extent that the physics of interest does not depend on this boundary data (such as if there are no boundaries, or if spatial infinity is the 'boundary' and all fields fall off sufficiently quickly at infinity), such total derivatives can be dropped.

Of course, no mistakes are actually made by keeping redundant interactions in a calculation; one just works unnecessarily hard. This is because couplings like g in the example above do not in any case appear in physical observables. To see how this can happen in detail, consider the example of the momentum–space Feynman rule computed for the 3-point vertex described by Eq. (2.90):

$$g(p_1 + p_2 + p_3) \cdot p_3 \left[i(2\pi)^4 \delta^4(p_1 + p_2 + p_3) \right]. \tag{2.92}$$

Here, $p_i \cdot p_j = \eta_{\mu\nu} p_i^\mu p_j^\nu$ where p_i^μ denote the inward-pointing 4-momenta for each of the lines attached to the vertex. Due to the presence of the energy–momentum conserving delta function, this has the form $x\,\delta(x) = 0$, and so identically vanishes. Consequently, this kind of interaction cannot contribute to any order in a perturbative expansion organized in terms of Feynman graphs.

The neglect of total-derivative terms must be re-examined in situations with boundaries or where the asymptotic behaviour of fields cannot be ignored. This can happen either when there really are physical boundaries or when there are fields that are sensitive to nontrivial topology.[12] When boundaries are present the above discussion continues to apply provided that the appropriate boundary interactions are also included (see §5 and §7.4 and §15.3 for examples along these lines).

[12] Topology enters if there are terms in the lagrangian that are locally total derivatives, but cannot be globally written this way throughout all of space.

Field Redefinitions

A second type of ignorable interaction is one that can be removed by performing a local field redefinition. Since physical quantities cannot depend on the particular choice of variables used by specific physicists for their description,[13] anything that can be removed by a nonsingular change of variables cannot contribute to any physical observables.

But how does one decide if a particular interaction can be removed in this way? It turns out there is a very simple criterion that works for any situation where the action is given as a series in a small quantity, ϵ:

$$S[\phi] = S_0[\phi] + \epsilon\, S_1[\phi] + \epsilon^2\, S_2[\phi] + \cdots . \tag{2.93}$$

This form always applies for the Wilson action in particular, where the corresponding expansion would be the low-energy approximation.

Imagine performing a generic infinitesimal field redefinition of the form

$$\delta\phi = \epsilon\, f_1[\phi] + \epsilon^2\, f_2[\phi] + \cdots , \tag{2.94}$$

for some arbitrary local functions, $f_i = f_i(\phi, \partial\phi, \cdots)$, of the fields and their derivatives. The change in Eq. (2.93) is

$$\begin{aligned}\delta S &= \int d^4x \left\{ \frac{\delta S_0}{\delta\phi(x)} + \epsilon\,\frac{\delta S_1}{\delta\phi(x)} + \epsilon^2\,\frac{\delta S_2}{\delta\phi(x)} + \cdots \right\} \delta\phi \\ &= \int d^4x \left\{ \epsilon\left[\frac{\delta S_0}{\delta\phi(x)}\, f_1\right] + \epsilon^2 \left[\frac{\delta S_0}{\delta\phi(x)}\, f_2 + \frac{\delta S_1}{\delta\phi(x)}\, f_1\right] + \cdots \right\}.\end{aligned} \tag{2.95}$$

This shows that the function f_1 can be used to remove any interaction in S_1 that is proportional to $\delta S_0/\delta\phi$ – and so vanishes when its lowest-order equations of motion are used[14] – up to terms that are $O\left(\epsilon^2\right)$. For instance, a term in S_i of the form

$$S_i \supset \int d^4x\, \frac{\delta S_0}{\delta\phi(x)}\, B[\phi], \tag{2.96}$$

(for some local function $B[\phi]$) is removed using the choice $f_i = -B$. That is, the quantities f_1 through f_{n+1} can be used to remove any interaction in $S - S_0$ that vanishes when $\delta S_0/\delta\phi = 0$, order-by-order in ϵ. Notice also that (2.95) also shows that the redefinition that removes terms from S_i can do so at the expense of also introducing new terms into S_j for $j > i$.

Example: The Toy Model

As usual, it is useful to make things concrete with an explicit example, for which the toy model of Chapter 1 is again pressed into service. To apply these ideas to the toy model, recall what has been found for its Wilson action so far. The calculations performed in the previous sections show it to have the form

[13] This is a theorem for scattering amplitudes [8], but it also applies to more general observables.

[14] This observation seems to have been general knowledge back into the mists of time, but the earliest explicit mention of it in the literature I have found is [20] (see also footnote 9 of [2]).

$$S_W[\xi] = -\int d^4x \left[\frac{1}{2}\partial_\mu\xi\partial^\mu\xi - a(\partial_\mu\xi\partial^\mu\xi)^2 - a'(\partial_\mu\xi\partial^\mu\xi)\Box(\partial_\mu\xi\partial^\mu\xi) - b(\partial_\mu\xi\partial^\mu\xi)^3 - b'(\partial_\mu\xi\partial^\mu\xi)\Box(\partial_\mu\xi\partial^\mu\xi)^3 + \cdots\right], \tag{2.97}$$

where

$$a = \frac{1}{m_R^4}\left[\frac{\lambda}{4} + O(\lambda^2)\right], \quad a' = \frac{1}{m_R^6}\left[\frac{\lambda}{4} + O(\lambda^2)\right], \quad b = \frac{1}{m_R^8}\left[0 + O(\lambda^3)\right] \tag{2.98}$$

and so on. The question is: are these the most general kinds of interactions possible? In particular, are there terms suppressed by only two powers of $1/m_R$? If not, why not?

Of course, symmetries restrict the form of S_W, and for the toy model symmetry under the shift $\xi \to \xi + \sqrt{2}\, v\, \omega$ requires ξ always to appear in S_W differentiated, so all interactions must involve at least as many derivatives as powers of ξ. Furthermore, to be a Lorentz scalar it must involve an even number of derivatives, so that all Lorentz indices can be contracted. But these conditions allow interactions that do not appear in Eq. (2.97). For instance, they allow the following effective interactions with dimension (energy)^6,

$$\begin{aligned}\mathfrak{L}_6 &= \frac{a_1}{m_R^2}(\partial_\mu\xi\Box\partial^\mu\xi) + \frac{a_2}{m_R^2}(\partial_\mu\partial_\nu\xi\,\partial^\mu\partial^\nu\xi) \\ &= \frac{(a_1 - a_2)}{m_R^2}(\partial_\mu\xi\Box\partial^\mu\xi) + \frac{a_2}{m_R^2}\partial_\nu(\partial_\mu\xi\,\partial^\mu\partial^\nu\xi),\end{aligned} \tag{2.99}$$

where (given the explicit dimensional factor of m_R^{-2}) a_1 and a_2 are dimensionless effective couplings.

The point is that both of these interactions are redundant, in the sense outlined above. The second line shows that one combination can be regarded as a total derivative, and so it is redundant to the extent that boundaries (or topology) do not play an important role in the physics of interest. The remaining term, involving the combination $\partial_\mu\xi\Box\partial^\mu\xi$, vanishes once evaluated using the equations of motion, $\Box\xi = 0$, for the lowest-order action. It can therefore be removed using the field redefinition

$$\xi \to \xi + \frac{a_2 - a_1}{m_R^2}\Box\xi, \tag{2.100}$$

since in this case

$$-\frac{1}{2}\partial_\mu\xi\partial^\mu\xi \to -\frac{1}{2}\partial_\mu\xi\partial^\mu\xi + \frac{a_2 - a_1}{m_R^2}(\partial_\mu\xi\Box\partial^\mu\xi), \tag{2.101}$$

up to terms of order $1/m_R^4$. This shows that (in the absence of boundaries) the first low-energy effects of virtual heavy particles arise at order $1/m_R^4$ rather than $1/m_R^2$.

What about interactions with dimension (energy)^8: is $(\partial_\mu\xi\,\partial^\mu\xi)^2$ the only allowed dimension-8 interaction? Since total derivatives are dropped, integration by parts can be used freely to simplify any candidate interactions. The most general possible Lorentz-scalar interactions invariant under constant shifts of ξ then are

$$\mathfrak{L}_8 = -a(\partial_\mu\xi\partial^\mu\xi)^2 - \frac{a_3}{m_R^4}(\partial_\mu\xi\Box^2\partial^\mu\xi), \tag{2.102}$$

where a_3 is a new dimensionless coupling and the freedom to integrate by parts is used (for the terms quadratic in ξ) to ensure all of the derivatives but one act on only one of the fields. The second term in (2.102) can be removed using the field redefinition

$$\xi \to \xi + \frac{a_3}{m_R^4} \Box^2 \xi, \tag{2.103}$$

without changing the coefficient a (or coefficients of any lower-dimension interactions), showing that a captures all of the effects that can arise at order $1/m_R^4$.

2.6 Summary

This chapter lays one of the cornerstones for the rest of the book; laying out how effective lagrangians fit into the broader context of generating functionals and the quantum (1PI) action.[15] By doing so, it provides a constructive framework for defining and explicitly building effective actions for a broad class of physical systems.

The 1PI action is a useful starting point for this purpose because it already plays a central role in quantum field theory. It does so partly because it is related to the full correlation functions and the energetics of field expectation values in the same way that the classical action is related to the classical correlation functions and the energetics that fixes the values of classical background fields. The low-energy 1LPI action is the natural generalization of the 1PI action because it is constructed in precisely the same way, but with the proviso that it only samples slowly varying field configurations. As such, it contains all the information needed to construct any observable that involves only low-energy degrees of freedom.

In this chapter, the Wilson action, S_W, emerges as the minimal object for capturing the implications of high-energy degrees of freedom for the low-energy theory. The Wilson action is related to the 1LPI action in the low-energy theory in precisely the same way that the classical action is related to the 1PI generator in the full theory. Because S_W is obtained by integrating out only high-energy states, its interactions efficiently encode their low-energy implications. And because knowledge of S_W allows the calculation of the 1LPI action it contains all of the information required to compute any low-energy observable.

The chapter concludes with a few tools that will prove useful in later chapters when computing and using the Wilson action. The first tool is simply dimensional analysis, which classifies effective interactions based on their operator dimension (in powers of energy). More and more interactions exist with larger operator dimension, but it is the relatively few lower-dimension interactions in this classification that are more important at low energies. This chapter also describes the related renormalization-group scaling satisfied by the effective couplings. These express how the effective couplings differentially adjust as more and more modes are integrated out, lowering the energy scale Λ that differentiates low energies from high. (The next chapter also has more to say about this dimensional scaling and its utility for identifying which interactions are important at low energies.)

The second tool described in this chapter identifies classes of effective interactions that are redundant in the sense that they do not contribute at all to physical processes. They do not contribute for one of two

[15] In some of the earlier literature the quantum action is also called the effective action, unlike the modern usage, where effective action usually means the Wilson action.

reasons: either they are total derivatives and so are only sensitive to physics that depends in detail on the information at the system's boundaries; or because a change of variables exists that allows them to be completely removed.

Exercises

Exercise 2.1 Prove Eq. (2.6) starting from Eqs. (2.2) and (2.4).

Exercise 2.2 Draw all possible two-loop vacuum diagrams that contribute to $Z[J]$ in a theory involving both cubic and quartic interactions (such as a scalar potential $V(\phi) = g\phi^3 + \lambda\phi^4$ for scalar-field self-interactions). Which of these diagrams contribute to $W[J]$ and to $\Gamma[\varphi]$?

Exercise 2.3 For a scalar field self-interacting through the potential $V = g\phi^3 + \lambda\phi^4$ express Eq. (2.15) as a sum of Feynman graphs with two external lines. Draw all graphs that contribute out to two-loop order. Show how the disconnected graphs cancel in the result.

Exercise 2.4 Prove that the graphical expansion of $W[J]$, defined by $Z[J] = \exp\{iW[J]\}$, is obtained by simply omitting any disconnected graphs that contribute to $Z[J]$. Do so by showing that the exponential of the sum of all connected graphs reproduces all of the combinatorial factors in the sum over all (connected and disconnected) graphs. For this argument it is not necessary to assume that only cubic or quartic interactions arise in $S_{\rm int}$.

Exercise 2.5 Consider a single scalar field, φ, self-interacting through a scalar potential $U(\varphi)$. Evaluate Eq. (2.23) in one-loop approximation for φ specialized to a constant spacetime-independent configuration. To do so, use the identity $\ln\det\Delta = \text{Tr}\ln\Delta$ and work in momentum space, for which $\Delta(p,p') = (p^2 + m^2 - i\epsilon)\,\delta^4(p-p')$, where $m^2 := U'' := \partial^2 U/\partial\varphi^2$. Evaluate the trace explicitly and Wick rotate to Euclidean signature ($p^0 = ip_E^4$) to derive the following expression

$$V_{\rm q}(\varphi) = U(\varphi) + \frac{1}{2}\int \frac{d^4 p_E}{(2\pi)^4}\,\ln(p_E^2 + m^2) = U_\infty(\varphi) + \frac{1}{64\pi^2}\, m^4 \ln\left(\frac{m^2}{\mu^2}\right),$$

for the quantum effective potential. Regulate the UV divergences using dimensional regularization (for which μ is the arbitrary scale: see Appendix A.2.4), and show that $U_\infty(\varphi) = U(\varphi) + A + B\,m^2(\varphi) + Cm^4(\varphi)$, where A, B and C are divergent constants in four spacetime dimensions. Show that if $U(\varphi) = U_0 + U_2\varphi^2 + U_4\varphi^4$ is quartic (and so renormalizable) then all divergences can be absorbed into the constants U_0, U_2 and U_4.

Exercise 2.6 Prove that the quantum effective potential is always convex [16, 21] when constructed about a stable vacuum. That is, show that for $0 \le s \le 1$

$$V_{\rm q}[s\varphi_1 + (1-s)\varphi_2] \le sV_{\rm q}(\varphi_1) + (1-s)V_{\rm q}(\varphi_2).$$

Exercise 2.7 Suppose the action $S[\phi]$ for a field theory is invariant under a symmetry transformation of the form $\delta\phi^a = \omega\,\zeta^a(\phi)$, where $\zeta^a(\phi)$ is a possibly nonlinear function and ω is an infinitesimal symmetry parameter. Show that the 1PI generator, $\Gamma[\varphi]$, is invariant under the symmetry $\delta\varphi^a = \omega\,\langle\zeta^a\rangle_J$, where the matrix element is taken in the adiabatic vacuum in the presence of the current $J_a(\varphi)$ defined by Eq. (2.16). In the special case of a linear transformation, with $\zeta^a(\phi) = M^a{}_b\,\phi^b$ the condition $\langle\phi^a\rangle_J = \varphi^a$ implies that both $\Gamma[\varphi]$ and $S[\phi]$ share a symmetry with the same functional form.

The invariance condition $\delta\Gamma = 0$ for this transformation can be expressed as

$$\int \mathrm{d}^4x\,\langle\zeta^a(x)\rangle_J\,\frac{\delta\Gamma}{\delta\varphi^a(x)} = 0.$$

Since this is true for all $\varphi^a(x)$ repeated functional differentiation leads to a sequence of relations – called Taylor–Slavnov identities [22, 23] – amongst the 1PI correlation functions obtained by differentiating $\Gamma[\varphi]$.

Exercise 2.8 For the toy model of §1 draw all tree-level (no loops) Feynman graphs that can contribute to the effective interaction $\mathfrak{L}_W \supset c\,(\partial_\mu\,\xi\,\partial^\mu\,\xi)^4$ within the Wilson action. Evaluate these graphs and compute the effective coupling c at tree level.

Exercise 2.9 Construct the most general possible renormalizable relativistic interactions for a single real scalar field ϕ in $D = 4$ spacetime dimensions. Repeat this exercise for $D = 6$ spacetime dimensions. For $D = 4$ find the most general possible renormalizable relativistic interactions for a real scalar field coupled to a spin-half Dirac field ψ.

Exercise 2.10 For a real scalar field, ξ, subject to a shift symmetry, $\xi \to \xi + \text{constant}$, every appearance of ξ in the Wilson action must be differentiated at least once. Show that the most general effective interactions possible for such a field involving six or fewer derivatives is

$$\begin{aligned}\mathfrak{L}_W = &-\frac{1}{2}\,(\partial_\mu\xi\,\partial^\mu\xi) + a\,(\partial_\mu\xi\,\partial^\mu\xi)^2 + b\,(\partial_\mu\xi\,\partial^\mu\xi)^3 \\ &+ c\,(\partial_\mu\xi\,\partial^\mu\xi)(\partial_\lambda\partial_\rho\xi)\,(\partial^\lambda\partial^\rho\xi),\end{aligned}$$

up to redundant interactions, for effective couplings a, b and c.

3 Power Counting and Matching

The previous chapter argues that the Wilson action captures the influence of virtual high-energy states on all low-energy observables, but a good number of questions remain to be addressed before it becomes a tool of practical utility. In particular, the Wilson action in principle contains an infinite number of interactions of various types, and although these are local (once expanded in inverse powers of the heavy scale) they ultimately involve arbitrarily many powers of the low-energy fields and their derivatives. What is missing is a simple way to identify systematically which interactions are required to calculate any given observable to a given order in the low-energy (and any other) expansions.

In principle, as argued in §2.4, what makes the Wilson action useful is dimensional analysis, which shows that more complicated interactions (with more derivatives or powers of fields) have coupling constants more suppressed by inverse powers of the physical heavy mass scale, M (like m_R in the toy model). This suggests that a dimension-Δ interaction can be ignored at low energies, E, provided effects of order $(E/M)^p$, with $p = \Delta - 4$, are negligible.

Sounds simple. Unfortunately, there is a confounding factor that complicates the simple dimensional argument. Although each use of an effective interaction within a Feynman graph costs inverse powers of a heavy scale like M, it is also true that the 4-momenta of virtual particles circulating within loops can include energies that are not small. This means that heavy scales can appear in numerators of calculations as well as in denominators, making it trickier to quantify the size of higher-order effects. Power counting – the main subject of this chapter – makes this argument more precise, and is the tool with which to identify which effective interactions are relevant to any particular order in the low-energy expansion.

To see how scales appear in calculations, for some purposes it is useful to explicitly track cutoffs, like Λ, that label the highest energies allowed to circulate within loops. Depending on the relative size of scales like M and Λ, it can happen that loop effects can cause effective couplings to acquire coefficients $c_n \propto \Lambda^{-p}$ rather than $c_n \propto M^{-p}$. For $p > 0$ these naively dominate because $\Lambda \ll M$. Estimates of the size of such corrections are discussed in this chapter in the section devoted to the 'exact renormalization group' (or 'exact RG').

But it is also true that Λ ultimately drops out of physical quantities, making its presence an unnecessary complication when formulating dimensional arguments. (Λ drops out of physical quantities because the precise split between low- and high-energy quantities is ultimately a book-keeping device for making calculations convenient, so Λ is not a physical scale. As this chapter shows, the disappearance of Λ in physical predictions happens in detail because the explicit Λ-dependence of the effective couplings in S_W cancels the Λ-dependence implicit in the definition of the low-energy path integral in which S_W is used.)

These observations suggest it is likely to be more efficient to formulate low-energy quantities in a cutoff-independent way. In particular, power counting is most efficiently formulated when dimensional regularization is used to define the Wilson action, rather than a floating cutoff like Λ. Ultimately, power counting turns the Wilson action from a chain-saw into a scalpel, making it a tool for making precision calculations. Along the way, it shows why UV divergences are only nuisances that are not fatal complications to the formulation of the low-energy theory, and provides deep conceptual insights into the physical meaning of renormalizability.

3.1 Loops, Cutoffs and the Exact RG ♠

The goal of this section is to explicitly track how the scales appearing in the Wilson action propagate into low-energy observables.[1] To this end, suppose the Wilson action computed from a specific underlying theory has the form

$$S_W = S_{W,0} + S_{W,\text{int}}, \tag{3.1}$$

with

$$\begin{aligned} S_{W,0} &= -\frac{\mathfrak{f}^4}{M^2 v^2} \int d^4x \left[\partial_\mu \phi \, \partial^\mu \phi + m^2 \phi^2 \right] \\ S_{W,\text{int}} &= -\mathfrak{f}^4 \sum_n \frac{\hat{c}_n}{M^{d_n} v^{f_n}} \int d^4x \, O_n(\phi), \end{aligned} \tag{3.2}$$

where ϕ denotes a generic low-energy field which the form for S_0 assumes to be bosonic, with $[\phi] = 1$. For simplicity only one field is kept here, though the dimensional arguments to follow remain unchanged if ϕ instead represents a collection of fields. The index n runs over a complete set of labels for all possible interactions, for each of which there are two non-negative integers, d_n and f_n, that respectively count the number of powers of derivatives and fields that appear in the interaction O_n. For example, for an effective interaction like $O = \phi^2 \partial_\mu \phi \, \partial^\mu \phi$ these constants are $d_n = 2$ and $f_n = 4$.

The three dimensionful quantities $\mathfrak{f}$, v and M are all energy scales much larger than the light-particle mass, m, that can be (but need not be) independent of one another or of Λ. Roughly speaking, this writes the action as an expansion in powers of fields, ϕ/v, and derivatives, ∂/M, with no *a-priori* requirement that the two comparison scales v and M be similar. The scale $\mathfrak{f}^4$ gives the rough energy density associated with $\phi \sim v$ and $\partial \sim M$. The goal is to track how these scales appear in physical quantities once S_W is used to compute them.

The kinetic term coming from $S_{W,0}$ carries the factor $\mathfrak{f}^4/M^2v^2$ so that it scales with these parameters in the same way as do the interaction terms. Although this means ϕ is not canonically normalized (unless $\mathfrak{f}^4 = \frac{1}{2} M^2 v^2$), the discussion is nonetheless kept general by working with Feynman rules that do not assume canonical normalization.

[1] This and later sections broadly follow the logic of [2], though details and notation used follow that of [24].

With these definitions, to leading order the total dimension of an interaction having d_n derivatives and f_n powers of ϕ is therefore

$$\Delta_n := [O_n] = d_n + f_n, \tag{3.3}$$

so the powers of $\mathfrak{f}$, M and v ensure that the coefficients $\hat{c}_n$ are dimensionless to leading order (assuming $[\phi] = 1$).

3.1.1 Low-Energy Amplitudes

Imagine now using these effective interactions in a path integral to compute a low-energy observable. Working perturbatively in $S_{w,\text{int}}$ amounts to evaluating various Feynman graphs using $S_{w,0}$ to define the internal lines and $S_{w,\text{int}}$ to define the effective interaction vertices.

Suppose $\mathcal{A}_{\mathcal{E}}(q)$ denotes the result of evaluating an amplitude involving $\mathcal{E}$ external lines, regarded as a function of a collection of external kinetic variables, q, sharing a common low-energy scale $E \ll \mathfrak{f}, M, v, \Lambda$. $\mathcal{A}_{\mathcal{E}}$ could be a scattering amplitude among low-energy particles, or might be a contribution to the generating functional[2] Γ_{le}. The goal is to determine the systematics of how $\mathcal{A}_{\mathcal{E}}(q)$ depends on the various energy scales as a function of $\mathcal{E}$ and q. But, in general, different Feynman graphs involving different numbers of internal lines and vertices can depend on these variables differently, so it is also worth tracking the dependence on other quantities like the number, $\mathcal{I}$, of internal lines and the number, $\mathcal{V}_n$, of vertices coming from the interaction O_n. Since O_n involves f_n fields, there are f_n lines that converge at the corresponding vertex, while d_n counts the number of derivatives appearing in the corresponding Feynman rule.

Some Useful Identities

The first observation is that the positive integers, $\mathcal{I}$, $\mathcal{E}$ and $\mathcal{V}_n$, that characterize any particular graph are not all independent. Rather, they are related by the rules for constructing graphs from lines and vertices.

One such relation is obtained by equating the two equivalent ways of counting the number of ends of internal and external lines in a graph:

- On one hand, since all lines end at a vertex, the number of ends is given by summing over all of the ends appearing in all of the vertices: $\sum_n f_n \mathcal{V}_n$;
- On the other hand, there are two ends for each internal line, and one end for each external line in the graph: making a total of $2\mathcal{I} + \mathcal{E}$ ends.

Equating these two ways of counting gives the identity expressing the 'conservation of ends':

$$2\mathcal{I} + \mathcal{E} = \sum_n f_n \mathcal{V}_n, \qquad \text{(conservation of ends)}. \tag{3.4}$$

[2] For applications to scattering amplitudes and generating functionals external lines are amputated from Feynman graphs, which matters when applying dimensional analysis to the result. For this reason a minor modification is required to apply the power-counting results here to correlation functions – see *e.g.* the discussion of §10.2.1.

A second useful identity defines the number of loops, $\mathcal{L}$, for each (connected) graph:

$$\mathcal{L} = 1 + \mathcal{I} - \sum_n \mathcal{V}_n, \qquad \text{(definition of } \mathcal{L}\text{)}. \tag{3.5}$$

As mentioned around Eq. (2.24), this definition is motivated by the topological identity that applies to any graph that can be drawn on a plane, that states that $\mathcal{L} - \mathcal{I} + \sum_n \mathcal{V}_n = 1$ (which is the Euler number of a disc). In what follows, Eqs. (3.4) and (3.5) are used to eliminate $\mathcal{I}$ and $\sum_n f_n \mathcal{V}_n$.

Feynman Rules

The next step is to use the action of Eqs. (3.2) and (3.55) to construct the Feynman rules for the graph of interest. This is done here in momentum space, but since the argument to be made is in essence a dimensional one, it could equally well be made in position space.

Schematically, in momentum space the product of all of the vertices contributes the following factor to the amplitude:

$$\text{(Vertices)} = \prod_n \left[i(2\pi)^4 \delta^4(p)\, \mathfrak{f}^4 \left(\frac{p}{M}\right)^{d_n} \left(\frac{1}{v}\right)^{f_n} \right]^{\mathcal{V}_n}, \tag{3.6}$$

where p generically denotes the various momenta running through the vertex. The product of all of the internal line factors gives the additional contribution:

$$\text{(Internal Lines)} = \left[-i \int \frac{d^4 p}{(2\pi)^4} \left(\frac{M^2 v^2}{\mathfrak{f}^4}\right) \frac{1}{p^2 + m^2} \right]^{\mathcal{I}}, \tag{3.7}$$

where p again denotes the generic momentum flowing through the lines. m is the mass of the light particle (or their generic order of magnitude – for simplicity taken to be similar in size – should there be more than one light field) coming from the unperturbed term, Eq. (3.2). For the 'amputated' Feynman graphs relevant to scattering amplitudes and contributions to effective couplings in Γ_{le} the external lines are removed, and so no similar factors are included for external lines.

The momentum-conserving delta functions appearing in (3.6) can be used to perform many of the integrals appearing in (3.7) in the usual way. Once this is done, one delta function remains that depends only on external momenta, $\delta^4(q)$, and so cannot be used to perform additional integrals. This is the delta function that enforces the overall conservation of energy and momentum for the amplitude. It is useful to extract this factor once and for all, by defining the reduced amplitude, $\mathcal{A}_{\mathcal{E}}$, by

$$A_{\mathcal{E}}(q) = i(2\pi)^4 \delta^4(q)\, \mathcal{A}_{\mathcal{E}}(q). \tag{3.8}$$

The total number of integrations that survive after having used all of the momentum-conserving delta functions is then $\mathcal{I} - \sum_n \mathcal{V}_n + 1 = \mathcal{L}$. This last equality uses the definition, Eq. (3.5), of the number of loops, $\mathcal{L}$.

3.1.2 Power Counting Using Cutoffs

The hard part in computing $\mathcal{A}_{\mathcal{E}}(q)$ is to evaluate the remaining multi-dimensional integrals. Things are not so bad if the only goal is to track how the result depends on the scales $\mathfrak{f}$, M and v, however, since then it suffices to use dimensional arguments to estimate the size of the result. Since the integrals typically diverge in the ultraviolet, they are most sensitive to the largest momenta in the loop, and according to Eq. (2.59) this is set by the cutoff Λ. (The contributions of loops having momenta higher than Λ are the ones used when computing S_w itself from the underlying theory.)

This leads to the following dimensional estimate for the result of the integration

$$\int \cdots \int \left[\frac{\mathrm{d}^4 p}{(2\pi)^4}\right]^A \frac{p^B}{(p^2+m^2)^C} \sim \left(\frac{1}{4\pi}\right)^{2A} \Lambda^{4A+B-2C}. \tag{3.9}$$

For the purposes of counting 2π's, a factor of π^2 is included for each $\mathrm{d}^4 p$ integration corresponding to the result of performing the three angular integrations.[3]

The idea is to Taylor expand the amplitude $\mathcal{A}_{\mathcal{E}}(q)$ in powers of external momentum, q, using Eq. (3.9) to estimate the size of the coefficients. Schematically,

$$\mathcal{A}_{\mathcal{E}}(q) \simeq \sum_{\mathcal{D}} \mathcal{A}_{\mathcal{E}\mathcal{D}}\, q^{\mathcal{D}}, \tag{3.10}$$

where the coefficients require an estimate for the following integral

$$\begin{aligned} \mathcal{A}_{\mathcal{E}\mathcal{D}}\, q^{\mathcal{D}} &\propto \int \cdots \int \left[\frac{\mathrm{d}^4 p}{(2\pi)^4}\right]^{\mathcal{L}} \frac{1}{(p^2+m^2)^{\mathcal{I}}} \left(\frac{q}{p}\right)^{\mathcal{D}} \prod_n p^{d_n \mathcal{V}_n} \\ &\sim \left(\frac{1}{4\pi}\right)^{2\mathcal{L}} \left(\frac{q}{\Lambda}\right)^{\mathcal{D}} \Lambda^{4\mathcal{L}-2\mathcal{I}+\sum_n d_n \mathcal{V}_n}. \end{aligned} \tag{3.11}$$

Combining this with the powers of $\mathfrak{f}$, M and v given by the Feynman rules then gives, after using identities (3.4) and (3.5),

$$\mathcal{A}_{\mathcal{E}\mathcal{D}}\, q^{\mathcal{D}} \sim \mathfrak{f}^4 \left(\frac{1}{v}\right)^{\mathcal{E}} \left(\frac{q}{\Lambda}\right)^{\mathcal{D}} \left(\frac{M\Lambda}{4\pi\,\mathfrak{f}^2}\right)^{2\mathcal{L}} \left(\frac{\Lambda}{M}\right)^{2+\sum_n (d_n-2)\mathcal{V}_n}. \tag{3.12}$$

This is the main result of this section, whose properties are now explored.

A reality check for this formula comes if it is applied to the simplest graph of all (see Fig. 3.1): one including no internal lines (so $\mathcal{L} = 0$) and only a single vertex, $n = n_0$, with $f_{n_0} = \mathcal{E}$ external lines and $d_{n_0} = \mathcal{D}$ derivatives (so $\sum_n \mathcal{V}_n = 1$ and $\sum_n d_n \mathcal{V}_n = \mathcal{D}$). In this case, (3.12) implies that the amplitude depends on the scales M, Λ and $\mathfrak{f}$ in precisely the same way as does the starting lagrangian (3.2): $\mathcal{A}_{\mathcal{E}\mathcal{D}}\, q^{\mathcal{D}} \sim \mathfrak{f}^4 (1/v)^{\mathcal{E}} (q/M)^{\mathcal{D}}$.

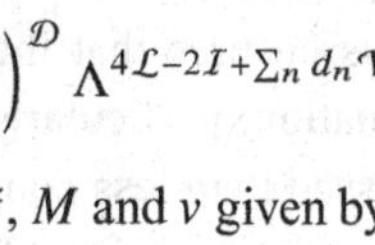

Fig. 3.1 The graph describing the insertion of a single effective vertex with $\mathcal{E}$ external lines and no internal lines.

[3] The factor of π^2 is clearest to see if momenta are Wick rotated to euclidean signature, since it there represents the volume of the unit 3-sphere corresponding to the integration over the three directions taken by a 4-vector.

A second reality check applies (3.12) to the special case where all scales are set by Λ (*i.e.* $\mathfrak{f} = M = v = \Lambda$) since this corresponds to the choices made in the dimensional arguments of §2.4.1. In this limit (3.12) becomes

$$\mathcal{A}_{\mathcal{E}\mathcal{D}}\, q^{\mathcal{D}} \sim \Lambda^4 \left(\frac{1}{\Lambda}\right)^{\mathcal{E}} \left(\frac{q}{\Lambda}\right)^{\mathcal{D}} \left(\frac{1}{4\pi}\right)^{2\mathcal{L}}. \tag{3.13}$$

Since $\mathcal{E}$ and $\mathcal{D}$ are fixed by external characteristics (the total number of external legs and power of q in the final answer), the last factor is the only part of (3.13) that changes for more and more complicated diagrams that share these external properties. This factor simply says that $1/(4\pi)^2$ is the price for each additional loop; that is to say, all graphs with a fixed number of loops are similar in size assuming the (unwritten) dimensionless couplings – *i.e.* the $\hat{c}_n$ of (3.2) – are also similar in size. Furthermore, perturbation theory in this regime is ultimately controlled by the ratio of the dimensionless $\hat{c}_n$ compared with $16\pi^2$. The statement that perturbation theory applies for small enough $\hat{c}_n$ agrees with (and refines) the simple estimate of §2.4.1.

Validity of the Perturbative Expansion

More broadly, Eq. (3.12) outlines the domain of validity of the perturbative expansion itself. If the contribution estimated in Eq. (3.12) is small for all choices of $\mathcal{D}$, $\mathcal{E}$, $\mathcal{L}$ and $\mathcal{V}_n$, then this ensures that the perturbative expansion used in its derivation is a good approximation, particularly if more complicated graphs (higher $\mathcal{L}$ and $\mathcal{V}_n$) are more suppressed than less complicated ones. Conversely, if there is a choice for $\mathcal{D}$, $\mathcal{E}$, $\mathcal{L}$ and $\mathcal{V}_n$ for which Eq. (3.12) is not small, then the perturbative expansion fails unless some other small parameter – such as the dimensionless couplings $\hat{c}_n$ of (3.2) – can be found that can systematically suppress more complicated graphs. Furthermore, since the semiclassical expansion is an expansion in loops,[4] the perturbative expansion becomes a semiclassical expansion when it is the $\mathcal{L}$-dependent factor that controls perturbation theory.

Eq. (3.12) shows that there are three small quantities whose size can help control perturbative corrections: q/Λ, Λ/M and $\Lambda M/4\pi\mathfrak{f}^2$. Some remarks are in order for each of these.

Derivative Expansion

Consider first the factor q/Λ, which controls the suppression of higher powers of external momenta. There is no question that $q/\Lambda \ll 1$ since the entire construction of the low-energy theory presupposes Λ can be chosen much smaller than the scale of the heavy physics that is being integrated out, but much larger than the low energies, $q \simeq E$ of applications. But when $\Lambda \ll M$ the ratio q/Λ is much bigger than q/M, even if both are separately very small. Eq. (3.12) therefore shows that it could happen that the derivative expansion in physical quantities (like scattering amplitudes) and in quantities like S_w ends up being controlled by powers of q/Λ rather than the powers of q/M assumed for the original action, (3.2). This point is returned to in §3.1, but it means that (all other things being equal) derivatives like to be suppressed by the

[4] The connection between loops and the semiclassical expansion is established in the discussion surrounding (2.24). In essence, the semiclassical expansion counts loops because it is an expansion in powers of $\hbar$, which appears as an overall factor in the quantity $S/\hbar$ within the path integral.

lowest possible UV scale available: in this case Λ. The case $M \simeq \Lambda$ is one that should be taken seriously in what follows.

Field Expansion

Notice that the same thing does *not* happen for the expansion in powers of ϕ/v, since the factor $(1/v)^{\mathcal{E}}$ assumed to appear in S_w does not get converted into a power like $(1/\Lambda)^{\mathcal{E}}$ or $(1/M)^{\mathcal{E}}$ in $\mathcal{A}_{\mathcal{E}}$. This means that it is consistent to have the scale v that controls the field expansion be systematically different from the scales Λ or M that control the derivative expansion. These scales are logically independent because, in general, large fields need not imply large energies, so small-field expansions are not necessarily required in a low-energy limit.

Loop Expansion

Next consider the suppressions coming from loops and vertices. Notice first that if $M \simeq \Lambda$ then the only systematic perturbative suppression in (3.12) comes from loops, due to the factor $(\Lambda^2/4\pi\mathfrak{f}^2)^{2\mathcal{L}}$. If all dimensionless couplings are order unity then perturbation theory in this limit is revealed to be a semiclassical expansion (*i.e.* controlled purely by the number of loops) whose validity rests on the assumption $4\pi\mathfrak{f}^2 \gg \Lambda^2$.

This condition is automatically satisfied in the regime $\Lambda \ll M$ provided that $\mathfrak{f} \gtrsim M$ is also true. It is a much stronger condition on Λ, however, if $\mathfrak{f}$ should be much smaller than M. In the particularly interesting case where $\mathfrak{f}^2 \simeq Mv$ (corresponding to canonical normalization in (3.2)) the loop-suppression factor becomes $(M\Lambda/4\pi\mathfrak{f}^2)^{2\mathcal{L}} \simeq (\Lambda/4\pi v)^{2\mathcal{L}}$.

Dangerous Non-Derivative Interactions

Finally, consider the final factor in (3.12). If $\Lambda \lesssim M$ the power of $(\Lambda/M)^{\mathcal{P}}$ appearing in Eq. (3.12) represents a suppression rather than an enhancement, provided the power

$$\mathcal{P} := 2 + \sum_n (d_n - 2)\mathcal{V}_n, \tag{3.14}$$

is non-negative. It is this factor that expresses the suppression of the effects of interactions involving three or more derivatives.

Lorentz invariance often requires d_n to be even (*e.g.* for scalar fields), and in this case it is only interactions with no derivatives at all ($d_n = 0$) for which $\mathcal{P}$ can be negative. These interactions are potentially dangerous in that they can in principle allow an *enhancement* in $\mathcal{A}_{\mathcal{E}}$ when $\Lambda \ll M$. When such interactions exist a more detailed estimate is required to see whether higher-order effects really are suppressed.

As an example of non-derivative interactions, imagine the low-energy field, ϕ, self-interacts through a scalar potential,

$$S_W \supset -\int d^4x\, V(\phi), \tag{3.15}$$

where

$$V(\phi) := \mathfrak{f}_v^4 \sum_r g_r \left(\frac{\phi}{v}\right)^r. \tag{3.16}$$

Here, g_r are dimensionless couplings and $\mathfrak{f}_v^4$ is the typical potential energy density associated with fields of order $\phi \simeq v$. If $\mathfrak{f}_v \neq \mathfrak{f}$ then repeating the above power counting argument shows that each appearance of a vertex drawn from $V(\phi)$ contributes an additional factor $g_r(\mathfrak{f}_v/\mathfrak{f})^4$, modifying Eq. (3.12) to

$$\mathcal{A}_{\mathcal{E}\mathcal{D}}\, q^{\mathcal{D}} \sim \frac{\Lambda^2 \mathfrak{f}^4}{M^2} \left(\frac{1}{v}\right)^{\mathcal{E}} \left(\frac{q}{\Lambda}\right)^{\mathcal{D}} \left(\frac{M\Lambda}{4\pi\, \mathfrak{f}^2}\right)^{2\mathcal{L}} \times \left\{ \prod_r \left[g_r \left(\frac{\mathfrak{f}_v^4 M^2}{\mathfrak{f}^4 \Lambda^2}\right)\right]^{\mathcal{V}_{0,r}} \right\} \left\{ \prod_{d\geq 2} \prod_{i_d} \left(\frac{\Lambda}{M}\right)^{(d-2)\mathcal{V}_{d,i_d}} \right\}, \tag{3.17}$$

where the product over vertex labels is now subdivided into groups involving precisely d derivatives: $\{n\} = \{d, i_d\}$, with $i_0 = r$.

This last expression shows that the potentially hazardous enhancement factor, $(M/\Lambda)^{2\mathcal{V}_{0,r}}$, need not be dangerous if the potential energy density in the low-energy theory is sufficiently small relative to the generic energy density, $\mathfrak{f}_v^4/\mathfrak{f}^4 \lesssim \Lambda^2/M^2$. But if this is not so, generic non-derivative interactions can be legitimate obstructions to having a well-behaved low-energy limit, a point that must be checked on a case-by-case basis.

Example: The Toy Model

The Wilson action for the toy model, Eq. (2.97), is a special case of the general form assumed in Eqs. (3.2), with $M = m_R$ and $\mathfrak{f}^2 = m_R v$ and no zero-derivative interactions for the low-energy field ξ. For this special case the estimate Eq. (3.12) becomes

$$\mathcal{A}_{\mathcal{E}\mathcal{D}}\, q^{\mathcal{D}} \sim v^2 \Lambda^2 \left(\frac{1}{v}\right)^{\mathcal{E}} \left(\frac{q}{\Lambda}\right)^{\mathcal{D}} \left(\frac{\Lambda}{4\pi v}\right)^{2\mathcal{L}} \left(\frac{\Lambda}{m_R}\right)^{\sum_n (d_n - 2)\mathcal{V}_n}, \tag{3.18}$$

which neglects dimensionless factors that come as a series in powers of the coupling λ.

Notice that Eq. (3.18) agrees with the calculations of the previous chapter for the size of tree and loop contributions to the effective vertex, $a_{\rm le}(\partial_\mu \xi\, \partial^\mu \xi)^2$ appearing in $\Gamma_{\rm le}$, for which $\mathcal{E} = \mathcal{D} = 4$. For instance, consider the three contributions of Fig. 2.5. Figure (a) has $\mathcal{L} = 0$ and $\mathcal{V}_{4,4} = 1$, and so Eq. (3.18) gives

$$\delta a_{\rm le} \sim v^2 \Lambda^2 (1/v)^4 (1/\Lambda)^4 (\Lambda/m_R)^2 \sim 1/(v^2 m_R^2) = \lambda / m_R^4\,; \tag{3.19}$$

Figure (b) has $\mathcal{L} = 1$ and $\mathcal{V}_{4,4} = 2$, and so Eq. (3.18) gives

$$\delta a_{\rm le} \sim v^2 \Lambda^2 (1/v)^4 (1/\Lambda)^4 (\Lambda/4\pi v)^2 (\Lambda/m_R)^4 \sim [1/(16\pi^2 v^4)](\Lambda/m_R)^4\,; \tag{3.20}$$

Figure (c) has $\mathcal{L} = 1$ and $\mathcal{V}_{6,6} = 1$, and so Eq. (3.18) gives

$$\delta a_{\rm le} \sim v^2 \Lambda^2 (1/v)^4 (1/\Lambda)^4 (\Lambda/4\pi v)^2 (\Lambda/m_R)^4 \sim [1/(16\pi^2 v^4)](\Lambda/m_R)^4. \tag{3.21}$$

These all agree with the estimates performed in §2.3.

But Eq. (3.18) contains much more information than just this. Most importantly, since the symmetry $\xi \to \xi +$ constant implies that there are no interactions with

$d < 4$, it follows that higher-order graphs are always suppressed by positive powers of the small ratios $\Lambda/4\pi v$ or Λ/m_R. The small size of these ratios is what is responsible for weak coupling in the low-energy theory, showing that it is the derivative coupling of Goldstone bosons at low energies that allows S_W to be treated perturbatively (and *not* the size of the coupling λ in the underlying theory).

The toy model also provides insight into the relationship between the scales M and v that respectively control the derivative and field expansions in the Wilson action. To see why, recall that $m_R^2 = \lambda v^2$, so the perturbative semiclassical regime $\lambda \ll 1$ is where the scales $M = m_R$ and v are very different from one another. Yet even so, Eq. (3.18) demonstrates that each external line (which, for calculations of $\Gamma_{\rm le}$, counts the power of ϕ) is accompanied by at least one power of $1/v$. Quantum corrections do not change the fact that it is always the size of ϕ/v that controls the field expansion, regardless of the scale appearing in the low-energy expansion.

3.1.3 The Exact Renormalization Group

Another instructive use of (3.12) is to estimate how the couplings in S_W themselves evolve as Λ changes. At first sight, this might seem surprising, since the evolution of couplings in S_W are obtained by starting from the underlying high-energy theory and integrating out all physics at energy scales above Λ. For this calculation, however, Λ is the *lowest* scale in the integration rather than the highest scale, so naively the dimensional arguments leading to (3.12) (which take Λ to be the *largest* scale in all integrals) might seem not to apply.

The estimate (3.12) is nevertheless useful because any Λ-dependence acquired by S_W when integrating out modes with energies larger than Λ must ultimately be cancelled when S_W is used to integrate out the remaining modes with energies smaller than Λ – to which (3.12) does apply.

This can be formalized by comparing the result obtained when computing something like the 1LPI generator, $\Gamma_{\rm le}[\phi;\Lambda]$, using the Wilson action defined at a cutoff scale Λ with that obtained with the Wilson action defined at a slightly lower cutoff $S_W[\phi,\Lambda']$, with $\Lambda' = \Lambda - {\rm d}\Lambda$. Either of these is an equally good starting point for computing $\Gamma_{\rm le}$, since this is Λ-independent. (It could, after all, have been computed for the full theory without ever dividing the problem into a contribution from above and below the scale Λ).

For example, for scalar fields when used with (2.59) this implies that

$$0 = \Lambda \frac{{\rm d}}{{\rm d}\Lambda} e^{i\Gamma_{\rm le}[\varphi]} = \Lambda \frac{{\rm d}}{{\rm d}\Lambda} \int \mathcal{D}\phi \; e^{iS_W[\varphi+\phi;\Lambda]+\int {\rm d}^4x \, j_a \phi^a}, \tag{3.22}$$

where $j_a = -\delta\Gamma_{\rm le}[\varphi]/\delta\varphi^a$ has support only for modes well below Λ. In detail, the full result is independent of Λ because the Λ-dependence of S_W cancels the Λ-dependence implicit in the measure $\mathcal{D}\phi$, since this includes a functional integration only over modes with energies below Λ.

In practice, it is useful to implement the cutoff excluding modes larger than Λ from the path integral by suitably modifying the Wilson action.[5] For instance, in perturbation theory high-energy modes can be suppressed in internal lines by Wick

[5] If cutoffs are instead implemented directly for the integrations in Feynman graphs the results can depend on the way virtual momentum is routed through the graph.

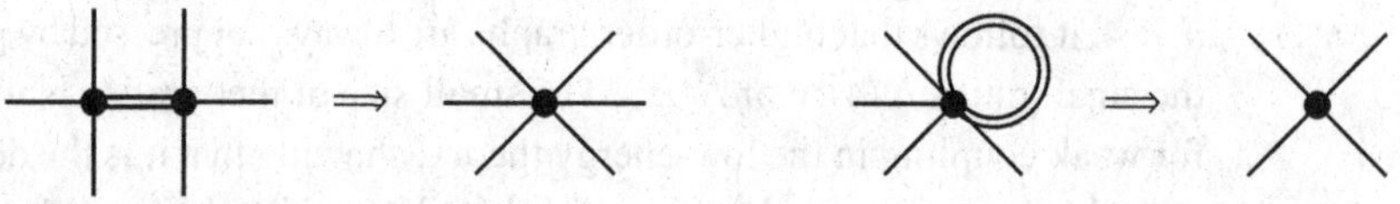

Fig. 3.2 Graphs illustrating the two effects that occur when an internal line is contracted to a point, depending on whether or not the propagator connects distinct vertices (left two figures) or ties off a loop on a single vertex (right two figures). In both cases, a double line represents the differentiated propagator. The two options respectively correspond to the terms $[\delta S_{w,\text{int}}/\delta\phi(p)][\delta S_{w,\text{int}}/\delta\phi(-p)]$ and $\delta^2 S_{w,\text{int}}/\delta\phi(p)\delta\phi(-p)$ appearing in the Wilson–Polchinski relation, Eq. (3.25) of the text.

rotating to imaginary time and introducing a cutoff function into the unperturbed action. Writing $S_w = S_{w0} + S_{w,\text{int}}$ one takes

$$S_{w0} = -\frac{1}{2}\int \mathrm{d}^4x\, \mathrm{d}^4x'\, \mathcal{K}_\Lambda(x-x')\left[\partial_\mu\phi(x)\,\partial'^\mu\phi(x') + m^2\phi(x)\,\phi(x')\right]$$
$$= -\frac{1}{2}\int \frac{\mathrm{d}^4p}{(2\pi)^4}\,\phi(p)\phi(-p)(p^2+m^2)K^{-1}(p^2/\Lambda^2), \tag{3.23}$$

where the kernel $\mathcal{K}_\Lambda(x-x')$ is defined by its Fourier transform

$$\mathcal{K}_\Lambda(x-x') = \int \frac{\mathrm{d}^4p}{(2\pi)^4}\,K^{-1}(p^2/\Lambda^2)\,e^{ip\cdot(x-x')}. \tag{3.24}$$

$K(p^2/\Lambda^2)$ is a smooth step-like cutoff function that satisfies $K(u) = 1$ for $u \ll 1$ and $K(u) \to 0$ for $u \gg 1$.

Once this is done, Eq. (3.22) can be read as a differential equation governing how $S_w[\phi;\Lambda]$ depends on Λ, called the 'exact renormalization group'. When the derivative hits the factor $K(p^2/\Lambda^2)$ in a propagator $K(p^2/\Lambda^2)/(p^2+m^2)$, the result has support only for $p^2 \simeq \Lambda^2$, effectively removing the corresponding internal line from the given Feynman graph.

The change wrought by this in the path integral must be compensated by appropriately modifying the interactions, and this is what defines the flow of effective couplings with Λ (along the lines illustrated in Fig. 3.2). This turns out to imply an evolution equation for the interaction lagrangian density, $S_{w,\text{int}}$, of the form [25–27]

$$\Lambda\frac{\mathrm{d}S_{w,\text{int}}}{\mathrm{d}\Lambda} = -\frac{1}{2}\int \mathrm{d}^4p\left[\frac{(2\pi)^4}{p^2+m^2}\right]\Lambda\frac{\partial K}{\partial\Lambda}\left[\frac{\delta S_{w,\text{int}}}{\delta\phi(p)}\,\frac{\delta S_{w,\text{int}}}{\partial\phi(-p)} + \frac{\delta^2 S_{w,\text{int}}}{\delta\phi(p)\,\delta\phi(-p)}\right]. \tag{3.25}$$

The size of the resulting changes to the effective couplings in $S_{w,\text{int}}$ can be estimated using (3.12). To this end, suppose the interaction terms can be written as

$$S_{w,\text{int}} = -\Lambda^2 v^2 \sum_n \frac{\hat{c}_n}{\Lambda^{d_n} v^{f_n}}\int \mathrm{d}^4x\, O_n^{(d_n,f_n)}(\phi) \tag{3.26}$$
$$= -\int \mathrm{d}^4x\left[\frac{\hat{c}_{4,0}\Lambda^2}{v^2}\phi^4 + \frac{\hat{c}_{4,2}}{v^2}\phi^2(\partial\phi)^2 + \cdots\right],$$

where the $O_n^{(d_n,f_n)}$ describe all possible local interactions involving f_n powers of the fields and d_n derivatives, and the last line specializes to a single scalar field for concreteness' sake.

The power counting result, (3.12), provides an estimate of the size of amputated Feynman graphs built using these interactions, involving fields defined below the cutoff Λ. But this also determines the Λ-dependence of perturbative corrections to the couplings in $S_{w\text{int}}$ because the direct contribution of interactions represented by graphs like Fig. 3.1 must precisely cancel the Λ-dependence coming from loop graphs, as estimated by Eq. (3.12).

For the action of (3.26) the estimate (3.12) gives the contribution to the effective coupling of a term in S_w involving $\mathcal{E}$ powers of ϕ and $\mathcal{D}$ derivatives to be

$$\delta\left[\hat{c}_n v^2\Lambda^2\left(\frac{1}{v}\right)^{\mathcal{E}}\left(\frac{1}{\Lambda}\right)^{\mathcal{D}}\right] \sim v^2\Lambda^2\left(\frac{1}{v}\right)^{\mathcal{E}}\left(\frac{1}{\Lambda}\right)^{\mathcal{D}}\left(\frac{\Lambda}{4\pi v}\right)^{2\mathcal{L}}, \tag{3.27}$$

and so $\delta\hat{c}_n$ acquires corrections from $\mathcal{L}$-loop graphs that are of order

$$\delta\,\hat{c}_n \sim \left(\frac{\Lambda}{4\pi v}\right)^{2\mathcal{L}} \times \quad \text{(combinations of other } \hat{c}_n\text{'s)}. \tag{3.28}$$

If $v \gtrsim \Lambda$, this shows that it is consistent to have the $\hat{c}_n$'s all be generically $\lesssim 1$ for all Λ. Some couplings can be much smaller than this if $v \gg \Lambda$ (or other hierarchies like powers of Λ/M or small dimensionless couplings are buried in the $\hat{c}_n$'s), provided these additional suppressions preserve any initially small values.

Log Running vs Power-Law Running

The exact cutoff-dependent renormalization group is not pursued further in this book, since the focus here is instead on more practical methods of approximate calculation. Before leaving the subject, though, it is useful to address a conceptual question and by so doing contrast the implications of logarithmic and power-law running of effective couplings, $\hat{c}_n(\Lambda)$, as Λ is varied.

The conceptual question is this: why does one care how couplings run with Λ if Λ itself ultimately does not appear in any physical results? It is emphasized many times in this book that Λ enters calculations purely as a convenient book-keeping device: it is useful to organize calculations by scales and integrate out physics one scale at a time. But in the end, physical quantities are obtained only after *all* scales are integrated out, after which the arbitrary separations between these scales disappear.

This section follows [28] to argue that understanding the running of couplings in S_w is useful to the extent that it helps track the dependence of physical quantities on large physical ratios of scale, M/m. In particular, there is often a precise connection between logarithmic dependence of low-energy quantities on the cutoff Λ and a logarithmic dependence on physical scales. The analogous connection is only qualitative for power-law dependence, however, and so is usually less useful.

To see why this is so, imagine a system characterized by two very different scales, $m \ll M$, such as the masses of two different particles. Further imagine that there is a physical quantity, A, whose dependence on M/m happens to be logarithmic, so

$$A = a_0 \ln\left(\frac{M}{m}\right) + a_1, \tag{3.29}$$

for some calculable constants a_0 and a_1. If both of these constants are similar in size, then the value of a_0 can be important in practice since the large logarithm can make it dominate numerically in the total result.

Next, suppose a Wilsonian calculation is performed that divides the contributions to A coming from physics above and below the scale Λ, with $m \ll \Lambda \ll M$. How does the large logarithm get into the low-energy part of the theory, given that it depends on scales that lie on opposite sides of Λ? Typically, this happens as follows:

$$\begin{aligned} A_{\rm le} &= a_{0\,\rm le} \ln\left(\frac{\Lambda}{m}\right) + a_{1\,\rm le} \\ A_{\rm he} &= a_{0\,\rm he} \ln\left(\frac{M}{\Lambda}\right) + a_{1\,\rm he} \\ \text{so that} \quad A = A_{\rm le} + A_{\rm he} &= a_0 \ln\left(\frac{M}{m}\right) + a_1. \end{aligned} \tag{3.30}$$

The requirement that Λ cancels implies that $a_{0\,\rm le} = a_{0\,\rm he}$ and then having the results agree with the full theory implies that $a_{0\,\rm le} = a_{0\,\rm he} = a_0$ and $a_1 = a_{1\,\rm le} + a_{1\,\rm he}$. What is significant is that the coefficient, a_0, of the logarithm in the full answer is calculable purely within the low-energy theory because Λ-cancellation dictates that $a_{0\,\rm le} = a_0$.

The same is not so for power-law dependence. Suppose, for example, that another observable, B, is computed that depends quadratically on masses, so

$$B = b_0\, M^2 + b_1\, m^2. \tag{3.31}$$

Again the coefficient b_0 is of practical interest since the large size of M can make this term dominate numerically. In this case, the low- and high-energy parts of the calculations instead are

$$\begin{aligned} B_{\rm le} &= b_{0\,\rm le}\, \Lambda^2 + b_{1\,\rm le}\, m^2 + \cdots \\ B_{\rm he} &= b_{0\,\rm he}\, M^2 + b_{1\,\rm he}\, \Lambda^2 + \cdots \\ \text{so that} \quad B = B_{\rm le} + B_{\rm he} &= b_0\, M^2 + b_1\, m^2, \end{aligned} \tag{3.32}$$

with $b_0 = b_{0\,\rm he}$ and $b_1 = b_{1\,\rm le}$, while Λ-cancellation requires $b_{0\,\rm le} + b_{1\,\rm he} = 0$.

Evidently, the b_0 term cannot be computed purely within the low-energy part of a Wilsonian calculation simply by tracking the dependence on Λ^2, unlike the way in which the $\ln \Lambda$ terms reproduce the value for a_0. This is a fairly generic result: quantitative predictions for quantities like b_0 really require detailed knowledge of the UV theory and cannot be computed using the low-energy Wilsonian theory alone. But logarithms can often be inferred purely from within the low-energy Wilsonian perspective. For this reason considerable attention is given to renormalization-group methods that allow efficient extraction of large logarithms using Wilsonian EFTs.

Method of Regions I

The cancellations of powers of cutoff and the utility of logarithms can be promoted to a useful tool – sometimes called the *method of regions* [29] – for estimating

Feynman integrals in situations where different integration regions compete in their contributions to the final result.

To illustrate the method, follow [30] and consider the integral

$$\int_0^\infty \frac{k\,dk}{(k^2+m^2)(k^2+M^2)} = \frac{\ln(M/m)}{M^2-m^2} \simeq \frac{\ln(M/m)}{M^2}\left[1+\frac{m^2}{M^2}+\cdots\right], \tag{3.33}$$

and imagine trying to extract the dominant small-m/M expansion without first evaluating the full integral. Naively, one simply Taylor expands the integrand in powers of m and integrates term-by-term, but this has the problem that each term involves an integral that diverges in the infrared

$$\int_0^\infty \frac{k\,dk}{k^2(k^2+M^2)}\left[1-\frac{m^2}{k^2}+\cdots\right], \tag{3.34}$$

as might be expected given that the full result (3.33) is not analytic at $m=0$.

A better procedure instead separates the integral into two regions, an IR region $0<k<\Lambda$ and a UV region $k>\Lambda$, and expands the integrand differently in each. For the IR the integrand is expanded with $k^2 \sim m^2 \ll M^2$, while in the UV one takes instead $m \ll k \sim M$. This leads to the result $\mathcal{I}(m,M) = \mathcal{I}^{IR}(m,M,\Lambda) + \mathcal{I}^{UV}(m,M,\Lambda)$, with

$$\mathcal{I}^{IR} = \int_0^\Lambda \frac{k\,dk}{(k^2+m^2)M^2}\left[1-\frac{k^2}{M^2}+\cdots\right] \simeq \frac{\ln(1+\Lambda^2/m^2)}{2M^2} - \frac{\Lambda^2}{2M^4}+\cdots, \tag{3.35}$$

and

$$\mathcal{I}^{UV} = \int_\Lambda^\infty \frac{k\,dk}{k^2(k^2+M^2)}\left[1-\frac{m^2}{k^2}+\cdots\right] \simeq \frac{\ln(1+M^2/\Lambda^2)}{2M^2} + O(m^2). \tag{3.36}$$

Once these last two formulae are summed, all Λ-dependence cancels (as it must), leaving residual logarithms of M/m as outlined above that reproduce the expansion of (3.33). Large logarithms like $\ln(M/m)$ are ultimately leftovers from the cancellation between the IR divergence of $\mathcal{I}^{UV}$ and the UV divergence of $\mathcal{I}^{IR}$.

3.1.4 Rationale behind Renormalization ◇

The above discussion about integrating out high-energies also provides physical insight into the entire framework of renormalization. This is because a central message is that the scale Λ is ultimately a calculational convenience that drops out of all physical quantities. In detail, Λ drops out because of a cancellation between: (*i*) the explicit Λ-dependence of the cutoff on the limits of integration for virtual low-energy states in loops, and (*ii*) the cutoff-dependence that is implicitly contained within the effective couplings of $\mathfrak{L}_W$.

But this cancellation is eerily reminiscent of how UV divergences are traditionally handled within *any* renormalizable theory, and in particular for the underlying UV theory from which S_W is calculated. The entire renormalization program relies on any UV-divergent cutoff-dependence arising from regulated loop integrals being cancelled by the regularization dependence of the counterterms of the renormalized lagrangian. There are, however, the following important differences.

1. The cancellations in the effective theory occur even though Λ is not sent to infinity, and even though $\mathfrak{L}_W$ contains arbitrarily many terms that are not renormalizable in the traditional sense.
2. The cancellation of regularization dependence in the traditional picture of renormalization appears completely ad-hoc and implausible, while the cancellation of Λ from observables within the effective theory is essentially obvious. It is obvious due to the fact that Λ was only introduced as an intermediate step in a calculation, and so cannot survive uncancelled in the answer.

This resemblance is likely not accidental. It suggests that rather than considering a model's classical lagrangian as something pristine or fundamental, it is better regarded as an effective lagrangian obtained by integrating out still-more-microscopic degrees of freedom. The cancellation of ultraviolet divergences within the renormalization program is, within this interpretation, simply the usual removal of an intermediate step in a calculation to whose microscopic part we are not privy.

This is the modern picture of what renormalization really means. When discovering successful theories, what is found is not a 'classical' action, to be quantized and compared with experiment. What is found is really a Wilsonian action describing the low-energy limit obtained by integrating out high-energy degrees of freedom in some more fundamental theory that describes what is really going on at much, much higher energies.

It is this Wilsonian theory, itself potentially already containing many high-energy quantum effects, whose low-energy states are quantized and compared with observations. Physics progresses by successively peeling back layer after layer of structure in nature, and our mathematics describes this through a succession of Wilsonian descriptions with ever-increasing accuracy.

This is how real progress often happens in science. Efforts to solve concrete practical questions – in this case, a desire to exploit hierarchies of scale as efficiently as possible – can ultimately provide deep insights about foundational issues – in this case, about what it is that is really achieved when new fundamental theories (be it Maxwell's equations, General Relativity or the Standard Model) are discovered.

3.2 Power Counting and Dimensional Regularization ◊

As previous sections make clear, there is a lot of freedom of definition when setting up a Wilson action: besides the freedom to make field redefinitions, there are also all the details of precisely how to differentiate between scales above and below Λ. Physical results do not depend on any of these choices at all since observables are independent of field redefinitions and are blind to the details of a regularization scheme. This freedom should be exploited to make the Wilson action as useful as possible for practical calculations. In particular, it should be used to optimize the efficiency with which effective interactions and Feynman graphs can be identified that completely capture the contributions to low-energy processes at any fixed order in low-energy expansion parameters like q/M.

Though instructive, the power counting analysis of the previous section does not yet do this, due to the appearance in all estimates of the cutoff Λ. Since Λ ultimately cancels in all physical quantities, it is inconvenient to have to rely on it when estimating the size of contributions from different interactions in the low-energy Wilson action. For this reason most practical applications (and most of the rest of this book) define the Wilson action using dimensional regularization rather than cutoffs [31, 32]. Dimensional regularization is useful because it is both simple to use and preserves more symmetries than do other regularization schemes. This section explores how this can be done.

3.2.1 EFTs in Dimensional Regularization

At first sight, it seems impossible to define a Wilson action in terms of dimensional regularization at all. After all, the entire purpose of the Wilson action is to efficiently encapsulate the high-energy part of a calculation, for later use in a variety of low-energy applications. This seems to *require* something like Λ to distinguish high energies from low energies. By contrast, although dimensional regularization is designed to regulate UV-divergent integrals, it does not do so by cutting them off at large momenta and energies. The regularization is instead provided by defining the integral (including contributions from arbitrarily large momenta) for complex dimension, D, taking advantage of the fact that the integral converges in the ultraviolet if D is sufficiently small or negative. The result still diverges in the limit $D \to 4$, but usually as a pole or other type of isolated singularity when D is a positive integer. The limit $D \to 4$ is taken at the end of a calculation, after any singularities are absorbed into the renormalization of the appropriate couplings.

This section describes how dimensional regularization can nonetheless be used to define a Wilson action, despite it not seeming to explicitly separate high from low energies. This is done first by briefly describing dimensional regularization itself, followed by a presentation of the logic of constructing an effective theory using it. (See also Appendix §A.2.4 for more details about dimensional regularization.)

What is Dimensional Regularization?

Consider the following integral over D-dimensional Euclidean momentum p^μ, where $p^2 = \delta_{\mu\nu} p^\mu p^\nu$ (and similarly for q^2),[6] [33, 34]

$$I_D^{(A,B)}(q) := \int \frac{\mathrm{d}^D p_E}{(2\pi)^D} \left[\frac{p^{2A}}{(p^2+q^2)^B}\right] = \frac{1}{(4\pi)^{D/2}} \left[\frac{\Gamma\left(A+\frac{D}{2}\right)\Gamma\left(B-A-\frac{D}{2}\right)}{\Gamma(B)\,\Gamma\left(\frac{D}{2}\right)}\right] \left(q^2\right)^{A-B+D/2}, \tag{3.37}$$

where $\Gamma(z)$ is Euler's Gamma function, defined to satisfy $z\,\Gamma(z) = \Gamma(z+1)$ with $\Gamma(n+1) = n!$ when restricted to positive integers, n. This integral converges in the

[6] Such Euclidean expressions are given in Euclidean signature, obtained by Wick rotating with $\mathrm{d}^4 p = \mathrm{i}\,\mathrm{d}^4 p_E$, meaning that the Minkowski-signature result has an additional factor of i. Notice that there are no additional explicit signs in continuing positive q^2 from Euclidean to Minkowski signature because of the wisdom of using conventions with a $(-+++)$ metric.

UV if the real part of $2(B-A)$ is bigger than D, and the right-hand side is the result of explicit evaluation.

Dimensional regularization uses the right-hand side to *define* this integral for values of A, B and D for which it does not converge, even when D is not an integer. In dimensional regularization D is regarded as complex during intermediate steps, with $D \to 4$ taken at the end of the calculation. It happens that $\Gamma(z)$ is analytic for all complex z apart from poles at the non-positive integers, and so when A and B are positive integers, Eq. (3.37) provides a definition of $I_D(q)$ that is finite for all complex D apart from possible poles when D is a positive even integer.

In particular, integrals defined in this way typically have divergences that arise as poles at $D = 4$. For instance, a useful example encountered earlier is the case $A = 0$ and $B = 2$,

$$\begin{aligned}\int \frac{d^D p}{(2\pi\mu)^D}\left[\frac{\mu^4}{(p^2+q^2)^2}\right] &= \frac{\Gamma\left(2-\frac{D}{2}\right)}{(4\pi)^{D/2}}\left(\frac{q^2}{\mu^2}\right)^{(D-4)/2} \\ &= \frac{1}{16\pi^2}\left[\frac{2}{4-D} - \gamma + \frac{1}{2}\ln\left(\frac{q^2}{4\pi\mu^2}\right) + \cdots\right],\end{aligned} \tag{3.38}$$

where the ellipses represent terms that vanish when $D \to 4$; μ is an arbitrary scale included on dimensional grounds and $\gamma \simeq 0.577215664901532\ldots$ is the Euler–Mascheroni constant. The pole as $D \to 4$ reflects the logarithmic divergence that would have been present if the integral were to be defined with $D = 4$ from the get-go.

Similarly, the integral with $A = B = k$ gives

$$\begin{aligned}\int \frac{d^D p}{(2\pi\mu)^D}\left[\frac{p^{2k}}{(p^2+q^2)^k}\right] &= \frac{\Gamma\left(k+\frac{D}{2}\right)\Gamma\left(-\frac{D}{2}\right)}{\Gamma(k)\Gamma\left(\frac{D}{2}\right)}\left(\frac{q^2}{4\pi\mu^2}\right)^{D/2} \\ &= \left(\frac{q^2}{4\pi\mu^2}\right)^2\left[\frac{2}{4-D} - \gamma_k + \frac{1}{2}\ln\left(\frac{q^2}{\mu^2}\right) + \cdots\right],\end{aligned} \tag{3.39}$$

where γ_k is a k-dependent pure number, whose precise value is not important for later purposes. This is an example of an integral that would diverge like a power of the cutoff if directly evaluated at $D = 4$. Notice that the dimensionally regularized version vanishes as $q^2 \to 0$, because the integrand has no other scale to which the result can be proportional.[7]

Because momenta get integrated from $-\infty$ to ∞ in dimensional regularization, both high and low energies are explicitly included. This makes its use seem contrary to the entire philosophy of defining a low-energy effective theory. But the utility of effective field theories is founded on the observation that *any* contribution of generic high-energy dynamics to low-energy amplitudes can be captured within the low-energy sector by an appropriate collection of local effective interactions. Since the error made in dimensional regularization by keeping all modes up to infinite energies is itself a particular choice of high-energy physics, any damage done can also be undone using an appropriate choice of effective couplings in the low-energy theory.

[7] The result cannot be proportional to μ when $D = 4$ since this scale is introduced in such a way as to ensure that the integral is proportional to μ^{D-4}.

To see how the Wilson action is defined in dimensional regularization, consider a theory containing a light field, ϕ, of mass m, and a heavy field, ψ, of mass $M \gg m$. For the purposes of argument the full theory can be imagined to be a renormalizable theory coupling ϕ to ψ, $S = S[\phi, \psi]$, regularized using dimensional regularization and then renormalized in any convenient way (such as with the modified minimal subtraction – or $\overline{\text{MS}}$ – scheme).[8]

The low-energy applications of interest in this model are to $E \ll M$. The effective Wilsonian action in this regime contains only the light field, $S_W = S_W[\phi]$. Just like the full theory this effective theory is also dimensionally regularized (and renormalized in a way specified below). In practice, this means that the dimensionally regularized effective theory is not obtained by explicitly integrating out successively higher-energy modes of all the fields in the underlying theory. Instead, the dimensionally regularized effective theory simply omits the heavy field ψ.

A convenient renormalization choice for the effective couplings in S_W demands that the predictions of $S_W[\phi]$ agree with the low-energy predictions of $S[\phi, \psi]$ order-by-order in $1/M$. That is, the renormalized effective couplings of the low-energy theory are obtained by performing a *matching* calculation, whereby the couplings of the low-energy effective theory are chosen to reproduce scattering amplitudes or Greens functions of the underlying theory order-by-order in powers of the inverse heavy scale, $1/M$. Once the couplings of the effective theory are determined in this way in terms of those of the underlying fundamental theory, they may be used to compute any other purely low-energy observable.

Method of Regions II

Before seeing how this all works in detail, it is worth first pausing here to see how dimensional regularization works when exploring how integrals depend on large hierarchies of parameters. To this end, consider again the sample integral, (3.33), examined above when discussing the method of regions. In dimensional regularization this integral would have the form

$$\mathcal{I}_\epsilon(m, M) := \int_0^\infty \left(\frac{\mu}{k}\right)^\epsilon \frac{k\,dk}{(k^2+m^2)(k^2+M^2)}$$
$$= \Gamma\left(1-\frac{\epsilon}{2}\right)\Gamma\left(\frac{\epsilon}{2}\right) \frac{(m/\mu)^{-\epsilon} - (M/\mu)^{-\epsilon}}{M^2 - m^2}, \tag{3.40}$$

where $\epsilon \propto D - 4$ is a parameter that is taken to zero at the end. The limit $\epsilon \to 0$ is nonsingular in this case because the integral of (3.33) converges when $D = 4$, with $\mathcal{I}_\epsilon = \ln(M/m)/(M^2 - m^2) + O(\epsilon)$ near $\epsilon = 0$.

In the spirit of the method of regions consider regarding $\mathcal{I}_\epsilon$ again as the sum of an IR contribution, for which $k^2 \sim m^2 \ll M^2$, plus a UV contribution with $m \ll k \sim M$:

$$\mathcal{I}_\epsilon^{IR} := \int_0^\infty \left(\frac{\mu}{k}\right)^\epsilon \frac{k\,dk}{(k^2+m^2)M^2}\left[1 - \frac{k^2}{M^2} + \cdots\right] \tag{3.41}$$
$$\simeq \frac{(m/\mu)^{-\epsilon}}{M^2}\left[\frac{1}{\epsilon} + O(\epsilon)\right] = \frac{1}{M^2}\left[\frac{1}{\epsilon} - \ln\left(\frac{m}{\mu}\right) + O(\epsilon)\right],$$

[8] As described in Appendix A.2.4, 'minimal subtraction' simply drops the $(4 - D)^{-1}$ term in divergent quantities, while 'modified minimal subtraction' drops both the $(4 - D)^{-1}$ and the constants γ and $\ln(4\pi)$ [35–37].

and

$$\mathcal{I}^{UV}_\epsilon := \int_0^\infty \left(\frac{\mu}{k}\right)^\epsilon \frac{k\,\mathrm{d}k}{k^2(k^2+M^2)}\left[1-\frac{m^2}{k^2}+\cdots\right] \qquad (3.42)$$
$$\simeq \frac{(M/\mu)^{-\epsilon}}{M^2}\left[-\frac{1}{\epsilon}+O(\epsilon)\right] = \frac{1}{M^2}\left[-\frac{1}{\epsilon}+\ln\left(\frac{M}{\mu}\right)+O(\epsilon)\right].$$

The approximate equality starting the second line for both of these formulae drops all subdominant terms in powers of m/M. Notice that the pole $1/\epsilon$ arises due to a UV divergence in $\mathcal{I}^{IR}$ as $\epsilon \to 0$, while it corresponds to the IR divergence in $\mathcal{I}^{UV}$ as $\epsilon \to 0$. Notice also that these poles cancel in the sum $\mathcal{I} = \mathcal{I}^{UV} + \mathcal{I}^{IR}$, after which the limit $\epsilon \to 0$ can be taken, revealing agreement with (3.40).

The surprise in this exercise was that it was not important to explicitly separate out the region with $k < \Lambda$ and $k > \Lambda$ when defining $\mathcal{I}^{UV}$ and $\mathcal{I}^{IR}$, which nevertheless reproduce the correct dependence on m/M once summed to give the full integral. This works because any cutoff-dependence in the definition of these integrals is guaranteed to cancel in any case, and so although including the cutoff-dependence could be done, it is wasted effort.

Several concrete examples of the use of dimensional regularization when matching between underlying and effective theories are examined in more detail (for relativistic theories) in Chapter 7, which also explores a modification to minimal subtraction called 'decoupling subtraction' that proves useful when matching is done at or above one-loop accuracy. Nonrelativistic examples of beyond-leading-order matching are similarly studied in §12 and §15.

3.2.2 Matching vs Integrating Out

Matching – the fixing of low-energy couplings by comparing the predictions of the full theory with the predictions of its low-energy Wilsonian approximation – is often much easier to carry out than is the process of explicitly integrating out a heavy state using a cut-off path integral. This is partly because the comparison can be made for *any* physical quantity, and, in particular, this quantity can be chosen to make the comparison as simple as possible. Furthermore, because the comparison is made at the level of renormalized interactions, for both the full and low-energy theories, there are no UV divergences to worry about.

Example: The Toy Model

As usual, the toy model helps make the above statements more concrete. For the toy model the heavy mass scale is m_R and the full theory describing the physics of the two fields χ and ξ above this scale is given by the action $S[\chi, \xi]$ of Eq. (1.24) and (1.25) (repeated for convenience here):

$$S = -\int \mathrm{d}^4x \left[\frac{1}{2}\partial_\mu\chi\partial^\mu\chi + \frac{1}{2}\left(1+\frac{\chi}{\sqrt{2}v}\right)^2 \partial_\mu\xi\partial^\mu\xi + V(\chi)\right], \qquad (3.43)$$

with

$$V(\chi) = \frac{\lambda v^2}{2}\chi^2 + \frac{\lambda v}{2\sqrt{2}}\chi^3 + \frac{\lambda}{16}\chi^4. \qquad (3.44)$$

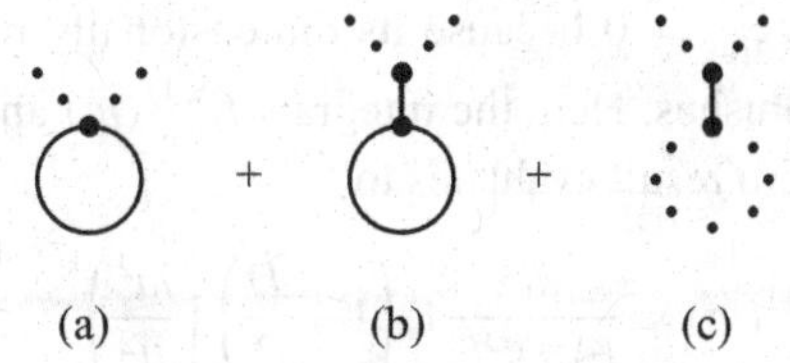

Fig. 3.3 One-loop graphs that contribute to the $\partial_\mu \xi \partial^\mu \xi$ kinetic term in the Wilson and 1LPI actions using the interactions of Eqs. (3.43) and (3.44). Solid (dotted) lines represent χ (and ξ) fields.

One could equivalently use the fields $\hat{\phi}_R$ and $\hat{\phi}_I$ with the action $S[\hat{\phi}_R, \hat{\phi}_I]$ of Eqs. (1.1) and (1.2), and renormalization is actually easier using these variables. This chapter sticks to χ and ξ to keep the symmetries of the problem more manifest. UV divergences in this theory are handled using dimensional regularization, and where necessary divergences are renormalized using modified minimal subtraction (see Appendix A.2.4 for details).

The low-energy Wilson action in this case is $S_W[\xi]$ (or $S_W[\hat{\phi}_I]$), depending only on the single light field, with UV divergent integrals again defined using dimensional regularization. Renormalization is again based on minimal subtraction, though with the difference that the finite part of the coupling is fixed by matching to predictions of the full theory (rather than again using modified minimal subtraction).

For present purposes the first step when matching is to write down all possible interactions in S_W up to some order in $1/m_R$, since this identifies the effective couplings whose value matching is meant to determine. For the toy model we know from §2.5 that all $1/m_R^2$ interactions are redundant, and the most general interactions (consistent with Lorentz invariance and the symmetry under constant shifts in ξ) are given to order $1/m_R^4$ by:

$$S_W = -\int d^4x \left[\frac{z_W}{2}\, \partial_\mu \xi\, \partial^\mu \xi - a_W (\partial_\mu \xi\, \partial^\mu \xi)^2 + \cdots \right], \tag{3.45}$$

where on dimensional grounds z_W is dimensionless while $a_W \propto 1/m_R^4$ and terms not explicitly written are suppressed by at least $1/m_R^6$ (see §2.5).

Whereas earlier sections use the freedom to rescale ξ to set $z_W = 1$ (*i.e.* to canonically normalize ξ), writing (3.45) recognizes this has only been done at the classical level and not exactly, so at one loop $z_W = 1 + z_W^{(1)}$ with $z_W^{(1)} \simeq O(\lambda/16\pi^2)$ due to graphs like those of Fig. 3.3.

The contributions to $z_W^{(1)}$ found by Wick rotating these graphs, evaluating them in dimensional regularization and matching them to the contribution of (3.45) are

$$\mathrm{i} z^{(1)}_{W\,3.3(a)} = \mathrm{i}\left(-\frac{1}{4v^2}\right) \int \frac{\mathrm{i}\, d^4 p_E}{(2\pi)^4}\, \frac{-\mathrm{i}}{p^2 + m_R^2} = -\frac{\mathrm{i}}{4v^2}\, I_D^{(0,1)}(m_R) \tag{3.46}$$

and

$$\mathrm{i} z^{(1)}_{W3.3(b)} \simeq 3\left[2 \times \frac{\mathrm{i}^2}{2!}\right] \left(-\frac{1}{\sqrt{2}\, v}\right) \left(-\frac{\lambda v}{2\sqrt{2}}\right) \left(-\frac{\mathrm{i}}{m_R^2}\right)$$
$$\times \int \frac{\mathrm{i}\, d^4 p_E}{(2\pi)^4}\, \frac{-\mathrm{i}}{p^2 + m_R^2} = \frac{3\mathrm{i}\lambda}{4 m_R^2}\, I_D^{(0,1)}(m_R) \tag{3.47}$$

while $z^{(1)}_{W3.3(c)} = 0$ because its dimensionally regularized loop evaluates to $I^{(1,1)}_D(0)$ and so vanishes. Here the integrals $I^{(0,1)}_D(m)$ and $I^{(1,1)}_D(m)$ are defined in (3.37), and the nonzero result evaluates to

$$I^{(0,1)}_D(m) = \frac{m^2}{(4\pi)^{D/2}}\Gamma\left(1-\frac{D}{2}\right)\left(\frac{m^2}{\mu^2}\right)^{(D-4)/2}$$
$$= \frac{m^2}{16\pi^2}\left[\frac{1}{(D/2)-2}+\gamma-1+\ln\left(\frac{m^2}{4\pi\mu^2}\right)+O(D-4)\right]. \tag{3.48}$$

Summing these contributions (using $m_R^2 = \lambda v^2$) gives the one-loop prediction

$$z^{(1)}_W = \frac{I^{(0,1)}_D(m_R)}{2v^2} = \frac{\lambda}{16\pi^2}\left[\frac{1}{D-4}+\frac{1}{2}(\gamma-1)+\frac{1}{2}\ln\left(\frac{m_R^2}{4\pi\mu^2}\right)+O(D-4)\right]. \tag{3.49}$$

These corrections to z_W can again be absorbed into a rescaling of ξ – *i.e.* ξ is 're-normalized' by defining $\xi \to z_W^{-1/2}\xi$ – leading to the following rescaled version of (3.45):

$$S_W = -\int d^4x\left[\frac{1}{2}\partial_\mu\xi\,\partial^\mu\xi - \frac{a_W}{z_W^2}(\partial_\mu\xi\,\partial^\mu\xi)^2 + \cdots\right]. \tag{3.50}$$

The remainder of the matching calculation computes another observable using (3.50) and (3.49) and compares the result to the calculation of the same quantity in the full theory to read off the coefficient a_W. It is relatively simple to do this with the same quantity as used at lowest order in §1.2.1: the amplitude for $\xi(p)+\xi(q)\to\xi(p')+\xi(q')$ scattering, keeping only terms up to order $1/m_R^4$. Since this calculation is already performed in chapter 1 at leading order in λ, it suffices here to sketch how things change once subdominant contributions are included.

To this end, write the coefficients in S_W as a series in λ,

$$a_W = a^{(0)}_W + a^{(1)}_W + \cdots \qquad \text{and} \qquad z_W = 1 + z^{(1)}_W + \cdots, \tag{3.51}$$

for which $z^{(1)}_W$ is given in (3.49). Starting on the Wilson side of the calculation, for $\xi-\xi$ scattering the required graphs up to one loop order are given by (a), (b) and (c) of Fig. 2.5. Of these, graphs (b) and (c) and their crossed counterparts both evaluate to give a contribution to $\xi\xi\to\xi\xi$ scattering that is suppressed by more than just four powers of $1/m_R$. This is easy to see in dimensional regularization because the coefficients of the interactions are suppressed by more than $1/m_R^4$ and the loop integrals only involve massless states and so cannot introduce compensating factors of m_R into the numerator. If one wishes to work only to lowest order in $1/m_R$ but to higher order in λ, it suffices to work with the tree contribution, graph (a), within the Wilsonian theory, but with λ-corrected effective coefficients $a^{(1)}_W$ and $z^{(1)}_W$.

Evaluating graph (a) using the Wilsonian coupling a_W/z_W^2 expanded out to subdominant order in λ, $a^{(1)}_W - 2a^{(0)}_W z^{(1)}_W$, then gives:

$$\mathcal{A}^{W(a)}_{\xi\xi\to\xi\xi} = 8i\left(a^{(0)}_W + a^{(1)}_W - 2a^{(0)}_W z^{(1)}_W\right)$$
$$\times\left[(p\cdot q)(p'\cdot q') + (q\cdot q')(p\cdot p') + (p\cdot q')(p'\cdot q)\right] + \cdots. \tag{3.52}$$

Field-theory aficionados will recognize the $z_W^{(1)}$ term as the wave-function renormalization counter-term that cancels UV divergences due to the loops of Fig. 3.3 inserted into the external lines of the tree-level scattering graphs. (These graphs are not drawn explicitly in Fig. 2.4.)

This is to be compared with the one-loop contributions computed within the UV theory, working out to subdominant order in λ. The leading contribution comes from the tree graphs of Fig. 1.3, which evaluate to the result given in Eq. (1.28):

$$\mathcal{A}^{\rm full,tree}_{\xi\xi\to\xi\xi} = \frac{2i\lambda}{m_R^4}\Big[(p\cdot q)(p'\cdot q') + (q\cdot q')(p\cdot p') + (p\cdot q')(p'\cdot q)\Big] + \cdots, \quad (3.53)$$

where the ellipses contain terms of higher order than $1/m_R^4$. Equating this to the lowest-order part of (3.52) then gives the previously obtained tree-level result: $a_W^{(0)} = \lambda/(4m_R^4) = 1/(4\lambda v^4)$.

Repeating this procedure including one-loop $O(\lambda/16\pi^2)$ corrections in the UV theory is less trivial, but in principle proceeds in precisely the same manner. This involves evaluating the graphs of Fig. 2.4, plus the 'wave-function renormalization' graphs obtained by inserting Fig. 3.3 into the external lines of the tree-level scattering graphs of Fig. 1.3. The $z_W^{(1)}$ contributions of (3.52) are important for reproducing the effects of these latter graphs in the full theory. The final result is a prediction for $a_W^{(1)}$ that is of order $\lambda^2/(16\pi^2 m_R^4) = 1/(4\pi v^2)^2$ in size.

This example shows how loops in the Wilsonian theory are not counted in the same way as are loops in the UV theory. Loops in the Wilsonian theory necessarily involve higher powers of E/m_R (more about this below), while loops in the UV theory are suppressed by factors of $\lambda/(16\pi^2)$ only, with all powers of E/m_R appearing at each loop order.

3.2.3 Power Counting Using Dimensional Regularization

The previous section made assertions about the size of the contributions of loop graphs – like graphs (b) and (c) of Fig. 2.5 in the toy model – which this section explores more systematically. More generally, this section's goal is to track how a generic Feynman graph computed using the Wilsonian action depends on a heavy scale like $1/m_R$, given that this scale does not appear in the same way for all interactions within $\mathfrak{L}_W$.

The logic here is much as used in §3.1.1, where dimensional analysis was employed to track how the cutoff Λ appears in amplitudes. The only difference now is to regulate the UV divergences in these graphs with dimensional regularization, since the size of a dimensionally regulated integral is set by the physical scales (light masses or external momenta) that appear in the integrand (rather than Λ). The power counting rules obtained in this way are much more useful since they directly track how amplitudes depend on physical variables, rather than unphysical quantities like Λ that in any case cancel from physical quantities.

The basic observation is that dimensional analysis applied to a dimensionally regulated integral estimates its size as

$$\int \cdots \int \left(\frac{d^D p}{(2\pi)^D} \right)^A \frac{p^B}{(p^2 + q^2)^C} \sim \left(\frac{1}{4\pi} \right)^{DA} q^{DA+B-2C}, \tag{3.54}$$

with a dimensionless prefactor that depends on the dimension, D, of spacetime, and which may well be singular in the limit that $D \to 4$. Here, q represents the dominant scale appearing in the integrand of the momentum integrations. If the light particles appearing as external states in $A_E(q)$ should be massless, or highly relativistic, then the typical external momenta are much larger than their masses and q in the above expression represents these momenta.[9] If all masses and momenta are comparable, then q is their common value. The important assumption is that there is only one low-energy scale (the more complicated case of multiple hierarchies is examined in later chapters, in particular in the nonrelativistic applications of Part III for which small speed, $v \sim E_{\rm kin}/p$, can be regarded as a ratio of two separate low-energy scales).

With this in mind, the idea is to repeat the steps of §3.1.1 and use the effective action, $S_W = S_{W,0} + S_{W,\rm int}$, of (3.2) – repeated here for ease of reference:

$$S_{W,0} = -\frac{\mathfrak{f}^4}{M^2 v^2} \int d^4x \left[\partial_\mu \phi \, \partial^\mu \phi + m^2 \phi^2 \right]$$
$$S_{W,\rm int} = -\mathfrak{f}^4 \sum_n \frac{\hat{c}_n}{M^{d_n} v^{f_n}} \int d^4x \, O_n(\phi), \tag{3.55}$$

to compute amputated Feynman amplitudes, $\mathcal{A}_\mathcal{E}(q)$, having $\mathcal{E}$ external lines, $\mathcal{I}$ internal lines, $\mathcal{L}$ loops and $\mathcal{V}_n$ vertices coming from the effective interaction with label 'n'. Respectively denoting (as before) the number of derivatives and fields appearing in this interaction as d_n and f_n, the amplitude becomes proportional to the following multiple integral:

$$\int \cdots \int \left(\frac{d^D p}{(2\pi)^D} \right)^\mathcal{L} \frac{p^\mathcal{R}}{(p^2 + q^2)^\mathcal{I}} \sim \left(\frac{1}{4\pi} \right)^{2\mathcal{L}} q^{4\mathcal{L}-2\mathcal{I}+\mathcal{R}}, \tag{3.56}$$

where $\mathcal{R} = \sum_n d_n \mathcal{V}_n$ and the final estimate takes $D \to 4$. Liberally using the identities (3.4) and (3.5) then gives the following order of magnitude for $\mathcal{A}_\mathcal{E}(q)$:

$$\mathcal{A}_\mathcal{E}(q) \sim \mathfrak{f}^4 \left(\frac{1}{v} \right)^\mathcal{E} \left(\frac{Mq}{4\pi \mathfrak{f}^2} \right)^{2\mathcal{L}} \left(\frac{q}{M} \right)^{2+\sum_n (d_n-2)\mathcal{V}_n}. \tag{3.57}$$

This last formula is the main result, used extensively in many applications considered later. Its utility lies in the fact that it links the contributions of the various effective interactions in the effective lagrangian, (3.55), with the dependence of observables on small ratios of physical scales such as q/M. Notice in particular that more and more complicated graphs – for which $\mathcal{L}$ and $\mathcal{V}_n$ become larger and larger – are generically suppressed in their contributions to the graphical expansion if q is much smaller than the other scales M and $\mathfrak{f}^2/M$. This suppression assumes only that the powers appearing in (3.57) are all non-negative, and this is true so long as $d_n \geq 2$. The special cases where $d_n = 0, 1$ are potentially dangerous in this context, and require examination on a case-by-case basis.

[9] Any logarithmic dependence on q and infrared mass singularities that might arise in this limit are ignored here (for now), since the main interest is in following *powers* of ratios of the light and heavy mass scales.

Example: The Toy Model

The toy model Wilsonian action has the form of (3.55) provided we take $M = m_R$ and $\mathfrak{f}^2 = m_R v$, in which case (3.57) becomes

$$\mathcal{A}_{\mathcal{E}}(q) \sim q^2 v^2 \left(\frac{1}{v}\right)^{\mathcal{E}} \left(\frac{q}{4\pi v}\right)^{2\mathcal{L}} \left(\frac{q}{m_R}\right)^{\sum_n (d_n-2)\mathcal{V}_n}, \tag{3.58}$$

underlining that low-energy amplitudes come as a double expansion, in powers of both q/m_R and $q/4\pi v$. Furthermore, the symmetry $\xi \to \xi + \text{constant}$ implies that all interactions in $S_{W,\,\text{int}}$ have $d_n \geq 4$ and so interactions involving larger powers of $\mathcal{L}$ and $\mathcal{V}_n$ are always suppressed by one or both of these small parameters.

Recalling that the validity of the UV theory's loop expansion — $\lambda/(16\pi^2) \ll 1$ — implies that $m_R = \sqrt{\lambda}\, v \ll 4\pi v$, it follows that the expansion in q/m_R converges more slowly in this regime than does the expansion in $q/4\pi v$. Both expansion parameters in the low-energy theory become similar in size just at the border of the domain of perturbative validity in the UV theory. In no way does the effective theory require small λ in order to be predictive, and the diagnostic for weak coupling in the underlying UV theory is the existence of two separate scales, m_R and $4\pi v$, against which q becomes compared.

Applying the estimate of (3.58) to the graphs of Fig. 2.5 requires specializing to $\mathcal{E} = 4$, with graph (a) having $\mathcal{L} = 0$ and $\sum_n (d_n - 2)\mathcal{V}_n = 2$ and so

$$\mathcal{A}_4^{(a)}(q) \sim \frac{q^2}{v^2} \left(\frac{q}{m_R}\right)^2, \tag{3.59}$$

in agreement with the explicit tree-level formulae found earlier.

The one-loop contributions coming from graphs (b) and (c) similarly have $\mathcal{L} = 1$ and $\sum_n (d_n - 2)\mathcal{V}_n = 4$ and so both satisfy

$$\mathcal{A}_4^{(b)}(q) \sim \mathcal{A}_4^{(c)}(q) \sim \frac{q^2}{v^2} \left(\frac{q}{4\pi v}\right)^2 \left(\frac{q}{m_R}\right)^4. \tag{3.60}$$

Both are similar in size and are suppressed relative to the tree-level result by the small factor $q^4/(4\pi v\, m_R)^2$.

Notice that the above power counting estimates apply equally well for *any* theory of an abelian Goldstone boson subject to the shift symmetry $\xi \to \xi + \text{constant}$, provided that scales like v and M are regarded as both being large compared with q and that dimensionless parameters $\hat{a}$ and $\hat{b}$ in effective couplings like $a = \hat{a}\, \mathfrak{f}^4/(Mv)^4$ and $b = \hat{b}\, \mathfrak{f}^4/(Mv)^6$ in Eq. (2.97) are regarded as being independent and not systematically large. The low-energy limit for the specific toy model of §1.1 is more predictive than is the generic low-energy Goldstone-boson theory, because it predicts all of these parameters in terms of two fundamental ones: λ and v, say. The predictiveness enters both because of the relationship implied amongst the generic scales – *e.g.* $\mathfrak{f}^2 = Mv$ and $M^2/v^2 = \lambda$ – and the inferences it allows for the values of coefficients like $\hat{a}$ and $\hat{b}$, given as a series in powers of λ. What is usually informative in any application of EFT methods is therefore the comparison between the generic expectations for the assumed low-energy field content and the more detailed predictions allowed by specific UV theories that can give rise to this field content.

3.2.4 Power Counting with Fermions

It is straightforward to extend these results to include light fermions in the effective theory. The complication here is not really the statistics of the fields; it is the different momentum dependence of the propagators and the related canonical dimension of the fields.

To sort this out, first generalize the starting form assumed for the lagrangian to include fermion fields, ψ, in addition to boson fields, ϕ:

$$\mathfrak{L}_{\rm eff} = \mathfrak{f}^4 \sum_n \frac{c_n}{M^{d_n}}\, O_n\left(\frac{\phi}{v_B}, \frac{\psi}{v_F^{3/2}}\right). \tag{3.61}$$

Fermions and bosons come with different powers of the scales v_B and v_F because their kinetic terms (that dominate the unperturbed action in a weak-coupling perturbative analysis) involve different numbers of derivatives: only one for fermions but two for bosons. This also implies that fermion and boson propagators fall off differently for large momenta, with bosonic propagators varying like $1/p^2$ for large p and fermion propagators only falling like $1/p$. The resulting differences in contributions to the power counting of Feynman graphs makes it important to keep separate track of the number of fermion and boson lines.

To this end, it is useful to choose to label vertices using *three* indices: d_n, b_n and f_n. As before, d_n labels the numbers of derivatives in the interaction, but now b_n and f_n separately count the number of bose and fermi lines terminating at the vertex of interest. The number of vertices in a graph carrying a given value for d_n, b_n and f_n is, as before, labeled by $\mathcal{V}_n$.

Consider now computing an amplitude with $\mathcal{E}_B$ external bosonic lines, $\mathcal{E}_F$ external fermion lines and $\mathcal{I}_B$ and $\mathcal{I}_F$ internal bose and fermi lines. The constraints of graph-making relate these in three ways. First, the definition of the number of loops generalizes (3.5) to

$$\mathcal{L} = 1 + \mathcal{I}_B + \mathcal{I}_F - \sum_n \mathcal{V}_n. \tag{3.62}$$

Similarly, 'conservation of ends' now holds separately for both bosonic and for fermionic lines and so implies that (3.4) is replaced by the two separate conditions

$$2\mathcal{I}_B + \mathcal{E}_B = \sum_n b_n \mathcal{V}_n \quad \text{and} \quad 2\mathcal{I}_F + \mathcal{E}_F = \sum_n f_n \mathcal{V}_n. \tag{3.63}$$

Repeating, with the lagrangian of Eq. (3.61), the power counting argument which led (using dimensional regularization) to Eq. (3.57) now gives the following result instead:

$$\mathcal{A}_{\mathcal{E}_B\mathcal{E}_F}(q) \sim \mathfrak{f}^4 \left(\frac{1}{v_B}\right)^{\mathcal{E}_B} \left(\frac{1}{v_F}\right)^{3\mathcal{E}_F/2} \left(\frac{Mq}{4\pi\,\mathfrak{f}^2}\right)^{2\mathcal{L}} \left(\frac{q}{M}\right)^{\mathcal{P}}, \tag{3.64}$$

where the power $\mathcal{P}$ can be written

$$\mathcal{P} = 2 + \mathcal{I}_F + \sum_n (d_n - 2)\,\mathcal{V}_n = 2 - \frac{1}{2}\,\mathcal{E}_F + \sum_n \left(d_n + \frac{1}{2}\,f_n - 2\right)\mathcal{V}_n. \tag{3.65}$$

Since $\mathcal{I}_F \geq 0$, the first equality shows (3.64) is suppressed relative to the corresponding term in the purely bosonic result (3.57), as makes sense since each fermionic propagator is order $1/q$, and so is suppressed by q relative to the bosonic propagator $1/q^2$. The second equality trades the dependence on $\mathcal{I}_F$ for $\mathcal{E}_F$ using (3.63).

The factor $(q/M)^{-\mathcal{E}_F/2}$ in (3.64) might also seem problematic at low energies, indicating as it does that more external lines necessarily imply more factors of the large ratio M/q. However, because such factors are fixed for all graphs contributing to any explicit process with a given number of external legs they usually do not in themselves undermine the validity of a perturbative expansion.

Furthermore, for the specific case where $\mathcal{A}_{\mathcal{E}_B\mathcal{E}_F}(q)$ represents a scattering amplitude each external fermion line corresponds to an initial-state or final-state spinor – $u_{q\,\sigma}$ or $\overline{u}_{q\,\sigma}$ – or the corresponding antiparticle spinor – $v_{q\,\sigma}$ or $\overline{v}_{q\,\sigma}$ – labelled by the corresponding state's momentum and spin. But each of these is itself proportional to an external particle – and so low-energy, $O(q)$ – scale, as can be seen from their appearance in spin-averaged expressions like $\sum_\sigma u_{q\,\sigma}\,\overline{u}_{q\,\sigma} = m - i\not{q}$ and $\sum_\sigma v_{q\,\sigma}\,\overline{v}_{q\,\sigma} = -m - i\not{q}$. This $q^{1/2}$ scaling of each external fermion line systematically cancels the factor $q^{-\mathcal{E}_F/2}$ in the amplitude $\mathcal{A}_{\mathcal{E}_B\mathcal{E}_F}(q)$, leading to non-singular predictions for scattering process at low energies. The same is true for effective couplings in S_W if these are obtained by matching to scattering processes.

Dangerous Interactions

As usual, interactions with the same number of fields and derivatives as the kinetic terms – either $f_n = 0$ and $d_n = 2$ (for bosons) or $f_n = 2$ and $d_n = 1$ (for fermions) – are unsuppressed by powers of q/M, beyond the usual loop factor. Interactions with more fields or derivatives than the kinetic terms additionally suppress a graph each time they are used. But interactions with no derivatives and two or fewer fermions can be potentially dangerous at low energies, introducing as they do negative powers of the small ratio q/M.

The kinds of interactions that are dangerous in this way are terms in a scalar potential ($d_n = f_n = 0$) and Yukawa couplings ($d_n = 0$ and $f_n = 2$). In principle, these kinds of interactions can be genuine threats to the consistency of the low-energy expansion, and whether such interactions are consistent with low-energy physics depends on the details.

What can make these interactions benign at low energies is if they do not carry too much energy for generic field configurations, $\phi \sim v_B$ and $\psi \sim v_F^{3/2}$. For instance, suppose, following the discussion around Eq. (3.17), that the scalar potential only carries energy density $\mathfrak{f}_V^4 \ll \mathfrak{f}^4$ when fields are order $\phi \sim v_B$ in size, such as if

$$V(\phi) \sim \mathfrak{f}_V^4 \sum_r g_r \left(\frac{\phi}{v_B}\right)^r . \tag{3.66}$$

In particular, the $r = 2$ term represents a mass for the field ϕ of order $m_B^2 \sim \mathfrak{f}_V^4/v_B^2$, so a natural criterion for ϕ to survive into the low-energy theory might be that $\mathfrak{f}_V^4 \sim m_B^2 v_B^2$ with $m_B \lesssim q$ for q a typical (possibly relativistic) momentum in the low-energy sector.

If this is the case then – assuming the couplings g_r are order unity – all the dimensionless couplings c_n of Eq. (3.61) for these particular interactions are secretly suppressed, with $c_n(d_n = 0) \sim g_r\,(\mathfrak{f}_V^4/\mathfrak{f}^4) \sim g_r\,(m_B^2 v_B^2/\mathfrak{f}^4)$. The contributions of these particular $d_n = f_n = 0$ interactions to (3.64) then become

$$\prod_{d_n=f_n=0}\left[c_n\left(\frac{q}{M}\right)^{-2}\right]^{\mathcal{V}_n} \sim \prod_r\left[g_r\left(\frac{\mathfrak{f}_V^4 M^2}{\mathfrak{f}^4 q^2}\right)\right]^{\mathcal{V}_r} \sim \prod_r\left[g_r\left(\frac{m_B^2}{q^2}\right)\left(\frac{v_B^2 M^2}{\mathfrak{f}^4}\right)\right]^{\mathcal{V}_r}, \tag{3.67}$$

which is no longer enhanced because $m_B \lesssim q$.

A similar story goes through for Yukawa interactions for which an interaction like

$$\mathfrak{L}_{\text{yuk}} \sim \mathfrak{f}_Y^4 \sum_r h_r \left(\frac{\phi}{v_B}\right)^r \left(\frac{\overline{\psi}\psi}{v_F^3}\right), \tag{3.68}$$

would contribute a fermion mass of order $m_F \sim \mathfrak{f}_Y^4/v_F^3$ for fields $\phi \sim v_B$ and couplings $h_r \sim O(1)$. This can remain a genuine low-energy mass satisfying $m_F \lesssim q$ if $\mathfrak{f}_Y^4 \sim m_F v_F^3$ is systematically small relative to $\mathfrak{f}^4$. In this case, the contribution of these $d_n = 0$ but $f_n = 2$ terms to (3.64) (assuming the h_r are order unity) then become

$$\prod_{d_n=0,\,f_n=2}\left[c_n\left(\frac{q}{M}\right)^{-1}\right]^{\mathcal{V}_n} \sim \prod_r\left[h_r\left(\frac{\mathfrak{f}_Y^4 M}{\mathfrak{f}^4 q}\right)\right]^{\mathcal{V}_r} \sim \prod_r\left[h_r\left(\frac{m_F}{q}\right)\left(\frac{v_F^3 M}{\mathfrak{f}^4}\right)\right]^{\mathcal{V}_r}, \tag{3.69}$$

which again would not be enhanced.

3.3 The Big Picture ◇

So far, this chapter has examined a variety of types of effective actions, 1PI, 1LPI and Wilsonian, and explored how dimensional analysis constrains how their effective couplings contribute to low-energy observables (both using a cutoff and in dimensional regularization). It is easy at this point to lose sight of the forest for the trees, so this section aims to reiterate some broader features of the overall EFT program.

3.3.1 Low-Energy Theorems

Given the huge number of effective couplings available in the Wilson action, it is tempting to conclude that there is nothing robust that can be said about the low-energy world. But this is misleading, because there are some predictions that can be made very robustly without running into renormalization ambiguities. These predictions are generally known as an EFT's 'low-energy theorems', and consist of those predictions that depend only on the low-energy theory's leading non-renormalizable interaction and the number and type of low-energy states present. In particular, they are insensitive to the values of the long list of higher-dimension effective couplings in the Wilsonian EFT.

The main observation underlying the construction of these theorems is that locality implies that the higher-dimensional effective interactions all depend only on polynomials of momenta. They do so because they are built using only powers of fields and their derivatives. As a result, they do not in themselves give rise to non-analytic contributions to scattering amplitudes at vanishing external momenta.

This observation may not seem so remarkable given that the power counting arguments of the previous sections also only involve powers of the external energy scale q. But (as remarked in footnote [9]) these power counting arguments focus exclusively on powers of q and explicitly ignore quantities like $\ln(q^2/\mu^2)$ that could make the full result non-analytic at $q = 0$. Indeed, such non-analytic contributions do arise in scattering amplitudes, with their presence being guaranteed by general principles [38, 39].

In fact, direct evaluation of the scattering amplitude for $\xi\xi \to \xi\xi$ scattering at one-loop order in the Toy Model (in particular graph (b) of Fig. 2.5) gives precisely this kind of contribution, behaving as $q^8 \ln(q^2/m_R^2)$. The coefficient of this dependence is proportional to $a_{\rm eff}^2/(4\pi)^2$ (consistent with (3.60)), with $a_{\rm eff} = \lambda/(4m_R^4) = 1/(4v^2m_R^2)$ the lowest-order effective coupling found from tree-level scattering in Eqs. (1.15) and (1.16).

Here is the point: although loop graphs like this arise potentially at the same order as tree graphs built from higher-derivative terms (whose coefficients renormalize the UV divergences in the loops), the higher-derivative interactions only contribute to the analytic part of the amplitude at $q = 0$ and so *cannot* contribute to the coefficient of the non-analytic $q^8 \ln q^2$ term. The leading coefficients of such non-analytic terms are therefore absolute predictions, given only the lowest-order non-renormalizable effective coupling (in the Toy Model case the coupling $a_{\rm eff}$). These predictions, collectively with the other predictions depending only on $a_{\rm eff}$, are called the EFT's low-energy theorems. (The applications of EFT methods in later chapters give practical examples of these theorems for the strong interactions §8 and for General Relativity §10.)

High-energy physics cannot contribute to non-analytic behaviour for q in a low-energy regime because the general arguments of [38, 39] imply this non-analytic behaviour arises only at the threshold for the particle whose circulation within loops generates the non-analytic behaviour. Non-analytic behaviour at $q^2 = 0$ comes from massless particles and so does not arise when heavy particles are integrated out. (This is also why such non-analytic dependence on derivatives does not arise in the Wilson action itself, for which the light states responsible for singularities at $q^2 = 0$ are not yet integrated out.)

3.3.2 The Effective-Action Logic $^\diamond$

Historically, what made theories with non-renormalizable interactions daunting was the seeming necessity to include an infinite number of interactions. This is partly because there are only a finite number of renormalizable (and super-renormalizable) interactions, but an infinite number of non-renormalizable ones. But it is also true that one usually cannot cherry-pick amongst non-renormalizable interactions because they are all needed to absorb the UV divergences appearing at higher loop orders.[10]

The utility of power counting formulae like (3.57) lies in their ability to cut through the conundrum of how to deal with so many interactions, and so to organize how

[10] It is this need for a nominally infinite number of couplings to renormalize UV divergences that underlies the name 'non-renormalizable'.

to calculate predictively (including quantum effects). The key observation is that non-renormalizable theories always implicitly involve a low-energy expansion, so predictivity must be assessed order-by-order in this expansion. The logic for doing so unfolds with the following steps:

[1] Choose the accuracy desired in the answer. (For instance an accuracy of 1% might be required for a particular scattering amplitude.)

[2] Determine the order in the small ratio of scales q/M (*e.g.* q/m_R in the toy model) required to achieve the desired accuracy. (For instance, if $q/M = 0.1$, then order $(q/M)^2$ would be required to achieve 1% accuracy.)

[3] Use the power counting results to identify which terms in $\mathfrak{L}_W$ can contribute to the observable of interest to the desired order in q/M. At any fixed order only a finite number (say, N) of terms in $\mathfrak{L}_W$ can contribute.

[4a] If the underlying theory is known and is calculable, then compute the coefficients of the N required effective interactions to the needed accuracy.

[4b] If the underlying theory is unknown, or is too complicated to permit explicit *ab initio* calculation of $\mathfrak{L}_W$, then treat the N required coefficients as free parameters. This is nonetheless predictive if more than N observables can be identified whose predictions depend only on these parameters.

It is in the spirit of step [4a] that EFT methods are developed for the toy model in previous sections: the full theory is explicitly known and parameters are within a calculable regime. (For the toy model this corresponds to using the full theory within the perturbative small-λ regime.) In this case, EFT methods simply provide an efficient means to identify and calculate the combination of parameters on which low-energy observables depend.

It is option [4b], however, that is responsible for the great versatility of EFT methods, because it completely divorces the utility of the low-energy theory from the issue of whether or not the underlying theory is understood. It allows EFT methods to be used for *any* low-energy situation, regardless of whether the underlying theory is completely unknown or is known but too complicated to allow reliable predictions. (Examples of both of these cases are considered in subsequent sections.)

From this point of view, the conundrum of dealing with an infinite number of interactions is really only a problem in the limit where infinite precision is required of the answer, since it is only then that one must work to all orders in the low-energy expansion.

Within this point of view, traditional renormalizable theories are simply the special case where the above logic is invoked, but with accuracy that only requires working at zeroth order in q/M. This is equivalent to renormalizability because in this case *all* interactions suppressed by $1/M$ can be dropped, which amounts to dropping all interactions whose couplings have dimensions of negative power of mass (*i.e.* all non-renormalizable interactions).

Renormalizable theories are revealed in this way to be the ones that should always dominate in the limit that the light scales q and the heavy scales M are so widely separated as to allow the complete neglect of heavy-particle effects. This is the beginnings of an explanation of *why* renormalizable interactions turn out to play such ubiquitous roles throughout physics: the message behind their success is that any 'new' physics not included in them involves scales too high to be relevant in practice.

3.4 Summary

The main topic of this chapter is power counting: the tool that makes the Wilson action a precise instrument. The purpose of this tool is to identify the effective interactions and Feynman graphs that are required to make physical predictions to any given order in the low-energy expansion, E/M.

At face value, this counting seems like it should be easy: the power of $1/M$ for any graph is found simply by collecting all such factors from the coefficients of a graph's effective interactions. What complicates this argument is the extreme sensitivity of some loop integrals to short wavelengths, which leads to the appearance of large scales like M in numerators rather than denominators. Successful tracking of high-energy scales in loops therefore requires handling their ultraviolet divergences.

This chapter provides two kinds of power-counting estimates. One of these regulates UV divergences with explicit cutoffs, and because short wavelengths often dominate in loops, what is mostly learned is how a graph depends on this cutoff regulator. This can be useful, particularly when asking how effective couplings flow as successive scales are integrated out. This flow is described by exact renormalization-group methods, culminating with Wilson–Polchinski type evolution equations.

The second kind of power-counting estimate this chapter provides regulates divergences using dimensional regularization. This is usually more convenient for practical applications, partly because dimensional regularization allows useful symmetries to be kept explicit. For EFTs, dimensional regularization also proves useful because the absence of an explicit cutoff scale simplifies dimensional power-counting arguments revealing how graphs depend on low- and high-energy scales. These arguments culminate in the very useful expressions (3.64) and (3.65), appropriate for low-energy theories dominated by a single low-energy scale.

Power counting has all of the utility and glamour of accounting: it is both crucial and not that exciting to do. The payoff for understanding it in detail is the power of the overall perspective it provides. One such insight is about what the cancellation of divergences during renormalization really means. In this new picture renormalization stops being a miraculous cancellation between divergences and counterterms and starts being an obvious cancellation of a scale, Λ, that is not really in the original problem.

A second insight concerns the utility of both renormalizable theories and non-renormalizable theories. Non-renormalizable theories are not daunting in themselves once it is recognized that they can be predictive to the extent that they only hope to capture low orders in a low-energy expansion. The enormous predictive success of renormalizable theories similarly emerges as a special case of this general low-energy predictiveness when the UV scale M is so high that it suffices to work to zeroth order in the ratio E/M.

Exercises

Exercise 3.1 Derive the cut-off dependent power-counting result of Eq. (3.12), starting from the effective lagrangian of Eq. (3.2).

Exercise 3.2 Extend the result of Exercise 3.1 to see how Eq. (3.12) changes when graphs are built using fermions as well as bosons.

Exercise 3.3 For Quantum Electrodynamics there is only a single type of interaction, $\mathfrak{L}_{\rm int} = ieA_\mu\overline{\psi}\gamma^\mu\psi$, for which $d = 0$ (no derivatives), $f = 2$ (two fermion fields) and $b = 1$ (one bosonic field). Show that any Feynman graph built with only this vertex (and the Dirac and electromagnetic propagators) satisfies

$$\mathcal{I}_B = \frac{\mathcal{E}_F}{2} + \mathcal{L} - 1, \quad \mathcal{I}_F = \frac{\mathcal{E}_F}{2} + \mathcal{E}_B + 2(\mathcal{L} - 1) \quad \text{and}$$
$$\mathcal{V} = \mathcal{E}_F + \mathcal{E}_B + 2(\mathcal{L} - 1),$$

for the number of internal fermionic lines, bosonic lines and vertices. Here, $\mathcal{E}_F$ and $\mathcal{E}_B$ are the number of external fermionic and bosonic lines and $\mathcal{L}$ is the number of loops, defined by (3.62).

Apply the power counting arguments leading to (3.12) to QED and show that they predict that a generic amputated Feynman graph evaluated at zero external momentum varies as $\mathcal{A}_{\mathcal{E}_F\mathcal{E}_B} \propto \Lambda^{\mathcal{S}}$ where $\mathcal{S} = 4 - \frac{3}{2}\mathcal{E}_F - \mathcal{E}_B$ is called the graph's superficial degree of divergence. Show that $\mathcal{S}$ is only non-negative for configurations for which tree-level vertices exist in the action.

Show that the power of electromagnetic coupling appearing in any graph is

$$\mathcal{A}_{\mathcal{E}_F\mathcal{E}_B} \propto e^{\mathcal{E}_F+\mathcal{E}_B-2}\left(\frac{e^2}{16\pi^2}\right)^{\mathcal{L}},$$

and thereby that it is $\alpha/4\pi$, where $\alpha = e^2/4\pi$ is the fine-structure constant, that controls the loop expansion.

Exercise 3.4 The Fermi theory of weak interactions involves only fermions and has only a single interaction, for which $d = 0$ (no derivatives) and $f = 4$ (four fermion fields). Show that any Feynman graph built using only this vertex (and the Dirac propagators) satisfies $\mathcal{I} = 2(\mathcal{L} - 1) + \frac{1}{2}\mathcal{E}$ and $\mathcal{V} = (\mathcal{L} - 1) + \frac{1}{2}\mathcal{E}$, where $\mathcal{E}$ denotes the number of external lines and $\mathcal{L}$ is the number of loops, defined by (3.5). Show that power counting predicts a generic amputated Feynman graph evaluated at zero external momentum varies as $\mathcal{A}_{\mathcal{E}} \propto \Lambda^{\mathcal{S}}$ where $\mathcal{S} = 2(\mathcal{L}+1) - \frac{1}{2}\mathcal{E}$. Unlike for QED (see Exercise 3.3), this eventually becomes positive (and so the graph becomes UV divergent) for large enough $\mathcal{L}$, regardless of the number of external lines involved.

Exercise 3.5 Complete the calculation of this chapter and use the toy model of §1.1 to compute the order λ^2/m_R^4 contribution to the effective coupling a appearing in the interaction $a\,(\partial_\mu\xi\,\partial^\mu\xi)^2 \in \mathfrak{L}_W$.

Exercise 3.6 Derive the central power counting result, Eqs. (3.64) and (3.65), starting from the lagrangian (3.61) and regulating UV divergences using dimensional regularization.

Exercise 3.7 Consider a Goldstone boson (or axion) with shift symmetry $\xi \to \xi$ + constant coupled to a fermion ψ at low energies. What are the lowest-dimension couplings possible involving these fields within the Wilsonian effective lagrangian? Assume the couplings are such that the lagrangian has the form (3.61) with two independent scales, M and v, with $v_B = v$, $v_F = M$ and $\mathfrak{f}^2 = Mv$. Use the power counting result of (3.64) to derive the leading dependence on these scales of the reduced amplitude $\mathcal{A}_4(q)$ for $2 \to 2$ fermion scattering in this theory. Draw the Feynman graphs that provide this leading contribution. Identify and draw the Feynman graphs that provide the next-to-leading contribution, both in the case where $M \sim 4\pi v$ and when $M \ll 4\pi v$. How suppressed are these subleading contributions relative to the leading order contribution? Repeat these leading and subleading power counting estimates for $2 \to 2$ axion-fermion scattering rather than fermion-fermion scattering.

Exercise 3.8 Compute the graphs of Fig. 2.5 using the Toy Model's low-energy Wilsonian action to evaluate the one-loop contribution to the $\xi(q_1)\xi(q_2) \to \xi(q_3)\xi(q_4)$ scattering amplitude, $\mathcal{A}_4$, as a function of the Mandelstam variables $s = -(q_1 + q_2)^2$, $t = (q_1 - q_3)^2$ and $u = (q_1 - q_4)^2$. Use dimensional regularization to evaluate any UV divergences you encounter. Perform the same calculation for $\phi_I\phi_I \to \phi_I\phi_I$ using the renormalizable Feynman rules of the underlying full theory, also to one-loop order using dimensional regularization. Show that these calculations agree on the leading low-energy limit of the amplitude, and in particular on the form and coefficient of the logarithmic term (which in the full theory requires a cancellation of the first few powers of q^2). Which Feynman graphs are responsible for the cancellations in the full theory?

Exercise 3.9 For the discussion of the 'method of regions' introduce a cutoff into the integrals $\mathcal{I}_\epsilon^{IR}$ and $\mathcal{I}_\epsilon^{UV}$ by defining

$$\mathcal{I}_\epsilon^{IR} := \int_0^\Lambda \left(\frac{\mu}{k}\right)^\epsilon f(k, M, m) \quad \text{and} \quad \mathcal{I}_\epsilon^{UV} := \int_\Lambda^\infty \left(\frac{\mu}{k}\right)^\epsilon f(k, M, m),$$

with integrands as given in (3.41) and (3.42). Evaluate the first few terms in the $1/M$ expansions explicitly as functions of m, M and Λ. Show that the sum $\mathcal{I}_\epsilon = \mathcal{I}_\epsilon^{IR} + \mathcal{I}_\epsilon^{UV}$ is given by the same result as in (3.40). This shows that there is no loss in removing the cutoff when defining $\mathcal{I}_\epsilon^{IR}$ and $\mathcal{I}_\epsilon^{UV}$, as done in the main text.

4 Symmetries

The previous sections involving the toy model show that the low-energy implications of an EFT – such as the suppression of scattering amplitudes by powers of energy – can be much more transparent when some fields (*e.g.* the field ξ in the toy model) are used to represent the light particles in the Wilsonian theory than they are when expressed in terms of others (such as ϕ_I in the toy model). Why should this be?

Notice that the issue here is *not* that different variables imply different predictions, since the calculations of §1 reveal the low-energy suppression of scattering amplitudes to be precisely the same when computed using either ξ or ϕ_I. The issue instead is why this suppression is manifest at every step when using ξ, whereas with ϕ_I the suppression emerges quite mysteriously only at the end of the calculation due to cancellations amongst the contributing Feynman graphs.

This section argues that the main difference between these variables is the way they realize the symmetries of the system. How they are realized is relevant because symmetries (by way of Goldstone's theorem – see below) are ultimately the origin of the low-energy suppression seen in scattering amplitudes. As is so often the case, the lesson is: although predictions for physical quantities can be made using any variables you like, if you use the wrong ones you will be sorry.[1]

To make this point it is first good to step back and summarize some implications of symmetries more generally. In particular, from an EFT perspective the discussion divides into two cases, depending on whether or not particles related to one another by a symmetry all lie within the low-energy effective theory or if some symmetries relate low-energy states to high-energy states. This latter situation can happen in particular when the relevant symmetry is spontaneously broken. From the point of view of EFTs, the main observation is that the nature of the description necessarily changes if the energy scale, v, associated with symmetry breaking becomes much larger than the scale, M, associated with any heavy states that have been integrated out.

4.1 Symmetries in Field Theory ♡

The first step is to review how symmetries act within quantum field theory, and more generally within quantum mechanics.

[1] This paraphrases one of Steven Weinberg's three laws of theoretical physics [40].

As reviewed in Appendix C.4, a symmetry is described in quantum mechanics by a unitary transformation,[2] $|\psi\rangle \to |\psi'\rangle = U|\psi\rangle$ with $U^*U = UU^* = I$, within Hilbert space, that leaves the system's Hamiltonian unchanged:

$$H \to H' = UHU^* = H. \tag{4.1}$$

Such transformations are important for (at least) two reasons:[3]

- *Spectral degeneracy:* Because (4.1) implies that $HU = UH$, a symmetry can only relate distinct energy eigenstates to one another if they share exactly the same energy eigenvalue. That is, if $H|\psi_i\rangle = E_i|\psi_i\rangle$ and $H|\psi_j\rangle = E_j|\psi_j\rangle$ and $|\psi_j\rangle = U|\psi_i\rangle$ are all true for some i and j, then $E_i = E_j$. This ensures that energy eigenspaces can all be organized into linear representations of the symmetry group: $U|\psi_i\rangle = \mathcal{U}^j{}_i|\psi_j\rangle$ for some coefficients $\mathcal{U}^j{}_i$, where $U_1U_2 = U_3$ implies that $(\mathcal{U}_1)^i{}_j(\mathcal{U}_2)^j{}_k = (\mathcal{U}_3)^i{}_k$.
- *Conservation laws:* If a symmetry is labelled by a continuous parameter, θ (such as is true for rotations in space, or the symmetry (1.21) of the toy model), then it can be written $U(\theta) = \exp[\mathrm{i}\theta Q]$, for some hermitian operator Q (unitarity of U implies that Q is hermitian). Because Q is hermitian, it is an observable, and because $[U, H] = 0$ implies that $[Q, H] = 0$, the quantity Q is conserved. That is, if $Q|\psi(t = 0)\rangle = \mathfrak{q}\,|\psi(t = 0)\rangle$ for some real eigenvalue $\mathfrak{q}$ at a particular time, then $Q|\psi(t)\rangle = \mathfrak{q}\,|\psi(t)\rangle$ for all t. This follows because $[Q, H] = 0$ implies that $Q|\psi(t)\rangle = Q\exp[-\mathrm{i}Ht]|\psi(0)\rangle = \exp[-\mathrm{i}Ht]\,Q|\psi(0)\rangle$.

Similar implications also apply in quantum field theory, though with important qualifications. The difference arises because in essence field theory makes quantum mechanics local by assigning different operators at different spacetime points. As a result, symmetries in field theory are usually defined in terms of local transformations amongst fields that leave the action invariant – such as, for the toy model, the transformation (1.21), whose action leaves the lagrangian density $\mathfrak{L}$ of (1.1) unchanged.

This difference in starting point sometimes leads to symmetries being realized differently in field theory than in ordinary quantum mechanics, as this and the following sections describe. In particular, for reasons explained below, it turns out that having a symmetry act linearly on fields need not imply that particles fall into linear representations of the symmetry with identical energies.

A second issue raised by the local framework of field theory is the possibility that continuous symmetries might have position-dependent symmetry parameters: $\theta = \theta(x)$. Spacetime dependent symmetry transformation rules are called *local* or *gauge* symmetries, in contrast with *global* symmetries – for which θ and $U(\theta)$ do not depend on spacetime position. The focus of this chapter is mostly on global symmetries, though some of the issues arising for gauge symmetries are also discussed (both here and in Appendices C.3.3 and C.5).

2 Except for time-reversal, which is described by an anti-unitary transformation. The discussion of symmetries in quantum mechanics goes back to [41].

3 Here and throughout, the Einstein summation convention is used, for which repeated indices are implicitly summed over their entire range: $\mathcal{U}^j{}_i|\psi_j\rangle := \sum_j \mathcal{U}^j{}_i|\psi_j\rangle$.

A further distinction amongst symmetries is between internal symmetries and spacetime symmetries. These differ by whether or not they act on spacetime position, with, for example,

$$\phi^a(x) \to U\phi^a(x)U^* = \mathcal{U}_b{}^a\, \phi^b(x), \tag{4.2}$$

being an example of an internal symmetry (because the spacetime coordinate x is unchanged), while a Lorentz transformation like

$$V^\mu(x) \to UV^\mu(x)U^* = \Lambda_\nu{}^\mu V^\nu(x') \quad \text{with} \quad x'^\mu = \Lambda_\nu{}^\mu x^\nu, \tag{4.3}$$

is a representative spacetime symmetry. Both internal and spacetime symmetries can arise in global or gauged varieties, with the local spacetime symmetries leading to the diffeomorphism invariance of general-covariant theories like General Relativity. Unless otherwise stated, most of this chapter restricts to internal symmetries, which is not too restrictive in practice because it includes the majority of the symmetries of practical interest in later applications.

4.1.1 Unbroken Continuous Symmetries

Suppose, then, that a field theory enjoys a symmetry defined as some action-preserving infinitesimal continuous global transformation of the fields of the form $\phi^i \to \phi^i + \delta\phi^i$ with

$$\delta\phi^i := \omega^a \Sigma^i_a(\phi), \tag{4.4}$$

where ϕ^i denote a generic collection of fields and ω^a denote a set of independent and spacetime-independent symmetry parameters, while $\Sigma^i_a(\phi)$ are a given (possibly nonlinear) collection of functions of the fields at a specific spacetime point.

What makes this a symmetry is the requirement that it leaves invariant the system's action: $\delta S = \int d^4x\ \delta\mathfrak{L} = 0$. For internal symmetries the transformation satisfies the stronger condition $\delta\mathfrak{L} = 0$ separately at each point in spacetime, while for spacetime transformations the invariance of the action only requires the weaker condition

$$\delta\mathfrak{L} = \partial_\mu \left(\omega^a V^\mu_a\right) = \omega^a \partial_\mu V^\mu_a, \tag{4.5}$$

for some quantities $V^\mu_a(\phi)$, since in this case $\delta S = \int d^4x\ \delta\mathfrak{L}$ can still vanish.

Noether's Theorem

In field theory, the existence of a continuous class of action-preserving field transformations guarantees the existence of a conserved current, j^μ; with there typically being one current for every global continuous symmetry of the action [42]. To low orders in the derivative expansion it is usually enough to work with actions that depend only on the fields and their first derivatives, so for simplicity the rest of the argument deriving these currents is restricted to this case.

Consider therefore an action $S = \int d^4x\ \mathfrak{L}(\phi, \partial_\mu\phi)$, that by assumption is invariant under the transformations (4.4). Including global spacetime symmetries in the discussion, this invariance implies that $\mathfrak{L}$ must vary at most into a total derivative, so (4.5) is satisfied with $\delta\mathfrak{L}$ on the left-hand side found by directly varying the fields and their derivatives. Equating to zero the coefficient of the arbitrary constant ω^a in the result then gives:

$$\begin{aligned}\partial_\mu V_a^\mu &= \frac{\partial \mathfrak{L}}{\partial \phi^i}\,\Sigma_a^i + \frac{\partial \mathfrak{L}}{\partial(\partial_\mu \phi^i)}\,\partial_\mu \Sigma_a^i \\ &= \left[\frac{\partial \mathfrak{L}}{\partial \phi^i} - \partial_\mu\left(\frac{\partial \mathfrak{L}}{\partial(\partial_\mu \phi^i)}\right)\right]\Sigma_a^i + \partial_\mu\left(\frac{\partial \mathfrak{L}}{\partial(\partial_\mu \phi^i)}\,\Sigma_a^i\right).\end{aligned} \tag{4.6}$$

This equation holds as an identity, both for arbitrary field configurations, ϕ^i, and for arbitrary (though spacetime-independent) symmetry parameters, ω^a.

The theorem follows from this last equation, which says that the definitions

$$j_a^\mu := \frac{\partial \mathfrak{L}}{\partial(\partial_\mu \phi^i)}\,\Sigma_a^i - V_a^\mu, \tag{4.7}$$

automatically satisfy the property

$$\partial_\mu j_a^\mu = 0, \tag{4.8}$$

whenever they are evaluated at any solution to the equations of motion for ϕ^i – *i.e.* on fields satisfying $\delta S = 0$, which is equivalent to the field equation

$$\frac{\partial \mathfrak{L}}{\partial \phi^i} - \partial_\mu\left(\frac{\partial \mathfrak{L}}{\partial(\partial_\mu \phi^i)}\right) = 0. \tag{4.9}$$

The conclusion, Eq. (4.8), is more general than the relativistic notation being used here seems to suggest, since it holds also for nonrelativistic systems. For these systems write $\rho_a = j_a^0$ for the time component of j_a^μ, and denote its spatial components by the three-vector $\mathbf{j}_a$. Then current conservation – Eq. (4.8) – becomes the continuity equation

$$\frac{\partial \rho_a}{\partial t} + \nabla\cdot\mathbf{j}_a = 0. \tag{4.10}$$

Eqs. (4.8) and (4.10) are conservation laws because they guarantee that the *charges*, Q_a, defined by

$$Q_a(t) = \int_{\text{fixed } t} \mathrm{d}^3x\,\rho_a(\mathbf{r},t) = \int_{\text{fixed } t} \mathrm{d}^3x\, j_a^0(x), \tag{4.11}$$

are conserved in the sense that they are independent of t. This t-independence may be seen by using Stokes' theorem to infer

$$\partial_t Q_a = \int \mathrm{d}^3x\,\partial_t\rho_a = -\int \mathrm{d}^3x\,\nabla\cdot\mathbf{j}_a = -\oint_{r\to\infty} \mathrm{d}^2\Omega\,\mathbf{e}_r\cdot\mathbf{j}_a = 0, \tag{4.12}$$

where $\mathbf{e}_r = \mathbf{r}/r$ is the unit vector in the radial direction, and the last equality assumes boundary conditions are such that the net flux of the current $\mathbf{j}_a$ through a sphere at spatial infinity vanishes.

Representation on Particle States

The charges Q_a provide the link back to the usual description of symmetries in quantum mechanics because their commutator with the fields gives the symmetry transformation itself

$$i\omega^a \left[Q_a, \phi^i(x)\right] = \mathrm{i}\omega^a \int_{y^0=x^0} \mathrm{d}^3y \left[\rho_a(y), \phi^i(x)\right] = \omega^a \, \Sigma^i_a[\phi(x)] = \delta\phi^i(x). \tag{4.13}$$

The first equality here uses conservation of Q_a to choose the time at which $j^0_a(x)$ is evaluated to agree with the time appearing in $\phi^i(x)$. The second equality then uses the definition (4.7), written as

$$\rho_a = j^0_a = \frac{\partial \mathfrak{L}}{\partial \dot{\phi}^i} \Sigma^i_a - V^0_a = \Pi_j \, \Sigma^i_a + V^0_a, \tag{4.14}$$

where overdots denote differentiation with respect to time and $\Pi_j(x) = \delta S/\delta\dot{\phi}^j(x)$ is the canonical momentum for the field $\phi^i(x)$. The second equality of (4.13) then follows from these definitions, together with the equal-time canonical commutation relations

$$\left[\Pi_j(\mathbf{x},t), \phi^i(\mathbf{y},t)\right]_{x^0=y^0} = -\mathrm{i}\,\delta^3(\mathbf{x}-\mathbf{y})\delta^i_j \quad \text{and} \quad \left[\phi^i(\mathbf{x},t), \phi^j(\mathbf{y},t)\right] = 0. \tag{4.15}$$

This derivation also assumes both $\Sigma^i_a(\phi)$ and V^0_a depend only on ϕ^j and not also on the canonical momenta.

Eq. (4.13) makes a connection to the usual story of symmetries in quantum mechanics because it ensures that $U = \exp[\mathrm{i}\omega^a Q_a]$ (if well-defined – see below for when it is not) would be the unitary operator that implements the action of a finite symmetry transformation within the Hilbert space:

$$\phi^i \to \tilde{\phi}^i = U\phi^i U^*. \tag{4.16}$$

To see how this works, consider a weakly coupled system of particles, for which interactions can be treated perturbatively. Working within the interaction picture means that the fields satisfy the free-field equations and so can be expanded in a complete set of free single-particle modes, $u^i_n(x)$ (see Appendix C for details)

$$\phi^i(x) = \sum_n \left[u^i_n(x)\mathfrak{a}_n + \text{c.c.}\right], \tag{4.17}$$

where $\mathfrak{a}_n$ is the destruction operator of a particle with label 'n,' $\mathfrak{a}_n|m\rangle = \delta_{mn}|0\rangle$, whose adjoint is the particle creation operator: $|n\rangle = \mathfrak{a}^*_n|0\rangle$. Here $|0\rangle$ denotes the usual no-particle ground state defined by $\mathfrak{a}_n|0\rangle = 0$.

If the no-particle state is invariant under the symmetry (as would be automatic if it were non-degenerate and separated from all other energy eigenstates by an energy gap – as is often true for simple systems – then $U|0\rangle = |0\rangle = U^*|0\rangle$, and so single-particle states are related to one another by the symmetry just like in ordinary quantum mechanics. That is, because (4.16) implies that $\tilde{\mathfrak{a}}_n = U\mathfrak{a}_n U^*$ (and its adjoint) it follows that

$$|\tilde{n}\rangle = \tilde{\mathfrak{a}}^*_n|0\rangle = (U\mathfrak{a}^*_n U^*)|0\rangle = U\,\mathfrak{a}_n{}^*|0\rangle = U|n\rangle. \tag{4.18}$$

In particular, this ensures all the usual consequences of symmetries within quantum mechanics; in particular, that the single-particle states $|\tilde{n}\rangle$ and $|n\rangle$ must have the same energy, and so on. Since internal symmetries commute with spacetime translations, they do not change particle momenta and so particles related by a symmetry must also have the same rest mass.

The above arguments show that when the ground state is invariant under a symmetry then all of the usual implications of the symmetry in quantum mechanics go through as usual. In particular, there is no loss of generality in representing the symmetry linearly on single-particle states and so also using linear transformations amongst the creation and annihilation operators and fields: $\Sigma^i_a(\phi) = S_j{}^i\,\phi^j$ with $S_j{}^i$ field-independent. Such a symmetry is said to be *linearly realized*.

The toy model provides an explicit example of this type of symmetry, but only in the special case that $v = 0$ since this choice makes the classical ground state $\phi = v$ invariant under the symmetry $\phi \to e^{i\omega}\phi$ – or (1.21). As expected, in this case both fields ϕ_R and ϕ_I represent particles with exactly the same mass (both are massless in this limit). More generally, if the toy model were instead to have a scalar potential with the sign of the $\phi^*\phi$ term reversed, as in $V = V_0 + m^2\phi^*\phi + \frac{1}{4}\lambda\,(\phi^*\phi)^2$, then the invariant configuration $\phi = 0$ remains the ground state also for nonzero masses, and in this case both ϕ_R and ϕ_I share the common nonzero mass m.

4.1.2 Spontaneous Symmetry Breaking

The key assumption in the previous section is that the ground state is invariant, and this is absolutely crucial for the above arguments to go through. Furthermore, this is not an empty exception: for field theories the ground state can fail to be invariant.[4] When the ground state of a system is not invariant under a symmetry of its action the symmetry is said to be *spontaneously broken*.

If a symmetry is spontaneously broken, then (by assumption) another state is produced once a transformation is applied to the ground state. Since the transformation is a symmetry, this new state must have the same energy and so also be a candidate ground state. For a continuous symmetry one expects a continuous family of vacua, all sharing the same energy. This is indeed what happens for the toy model, for which the semiclassical vacuum corresponds to any spacetime-independent configuration satisfying $\phi^*\phi = v^2$. The one-parameter family of ground states is parameterized by $\phi = v\,e^{i\xi}$ for any constant ξ. From here on a ground state of such a system is denoted by $|\Omega\rangle$ rather than $|0\rangle$, to emphasize the fact that it can be more complicated than the single no-particle state of a simple harmonic Fock space.

Whenever $U|\Omega\rangle \neq |\Omega\rangle$ the line of argument given above that says single-particle states, $|n\rangle = \mathfrak{a}^*_n|\Omega\rangle$, must be linearly related by the symmetry, $|\tilde{n}\rangle = U|n\rangle$, also fails.[5] It fails because the symmetry changes the ground state, $|\Omega\rangle \to |\widetilde{\Omega}\rangle$, in addition to acting on the single-particle states that are built from them. As a result, fields related by the symmetry (like ϕ_R and ϕ_I of the toy model) need no longer correspond to particles with equal masses, unlike what happens when the vacuum is invariant. This is seen explicitly in the toy model spectrum when $v \neq 0$, since in this case the fields ϕ_R and ϕ_I represent particles with masses $m_R = \sqrt{\lambda}\,v$ and $m_I = 0$, respectively, despite being linearly related by the symmetry (1.21).

[4] This is unlike what happens for the quantum mechanics of a small number of degrees of freedom, for which the ground state tends to be unique (and so therefore is also invariant under a symmetry transformation).

[5] For field theories there can be problems even defining operators like U for spontaneously broken symmetries [43] (see also Appendix C.5.1).

Although traditional implications (like equality of particle masses) can break down for spontaneously broken symmetries, these symmetries nonetheless do come with consequences, as is now described [3–5].

Goldstone's Theorem

Whenever the ground state of a system does not respect one of the system's global continuous symmetries, there are very general implications for the low-energy theory that are summarized by *Goldstone's theorem* [4].

Goldstone's theorem states that any system for which a continuous, global symmetry is spontaneously broken must contain a state, $|G\rangle$ – called a *Goldstone mode*, or *Nambu-Goldstone boson* [6] – with the defining property that it is created from the ground state by a spacetime-dependent symmetry transformation.

In equations, $|G\rangle$ is defined by the condition that the following matrix element cannot vanish:

$$\langle G|\, \rho(\mathbf{r},t)|\Omega\rangle \neq 0, \tag{4.19}$$

where $|\Omega\rangle$ represents the ground state and[7] $\rho = j^0$ is the density for the symmetry's conserved charge, as is guaranteed to exist by Noether's theorem.

All of the properties of a Goldstone state follow from the definition (4.19), but before turning to them it is worth first sketching why this equation is true. The starting point is the assumption of the existence of a *local order parameter*. This is a field, $\phi(x)$, in the problem satisfying two defining conditions:

1. ϕ must transform nontrivially under the symmetry in question: *i.e.* there is another field, $\psi(x)$, for which:
$$\delta\psi \equiv \mathrm{i}[Q, \psi(x)] = \phi(x), \tag{4.20}$$
where Q is the conserved charge defined by integrating the current density, $\rho(\mathbf{r},t)$, throughout all of space.
2. The field ϕ must have a nonzero expectation in the ground state:
$$\big\langle\phi(x)\big\rangle := \langle\Omega|\phi(x)|\Omega\rangle = v \neq 0. \tag{4.21}$$
This last condition would be inconsistent with Eq. (4.20) if the ground state were invariant under the symmetry of interest, since invariance means $Q|\Omega\rangle = 0$, and this would mean the expectation value of Eq. (4.20) must vanish.

To see why (4.19) follows from (4.20) and (4.21) use the following steps. First substitute (4.20) into Eq. (4.21). Second, use $Q = \int \rho \; \mathrm{d}^3x$ in the result, as is guaranteed to be possible by Noether's theorem. Third, insert a partition of unity, $1 = \sum_n |n\rangle\langle n|$, between the operators ρ and ψ. The resulting expression shows that if no state exists satisfying the defining condition, Eq. (4.19), then the right-hand side of Eq. (4.21) must vanish, in contradiction with the starting assumptions.

[6] For internal symmetries this state must be a boson, but for graded symmetries like supersymmetry spontaneous breakdown ensures the existence of a Goldstone fermion, the goldstino.

[7] The nonrelativistic notation $\rho = j^0$ is used to emphasize that the conclusions presented are not specific to relativistic systems.

Goldstone's theorem states that the consistency of the matrix element, Eq. (4.19), with the conservation law, Eq. (4.10), requires the state $|G\rangle$ to have a number of important properties. Two of these are the state's spin and statistics. Because $\rho(x)$ transforms under rotations as a scalar and because $|\Omega\rangle$ is rotationally invariant, it follows that $|G\rangle$ must have spin zero and so (from the spin-statistics theorem) must also be a boson.

Furthermore, the state $|G\rangle$ must also be *gapless*, in that its energy must vanish in the limit that its (three-) momentum vanishes:

$$\lim_{p\to 0} E(p) = 0. \tag{4.22}$$

In relativistic systems, for which $E(p) = \sqrt{p^2 + m^2}$ with m being the particle's rest mass, the gapless condition is equivalent to the masslessness of the Goldstone particle. This gaplessness follows by using the fact that $|G\rangle$ and $|\Omega\rangle$ are both energy and momentum eigenstates (with $E_\Omega = \mathbf{p}_\Omega = 0$) to write (see *e.g.* Appendix C.5.1)

$$\langle G|\rho(\mathbf{r}, t)|\Omega\rangle = e^{iE_G t - i\mathbf{p}_G\cdot\mathbf{r}}\langle G|\rho(0)|\Omega\rangle. \tag{4.23}$$

Using this (and a similar expression for $\langle G|\mathbf{j}(\mathbf{r}, t)|\Omega\rangle$) when taking the matrix element of the conservation equation, (4.10), between $\langle G|$ and $|\Omega\rangle$, leads to

$$0 = \langle G|\left(\partial_t\rho + \nabla\cdot\mathbf{j}\right)|\Omega\rangle = \mathrm{i}\left[E_G\langle G|\rho(\mathbf{r}, t)|\Omega\rangle - \mathbf{p}_G\cdot\langle G|\mathbf{j}(\mathbf{r}, t)|\Omega\rangle\right]. \tag{4.24}$$

Since the last term vanishes as $\mathbf{p}_G \to 0$ and we know $\langle G|\rho(\mathbf{r}, t)|\Omega\rangle \neq 0$, it follows that E_G also vanishes in this limit.

More generally, the Goldstone boson must completely decouple from all of its interactions in the limit that its momentum vanishes. This is because Eq. (4.19) states that in the zero-momentum limit the Goldstone state literally is a symmetry transformation of the ground state. As a result, it is *completely indistinguishable* from the vacuum in this limit.

These properties say a lot about the low-energy behaviour of any system that satisfies the assumptions of the theorem. Gaplessness guarantees that the Goldstone boson must itself be one of the light states of the theory, and so it must be included in any effective lagrangian analysis of this low-energy behaviour. Low-energy decoupling ensures that the Goldstone mode must be very weakly coupled in the low-energy limit, and strongly limits the possible form its interactions can take.

The toy model again provides a simple example of all of these consequences. The symmetry of this model is spontaneously broken whenever $v \neq 0$, and there is certainly a gapless state in the spectrum whenever this is true: the state represented by the field ϕ_I or ξ. To see that this state satisfies (4.19) requires constructing the Noether current for the symmetry (1.21). For the toy model this is

$$\begin{aligned} j_\mu = \mathrm{i}(\phi\,\partial_\mu\phi^* - \phi^*\partial_\mu\phi) &= \sqrt{2}\left(v + \frac{\hat\phi_R}{\sqrt{2}}\right)\partial_\mu\hat\phi_I - \hat\phi_I\,\partial_\mu\hat\phi_R \\ &= \sqrt{2}\left(v + \frac{\chi}{\sqrt{2}}\right)^2 \frac{\partial_\mu\xi}{v} \end{aligned} \tag{4.25}$$

and so its single-particle matrix element is $\langle\xi(p)|j^\mu(x)|\Omega\rangle \propto \sqrt{2}\,vp^\mu\, e^{-\mathrm{i}p\cdot x}$, which is nonzero whenever $v \neq 0$. Furthermore, conservation of the Noether current also implies that this particle is massless because

$$0 = \langle \xi(p)|\partial_\mu j^\mu(x)|\Omega\rangle \propto \partial_\mu\left(\sqrt{2}\,v p^\mu\, e^{-ip\cdot x}\right) = -i\sqrt{2}\,v\, p_\mu p^\mu e^{-ip\cdot x} = i\sqrt{2}\,v\, m^2 e^{-ip\cdot x}. \tag{4.26}$$

The more general statement that interactions for the Goldstone particle must turn off at zero momentum is also clear for the toy model, since this is the property (much discussed in §1) that the scattering amplitudes for toy-model massless states approach zero as the scattering energy goes to zero.

4.2 Linear vs Nonlinear Realizations ◇

With the above discussion in mind, we are in a position to formalize more explicitly how symmetries are realized within a Wilsonian low-energy effective theory. The most basic statement is that the low-energy theory must share the symmetry properties of the full UV theory, both for the symmetries of the action and the symmetries of the ground state. That is, it should be possible to read off the symmetry properties of the system directly from the EFT at any scale one chooses.[8]

The discussion of the previous section shows that how this is done depends on whether or not the symmetry of interest is spontaneously broken. If the symmetry is unbroken, then it can without loss of generality be realized to act linearly on the fields. Since in this case all particles related by the symmetry share the same mass, if any of them is light enough to be in the low-energy theory then all of them are. But the same need not be true if the symmetry is spontaneously broken.

Whether a linear realization is possible for a spontaneously broken symmetry depends on the relative size of two important scales: the scale M of the UV physics whose integrating out led to the EFT in question, and the scale v of the expectation value responsible for the symmetry's spontaneous breaking (see Fig. 4.1). This section argues that a linear realization can continue to be useful when $v \ll M$, but

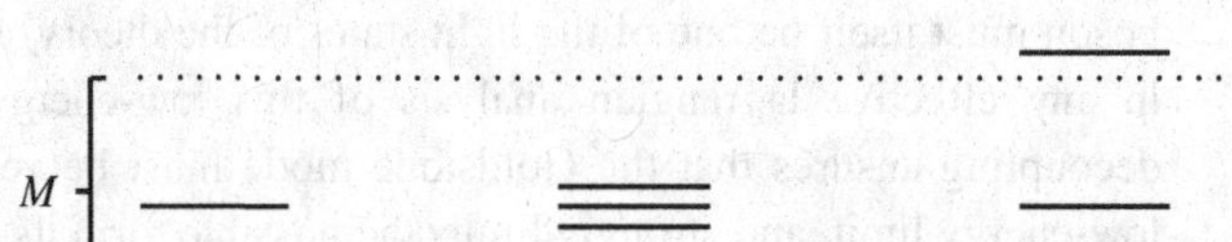

Fig. 4.1 A sketch of energy levels in the low-energy theory relative to the high-energy scale, M, and the relative splitting, v, within a global 'symmetry' multiplet. Three cases are pictured: panel (a) unbroken symmetry (with unsplit multiplets); panel (b) low-energy breaking ($v \ll M$) and panel (c) high-energy breaking (with $v \gtrsim M$). Symmetries are linearly realized in cases (a) and (b) but not (c). If spontaneously broken, symmetries in case (c) are nonlinearly realized in the EFT below M. (If explicitly broken in case (c) there is little sense in which the effective theory has approximate symmetry at all.)

[8] Anomalies – the failure of a classical symmetry to survive quantization – can complicate this statement slightly, inasmuch as what can look like an anomalous symmetry at some scales can look like a classical breaking of the symmetry at other scales. More about this in §4.3.

necessarily breaks down if M is too small relative to v. If the symmetry cannot be realized linearly, there turns out to be an alternative standard realization that is always possible for a broad class of symmetry-breaking patterns.

The toy model provides an example where linear realization remains useful even though a symmetry spontaneously breaks. That is, suppose the entire toy model were itself regarded as being the low-energy limit of some larger UV completion; being obtained by integrating out some new states, ψ^i, with masses $M_\psi \gg v \gtrsim m_R = \sqrt{\lambda}\, v$. In this case, the toy model (plus all possible higher-dimensional interactions built from ϕ) is the EFT for energies in the regime $v \ll E \ll M_\psi$, and includes both fields ϕ_R and ϕ_I, with the $U(1)$ toy model symmetry realized linearly as in (1.21).

But this linear realization is no longer possible for an EFT aimed at the regime $E \ll m_R$. Linear realization is not possible in this regime because one of the two fields required for its existence is now integrated out. Notice that the possibility of integrating out part of a symmetry multiplet only arises if the symmetry is spontaneously broken because only then can some particles within a symmetry multiplet differ in mass. In the toy model what is required in this regime is a way to realize the symmetry using the EFT's only low-energy field. The required realization is provided by the symmetry under which the field ξ shifts, as in (1.23). The inhomogeneous nature of this symmetry is characteristic of a Goldstone boson, since the symmetry is necessarily inconsistent with choosing a specific vacuum value for the field ξ.

4.2.1 Linearly Realized Symmetries

The simplest situation is when $v \ll M$, which includes the case $v = 0$, where the symmetry is not broken at all. In this case, the low-energy theory contains the right number of particles to fill out linear representations of the symmetry, and so fields can be chosen in such a way as to represent the symmetry linearly. For an internal symmetry this means we can take:[9]

$$\phi^i(x) \to \tilde{\phi}^i(x) = \mathcal{M}^i{}_j\, \phi^j(x), \tag{4.27}$$

for some choice of matrices $\mathcal{M}^i{}_j$.

Not much need be said in this case, which is the one most commonly used in practice. Because all of the fields needed for a linear realization are present, the potential order parameters are also present as fields within the low-energy theory. The only difference between unbroken symmetry and spontaneous breaking therefore lies in the choice of action and the energy it assigns to a nonzero value for the order parameters.

Explicit Symmetry Breaking

An important variation on this chapter's theme is the situation where symmetries are only approximate rather than exact. In this case, the action for the system at any scale is assumed to have the form

[9] The attentive reader may notice a difference in the index ordering between this equation and (4.2). These representations are conjugates of one another, with the choice of (4.2) designed to ensure that if $U_3 = U_1 U_2$ then $(\mathcal{U}_3)_i{}^j = (\mathcal{U}_1)_i{}^k (\mathcal{U}_2)_k{}^j$.

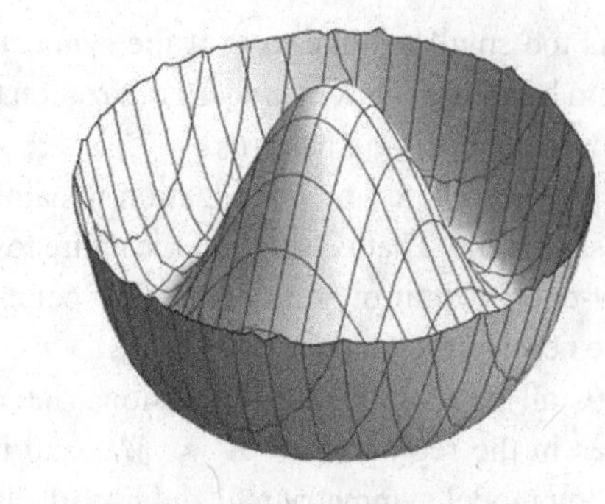

(a) The shape of the potential $V(\phi_R, \phi_I)$, in the limit of no explicit symmetry breaking, showing its sombrero shape and the circular line of minima.

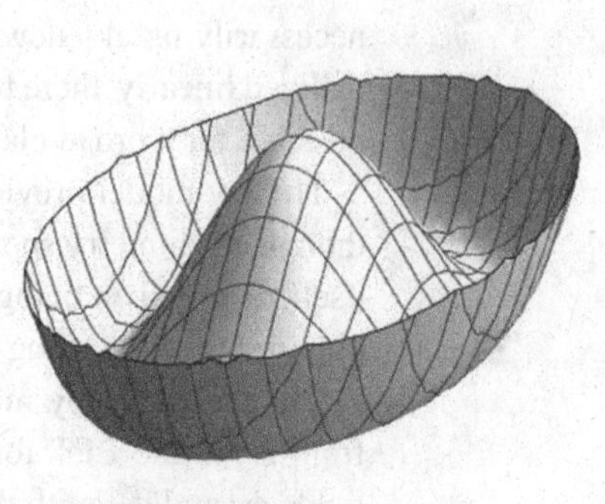

(b) The potential $V(\phi_R, \phi_I)$, with an explicit symmetry-breaking term added that is linear in ϕ_R, showing how the degeneracy of the minimum gets lifted.

$$S = S_{\rm inv} + \epsilon\, S_1 + \epsilon^2 S_2 + \cdots , \tag{4.28}$$

where $S_{\rm inv}$ is invariant under some group of symmetries while the S_i are not. Some small dimensionless parameter ϵ is assumed to be present to quantify the notion that the symmetry breaking is 'small'. In particular, if an expansion like (4.28) is possible in the full high-energy theory, it must also be possible in the low-energy theory (which can be a useful observation when explicit *ab-initio* calculations are not possible in the high-energy theory – an important example of which is described in §8).

An explicit example helps make things concrete. The toy model's $U(1)$ symmetry $\phi \to e^{i\theta}\phi$ is responsible for many of its predictions, such as the equality of the two particle masses when $v = 0$ and the elimination of the Goldstone boson from the scalar potential when $v \neq 0$. Both of these properties are easily seen to fail once interactions breaking the $U(1)$ symmetry are included. Approximate symmetries remain useful, however, because any failure of symmetry relations tends to zero as the symmetry breaking turns off (*i.e.* as $\epsilon \to 0$), allowing predictions to be given perturbatively, in powers of ϵ.

The simplest example of explicit symmetry breaking in this model is to add a term linear in ϕ that tilts the potential, such as by adding to (1.1) the symmetry-breaking term

$$\mathfrak{L}_{\rm sb} = \mu^3\,(\phi + \phi^*), \tag{4.29}$$

where μ is a symmetry-breaking parameter with dimensions of mass. Writing $\phi = \frac{1}{\sqrt{2}}\,(\phi_R + i\phi_I)$, the modified potential becomes

$$V(\phi_R, \phi_I) = \frac{\lambda}{16}\left(\phi_R^2 + \phi_I^2 - 2v^2\right)^2 - \sqrt{2}\,\mu^3\phi_R, \tag{4.30}$$

which has extrema that are given to leading order in $0 \le \mu^3/(\lambda v^3) \ll 1$ by

$$\phi_0 \simeq -\frac{2\mu^3}{\lambda v^2} \quad \text{and} \quad \phi_\pm \simeq \pm v + \frac{\mu^3}{\lambda v^2}. \tag{4.31}$$

The degenerate circular minimum present for the sombrero-shaped potential when $\mu = 0$ is now tilted, leaving a unique real minimum at ϕ_+ (assuming $\mu > 0$) with a saddle point on the opposite side of the circle at ϕ_-. ϕ_0 turns out to be the shifted local maximum at the centre of the Mexican hat. The mass eigenvalues for the two fields, found as before by evaluating the potential's second derivatives at $\phi_\pm$, then become

$$m^2_{R\pm} = \lambda v^2 \pm \frac{3\mu^3}{v} \quad \text{and} \quad m^2_{I\pm} = \pm\frac{\mu^3}{v}. \tag{4.32}$$

In particular, this shows how explicit symmetry breaking gives the would-be Goldstone boson state – commonly called a 'pseudo-Goldstone' boson [44] – a nonzero mass $m_G^2 = m^2_{I+} \simeq \mu^3/v$, that vanishes (as it must) when $\mu \to 0$.

Explicit symmetry breaking also interferes with the other low-energy Goldstone properties, as can be seen by expressing the toy model in terms of the fields χ and ξ using (1.22). With these variables, the symmetry-breaking term (4.29) modifies the potential from (1.25) to

$$V(\chi, \xi) = \frac{\lambda}{4}\left(\sqrt{2}\, v\, \chi + \frac{\chi^2}{2}\right)^2 - 2\mu^3\left(v + \frac{\chi}{\sqrt{2}}\right)\cos\left(\frac{\xi}{\sqrt{2}\, v}\right). \tag{4.33}$$

The would-be Goldstone particle, ξ, no longer drops out of the potential, and as a result its scattering amplitudes no longer need vanish at low energies, by a calculable amount in powers of μ^3. The low-energy Wilson action for ξ then also acquires a scalar potential whose leading contribution is proportional to μ^3, in order to capture the low-energy limit of the full theory's symmetry-breaking behaviour,

$$V_W(\xi) \simeq -2\mu^3 v \cos\left(\frac{\xi}{\sqrt{2}\, v}\right) + \cdots. \tag{4.34}$$

Here, the ellipses represent terms suppressed by additional powers of $1/m_R$ and/or μ^3.

The toy model illustrates what is also true in the general case: what were exact symmetry implications survive in an approximate form in the limit that the symmetry-breaking physics is small. Deviations from these predictions are then obtained by expanding systematically in powers of the small symmetry-breaking parameter ϵ. In particular, although the light particle ξ is no longer massless, it remains much lighter than the heavy χ particle provided the symmetry-breaking physics is small: $m_G^2/m_R^2 = \mu^3/(\lambda v^3)$. Although nonzero, its low-energy couplings at zero momentum are similarly suppressed. Chapter 8 includes a very practical example along these lines that was very influential for the development of effective field theories.

4.2.2 Nonlinearly Realized Symmetries

An important constraint on any low-energy EFT is the requirement that it shares, all the symmetry properties of its underlying UV completion. This is fairly straightforward to do when the symmetry is linearly realized since the symmetry then groups particles into multiplets with similar masses and couplings, both at high and low energies.

How spontaneously broken symmetries get expressed in the low-energy theory is more subtle, particularly when the symmetry-breaking scale, v, is bigger than the UV scalex M – *i.e.* panel (c) of Fig. 4.1. In the regime $E \ll M \ll v$ the low-energy field content can be inconsistent with realizing the symmetry linearly, since some members of multiplets in the UV theory might be heavy enough to have been integrated out while others remain light enough to appear in the effective theory.

This section explores how the symmetries of the underlying UV theory manifest themselves at low energies in this case.

Abelian Case

Once again, the toy model is informative, since it enjoys an abelian $U(1)$ global symmetry and its best-understood parameter regime (that of small λ) satisfies $m_R \lesssim v$. The low-energy theory described in earlier sections for $E \ll m_R$ therefore falls precisely into the regime of interest. With only one low-energy field present at these energies it is impossible to realize the $U(1)$ linearly, and it is instead realized as an inhomogeneous shift symmetry on the one low-energy field: $\xi \to \xi + \sqrt{2}\,\omega v$ (where ω is the symmetry parameter).

In §1 this formulation was found by redefining the fields $\phi = F(\chi)\, e^{i\xi/f}$ – with the specific form $F(\chi) = v + (\chi/\sqrt{2})$ not important and $f = \sqrt{2}\,v$ if ξ is to be canonically normalized. The significance of this choice is that ξ appears as would the parameter of a symmetry transformation, $\phi \to \phi\, e^{i\theta}$, with θ replaced by $\xi(x)/(\sqrt{2}\,v)$. Because ξ, when spacetime-independent, appears as does a symmetry parameter, constant configurations of ξ must drop out of the action (which is, after all, invariant under the symmetry). Consequently, the action can only depend on ξ through its derivative, $\partial_\mu \xi$. The heavy field χ is then found by identifying that part of ϕ that is in some sense orthogonal to this symmetry direction.

Two things are instructive about this construction. First, as the next section shows, it can be extended to more general (and in particular non-abelian) symmetries. Second, the rules for constructing the most general $U(1)$-invariant lagrangian in the EFT for the toy model are fairly simple, and also generalize to more complicated groups.

If written out pedantically, the instructions for building a general interaction for a $U(1)$ Goldstone boson (such as ξ in the toy model) go as follows. First build a generic local action using a vector field V_μ and its derivatives, plus any other fields ψ^m that happen to be in the low-energy theory, constrained only by Lorentz-invariance (or whatever other spacetime and unbroken internal symmetries are relevant). In particular, do so without making any reference at all to a global $U(1)$ symmetry. The terms involving the fewest V_μ fields and the fewest derivatives then have the general form

$$\mathfrak{L}(\psi, V_\mu) = \mathfrak{L}_0(\psi) - j^\mu(\psi) V_\mu - \frac{1}{2}\, m^{\mu\nu}(\psi)\, V_\mu V_\nu - n^{\mu\nu}(\psi)\, \partial_\mu V_\nu + \cdots\,, \tag{4.35}$$

where $\mathfrak{L}_0$, j^μ, $m^{\mu\nu}$ and $n^{\mu\nu}$ are functions of the ψ^m and their derivatives. Then *any* such lagrangian is automatically promoted to a $U(1)$-invariant one simply by replacing everywhere $V_\mu \to \partial_\mu \xi$.

The lesson is this: whereas a linearly realized symmetry restricts the kinds of terms that are allowed in a lagrangian (such as by forbidding interactions with unequal powers of ϕ and ϕ^* in the toy model lagrangian), the spontaneously broken $U(1)$ symmetry does not directly restrict the kinds of terms that can be written in (4.35). What the spontaneously broken $U(1)$ symmetry instead does is dictate how the Goldstone boson must couple to other fields at low energies, given a generic lagrangian, like (4.35), constrained only by the *unbroken* symmetries. It is only through the couplings of the Goldstone field that the low-energy EFT 'learns' about the existence of the broken $U(1)$ theory in its UV completion.

In particular, it is ultimately the shift-symmetry realization of the $U(1)$ symmetry in the low-energy EFT (with $v \gg M$) that forces the Goldstone boson to couple to other fields only through derivatives. In this way the implications of Goldstone's theorem emerge as automatic symmetry consequences when constructing the Wilson action. Properly realizing symmetries in low-energy effective actions means not having to be clever in order to extract their low-energy consequences, with no need for fancy 'current algebra' operator arguments (*e.g.* compare [11] with [45, 180, 181] or [48] to [49]).

Non-abelian Case ♠

This section sketches how nonlinear realizations work for more general patterns of spontaneous symmetry breaking, for which the action has a symmetry group G but the ground state is invariant only under a subgroup $H \subset G$. The treatment here is meant mainly to summarize the main results, but since this is an important EFT topic, more details are presented in Appendix C.6 about the motivation for and derivation of the results given here – the so-called standard realization, as well as a lightning review of the main properties of Lie groups and Lie algebras that are needed to do so.

There are many ways to represent the Goldstone bosons for a generic symmetry-breaking pattern, $G \to H$, but these are all related by field redefinitions to (and so are equivalent to) a standard one, whose properties are summarized here. (Geometrically, the Goldstone boson fields can be regarded as being coordinates on the coset space G/H; see Appendix C.6.) Before describing this, though, there are a few group-theoretical facts worth collecting.

Group-Theoretic Aside

As described in Appendix C.6, it is useful to work with the Lie algebra of G rather than in terms of the group itself. Any group element connected to the identity element, $g = 1$, can be written as a matrix exponential:

$$g = \exp\left[i\omega^a T_a\right], \tag{4.36}$$

where for a p-parameter group the p generators, T_a, $a = 1, \ldots, p$, form a basis of the Lie algebra of G. As is often the case in physics, in the first instance the interest is often in a specific matrix representation of the group elements for $g \in G$, perhaps acting on the fields of interest in the UV theory for which the symmetries are linearly realized, rather than something more abstract. For N fields the matrices g would be $N \times N$. As discussed earlier, these specific matrix representations are unitary and they can be also chosen to be real. (Complex fields, like ϕ in the toy model, can always be broken into their real and imaginary parts, like ϕ_R and ϕ_I.) When this is done the matrices T_a are hermitian and imaginary:

$$T_a = T_a^\dagger = -T_a^* = -T_a^T, \tag{4.37}$$

where the superscript 'T' denotes transpose.

The properties of the group G are encoded in its group multiplication law, and this also has implications for the matrices T_a. In particular, the closure property of the group multiplication law for G implies that the generators satisfy commutation relations

$$T_a T_b - T_b T_a = \mathrm{i}\, c_{abd}\, T_d, \tag{4.38}$$

with an implied sum on the index 'd'. Like the group multiplication law, the constant coefficients, c_{abd}, are characteristic of the group involved.

When describing a symmetry-breaking pattern where G breaks to H it is convenient to choose the basis of generators to include the generators of H as a subset:

$$\{T_a\} = \{t_i, X_\alpha\}, \tag{4.39}$$

where the t_i's generate the Lie algebra of H and the X_α's constitute the rest. The broken generators X_α typically do not also generate a group, since the commutator of two X_α's need not involve only X_α's with no t_i's. They instead can be regarded as generating[10] the space of 'cosets', G/H.

The closure of H under multiplication ensures that

$$t_i\, t_j - t_j\, t_i = \mathrm{i}\, c_{ijk}\, t_k, \tag{4.40}$$

with no X_α's on the right-hand side, so $c_{ij\,\alpha} = 0$. Under broad assumptions it is also possible to choose a basis of generators so that $c_{i\alpha j} = 0$, so

$$t_i\, X_\alpha - X_\alpha\, t_i = \mathrm{i}\, c_{i\alpha\beta} X_\beta, \tag{4.41}$$

with no t_j's on the right-hand side. This implies that the X_α's fall into a (possibly reducible) representation of H, which when exponentiated to a finite transformation implies that

$$h\, X_\alpha\, h^{-1} = L_\alpha{}^\beta X_\beta \tag{4.42}$$

for some coefficients, $L_\alpha{}^\beta$, and for any $h = \exp[\mathrm{i}\omega^i t_i] \in H$.

The Realization

For internal symmetries, with the symmetry-breaking pattern $G \to H$, the low-energy theory contains a Goldstone boson, ξ^α, for each broken generator, X_α (the counting can be different for other situations, like spacetime symmetries – see §14.3). The low-energy theory might also contain a collection of other non-Goldstone low-energy fields, χ^n, depending on the particular system of interest. The rest of this section provides an explicit nonlinear realization of the symmetry group G on the collection $\{\xi^\alpha\}$ and separately on the collection $\{\chi^n\}$.

As motivated in Appendix C.6, it is always possible to perform a field redefinition so that the fields ξ^α and χ^n transform according to [12, 13]

$$\xi^\alpha \to \tilde{\xi}^\alpha(\xi, g) \qquad \text{and} \qquad \chi^n \to \tilde{\chi}^n(\xi, g, \chi), \tag{4.43}$$

[10] Formally, a coset G/H is an equivalence class wherein two elements $g_1, g_2 \in G$ are regarded as equivalent if g_1 can be obtained from g_2 by multiplying by some $h \in H$.

where $\tilde{\xi}^\alpha$ and $\tilde{\chi}^n$ are defined by the relations

$$g\, e^{i\xi^\alpha X_\alpha} = e^{i\tilde{\xi}^\alpha X_\alpha}\, e^{iu^i t_i} \quad \text{and} \quad \widetilde{X} = e^{iu^i t_i}\, X. \tag{4.44}$$

Here, X (or $\widetilde{X}$) denotes a column vector whose entries are the fields χ^n (or $\tilde{\chi}^n$).

The first of Eqs. (4.44) should be read as defining the nonlinear functions $\tilde{\xi}^\alpha(\xi, g)$ and $u^i(\xi, g)$. Starting with $e^{i\xi \cdot X}$ one multiplies through on the left by $g \in G$ to construct a new element of G: $g\, e^{i\xi \cdot X}$. The functions $\tilde{\xi}^\alpha$ and u^i are then defined by decomposing this new matrix into the product of a factor, $e^{i\tilde{\xi} \cdot X}$, lying in G/H times an element, $e^{iu \cdot t}$, in H. The second of Eqs. (4.44) then defines the transformation rule for the non-Goldstone fields, χ^n.

These transformations simplify in the special case where $g = h$ lies in H, in which case both χ^n and ξ^α turn out to transform *linearly* under the unbroken symmetry transformations of H. The simplification happens because the above definitions in this case reduce to:

$$\xi^\alpha X_\alpha \to \tilde{\xi}^\alpha X_\alpha = h(\xi^\alpha X_\alpha)h^{-1} = \xi^\alpha L_\alpha{}^\beta X_\beta,$$
$$X \to \widetilde{X} = hX, \tag{4.45}$$

where the last equality in the first line uses (4.42).

More generally, the transformation laws are both inhomogeneous and nonlinear in the Goldstone fields, ξ^α. Explicit closed-form expressions can be found for infinitesimal transformations – c.f Eqs. (C.124) and (C.125) – which when expanded in powers of fields give[11]

$$u^i \approx -c^i{}_{\alpha\beta}\omega^\alpha \xi^\beta + O(\omega\xi^2) \quad \text{and} \quad \delta\xi^\alpha = \omega^\alpha - c^\alpha{}_{\beta\gamma}\omega^\beta \xi^\gamma + O(\omega\xi^2). \tag{4.46}$$

Because these transformations are nonlinear, they are effectively spacetime-dependent due to their dependence on the field $\xi^\alpha(x)$. This complicates the algorithm for finding the general form for invariant lagrangians, as is now briefly described.

Invariant Actions

The field-dependent matrix-valued quantity $U(\xi) := \exp[i\xi^\alpha(x)X_\alpha]$ provides the starting point for constructing actions that are invariant under transformations (4.43) and (4.44). To see why, notice that U transforms under the above rules as $U(\xi) \to U(\tilde{\xi})$, where $gU(\xi) = U(\tilde{\xi})\,\text{h}$, with

$$\text{h} := \exp\left[iu^i(\xi, g)t_i\right] \in H, \tag{4.47}$$

being the matrix appearing in (4.44). The key observation is that this implies that the combination[12] $U^{-1}\partial_\mu U$ transforms like a gauge-potential (compare this with Eq. (C.72), keeping in mind the footnote immediately after (C.130))

$$U^{-1}\partial_\mu U \to \tilde{U}^{-1}\partial_\mu \tilde{U} = \text{h}\,(U^{-1}\partial_\mu U)\,\text{h}^{-1} - \partial_\mu \text{h}\,\text{h}^{-1}. \tag{4.48}$$

---

[11] Indices on structure constants are raised and lowered using the Killing metric defined in Appendix C.4.1.
[12] This useful combination is called a Maurer-Cartan form [50].

Separating $U^{-1}\partial_\mu U$ into a piece proportional to X_α plus one proportional to t_i defines two important quantities, $\mathcal{A}^i_\mu(\xi)$ and $e^\alpha_\mu(\xi)$, according to

$$U^{-1}\partial_\mu U = -\mathrm{i}\mathcal{A}^i_\mu t_i + \mathrm{i}e^\alpha_\mu X_\alpha. \tag{4.49}$$

Extracting an overall factor of $\partial_\mu \xi^\alpha$, so that $\mathcal{A}^i_\mu = \mathcal{A}^i_\alpha(\xi)\ \partial_\mu\xi^\alpha$ and $e^\alpha_\mu = e^\alpha{}_\beta(\xi)\ \partial_\mu\xi^\beta$, then the explicit expressions for the small-ξ expansion of these quantities become (see Appendix C.6.3)

$$\mathcal{A}^i_\alpha(\xi) = -\int_0^1 ds\ \mathrm{Tr}\left[t^i e^{-\mathrm{i}s\,\xi\cdot X} X_\alpha\, e^{\mathrm{i}s\,\xi\cdot X}\right] \simeq \frac{1}{2}\, c^i{}_{\alpha\beta}\xi^\beta + O(\xi^2), \tag{4.50}$$

and

$$e^\alpha{}_\beta(\xi) = \int_0^1 ds\ \mathrm{Tr}\left[X^\alpha e^{-\mathrm{i}s\,\xi\cdot X} X_\beta\, e^{\mathrm{i}s\,\xi\cdot X}\right] \simeq \delta^\alpha{}_\beta - \frac{1}{2}\, c^\alpha{}_{\beta\gamma}\xi^\gamma + O(\xi^2). \tag{4.51}$$

Their infinitesimal transformation rules similarly are

$$\delta\mathcal{A}^i_\mu(\xi) = \partial_\mu u^i(\xi,\omega) - c^i{}_{jk}\, u^j(\xi,\omega)\mathcal{A}^k_\mu(\xi), \tag{4.52}$$

and

$$\delta e^\alpha_\mu(\xi) = -c^\alpha{}_{i\beta}\, u^i(\xi,\omega)\, e^\beta_\mu(\xi). \tag{4.53}$$

In this last expression, the structure constants themselves define representation matrices, $(\mathcal{T}_i)^\alpha{}_\beta = c^\alpha{}_{i\beta}$, of the Lie algebra of H, whose exponentials appear in (4.42).

To build self-interactions for the Goldstone bosons using these tools one combines the covariant quantity, $e^\alpha_\mu = e^\alpha{}_\beta\,\partial_\mu\xi^\beta$ in all possible H-invariant ways. This is simple to do since this quantity transforms very simply under G: $e_\mu \cdot X \to \mathtt{h}\,(e_\mu\cdot X)\,\mathtt{h}^{-1}$. Derivatives of e^α_μ are then included by using the covariant derivative constructed from $\mathcal{A}^i_\mu t_i$:

$$(D_\mu e_\nu)^\alpha = \partial_\mu e^\alpha_\nu + c^\alpha{}_{i\beta}\mathcal{A}^i_\mu\, e^\beta_\nu, \tag{4.54}$$

which transforms in the same way as does e^α_μ: $\delta(D_\mu e_\nu)^\alpha = -c^\alpha{}_{i\beta}u^i(D_\mu e_\nu)^\beta$.

The invariant lagrangian then is $\mathfrak{L}(e_\mu, D_\mu e_\nu, \dots)$, where the ellipses denote terms involving higher covariant derivatives and the lagrangian is constrained to be globally H invariant:

$$\mathfrak{L}(he_\mu h^{-1}, hD_\mu e_\nu h^{-1}, \dots) \equiv \mathfrak{L}(e_\mu, D_\mu e_\nu, \dots). \tag{4.55}$$

Whenever $\mathfrak{L}$ satisfies (4.55) for constant h, the definitions of $e^\alpha{}_\beta$ and $\mathcal{A}^i_\alpha$ ensure it is also *automatically* invariant under global G transformations.

For a Poincaré-invariant system, this leads to the following terms involving the fewest derivatives

$$\mathfrak{L}_{GB} = -\frac{1}{2}\, g_{\alpha\beta}(\xi)\,\partial^\mu\xi^\alpha\partial_\mu\xi^\beta + \text{(higher-derivative terms)}, \tag{4.56}$$

with $g_{\alpha\beta}(\xi) = f_{\gamma\delta}\, e^\gamma{}_\alpha\, e^\delta{}_\beta$ where $f_{\alpha\beta}$ is a constant positive-definite matrix that must satisfy

$$f_{\lambda\beta}c^\lambda{}_{i\alpha} + f_{\alpha\lambda}c^\lambda{}_{i\beta} = 0, \tag{4.57}$$

in order for the lagrangian of (4.56) to be G-invariant. In many situations the representation matrices $(\mathcal{T}_i)^\alpha{}_\beta$ form an irreducible representation, in which case

Schur's lemma implies that $f_{\alpha\beta}$ must be proportional to the unit matrix $f_{\alpha\beta} = F^2\delta_{\alpha\beta}$ where F is a constant parameter.

The action for the other matter fields is similarly constructed by using $\mathcal{A}^i_\mu(\xi)$ to build covariant derivatives for the χ^n: $D_\mu\mathcal{X} = \partial_\mu\mathcal{X} - i\mathcal{A}^i t_i\mathcal{X}$. Because the symmetry H is unbroken, these fields must all transform linearly under H: $\mathcal{X} \to h\mathcal{X}$, for some representation matrices, h, of H. This gets promoted to a nonlinearly realized G-transformation because the transformation law for the χ^n is $\mathcal{X} \to \mathrm{h}\mathcal{X}$, with $\mathrm{h} = \mathrm{h}(\xi, g) \in H$, as defined in (4.47). The covariant derivative $D_\mu\mathcal{X}$ is defined so that it also transforms in the same way as does $\mathcal{X}$ itself, $D_\mu\mathcal{X} \to \mathrm{h}(\xi, g)\, D_\mu\mathcal{X}$, under the nonlinearly realized G-transformations.

With these rules, any old globally H-invariant lagrangian for $\mathcal{X}$ automatically becomes promoted to a G invariant lagrangian once all derivatives are replaced by the ξ-dependent covariant derivatives. This works because global H-invariance of the original lagrangian means it satisfies

$$\mathfrak{L}(he_\mu h^{-1}, h\mathcal{X}, h\partial_\mu e_\nu h^{-1}, h\,\partial_\mu\mathcal{X}, \dots) \equiv \mathfrak{L}(e_\mu, \mathcal{X}, \partial_\mu e_\nu, \partial_\mu\mathcal{X}, \dots), \tag{4.58}$$

for any $h \in H$. But the above constructions are designed to ensure that each covariant quantity transforms under G as $e_\mu \to \mathrm{h}\, e_\mu\, \mathrm{h}^{-1}$, $\mathcal{X} \to \mathrm{h}\mathcal{X}$, $\mathcal{D}_\mu\mathcal{X} \to \mathrm{h}\,\mathcal{D}_\mu\mathcal{X}$ and so on, so the condition for G invariance

$$\mathfrak{L}(\mathrm{h}e_\mu\mathrm{h}^{-1}, \mathrm{h}\mathcal{X}, \mathrm{h}\,D_\mu e_\nu\mathrm{h}^{-1}, \mathrm{h}\,D_\mu\mathcal{X}, \dots) \equiv \mathfrak{L}(e_\mu, \mathcal{X}, D_\mu e_\nu, D_\mu\mathcal{X}, \dots), \tag{4.59}$$

becomes an automatic consequence of (4.58). As shown in Appendix C.6, this construction of the invariant lagrangian is also unique, given the assumed transformation rules for the fields.

4.2.3 Gauge Symmetries

Up to this point, the discussion has been completely aimed at global symmetries, for which the symmetry parameter is (by definition) independent of spacetime position. At this point, a brief aside is warranted on how local (or gauge) symmetries are nonlinearly realized within the low-energy Wilsonian EFT, where the parameters ω^a are no longer required to be constants.

The motivation for doing so is because this situation is not hypothetical. The twin constraints of Lorentz invariance and unitarity in quantum mechanics dictate that the couplings of *any* massless spin-one particle must be gauge invariant [51, 54] (see Appendix C.3.3), to the extent that their coupling to other matter at very low energies (*i.e.* their renormalizable interactions) is only possible if this other matter enjoys some sort of gauge symmetries. By extension, low-energy couplings of massive spin-one particles (whose masses are nonzero but very small compared with other scales) are only possible to matter that enjoys a spontaneously broken gauge symmetry [55–60]. This framework includes, in particular, all presently known fundamental[13] massive spin-one particles in nature [61].

[13] As this book makes clear, fundamental here simply means point-like in the best effective description known to date, and does not exclude the possibility of their being found to have substructure as knowledge improves.

Just like for global symmetries, the discussion naturally breaks up into linearly realized and nonlinearly realized symmetries, so each is considered in turn. For simplicity this section restricts itself to abelian symmetries, though the conclusions drawn apply more generally (see Appendix C.5).

At face value, it is simple to construct lagrangians invariant under linearly realized gauge symmetries along standard lines. One starts with a lagrangian that is invariant under a global symmetry and promotes the global symmetry to a gauge symmetry by combining all derivatives of the fields (like $\partial_\mu \psi^m$) with a gauge potential, A_μ, to make gauge-covariant derivatives (denoted $D_\mu \psi^m$). For abelian symmetries the gauge potential transforms as $A_\mu \to A_\mu + \partial_\mu \omega$. These covariant derivatives are designed so that $D_\mu \psi^m$ transforms under the position-dependent symmetry in precisely the same way as $\partial_\mu \psi^m$ did when the symmetry parameter was constant.

For example, in the UV version of the toy model, the symmetry transformation is $\phi \to e^{i\omega}\phi$, and so if ω were not a constant then $\partial_\mu \phi \to e^{i\omega}(\partial_\mu \phi + i\partial_\mu \omega\, \phi)$. The corresponding covariant derivative then is $D_\mu \phi = \partial_\mu \phi - iA_\mu \phi$ since this transforms like $D_\mu \phi \to e^{i\omega} D_\mu \phi$ even for spacetime-dependent ω. No additional symmetry restrictions are placed on the lagrangian itself beyond the requirements already imposed by global invariance.

The gauge-invariant version of the theory is then found by replacing $\partial_\mu \psi^m \to D_\mu \psi^m$ everywhere, and supplementing the result with a dependence on derivatives of A_μ, which appear through the gauge-invariant field strength $F_{\mu\nu} = \partial_\mu A_\nu - \partial_\mu A_\nu$. For instance, the gauge-invariant version of the toy model replaces (1.1) by

$$S := -\int d^4x \left[\frac{1}{4g^2} F^{\mu\nu} F_{\mu\nu} + D_\mu \phi^* D^\mu \phi + V(\phi^* \phi) \right], \tag{4.60}$$

where $D_\mu \phi = \partial_\mu \phi - iA_\mu \phi$ as before, and only renormalizable interactions are kept (*i.e.* those with couplings having non-negative dimension in powers of mass).

This describes the particles of the toy model interacting with a spin-one particle described by the field A_μ. The field A_μ can be canonically normalized (see Appendix C.3.3) by rescaling $A_\mu \to gA_\mu$. When this is done the covariant derivative becomes $D_\mu \phi = \partial_\mu \phi - igA_\mu \phi$, revealing g to be the gauge coupling whose value controls how strongly A_μ couples to ϕ. If the scalar potential is minimized at $\phi^* \phi = v^2$, expanding about $\phi = v$ shows that the quadratic terms in A_μ become

$$-\frac{1}{4} F_{\mu\nu} F^{\mu\nu} - D_\mu \phi^* D^\mu \phi \supset -\frac{1}{4} F_{\mu\nu} F^{\mu\nu} - g^2 v^2 A_\mu A^\mu, \tag{4.61}$$

and so spontaneous symmetry breaking gives the spin-one particle a mass $M_A^2 = 2g^2 v^2$.

So far, so standard. The next step is to ask how a gauge symmetry is manifest in a low-energy EFT in the limit $v \gg M$ and so for which the symmetry is nonlinearly realized.

Explicit vs Spontaneous Breaking for Gauge Symmetries

In principle, the procedure for gauging a nonlinearly realized symmetry remains the same as before: combine the derivatives of the Goldstone field, $\partial_\mu \xi$, (which is the only one that transforms in the EFT under the spontaneously broken global abelian symmetry) with the gauge potential A_μ to build a covariant derivative. The only difference is that ξ transforms as a shift under the symmetry, $\xi \to \xi + \sqrt{2}\, v\, \omega$, rather than being multiplied by a phase or a matrix. (A field that shifts as ξ does under a gauge transformation is often called a 'Stueckelberg' field [63].) Consequently, the required covariant derivative this time is $D_\mu \xi = \partial_\mu \xi - \sqrt{2}\, v\, A_\mu$, if $A_\mu \to A_\mu + \partial_\mu \omega$. For canonically normalized A_μ the covariant derivative is instead $D_\mu = \partial_\mu - \sqrt{2}\, g v\, A_\mu$.

Notice in particular that because ω is an arbitrary function it can be used to completely remove ξ by setting it to zero everywhere. This choice is called 'unitary gauge' [64, 65] and in this gauge $D_\mu \xi = -\sqrt{2}\, g v A_\mu$, revealing how ξ can be completely absorbed into the spin-one field A_μ. In this gauge the canonically normalized kinetic terms for ξ and A_μ become

$$-\frac{1}{4} F_{\mu\nu} F^{\mu\nu} - \frac{1}{2} D_\mu \xi D^\mu \xi = -\frac{1}{4} F_{\mu\nu} F^{\mu\nu} - g^2 v^2 A_\mu A^\mu, \tag{4.62}$$

revealing the spin-one particle to have mass $M_A^2 = 2 g^2 v^2$, in agreement with the UV theory (*c.f.* Appendix C.3.3). This absorption of ξ to give the spin-one field a mass is the usual Higgs mechanism in action, with ξ providing the missing degrees of freedom required to convert the two spin states of a massless spin-one particle to the three spin states of massive spin one.

What is interesting is that there are two ways to regard the resulting low-energy EFT. The first way is to think of it in the way just described: it is a massless spin-one field coupled to a nonlinearly realized abelian symmetry. The other way to think of it is simply as a generic theory of a massive spin-one vector field, V_μ, coupled to other particles in an arbitrary way with no reference made to gauge invariance at all; that is to say, the gauge symmetry is explicitly broken.

These two ways of thinking lead to precisely the same lagrangian, since the discussion surrounding Eq. (4.35) shows that the most general nonlinearly realized lagrangian is built using arbitrary combinations of a vector field V_μ – with no constraints on the lagrangian coming from the abelian symmetry – followed by the replacement $V_\mu \to \partial_\mu \xi$. In the gauge-invariant construction this vector is simply $V_\mu = D_\mu \xi / (\sqrt{2}\, v)$, and so is simply A_μ written in unitary gauge.

The upshot is this: from a low-energy perspective there is operationally no difference between a nonlinearly realized gauge symmetry and the complete absence of a gauge symmetry (or an explicitly broken gauge symmetry).[14] There is no royal road that allows an observer to learn about broken high-energy gauge symmetries using only low-energy methods. The same is not true for global symmetries because for these the physical Goldstone mode is always present at low energies to bring the news about the UV theory's broken symmetries.

Of course, this doesn't mean that there is no utility in sometimes using unitary gauge (and thereby ignoring the symmetries) and sometimes using a more general covariant gauge (for which the Stueckelberg field is kept and the gauge symmetry is nonlinearly realized). Unitary gauge is usually more useful at tree level, since it makes the physical particle spectrum more transparent. Covariant gauges are more

[14] A similar statement applies for non-abelian symmetries [28].

convenient when computing loops (or power counting in general), for reasons that become clear once the massive spin-one propagator is written.

For a one-parameter family of covariant gauges [66] the massive spin-one propagator has the form

$$G_{\mu\nu}(x-y) = -\mathrm{i}\int \frac{\mathrm{d}^4 p}{(2\pi)^4} \frac{1}{p^2+m^2-\mathrm{i}\epsilon}\left[\eta_{\mu\nu} + (\zeta - 1)\frac{p_\mu p_\nu}{p^2+\zeta m^2}\right] e^{\mathrm{i}p\cdot(x-y)}, \tag{4.63}$$

where the real parameter ζ labels the gauge choice. Popular gauge choices within this class are Feynman gauge ($\zeta = 1$), Landau gauge ($\zeta = 0$) and unitary gauge, which corresponds to the limit $\zeta \to \infty$. What is inconvenient about loops in unitary gauge is the propagator's large-momentum limit, since it does not fall off quadratically with momentum as the components of p_μ get large. Naive use of unitary gauge in the power-counting estimates of §3 leads to completely misleading results.[15]

Unitarity Bound: The Gauged Toy Model

How can describing a spin-one particle without using gauge invariance be consistent with the statement made at the beginning of this section that relativity and unitarity in quantum mechanics require a massless spin-one particle to be associated with a gauge symmetry? To explore this it is again instructive to ask the question for the toy model of §1; or rather for the gauged version of the toy model for which a gauge potential A_μ 'gauges' the toy model's $U(1)$ symmetry (*i.e* promotes it from a global symmetry to a local one). The renormalizable lagrangian for the UV version of this model is given by (4.60) instead of (1.1), though with scalar potential still given by (1.2).

For nonzero v both the scalar ϕ_R (or χ) and the spin-one field A_μ acquire a mass, with $m_R^2 = \lambda v^2$ and $m_A^2 = 2g^2v^2$. In the regime where the gauge coupling, g, and the scalar self-coupling, λ, are both small (to justify semiclassical methods) but with $g^2 \ll \lambda$ these satisfy $m_A^2 \ll m_R^2$. In this case, the low-energy spectrum for energies below m_R consists only of the massive spin-one particle, whose lagrangian is not constrained by gauge invariance apart from the observation that it can be built using arbitrary powers of the invariant field $V_\mu = \partial_\mu \xi - \sqrt{2}\, gvA_\mu$.

With these choices (and in a covariant gauge[16]), the power-counting arguments of §3 go through in the Wilsonian theory of V_μ interactions with only minor modifications due to the presence of the gauge field A_μ. Writing the basic action in the form

$$\mathcal{L}_W = \mathfrak{f}^4 \sum_n c_n\, \mathcal{O}\left(\frac{\partial}{M}, \frac{\xi}{v}, \frac{A}{v_A}\right), \tag{4.64}$$

for dimensionless couplings c_n, and repeating the dimensional power-counting arguments of §3.2.3 leads to the following minor generalization of (3.57),

[15] This is another example where if you use the wrong variables you will be sorry. In principle, once everything is included all gauges give the same answer to physical (and so gauge-invariant) quantities. But results that are manifest at every step in covariant gauges emerge in unitary gauge only after obscure cancellations.

[16] For aficianados: including 'ghosts', though these do not play an important role for the present purposes.

$$\mathcal{A}_{\mathcal{E}_\xi \mathcal{E}_A}(q) \sim \frac{q^2 \mathfrak{f}^4}{M^2} \left(\frac{1}{v}\right)^{\mathcal{E}_\xi} \left(\frac{1}{v_A}\right)^{\mathcal{E}_A} \left(\frac{Mq}{4\pi \mathfrak{f}^2}\right)^{2\mathcal{L}} \prod_n \left[c_n \left(\frac{q}{M}\right)^{d_n - 2}\right]^{\mathcal{V}_n}, \qquad (4.65)$$

as an estimate for a graph with $\mathcal{E}_\xi$ and $\mathcal{E}_A$ external ξ and A_μ legs, $\mathcal{L}$ loops and $\mathcal{V}_n$ vertices involving d_n derivatives each. As usual, q denotes here the generic size of the external energies flowing through the graph and assumes this is the only scale relevant when making a dimensional estimate of the result's size.

From §3.2.3 it is known that using the choices $\mathfrak{f}^2 = m_R v$ and $M = m_R$ in (4.64) correctly captures the dependence of $\mathcal{L}_W$ on v and m_R, at least for the ξ-dependent terms. With these choices choosing $v_A = M/g = m_R/g$ in (4.64) also ensures that both ∂_μ and gA_μ enter with the same dimensional factor of $1/M$ in $\mathfrak{L}_W$, thereby ensuring that all appearances of the covariant derivative appearing in the combination

$$\frac{D_\mu \xi}{vM} = \frac{\partial_\mu \xi - \sqrt{2}\, gvA_\mu}{vM} = \frac{\partial_\mu \xi}{vM} - \sqrt{2}\,\frac{A_\mu}{v_A} \qquad (4.66)$$

appear consistent with the assumption made in (4.64).

What remains is to track the remaining factors of the gauge coupling, g, and to see whether or not there are any systematic powers of m_R and v hidden within the dimensionless coefficients, c_n. The above choices ensure there are none in any terms that explicitly involve the field ξ. What they do not get right are the terms that involve A_μ without ξ, such as terms built using the covariant field strength $F_{\mu\nu} = \partial_\mu A_\nu - \partial_\nu A_\mu$.

In particular, using the above rules in (4.64) would predict such terms appear in $\mathcal{L}_W$ proportional to

$$c_n\, \mathfrak{f}^4\, O_n\left(\frac{\partial}{M}, \frac{F_{\mu\nu}}{Mv_A}\right) = c_n\, m_R^2 v^2\, O_n\left(\frac{\partial}{m_R}, \frac{gF_{\mu\nu}}{m_R^2}\right). \qquad (4.67)$$

This gets two things wrong when compared with what is obtained by integrating out the heavy χ scalar from the UV theory using (4.60). First, it predicts a kinetic Maxwell action for the gauge field of size

$$\frac{\mathfrak{f}^4}{M^2 v_A^2}\, F_{\mu\nu}F^{\mu\nu} = \frac{g^2 v^2}{M^2}\, F_{\mu\nu}F^{\mu\nu}, \qquad (4.68)$$

rather than the standard Maxwell action inherited from (4.60). This shows that the estimate (4.65) misses a factor of g^2v^2/M^2 from each internal gauge field line, requiring it to be corrected by the factor

$$\text{Correction factor} = \left(\frac{g^2 v^2}{M^2}\right)^{\mathcal{I}_A} = \left(\frac{gv}{M}\right)^{-\mathcal{E}_A + \sum_n f_n^A \mathcal{V}_n}, \qquad (4.69)$$

where $\mathcal{I}_A$ is the number of internal gauge-field lines in the graph, and f_n^A is the number of gauge lines that meet at the vertex labelled by interaction 'n'. This expression uses 'conservation of ends' for gauge field lines, $\mathcal{E}_A + 2\mathcal{I}_A = \sum_n f_n^A \mathcal{V}_n$. This correction factor says that each internal gauge-boson line brings an additional suppression by a power of $g^2v^2/M^2 \simeq m_A^2/m_R^2 = g^2/\lambda$, which is small because of the assumption that the spin-one particle is light enough to be in the low-energy theory.

Similarly, integrating out the χ field in the full theory also gives higher powers of $F_{\mu\nu}$ (and its derivatives) that come with a factor of g for each $F_{\mu\nu}$ and with dimensions set by powers of m_R, as in

$$h_n \, m_R^4 \, O\left(\frac{\partial}{m_R}, \frac{gF}{m_R^2}\right), \tag{4.70}$$

where h_n is a pure number (that contains a factor of $1/(16\pi^2)$ for each loop in the full theory required to generate the operator in question). Comparing this with (4.67) shows that the dimensionless couplings c_n and h_n are related by

$$c_n = h_n \left(\frac{m_R^4}{\mathfrak{f}^4}\right) = h_n \left(\frac{m_R^2}{v^2}\right) = \lambda h_n, \tag{4.71}$$

(where the last equality uses $m_R^2 = \lambda v^2$), but only for those interactions that involve only $F_{\mu\nu}$ and not ξ.

Physically, interactions in $\mathfrak{L}_W$ that are gauge-invariant and independent of ξ involve only the transverse polarizations of the spin-one particle, while ξ itself plays the role of the massive spin-one particle's longitudinal spin state. The above estimates show that (when $m_R^2 \ll v^2$) the interactions of the transverse states are over-estimated at low-energies when using (4.64) and (4.65), though these estimates do get the size of the contribution of the longitudinal state right.

Now comes the main point: concentrating exclusively on the interactions of the longitudinal state, the dominant size of an $\mathcal{E}$-point amplitude is obtained from the estimate (4.65) using only ξ external and internal lines, leading to the estimate (3.58) (or, equivalently, the estimate (3.57)) for a graph with $\mathcal{E}$ external legs, $\mathcal{L}$ loops and $\mathcal{V}_n$ vertices, each of which involves d_n derivatives:

$$\mathcal{A}_{\mathcal{E}}(q) \sim q^2 v^2 \left(\frac{1}{v}\right)^{\mathcal{E}} \left(\frac{q}{4\pi v}\right)^{2\mathcal{L}} \prod_n \left[\left(\frac{q}{m_R}\right)^{(d_n-2)}\right]^{\mathcal{V}_n}. \tag{4.72}$$

Not surprisingly, this says that the low-energy expansion is the key to the validity of the loop expansion in the Wilsonian theory. For $q \ll m_R \ll 4\pi v$ all graphs are suppressed and the expansion is controlled. When $q \sim m_R \ll 4\pi v$ multiple insertions of vertices at fixed loop order can become unsuppressed, but need not represent loss of control provided there are fixed numbers of graphs possible at any loop order. But the low-energy expansion underlying the Wilsonian lagrangian necessarily requires

$$q \ll 4\pi v \sim \frac{4\pi m_A}{g}. \tag{4.73}$$

This kind of restriction for an EFT is often called a 'unitarity bound' [67], because it is often identified by asking when a cross section computed within the low-energy theory becomes inconsistent with the energy dependence required at high energy by unitarity [68]. It would, of course, be misleading to regard inconsistency as a *bona fide* loss of unitarity in the low-energy theory (which nobody does), since it is hard to see how unitarity can be lost if the Hamiltonian remains hermitian (as is typically the case). What is really failing is the validity of the low-energy expansion used to infer the perturbative cross-section in the low-energy regime, and this failure is even more systematically revealed through power-counting estimates like (4.72).

It is now possible to circle back to the question that started this section: how can the necessity for gauge invariance for massless spin-one particles be consistent with the observation that nonlinearly realized gauge symmetry is operationally the same as explicit breaking of the gauge symmetry in the EFT for massive spin-one states?

As is seen above, the low-energy description of a massive spin-one particle (either without gauge invariance or with a nonlinearly realized gauge symmetry) always breaks down at energies of order $q \sim 4\pi m_A/g$, by which point some new UV description must necessarily intervene. (Usually what intervenes is a description involving linearly realized gauge invariance.) Although a non-gauge invariant description of a massive spin-one particle can make sense within an EFT, this description cannot work up to energies that are hierarchically large compared to its mass, and cannot work at all for nonzero g if the particle is massless.[17] Similar restrictions do *not* apply for linearly realized gauge symmetries, since these can be renormalizable and so be valid up to energies much higher than any of those that appear explicitly in the low-energy theory itself (indeed this is one way that they can be derived [69]).

4.3 Anomaly Matching ♠

The previous sections discuss symmetries as if their existence is established by showing the invariance of the *classical* action and so ignore the possibility that classical symmetries might not survive quantization. Traditionally, when a classical symmetry fails to survive quantization it is known as an 'anomalous' symmetry [70–73, 75].

This kind of separate treatment of the classical action and its quantum corrections is a bit too old-school within an EFT framework, because what one naively calls the 'classical' action is really better understood as the Wilsonian action obtained by integrating out higher-energy degrees of freedom. As this section now argues, it is more useful to organize one's thinking in terms of the scales involved than to divide the world artificially into a quantum and classical part. From this point of view, an anomalous symmetry is a particular instance of a transformation that is simply not a symmetry, but under which the action transforms in a specific way.

4.3.1 Anomalies♡

One way to characterize the failure of a classical symmetry at the quantum level is if the system's 1PI action is not invariant under the transformations in question, even though the classical (or Wilson) action is. To see how this might happen recall the relation between the 1PI and classical actions, given by (2.19), reproduced here for convenience of reference:

$$\exp\Big\{\mathrm{i}\Gamma[\varphi\,]\Big\} = \int \mathcal{D}\hat{\phi}\ \exp\left\{\mathrm{i}S[\varphi+\hat{\phi}\,]+\mathrm{i}\int \mathrm{d}^4x\,\hat{\phi}^a J_a(\varphi)\right\}, \tag{4.74}$$

where $J_a(\varphi) = -\delta\Gamma/\delta\varphi^a$.

[17] A loophole in this argument arises for massive abelian gauge bosons, since for these interactions for the would-be Goldstone field ξ need not exist at all (in which case, the massive Stueckelberg description can be renormalizable). (This is only an option for abelian bosons because – as §4.2.2 shows – non-renormalizable self-interactions are compulsory for non-abelian Goldstone bosons.) As the toy model example shows, the absence of interactions for ξ is a strong condition even for abelian theories, and is not generic in any particular UV completion.

In this expression, consider transforming the argument of $\Gamma[\varphi]$ under a symmetry transformation, which for simplicity's sake[18] is taken to act linearly on the fields: $\varphi^a \to (U\varphi)^a = U^a{}_b\, \varphi^b$. This gives

$$\exp\{i\Gamma[U\varphi]\} = \int \mathcal{D}\hat{\phi}\, \exp\left\{iS[U\varphi + \hat{\phi}] + i\int d^4x\, \hat{\phi}^a J_a\right\} \tag{4.75}$$
$$= \int \mathcal{D}\hat{\phi}_u\, \mathfrak{J}(\varphi, \hat{\phi}_u)\, \exp\left\{iS[\varphi + \hat{\phi}_u] + i\int d^4x\, \hat{\phi}^a_u \mathcal{J}_a\right\},$$

where the second line performs the change of integration variable $\hat{\phi} \to \hat{\phi}_u$ where $\hat{\phi} = U\hat{\phi}_u$ (for which $\mathfrak{J}(\varphi, \hat{\phi}_u)$ is the Jacobian – more about which below) and uses the invariance of the classical action to write $S[U(\varphi + \hat{\phi}_u)] = S[\varphi + \hat{\phi}_u]$. Also used is the definition of the current, $J_a(\varphi) = -\delta\Gamma[\varphi]/\delta\varphi^a$, which implies that $J_a(U\varphi) = -(\delta\Gamma[\psi]/\delta\psi^a)_{\psi=U\varphi}$, while $\mathcal{J}_a := -\delta\Gamma[U\varphi]/\delta\varphi^a = U^b{}_a J_b(U\varphi)$.

This manipulation shows that it is consistent to have $\Gamma[U\varphi] = \Gamma[\varphi]$ if both the classical action is invariant *and* the path integral measure is invariant – *i.e.* the Jacobian is trivial: $\mathfrak{J} = 1$. One way to think about anomalies is that they are the situation where there is an obstruction to constructing this type of invariant measure for the path integral [76]. Although it goes beyond the scope of this book to derive the conditions for anomalies in great detail (see the bibliography, §D.2, for further reading), suffice it to say that obstructions arise when a system is 'chiral' in the sense that its interactions treat left- and right-handed particles differently. In four dimensions this boils down to systems with chiral fermions.

A concrete way to identify when there is an anomaly is to evaluate the matrix elements of the conserved Noether current, regarded as a quantum operator. Recall that for each classical symmetry Noether's theorem ensures the existence of a current, J^μ, that is locally conserved inasmuch as the field equations imply $\partial_\mu J^\mu = 0$. Explicit calculations of matrix elements like $\langle f|J^\mu|\Omega\rangle$, where $|\Omega\rangle$ is the ground state and $|f\rangle = |A(k), A(q)\rangle$ is a state involving two spin-one particles, show that the matrix element $\langle f|\partial_\mu J^\mu|\Omega\rangle$ cannot be zero, so local conservation fails as an operator statement.

Evaluating the graph of Fig. 4.2 gives the following result for the conservation of the operator current [70–73][19]

$$\partial_\lambda J^\lambda_a = A_{abc}\, \frac{g_b g_c}{64\pi^2}\, \epsilon^{\mu\nu\alpha\beta} F^b_{\mu\nu} F^c_{\alpha\beta} \tag{4.76}$$

with $F^b_{\mu\nu}$ the field strength corresponding to a gauge symmetry generator T_b, and g_b its associated gauge coupling. The quantities A_{abc} are called *anomaly coefficients*, and they are given in terms of the symmetry generators (acting on left-handed spin-half fields) by

[18] A similar argument for nonlinearly realized symmetries couples the current to a combination $\sigma^a(\varphi, \hat{\phi})$ that transforms more covariantly under field redefinitions. The functional form of the transformation rule can also evolve with scale, and so differ between the microscopic fields ϕ^a and $\varphi^a = \langle\phi^a\rangle$ (see also Exercise 2.7).

[19] Non-invariance of $\Gamma[\phi]$ is related to the failure of the Noether current to be conserved, as can be seen by performing a *local* symmetry transformation, under which (for a global symmetry) the classical action (4.81) is not invariant. Evaluating explicitly how it transforms shows: $\delta\Gamma = -\int d^4x\, \partial_\mu\omega^a J^\mu_a$, which after integration by parts gives $\delta\Gamma = \int d^4x\, \omega^a\, \partial_\mu J^\mu_a$.

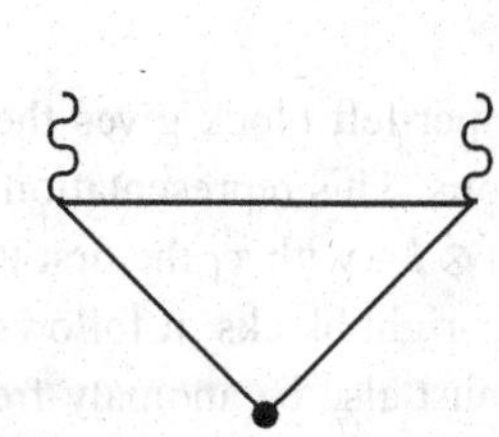

Fig. 4.2 The triangle graph that is responsible for anomalous symmetries (in four spacetime dimensions). The dot represents the operator J^μ and the external lines represent gauge bosons in the matrix element $\langle gg|J^\mu|\Omega\rangle$, where $|\Omega\rangle$ is the ground state.

$$A_{abc} = \text{tr}\Big(T_a\big\{T_b, T_c\big\}\Big) \tag{4.77}$$

where the curly brackets denote the anticommutator, $\big\{T_b, T_c\big\} := T_bT_c + T_cT_b$. As defined, A_{abc} is completely symmetric under the interchange of any pair of indices, and it is real because the generators T_a are hermitian. The trace is over the reducible representation of the symmetry acting on the complete set of left-handed fermions.

A classical gauge symmetry survives quantization – and is said to be 'anomaly free' – if $A_{abc} = 0$ for all T_a, T_b and T_c in the symmetry's Lie algebra. Because gauge invariance ultimately is required by the interplay between Lorentz invariance and unitarity, gauge symmetries must be anomaly-free to be consistent. (In §9.1.2 it is shown that anomalies cancel in an interesting way for the symmetry group $SU_C(3) \times SU_L(2) \times U_Y(1)$ using a single 'generation' of fermion content from the Standard Model.)

An important sufficient condition for the absence of anomalies is simple to state. Any symmetry group must be anomaly-free if its representation on left-handed fermions is *real* (or pseudoreal). A representation is real if its group-representation matrices $\exp(i\omega^a T_a)$ are real and so the matrices T_a are imaginary. It is pseudo-real if the matrices T_a are imaginary up to a similarity transformation: $T_a^* = -S\,T_aS^{-1}$ for some invertible matrix S.

To see why pseudo-reality ensures freedom from anomalies, notice that because the generators T_a are in general hermitian, it follows that $T_a^T = T_a^*$. Because the trace of a matrix equals the trace of its transpose it follows that

$$\begin{aligned} A_{abc} &= \text{tr}\,[(T_a\{T_b, T_c\})^T] = \text{tr}\,(\{T_c^T, T_b^T\}T_a^T) \\ &= \text{tr}\,(\{T_c^*, T_b^*\}T_a^*) = -\,\text{tr}\,(S\{T_c, T_b\}T_aS^{-1}) \\ &= -\text{tr}\,(\{T_c, T_b\}T_a) = -A_{abc}, \end{aligned} \tag{4.78}$$

and so $A_{abc} = 0$.

A special case of this last result shows why only chiral symmetries are anomalous. To see why, imagine that fermion number is conserved (so that fermions and antifermions are distinguishable) and further assume the symmetry is *not* chiral, in that left- and right-handed fermions (as opposed to antifermions) transform in the same representation, t_a say, of the group. In this case, the group generators acting on *all* the left-handed fermions – for fermions *and* antifermions – can be written in the block-diagonal form

$$T_a = \begin{pmatrix} t_a & 0 \\ 0 & -t_a^* \end{pmatrix} \tag{4.79}$$

where the upper-left block gives the action on fermions and the lower-right block on antifermions. This representation is manifestly pseudoreal since $T_a^* = -S\,T_a S^{-1}$, where $S = \tau_1 \otimes I$ – with τ_1 the first Pauli matrix – is the matrix that swaps the upper-left and lower-right blocks. It follows that any symmetry that is left–right symmetric in this way must also be anomaly-free. This is, in particular, why anomalies are not an issue for either Quantum Electrodynamics or Quantum Chromodynamics.

An important property of the definition of an anomaly is the inability to remove it (and so to restore the symmetry) simply by appropriately adding non-invariant local counter-terms to the lagrangian density. This in itself means that anomalies must have their origins in the low-energy part of the theory, rather than the high-energy part. (After all, the EFT program argues that *any* high-energy physics can be described by some choice for local interactions within an effective theory.) It is for this reason that anomalies are relevant when setting up the Wilsonian description of the low-energy sector [74].

The observation that anomalies cannot (by definition) be canceled by local counter-terms also reveals the difference between an honest-to-God anomaly and just regularizing in a silly way. Any damage done by using an ill-conceived regularization procedure – such as one that does not preserve a system's symmetries – can be undone by renormalizing parameters appropriately, but this is *not* possible if a symmetry is anomalous. Many simple regularization schemes (like explicit cut-offs in momentum integrals or point-splitting techniques) break symmetries (like Lorentz invariance or gauge invariance), but their use does not mean that the physics being described must break these symmetries. This is why it can make sense (though is not normally convenient) to define effective Wilson actions with cutoffs, even for Lorentz-invariant systems or systems with gauge symmetries. The implicit choice made in such cases is to undo any regulatory damage by appropriately renormalizing the theory to restore these symmetries.

A sufficient condition for a renormalization scheme to exist that preserves a symmetry is the existence of a regularization scheme that explicitly preserves it (such as is often true with dimensional regularization, for example). Since the difference between any two regularization schemes lies purely at high energies, it must be captured by some choice of effective local couplings. But while this shows that invariant regularizations are not possible for anomalous symmetries,[20] the absence of a known invariant regularization does not necessarily imply the existence of an anomaly.

4.3.2 Anomalies and EFTs

This path integral way of formulating things shows how the classical/quantum split is more subtle when phrased in terms of the Wilsonian effective action. Given a hierarchy of scales with light, l, and heavy, $\mathfrak{h}$, degrees of freedom, the definition (2.58) of the Wilsonian action (reproduced here)

[20] Sometimes regularizations superficially appear to preserve an anomalous symmetry (such as the anomaly in Weyl invariance when regularized in $D \neq 4$ dimensions), but when this happens the regularization scheme introduces new light degrees of freedom (such as the $D-4$ components of D-dimensional tensor fields) [75].

$$\exp\left\{iS_W[l]\right\} := \int \mathcal{D}\mathfrak{h}\, \exp\left\{iS[l+\mathfrak{h}]\right\}, \tag{4.80}$$

shows that a nontrivial Jacobian potentially moves the high-energy part of the anomaly from the measure into the (Wilsonian) action itself.

As ever, the basic symmetry statement is – at *any* scale – that the transformation properties of the Wilsonian action are whatever they must be to reproduce the correct transformation properties of quantities like the 1PI action for the full microscopic theory. This means invariance at all scales for honest-to-God symmetries, and it means reproducing the nontrivial transformation properties of $\delta\Gamma[\varphi]$ for broken symmetries. (For anomalous symmetries, this condition that the Wilsonian action reproduce the anomalies of the full theory is called 'anomaly matching', and its power lies in the fact that for anomalies $\delta\Gamma[\varphi]$ takes a restricted form [78, 80].)

Gauge Symmetries

For these purposes it is important to distinguish between local (or gauge) symmetries and global symmetries. Linearly realized gauge symmetries are central to the consistency of the coupling of light spin-one particles, since their interactions are only Lorentz invariant and unitary if they are also invariant under local gauge transformations. Consequently, gauge symmetries cannot be anomalous, and this must be true for the Wilsonian theory at any energy for which one cares to ask the question.

Absence of anomalies usually means invariance for both S_W and the Jacobian $\mathfrak{J}$, and this is how things turn out to transpire for the Standard Model (see §9.1.2). This need not necessarily be so, however, since in principle both δS_W and $\mathfrak{J}-1$ can be nonzero, so long as the total combination $\mathcal{D}\hat{\phi}\, e^{iS_W[\varphi+\hat{\phi}]}$ is invariant. This is not just an academic observation because this is the way gauge symmetries in many string theories (see §10.3) turn out to be anomaly free. In this context, the cancellation between the variation of the action and the Jacobian is called 'Green-Schwarz' anomaly cancellation [77].

What makes a cancellation between the variation of $\mathcal{D}\hat{\phi}$ and e^{iS_W} tricky is the fact that $S_W = \int d^4x\, \mathfrak{L}_W$ is local. In fact, the locality of the Wilson action means the idea that terms in S_W can cancel an anomaly (regarded as the variation of $\mathcal{D}\hat{\phi}$) needs some clarification. This is because – as stated explicitly above – an anomaly is *defined* as a nontrivial transformation of $\mathcal{D}\hat{\phi}$ that cannot be removed by adding local counter-terms to the action.

The main issue here is semantic. Strictly speaking, an 'anomalous' symmetry is not really anomalous if it can be cancelled by terms in S_W, as in Green–Schwarz anomaly cancellation. It is not anomalous precisely because anomalies are defined modulo the variation of local terms in the action. When one speaks of Green–Schwarz cancellations the anomalies in question arise from a particular sector of the theory, usually chiral fermions. The corresponding symmetries are anomalous in the sense that they cannot be cancelled by local counter-terms *purely within this sector*. Green–Schwarz anomaly cancellation becomes possible once the rest of the fields from other sectors are also included.

An example might be helpful here. Consider, therefore, Quantum Electrodynamics (QED) in the limit of vanishing fermion mass. In this case, the fields involved are the electromagnetic potential, A_μ, and the fermion's spinor field, ψ, and the leading, renormalizable, terms in the (Wilsonian) lagrangian are

$$\mathfrak{L}_W = -\frac{1}{4} F_{\mu\nu}F^{\mu\nu} - \overline{\psi}\not{D}\psi. \tag{4.81}$$

Here, $\not{D}\psi = \gamma^\mu D_\mu \psi = \gamma^\mu(\partial_\mu - iqA_\mu)\psi$, where q is the fermion's charge ($q = -e$ for an electron, say) and $F_{\mu\nu} = \partial_\mu A_\nu - \partial_\nu A_\mu$.

This lagrangian enjoys a classical $U_V(1) \times U_A(1)$ symmetry where $U_V(1)$ is the electromagnetic gauge symmetry: $\delta A_\mu = \partial_\mu \zeta$ with $\delta\psi = iq\zeta\,\psi$, where $\zeta(x)$ is an arbitrary infinitesimal real local symmetry parameter. Unlike the gauge symmetry, the global $U_A(1)$ symmetry – $\delta\psi = i\omega\,\gamma_5\psi$ for constant, real infinitesimal symmetry parameter ω – is only present due to the absence of a mass term.

The axial symmetry, $U_A(1)$, in this theory proves to have an anomaly[21] under which the 1PI action transforms as

$$\delta\Gamma = \frac{\omega\, q^2}{16\pi^2} \int d^4x\; \epsilon^{\mu\nu\lambda\rho} F_{\mu\nu} F_{\lambda\rho}, \tag{4.82}$$

where indeed the right-hand side cannot be written as the variation, δS, of some local functional of A_μ and ψ. This is not inconsistent with (4.82) because – unlike the Wilson action – the Γ that satisfies (4.82) is not local.

But this anomaly *can* be the variation of a local action (and so amenable to Green–Schwarz anomaly cancellation) once other fields are added. In particular, adding a real scalar field, ϕ, transforming inhomogeneously under $U_A(1)$ as a Goldstone field, $\delta\phi = \omega$, allows (4.82) to be cancelled by a contribution to the Wilson action of the form

$$\mathfrak{L}_{GS} = -\frac{q^2}{16\pi^2} \int d^4x\; \phi\, \epsilon^{\mu\nu\lambda\rho} F_{\mu\nu} F_{\lambda\rho}. \tag{4.83}$$

Global Symmetries and Anomaly Matching

Anomaly matching enters for global symmetries, since these can be anomalous within a consistent theory. The presence of anomalies for global symmetries can make the difference between having a theory agree with experiment or not.

A famous practical example of this arises in Quantum Chromodynamics (QCD) (discussed in more detail in §8), wherein anomalies prove to be crucial for describing the decay rate for π^0 mesons. Unlike most mesons, π^0 mesons are seen to decay electromagnetically through the decay into two photons, $\pi^0 \to \gamma\gamma$. Its decays are well-described by an interaction term involving pions and photons of the form

$$\mathfrak{L}_{\rm decay} = \frac{e^2}{32\pi^2 F_\pi}\, \pi^0\, \epsilon^{\mu\nu\lambda\rho} F_{\mu\nu} F_{\lambda\rho}, \tag{4.84}$$

where e is the electromagnetic coupling and $F_\pi = 92$ MeV is a parameter discussed in some detail in §8.

[21] Strictly speaking, there is an anomaly in the full $U_V(1) \times U_A(1)$ symmetry, but precisely which factor is anomalous can be chosen by adding appropriate counterterms. Requiring Lorentz invariance and unitarity precludes letting the $U_V(1)$ factor be anomalous and so forces the anomaly onto the axial transformation, $U_A(1)$.

The interaction (4.84) was a puzzle in QCD before the role played by anomalies was appreciated [81–86]. This is because in QCD the π^0 is understood to be a bound state consisting of a quark–antiquark combination, where quarks (and their antiparticles) are bound together by the strong interactions. The quarks involved in the π^0 are the up and down quarks, each of which comes in $N_c = 3$ copies (or colours) and whose charges are, respectively, $q_u = \frac{2}{3}\,e$ and $q_d = -\frac{1}{3}\,e$. The strong force in QCD couples to colour in much the way that electromagnetism couples to electric charge in QED, and this binds the quarks together into mesons with binding energies of order $4\pi F_\pi \sim 1$ GeV or so. In a Wilsonian picture the EFT appropriate at energies higher than this is built using the quarks, while at energies well below $4\pi F_\pi \sim 1$ GeV the effective action instead directly involves bound-states like π^0 (for more details see §8).

The puzzle arises because the π^0 proves to be a Goldstone boson for a global symmetry, $\delta\pi^0 = \omega F_\pi$, of the strong and electromagnetic interactions, but this seems at first sight to be inconsistent with its appearing undifferentiated in the sub-GeV EFT in a term like (4.84). Instead of being invariant, Eq. (4.84) predicts

$$\delta\mathfrak{L}_{\rm decay} = \frac{\omega\, e^2}{32\pi^2}\,\epsilon^{\mu\nu\lambda\rho}F_{\mu\nu}F_{\lambda\rho}. \tag{4.85}$$

The resolution of the puzzle lies in the observation that QCD predicts there is an anomaly in the underlying symmetry for which π^0 is a Goldstone boson. It is the anomaly that allows terms like (4.84), and it is anomaly matching that predicts the size of its coefficient.

To see how this works, it is convenient to write the action on the u and d quarks of the symmetry for which π^0 is the Goldstone boson action as

$$\delta\begin{pmatrix} u \\ d \end{pmatrix} = \mathrm{i}T_{A}\gamma_5\begin{pmatrix} u \\ d \end{pmatrix} \quad \text{with} \quad T_{A} = \begin{pmatrix} \frac{1}{2} & \\ & -\frac{1}{2} \end{pmatrix}. \tag{4.86}$$

In this same notation the electric charge of these quarks has the form

$$Q_{\rm em} = \begin{pmatrix} \frac{2}{3}\,e & \\ & -\frac{1}{3}\,e \end{pmatrix}. \tag{4.87}$$

Since (4.86) is an axial symmetry, it has an anomaly of the form given in (4.82), which when summed over all the colours of the two types of quarks gives

$$\delta\Gamma = \frac{\omega\,\mathcal{A}}{16\pi^2}\int \mathrm{d}^4x\;\epsilon^{\mu\nu\lambda\rho}F_{\mu\nu}F_{\lambda\rho}, \tag{4.88}$$

with anomaly coefficient that counts the number of quarks, weighted by their electric charges

$$\mathcal{A} = \mathrm{tr}\left[T_{A}Q_{\rm em}^2\right] = \frac{N_c}{2}\left[\left(\frac{2}{3}\right)^2 - \left(-\frac{1}{3}\right)^2\right] = \frac{N_c}{6}. \tag{4.89}$$

The success of (4.84) in describing π^0 decays provides one of the experimental confirmations that $N_c = 3$ is the number of colours in QCD.

This success is a special case of anomaly matching for a larger group of approximate global symmetries in low-energy QCD, which includes an entire $U_L(3) \times U_R(3)$

invariance, associated with separate unitary rotations amongst the left- and right-handed parts of the three lightest quarks: u, d and s. Some of the broader implications of these symmetries are described in §8.

In particular, all evidence indicates that the QCD vacuum spontaneously breaks the axial combination of these symmetries, giving rise to eight pseudo-Goldstone bosons.[22] These symmetries experience several anomalies when combined with various Standard Model gauge symmetries, and their existence implies that the existence in the low-energy meson EFT of a specific kind of self-interaction amongst the eight would-be Goldstone particles. It goes beyond the scope of the book to work out this anomaly-matching lagrangian – called a *Wess-Zumino* action – for the entire anomalous action, but the leading term it generates once it is expanded in powers of the 3×3 hermitian, traceless meson field, $\mathcal{M}$, has the form

$$\mathcal{L}_{WZW} = \frac{N_c}{240\pi^2 F_\pi^5} \epsilon^{\mu\nu\lambda\rho} \operatorname{tr} \left[\mathcal{M}\, \partial_\mu \mathcal{M}\, \partial_\nu \mathcal{M}\, \partial_\lambda \mathcal{M}\, \partial_\rho \mathcal{M} \right] + \cdots \tag{4.90}$$

where the ellipses represent terms involving more powers of $\mathcal{M}$ and $N_c = 3$ here denotes the number of quark colours.

As above, the coefficient is fixed by demanding that its transformation under $SU_L(3) \times SU_R(3)$ reproduces the anomalies of the underlying quarks, and the resulting value is successful in describing low-energy meson properties.

Anomaly matching can also provide a powerful constraint for theories where it is different species of chiral fermions that contribute to the anomalies at high energies and in the low-energy theory. Particularly interesting models of this form are those where chiral elementary fermions get bound into composite fermions that are also chiral. Such theories arise when contemplating whether or not quarks or leptons might be built from smaller constituents, much as protons and neutrons (once considered to be fundamental) are built from up and down quarks.

The physics involved in such models is often chiral because the puzzle such theories raise is why the bound-state masses (*i.e.* the mass of the ordinary quark or lepton) should be so much smaller than the typical energy $E \sim 1/\ell$ associated with the size of the bound object. Experimental searches for compositeness already tell us that if quarks or leptons are composite then the size of the associated bound state must be extremely small, $m \ll 1/\ell$. Chiral theories can help with this because chiral symmetries can allow such theories to have states whose binding energies are much smaller than their size.

For any such model the total pattern of anomalies carried by the constituent fermions at high energies must also be reflected in the spectrum of particles at lower energies, either by having the composite fermions produce the same anomalies or by having composite Goldstone bosons arise (much like the π^0 meson in the QCD example described above). See Exercise 4.6 for a more explicit example of anomaly matching with composite fermions.

[22] There are only eight rather than the nine Goldstone bosons expected for $U_A(3)$ because the overall rotation of all three quarks by a common axial phase is also anomalous and so is broken by the strong interactions.

4.4 Summary

Symmetries play a central role in modern physics, and effective theories are no different in this regard. This chapter opens with a section that recaps the various roles that symmetries play in quantum mechanics and in quantum field theory. The main new ingredient that the locality of quantum field theory introduces is the possibility that symmetries can be spontaneously broken: the ground state might not be invariant under some of the symmetries of the action (or equations of motion).

If the symmetry that breaks spontaneously is both a continuous symmetry and a global symmetry then spontaneous breaking requires the existence of gapless (or, in a relativistic context, massless) Goldstone states. This makes them card-carrying members of the low-energy sector, whose properties are largely dictated on symmetry grounds.

The main message of this chapter is that any symmetry properties of the full UV theory must also be reflected in any Wilsonian description of its low-energy sector. Much of the discussion is devoted to identifying Goldstone boson properties as a function of the assumed symmetry breaking pattern. This is done by identifying the general nonlinear realization of the broken global symmetries that the Goldstone bosons carry since that is how the news of these symmetries gets brought to the low-energy theory. Many examples of the structures found here arise in later chapters on applications, such as §8, §13 and §14.

Finally, the latter sections of this chapter examine related issues, such as how a nonlinear realization goes through when the spontaneously broken symmetry is local rather than global (*i.e.* is a gauge symmetry). The main new feature is that spontaneously broken gauge symmetries have no gap, inasmuch as the would-be Goldstone bosons get 'eaten' (through the Higgs mechanism) to provide the longitudinal spin state required for a massive spin-one particle.

As a result, the low-energy theory loses the information about the existence of the symmetry in the high-energy sector. For the Wilson action there is operationally no difference at all between a nonlinearly realized gauge symmetry and no gauge symmetry at all. The consistency of this observation with the requirement of gauge symmetries for massless spin-one particles is explored, including the associated breakdown of the low-energy EFT at scales not higher than of order $4\pi m_A/g$, if m_A and g are the spin-one particle's mass and coupling constant.

The final section provides a superficial description of anomalies – the failure of a classical symmetry to survive quantization – as a lead-in to a discussion of anomaly matching. From an EFT perspective anomalous symmetries are not symmetries at all, since for them the 1PI and 1LPI actions are not invariant, $\delta\Gamma[\varphi] \neq 0$, even if the classical action might be. What is special about anomalies is that $\delta\Gamma[\varphi]$ is quite constrained in form, so there can be content in requiring the Wilsonian action to reproduce the transformation properties of the underlying theory.

Exercises

Exercise 4.1 Consider the Goldstone bosons for the symmetry-breaking pattern where the group $G = SU(2)$ breaks down to $H = U(1)$. Take the generators of G to be the 2×2 Pauli matrices $T_a = \frac{1}{2}\tau_a$, with (as usual)

$$\tau_1 = \begin{pmatrix} 0 & 1 \\ 1 & 0 \end{pmatrix}, \quad \tau_2 = \begin{pmatrix} 0 & -i \\ i & 0 \end{pmatrix} \quad \text{and} \quad \tau_3 = \begin{pmatrix} 1 & 0 \\ 0 & -1 \end{pmatrix},$$

and take the generator of H to be T_3. Using the standard realization compute explicit formulae for the two Goldstone fields, ξ^1 and ξ^2, under arbitrary infinitesimal G transformations. Compute the Maurer–Cartan form and the associated quantities $\mathcal{A}_\alpha(\xi)$ and $e^\alpha{}_\beta(\xi)$, and their transformation properties under G. Write down the most general lagrangian up to two derivatives describing the self-couplings of these Goldstone fields, ξ^a, and compute the Noether currents implied by this action for the symmetry group G.

Show that a change of variables $(\xi^1, \xi^2) \to (\vartheta, \varphi)$ exists that turns your result into the Goldstone fields for a target space that is a 2-sphere:

$$\mathfrak{L}_W = -\frac{F^2}{2}\left[(\partial_\mu \vartheta\, \partial^\mu \vartheta) + \sin^2 \vartheta\, (\partial_\mu \varphi\, \partial^\mu \varphi)\right], \tag{4.91}$$

with F^2 an arbitrary positive real constant.

Exercise 4.2 Consider the Goldstone bosons for the symmetry-breaking pattern where the group $G = SU(2) \times SU(2)$ breaks down to $H = SU(2)$ corresponding to the diagonal subgroup (for which both $SU(2)$ factors rotate in the same way rather than independently). How many Goldstone bosons are there for this pattern?

Using the standard realization compute explicit formulae for the Goldstone fields under arbitrary infinitesimal G transformations. Compute the Maurer–Cartan form and the associated quantities $\mathcal{A}_\alpha(\xi)$ and $e^\alpha{}_\beta(\xi)$, and their transformation properties under G. Write down the most general two-derivative self-couplings for the Goldstone fields ξ^α, and compute its Noether currents for the symmetry group G. This action describes the low-energy interactions of pions.

Show that there is a change of variables that allows your result to be rewritten in the 'nonlinear σ-model' form

$$\mathfrak{L}_W = -\frac{1}{2}\frac{\partial_\mu \vec{\pi} \cdot \partial^\mu \vec{\pi}}{(1 + \vec{\pi}\cdot\vec{\pi}/F^2)^2}. \tag{4.92}$$

Exercise 4.3 Derive the useful identity, Eq. (C.133), that is used when proving formulae (4.50) and (4.51) of the main text.

Exercise 4.4 For the symmetry breaking pattern of Exercise 4.1 suppose that the group G is gauged. Show that the low-energy nonlinear realization is equivalent to the theory of a massive charged complex vector field W_μ coupled to a single unbroken $U(1)$ gauge boson, subject only to the constraints of the $U(1)$ invariance. Compute the most general interactions for this theory involving up to four fields and at most two derivatives.

Exercise 4.5 Explicitly evaluate the Feynman graph of Fig. 4.2 and derive the anomaly equation, Eq. (4.76).

Exercise 4.6 The strong interactions have a gauge group $SU(3)_c$ (where 'c' stands for 'colour' – for more details see §8). Suppose there are three types of left-handed massless spin-half quarks, q, that each transform under $SU(3)_c$ as a triplet (**3**) as well as three types of left-handed spin-half anti-quarks, $\bar{q}$, that each transform as an anti-triplet ($\bar{\mathbf{3}}$). It happens that the strong dynamics preserves a global 'flavour' symmetry group $G_f := SU(3)_L \times SU(3)_R \times U(1)$ that commutes with $SU(3)_c$, under which the q's transform as $(\mathbf{3}, \mathbf{1})_1$ while the $\bar{q}$ transform as $(\mathbf{1}, \bar{\mathbf{3}})_{-1}$, where the subscript gives the charge of the field for the $U(1)$ generator. Evaluate the anomaly coefficients A_{abc} for the generators of G_f using the generators T_a acting on the left-handed quarks and anti-quarks.

It is believed that the strong interactions form bound states that are singlets under $SU(3)_c$. For this quark content these include fermionic bound states (or 'baryons') in the completely antisymmetric colour combination: $B = \epsilon_{abc} q^a q^b q^c$ as well as its conjugate (or 'anti-baryon') $\overline{B} = \epsilon^{abc} \bar{q}_a \bar{q}_b \bar{q}_c$. What are the possible representations that B and $\overline{B}$ can transform in under the flavour group G_f?

Evaluate the anomaly coefficients A_{abc} for the generators of G_f acting on the baryons in each of these representations allowed for the bound-state baryons. Prove that it is impossible to choose the number of types of these representations in the bound-state spectrum in such a way that the A_{abc} for the baryons agree with those obtained from the quarks. The impossibility of doing so provides an argument that for these choices of quantum numbers the strong interactions must spontaneously break the flavour group G_f.

5 Boundaries

The presence of boundaries modifies the previous discussion in several ways, such as by removing the freedom to drop total derivative effective interactions in the low-energy action when identifying redundant terms. This short chapter sketches in some of the details of the new features that boundaries bring to low-energy theories. It is short because this is an area for which Wilsonian methods remain relatively poorly developed. It is nonetheless included because it provides a useful starting point for later sections, such as §7.4 and §13, of this book.

In general, spacetime is regarded to be a manifold $\mathcal{M}$ with boundary $\partial\mathcal{M}$, and the action must be specified on both of these regions to completely specify the problem:

$$S = S_B(\phi) + S_b(\phi, \psi) = \int_{\mathcal{M}} \mathrm{d}^4x\, \mathfrak{L}_B + \int_{\partial\mathcal{M}} \mathrm{d}^3x\, \mathfrak{L}_b, \tag{5.1}$$

where S_B (or the 'bulk' action) describes the dynamics of a collection of fields, $\phi(x)$, in the interior of $\mathcal{M}$ while S_b (or the 'boundary' action) describes how these bulk fields couple to the boundary, possibly including to any boundary-localized dynamical degrees of freedom (such as the boundary position, $y^\mu(t)$, itself, if it is free to move). Unless stated otherwise, the boundary of interest is timelike, consisting of a boundary to space at a given time (in some preferred frame) whose world-volume sweeps out the spacetime boundary as time evolves.

The division of interactions between the bulk and boundary is somewhat fluid since Stokes' theorem can be used to rewrite total derivatives in $\mathfrak{L}_B$ as a contribution to $\mathfrak{L}_b$. That is, if $\mathfrak{L}_B \supset \mathfrak{L}_{\text{tot deriv}} = \partial_\mu V^\mu$ for some V^μ, then the corresponding contribution to the action is

$$\int_{\mathcal{M}} \mathrm{d}^4x\, \mathfrak{L}_{\text{tot deriv}} = \int_{\mathcal{M}} \mathrm{d}^4x\, \partial_\mu V^\mu = \int_{\partial\mathcal{M}} \mathrm{d}^3x\, n_\mu V^\mu, \tag{5.2}$$

where $n_\nu = \{0, \mathbf{n}\}$ is a normal vector on the surface, conventionally chosen to point *out* of the bulk. In the absence of boundaries total derivatives are redundant because they can be simply dropped from the action with no physical consequences. With boundaries these same effective couplings carry consequences, but remain redundant in as much as their consequences are not distinct from those of interactions within the boundary action.

5.1 'Induced' Boundary Conditions

What boundary conditions should be imposed at the boundary $\partial\mathcal{M}$? Since quantum field theory is the business of evaluating path integrals over fields, one way to approach boundary conditions is to imagine formulating the path integral itself to be over a space of fields all of which satisfy some condition on the boundary – perhaps the fields or their derivatives or some combination of these vanish. The classical limit of such a problem then involves expanding about a saddle point defined by solving the classical field equations subject to the assumed boundary conditions on $\partial\mathcal{M}$. What is unsatisfying about such problems is the arbitrariness of the boundary conditions, which are simply handed down by God when formulating the problem.

Less arbitrary are systems for which the boundary conditions on $\partial\mathcal{M}$ are 'induced' inasmuch as they can be derived from the form of the action, with the action itself acquiring a new contribution specifically associated with the boundary. These kinds of boundary conditions typically arise when the path integral runs over arbitrary field configurations, both within the interior of $\mathcal{M}$ and on its boundary, as is very often the case in real systems.

Induced boundary conditions are of physical interest because in real applications the boundary is usually not a physical ending of spacetime on which fields (and/or their derivatives) are specified once and for all. Instead, the boundary usually arises as an approximate description of a place where there is a very rapid change of background properties; for electromagnetic applications perhaps it is the edge of a conducting region or a dielectric object beyond which one chooses not to track field behaviour (see, for example, Exercise 5.4). Deep down, boundary physics is in such cases no different from bulk physics, and one should imagine integrating over all possible values of both bulk and boundary fields when performing the path integral.

At the classical level this means that saddle points are chosen by demanding that the action is stationary against variations of the fields both in the bulk and on the boundary. It is stationarity against variation of fields on the boundary that dynamically dictates the classical boundary conditions that hold on $\partial\mathcal{M}$. Most importantly, tying boundary conditions to an action in this way ultimately allows all of the EFT reasoning described in this book to be brought to bear when deciding which boundary conditions should arise in any given situation.

The Toy Model

To make the issues concrete, return to the toy model of §1.1: the self-interactions of a complex field to which is now added a boundary term:

$$S = -\int_{\mathcal{M}} \mathrm{d}^4x \left[\partial_\mu\phi^*\partial^\mu\phi + V(\phi^*\phi)\right] + \int_{\partial\mathcal{M}} \mathrm{d}^3x \, \mu\,\phi^*\phi. \tag{5.3}$$

Here μ is a new parameter with dimensions of mass, and the above choice for S_b is the lowest-dimension possibility involving ϕ that is local and invariant under the symmetry $\phi \to e^{i\theta}\phi$. The path integral over ϕ is unconstrained both throughout the interior and boundary of $\mathcal{M}$.

In the classical approximation the path integral is computed as an expansion about a saddle point, $\phi_c(x)$, defined as the configuration where $\delta S(\phi = \phi_c)$ vanishes. Writing $\phi \to \phi + \delta\phi$ and linearizing the action in $\delta\phi$ then leads to the expression

$$\delta S = \int_{\mathcal{M}} \mathrm{d}^4x \left[\left(\Box\phi - \frac{\partial V}{\partial\phi^*}\right)\delta\phi^* + \text{c.c.}\right] + \int_{\partial\mathcal{M}} \mathrm{d}^3x \left[\left(-\partial_n\phi + \mu\phi\right)\delta\phi^* + \text{c.c.}\right], \tag{5.4}$$

where the first term in the second integral comes from an integration by parts in the bulk, with $\partial_n := n_\mu \partial^\mu$ denoting the normal derivative at the boundary.

Because the path integral is over arbitrary fields the saddle point must be stationary against arbitrary variations $\delta\phi$ everywhere within $\mathcal{M}$ and $\partial\mathcal{M}$. Restricting first to those variations that vanish on the boundary shows that $\phi_c(x)$ must satisfy the usual classical field equations throughout $\mathcal{M}$:

$$\left(\Box\phi - \frac{\partial V}{\partial \phi^*} \right)_{\phi=\phi_c} = 0. \tag{5.5}$$

Stationarity of the action against arbitrary variations on the boundary then shows the saddle point must satisfy the induced boundary condition

$$\partial_n \phi_c = \mu \, \phi_c \qquad \text{on } \partial\mathcal{M}. \tag{5.6}$$

Several things about this boundary-value problem are noteworthy. First, the boundary condition (5.6) is linear. Because the boundary condition is derived from the action, this is not automatic, and in this particular example it is a consequence of using only the lowest-dimension term (which is quadratic in ϕ) for the boundary action, S_b, in (5.3). As for any Wilsonian action, ultimately the justification for using low-dimension terms in S_b will rely on the low-energy approximation, and in this lies the seeds of an explanation as to why linear boundary conditions so often play a role throughout physics.

Second, notice that the boundary condition (5.6) forbids the vanilla vacuum solution of constant field, $\phi = v$, which minimizes V. In general, the coupling to the boundary causes a trade-off between trying to minimize the scalar potential throughout the bulk and paying some gradient energy to satisfy (5.6) on the boundary.

For instance, suppose the boundary is the $x - y$ plane (at $z = 0$) and the bulk is the region $z > 0$. Then, neglecting the interactions of the potential that are cubic and quartic in $\psi = \phi - v$ implies that a bulk solution of the form $\phi_c = v + \psi_c(z)$ satisfies $\psi_c'' - m_R^2 \psi_c \simeq 0$, where primes denote d/dz and (as before) $m_R^2 = \lambda v^2$. Requiring $\phi \to v$ as $z \to \infty$ and satisfying the boundary condition (5.6) implies the approximate saddle point solution

$$\phi_c(z) \simeq v \left(1 - \frac{\mu}{\mu + m_R} \, e^{-m_R z} \right). \tag{5.7}$$

Eq. (5.7) is consistent with the neglect of ψ^3 and ψ^4 in V for the regime $\mu \ll m_R$ since in this case $|\phi_c - v| \sim \mathcal{O}(\mu v / m_R) \ll v$.

The bulk energy cost of interacting with the boundary can be estimated by evaluating the classical energy at the solution (5.7). Dropping subdominant powers of μ/m_R, the resulting classical bulk energy-per-unit-area is

$$\frac{E_B}{A} \simeq \int_0^\infty \mathrm{d}z \, \left[|\phi'|^2 + V(\phi^*\phi) \right] \simeq \frac{\mu^2 v^2}{m_R} = \frac{\mu^2 v}{\sqrt{\lambda}}. \tag{5.8}$$

This expression drops the cubic and quartic terms of V, since these are also down relative to (5.8) by at least one power of μ/m_R. This bulk energy cost is more than compensated by the boundary contribution to the energy-per-unit-area, which is

$$\frac{E_b}{A} = -\mu|\phi(0)|^2 \simeq -\mu v^2 + \frac{2\mu^2 v^2}{m_R} + \cdots]. \tag{5.9}$$

Semiclassical quantum corrections to this classical result are computed using the same steps as used without a boundary: expand all fields about the background $\phi = \phi_c(z) + \hat{\phi}$ and quantize the fluctuations $\hat{\phi}$. The main difference is that any expansion in terms of modes $\hat{\phi}(x) = \sum_n [u_n(x)\, \mathfrak{a}_n + \text{c.c.}]$ involves modes defined in the presence of the background. In the present instance this means that they are not eigenstates of the z-component of momentum, due to the breaking of translation invariance in this direction by the background $\phi_c(z)$. See §13.1 for more about semiclassical expansions about position-dependent classical background fields.

5.2 The Low-Energy Perspective

Any boundary physics of the full theory that persists to low energies should be directly describable in terms of a low-energy EFT. For induced boundary conditions derivable from a boundary action this means the Wilsonian action should also have a boundary component from which the low-energy boundary physics can be inferred. As always with a Wilsonian action the form of the low-energy boundary action is obtained by matching, inasmuch as it is defined by the requirement that it reproduces the boundary physics of the full theory order by order in the low-energy expansion.

This is all made more concrete using the toy model example just described. For fields varying slowly compared with the length scale m_R^{-1} the effects of the boundary should also be calculable within the low-energy EFT appropriate below the mass scale m_R, for which only the Goldstone field ξ survives. This EFT is the one encountered in earlier sections, with a shift symmetry $\xi \to \xi + c$, but now including boundary interactions that also respect this symmetry (because the boundary term in (5.3) respects the $U(1)$ symmetry under rephasings of ϕ).

At the classical level, the physics of the UV system that the boundary part of the Wilson action captures is the boundary condition satisfied at $\partial\mathcal{M}$ by ξ. In the example above, the UV boundary condition is $\partial_n \phi = \mu\phi$, and the implications of this condition for the light fields can be found in the full theory by using $\phi = \varrho \exp\left[i\xi/\sqrt{2}\,v\right]$, with $\varrho := v + \chi/\sqrt{2}$. For real μ the real and imaginary parts of the boundary condition for ϕ give the two separate real boundary conditions for χ and ξ (or, equivalently, for ϱ and ξ)

$$\partial_n \varrho = \mu\varrho \qquad \text{and} \qquad \varrho\, \partial_n \xi = 0 \quad \text{on } \partial\mathcal{M}. \tag{5.10}$$

Assuming $\varrho \neq 0$ at $\partial\mathcal{M}$, at leading (classical) order in λ the low-energy boundary action at $\partial\mathcal{M}$ should imply $\partial_n \xi = 0$ there. More generally, S_b in the effective theory captures the dependence of the ξ boundary physics order by order in the low-energy expansion.

What does this imply explicitly for S_b in the toy model? Consider first the self-interactions of ξ involving the smallest mass dimension. The most general possible local bulk and boundary interactions consistent with the symmetries[1] are

[1] In this example, $\partial\mathcal{M}$ is chosen to be flat and Poincaré invariant along the directions parallel to the boundaries, although dependence on the boundary's local geometry – such as its curvature – is in general present if the physics of the boundary is more complicated.

$$\mathfrak{L}_B = -\frac{Z_1}{2}\,\partial_\mu\xi\,\partial^\mu\xi + \frac{Z_2}{2}\,\xi\Box\xi - m_B\,\Box\xi + \cdots$$
$$\mathfrak{L}_b = -w^3 - m_b\,\partial_n\xi - \frac{Z_2}{2}\,\xi\,\partial_n\xi - c_1\,\partial_n^2\xi - c_2\,\partial^a\partial_a\xi - \frac{h_1}{2}(\partial_n\xi)^2 + \cdots, \quad (5.11)$$

where the effective couplings w, m_B and m_b have dimension mass, the couplings, Z_1, Z_2, c_1, c_2 and c_3 are dimensionless and h_1 has dimension (mass)$^{-1}$. (This list does not exhaust the possibilities for the dimensions shown.) Here, ∂_n denotes the normal derivative $n_\mu\partial^\mu$, while ∂_a denotes derivatives only along directions parallel to $\partial\mathcal{M}$ (as opposed to ∂_n, which is in the direction perpendicular to $\partial\mathcal{M}$, and ∂_μ, which indicates differentiation in all of the directions within $\mathcal{M}$). The coefficients of the $\xi\Box\xi$ term in $\mathfrak{L}_B$ and the $\xi\partial_n\xi$ term in $\mathfrak{L}_b$ must be related in the way indicated in order to preserve the invariance of the total action, $S = S_B + S_b$, under the shift symmetry.

Some of these interactions are redundant, for both of the reasons discussed in §2.5. An important difference from this earlier discussion is that total derivatives in $\mathfrak{L}_B$ can no longer simply be dropped. Instead, Stokes' theorem relates such terms to terms in the boundary action. For example, the $\Box\xi$ term in $\mathfrak{L}_B$ can be converted in this way to the $\partial_n\xi$ term on $\mathfrak{L}_b$, showing that physical quantities can only depend on the combination $\tilde{m} := m_b + m_B$ rather than either m_b or m_B separately. Similarly, integrating by parts either the Z_1 or Z_2 terms in $\mathfrak{L}_B$ shows that these parameters can only contribute as the sum $Z := Z_1 + Z_2$. Furthermore, the combination $\partial^a\partial_a\xi$ is a total derivative within $\partial\mathcal{M}$, and can always be dropped given that the boundary itself has no boundary.

Using this freedom allows the above action to be rewritten as

$$\mathfrak{L}_B = -\frac{Z}{2}\,\partial_\mu\xi\,\partial^\mu\xi \quad \text{and} \quad \mathfrak{L}_b = -w^3 - \tilde{m}\,\partial_n\xi - c_1\,\partial_n^2\xi - \frac{h_1}{2}(\partial_n\xi)^2, \quad (5.12)$$

and rescaling the field, $\xi \to \xi/\sqrt{Z}$, shows the four parameters Z, $\tilde{m}$, c_1 and h_1 only enter physical quantities through the three combinations $\tilde{m}/\sqrt{Z}$, $c_1/\sqrt{Z}$ and h_1/Z. This freedom is now used to set $Z = 1$ (*i.e.* to 'canonically normalize' the field), leaving only three potentially independent parameters $\tilde{m}$, c_1 and h_1 of the terms considered in (5.11).

But even these parameters need not be independent (or present at all). To see why, recall this effective theory arises from UV physics where all fields are integrated freely both within the bulk and on the boundary. Consequently, the functional integral over ξ in the low-energy theory is also unconstrained in both the bulk and on the boundary. The classical limit for such a free integration is then found by evaluating the path integral at a classical path chosen to make the action stationary against arbitrary variations of ξ both in the interior of $\mathcal{M}$ *and* throughout $\partial\mathcal{M}$, and this should be consistent with what is found for the UV completion.

To see what this requires write $\xi \to \xi + \delta\xi$ and linearize the action in $\delta\xi$, to find

$$\delta S = -\int_{\mathcal{M}} d^4x\;\partial_\mu\xi\,\partial^\mu\delta\xi - \int_{\partial\mathcal{M}} d^3x\left[\tilde{m}\,\partial_n\delta\xi + c_1\,\partial_n^2\delta\xi + h_1\partial_n\xi\,\partial_n\delta\xi\right] \quad (5.13)$$
$$= \int_{\mathcal{M}} d^4x\,(\Box\xi)\,\delta\xi - \int_{\partial\mathcal{M}} d^3x\left[\partial_n\xi\,\delta\xi + (\tilde{m} + h_1\partial_n\xi)\partial_n\delta\xi + c_1\,\partial_n^2\delta\xi\right].$$

Since the action must be stationary against *arbitrary* $\delta\xi$, first choose $\delta\xi$ and its derivatives to be only nonzero away from the boundary. As usual, this implies that the saddle point $\xi_c(x)$ must satisfy the classical field equations $\Box\xi_c = 0$ everywhere in $\mathcal{M}$.

Next, demand also $\delta S = 0$ for arbitrary variations of ξ on $\partial\mathcal{M}$. First, do so by choosing the variations so that $\partial_n \delta\xi = \partial_n^2 \delta\xi = 0$ but with $\delta\xi \neq 0$ on $\partial\mathcal{M}$. Requiring $\delta S = 0$ for all such $\delta\xi$ implies that the saddle point must satisfy *Neumann* boundary conditions: $\partial_n \xi_c(x) = 0$ on $\partial\mathcal{M}$. But now requiring $\delta S = 0$ for variations with $\partial_n^2 \delta\xi = 0$ but $\partial_n \delta\xi \neq 0$ requires $\tilde{m} + h_1 \partial_n \xi_c = 0$ on $\partial\mathcal{M}$; a result inconsistent with Neumann boundary conditions unless $\tilde{m} = 0$. Similarly, variations with $\partial_n^2 \xi \neq 0$ imply further conditions ($c_1 = 0$ if only the displayed terms are kept).

The very presence of nonzero couplings $\tilde{m}$ and c_1 presents an obstruction to being able to find a consistent boundary condition at $\partial\mathcal{M}$, and thereby also obstructs there being a saddle point for which $\delta S = 0$ when ξ varies arbitrarily on the boundary. Since this obstruction did not arise in the UV theory, the appropriate matching condition must be that $\tilde{m} = c_1 = 0$. The resulting boundary condition at this order is then $\partial_n \xi_c = 0$ on $\partial\mathcal{M}$, agreeing with the result found in the full theory in (5.10). Nontrivial effective coupling can arise in S_b at higher orders in the semiclassical expansion to the extent that they are required in order to reproduce modifications to the UV physics there.

This example shows that terms in $\mathfrak{L}_b$ involving normal derivatives of the fields can (in general) over-determine the boundary conditions obtained by varying the action freely on the boundary. Normal derivatives are special in this way because they cannot be integrated by parts on the boundary, making it impossible to rewrite their variation in terms only of $\delta\xi$ (as opposed to its derivatives). Because of this, it is generic that the effective couplings for interactions involving normal derivatives in the boundary action are completely determined by the effective couplings for terms in $\mathfrak{L}_B$. (For example, in the example above the coefficient of $\xi\,\partial_n\xi \in \mathfrak{L}_b$ is not independent of the coefficient of $\xi\,\Box\xi \in \mathfrak{L}_B$.) It is only the couplings for the rest of the effective interactions that represent independent parameters describing low-energy properties of the boundary.

The other way that interactions can be redundant is if they can be removed using a local field redefinition, again following the arguments of §2.5. To see how this works for terms on the boundary suppose the bulk action is dominated by the kinetic term, $\mathfrak{L}_{B0} = -\frac{1}{2}(\partial_\mu\xi\,\partial^\mu\xi)$, in the regime of semiclassical perturbation theory (as it is for the toy model example). Performing a change of variables $\xi \to \xi + \epsilon\zeta(\xi)$ – with ϵ a small perturbation parameter – then use of Stoke's theorem leads to the following change in the bulk action

$$\delta S_{B0} = -\epsilon \int_{\mathcal{M}} \mathrm{d}^4x\; \partial_\mu\xi\,\partial^\mu\zeta = -\epsilon \int_{\partial\mathcal{M}} \mathrm{d}^3x\; \partial_n\xi\,\zeta + \epsilon \int_{\mathcal{M}} \mathrm{d}^4x\; \zeta\,\Box\xi. \tag{5.14}$$

For generic $\zeta(\xi)$ it is the last term of (5.14) that was used in §2.5 to argue that terms proportional to $\Box\xi$ can be removed in an order-ϵ term of the bulk action. Eq. (5.14) then shows that the change of variables also removes the order-ϵ terms in S_b whose effective coupling is tied to the removed bulk term. For example, a transformation with $\zeta = c_{\text{eff}}\partial_\mu\xi\,\partial^\mu\xi$ that removes a term $-\epsilon c_{\text{eff}}(\partial_\mu\xi\,\partial^\mu\xi)\Box\xi$ in $\mathfrak{L}_B$ also adds a term $-\epsilon c_{\text{eff}}(\partial_\mu\xi\,\partial^\mu\xi)\partial_n\xi$ to $\mathfrak{L}_b$.

A coupling in (5.11) not determined by boundary conditions in this way is the parameter w^3. At first sight, this coupling might be thought to be redundant inasmuch as the corresponding term in the effective action does not depend on the low-energy field ξ. This parameter nonetheless carries physical content since it contributes to the low-energy stress energy[2] and so also to the energy-per-unit-area of the boundary. Matching this to the result, (5.8), obtained in the UV theory for the toy model with a boundary at $z = 0$ then implies that

$$w^3 \simeq -\mu v^2 + \frac{3\mu^2 v^2}{m_R} = -\mu v^2 + \frac{3\mu^2 v}{\sqrt{\lambda}}. \tag{5.15}$$

5.3 Dynamical Boundary Degrees of Freedom

One thing not yet captured by this chapter's discussion is the possible existence of fields, ψ, that appear only in the boundary action, S_b, and not at all in the bulk, S_B, so $S[\phi, \psi] = S_B[\phi] + S_b[\phi, \psi]$. Such fields capture the low-energy physics of states in the UV theory whose mode-functions have support only in the immediate vicinity of the boundary, and so are said to be 'localized' at the boundary. These could be anything from surface charges on a conductor or boundary states at the interface between materials to the motion of p-branes [87] in supergravity and open strings attached to D-branes [88] in string theory.

Boundary-localized fields depend only on the three coordinates, σ^a, that label position on $\partial\mathcal{M}$. For example, for a boundary consisting of the $x - y$ plane at $z = 0$ in a flat cartesian space these three coordinates might be $\{\sigma^a\} = \{t, x, y\}$.

Perhaps the simplest such localized field describes the position of the boundary itself, $y^\mu(\sigma)$, where the boundary's position in spacetime is denoted $x^\mu = y^\mu(\sigma)$. This is a dynamical field if this boundary position is free to move at low energies.[3] In the presence of such fields the boundary action, S_b, does double duty: it both determines the dynamics of y^μ given the presence of any bulk 'background' fields, $\phi(x)$, and it determines how boundaries source these same bulk fields.

For example, consider an ordinary real scalar field, ϕ, coupled to a dynamical but slowly moving boundary $y^\mu(\sigma) = \{t, x, y, \mathfrak{z}(x, y, t)\}$, located at $z = \mathfrak{z}(x, y, t)$ with $\mathfrak{z}(x, y, t)$ a single-valued function whose derivatives are small: *e.g.* $|\dot{\mathfrak{z}}| \ll 1$ with over-dots representing $\mathrm{d}/\mathrm{d}t$. Then, an expansion of the boundary action in powers of $\dot{\mathfrak{z}}$ might take the form

$$S_b[\phi, y] = -\int \mathrm{d}t\, \mathrm{d}x\, \mathrm{d}y \left[W[\phi(z = \mathfrak{z})] + \frac{1}{2} K[\phi(z = \mathfrak{z})]\, \dot{\mathfrak{z}}^2 + \cdots \right], \tag{5.16}$$

where $W(\phi)$ and $K(\phi)$ are specified functions that are characteristic of the surface. (Spatial derivatives of $\mathfrak{z}$ can also be considered but are dropped here for simplicity.) The evolution of the bulk field, ϕ, is for simplicity imagined to have a bulk lagrangian dominated by

$$S_B[\phi] = -\int \mathrm{d}^4x \left[\frac{1}{2}\, \partial_\mu\phi\, \partial^\mu\phi + \frac{1}{2}\, m_\phi^2\, \phi^2 + \cdots \right]. \tag{5.17}$$

[2] That is to say, it does couple to a low-energy field: the spacetime metric.

[3] §6.3.1, §13.1.2 and §14.3.1 argue why these fields often behave like Goldstone modes for spacetime symmetries, and as such naturally appear in the low-energy sector relevant for EFT methods.

With these choices the motion of the boundary in the presence of a given bulk field configuration, $\phi(x)$, is found at the classical level by varying S_b with respect to $\mathfrak{z}$, and gives

$$\left[K\ddot{\mathfrak{z}} + \left(\frac{1}{2}\frac{\partial K}{\partial\phi}\dot{\mathfrak{z}}^2 - \frac{\partial W}{\partial\phi}\right)\frac{\partial\phi}{\partial z}\right]_{z=\mathfrak{z}(x,y,t)} = 0, \tag{5.18}$$

as the equation of motion governing the time-dependence of $\mathfrak{z}(x, y, t)$. The classical field equations for ϕ obtained by varying $S_B + S_b$ in the bulk similarly give

$$(\Box - m_\phi^2)\phi = 0 \qquad \text{for } z > \mathfrak{z}(x, y, t), \tag{5.19}$$

while variations on the boundary give the condition

$$\partial_n\phi + \left(\frac{\partial W}{\partial\phi} + \frac{1}{2}\frac{\partial K}{\partial\phi}\dot{\mathfrak{z}}^2 + \cdots\right) = 0 \quad \text{for } z = \mathfrak{z}(x, y, t). \tag{5.20}$$

A slightly different but related picture can arise if the boundary is instead regarded as a thin surface (*i.e.* membrane – or in its relativistic incarnations 'brane' [87]) with two sides, rather than effectively being the edge of spacetime. In this case, the boundary action (5.16) can instead be written more usefully as the two-sided brane action

$$S_b[\phi, y] = -\int d^4x \left[W[\phi(x)] + \frac{1}{2}K[\phi(x)]\dot{\mathfrak{z}}^2 + \cdots\right]\delta[z - \mathfrak{z}(x, y, t)], \tag{5.21}$$

leading to a ϕ equation of the form

$$(\Box - m_\phi^2)\phi - \left[\frac{\partial W}{\partial\phi} + \frac{1}{2}\frac{\partial K}{\partial\phi}\dot{\mathfrak{z}}^2 + \cdots\right]\delta[z - \mathfrak{z}(x, y, t)] = 0. \tag{5.22}$$

Here, the boundary condition becomes a 'jump' condition obtained by integrating (5.22) over an infinitesimal region $\mathfrak{z} - \epsilon < z < \mathfrak{z} + \epsilon$ that includes the delta function:

$$\Big[\partial_n\phi\Big]_{\mathfrak{z}} + \left(\frac{\partial W}{\partial\phi} + \frac{1}{2}\frac{\partial K}{\partial\phi}\dot{\mathfrak{z}}^2 + \cdots\right)_{z=\mathfrak{z}(x,y,t)} = 0, \tag{5.23}$$

where the square bracket denotes the jump in a quantity across $z = \mathfrak{z}$, as in $\Big[\partial_n\phi\Big]_{\mathfrak{z}} = \partial_n\phi(z = \mathfrak{z} + \epsilon) - \partial_n\phi(z = \mathfrak{z} - \epsilon)$ with $\epsilon \to 0$ at the end.

5.4 Summary

Boundaries do not change the EFT story in a dramatic way. This chapter maps out the various small ways that boundaries do change low-energy dynamics focussing on 'induced' boundary conditions, defined as those that are obtained by extremizing an action with respect to field variations on the boundary. Such boundary conditions arise naturally in situations where the path integral is over all fields in an unconstrained way, both in the bulk and on the boundary.

Like the devil, the main differences associated with boundaries are in the details. The central new feature is the addition of a local boundary component to the Wilsonian action. Its effective couplings are (as usual) obtained by demanding that the low-energy theory reproduces the full theory's boundary physics order-by-order in the low-energy expansion. The precise ways that redundant interactions are identified in the Wilsonian action are slightly modified due to the ability to swap terms between the bulk and boundary actions by integrating by parts.

Terms in the boundary action involving normal derivatives play a special role because their presence can over-determine the boundary conditions when extremizing against arbitrary field variations. Consequently, the matching process often ends up fixing their effective couplings in terms of the values of couplings appearing in the bulk lagrangian (or makes them vanish). When this happens they do not represent independent parameters associated with the physics of the boundary.

An attractive feature about tying boundary conditions to boundary components of a low-energy Wilsonian action is the context it provides for understanding why some boundary conditions arise more often in physical situations than do others. The leading low-energy contributions to boundary terms in the Wilson action are usually not complicated for the same reasons that terms in the bulk are not: low energy often puts a premium on having few fields and derivatives and not many terms are possible involving the fewest of each. This suggests an underlying reason why relatively simple boundary conditions (like those linear in the fields) arise so frequently in practice.

A final qualitatively new feature that boundaries can introduce are localized degrees of freedom that live only at the boundary. When these exist, their interactions with bulk fields are governed by the boundary action, and their semiclassical treatment goes through much the same as for bulk fields alone.

Exercises

Exercise 5.1 Derive the approximate classical solution Eq. (5.7) in the bulk for a field satisfying the field equation (5.5) in the regime $z \geq 0$ subject to the boundary condition $\phi \to v$ as $z \to \infty$ and Eq. (5.6) at $z = 0$.

Evaluate the classical energy per unit area, E/A, and verify (5.8) holds when the cubic and quartic terms in $(\phi - v)$ are neglected in the potential V. Include the cubic and quartic terms in the energy when evaluating E/A at the solution of (5.7), and thereby quantify how suppressed they are as a function of the small dimensionless parameters μ/m_R and $\sqrt{\lambda} = m_R/v$.

Exercise 5.2 For the same bulk and boundary action as in Exercise 5.1 write the full quantum field as $\phi = \phi_c + \hat{\phi}$ where ϕ_c is the classical solution (5.7). What are the boundary conditions for the quantum fluctuation field $\hat{\phi}$ at the boundary at $z = 0$? Using this boundary condition compute the mode functions, $u_n(z)e^{i(k_x x + k_y y - \omega t)}$, appearing in the expansion of $\hat{\phi}$ in terms of creation and annihilation operators. (Neglect the cubic and quartic interaction terms in the bulk scalar potential when doing so.) Are any of these modes bound states localized near the boundary? (Bound states have energies $\omega^2 < k_x^2 + k_y^2 + m_R^2$ and so have wave-functions that are normalizable in the z-direction.) If so, what is the mode profile $u_n(z)$ and energy ω_n?

Exercise 5.3 Repeat Exercise 5.1, but with a flat mobile brane held stationary at $z = z_b$ with $z_b > 0$. Compute the approximate field profile for $\phi(z)$ (as a function of z, z_b, μ and m_R) on both sides of the mobile brane (*i.e.* for both $0 \leq z < z_b$ and $z_b < z < \infty$), using the same boundary conditions as before at $z = 0$ and $z \to \infty$. Neglect the cubic and quartic interactions in the bulk potential when doing so, and suppose the action for the mobile brane is given by (5.21) with $K = 1$ and $W = g\phi^*\phi$. Use the jump condition (5.23) (as well as continuity of ϕ itself) to evaluate the boundary conditions at $z = z_b$.

Use Eq. (5.18) to evaluate the acceleration of the mobile brane, $\ddot{z}_b$, if it were free to move. Which direction does the mobile brane go once it is released from rest at $z = z_b$?

Exercise 5.4 Consider two bulk regions, R_1 and R_2, on either side of an interface, F, with the interface regarded as a common boundary shared by the two regions. Suppose the bulk action for the electromagnetic field in each region is that of a dielectric with differing dielectric constants, corresponding to

$$S_{R_i} = \frac{1}{2} \int_{R_i} d^4x \left[\varepsilon_i \, \mathbf{E} \cdot \mathbf{E} - \frac{\mathbf{B} \cdot \mathbf{B}}{\mu_i} \right],$$

where ε_i is the dielectric constant and μ_i the magnetic permeability for each bulk region. Take the boundary action for the common interface to be

$$S_F = - \int_F d^3x \; A_\mu J^\mu$$

where $A^0 = \Phi$ is the electrostatic potential and $\mathbf{A}$ is the vector potential, while $J^0 = \rho$ is the interface's surface charge density and $\mathbf{J}$ is its surface current (satisfying $\partial_t \rho + \nabla \cdot \mathbf{J} = 0$). By varying the electromagnetic potentials Φ and $\mathbf{A}$ derive the dielectric Maxwell equations in each of the bulk regions R_i as well as the boundary conditions obtained by demanding that the action $S_{R_1} + S_{R_2} + S_F$ is stationary under arbitrary variations δA_μ on the interface. Show that your results reproduce the standard ones: the jump across of the interface of the normal component of $\mathbf{D} = \varepsilon \, \mathbf{E}$ is given by the charge density; the jump in the tangential component of $\mathbf{H} = \mathbf{B}/\mu$ is given by the surface current; while the other components of $\mathbf{E}$ and $\mathbf{B}$ are continuous at F.

6 Time-Dependent Systems

Up to this point in the discussion, many of the EFT applications have been to scattering problems for low-energy states arising as fluctuations about a stationary ground state. This really only scratches the surface of the utility of effective field theories, as this chapter hopes to convey. This chapter asks how to apply EFT methods to systems involving background fields that evolve in time. This kind of problem arises throughout physics, including (but not limited to) atomic interactions with time-dependent electromagnetic fields, particle motion through inhomogeneous media and the time-varying fields of early-universe cosmology.

For the purposes of argument in this chapter the time-varying background field is taken to be a scalar, in order to make better contact with the toy model. But examples could equally well be considered using background electromagnetic or gravitational fields, some of which are considered amongst the examples examined in later sections.

A number of new conceptual issues arise when setting up an effective description of systems with time-dependent backgrounds. One such asks why 'low-energy' and 'high-energy' remain useful as criteria for splitting up the space of states given that the breaking of time-translation invariance means fluctuation energy is not strictly conserved. Another asks whether all solutions to the full theory's field equations have counterpart solutions in the effective theory, and vice versa. A third asks what the correct number of initial conditions should be in an effective theory, given that the low-energy field equations can involve more than two time derivatives.

These issues do not arise in simpler static settings, and although none of them need preclude using low-energy techniques, the validity of EFT methods sometimes involves additional criteria that must be checked explicitly for any particular application. Most notable among these new conditions is the requirement that any background evolution be slow enough to be adiabatic (in a sense that is further elaborated below).

6.1 Sample Time-Dependent Backgrounds ◇

Just like in earlier sections it is useful to ground a general discussion of issues by having a concrete example in mind. To this end, this chapter starts with an example of time-dependent backgrounds within the toy model introduced in §1.1, using it to illustrate how EFT methods work for time-dependent settings and why they can sometimes fail.

First, we present a brief reminder of the main features of the toy model, for ease of reference. Its lagrangian density is given by

$$\mathfrak{L} = -\partial_\mu \phi^* \partial^\mu \phi - V(\phi^* \phi), \tag{6.1}$$

where the complex field ϕ is written in terms of two real fields using either $\phi = \frac{1}{\sqrt{2}}(\phi_R + i\phi_I)$ or $\phi = \varrho\, e^{i\vartheta}$, where these are related to the variables used in previous sections by $\phi_R = \sqrt{2}\, v + \tilde{\phi}_R$, $\phi_I = \tilde{\phi}_I$, $\xi = \sqrt{2}\, v\, \vartheta$ and $\chi = \sqrt{2}(\varrho - v)$. Furthermore, the potential has the explicit 'Mexican hat' or 'wine-bottle' form

$$V(\phi^*\phi) = \frac{\lambda}{4}\left(\phi^*\phi - v^2\right)^2, \tag{6.2}$$

which at low energies has a level, circular trough with a bottom at $V = 0$ along the curve $\sqrt{2}\, \varrho = \sqrt{\phi_R^2 + \phi_I^2} = \sqrt{2}\, |\phi| = \sqrt{2}\, v$.

The model's Noether current for the $U(1)$ symmetry is given by (4.25),

$$j_\mu = i(\phi\, \partial_\mu \phi^* - \phi^* \partial_\mu \phi) = \phi_R \partial_\mu \phi_I - \phi_I\, \partial_\mu \phi_R = 2\varrho^2 \partial_\mu \vartheta, \tag{6.3}$$

and Noether's theorem implies that this satisfies $\partial_\mu j^\mu = 0$ whenever the field equations,

$$\Box \phi = -\partial_t^2 \phi + \nabla^2 \phi = \frac{\lambda}{2}\left(\phi^*\phi - v^2\right)\phi, \tag{6.4}$$

are satisfied.

Slow-Roll Backgrounds

Until now the only background solution to (6.4) to be considered has been the static vacuum solution $\phi = v$. Consider instead the time-dependent background corresponding to the scalar field homogeneously rolling around the bottom of its potential [89]:

$$\varrho = \varrho_0 \qquad \text{and} \qquad \vartheta(t) = \vartheta_0 + \omega t, \tag{6.5}$$

for constants ϱ_0, ϑ_0 and ω. Eq. (6.4) implies that these constants must satisfy

$$\left[\frac{\lambda}{2}\left(\varrho_0^2 - v^2\right) - \omega^2\right]\varrho_0 = 0, \tag{6.6}$$

so the only solution with $\varrho_0 > 0$ is

$$\varrho_0 = \sqrt{v^2 + \frac{2\omega^2}{\lambda}}. \tag{6.7}$$

This shows how the force due to the scalar potential's gradient competes with the centripetal acceleration due to the circular motion to drive the radial field ϱ slightly away from the trough's bottom.

The density of the conserved Noether charge evaluated at this solution is

$$j^\mu = -2\omega\, \varrho_0^2\, \delta_0^\mu = -2\omega\left(v^2 + \frac{2\omega^2}{\lambda}\right)\delta_0^\mu, \tag{6.8}$$

and its energy density is

$$\varepsilon = \dot{\phi}^*\dot{\phi} + \nabla\phi^* \cdot \nabla\phi + \frac{\lambda}{4}(\phi^*\phi - v^2)^2 = \omega^2 \varrho_0^2 + \frac{\lambda}{4}(\varrho_0^2 - v^2)^2$$
$$= \omega^2 \left(v^2 + \frac{3\omega^2}{\lambda} \right), \tag{6.9}$$

where over-dots denote differentiation with respect to t. For later purposes notice that the appearance of a 3 (instead of a 2) in the last term of the last line of this last equation can be traced to the contribution of the scalar potential to ε in the second-last line. The potential contributes because the motion displaces the field away from the potential's minimum by the amount given in (6.7).

6.1.1 View from the EFT

This section now asks how the above rolling solution looks from the point of view of the low-energy EFT appropriate at energies well below m_R, which should be a valid regime for a sufficiently slowly moving background. In particular, how does the EFT know about the energy increase of (6.9) due to the field ϱ climbing part way up the potential if there is no field ϱ left in the effective theory to adjust to balance centrifugal forces, and no scalar potential (or indeed notion of centripetal acceleration) within the EFT.

Previous sections show that the leading approximation to the low-energy EFT for this model is given by (2.97), which in the classical approximation (using $m_R^2 = \lambda v^2$) is

$$S_W[\xi] = -\int d^4x \left[\frac{1}{2}\,\partial_\mu\xi\,\partial^\mu\xi - \frac{\lambda}{4m_R^4}(\partial_\mu\xi\,\partial^\mu\xi)^2 + \cdots \right]$$
$$= -\int d^4x \left[v^2\partial_\mu\vartheta\,\partial^\mu\vartheta - \frac{v^2}{m_R^2}(\partial_\mu\vartheta\,\partial^\mu\vartheta)^2 + \cdots \right]. \tag{6.10}$$

The field equations for ϑ predicted by this action are

$$\partial_\mu \left\{ \partial^\mu\vartheta \left[1 - \frac{2}{m_R^2}(\partial_\nu\vartheta\,\partial^\nu\vartheta) + \cdots \right] \right\} = 0, \tag{6.11}$$

which admits the solution $\vartheta = \vartheta_0 + \omega t$ for which $\partial_\mu\vartheta = \omega\,\delta^0_\mu$ is constant.

Applying Noether's theorem to this action with the low-energy shift symmetry $\vartheta \to \vartheta + c$ ensures the existence of the conserved current,

$$j^\mu_{\rm eff} = 2v^2\partial^\mu\vartheta \left[1 - \frac{2}{m_R^2}(\partial_\nu\vartheta\,\partial^\nu\vartheta) + \cdots \right], \tag{6.12}$$

for which the equations of motion (6.11) clearly imply $\partial_\mu j^\mu_{\rm eff} = 0$. Evaluating this at the rolling solution, $\vartheta = \omega t$ then gives

$$j^\mu_{\rm eff} = -2v^2\omega \left(1 + \frac{2\omega^2}{m_R^2} + \cdots \right) \delta^\mu_0 = -2\omega \left(v^2 + \frac{2\omega^2}{\lambda} + \cdots \right) \delta^\mu_0, \tag{6.13}$$

in agreement with (6.8).

To calculate the energy density of this solution in the effective theory compute the effective Hamiltonian density,

$$\mathcal{H}_{\rm eff} = \pi_{\rm eff}\,\dot{\vartheta} - \mathfrak{L}_{\rm eff}, \tag{6.14}$$

where the canonical momentum is defined by

$$\pi_{\rm eff} := \frac{\delta S_{\rm eff}}{\delta \dot{\vartheta}} = 2v^2 \left(\dot{\vartheta} + \frac{2\dot{\vartheta}^3}{m_R^2} + \cdots \right). \tag{6.15}$$

Using this, the Hamiltonian density becomes

$$\mathcal{H}_{\rm eff} = v^2 \dot{\vartheta}^2 + v^2 \nabla\vartheta \cdot \nabla\vartheta + \frac{3\lambda v^4\, \dot{\vartheta}^4}{m_R^4} + \cdots, \tag{6.16}$$

and so the energy density obtained by evaluating this at $\dot{\vartheta} = \omega$ is

$$\varepsilon_{\rm eff} = v^2\omega^2 + \frac{3\lambda v^4\, \omega^4}{m_R^4} + \cdots = v^2\omega^2 + \frac{3\omega^4}{\lambda} + \cdots, \tag{6.17}$$

in agreement with (6.9), including the last term's factor of 3.

These calculations reveal that it is the first subleading term, $(\partial_\mu \xi\, \partial^\mu \xi)^2$, in $\mathfrak{L}_W$ that brings the news to the EFT about the adjustment of ϱ_0 and the centripetal acceleration in the UV theory. Furthermore, the matching performed in previous sections gives precisely the value for the effective coupling needed to get the answer right. This despite the fact that these earlier matching calculations obtain the coupling's value using scattering amplitudes, rather than classical background evolution.

Because it is the higher-derivative terms that carry the information about the ω-dependent response of the system in the EFT, it is also clear that a purely EFT description of this response assumes $\omega \ll m_R$ if it is to neglect the contributions of higher-dimension interactions.

6.2 EFTs and Background Solutions ◇

The toy model example just considered shows that the field equations of the Wilsonian effective theory properly capture the time-dependence of slowly evolving classical background solutions of the full UV theory (see §6.3 for the analogous story about fluctuations about such backgrounds). This section asks more generally when the background solutions within an EFT should (and shouldn't) be expected to reproduce the solutions of the underlying UV theory.

The main message is that the space of solutions solving the field equations of an EFT overlaps with (but neither contains nor is contained within) the space of solutions for the UV theory. Not surprisingly, solutions to the EFT's field equations do include those of the full theory that evolve adiabatically but not those that evolve too quickly. But the EFT field equations also have solutions that are not related to those of the full theory (and are typically singular in the limit that the UV scale goes to infinity). To justify these statements (and make them more precise), and to see how to identify which EFT solutions are relevant to the full theory's low-energy limit (and which are not), it is worth recalling some features of the discussion in §2.1 and §2.2.

6.2.1 Adiabatic Equivalence of EFT and Full Evolution

Why should background solutions for the full theory and low-energy theories agree with one another, and precisely which equations do backgrounds solve?

For the full theory the relation between the field expectation value and the action is given by (2.14) and (2.18), which (in the absence of an external current) state that

$$\varphi^i(x) = \langle \phi^i(x) \rangle = \frac{\langle \text{out}|\phi^i(x)|\text{in}\rangle}{\langle \text{out}|\text{in}\rangle} \tag{6.18}$$

satisfies

$$\left(\frac{\delta \Gamma[\phi]}{\delta \phi^i(x)}\right)_{\phi=\varphi} = 0, \tag{6.19}$$

where $\Gamma[\phi]$ is the generator of 1PI graphs, $|\text{in}\rangle$ is the vacuum state in the remote past and $|\text{out}\rangle$ is the vacuum state in the remote future (which need not be the same in the presence of time-dependent fields in between). Crucially, the derivation of this statement assumes $|\text{in}\rangle$ evolves adiabatically[1] as a function of the evolving background quantities as it eventually turns into $|\text{out}\rangle$ [17].

For a system whose fields divide into heavy and light degrees of freedom, $\{\phi^i(x)\} = \{h^a(x), \ell^\alpha(x)\}$, Eq. (6.19) holds both for $\langle h^a(x)\rangle$ and $\langle \ell^\alpha(x)\rangle$,

$$\left(\frac{\delta \Gamma[h,\ell]}{\delta h^a(x)}\right)_{\langle h\rangle,\langle \ell\rangle} = \left(\frac{\delta \Gamma[h,\ell]}{\delta \ell^\alpha(x)}\right)_{\langle h\rangle,\langle \ell\rangle} = 0. \tag{6.20}$$

The closest analogue of $\Gamma[h,\ell]$ for the low-energy part of this theory is the 1LPI generator, $\Gamma_{\text{le}}[\ell]$, introduced in §2.2.3, for which external currents are only turned on for the light fields. Chasing through the definitions implies that the relation between $\Gamma[h,\ell]$ and $\Gamma_{\text{le}}[\ell]$ is given by (2.45), which states

$$\Gamma_{\text{le}}[\ell] = \Gamma[h_{\text{le}}(\ell),\ell] \quad \text{where} \quad \left(\frac{\delta \Gamma}{\delta h^a}\right)_{h=h_{\text{le}}(\ell)} = 0, \tag{6.21}$$

and so $h_{\text{le}}(\ell) = \langle h\rangle$ is regarded as a function of the specified value for the light field.

Varying this expression with respect to $\ell^\alpha(x)$ – keeping in mind (6.20) – then shows that $\langle \ell^\alpha(x)\rangle$ satisfies the purely low-energy condition

$$\left(\frac{\delta \Gamma_{\text{le}}[\ell]}{\delta \ell^\alpha(x)}\right)_{\langle \ell\rangle} = \left(\frac{\delta \Gamma[h,\ell]}{\delta \ell^\alpha(x)}\right)_{\langle h\rangle,\langle \ell\rangle} = 0. \tag{6.22}$$

This shows that $\langle \ell^\alpha(x)\rangle$ can equally well be computed by extremizing $\Gamma[h,\ell]$ in the full theory or by extremizing $\Gamma_{\text{le}}[\ell]$ of the low-energy sector alone. But a central part of this argument is the adiabatic assumption that underpins the starting point, (6.19).

Classical Limit

Time-dependent backgrounds are commonly encountered within a semiclassical approximation, wherein the time-dependent background is the dominant, classical, configuration: $\langle \phi^i(x)\rangle \simeq \phi^i_c(x)$. In this case, the above argument goes through order-by-order in the semiclassical expansion, with the leading (classical) contribution being given by

$$\Gamma[h,\ell] \simeq S[h,\ell] \quad \text{and} \quad \Gamma_{\text{le}}[\ell] \simeq S_w[\ell]. \tag{6.23}$$

[1] Adiabatic evolution here means the solutions are solved as if they are static functions of external parameters, like the currents J, but these external parameters are themselves allowed to evolve very slowly with time. This is as opposed, for instance, to having levels cross or some other drama between $t = \pm\infty$.

Here, S is the classical action for the full theory and S_W is the Wilson action defined in §2.3 and given in the classical approximation by $S_W[\ell] \simeq S[h_c(\ell), \ell]$, where $h_c(\ell)$ is found by solving $\delta S[h, \ell]/\delta h = 0$ as a function of a specified light field ℓ (*c.f.* Eq. (6.21)). Then (6.22) becomes a statement relating the classical solutions for these two actions:

$$\left(\frac{\delta S_W[\ell]}{\delta \ell^a(x)}\right)_{\ell_c} = \left(\frac{\delta S[h, \ell]}{\delta \ell^a(x)}\right)_{h_c, \ell_c} = 0, \tag{6.24}$$

Again it seems clear that solutions to the equations of motion for the classical Wilson action reproduce the light-field part of the solutions to the classical equations of the full theory. But how does the adiabatic requirement arise in this purely classical argument? To understand this it helps to consider an example, and our stalwart toy model once more comes in useful.

Example: The Toy Model

To this end, revisit the derivation given in §2.2.3, starting with Eq. (2.47) (reproduced here),

$$S[\xi, \chi] = -\int \mathrm{d}^4x \left[\frac{1}{2}\,\partial_\mu \chi \partial^\mu \chi + \frac{1}{2}\left(1 + \frac{\chi}{\sqrt{2}\,v}\right)^2 \partial_\mu \xi \partial^\mu \xi + V(\chi)\right], \tag{6.25}$$

in which the heavy field χ is explicitly integrated out within the classical approximation, using the classical potential

$$V(\chi) = \frac{m_R^2}{2}\,\chi^2 + \frac{\lambda\, v}{2\sqrt{2}}\,\chi^3 + \frac{\lambda}{16}\,\chi^4. \tag{6.26}$$

To compute $S_W[\xi] \simeq S[\xi, \chi_c(\xi)]$ classically requires solving for $\chi_c(\xi)$ using Eq. (2.49), which to leading nontrivial order is approximately

$$\left(-\Box + m_R^2\right)\chi_c \simeq -\frac{1}{\sqrt{2}\,v}\,\partial_\mu \xi \partial^\mu \xi + \cdots, \tag{6.27}$$

where the higher-order terms are not required to make the point soon to follow. The solution to this equation used in §2.2.3 is

$$\chi_c \simeq -\frac{1}{\sqrt{2}\,v m_R^2}\,(\partial_\mu \xi \partial^\mu \xi) + \cdots, \tag{6.28}$$

and substituting this into $S[\chi, \xi]$ leads to the $(\partial_\mu \xi\, \partial^\mu \xi)^2$ interaction found earlier for S_W.

Now comes the main point: the transition from (6.27) to (6.28) given in §2.2.3 proceeds as if the solution to (6.27) were *unique*. But we know the general solution to (6.27) is really given by the sum of (6.28) and an arbitrary solution, χ_h, to the homogenous equation $(-\Box + m_R^2)\chi_h = 0$.

It is the adiabatic approximation that dictates choosing the solution $\chi_h = 0$, and it does so because χ_h is generally a sum of modes whose time-dependence is given by e^{-iEt} where $E \geq m_R$. Such modes could be excited if the time evolution of the background were sufficiently rapid, but are not directly excited for slow adiabatic evolution. Of course, interactions can also introduce modes with higher mode energies starting only from those at lower energies, because once interactions are included, only the total energy (including interactions) is strictly conserved, and not just the energy of isolated linearized modes (more about this later).

6.2.2 Initial Data and Higher-Derivative Instabilities *

The previous section's observation that classical solutions to the Wilsonian equations of motion do not precisely overlap with those of the underlying UV theory is actually an EFT feature rather than a bug. It is, of course, reasonable that the UV theory should contain solutions not in the low-energy theory, since the latter cannot capture those solutions of the full theory that evolve rapidly (and so do not exclusively involve low-energy modes). This section argues that there are also solutions to the low-energy equations that do not correspond to solutions of the full UV theory, and that this is also a good thing.

Extra unwanted solutions arise in the low-energy theory because the Wilson action generically contains all possible interactions allowed by the low-energy field content and symmetries. As a result, it usually contains terms for which the fields appear multiply differentiated. For instance, at the six-derivative level for the toy model one can have

$$\mathfrak{L}_{\rm hd} = -c_{61}\, X^3 - c_{62}\, X\, \vartheta_{\mu\nu}\, \vartheta^{\mu\nu}, \tag{6.29}$$

where

$$X := -\partial_\lambda \vartheta\, \partial^\lambda \vartheta \quad \text{and} \quad \vartheta_{\mu\nu} := \partial_\mu \partial_\nu \vartheta, \tag{6.30}$$

while c_{6n} denotes the relevant effective coupling.

What is important for the purposes of counting solutions is that the last term of (6.29) involves more than two time derivatives, as is most easily seen by temporarily ignoring all spatial derivatives. In this case, (6.29) contains the term $\mathfrak{L}_{62} = c_{62}\, \ddot{\vartheta}^2 \dot{\vartheta}^2$, whose variation is

$$\frac{\delta \mathfrak{L}_{62}}{2c_{62}} = \left[\ddot{\vartheta}\dot{\vartheta}^2\right]\delta\ddot{\vartheta} + \left[\dot{\vartheta}\ddot{\vartheta}^2\right]\delta\dot{\vartheta} = \left[\ddddot{\vartheta}\,\dot{\vartheta}^2 + 4\,\dddot{\vartheta}\ddot{\vartheta}\,\dot{\vartheta} + \ddot{\vartheta}^3\right]\delta\vartheta, \tag{6.31}$$

and the last equality drops surface terms coming from several integrations by parts. Because the field equation obtained from $\delta S/\delta\vartheta = 0$ involves fourth derivatives of ϑ its integration requires more initial data than usual (it requires initial values for $\ddot{\vartheta}$ and $\dddot{\vartheta}$ in addition to the usual initial values for ϑ and $\dot{\vartheta}$).

Related to the requirement for more initial data is the observation that the general solutions to higher-derivative field equations involve more integration constants and so involve more than the usual two-parameter class of solutions appropriate to second-order field equations.

What is more troubling is that these new solutions almost always include unstable runaway solutions. The generic appearance of instability is most easily seen from the canonical formulation [90–92], for which all field equations are written in terms of single time derivatives by introducing new canonical 'momenta'. The argument is made here for lagrangians of the form $\mathfrak{L} = \mathfrak{L}(\phi, \dot{\phi}, \ddot{\phi})$, but generalizes to the inclusion in $\mathfrak{L}$ of still higher derivatives as well.

The lagrangian $\mathfrak{L} = \mathfrak{L}(\phi, \dot\phi, \ddot\phi)$ has higher-order equations of motion given by

$$\frac{d^2}{dt^2}\left[\frac{\partial \mathfrak{L}}{\partial \ddot\phi}\right] - \frac{d}{dt}\left[\frac{\partial \mathfrak{L}}{\partial \dot\phi}\right] + \frac{\partial \mathfrak{L}}{\partial \phi} = 0. \tag{6.32}$$

To set up a canonical formulation for these equations define the new variable $\psi = \dot\phi$ so that $\mathfrak{L} = \mathfrak{L}(\phi, \psi, \dot\psi)$ and define the standard (π) and new (ζ) canonical momenta by

$$\pi(\phi, \psi, \dot\psi) := \frac{\partial \mathfrak{L}}{\partial \dot\phi} = \frac{\partial \mathfrak{L}}{\partial \psi} \quad \text{and} \quad \zeta(\phi, \psi, \dot\psi) := \frac{\partial \mathfrak{L}}{\partial \ddot\phi} = \frac{\partial \mathfrak{L}}{\partial \dot\psi}, \tag{6.33}$$

of which it is assumed the defining equation for ζ can be solved for $\dot\psi$ to give an expression of the form $\dot\psi = \dot\psi(\phi, \psi, \zeta)$.

With these choices the Hamiltonian density

$$\begin{aligned}\mathcal{H}(\phi, \psi; \pi, \zeta) &:= \pi\,\dot\phi + \zeta\,\dot\psi - \mathfrak{L}(\phi, \psi, \dot\psi) \\ &= \pi\,\psi + \zeta\,\dot\psi(\phi, \psi, \zeta) - \mathfrak{L}[\phi, \psi, \dot\psi(\phi, \psi, \zeta)],\end{aligned} \tag{6.34}$$

generates the equations of motion (6.32) through the first-order system

$$\dot\phi = \frac{\partial \mathcal{H}}{\partial \pi}, \quad \dot\psi = \frac{\partial \mathcal{H}}{\partial \zeta}, \quad \dot\pi = -\frac{\partial \mathcal{H}}{\partial \phi} \quad \text{and} \quad \dot\zeta = -\frac{\partial \mathcal{H}}{\partial \psi}. \tag{6.35}$$

For stability arguments it is crucial that $\mathcal{H}$ also be conserved and bounded from below, since when these are both true the configuration minimizing $\mathcal{H}$ must be stable. In the present case, conservation goes through as usual (provided $\mathfrak{L}$ does not itself depend explicitly on t) because

$$\dot{\mathcal{H}} = \frac{\partial \mathcal{H}}{\partial \phi}\dot\phi + \frac{\partial \mathcal{H}}{\partial \psi}\dot\psi + \frac{\partial \mathcal{H}}{\partial \pi}\dot\pi + \frac{\partial \mathcal{H}}{\partial \zeta}\dot\zeta = 0 \tag{6.36}$$

with the last equality using (6.35). The generic problem with higher-derivative theories is that $\mathcal{H}$ is not bounded from below, as is seen because (6.34) shows $\mathcal{H}$ is *linear* in the variable π.

To obtain an intuition for how such an instability arises more concretely, consider the following quadratic (but higher-order) toy lagrangian [93]:

$$L = \frac{1}{2}\,\dot\vartheta^2 + \frac{1}{2M^2}\,\ddot\vartheta^2, \tag{6.37}$$

whose variation $\delta L = 0$ gives the linear equation of motion

$$-\ddot\vartheta + \frac{\ddddot\vartheta}{M^2} = 0. \tag{6.38}$$

The general solution to this equation is

$$\vartheta = A + Bt + Ce^{Mt} + De^{-Mt}, \tag{6.39}$$

where A, B, C and D are integration constants. This has an unstable runaway form apart from the special initial condition that chooses $C = 0$. The generic unstable mode encountered for higher-derivative theories is often called the *Ostrogradsky ghost*.

The question of why this issue is not a problem for the low-energy EFT is addressed below, after first a brief detour.

A Galileon Aside

Although the above arguments show that the introduction of higher-derivative interactions generically leads to instability, it is also true that not all higher-derivative effective interactions need do so. There are two kinds of relatively benign interactions of this type.

The first type of benign higher-derivative interaction consists of those that are redundant, in the sense made more precise in §2.5. As described there, interactions are redundant if they arise as a total derivative or if they can be removed through a local field redefinition. An example of these types of redundancy for the toy model would be a term like $\vartheta^{\mu\nu}\vartheta_{\mu\nu}$ – with $\vartheta_{\mu\nu}$ defined in (6.30) – since this can be rewritten using

$$\vartheta^{\mu\nu}\vartheta_{\mu\nu} = \partial_\mu\left(\vartheta^{\mu\nu}\partial_\nu\vartheta\right) - (\partial^\nu\Box\vartheta)\partial_\nu\vartheta = \partial_\mu\left(\vartheta^{\mu\nu}\partial_\nu\vartheta - \Box\vartheta\,\partial^\mu\vartheta\right) + (\Box\vartheta)^2. \tag{6.40}$$

The first term on the right-hand side is a total derivative and the second term vanishes when the lowest-order field equations, $\Box\vartheta = 0$, are used, showing that it can be removed to this order in the derivative expansion by performing a field redefinition of the form $\delta\vartheta \propto \Box\vartheta$.

But there is also a second way that nominally higher-derivative interactions can avoid introducing new solutions and instabilities. This is because there are a handful of higher-derivative lagrangian interactions for which the corresponding higher-derivative terms in the field equations happen to cancel. In four spacetime dimensions, using one scalar field ϕ, the most general such an interaction (up to total derivatives) turns out to be a linear combination of the following *Galileon* interactions [94–96]

$$\begin{aligned}
\mathfrak{L}_{G2} &:= G_2(\phi, X)\\
\mathfrak{L}_{G3} &:= G_3(\phi, X)\,\Box\phi\\
\mathfrak{L}_{G4} &:= G_4(\phi, X)\left[(\Box\phi)^2 - \phi_{\mu\nu}\phi^{\mu\nu}\right]\\
\mathfrak{L}_{G5} &:= G_5(\phi, X)\left[(\Box\phi)^3 - 3\phi_{\mu\nu}\phi^{\mu\nu}\,\Box\phi + 2\phi^{\mu\nu}\phi_{\nu\lambda}\phi^{\lambda}{}_{\mu}\right],
\end{aligned} \tag{6.41}$$

where, following earlier notation, these use the definitions $X := -\partial_\lambda\phi\,\partial^\lambda\phi$ and $\phi_{\mu\nu} := \partial_\mu\partial_\nu\phi$. Here, $G_i(\phi, X)$ with $i = 2, \dots 5$ are four arbitrary functions of two arguments. For generic G_i none of these is a total derivative and for all G_i they contribute only terms involving at most two time derivatives to the scalar field equations.

For low enough derivative order it sometimes happens that the most general form for the Wilsonian action is a special case of (6.41) [97]. For instance, for the toy model the most general terms arising out to four-derivative level can be written (up to a total derivative) as a linear combination of X^2, $X\Box\vartheta$ and $(\Box\vartheta)^2$. The last two of these vanish when $\Box\vartheta = 0$, and so can be removed by performing the field redefinition $\delta\vartheta = a\,\Box\vartheta + b\,X$ for appropriate choices for the constants a and b. The remaining term is a special case of the first of (6.41), with $G_2 = \frac{1}{2}X + c_4 X^2$ with c_4 the constant given in Eq. (1.12).

Similarly, the most general terms involving six derivatives are given by (6.29), once total derivatives and redundant interactions involving $\Box\vartheta$ are removed. Because these differ from the term in (6.41) just by $\Box\vartheta$ terms, at six-derivative level the shift-symmetric lagrangian also has the form of (6.41), up to field redefinitions. Up to six-derivative level the corresponding terms have $G_2 = \frac{1}{2}X + c_4 X^2 - c_{61} X^3$ and $G_4 = c_{62} X$.

Of course, the terms in (6.41) involve at most six derivatives not involved in a factor of X, and so eventually terms should arise with sufficient numbers of derivatives to preclude their being put into the Galileon form. And for theories with more general field content more structures are possible at each order, so there is no broad expectation that a generic system can always be written, at all orders in the derivative expansion, like (6.41) or its generalizations. How are the instabilities associated with higher-derivative interactions dealt with then?

A More General Argument

If EFTs generically involve higher-derivative effective interactions and if these interactions generically produce unstable solutions, how can a generic Wilson action hope to describe the time-evolution of a UV theory that is known to be stable (such as the toy model)?

A key step in the development of the Wilson action was the expansion in powers of $1/M$; in particular, it is only after this expansion that the EFT is described by a *local* lagrangian density. Because of this, a local Wilsonian action should only be expected to capture the properties of the underlying UV-completion order-by-order in powers of $1/M$. This is true in particular when seeking time-dependent solutions, which should only be trusted to the extent that they fall within the regime of the $1/M$ expansion.[2]

So a crucial feature of the 'new' solutions (including in particular the runaways) associated with the new higher-derivative terms is that they do *not* arise as a series in powers of $1/M$. They do not do so because they are singular perturbations of the zeroth-order differential equation (because it is the highest-derivative terms of the field equations that are multiplied with nonzero powers of $1/M$).

This is seen explicitly in the solution (6.39) and field equation (6.38) of the simple higher-derivative action given in (6.37). Only the two-parameter family of these solution obtained using $C = D = 0$ goes over to the solutions to the lowest-order field equation, obtained from the $M \to \infty$ lagrangian, $L_0 = \frac{1}{2}\dot{\vartheta}^2$; the other solutions are not captured at any finite order of $1/M$ because for them the $\dot{\vartheta}^2$ and $\ddot{\vartheta}^2$ terms are comparably large. This is manifest in exponential solutions like $\exp(\pm Mt)$ of Eq. (6.39), which have an essential singularity as $M \to \infty$ and are not described at any order by a series in $1/M$.

The lesson is this: a local Wilsonian EFT only aspires to capture the full theory order-by-order in $1/M$, and so any predictions it makes that fall outside of a $1/M$ expansion should be regarded as spurious. Such predictions should not be expected to capture properties of the underlying UV theory.

[2] This is one of those arguments that has been 'in the air' and widely known by those who know for decades, and because of that it has not been written down anywhere (almost; ref. [93] was written to record the argument, which at the time had not percolated into relatively new communities for EFT arguments).

Well-Posed Evolution

Just having equations of motion that are second-order in time does not mean one can relax, however, since in some circumstances time evolution can nonetheless be difficult to evaluate. This could be because an evolution equation's caustics begin to intersect or because short-wavelength modes grow too quickly even if not initially present. When this happens an initial configuration with small gradients can be driven into a regime of large derivatives, and so beyond the reach of EFT methods.

Sometimes this kind of behaviour is the right answer. The collection and focussing of light by lenses is an example where this is true, as is the phenomenon of gravitational collapse (for which an initially diffuse and low-energy cloud of dust becomes gravitationally compressed, possibly into a singularity with arbitrarily large derivatives). But there are also many other examples of this phenomenon throughout physics, such as the turbulent cascade of fluid energies down to small distances, or the development of caustics for the propagation of light in a medium, or the development of shock fronts within hot materials.[3]

In these situations energy conservation in itself does not prevent moving from long-wavelength initial conditions towards those with larger gradients, and so towards a breakdown of the low-energy description. This need not be a problem of principle for EFT methods (depending on how fast it happens) since nothing says that a system that starts in a long-wavelength regime must remain there. Indeed, if the underlying system moves from smooth configurations towards variations over microscopic scales then the EFT should be able to track the early part of this evolution before showing signs of breaking down.

Two features that lend themselves to this kind of breakdown are nonlinear field equations and the breaking of Lorentz invariance, features that are generic in real applications with time-dependent backgrounds. Both of these undermine the protection energy conservation naively gives against generating short-wavelength modes from long-wavelength initial data. Nonlinearities do so by allowing many low-energy modes to combine into a higher-energy one. Breaking Lorentz invariance can allow large mode momenta to coexist with low mode energies even without nonlinearities, and so can also interfere with the ability to discriminate against short-wavelength modes using only low energy as a criterion.

Studies of nonlinear classical field equations often frame the issue of the growth of small-wavelength modes in terms of the well-posedness of the initial-value problem [100]. An initial-value problem is said to be *locally well-posed* if, given suitable initial data, a unique solution of the equation of motion exists, and that the space of solutions depends continuously on the initial data. Well-posedness is local inasmuch as the solution is only required to exist for some nonzero, though possibly very small, time.

An example of the kind of thing that would make an initial-value problem ill-posed would be if modes of wave-number $\mathbf{k}$ were to grow in time as quickly as $\exp(+|\mathbf{k}|t)$, say. If the limit $|\mathbf{k}| \to \infty$ were allowed, this would represent an arbitrarily fast growth, undermining the continuity of the solutions regarded as functions of their initial data. From an EFT perspective things are never quite this bad, however, since within an

[3] See *e.g.* [98] and [99] for discussions of this issue with applications to gravity and fluids, respectively.

effective theory $|\mathbf{k}|$ is bounded to be smaller than some UV scale M. So, whereas mode growth can happen, the timeframe for catastrophic growth in an EFT is usually not arbitrarily short. But this might be cold comfort if it were instead to occur on a UV time-scale like M. As previous sections make clear, evolution over time-scales as short as M^{-1} lies beyond what an EFT can capture.

Well-posedness can also be an important issue even when only asking pragmatic questions well within the EFT regime (for which physical quantities do not evolve on microscopic time scales). This is because, in practice, evolution is calculated only approximately, perhaps numerically by breaking space and time into a discrete lattice. Such approximations necessarily introduce short-distance errors into the initial conditions and evolution equations, which for all intents and purposes play the role of unknown UV physics at the regulator scale Λ. If these regulation errors were to grow over time-scales as short as Λ^{-1} then this spurious growth could quickly swamp the much slower evolution of the physical system being modelled by the EFT description. It is the desire to integrate effective-field equations in nonlinear settings that makes discussions of well-posedness more than a purely mathematical exercise.

Although not a problem for well-posed evolution, these issues mean that approximate methods typically require some sort of smoothing procedure for ill-posed problems [102] to suppress spurious regulator-scale variations (for a discussion of these issues for the toy model considered here see [103]). Whether such smoothing is necessary requires a diagnosis of the well-posedness of EFT field equations.

Well-posedness for a nonlinear theory is ensured if its field equations are strongly hyperbolic.[4] Sadly, the field equations for many EFTs are known not to be strongly hyperbolic even if the underlying UV theory is. EFTs can run into trouble in this way – even those lying in the Galileon class discussed earlier [101] – because the derivatives appearing within effective interactions modify the character of the second-derivative terms on which hyperbolicity is based, perhaps as a function of the size of (or variation in) a background field.

Since any spurious regulator dependence is a special case of UV physics, in principle it can be absorbed into the values of an EFT's effective couplings. The problem is how to do this in practice, numerically and on the fly. As of this writing (2018), the issue of how to optimally simulate the time-evolution predicted order-by-order in an EFT's low-energy expansion is not yet settled, though is under active study.

6.3 Fluctuations about Evolving Backgrounds ♠

Earlier sections in this chapter mostly focus on how the background evolves and how this is captured by an effective Wilsonian description. But there can also be interest in the properties of fluctuations about non-static background configurations,

[4] A hyperbolic system is strongly hyperbolic if there is a norm for solutions whose behaviour at time t is bounded by the initial value of the same norm multiplied by a function of time that is independent of the initial data.

and this section explores some of the ways that fluctuations about time-evolving backgrounds differ from those about static vacua. The fluctuations of interest could either describe nearby solutions within a purely classical problem, or be full-on quantum fluctuations. Which one is relevant in any particular application can be determined using a power-counting analysis such as that given in §3.

Part of the practical interest in studying EFTs for fluctuations about evolving backgrounds (for relativistic systems) comes from cosmology [104]. As described in more detail in §10.2, quantitative predictions for fluctuations are pressing in cosmology because in the modern understanding the large-scale distribution of matter and radiation throughout the universe arises as the gravitational amplification of small primordial field fluctuations occurring within an expanding spacetime. This allows precise predictions of the properties of these fluctuations to be compared in detail with the wealth of modern cosmological observations.

6.3.1 Symmetries in an Evolving Background

Time-dependent backgrounds typically preserve fewer symmetries than do static vacua. For instance, for the toy model with a homogeneous time-dependent background, $\vartheta(t)$, the background breaks both time-translation invariance and Lorentz invariance, while preserving rotational symmetry and invariance under spatial translations. Since the symmetries are broken by a field configuration, the breaking can be considered to be spontaneous, though of spacetime symmetries rather than internal ones (for a recent systematic discussion of the issues see, for example, [105]).

The consequences of this symmetry-breaking pattern for fluctuations follow the general rules outlined earlier for spontaneous symmetry breaking. In particular, fluctuations fall into linear representations only of the unbroken subgroup of symmetries that leave the background invariant. For homogeneous time-dependent backgrounds this means that fluctuations can be labelled by their spin (*i.e.* representation of the field under rotations) and linear momentum (representation under translations). Total momentum and angular momentum are conserved by virtue of the background's invariance under spatial translations and rotations.

Other consequences of Poincaré invariance for static Lorentz-invariant vacua do not carry over to fluctuations about homogeneous time-dependent backgrounds. In principle, the breaking of time-translations by the background means that energy is not strictly conserved for the fluctuations. (That is to say: even if energy is conserved for the whole system – background plus fluctuations – in general there can be energy transfer between the two for time-dependent backgrounds, making the energy of fluctuations themselves not strictly conserved.)

Because Wilsonian actions only capture the time-dependence of adiabatic evolution in the UV theory, when EFT methods are useful it is possible to define a time-dependent energy satisfying

$$H(t)|\mathbf{k},\sigma\rangle = E(k,t)|\mathbf{k},\sigma\rangle, \tag{6.42}$$

acting on fluctuation states, where H might parametrically depend on time. Alternatively, the time-evolution of field mode functions can be approximately written as

$$u(t,\mathbf{x}) = v(\mathbf{x})\,\exp\left[-\mathrm{i}\int_{t_0}^{t} \mathrm{d}\tau E(k,\tau)\right]. \tag{6.43}$$

As described earlier, it is this energy that implicitly is used to distinguish low-energy from high-energy states for EFT applications with time-dependent backgrounds. These expressions use rotation invariance, which ensures a mode's energy eigenvalue (or dispersion relation) depends only on the magnitude $k = |\mathbf{k}|$.

Finally, for fluctuations about time-dependent backgrounds the dispersion relation $E(k)$ can differ from the Lorentz-invariant result $\sqrt{k^2 + m^2}$. For instance, the explicit calculations to follow for the time-dependent toy model example considered above show the Goldstone mode propagates with dispersion relation $E(k) = kc_s + O(k^2)$, with $1 - c_s = \delta_c$ being a calculable positive function of system parameters (like ω, λ and v in the toy model).

Fluctuations in the Toy Model

To make the story concrete, this section examines how fluctuations around a time-dependent background behave in the toy model of §1.1, both in the full theory and in its low-energy Wilsonian incarnation.

To this end, in the full theory expand $\phi = \varphi(t) + \tilde{\phi}$ where $\varphi(t) = \varrho_0\, e^{i\omega t}$ is the background solution considered above, with (6.7) implying $\varrho_0^2 = v^2 + 2\omega^2/\lambda$. With this choice the lagrangian can be expanded in powers of $\tilde{\phi}$, so $\mathfrak{L} = \mathfrak{L}^{(0)} + \mathfrak{L}^{(1)} + \mathfrak{L}^{(2)} + \mathfrak{L}^{(3)} + \mathfrak{L}^{(4)}$, where

$$
\begin{aligned}
\mathfrak{L}^{(0)} &= \omega^2 \varrho_0^2 - \frac{\omega^4}{\lambda} = \omega^2 v^2 + \frac{\omega^4}{\lambda} \\
\mathfrak{L}^{(1)} &= \sqrt{2}\, \varrho_0\, \omega\, \frac{\mathrm{d}}{\mathrm{d}t}\Big[-\tilde{\phi}_R s_t + \tilde{\phi}_I c_t\Big] \\
\mathfrak{L}^{(2)} &= -\frac{1}{2}\left(\partial_\mu \tilde{\phi}_R\, \partial^\mu \tilde{\phi}_R + \partial_\mu \tilde{\phi}_I\, \partial^\mu \tilde{\phi}_I\right) \\
&\quad - \frac{1}{2}\begin{pmatrix} \tilde{\phi}_R \\ \tilde{\phi}_I \end{pmatrix}^T \begin{pmatrix} \omega^2 + \lambda \varrho_0^2 c_t^2 & \lambda \varrho_0^2 c_t s_t \\ \lambda \varrho_0^2 c_t s_t & \omega^2 + \lambda \varrho_0^2 s_t^2 \end{pmatrix} \begin{pmatrix} \tilde{\phi}_R \\ \tilde{\phi}_I \end{pmatrix},
\end{aligned}
\tag{6.44}
$$

and so on for higher powers of $\tilde{\phi}$, where $\tilde{\phi} = \frac{1}{\sqrt{2}}(\tilde{\phi}_R + i\tilde{\phi}_I)$ while $c_t := \cos\omega t$ and $s_t := \sin\omega t$. The term linear in $\tilde{\phi}$ can be dropped for most purposes because it is a total derivative – as is always true when expanding about a classical solution.

Although the quadratic term seems to involve a standard kinetic piece plus lots of oscillatory time-dependence in the mass term, this is actually deceptive since the eigenvalues of the mass matrix are not time-dependent at all:

$$
m_+^2 := \omega^2 + \lambda \varrho_0^2 = 3\omega^2 + \lambda v^2 \quad \text{and} \quad m_-^2 := \omega^2. \tag{6.45}
$$

The oscillatory time-dependence seen in (6.44) and the tempting interpretation of (6.45) as nonzero masses can be misleading if they are used too naively when drawing physical consequences (such as the existence of an energy gap for fluctuations at zero momentum). They are misleading because the shorthand that allows a straightforward inference of physical quantities like masses from quadratic terms in an action breaks down in this particular choice of basis fields, $\tilde{\phi}_R$ and $\tilde{\phi}_I$. The problem arises because these basis fields are fixed in time while the physical basis of mass eigenstates rotates in field space with angular frequency ω.

Drawing inferences using the fields $\tilde{\phi}_R$ and $\tilde{\phi}_I$ might be warranted if there were a physical reason for choosing this basis – *e.g.* if other sectors of the theory were to break the $U(1)$ symmetry, such as if perhaps only ϕ_I were to couple to observable particles. Otherwise, performing the time-dependent rotation required to reach the mass basis transfers the effects of the time-dependent background into the kinetic part of the fluctuation fields, suggesting very different kinds of observable consequences.

Simpler than performing this time-dependent rotation is to directly use the fluctuation fields $\tilde{\chi}$ and $\tilde{\xi}$ defined by

$$\phi = \left(\varrho_0 + \frac{\tilde{\chi}}{\sqrt{2}}\right) \exp\left[\frac{i\xi}{\sqrt{2}\varrho_0}\right], \tag{6.46}$$

with $\xi = \sqrt{2}\,\varrho_0\,\omega t + \tilde{\xi}$, since in this case the lagrangian expansion becomes (*c.f.* Eqs. (1.24) and (1.25))

$$\mathfrak{L} = -\frac{1}{2}\,\partial_\mu\tilde{\chi}\partial^\mu\tilde{\chi} - \frac{1}{2}\left(1 + \frac{\tilde{\chi}}{\sqrt{2}\,\varrho_0}\right)^2 \partial_\mu\xi\partial^\mu\xi - V(\tilde{\chi}), \tag{6.47}$$

with

$$-\partial_\mu\xi\,\partial^\mu\xi = 2\omega^2\varrho_0^2 + 2\sqrt{2}\,\omega\,\varrho_0\,\partial_t\tilde{\xi} - \partial_\mu\tilde{\xi}\,\partial^\mu\tilde{\xi}, \tag{6.48}$$

and

$$V(\tilde{\chi}) = \frac{\lambda}{4}\left(\frac{2\omega^2}{\lambda} + \sqrt{2}\,\varrho_0\,\tilde{\chi} + \frac{\tilde{\chi}^2}{2}\right)^2. \tag{6.49}$$

Expanding this lagrangian in powers of $\tilde{\chi}$ and $\tilde{\xi}$ then gives the same expression as before for $\mathfrak{L}^{(0)}$; a total derivative for the linear terms; and the following quadratic term

$$\mathfrak{L}^{(2)} = -\frac{1}{2}\,\partial_\mu\tilde{\chi}\partial^\mu\tilde{\chi} - \frac{1}{2}\partial_\mu\tilde{\xi}\,\partial^\mu\tilde{\xi} + 2\omega\,\tilde{\chi}\,\partial_t\tilde{\xi} - \frac{1}{2}\lambda\varrho_0^2\,\tilde{\chi}^2. \tag{6.50}$$

Although not diagonal, this form does not have explicitly time-dependent coefficients.

To identify the dispersion relations of the propagating modes it is convenient to Fourier transform by switching to energy and momentum eigenstates, $\propto e^{i(-Et+\mathbf{k}\cdot\mathbf{x})}$, leading to a quadratic action proportional to

$$\begin{pmatrix} \tilde{\chi} \\ \tilde{\xi} \end{pmatrix}^\dagger \begin{pmatrix} E^2 - k^2 - \lambda\varrho_0^2 & -2i\omega E \\ 2i\omega E & E^2 - k^2 \end{pmatrix} \begin{pmatrix} \tilde{\chi} \\ \tilde{\xi} \end{pmatrix}, \tag{6.51}$$

where E and $k = |\mathbf{k}|$ are the energy and the magnitude of momentum for the corresponding mode. The dispersion relations, $E(k)$, for the propagating modes correspond to those choices that make the eigenvalues,

$$\Delta_\pm = E^2 - k^2 - \frac{1}{2}\lambda\varrho_0^2\left[1 \pm \sqrt{1 + \frac{16E^2\omega^2}{\lambda^2\varrho_0^4}}\right], \tag{6.52}$$

of this matrix vanish.

For $\omega E \ll \frac{1}{2}\lambda\varrho_0^2$ the corresponding dispersion relations, $E_\pm(k)$, therefore satisfy

$$E_-^2\left(1 + \frac{4\omega^2}{\lambda\varrho_0^2}\right) - k^2 \simeq E_+^2\left(1 - \frac{4\omega^2}{\lambda\varrho_0^2}\right) - k^2 - \lambda\varrho_0^2 \simeq 0. \tag{6.53}$$

These show that nonzero ω does not introduce an energy gap at zero momentum for the Goldstone boson, though such a gap does, of course, exist for the massive particle (though with v replaced with ϱ_0 when compared with the mass found in §1.1).

The main effect of nonzero ω for the Goldstone boson is to change its 'sound speed', c_s, defined by writing the small-k dispersion relation as $E^2 = k^2 c_s^2$. Comparing with the Goldstone mode relation, $E_-(k)$, implies that

$$c_{s-} \simeq \left(1 + \frac{4\omega^2}{\lambda \varrho_0^2}\right)^{-1/2} \simeq 1 - \frac{2\omega^2}{\lambda v^2}, \tag{6.54}$$

to leading order in ω^2/m_R^2.

As mentioned earlier, the result $c_s \neq 1$ never arises when expanding about a static background like $\phi = v$ because anything except $c_s = 1$ is in that case forbidden by Lorentz invariance. Nontrivial speed of sound arises for time-dependent backgrounds because these break the underlying Lorentz invariance of the action.

The Wilsonian Point of View

This same conclusion about the ω-dependence of the Goldstone-boson dispersion relation also follows directly from the toy model's Wilsonian EFT, given by (6.10) and repeated here:

$$\mathfrak{L}_W = -v^2 \partial_\mu \vartheta \partial^\mu \vartheta + \frac{v^2}{m_R^2}(\partial_\mu \vartheta\, \partial^\mu \vartheta)^2 + \cdots, \tag{6.55}$$

where $\phi = \varrho\, e^{i\vartheta}$. Expanding ϑ about the slowly rolling classical solution, $\vartheta = \omega t + \tilde{\vartheta}$ then implies that $-\partial_\mu \vartheta\, \partial^\mu \vartheta = \omega^2 + 2\omega\, \partial_t \tilde{\vartheta} + (\partial_t \tilde{\vartheta})^2 - \nabla\tilde{\vartheta} \cdot \nabla\tilde{\vartheta}$, so the quadratic part of the expanded action becomes

$$\mathfrak{L}_W^{(2)} = v^2\left[(\partial_t \tilde{\vartheta})^2 - \nabla\tilde{\vartheta} \cdot \nabla\tilde{\vartheta}\right] + \frac{\omega^2}{\lambda}\left[6(\partial_t \tilde{\vartheta})^2 - 2\nabla\tilde{\vartheta} \cdot \nabla\tilde{\vartheta}\right] + \cdots, \tag{6.56}$$

where ellipses denote terms involving higher powers of ω/m_R.

The field equations for ϑ predicted by this action therefore are

$$-\left(1 + \frac{6\omega^2}{\lambda v^2}\right) \partial_t^2 \tilde{\vartheta} + \left(1 + \frac{2\omega^2}{\lambda v^2}\right) \nabla^2 \tilde{\vartheta} \simeq 0, \tag{6.57}$$

which when compared to the wave equation $(-\partial_t^2 + c_s^2 \nabla^2)\tilde{\vartheta} = 0$ leads to a prediction

$$c_s \simeq \sqrt{\frac{1 + 2\omega^2/(\lambda v^2)}{1 + 6\omega^2/(\lambda v^2)}} \simeq 1 - \frac{2\omega^2}{\lambda v^2}, \tag{6.58}$$

that agrees to leading nontrivial order in ω^2/m_R^2 with (6.54). Notice that this ω-dependent Goldstone sound speed is less than the speed of light (*i.e.* $c_s < c = 1$) by virtue of the sign of the $(\partial_\mu \vartheta\, \partial^\mu \vartheta)^2$ term.

6.3.2 Counting Goldstone States and Currents *

Since time-dependent backgrounds spontaneously break spacetime symmetries, one might expect Goldstone's theorem to ensure the existence of new low-energy Goldstone degrees of freedom. Although it is sometimes true that each new broken symmetry generator implies a new Goldstone particle, the toy model shows that this naive counting of Goldstone states can be misleading, particularly for spacetime symmetries [230]. The discussion given here follows [107] (see also [108]), and is generalized to more general background fields in §14.3.1.

For example, for the toy model expanded about the time-dependent classical background $\vartheta_c = \omega t$, the background breaks both the internal $U(1)$ symmetry – for which $\vartheta \to \vartheta + c$ for constant c – and time-translation invariance: $t \to t + \tau$ for constant τ. Breaking two symmetries naively suggests there should be two Goldstone particles, yet the effective theory only contains the one low-energy state.

For these specific symmetries the real lesson of the toy model is this: with multiple symmetries one must be careful when counting how many symmetries are broken. That is, it is always possible to undo the action of time translation, $t \to t + \tau$, on the background by simultaneously performing a compensating $U(1)$ transformation, $\vartheta \to \vartheta - \omega\tau$, leaving ϑ_c invariant. Only one Goldstone particle arises because the background $\vartheta_c = \omega t$ really breaks only one combination of these two symmetries.

More generally, time-dependent backgrounds also break the six-dimensional group of Lorentz transformations down to the three-dimensional group of rotations. Why doesn't Goldstone's theorem imply there must also be Goldstone modes for these broken symmetries?

To see why, it is worth referring back to the derivation of Goldstone's theorem presented in §4.1.2. What matters for Goldstone's theorem is not the number of broken generators of the symmetry group. What matters instead is the number of independent conserved currents, $j^\mu(x)$, implied by the symmetry group, since for each independent current associated with a broken symmetry there should be a Goldstone state $|G\rangle$ satisfying the defining condition that

$$\langle G|j^0(x)|\Omega\rangle \neq 0, \tag{6.59}$$

where $|\Omega\rangle$ is the ground state. Furthermore, although Goldstone's theorem establishes the existence of such a state for each broken current, it *doesn't* require that a new state is required for each new current.

As reviewed in Appendix C.5.3, for spacetime symmetries there are only four independent conserved currents regardless of the dimension of the group of spacetime symmetries. This is because spacetime symmetries all have their roots in diffeomorphisms, $\delta x^\mu = V^\mu(x)$, for which the associated conserved current is the *stress-energy tensor*, $T_{\mu\nu}(x) = T_{\nu\mu}(x)$, defined in terms of the matter action by

$$T^{\mu\nu} = \frac{2}{\sqrt{-g}}\frac{\delta S_m}{\delta g_{\mu\nu}}, \tag{6.60}$$

where the spacetime metric is temporarily introduced for the purpose of performing the variation, before returning to the flat cartesian Minkowski metric of special relativity: $g_{\mu\nu} = \eta_{\mu\nu} = \text{diag}(-1, 1, 1, 1)$.

As also reviewed in Appendices C.5.2 and C.5.3, spacetime symmetries correspond to those diffeomorphisms that leave the background metric invariant, which for the Minkowski metric turns out to mean that V^μ must satisfy

$$\delta\eta_{\mu\nu} = \partial_\mu V_\nu + \partial_\nu V_\mu = 0. \tag{6.61}$$

This has as solutions $V_\mu = a_\mu + \omega_{\mu\nu}\, x^\nu$, with $\omega_{\mu\nu} = -\omega_{\nu\mu}$, corresponding to the usual translations in spacetime ($\delta x^\mu = a^\mu$) and Lorentz transformations ($\delta x^\mu = \omega^\mu{}_\nu\, x^\nu$).

For each solution to (6.61) a conserved current can be constructed using only the symmetric stress-energy tensor $T^{\mu\nu}$, since conservation $\partial_\mu T^{\mu\nu} = 0$ together with (6.61) implies that $\partial_\mu j^\mu_V = 0$, where

$$j^\lambda_V(x) := T^{\lambda\mu}(x)\, V_\mu(x). \tag{6.62}$$

The corresponding conserved charge (or generator) for this symmetry is constructed by integrating j^0_V over all of space.

With this in mind, the Goldstone states required by spontaneously broken spacetime symmetries are those for which the stress-energy matrix element,

$$\langle G|T^{0\mu}(x)|\Omega\rangle\, V_\mu(x) \neq 0, \tag{6.63}$$

is nonzero. As seen in §13.1 and §14.3, systems (such as solids or liquids) that spontaneously break Poincaré invariance typically do give rise to Goldstone modes of this type (corresponding to sound waves, or phonons). What is *not* in general guaranteed by Goldstone's theorem is that the state $|G\rangle$ appearing in (6.59) need be different than the state appearing in (6.63). The states appearing in these matrix elements can sometimes be different, but need not always be so.

The toy model provides an explicit example where both (6.59) and (6.63) are satisfied by the same state: the massless state described by the field ξ. Indeed, for weak coupling the low-energy sector only has a single state available to play both roles. To see this explicitly it is instructive to compute explicitly both the Noether current for the internal-$U(1)$ current and the stress energy.

Working to lowest order in the energy expansion the action is simply that of a massless free scalar field,

$$\mathfrak{L}_W \simeq -\frac{1}{2}\,(\partial^\mu\xi\,\partial_\mu\xi) + \frac{\lambda}{4m_R^4}(\partial^\mu\xi\,\partial_\mu\xi)^2 + \cdots, \tag{6.64}$$

for which the current predicted by (4.7) for the $U(1)$ symmetry $\xi \to \xi + \sqrt{2}\,cv$ (where c is the constant symmetry parameter) is

$$j_\mu = -\sqrt{2}\,v\,\partial_\mu\xi\left[1 - \frac{\lambda}{m_R^4}(\partial^\nu\xi\,\partial_\nu\xi) + \cdots\right]. \tag{6.65}$$

The stress energy predicted for a minimally coupled scalar is similarly given by

$$T_{\mu\nu} = \partial_\mu\xi\,\partial_\nu\xi - \frac{1}{2}(\partial^\lambda\xi\,\partial_\lambda\xi)\,\eta_{\mu\nu} + \cdots, \tag{6.66}$$

where the ellipses in both of these expression denote higher-derivative contributions than those written.

When expanded about a time-dependent solution, $\xi = \sqrt{2}\,v\,\omega t + \tilde{\xi}$, both j^0 and $T^0{}_\mu$ contain terms linear in the fluctuation $\tilde{\xi}$. Writing[5] $\langle p|\tilde{\xi}(x)|\Omega\rangle = F\,e^{ipx}$ (with nonzero F) for a single-particle momentum eigenstate $|p\rangle$ shows that the field $\tilde{\xi}$ plays the role of the Goldstone state for all of the broken symmetries, with

$$\langle p|j_\mu(x)|\Omega\rangle = -i\,\sqrt{2}\,v\,F p_\mu e^{ipx} + \cdots$$
$$\text{and}\quad \langle p|T^0{}_\mu(x)|\Omega\rangle = i\,\sqrt{2}\,v\,\omega\,F p_\mu\, e^{ipx} + \cdots, \tag{6.67}$$

[5] As written F contains factors of $\sqrt{E(p)}$ unless $|p\rangle$ is normalized covariantly (see *e.g.* Appendix B.1).

where ellipses denote terms of relative order ω^2/m_R^2 (or those suppressed by loop factors). In this sense, the low-energy sector of the toy model saturates the requirements of Goldstone's theorem in a minimal way.

6.4 Summary

To summarize this chapter, the bottom line is this: time-dependent evolution in the full theory (both of backgrounds and classical and quantum fluctuations about them) can be captured using time-dependent solutions to the low-energy effective theory, but only if the evolution of interest is sufficiently slow.

Generically, 'sufficiently slow' means demanding that $\mu_\phi := \dot{\phi}/\phi$ – for all choices of fields $\phi^i(x)$ in the problem – be much smaller than the UV scale M (*i.e.* in the toy model, m_R). This adiabatic condition is in addition to all the other requirements already needed when formulating a Wilsonian low-energy theory: such as that the energies of all fluctuation modes be much smaller than M.

This points to two kinds of generic new failure modes specific to EFTs applied to time-dependent problems. The first new failure mode arises if the background evolution itself should become too fast. In such a case, the transfer of energy between background and fluctuations (such as through particle production using energy extracted from the background) becomes too efficient, destroying the adiabatic approximation (and with it the approximately conserved notion of energy used to discriminate between low- and high-energy fluctuation modes).

A second type of new failure mode is simply the time-dependent version of the old failure mode: a nominally low energy, E, is not small enough to trust the E/M expansion. Time dependent drift of $E(t)$ and $M(t)$ means $E(t)/M(t)$ might eventually become large even if were small initially (such as occurs in level crossing, see panel (a) of Fig. 6.1).

Notice that level crossing – for which the EFT expansion in powers of E/M must fail – is different from having UV states simply evolve below some regulator scale Λ (panel (b) of Fig. 6.1). Nothing dramatic need happen as UV levels pass below a cutoff scale, provided the UV states evolve in their adiabatic vacua, since cutoff scales by construction do not appear in any physical quantities.

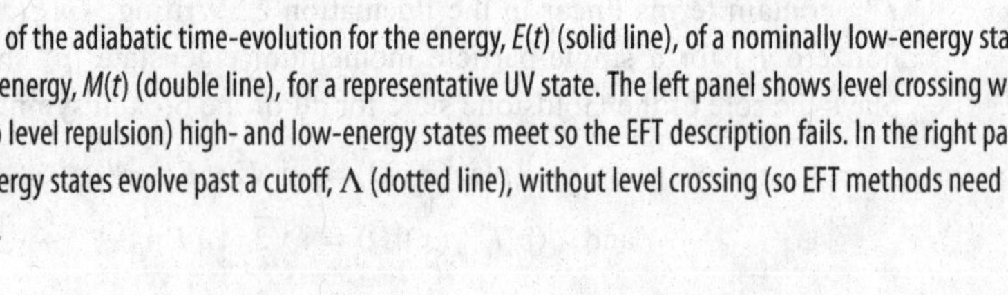

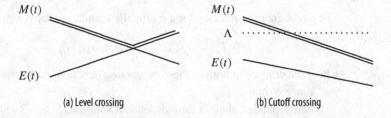

Fig. 6.1 A sketch of the adiabatic time-evolution for the energy, $E(t)$ (solid line), of a nominally low-energy state and the energy, $M(t)$ (double line), for a representative UV state. The left panel shows level crossing where (modulo level repulsion) high- and low-energy states meet so the EFT description fails. In the right panel high-energy states evolve past a cutoff, Λ (dotted line), without level crossing (so EFT methods need not fail).

Exercises

Exercise 6.1 Rederive Eq. (6.17) using the stress-energy tensor, $T^{\mu\nu}$, defined in (6.60) applied to the action built from the lagrangian density of Eq. (6.64). Doing so requires writing this matter action for a general metric:

$$S_m \simeq \int \mathrm{d}^4x\, \sqrt{-g}\left[\frac{1}{2}X + \frac{\lambda}{4m_R^4}X^2\right],$$

where $X := -g^{\mu\nu}\partial_\mu\xi\,\partial_\nu\xi$ and g denotes the determinant of (and $g^{\mu\nu}$ is the matrix inverse of) the covariant components of the metric, $g_{\mu\nu}$ (see Appendices A.2.1 and C.5.2 for more details). Once the stress-energy tensor is computed the energy density is given by $\varepsilon = T_{00}$.

Exercise 6.2 Too-rapid background time-dependence can ruin the low-energy approximation. Consider the toy model of §1.1 in the semiclassical regime, but instead of starting in the vacuum consider the background field configuration describing homogeneous heavy-field oscillations about its minimum: $\chi_c(t) = \chi_0\cos(m_R t)$, where $\chi_0 \ll v$ so that the cubic and quartic terms in the potential $V(\chi)$ can be neglected. Compute the energies of the ξ particles that are pair-produced by their interactions with this background oscillating field and calculate their production rate. Can the production of these ξ particles be described purely with a low-energy EFT description?

Exercise 6.3 As a toy model of level crossing (and repulsion) consider two real free scalar fields, ϕ_1 and ϕ_2, that mix with one another through the lagrangian density $\mathfrak{L} = -\frac{1}{2}\left[(\partial\phi_1)^2 + (\partial\phi_2)^2\right] + \mathfrak{L}_{\text{mix}}$ where

$$\mathfrak{L}_{\text{mix}} = -\frac{1}{2}\begin{pmatrix}\phi_1\\ \phi_2\end{pmatrix}^T\begin{pmatrix} gn(t) & \mu^2\\ \mu^2 & m^2\end{pmatrix}\begin{pmatrix}\phi_1\\ \phi_2\end{pmatrix}$$

where μ and m are positive and real mass parameters with $\mu \ll m$, g is a coupling constant and $n(t)$ is the density of particles in a medium within which the scalars are immersed. $n(t) = n_0\,e^{-t/\tau}$ is assumed to be monotonically decreasing, asymptoting to zero for large t. For any fixed t what are the eigenvalues and eigenvectors for the mass matrix? Assume the system is prepared in a state that is a ϕ_1 eigenstate with momentum $\mathbf{p}$ at $t = 0$ with $gn_0 > m^2$. After this, $n(t)$ falls slowly enough that the evolution is adiabatic – *i.e.* instantaneous energy eigenstates evolve with phase $\exp[-\mathrm{i}\int_{t_0}^t \mathrm{d}s\, E(s)]$. What is the likelihood that the state is measured at $t \to \infty$ to be in state ϕ_2?

Part II

Relativistic Applications

About Part II

The remaining three parts of this book aim to explore a variety of real-world applications of the principles developed in Part I. This one (Part II) is aimed at relativistic systems, many of which played a role in the development of EFT techniques. The remaining two parts explore applications for nonrelativistic and open systems.

The chapters in this section divide the applications into two main topics. The first examples are taken from systems for which the high-energy theory is well understood and calculable, such as for the charged-current weak interactions of the Standard Model, for Quantum Electrodynamics with electrons and muons, or for the interactions of the known elementary particles (gravitons, neutrinos and photons) at energies much below the electron mass. These examples allow low-energy methods to be explored in situations where the answer is already known from other methods. Included in these topics are examples of how renormalization-group (RG) techniques and EFTs are used to track large logarithms.

The second half of Part II then switches to cases for which the high-energy theory is either unknown (such as when the low-energy theory is General Relativity or the Standard Model itself) or when it is known but not understood well enough to allow precise calculations (such as the behaviour of pions in the low-energy limit of QCD).

7 Conceptual Issues (Relativistic Systems)

This chapter takes up the story with the simplest cases: situations where the full high-energy theory is well-understood and the low-energy EFT is explicitly calculable. Examples in this chapter are chosen primarily to illustrate useful conceptual points that arise more generically for effective theories.

7.1 The Fermi Theory of Weak Interactions ◊

We start with a classic example of an effective theory: the Fermi theory of (low-energy) weak interactions. For presentation purposes some facts about the charged-current weak interactions are required. The UV theory in this case is the Standard Model of particle physics [61, 109, 110], which describes them as the result of exchanges of W bosons.

7.1.1 Properties of the W Boson

Since the W-boson has spin one, it is represented by a gauge potential, W_μ, and because it also carries electric charge this gauge potential is complex rather than real. The free propagation of the W boson (within unitary gauge) is described by the lagrangian (see also §C.3.3 for more about massive spin-1 particles)

$$\mathfrak{L}_{\rm free} = -\frac{1}{2} W^*_{\mu\nu} W^{\mu\nu} - M_W^2 W^*_\mu W^\mu, \tag{7.1}$$

where $W_{\mu\nu} := \partial_\mu W_\nu - \partial_\nu W_\mu$, $M_W \simeq 80$ GeV is the mass of the W boson.[1]

The field equations obtained by varying the above action with respect to W^*_μ are

$$\Box W_\mu - \partial_\mu \partial^\nu W_\nu - M_W^2 W_\mu = 0, \tag{7.2}$$

for which the W-boson propagator is the Green's function, given by (see Eq. (4.63) for the massive spin-1 propagator in a more general gauge)

$$G^\mu{}_\nu(x,y) = -{\rm i} \int \frac{{\rm d}^4 k}{(2\pi)^4} \frac{e^{ik\cdot(x-y)}}{k^2 + M_W^2 - {\rm i}\epsilon} \left(\delta^\mu{}_\nu + \frac{k^\mu k_\nu}{M_W^2} \right). \tag{7.3}$$

Here, $k \cdot (x-y) := k_\mu (x-y)^\mu$ and ϵ is present to enforce Feynman boundary conditions, so it is a positive infinitesimal that is taken to zero at the end of any calculation.

[1] Unless otherwise stated, quoted experimentally measured values for parameters come from reference [62].

Table 7.1 The three generations $i = 1, 2, 3$ of fermion flavours

Leptons		Quarks	
ν-Type	ℓ-Type	u-Type	d-Type
ν_1	e	u	d
ν_2	μ	c	s
ν_3	τ	t	b

The Standard Model also endows the W boson with a number of interactions, such as the electromagnetic interactions that accompany its electric charge. These are incorporated by replacing

$$\partial_\mu W_\nu \to D_\mu W_\nu := (\partial_\mu + ieA_\mu) W_\nu$$
$$\partial_\mu W_\nu^* \to D_\mu W_\nu^* := (\partial_\mu - ieA_\mu) W_\nu^*, \tag{7.4}$$

where A_μ is the usual electromagnetic potential. Eqs. (7.4) show that the conventional choice has W_μ destroying particles with charge $-e$ and creating antiparticles with charge $+e$, where e is the charge of the proton, once W_μ is expanded in terms of creation and annihilation operators (see §C.3.3 for details).

Within this framework, the charged-current weak interactions are described by the interactions between W bosons and fermions, with the generic form[2] [61, 111]

$$\mathfrak{L}_{cc} = \frac{ig}{\sqrt{2}} \Big[U_{ja}^* \, W_\mu^* \, (\bar{\nu}^a \gamma^\mu \gamma_L \ell^j) + U_{ja} \, W_\mu \, (\bar{\ell}^j \gamma^\mu \gamma_L \nu^a) \Big] \tag{7.5}$$
$$+ V_{ij} \, W_\mu^* \, (\bar{u}^i \gamma^\mu \gamma_L d^j) + V_{ji}^* \, W_\mu \, (\bar{d}^j \gamma^\mu \gamma_L u^i) \Big],$$

where g is a fundamental coupling constant, related to the electromagnetic coupling e by $g = e/\sin\theta_w$, where the angle θ_w is called the weak-mixing angle (or Weinberg angle) and is a parameter of the theory that is measured to have size $\sin^2\theta_w \simeq 0.231$. The fermion fields ν^a, ℓ^j, u^i and d^j are spinors representing the various known elementary spin-half particles listed in Table 7.1. (See §A.2.3 and §C.3.2 for a refresher on spin-half fields, Dirac gamma-matrices and the spinor conventions used here.) An over-bar on a spin-half field denotes Dirac conjugation, the γ^μ denote the usual Dirac matrices, and the matrix $\gamma_L := \frac{1}{2}(1 + \gamma_5)$ projects onto left-handed Dirac spinors. The proportionality of interactions to γ_L expresses the experimental fact that only left-handed fermions couple to W bosons, making the W interactions *chiral* inasmuch as they treat left-handed and right-handed particles differently. The presence of both the matrices γ^μ and $\gamma^\mu\gamma_5$ shows that the weak interactions break parity invariance.

The indices i, j and a run over the labels $1, 2, 3$ corresponding to the three known generations of elementary fermions.[3] Every fermion is either 'u-type' (ν^a or u^i) or 'd-type' (ℓ^i and d^i). Fermions also split into two categories, called *leptons* (ν^a

[2] This expression modifies the Standard Model prediction slightly by introducing the matrix of coupling parameters $U_{ia} \neq \delta_{ia}$, as required since the discovery of neutrino oscillations.

[3] The neutrino index is labelled 'a' instead of 'i' because there may be more than three species of neutrinos.

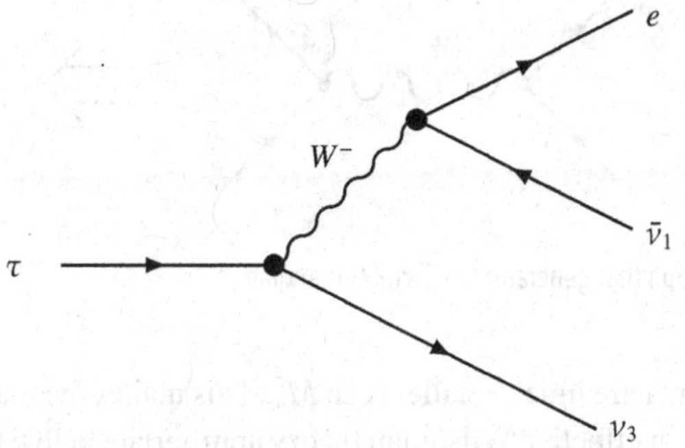

Fig. 7.1 The Feynman graph responsible for the decay $\tau \to e\nu_3\bar{\nu}_1$ at leading order in unitary gauge.

and ℓ^i), which do not take part in the strong interactions – *i.e.* do not carry 'colour' – or *quarks* (u^i and d^i), which do take part in the strong interactions – and so do carry colour.

There are two kinds of 3×3 matrices of coefficients, U_{ia} and V_{ij}, each of which is predicted (and measured) to be unitary: $U^\dagger U = V^\dagger V = 1$. Besides being unitary there is no other prediction for these matrices, and the values of each entry must be taken from experiment. The quark matrix (called the CKM matrix after physicists Cabbibo [112], Kobayashi and Maskawa [113]) is measured to be close to – but not exactly – diagonal, but measurements of the lepton matrix (the PMNS matrix, for Pontecorvo [114], Maki, Nakagawa and Sakata [115]) reveal it to be more complicated. (Notice that the conventions used to define U_{ia} relative to U^*_{ia} differ – for historical reasons – from those used for V_{ij} and V^*_{ij}.)

7.1.2 Weak Decays

Within the Standard Model the interactions between fermions and the W boson are special because they are the only ones that change fermion 'flavour' (or the basic type – ν_1 vs μ vs b *etc.*) of fermion. As a result, the emission and absorption of W bosons is responsible for all decays of elementary fermions.[4]

Consider, for instance, the decay $\tau^-(k) \to e^-(p)\nu_3(l)\bar{\nu}_1(q)$ of a τ lepton (with 4-momentum k^μ) into an electron (with 4-momentum p^μ), a ν_3 neutrino (4-momentum l^μ) and ν_1 anti-neutrino (4-momentum q^μ). In unitary gauge the leading contribution from the charged-current interaction (7.5) is found by evaluating the Feynman graph of Fig. 7.1, leading to a decay amplitude:

$$\begin{aligned}\mathcal{A}(\tau \to e\nu_3\bar{\nu}_1) &= \frac{g^2 U^*_{\tau 3} U_{e1}}{2}\Big[\bar{\mathrm{u}}_{\nu_3}(\mathbf{l})\gamma^\mu\gamma_L \mathrm{u}_\tau(\mathbf{k})\Big]\Big[\bar{\mathrm{u}}_e(\mathbf{p})\gamma^\nu\gamma_L \mathrm{v}_{\nu_1}(\mathbf{q})\Big] \\ &\quad\times \left[\frac{\eta_{\mu\nu} + (k-l)_\mu(k-l)_\nu/M_W^2}{(k-l)^2 + M_W^2 - \mathrm{i}\epsilon}\right].\end{aligned} \tag{7.6}$$

The small size of the mass ratios, $m_\tau^2/M_W^2 \approx 5\cdot 10^{-4}$ and $m_\mu^2/M_W^2 \approx 2\cdot 10^{-6}$, ensures that a frame exists (the rest frame of the decaying τ) for which all components of

[4] These decays are also often responsible for the decays of composite particles built from elementary fermions, such as nuclei or mesons, but these can also decay for other reasons.

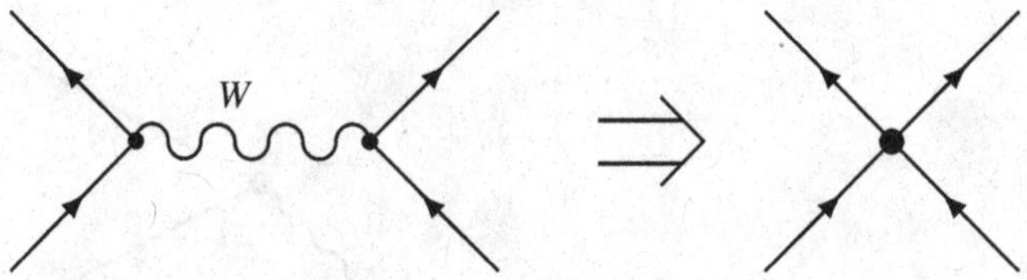

Fig. 7.2 The tree graph that generates the Fermi Lagrangian.

4-momenta are much smaller than M_W. This makes the reaction ripe for description in terms of an effective Wilsonian theory appropriate below the W boson mass (in which the W boson is integrated out). The required effective interaction should capture the above amplitude to lowest order in $1/M_W^2$, and thus can be determined by expanding the W-boson propagator in inverse powers of M_W^2

Keeping only the lowest order term gives the following result for the decay amplitude

$$\mathcal{A}(\tau \to e\nu_3\bar{\nu}_1) = \frac{G_F}{\sqrt{2}}\, U^*_{\tau 3} U_{e1} \Big[\bar{\mathfrak{u}}_{\nu_3}(\mathbf{l})\gamma^\mu(1+\gamma_5)\mathfrak{u}_\tau(\mathbf{k})\Big]\Big[\bar{\mathfrak{u}}_e(\mathbf{p})\gamma_\mu(1+\gamma_5)\mathfrak{v}_{\nu_1}(\mathbf{q})\Big], \tag{7.7}$$

which defines the *Fermi constant*, G_F, as the combination

$$\frac{G_F}{\sqrt{2}} := \frac{g^2}{8M_W^2}. \tag{7.8}$$

The key observation is that this is what would have been produced at lowest order in perturbation theory using the effective 4-fermion interaction, called the *Fermi Lagrangian*:

$$\mathfrak{L}'_{cc} = \frac{G_F}{\sqrt{2}}\, C^\mu C^*_\mu \tag{7.9}$$

in which the *charged current*, C^μ, is defined by

$$C^\mu = iU_{ia}\,\overline{\ell}^i\gamma^\mu(1+\gamma_5)\nu^a + iV^*_{ij}\,\overline{d}^j\gamma^\mu(1+\gamma_5)u^i, \tag{7.10}$$

That is to say, the lagrangian given by (7.9) and (7.10) reproduces *all* of the charged-current predictions of the Standard Model at energies well below M_W, to lowest nontrivial order in E^2/M_W^2. Pictorially, in any W-exchange Feynman graph it is effectively as if the W propagator were contracted to a point, as in Fig. 7.2.

Although the Fermi lagrangian is derived here as the low-energy limit of the Standard Model, historically it was the Fermi theory that was developed first.[5,6] Indeed, it is a great success of the Standard Model that it contains the Fermi theory of weak interactions as part of its low-energy EFT since the Fermi theory

[5] As the story goes, Fermi's paper was first rejected by the journal *Nature* (for being too speculative), and so it was published separately in Italian and German [116, 117].

[6] The version of the Fermi theory used here is actually the one given by Feynman & Gell-Mann [118] and separately by Sudarshan & Marshak [119], which differs from the original Fermi theory by the inclusion of parity violation among other features. The inclusion of quarks into the theory came somewhat later [111].

is known to describe well the many detailed measurements of decay properties of light elementary particles.

The emergence of the Fermi theory as the low-energy limit of the Standard Model also carries several conceptual insights. The Standard Model provides a more unified description of the electromagnetic and weak interactions, characterizing them in terms of two very similar couplings g and e, rather than the superficially very different historical couplings G_F and e. On dimensional grounds a weak-interaction decay computed at leading order using (7.9) that releases an energy Q to the decay daughters has a generic decay rate $\Gamma \propto G_F^2 Q^5$, showing that the weak interactions were historically regarded as being weak more because $Q \ll G_F^{-1/2} \simeq$ 300 GeV than because the underlying interaction is particularly weak compared to electromagnetism.

The experimental success of the Fermi theory also helped the particle-physics community to come to grips with using non-renormalizable interactions within a quantum context. Because the Fermi coupling G_F has negative mass-dimension – $G_F \propto (\text{energy})^{-2}$ – the Fermi theory is not renormalizable. Yet the development of a renormalizable theory like the Standard Model as its UV completion reveals very concretely that a low-energy non-renormalizable interaction need not preclude making sensible quantum predictions, even at low energies and including radiative corrections. What the UV theory makes clear is that more and more information about the high-energy sector must be known the more accurately the answer is required, in a way that is ultimately most efficiently captured at any order in terms of the low-energy Wilsonian EFT.

7.2 Quantum Electrodynamics

An important spinoff of the EFT point of view is a deeper understanding of why the theories we use have the form they do. In particular, it explains why renormalizable theories are so often found to be important in physics, as this section elaborates using Quantum Electrodynamics (QED) as an illustration.

A second, practical, reason for presenting QED in more detail here is to provide concrete (and useful) examples of more precise treatments – such as loop-level matching between the UV and low-energy theories – than were presented earlier for the toy model in Part I.

For example, the very lightest electromagnetically interacting elementary particles are the massless photon and the electron (whose mass is $m_e \simeq 0.511$ MeV). The next-lightest after these is the muon, which with $M_\mu \simeq 106$ MeV is about 200 times heavier. To the extent that one is happy to drop all corrections suppressed by powers of $1/M_\mu$, all electron-photon physics at energies $E \ll M_\mu$ should be described by a Wilsonian EFT with an action involving no couplings with negative mass-dimension; that is to say, by the most general renormalizable interactions built using only electrons and photons.

For a low-energy theory involving only electrons and photons the relevant field content is a Dirac spinor field, $\psi(x)$, for the electron plus the electromagetic gauge potential, $A_\mu(x)$. As discussed in §C.3.3, a massless spin-one particle (the photon) requires the existence of a linearly realized gauge symmetry (also called a field redundancy) under which ψ must also transform if it is electrically charged. For a single Dirac field – chosen to destroy particles with charge $-e$ – this implies that

$$\psi(x) \to \psi(x)\, e^{-ie\zeta(x)} \quad \text{and} \quad A_\mu(x) \to A_\mu(x) + \partial_\mu \zeta(x), \tag{7.11}$$

where $\zeta(x)$ is the symmetry's spacetime-dependent transformation parameter.

The most general renormalizable interactions consistent with this field content and electron charge assignment are then given by the lagrangian[7]

$$\mathfrak{L}_{\rm ren}(A_\mu, \psi) = -\frac{1}{4} F_{\mu\nu}F^{\mu\nu} - \overline{\psi}(\not{D} + m_e)\psi, \tag{7.12}$$

where $F_{\mu\nu} = \partial_\mu A_\nu - \partial_\nu A_\mu$ is the electromagnetic field strength, and both fields are rescaled to put their kinetic terms into standard form. Here $\not{D}$ denotes $\gamma^\mu D_\mu$ (as usual) with the covariant derivative defined by $D_\mu \psi = \partial_\mu \psi + ieA_\mu \psi$, with e the proton charge (or electromagnetic coupling constant). Any other local interactions that involve only these fields and are invariant under (7.11) necessarily involve couplings with negative mass-dimension (and so are not renormalizable), as can be seen from the field dimensions $[\psi] = 3/2$ and $[A_\mu] = 1$ (which follow from the dimension of the respective kinetic terms).

The remainder of this section uses QED to illustrate EFT methods in two ways. The first class of applications uses (7.12) as the UV theory, regarding the electron as the heavy particle whose removal generates the Wilsonian EFT. The second class regards (7.12) as part of the EFT obtained by integrating out the next-heaviest particle (the muon). Both categories of examples integrate out the 'heavy' fermion at the loop level, raising a number of conceptual issues not worked through in detail in the examples of Part I.

The ability to think in both of these ways also emphasizes how the notion of an EFT is a recursive one: having $\mathfrak{L}_1(\phi)$ as the Wilson action for a UV theory containing more fields, $\mathfrak{L}_2(\phi, \psi)$, does not preclude $\mathfrak{L}_2(\phi, \psi)$ from itself being the Wilson action for yet another underlying theory, $\mathfrak{L}_3(\phi, \psi, \chi)$, which applies at still higher energies (further into the UV).

7.2.1 Integrating Out the Electron

This section starts with (7.12) as the UV theory and integrates out the electron, leaving an effective theory involving only photons. Because a low-energy world containing only photons is a bit boring, in this section (7.12) is temporarily supplemented with a configuration of classical macroscopic currents.

There is also a physical motivation for adding classical currents to (7.12), since any practical applications of electromagnetism in real life usually involve the interaction of photons with large distributions of electric charges and currents that capture the bulk features of macroscopic collections of atoms (such as used to describe a wire

[7] This is precisely the lagrangian for Quantum Electrodynamics: the theory of photons [120, 121] coupled to relativistic electrons [122]. See also [123–126, 128].

or a capacitor in the laboratory).[8] In practice, such currents are of most interest at energies and momenta much smaller than m_e, so their study lends itself well to an EFT description below the electron mass.

To describe the currents add to (7.12) the interaction

$$\mathfrak{L}_J = e\,A_\mu\,\mathfrak{J}^\mu. \tag{7.13}$$

where the current $\mathfrak{J}^\mu$ is imagined to be an explicitly given function of spacetime. This coupling is only consistent with gauge invariance – *i.e.* the transformation (7.11) – if the current is identically conserved: $\partial_\mu \mathfrak{J}^\mu = 0$, independent of any equations of motion. It should also fall off sufficiently quickly to preclude there being a current flow at spatial infinity, and (of course) only vary slowly – over macroscopically large distances – so as to capture only currents appropriate for a low-energy analysis.

It often helps to have specific examples of such conserved configurations in mind. One such an example of practical interest might be a specified static charge distribution: $\mathfrak{J}^0 = \rho(\mathbf{r})$, $\mathfrak{J}^i = 0$, such as, in particular, for a point charge $\rho(\mathbf{r}) = Q\,\delta^3(\mathbf{r} - \mathbf{r}_0)$. A second example could be an electrical current with no electric charge density: $\mathfrak{J}^0 = 0$ and $\mathfrak{J}^i = j^i(\mathbf{r}, t)$, where $\mathbf{j}(\mathbf{r}, t)$ is a local current distribution satisfying $\nabla \cdot \mathbf{j} = 0$ at all times.

Lowest Dimension Effective Interactions

When constructing the Wilson action below m_e, the first step is to identify what kinds of effective lagrangian, $\mathfrak{L}_w(A\,, \mathfrak{J})$, can be envisioned consistent with $\partial_\mu \mathfrak{J}^\mu = 0$ and invariance under $A_\mu \to A_\mu + \partial_\mu \zeta$. When doing so, interactions are organized with higher-and-higher operator dimension, keeping in mind the field dimensions, $[A_\mu] = 1$ and $[\mathfrak{J}^\mu] = 3$ from which it also follows that $[F_{\mu\nu}] = 2$, for the field-strength tensor $F_{\mu\nu} = \partial_\mu A_\nu - \partial_\nu A_\mu$.

The Wilsonian action can be written (up to total derivatives[9]) as a sum over terms of increasing operator dimension, $\mathfrak{L}_w = \mathfrak{L}_4 + \mathfrak{L}_6 + \mathfrak{L}_8 + \cdots$, where $[\mathfrak{L}_d] = d$, and

$$\begin{aligned}
\mathfrak{L}_4 &= -\frac{Z}{4}\, F_{\mu\nu}F^{\mu\nu} + e\,A_\mu \mathfrak{J}^\mu, \\
\mathfrak{L}_6 &= \frac{a_1}{2m_e^2}\, e^2 \mathfrak{J}_\mu \mathfrak{J}^\mu + \frac{a_2}{4m_e^2}\, F_{\mu\nu}\Box F^{\mu\nu} + \frac{a_3}{2m_e^2}\, \partial_\mu F^{\mu\nu} \partial^\lambda F_{\lambda\nu}\,, \\
\mathfrak{L}_8 &= \frac{b_1}{m_e^4}\,(F_{\mu\nu}F^{\mu\nu})^2 + \frac{b_2}{m_e^4}\,(F_{\mu\nu}\widetilde{F}^{\mu\nu})^2 + \frac{b_3}{2m_e^4}\, e^2 \partial^\mu \mathfrak{J}^\nu \partial_\mu \mathfrak{J}_\nu + (\partial^4 F^2 \text{ terms}),
\end{aligned} \tag{7.14}$$

and so on. In this expression $\widetilde{F}_{\mu\nu} := \frac{1}{2}\,\epsilon_{\mu\nu\lambda\rho}F^{\lambda\rho}$ represents the 'dual' field-strength tensor and a term linear in $F^{\mu\nu}\widetilde{F}_{\mu\nu}$ is not written because – unlike 7.12 – it is not parity invariant (and in any case can locally be written as a total derivative). A power of $1/m_e$ is factored out of the coefficient of each term so that the constants, Z, a_i, b_i *etc.*, are dimensionless.

Notice that no $\partial^2 F^3$ terms appear in $\mathfrak{L}_8$, and they are not included because any terms involving an odd number of F's are forbidden by charge-conjugation (or C) invariance, which interchanges particles and anti-particles (*i.e.* interchanges their

[8] More detailed and systematic ways to arrive at macroscopic descriptions for large collections of atoms are described in Parts III and IV.

[9] Total derivatives are revisited in §7.4, where conducting or dielectric boundaries are present.

destruction operators $\mathfrak{a}_{k\lambda} \leftrightarrow \bar{\mathfrak{a}}_{k\lambda}$) and so takes $F_{\mu\nu} \to -F_{\mu\nu}$. (This is Furry's theorem in an EFT guise [127, 128].) Imposing invariance under C – or under parity, P (for which $\mathbf{x} \to -\mathbf{x}$), or their product CP – is appropriate to the extent that the UV theory also respects these symmetries. They are indeed symmetries if the underlying theory is given by (7.12), but invariance in the EFT below m_e should be revisited once non-renormalizable interactions – like those of the Fermi theory (7.9) – are added to the UV theory. (See Table C.1 for a summary of how quantities in QED transform under C, P and T.)

The final step performs field redefinitions to eliminate redundant operators from (7.14). As described in Part I, this in practice boils down to eliminating terms in $\mathfrak{L}_k$ (with $k \geq 6$) that vanish when evaluated at a solution to Maxwell's equations $\partial_\mu F^{\mu\nu} + e\mathfrak{J}^\nu = 0$ (which together with the Bianchi identity, $\partial_\mu F_{\nu\lambda} + \partial_\nu F_{\lambda\mu} + \partial_\lambda F_{\mu\nu} = 0$, also implies that $\Box F_{\mu\nu} + e(\partial_\mu \mathfrak{J}_\nu - \partial_\nu \mathfrak{J}_\mu) = 0$. Together with an integration by parts, this allows $\mathfrak{L}_6$ to be simplified to

$$\mathfrak{L}_6 = \frac{\tilde{a}_1}{2m_e^2}\, e^2 \mathfrak{J}_\mu \mathfrak{J}^\mu, \tag{7.15}$$

where $\tilde{a}_1 := a_1 - a_2 + a_3$ (or, equivalently, just setting $a_2 = a_3 = 0$). Identical arguments also allow the removal of the $\partial^4 F^2$ terms from $\mathfrak{L}_8$ in (7.14).

The leading non-redundant terms in the Wilsonian action below m_e therefore become

$$\begin{aligned}\mathfrak{L}_W = &-\frac{Z}{4} F_{\mu\nu}F^{\mu\nu} + eA_\mu \mathfrak{J}^\mu + \frac{\tilde{a}_1}{2m_e^2} e^2 \mathfrak{J}_\mu \mathfrak{J}^\mu \\ &+ \frac{b_1}{m_e^4}(F_{\mu\nu}F^{\mu\nu})^2 + \frac{b_2}{m_e^4}(F_{\mu\nu}\widetilde{F}^{\mu\nu})^2 + \frac{b_3}{2m_e^4} e^2 \partial^\mu \mathfrak{J}^\nu \partial_\mu \mathfrak{J}_\nu + \cdots\end{aligned} \tag{7.16}$$

with effective couplings to be obtained by matching to the UV theory.

Power Counting

The effective couplings Z, e, a_i and b_i are inferred by matching predictions for observables made using (7.16) to the low-energy limit of the full QED prediction. The first part of this requires power counting: to know precisely which Feynman graphs need evaluation and identifying which effective interactions must appear within them to capture all contributions to any order in $1/m_e$.

In this particular case, the general power-counting results of §3.2.3 can be carried over in whole cloth since the effective lagrangian of Eq. (7.14) is a special case of the form considered in Eq. (3.55), provided one makes the choices $\mathfrak{f} = M = v = m_e$. Directly using Eq. (3.57) for the $\mathcal{E}$-point scattering amplitude, $\mathcal{A}_\mathcal{E}(q)$, leads to:

$$\mathcal{A}_\mathcal{E}(q) \sim q^2 m_e^2 \left(\frac{1}{m_e}\right)^{\mathcal{E}} \left(\frac{q}{4\pi\, m_e}\right)^{2\mathcal{L}} \left(\frac{q}{m_e}\right)^{\sum_n (d_n - 2)\mathcal{V}_n}. \tag{7.17}$$

Gauge invariance also brings more specific information. In particular, since the gauge potential, A_μ, only appears in $\mathfrak{L}_W$ through its field strength, $F_{\mu\nu}$ (apart from the sole exception of the $A_\mu \mathfrak{J}^\mu$ term), all effective interactions in vertices must contain at least as many derivatives as powers of A_μ: *i.e.* $\mathcal{V}_n = 0$ unless $d_n \geq f_n$. Furthermore, charge-conjugation invariance implies that f_n must be even and so $\mathcal{V}_n = 0$ unless $d_n \geq f_n \geq 4$. (The exception to this counting would be if the Maxwell term,

Fig. 7.3 The Feynman graph contributing the leading contribution to photon-photon scattering in the effective theory for low-energy QED. The vertex represents either of the two dimension-eight interactions discussed in the text.

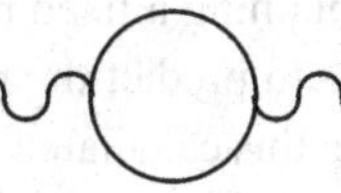

Fig. 7.4 The Feynman graph contributing the vacuum polarization. The circular line denotes a virtual electron loop, while the wavy lines represent external photon lines.

$ZF_{\mu\nu}F^{\mu\nu}$, were not all kept in the unperturbed lagrangian, but instead was partly written as a perturbative 'counter-term' interaction.)

As a result, vertices drawn from this particular EFT satisfy the inequality $\sum_n (d_n - 2)\mathcal{V}_n \geq 2$, implying $\mathcal{A}_\mathcal{E}(q)$ vanishes at least like q^4 for all $\mathcal{E}$. Furthermore, the only way to get $\mathcal{A}_E \propto q^4$ is if $\mathcal{L} = 0$ and $\sum_n (d_n - 2)\mathcal{V}_n = 2$, which can only happen if $\mathcal{V}_n = 0$ for all $d_n > 4$ but with $\mathcal{V}_n = 1$ for $d_n = 4$ (and so also $f_n = 4$). There is only one such graph, shown explicitly in Fig. 7.3 with the vertex corresponding to the effective couplings b_1 and b_2 of Eq. (7.16). Only $\mathcal{A}_4(q)$ can be proportional to q^4 while $\mathcal{A}_\mathcal{E}(q)$ is suppressed by at least q^6 for all $\mathcal{E} > 4$.

Dominant Low-Energy Behaviour

Power counting shows that the only part of the Wilsonian lagrangian, (7.14), that is completely unsuppressed by $1/m_e$ at low energies is

$$\mathfrak{L}_W \simeq \mathfrak{L}_4 = -\frac{Z}{4} F_{\mu\nu}F^{\mu\nu} + e\, A_\mu\, \mathfrak{J}^\mu, \tag{7.18}$$

so the dimensionless parameter, Z, contains all of the information about the leading contribution from virtual electrons.

Z may be explicitly computed in QED by evaluating the graph of Fig. 7.4, and keeping only the part quadratic in external momenta, $\mathcal{A}_{\mu\nu}(q) \propto q^2\eta_{\mu\nu} - q_\mu q_\nu$, leading to the expression (see §A.2.4)

$$Z = 1 - \frac{\alpha}{3\pi}\left[\frac{1}{\varepsilon} - \gamma_k + \log\left(\frac{m_e^2}{\mu^2}\right)\right], \tag{7.19}$$

where the integration over loop momentum is regulated using dimensional regularization, with $D = 4 - 2\varepsilon$ so $D \to 4$ corresponds to $\varepsilon \to 0$. The constant γ_k is the quantity encountered in Eq. (3.39), which appears universally with the divergence, $1/\varepsilon$, and μ is the usual (arbitrary) mass scale introduced in dimensional regularization to keep the coupling constant, e, dimensionless in D dimensions[10].

The physical interpretation of Z is found by performing the rescaling, $A_\mu = Z^{-1/2} A_\mu^{(R)}$, required to return the photon kinetic term to its canonical normalization – *i.e.* re-normalizing A_μ. This returns the effective theory

10 That is: $e_D = e\mu^{4-D}$.

$$\mathfrak{L}_4 = -\frac{1}{4}\, F^{(R)}_{\mu\nu} F^{\mu\nu}_{(R)} + e_{\rm phys}\, A^{(R)}_\mu \, \mathfrak{J}^\mu . \tag{7.20}$$

where $e := Z^{1/2}\, e_{\rm phys}$ and the label '(R)' is dropped from here on to avoid overly cluttering the notation. The charge $e_{\rm phys}$ appears in (7.20) in precisely the way that the proton charge would appear in the lagrangian density for classical electromagnetism. The subscript 'phys' emphasizes that this charge can be regarded as a physical observable whose value can in principle be experimentally determined. For instance, it could be measured by taking a static macroscopic distribution, $\mathfrak{J}^0$, of known total charge (perhaps containing a fixed number of protons, for example), and then using Maxwell's equations to predict the resulting flux of electric field at a large (known) distance. Comparing this calculated flux with the measured flux gives a measurement of $e_{\rm phys}$.

This fairly trivial effective theory reveals several things. First, it sharpens the notion of decoupling. Eq. (7.20) shows that to leading (zeroth) order in $1/m_e$ virtual electrons only affect low-energy photon properties through the *value* taken by the physical electric charge, $e_{\rm phys}$. So it is *not* true that integrating out the electron only produces effects at low energies that are suppressed by powers of $1/m_e$; this only becomes true after a suitable adjustment of low-energy parameters like e [1]. Decoupling only states that there exists a choice of low-energy couplings for which virtual high-energy physics is suppressed by powers of the high scale.

A second practical conclusion also follows from (7.20). Calculations of electromagnetic response to the macroscopic currents, $\mathfrak{J}^\mu$, do not get corrected at *any* order in α without also being suppressed by powers of $1/m_e$. Consequently, the justification of using Maxwell's equations to describe low-energy electromagnetic properties is the neglect of terms of order $1/m_e$, *not* neglect of powers of α. This is the root of the explanation of why Rutherford scattering [129] of nonrelativistic particles from a point charge distribution remains completely uncorrected by quantum corrections even though it is only usually derived in tree approximation within QED. The EFT explains the robustness of this result because any such correction must be additionally suppressed by powers of $p/m_e \propto v \ll 1$ and so get dropped in the nonrelativistic limit.[11]

At low energies, the sole effect of all higher-order corrections in α is just to renormalize the value of α in terms of which all observables are computed. As explained at length in §7.2.2, this renormalization actually matters once it is possible to measure the coupling e also at higher energies, since then the logarithmic running of couplings with scale captures the potential dependence of observables on logarithms of large mass ratios (and so encodes a real physical effect).

Scattering of Light by Light

Next examine the case $\mathcal{E} = 4$ (*i.e.* $2 \to 2$ photon scattering) in more detail, restricting for simplicity to a region where $\mathfrak{J}^\mu = 0$. For $\mathcal{E} = 4$ the power-counting arguments of the previous paragraphs give

$$\mathcal{A}_4(q) \sim \frac{q^2}{m_e^2} \left(\frac{q}{4\pi\, m_e}\right)^{2\mathcal{L}} \left(\frac{q}{m_e}\right)^{\sum_n (d_n - 2)\mathcal{V}_n}, \tag{7.21}$$

[11] A possibly apocryphal physics tale attributes to Rutherford great pride that his scattering formula (derived classically) survived unscathed after the invention of quantum mechanics.

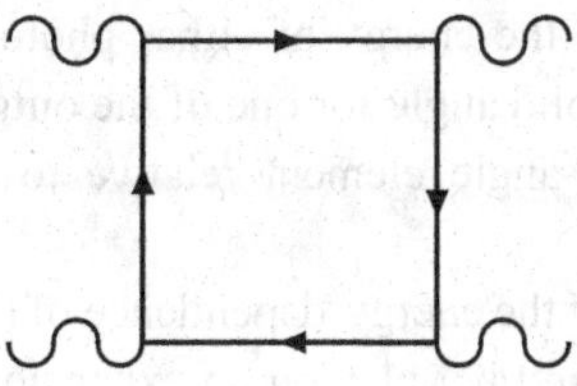

Fig. 7.5 The leading Feynman graphs in QED which generate the effective four-photon operators in the low-energy theory. Straight (wavy) lines represent electrons (photons).

with the leading contribution – given by Fig. 7.3 – being of order $\mathcal{A}_4 \sim q^4/m_e^4$. For photon-photon scattering the relevant value for q is the centre-of-mass energy $E_{\rm cm}$, and since the cross section – which has dimension (length)^2 – is an integral over final-state momenta of a transition amplitude proportional to $|\mathcal{A}_4|^2$, on dimensional grounds $\sigma \sim E_{\rm cm}^6/m_e^8$ for $E_{\rm cm} \ll m_e$.

The real power of the effective lagrangian comes when computing sub-leading contributions, as can be done using (7.21) to any order desired in q/m_e. Naively, the first subdominant term is suppressed by one more factor of q^2/m_e^2; however, this would require either $\mathcal{L} = 1$ and $\sum_n (d_n - 2)\mathcal{V}_n = 2$ or $\mathcal{L} = 0$ and $\sum_n (d_n - 2)\mathcal{V}_n = 4$. But $\sum_n (d_n - 2)\mathcal{V}_n = 2$ implies that $\mathcal{V}_n = 0$ for $d_n > 4$ and $\mathcal{V}_n = 1$ for $d_n = f_n = 4$, and there is no one-loop graph with four external lines built only using a single 4-point ($f_n = 4$) interaction. Similarly, no tree graphs with only four external lines can be built using vertices involving more than four fields ($f_n > 4$), so because gauge invariance implies that all vertices satisfy $d_n \geq f_n$ the only possible graph contributing at order q^6/m_e^6 to $\mathcal{A}_4$ would be the graph of Fig. 7.3 built using a single $d_n = 6$ and $f_n = 4$ (six-derivative four-field) vertex. And so on.

Returning to the leading term, $\mathcal{A}_4 \propto (q/m_e)^4$, the precise values for the couplings b_1 and b_2 are determined by matching to the full theory. This involves evaluating the 'box' graph of Fig. 7.5 in the full theory, expanded to lowest nontrivial order in powers of external momenta, and equating it to the implications of Eq. (7.16) in the Born approximation (*i.e.* Fig. 7.3). No counterterms or renormalizations are required in this calculation because the box graph is ultraviolet finite.

Agreement of the two calculations requires (see Exercise 7.2)

$$b_1 = \frac{4}{7}\, b_2 = \frac{\alpha^2}{90}, \tag{7.22}$$

so that the effective four-photon interaction becomes [130, 131]

$$\begin{aligned} \mathfrak{L}_{4\gamma} &= \frac{\alpha^2}{90\, m_e^4}\left[(F_{\mu\nu}F^{\mu\nu})^2 + \frac{7}{4}\,(F_{\mu\nu}\widetilde{F}^{\mu\nu})^2\right] \\ &= \frac{\alpha^2}{180\, m_e^4}\left[5(F_{\mu\nu}F^{\mu\nu})^2 - 14\, F_{\mu\nu}F^{\nu\lambda}F_{\lambda\rho}F^{\rho\mu}\right], \end{aligned} \tag{7.23}$$

and the cross-section for the light-by-light scattering is [132]

$$\frac{d\sigma_{\gamma\gamma}}{d\Omega} \simeq \frac{139}{4\pi^2}\left(\frac{\alpha^2}{90}\right)^2\left(\frac{E_{\rm cm}^6}{m_e^8}\right)\left(3 + \cos^2\theta\right)^2. \tag{7.24}$$

Here, $E_{\rm cm}$ is the energy of either photon in the CM frame, $d\Omega$ is the differential element of solid angle for one of the outgoing photons and θ is the angular position of this solid-angle element relative to the direction of (either of) the incoming photons.

Notice that the energy-dependence of (7.24) relies only on the expansion in powers of $E_{\rm cm}/m_e$, and not also on an expansion of the cross section in powers of α. It is only when using Eq. (7.22) that perturbation theory in α is first used.

7.2.2 $E \gg m_e$ and Large Logs *

The discussion now switches gears and regards the QED lagrangian (7.12) as the first terms in a Wilsonian effective theory rather than being the underlying UV completion. In this picture the electron is now a light particle and interest is extended to energies $q \gg m_e$. This regime allows an interpretation of the parameter Z as well as an examination of the leading effects due to integrating out the muon at energies $m_e \ll q \ll M_\mu$.

This section starts by exploring the implications of Z in this higher-energy regime, providing a concrete illustration of how matching can be done including loops, and so how to deal with both UV divergences and renormalization. This section shows how to perform these steps while both preserving the practical utility of dimensional regularization and keeping manifest how heavy particles decouple from low-energy observations.

To see the relevance of Z to large logarithms, it is instructive to contrast two useful renormalization schemes. The first is the one defined above, in which all of Z is completely absorbed into the fields and couplings:

$$A_\mu = Z_{\rm phys}^{-1/2}\, A_\mu^{\rm phys}, \qquad \text{and} \qquad e = Z_{\rm phys}^{1/2}\, e_{\rm phys},$$
$$\text{with} \qquad Z_{\rm phys} := Z \simeq 1 - \frac{\alpha}{3\pi}\left[\frac{1}{\varepsilon} - \gamma_k + \log\left(\frac{m_e^2}{\mu^2}\right)\right]. \tag{7.25}$$

The second renormalization scheme is the scheme of choice for most practical calculations. Called the 'modified minimal subtraction' (or $\overline{MS}$) scheme,[12] the renormalization is defined to subtract only the term $1/\varepsilon - \gamma_k$ in Z. That is:

$$A_\mu = Z_{\overline{MS}}^{-1/2}\, A_\mu^{\overline{MS}}, \qquad \text{and} \qquad e = Z_{\overline{MS}}^{1/2}\, e_{\overline{MS}},$$
$$\text{with} \qquad Z_{\overline{MS}} := 1 - \frac{\alpha}{3\pi}\left[\frac{1}{\varepsilon} - \gamma_k\right]. \tag{7.26}$$

In terms of this scheme the effective lagrangian becomes

$$\mathfrak{L}_{\rm eff} = -\frac{1}{4}\left[1 - \frac{\alpha}{3\pi}\log\left(\frac{m_e^2}{\mu^2}\right)\right] F_{\mu\nu}^{\overline{MS}} F^{\mu\nu}_{\overline{MS}} - e_{\overline{MS}}\, A_\mu^{\overline{MS}}\, \mathfrak{J}^\mu, \tag{7.27}$$

as opposed to (7.20), where α can be taken to be either $e^2_{\overline{MS}}/4\pi$ or $e^2_{\rm phys}/4\pi$ since the difference is higher order in α. To see this explicitly, notice that $\alpha_{\overline{MS}} = e^2_{\overline{MS}}/4\pi$ is related to $\alpha_{\rm phys} = e^2_{\rm phys}/4\pi$ by

[12] The awkward name arises historically because it came after the earlier 'minimal subtraction' (or MS) scheme, which subtracts only the pole: $1/\varepsilon$.

$$\alpha_{\overline{MS}} = \left(\frac{Z_{\rm phys}}{Z_{\overline{MS}}}\right) \alpha_{\rm phys} \simeq \left[1 - \frac{\alpha}{3\pi} \log\left(\frac{m_e^2}{\mu^2}\right)\right] \alpha_{\rm phys}. \tag{7.28}$$

No poles in $1/\varepsilon$ remain in this expression because both couplings $\alpha_{\overline{MS}}$ and $\alpha_{\rm phys}$ are renormalized quantities.

A key observation states that because $\alpha_{\rm phys}$ is a physical quantity, it cannot depend on the arbitrary scale μ. As a result, this last equation implies a μ-dependence for $\alpha_{\overline{MS}}$. It is only for $\mu = m_e$ that the two couplings agree:

$$\alpha_{\overline{MS}}(\mu = m_e) = \alpha_{\rm phys}. \tag{7.29}$$

There proves to be profit in re-expressing the μ dependence of $\alpha_{\overline{MS}}$ in Eq. (7.28) as a differential Callan–Symanzik [139, 140] relation,

$$\mu^2 \frac{{\rm d}\alpha_{\overline{MS}}}{{\rm d}\mu^2} = + \frac{\alpha_{\overline{MS}}^2}{3\pi}. \tag{7.30}$$

This is useful because (7.30) and (7.28) have different domains of validity. While the differential expression (7.30) requires only $\alpha_{\overline{MS}} \ll 1$ to be valid, Eq. (7.28) requires the stronger condition $\alpha_{\overline{MS}} \log(m_e^2/\mu^2) \ll 1$. One way to think of why there is an extended domain for (7.30) is that for any μ the evolution of the coupling can be computed for a small range of scales around $\mu = \mu_0$ using the analog of (7.28)

$$\alpha_{\overline{MS}}(\mu) \simeq \left[1 - \frac{\alpha}{3\pi} \log\left(\frac{\mu_0^2}{\mu^2}\right)\right] \alpha_{\overline{MS}}(\mu_0). \tag{7.31}$$

Although (7.31) is only valid for μ close enough to μ_0 to ensure $|\alpha \log(\mu^2/\mu_0^2)| \ll 1$, it is also true that μ_0 is arbitrary. So (7.30) can be derived on a sequence of overlapping domains centred about different values of μ_0, and what is important is that (7.30) is then valid on the *union* of all these overlapping domains [40, 133, 134].[13]

Integrating Eq. (7.30) allows an inference of the μ-dependence of $\alpha_{\overline{MS}}$ when $\alpha_{\overline{MS}} \ll 1$ but $\alpha_{\overline{MS}} \log(m_e^2/\mu^2)$ is not small:

$$\frac{1}{\alpha_{\overline{MS}}(\mu)} = \frac{1}{\alpha_{\overline{MS}}(\mu_0)} - \frac{1}{3\pi} \log\left(\frac{\mu^2}{\mu_0^2}\right) = \frac{1}{\alpha_{\rm phys}} - \frac{1}{3\pi} \log\left(\frac{\mu^2}{m_e^2}\right). \tag{7.32}$$

Called the 'renormalization-group' improved running, Eq. (7.32) is accurate to all orders in $(\alpha/3\pi)\log(\mu^2/m_e^2)$, so long as $(\alpha/3\pi) \ll 1$. Notice that Eq. (7.30) integrates so simply only because $\overline{MS}$ renormalization is a *mass-independent* scheme. That is, $\mu^2{\rm d}\alpha_{\overline{MS}}/{\rm d}\mu^2$ depends only on $\alpha_{\overline{MS}}$ and not also on ratios of mass scales like m_e/μ. ('On shell' renormalizations, such as where e is defined in terms of the value of a scattering amplitude at a specific energy threshold, furnish examples of schemes that are not mass-independent.)

Eq. (7.32) is ultimately useful because it provides a simple way to track how some large logarithms appear in physical observables. For instance, consider the cross section, σ, for the scattering of electrons with centre-of-mass energy E, plus an indeterminate number of soft photons, with energies up to $E_\gamma = fE$ with $1 > f \gg m_e/E$. On dimensional grounds one has

[13] This argument is met again in §16.4.1 and §16.4.2 when computing late-time evolution in perturbation theory.

$$\sigma(E, m_e, \alpha_{\text{phys}}) = \frac{1}{E^2}\, \mathcal{F}\left(\frac{m_e}{E}, \alpha_{\text{phys}}, f, \Theta_k\right), \tag{7.33}$$

where $\mathcal{F}$ is some calculable function and the Θ_k denote any number of dimensionless quantities (like angles) on which the observable depends.

Now comes the main point. Interest is often in the regime $E \gg m_e$, so it is tempting to Taylor expand σ in powers of m_e/E. Unfortunately, the function $\mathcal{F}$ proves to be singular when $m_e/E \to 0$ due to the appearance of large logarithms of the form $\log(E/m_e)$ that arise from infrared divergences that would be present in the limit $m_e \to 0$.

These divergences are not the usual 'Bloch-Nordsieck' infrared divergences [52, 53, 135–138] of QED that would arise if one were to take the unphysical limit $f \to 0$ (corresponding to not summing over the production of soft photons). Instead, the large logarithms of interest only appear after performing the 'on-shell' subtractions that renormalize the physical coupling e_{phys} (see, for example, the singularity as $m \to 0$ of the vacuum polarization computed in (A.53)). What is important is this: these singularities would not appear if $\mathcal{F}$ were instead to be expressed in terms of an 'off-shell' quantity like $\alpha_{\overline{MS}}$. This makes it convenient to compute σ using $\overline{MS}$ renormalization, and then use the absence of $m_e \to 0$ singularities to Taylor expand the result in powers of m_e/E, leading to

$$\sigma(E, m_e, \alpha_{\text{phys}}) = \frac{1}{E^2}\left[\mathcal{F}_0\left(\frac{E}{\mu}, \alpha_{\overline{MS}}(\mu), f, \Theta_k\right) + O(m_e/E)\right], \tag{7.34}$$

where μ is the arbitrary scale appearing in dimensional regularization.

What is important is that physical quantities like σ cannot depend on μ. Consequently, any explicit μ-dependence of $\mathcal{F}_0$ must precisely cancel the μ-dependence appearing implicitly through $\alpha_{\overline{MS}}(\mu)$. This allows the singular behaviour in σ to be identified – at all orders in $\alpha \log(E^2/m_e^2)$ – by using the convenient choice $\mu = E$ in (7.34). Making this choice and using Eq. (7.32) with Eq. (7.29) then gives

$$\sigma(E, m_e, \alpha_{\text{phys}}) = \frac{1}{E^2}\left[\mathcal{F}_0\left(1, \alpha_{\overline{MS}}(E), f, \Theta_k\right) + O(m_e/E)\right], \tag{7.35}$$

where

$$\alpha_{\overline{MS}}(E) = \frac{\alpha_{\text{phys}}}{1 - \frac{1}{3\pi}\,\alpha_{\text{phys}} \log\left(E^2/m_e^2\right)}. \tag{7.36}$$

Once the dependence of $\mathcal{F}_0$ on $\alpha_{\overline{MS}}$ is known in a simple (say, high-energy) regime, its dependence on $\alpha \log(E/m_e)$ is determined up to subdominant order $\alpha^2 \log(E/m_e)$ effects.

7.2.3 Muons and the Decoupling Subtraction Scheme ♠

Continuing to regard the original photon-electron theory as an EFT, it is useful now to be more explicit about what the UV theory is: the electrodynamics of photons, electrons and muons, with the photon-electron system obtained by integrating out the muons. The (renormalizable part of the) underlying UV theory then is

$$\mathfrak{L} = -\frac{1}{4}\, F_{\mu\nu}F^{\mu\nu} + eA_\mu \mathfrak{J}^\mu - \overline{\psi}(\not{D} + m_e)\psi - \overline{\chi}(\not{D} + M_\mu)\chi, \tag{7.37}$$

where χ is the Dirac spinor representing the muon and M_μ is the muon mass. For both fields the covariant derivative is as appropriate for a field with charge $q = -e$: $D_\mu = \partial_\mu + ieA_\mu$.

Integrating out the muon leads to a variety of effective interactions for the Wilson action below the muon mass. Those interactions involving only electromagnetic fields are identical to the ones obtained above when integrating out the electron. The key difference with the muon is that their effective couplings are suppressed by powers of M_μ rather than m_e. For instance, a muon in the graph of Fig. 7.5 generates the dimension-eight effective photon self-interactions of (7.16)

$$\mathfrak{L}_W \supset \frac{b_1}{M_\mu^4}(F_{\mu\nu}F^{\mu\nu})^2 + \frac{b_2}{M_\mu^4}(F_{\mu\nu}\tilde{F}^{\mu\nu})^2 + \cdots , \tag{7.38}$$

with b_1 and b_2 again as in (7.22).

These interactions are generally ignored when discussing photon-photon scattering because their $1/M_\mu^4$ suppression makes them much smaller at low energies than are the corresponding electron results. This is a general feature: non-renormalizable effective couplings generically arise as a series in powers of inverse masses, corresponding to the contribution from each threshold as heavy particles are successively integrated out. All other things being equal, it is the *smallest* mass that usually dominates the couplings of higher-dimensional interactions at low energies.

Integrating out the muon also generates a new class of non-renormalizable effective interactions not present in (7.16), involving the electron field ψ. Such terms get generated at one-loop initially as redundant interactions like (compare with Eq. (A.53) of §A.2.4)

$$\mathfrak{L}_W \supset \frac{a_2}{4M_\mu^2} F_{\mu\nu}\Box F^{\mu\nu} \quad \text{with} \quad a_2 = \frac{\alpha}{15\pi}, \tag{7.39}$$

once the lowest-order Maxwell equation is used: $\partial_\mu F^{\mu\nu} + e\mathfrak{J}^\mu - ie\overline{\psi}\gamma^\mu\psi = 0$, leading to the contact interaction (*c.f.* Eq. (7.15))

$$\begin{aligned}\mathfrak{L}_W &\supset -\frac{a_2}{2M_\mu^2}\, e^2\big(\mathfrak{J}_\mu - i\overline{\psi}\gamma_\mu\psi\big)\big(\mathfrak{J}^\mu - i\overline{\psi}\gamma^\mu\psi\big) \\ &= -\frac{2\alpha^2}{15M_\mu^2}\, e^2\big(\mathfrak{J}_\mu - i\overline{\psi}\gamma_\mu\psi\big)\big(\mathfrak{J}^\mu - i\overline{\psi}\gamma^\mu\psi\big).\end{aligned} \tag{7.40}$$

The terms completely unsuppressed by $1/M_\mu$ are precisely the terms found in (7.12) and (7.13), though with a non-canonical Maxwell action of the form of (7.18) with

$$Z = 1 - \frac{\alpha}{3\pi}\left[\frac{1}{\varepsilon} - \gamma_E + \log\left(\frac{M_\mu^2}{\mu^2}\right)\right]. \tag{7.41}$$

Once canonically normalized this is precisely the QED lagrangian, our starting point at the beginning of this section. In this observation lies the roots of an explanation of *why* QED is such a successful description of low-energy electron-photon interactions. Because QED contains the most general renormalizable couplings possible for a spin-half charged particle and a massless spin-one boson it is guaranteed to emerge as the dominant part of the Wilson action at sufficiently low energies.

This exercise of integrating out the muon to construct a Wilsonian EFT below the muon mass also shows how to efficiently track large logarithms by defining a renormalization scheme (decoupling subtraction) that keeps the simplicity of minimal subtraction without giving up on having heavy fields decouple from the running of low-energy couplings.

To see how this works, it is useful first to ask how couplings run in the full UV theory including both electrons and muons. Following the same steps as in the previous section leads to the following relation between couplings in the $\overline{MS}$ and physical renormalization schemes

$$\alpha_{\overline{MS}} = \left\{1 - \frac{\alpha}{3\pi}\left[\log\left(\frac{m_e^2}{\mu^2}\right) + \log\left(\frac{M_\mu^2}{\mu^2}\right)\right]\right\} \alpha_{\text{phys}}. \tag{7.42}$$

This relation includes a contribution from both electron and muon loops, and replaces Eq. (7.28) of the purely electron-photon theory. The physical coupling, α_{phys} (defined by a canonically normalized Maxwell action once both electrons and muons are integrated out), is now given by $\alpha_{\overline{MS}}(\mu = \sqrt{m_e M_\mu})$.

The corresponding RG equation for the running of $\alpha_{\overline{MS}}$ becomes

$$\mu^2 \frac{\mathrm{d}\alpha_{\overline{MS}}}{\mathrm{d}\mu^2} = +\frac{2\alpha^2}{3\pi}, \tag{7.43}$$

with solution

$$\frac{1}{\alpha_{\overline{MS}}(\mu)} = \frac{1}{\alpha_{\overline{MS}}(\mu_0)} - \frac{2}{3\pi}\log\left(\frac{\mu^2}{\mu_0^2}\right). \tag{7.44}$$

The coupling runs twice as fast as in (7.30) because both electrons and muons contribute.

Eqs. (7.42) through (7.44) reveal an inconvenience of the $\overline{MS}$ renormalization scheme: its mass-independence ensures that both the electron and the muon contribute equally to the running of $\alpha_{\overline{MS}}$ *at all scales*. This is true in particular for $\mu \ll M_\mu$, where the physical influence of the muon should decouple.

Of course, the physical effects of the muon indeed *do* decouple at scales well below the muon mass, with 'decoupling' meaning the existence of a choice of α for which all muonic effects are suppressed by powers of $1/M_\mu$. The presence in (7.44) of too-large running below the muon mass simply means that the $\overline{MS}$ coupling is not the one that makes decoupling manifest. Although muon decoupling is true for physical predictions, it is not manifest at intermediate steps when using the $\overline{MS}$ scheme.

The Wilsonian EFT suggests how to define a scheme that keeps the decoupling manifest without giving up the practical benefits of a mass-independent renormalization scheme. This is done by working with minimal subtraction, but only when running couplings in an energy range for which there are no particle masses. To track how couplings behave as energies pass below each particle mass, the trick is to switch to a new effective theory defined by integrating out this particle explicitly. The couplings in the new low-energy theory are then found by matching to the couplings defined in the underlying theory above the relevant mass scale. (This can be done as in §3.2.2 by computing a simple low-energy observable in both theories and choosing couplings in the low-energy theory so that they give the same result as does the high-energy theory for the observable in question.) The scheme defined by doing so through all particle thresholds is called the *decoupling subtraction* ($\overline{DS}$) renormalization scheme.

For instance, for the electrodynamics of electrons and muons, the coupling constant as defined in the $\overline{MS}$ and $\overline{DS}$ schemes is identical at energies above the muon mass: $\mu > M_\mu$. For $m_e < \mu < M_\mu$ the muon is integrated out to construct an effective theory involving only photons and electrons (as above), consisting of the usual QED lagrangian plus an infinite number of higher-dimension effective interactions encoding the low-energy implications of virtual muons. Within this effective lagrangian the electromagnetic coupling constant is again defined using minimal subtraction, but because there is no muon within this effective theory only the electron contributes to its running. This can be repeated as necessary to include the effects of any particles at still-higher energies.

Quantitatively, to one loop the RG equation for the $\overline{DS}$ scheme for the theory of electrons, muons and photons then becomes:

$$\mu^2 \frac{d\alpha_{\overline{DS}}}{d\mu^2} = \begin{cases} 2\alpha^2/3\pi & \text{if} \quad \mu > M_\mu \\ \alpha^2/3\pi & \text{if} \quad M_\mu > \mu > m_e \,, \\ 0 & \text{if} \quad m_e > \mu \end{cases} \tag{7.45}$$

with the boundary conditions at $\mu = M_\mu$ fixed by matching between the full and effective theories, and $\alpha_{\overline{DS}}(\mu = m_e) = \alpha_{\text{phys}}$ at $\mu = m_e$. At leading nontrivial order matching amounts to continuity of α, so

$$\begin{aligned} \frac{1}{\alpha_{\overline{DS}}(E)} &= \frac{1}{\alpha_{\text{phys}}} - \frac{1}{3\pi} \log\left(\frac{E^2}{m_e^2}\right) && \text{for } m_e < E < M_\mu, \\ &= \frac{1}{\alpha_{\text{phys}}} - \frac{1}{3\pi} \log\left(\frac{M_\mu^2}{m_e^2}\right) - \frac{2}{3\pi} \log\left(\frac{E^2}{M_\mu^2}\right) && \text{for } M_\mu < E. \end{aligned} \tag{7.46}$$

Using this last result in expressions for physical processes efficiently displays the large logarithms discussed in earlier sections, both for $m_e \ll E \ll M_\mu$ and for $E \gg M_\mu$. It has the virtue of running the coupling with the ease of a mass-independent scheme (for which equations like (7.45) are relatively easy to integrate), but with each particle explicitly decoupling from the running as the scale μ drops through the corresponding particle mass.

7.2.4 Gauge/Goldstone Equivalence Theorems

The Standard Model contains a myriad of other hierarchies of scale and so provides a great many other instructive applications of EFT techniques. This next example provides a simple illustration of why the study of Goldstone boson properties is not limited in its practical utility to cases where there are accidental global symmetries.

Goldstone bosons are also useful when studying properties of gauge bosons in a regime where the energies of interest satisfy $M_A \ll E \ll 4\pi v$, where v is the 'decay constant' (*i.e.* the scale at which the gauge symmetry is spontaneously broken) and $M_A \simeq gv$ is the gauge boson mass (with g being the gauge coupling). This can be a significant energy window when the gauge coupling is weak: $g \ll 4\pi$. For instance, for electroweak gauge bosons $M_A \sim 80 - 90$ GeV while $4\pi v \sim 3$ TeV.

When at rest, a gauge boson's three spin states are all alike since they are related to one another by a symmetry (rotations). This need no longer be true once

$E \gg M_A$, however, since the gauge bosons are then relativistic. In particular, although the assumption of small g ensures the couplings of transverse spin states to other particles is quite weak, the same need not be true of the gauge bosons' longitudinal spin states. After all, these started life as scalar fields and so can have couplings with other fields (or themselves) that remain unsuppressed[14] as $g \to 0$. When these couplings dominate, the dynamics of the gauge field is well approximated by the physics of its purely longitudinal mode and so is captured by the scalar interactions of the Goldstone bosons, which can be formulated quite generally whenever $E \ll 4\pi v$.

Probably the simplest way to see how this works is to use an explicit example. For this purpose consider the Standard model, but in an alternative world,[15] where the Higgs boson is heavy enough to allow the decays $h \to W^+W^-$ and $h \to ZZ$. The relevant interaction terms (to leading order, in unitary gauge) in the Standard Model are

$$\mathfrak{L}_{hVV} = -\frac{h}{v}\left(2M_W^2 W_\mu^* W^\mu + M_Z^2 Z_\mu Z^\mu\right). \tag{7.47}$$

This leads to the leading-order matrix element for $h(k) \to W^+(q)W^-(p)$,

$$\mathcal{A}[H(k) \to W^+(q)W^-(p)] = \frac{2M_W^2}{v}\epsilon_\mu^*(\mathbf{q},\zeta)\epsilon^{*\mu}(\mathbf{p},\sigma), \tag{7.48}$$

where ϵ_μ denotes the spin-one polarization vector. Squaring and summing over the final-state spins using

$$\sum_{\sigma=0,\pm1} \epsilon_\mu(\mathbf{p},\sigma)\epsilon_\nu^*(\mathbf{p},\sigma) = \eta_{\mu\nu} + \frac{p_\mu p_\nu}{M_W^2}, \tag{7.49}$$

leads to the unpolarized differential decay rate

$$d\Gamma = \left(\frac{M_W^2}{2\pi v}\right)^2 \left[2 + \frac{(p\cdot q)^2}{M_W^4}\right] \delta^4(p+q-k)\,\frac{d^3p\,d^3q}{2k^0p^0q^0}. \tag{7.50}$$

For the present purposes, what is important in this last expression is the square bracket, in which the factor 2 comes from the contribution of the two transverse polarization states of the W meson while the $(p\cdot q)^2$ term gives the momentum-dependent contribution of the longitudinal spin state. The dependence on the gauge coupling is buried in the pre-factor, but can be read off using $M_W = \frac{1}{2}gv$ so that $M_W^2/(2\pi v) = g^2v/(8\pi)$, showing how the decay into transverse polarizations vanishes when $g \to 0$ with v fixed. But the decay to longitudinal photon polarizations does *not* vanish in this limit, since for this the M_W factors cancel, leaving a result proportional to $(p\cdot q)^2/(2\pi v)^2$. Rather than being proportional to g^2 this is instead suppressed by the derivative coupling characteristic of a Goldstone boson.

Performing the final-state integrations gives the standard expression for the total decay rate in the Higgs rest frame

$$\begin{aligned}\Gamma(h \to W^+W^-) &= \frac{m_H^3}{16\pi v^2}\left[1-4\left(\frac{M_W^2}{m_H^2}\right)+12\left(\frac{M_W^2}{m_H^2}\right)^2\right]\sqrt{1-\frac{4M_W^2}{m_H^2}} \\ &\approx \frac{m_H^3}{16\pi v^2} \quad \text{for } M_W \ll m_H,\end{aligned} \tag{7.51}$$

[14] An example of this was encountered in §4.2.3, which showed that the most dangerous couplings in the low-energy expansion came from the longitudinal spin-one spin states.

[15] The threshold for W decays is 160 GeV and for decays into Z is 180 GeV, whereas the Higgs mass was recently measured to be 125 GeV.

revealing Γ/m_H to be controlled in the limit $M_W \ll m_H$ by m_H^2/v^2 rather than M_W^2/v^2. This makes it sensitive to the Higgs self-coupling, $m_H^2/(16\pi v^2) = \lambda/8\pi$ – with λ defined by the Standard Model Higgs potential, *c.f.* Eq. (9.5) below – rather than the gauge coupling. This is as expected given that the longitudinal gauge coupling started life as part of the Higgs doublet, Φ, and indeed the leading part of (7.51) could be written down directly using only the interactions coming from the scalar sector by computing the rate for decay into a pair of Goldstone bosons. It is this ability to compute the high-energy interactions of longitudinally polarized gauge bosons in terms of the scalars that they've eaten that is known as the 'gauge-Goldstone equivalence theorem' [69].

7.3 Photons, Gravitons and Neutrinos

Photons and gravitons (and possibly some[16] neutrinos) may be the only particles that are massless, or very nearly so. They would be the only degrees of freedom to arise at extremely low energies within the vacuum sector[17] (and possibly within other sectors) of the Standard Model.

This makes the study of very-low-energy physics largely an effective theory of gravitons, photons and neutrinos. It is perhaps no surprise in this context that much of macroscopic physics boils down to electromagnetic or gravitational interactions. Neutrinos can also have practical low-energy implications, because their small masses and weak interactions give them a unique role within very dense environments, such as those found in astrophysics and cosmology.

The remainder of this section examines a few of the features of this very-low-energy world, with two goals in mind. One goal is to illustrate how much of what we know about these particles follows very naturally from general EFT considerations. The second goal is to use this EFT to illustrate a conceptual point about how in some circumstances the mass scales suppressing specific effective interactions can be surprisingly different from naive expectations.

7.3.1 Renormalizable Interactions ◇

The goal is to describe the effective theory below the electron mass, involving only the known particles light enough to be present in this energy range: gravitons, photons and neutrinos. These particles are, respectively, represented by a symmetric-tensor field, $h_{\mu\nu}(x) = h_{\nu\mu}(x)$, for the graviton (see §C.3.4), a vector potential, $A_\mu(x)$, for the photon (see §C.3.3) and N spinor fields, $\nu^a(x)$, with $a = 1, \ldots, N$ running over all sufficiently light neutrino species. As described in §A.2.3, without loss of generality the fields ν^a can be taken to be Majorana spinors.

[16] At most, one neutrino can be massless, to the best present knowledge, if there are only three neutrino species light enough to appear in neutrino-oscillation experiments.

[17] Of more practical interest for the world around us is the low-energy theory also including massive (but long-lived and slowly moving) macroscopic objects like planets, or the Sun or everyday macroscopic objects, depending on the application. These are included by examining sectors of the theory that carry a conserved charge, like net baryon number. More about this in Chapters 8, 12, 13 and Part IV.

As ever, the dominant behaviour at low-energies is described by the renormalizable part of the EFT, since this is unsuppressed by any powers of heavier masses. This lagrangian must be built from the above fields and (in the vacuum sector) must be invariant under Lorentz transformations, spacetime translations and the gauge symmetries appropriate for massless spin-one particles,

$$A_\mu(x) \to A_\mu(x) + \partial_\mu \zeta(x), \tag{7.52}$$

and massless spin-two particles,

$$h_{\mu\nu}(x) \to h_{\mu\nu}(x) + \partial_\mu \zeta_\nu(x) + \partial_\nu \zeta_\mu(x). \tag{7.53}$$

Here, $\zeta(x)$ and $\zeta_\mu(x)$ are, respectively, arbitrary scalar and vector functions. The neutrino fields are deliberately not assigned a transformation under (7.52) because this is what it means for them to be electrically neutral.

With these fields and symmetries, the most general renormalizable lagrangian possible is simply the free lagrangian, $\mathfrak{L} = \mathfrak{L}_2 + \mathfrak{L}_1 + \mathfrak{L}_{1/2}$, where $\mathfrak{L}_s$ describes the free lagrangian for the spin-s field. Explicitly, the Dirac action for the neutrinos is[18]

$$\mathfrak{L}_{1/2} = -\bar{\nu}^a \left(\not{\partial} + m_a \right) \nu^a \tag{7.54}$$

where there is an implied sum over neutrino type 'a' and m_a is the mass for each type. The spin-one term is the Maxwell action

$$\mathfrak{L}_1 = -\frac{1}{4} F_{\mu\nu} F^{\mu\nu} = -\frac{1}{2} \partial_\mu A_\nu (\partial^\mu A^\nu - \partial^\nu A^\mu), \tag{7.55}$$

and the spin-two term is

$$\begin{aligned} \mathfrak{L}_2 &= h^{\mu\nu} \left(\mathcal{R}_{\mu\nu} - \frac{1}{2} \eta_{\mu\nu} \eta^{\alpha\beta} \mathcal{R}_{\alpha\beta} \right) \\ &= -\frac{1}{2} \left[\partial^\alpha h^{\mu\nu} \partial_\alpha h_{\mu\nu} - 2\, \partial^\alpha h_{\alpha\mu} \partial_\beta h^{\beta\mu} + 2\, \partial^\alpha h_{\alpha\mu} \partial^\mu h^\beta_\beta - \partial^\mu h^\alpha_\alpha \partial_\mu h^\beta_\beta \right], \end{aligned} \tag{7.56}$$

which uses integrations by parts, and the definition

$$\mathcal{R}_{\mu\nu} = \mathcal{R}_{\nu\mu} := \frac{1}{2} \eta^{\alpha\beta} \left(\partial_\alpha \partial_\beta h_{\mu\nu} - \partial_\mu \partial_\alpha h_{\beta\nu} - \partial_\nu \partial_\alpha h_{\beta\mu} + \partial_\mu \partial_\nu h_{\alpha\beta} \right). \tag{7.57}$$

Eq. (7.56) is invariant – up to a total derivative – under (7.53) because $\mathcal{R}_{\mu\nu}$ is invariant and also satisfies the identity $\partial^\mu \left(\mathcal{R}_{\mu\nu} - \frac{1}{2} \eta_{\mu\nu} \eta^{\alpha\beta} \mathcal{R}_{\alpha\beta} \right) = 0$.

There is an interaction linear in A_μ like $\mathcal{L}_{\rm int} \propto A_\mu J^\mu$ that is possible in principle, where J^μ is built from the other fields, but invariance under (7.52) requires $\partial_\mu J^\mu = 0$. Such a current is not possible for Majorana neutrinos, but $J^\mu = \bar{\nu}^a \gamma^\mu \nu^a$ is conserved if ν^a is a complex (Dirac) field, since J^μ is then the conserved current associated with rotating ν^a by a phase. (The same rotation is inconsistent with the reality condition for Majorana neutrinos.) But even for complex neutrinos, identifying phase rotations with the transformation (7.52) amounts to giving neutrinos an electric charge, which they do not have. A similar interaction like $h_{\mu\nu} T^{\mu\nu}$ is also ruled out on similar grounds since (7.53) requires $\partial_\mu T^{\mu\nu} = 0$, and the only conserved symmetric tensor built from neutrinos has dimension 4, making the coupling non-renormalizable.

[18] For the present vacuum-sector purposes it suffices to work in an expansion around flat Minkowski space.

The absence of possible renormalizable interactions partly explains why gravitons, photons and neutrinos are found to be so weakly interacting. At low energies there is no low-energy interaction available for them to have.

7.3.2 Strength of Non-renormalizable Interactions ◊

Since all interactions for these particles must be non-renormalizable, their strength is controlled by the size of the mass scale appearing in the effective couplings. The next question is what scale this should be.

For gravitons, photons and neutrinos it turns out that these scales are extremely different. Earlier sections show that photons experience interactions, like those of (7.16), suppressed by coefficients of order α^2/m_e^4 and α^2/M_μ^4 (and similar terms involving the masses of all possible charged particles whose integrating out might generate the loop). The appearance of $m_e \simeq 5\times10^{-4}$ GeV in the dominant term seems reasonable because this is the mass of the lightest such particle to be integrated out.

Neutrino interactions turn out generically to be much weaker than this since they are suppressed by powers of the Fermi constant, G_F, whose size is of order α/M_W^2, where $M_W \simeq 80$ GeV is the mass of the W boson. Gravitational interactions are even weaker, since they are suppressed by powers of the Planck mass:[19] $M_p \simeq 2\times10^{18}$ GeV. The Planck mass is related to Newton's constant of universal gravitation, G_N, by $8\pi G_N = 1/M_p^2$ in fundamental units.[20]

In the same way that $G_F^{-1/2} \sim 300$ GeV and $M_W \sim 80$ GeV are not exactly the same size (due to factors of α), $M_p = (8\pi G_N)^{-1/2}$ need not be the precise scale where new degrees of freedom enter that change how gravity behaves at high energies. It may instead be that $G_N \simeq g^2/M_g^2$ for some dimensionless coupling g and a new physical scale[21] M_g. But to the extent that $g \lesssim 1$ in this new sector then $M_g \lesssim M_p$, so the Planck mass is an upper bound for the scale where unknown physics is likely to intervene.

This hierarchy of scales illustrates an important point about effective field theories, and the low-energy limit of complicated systems. Earlier sections have argued that the generic size to be expected for non-renormalizable effective couplings is set by the mass of the lightest particle whose removal generates the EFT of interest. If this is so, why are low-energy graviton and neutrino interactions not swamped by contributions suppressed only by powers of m_e?

The key word in the previous paragraph's summary of earlier sections is 'generic'. *All other things being equal*, it is true that the size of a generic interaction in the EFT below the electron mass is set by the appropriate power of m_e (rather than a heavier mass, like M_μ or M_W, say). But in the case of neutrinos, we know that all other

19 On one hand, the observation that gravity is so weak explains why astrophysical objects are so large (they must be large in order to be massive enough for gravity to compete with other interactions). On the other hand, it is an unsolved problem *why* M_p should be so much larger than other known scales (more about this in Chapter 9.).

20 Strictly speaking, the factor of 8π makes this the 'reduced' Planck mass, as opposed to simply defining $G_N = 1/M_p^2$. The reduced Planck mass is more widely used by those toiling away in gravity sweatshops worldwide.

21 In string theory, for example, M_g would be the mass of the lightest new excitation mode of the fundamental string (called the string scale, M_s), as described in Chapter 10.

things are not equal: the only renormalizable couplings involving neutrinos in the Standard Model turn out to be its couplings to the W boson – as given in (7.5), for example – and a similar coupling to the electrically neutral Z boson (whose mass $M_Z \simeq 90$ GeV is not that different from $M_W \simeq 80$ GeV). Once the W and Z particles are integrated out, the neutrino only experiences non-renormalizable couplings, the largest of which are those of the weak interactions (whose strength is set by $G_F \sim \alpha/M_W^2$).

The suppression of neutrino interactions by factors of G_F then continues to lower and lower energies as lighter and lighter particles (such as the electron and muon) are integrated out. Although these lighter particles can generate non-renormalizable interactions for the photon (to which they have renormalizable couplings), they cannot change the fact that all neutrino interactions are suppressed by at least one power of G_F.

This understanding of the hierarchy between low-energy neutrino and photon interactions suggests a similar interpretation for the even greater suppression of graviton interactions. The hierarchy of interactions would make sense if the graviton also were to experience renormalizable interactions with other particles, but only at a scale M_g much larger than those now experimentally accessible (see Fig. 7.6). If this were true then the absence of renormalizable interactions involving gravitons

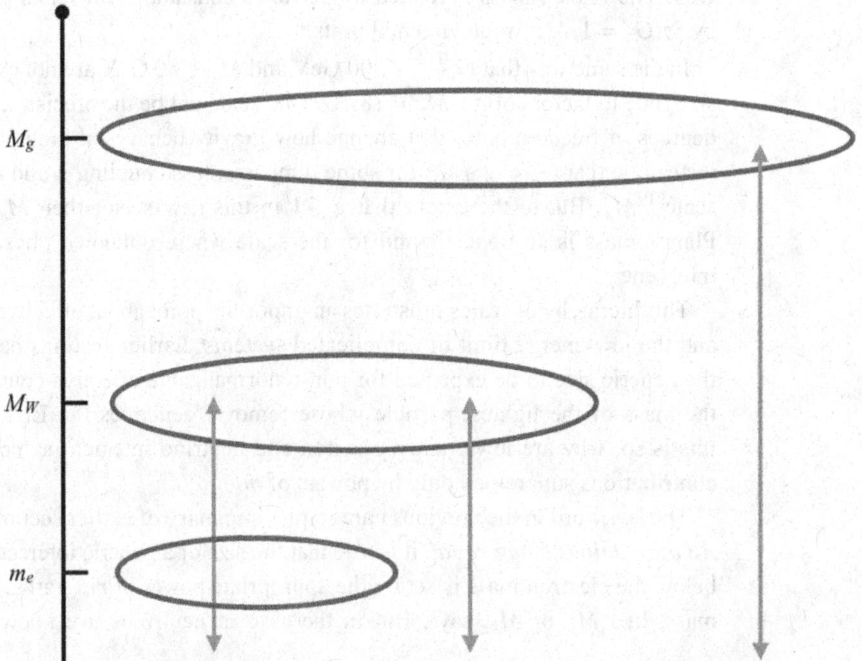

Fig. 7.6 Schematic of the energy scales and couplings responsible for the hierarchy of interactions among gravitons, photons and neutrinos. Here the ovals represent the collection of particles at a given energy that experience renormalizable interactions with one another. Three such circles are drawn, for energies at the electron mass, m_e, the W-boson mass, M_W, and a hypothetical scale, M_g, for whatever theory (perhaps string theory) describes gravity at very high energies.

at energies below M_g would mean all of their low-energy interactions must remain suppressed by at least one power of $g^2/M_g^2 \sim 1/M_p^2$, where g is the strength of the high-energy renormalizable couplings. This suppression would not be changed by the process of integrating out particles at still-lower energies. From this point of view it is perhaps less surprising that the low-energy field with the weakest couplings is also the one for which it is impossible to write down renormalizable interactions.

7.3.3 Neutrino-Photon Interactions ♣

Though it is an attractive (and basically correct) perspective, the above argument also turns out to be a bit too glib. This is because it comes with an important caveat: although it is true that *some* of the dimensions of low-energy effective neutrino couplings must be given by G_F, it is not true that *all* of them must come from powers of $1/M_W$. That is, an effective neutrino coupling of dimension (mass)$^{-n}$ need not be as small as M_W^{-n}; it could instead arise with size $M_W^{-2}m^{2-n}$ for $m \ll M_W$. The same is also true for gravitons: a higher-dimension effective graviton coupling of dimension (mass)$^{-n}$ can be order $M_p^{-2}m^{2-n}$ for $m \ll M_p$ rather than M_p^{-n} (examples of this for gravity are described in §10).

This section provides an explicit example of this phenomenon using the effective interactions of photons and neutrinos within the EFT below the electron mass. The example in question has operator dimension greater than six and so the effective coupling has dimension (mass)$^{-2(n+1)}$ with $n > 0$. Although two powers must come from G_F, the remaining scales are provided by the electron mass, leading to a coupling of order G_F/m_e^{2n} rather than the much smaller naive estimate, G_F^{n+1}.

The fact that low-energy neutrino interactions can be suppressed by powers of m_e rather than just M_W can lead to some surprises.[22] In particular, it turns out that $2 \to 2$ processes, like $\nu\overline{\nu} \to \gamma\gamma$ and $\nu\gamma \to \nu\gamma$ [141, 142], can be smaller than reactions involving more photons, like $\nu\overline{\nu} \to 3\gamma$ and $\nu\gamma \to \nu\gamma\gamma$ or $\overline{\nu}\nu \to 5\gamma$ [143, 144], even at the keV – MeV energies relevant to astrophysical applications. This is counter-intuitive because each additional photon costs a power of electromagnetic coupling and so, all other things being equal, the likelihood of coupling to three photons should be smaller than the likelihood of coupling just to two. Instead, the cross section for two-photon processes at low centre-of-mass energy E turns out to be of order $\sigma(2 \to 2) \sim G_F^4 E^6$, while those for three-photon processes instead behave as $\sigma(2 \to 3) \sim (\alpha/4\pi)^3 G_F^2 E^{10}/m_e^8$. At energies $E \sim 10$ keV the factor $(G_F E^2)^2 \simeq 10^{-30}$ makes the $2 \to 2$ processes much smaller than the $2 \to 3$ processes, which are suppressed only by $(\alpha/4\pi)^3 \sim 10^{-10}$ and $(E/m_e)^8 \sim 10^{-14}$.

EFT at the Weak Scale

To understand this low-energy dependence within the EFT first calculate the size of neutrino interactions just below the W-boson mass. At tree level the result is as found in §7.1, with both W and Z exchange combining to give the following photon and neutrino interactions

[22] These are conceptual surprises rather than practical ones; the cross sections involved are too small to be relevant in any known observable process.

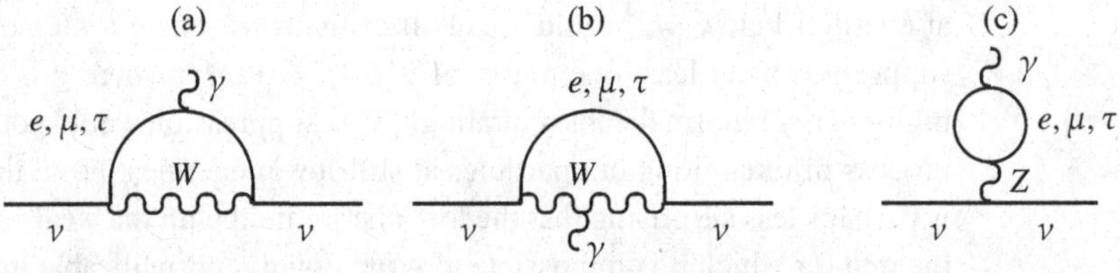

Fig. 7.7 Feynman graphs giving neutrino-photon interactions in the Standard Model. Graph (a) (left panel): contributions that can be regarded as low-energy renormalizations of the tree-level weak interaction. Graph (b) (middle panel): contributions generating higher-dimension interactions when integrating out the *W*. Graph (c) (right panel): contributions obtained when integrating out the *Z*. Although not labelled explicitly, quarks can also contribute to the loop in panel (c). Similar graphs with more photon legs contribute to neutrino/*n*-photon interactions.

$$\mathfrak{L}_{\rm wk}^{\rm tree}(\mu = M_W) = e\,A_\mu\,J^\mu_{\rm em} + \frac{G_F}{\sqrt{2}}\left(\mathrm{i}\overline{\nu}^a\gamma_\mu\gamma_L\nu^b\right)L^\mu_{ab} + O\left(\frac{1}{M_W^4}\right), \tag{7.58}$$

where there is an implied sum over the neutrino flavour indices, a, b, (and over charged-lepton indices $i, j = e, \mu, \tau$ in later expressions). The charged-lepton dependence in the above lagrangian is given by

$$L^\mu_{ab} = \mathrm{i}\overline{\ell}^i\gamma^\mu\left(v_{ab\,ij} + a_{ab\,ij}\gamma_5\right)\ell^j, \qquad J^\mu_{\rm em} = -\mathrm{i}\overline{\ell}^i\gamma^\mu\ell^i. \tag{7.59}$$

where the effective couplings, $v_{ab\,ij}$ and $a_{ab\,ij}$, as found from tree-level matching are given by:

$$v_{ab\,ij}(\mu = M_W) = U^*_{ja}\,U_{ib} + \delta_{ab}\,\delta_{ij}\left(-\frac{1}{2} + 2s_w^2\right) \tag{7.60}$$

$$\text{and} \qquad a_{ab\,ij}(\mu = M_W) = U^*_{ja}\,U_{ib} - \frac{1}{2}\,\delta_{ab}\,\delta_{ij},$$

where $s_w = \sin\theta_w$ is the sine of the weak mixing angle and U_{ia} denotes the PMNS matrix defined in §7.1.1. The terms involving U_{ia} in (7.60) arise due to W-boson exchange, while the others come from Z-boson exchange.

No direct couplings involving just neutrinos and photons arise in these tree-level effective interactions. Direct couplings do arise once loops are included, of which the leading graphs of interest for neutrino-photon scattering within the Standard Model are shown in Fig. 7.7. The next few paragraphs summarize their size, working in the limit of vanishing neutrino mass since the known neutrino masses cannot be larger than ~ 1 eV.

In the massless limit the graphs of Fig. 7.7 preserve neutrino helicity and so can only generate effective interactions for which an odd number of Dirac matrices appear between $\bar{\nu}^a$ and ν^b. Consider first panels (a) and (b) of Fig. 7.7. The lowest-dimension effective interactions with this Dirac-matrix structure and involving one or two electromagnetic fields (which must appear through the field strength $F_{\mu\nu}$ on grounds of gauge invariance, (7.52)) are

$$\mathfrak{L}^{\rm eff}_{\nu 2\gamma} = C^{(1)}_{ab}\,M^{ab}_{\mu\nu}\,F^{\mu\nu} + C^{(2)}_{ab}\,M^{ab}_{\mu\nu}\,F^{\mu\lambda}F_\lambda{}^\nu, \tag{7.61}$$

where $C^{(1)}_{ab}$ and $C^{(2)}_{ab}$ are dimensionless coefficients and

$$M^{ab}_{\mu\nu} := \mathrm{i}\overline{\nu}^a\gamma_\mu\gamma_L\partial_\nu\nu^b - \mathrm{i}\partial_\nu\overline{\nu}^a\gamma_\mu\gamma_L\nu^b. \tag{7.62}$$

The derivatives in (7.62) must act antisymmetrically because the symmetric combination is a total derivative, $\partial_\nu(\overline{\nu}^a\gamma_\mu\gamma_L\nu^b)$, for which both terms in (7.61) are redundant operators. Their redundancy can be seen by integrating by parts and using the Maxwell equations, $\partial_\mu F^{\mu\nu} = 0$ and $\partial^\mu F^{\nu\lambda} + \partial^\nu F^{\lambda\mu} + \partial^\lambda F^{\mu\nu} = 0$, together with $\not{\partial}\nu^a = 0$ – which also implies that $\partial^\mu(\bar{\nu}^a\gamma_\mu\gamma_L\nu^b) = 0$. The left-handed projection matrix, $\gamma_L = \frac{1}{2}(1+\gamma_5)$, appears in (7.62) because only left-handed fermions couple to the W boson in the Standard Model. The necessity of having the neutrino differentiated in this interaction – and the resulting suppression of low-energy neutrino-photon scattering – follows from neutrino helicity conservation for the first term, while for the second term it follows from Yang's theorem [145], which prohibits coupling two photons to a state of angular momentum one.

Chirality conservation implies that the leading contributions of panel (c) of Fig. 7.7 must also take the form of Eqs. (7.61) and (7.62), and as a result this graph does not contribute at all to the coefficients $C^{(1)}_{ab}$ and $C^{(2)}_{ab}$. This is because the neutrino fields in (c) necessarily arise in the neutral-current combination, $J^{ab}_\mu = i\overline{\nu}^a\gamma_\mu(A + B\gamma_5)\nu^b$ found in the tree level interactions, (7.58) and (7.59). But all effective operators at this dimension built from this current and its derivative plus one or two electromagnetic field strengths are redundant by the same arguments made in the previous paragraph.

In the limit where the mass of the charged lepton in the loop can be neglected (as appropriate for its UV contribution), the two-photon coefficient obtained by matching to the EFT using the graphs given above turns out to be

$$C^{(2)}_{ab}(\mu) = \frac{2\sqrt{2}\,\alpha\, G_F}{\pi M_W^2}\left[1 + \frac{4}{3}\log\left(\frac{M_W^2}{\mu^2}\right)\right]\delta_{ab}, \tag{7.63}$$

which uses the unitarity of the PMNS matrix: $\sum_i U^*_{ia}U_{ib} = \delta_{ab}$. This dimension-six interaction arises at order α^2/M_W^4 when the W boson is integrated out, as naively expected.

Integrating out the Charged Leptons

To identify where factors of lighter masses like $1/m_e$ or $1/M_\mu$ might enter into neutrino-photon scattering, consider integrating out these lighter fermions one by one, eventually obtaining the EFT below the electron mass. The corresponding graphs required to do so within the EFT below the W-boson mass are shown in Fig. 7.8. The sum of the graphs in Fig. 7.8 should agree with those of Fig. 7.7 to the first few orders in $1/M_W^2$, and the matching process chooses the coefficients of the effective interactions in the EFT below M_W to make this so.

In particular, the effective interactions shown as fat dots in panels (b) and (c) of Fig. 7.8 represent the leading direct neutrino-photon interactions described in (7.61) and (7.62), above. In particular, these come already suppressed by four powers of $1/M_W$ in the two-photon case due to the matching result (7.63) when integrating out the W boson. By contrast, the effective interaction appearing in the graph of panel (a) is simply the tree-level Fermi interaction, and so it is this graph that gives the contribution to neutrino-photon scattering suppressed by only one power of $1/M_W^2$.

What is important is that this graph – *i.e.* panel (a) of Fig. 7.8, with one or two external photons – does not contribute at all to the coefficients $C^{(1)}_{ab}$ or $C^{(2)}_{ab}$ of (7.61). It cannot do so because the neutrinos in this graph appear only through

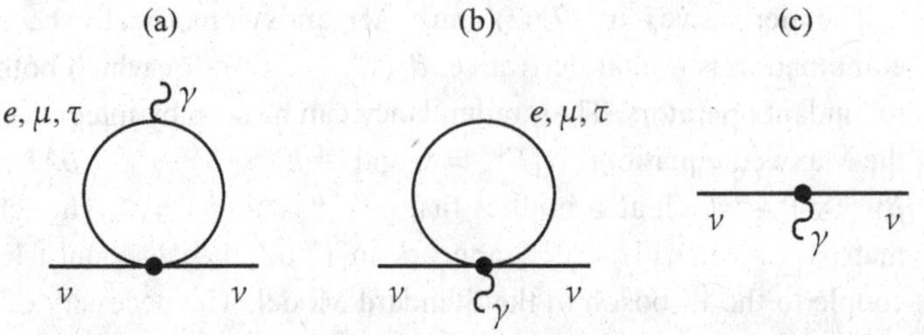

Fig. 7.8 Feynman graphs giving neutrino/single-photon interactions within the EFT below M_W. Graph (a) (left panel): loop corrections to the tree-level Fermi interaction. Graph (b) (middle panel): loop corrections to tree-level higher-dimension effective four-fermion/one-photon interactions. Graph (c) (right panel): loop-generated higher-dimension effective two-fermion/one-photon interactions. Similar graphs with more photon legs describe multiple-photon interactions.

their contribution to the Fermi current J^{ab}_μ and its derivative, much as was true for panel (c) of Fig. 7.7. Consequently, all its potential contributions to effective $\bar{\nu}\partial\nu F$ and $\bar{\nu}\partial\nu F^2$ interactions are redundant. As a result, the dominant contribution to low-energy $2 \to 2$ photon-neutrino scattering amplitudes remains suppressed by four powers of M_W, even after integrating out the charged leptons.

The next step is to identify what the dominant low-energy contribution is arising from panel (a) of Fig. 7.8 (and its counterparts with more external photons) since this can be suppressed only by two powers of $1/M_W$. To identify terms where the rest of the suppression comes from powers of $1/m_e$, the focus is on when it is electrons that circulate within the loop. This contribution to the effective neutrino-photon interaction lagrangian can be obtained by evaluating

$$\mathfrak{L}^{\rm eff}_{\nu n\gamma}(\mu = m_e) = \frac{G_F}{\sqrt{2}}\left(\mathrm{i}\bar{\nu}_a\gamma_\mu\gamma_L\nu_b\right)\left(v_{abee}\left\langle\bar{\psi}\,\gamma^\mu\,\psi\right\rangle + a_{abee}\left\langle\bar{\psi}\,\gamma^\mu\gamma_5\psi\right\rangle\right), \quad (7.64)$$

where ψ is the electron field and $\langle X^\mu\rangle$ represents the expectation of the operator X^μ, obtained by integrating out the electrons, weighted by the QED lagrangian

$$\langle X^\mu\rangle = \int \mathcal{D}\psi\mathcal{D}\bar{\psi}\; X^\mu\; \exp\left[\mathrm{i}\int d^4x\left(\mathfrak{L}_{\rm kin} - \mathrm{i}e\,A^\mu\,\bar{\psi}\,\gamma^\mu\,\psi\right)\right]. \quad (7.65)$$

Our interest is in how quantities like $\langle\bar{\psi}\gamma^\mu\psi\rangle$ and $\langle\bar{\psi}\gamma^\mu\gamma_5\psi\rangle$ depend on the external electromagnetic fields. Since the electromagnetic interactions preserve parity (P) and charge conjugation (C), these symmetries may be used to further organize the contributions to $\mathfrak{L}_W$. In particular, C and P invariance imply that any term in $\mathfrak{L}_W$ involving an odd power of $F_{\mu\nu}$ receives contributions only from the vector current, $\left\langle\bar{\psi}\gamma^\mu\,\psi\right\rangle$, while those involving even powers of $F_{\mu\nu}$ arise purely from the axial current, $\left\langle\bar{\psi}\gamma^\mu\gamma_5\,\psi\right\rangle$.

Since the leading low-energy couplings to two photons are suppressed by at least four powers of $1/M_W$, focus now on the low-energy couplings to three photons. This requires evaluating the expectation of the vector current, $\langle\bar{\psi}\gamma^\mu\,\psi\rangle$, at cubic order in the electromagnetic field. But because the vector current is also the electromagnetic current for the electron-photon effective theory, its expectation may be expressed in

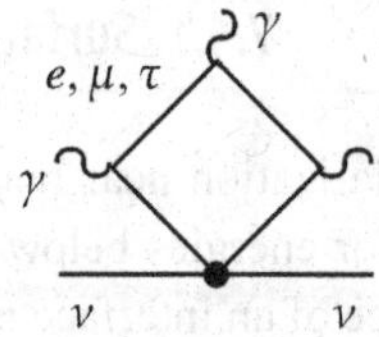

Fig. 7.9 Feynman graph showing how the light-by-light scattering box diagram appears in the 2 ⟶ 3 neutrino-photon scattering problem. The dot represents the tree-level Fermi coupling, though C and P invariance of electromagnetic interactions imply only the vector part need be used.

terms of the Euler-Heisenberg effective lagrangian, $W_{EH}[A]$, obtained in (7.23) for photon-photon scattering below m_e: $\left\langle \overline{\psi}\gamma^\mu \psi \right\rangle = (1/e) \left(\delta Z/\delta A_\mu \right)$ where

$$Z[A] = e^{iW_{EH}[A]} = \int \mathcal{D}\psi\mathcal{D}\overline{\psi} \; \exp\left[i \int d^4x \left(\mathfrak{L}_{\rm kin} - ie\, A^\mu\, \overline{\psi}\, \gamma^\mu\, \psi \right) \right]. \qquad (7.66)$$

As shown in (7.23), the leading term in $W[A]$ is quartic in $F_{\mu\nu}$ and so the leading contribution to $\langle \overline{\psi}\gamma^\mu\psi \rangle$ is cubic and given by:

$$\mathfrak{L}^{\rm eff}_{\nu 3\gamma} = \frac{e\, v_{ab}\, \alpha}{90\pi\, m_e^4} \left(\frac{G_F}{\sqrt{2}} \right) \left[5\, (N^{ab}_{\mu\nu}\, F^{\mu\nu})(F_{\lambda\rho}\, F^{\lambda\rho}) - 14\, (N^{ab}_{\mu\nu}\, F^{\nu\lambda}\, F_{\lambda\rho}\, F^{\rho\mu}) \right], \qquad (7.67)$$

with $N^{ab}_{\alpha\beta} = \partial_\alpha \left(\overline{\nu}^a \gamma_\beta \gamma_L \nu^b \right) - (\alpha \leftrightarrow \beta)$ and[23]

$$v_{ab} := v_{ab\,ee}(\mu = m_e) = U^*_{ea}\, U_{eb} + \delta_{ab} \left(-\frac{1}{2} + 2s_w^2 \right). \qquad (7.68)$$

This appearance of the light-by-light scattering lagrangian occurs because of the contribution of the box diagram that appears within panel (a) of Fig. 7.8, once three external photons are involved (as shown in Fig. 7.9).

As advertised, this particular effective interaction is suppressed by only two powers of $1/M_W$, with the rest of the dimensions of the effective coupling being filled by the factor of $1/m_e^4$. The non-existence of similar terms with fewer photons is seen to be related to the non-existence of photon-photon scattering interactions involving fewer than four powers of $F_{\mu\nu}$.

7.4 Boundary Effects

Low-energy electromagnetic interactions also provide simple practical examples where total derivatives (and boundary terms) in an EFT cannot be neglected. Two examples are explored here: surface polarizations (such as are experienced by electromagnetic fields in the presence of conducting or dielectric boundaries) and their implications for Casimir energies. (§15.3 describes the closely related phenomenon of boundary currents in the context of Quantum Hall materials.)

[23] This expression also runs the values taken by the effective couplings $v_{abee}(\mu = m_e)$ and $a_{abee}(\mu = m_e)$ down from $\mu = M_W$, although it happens that $v_{abee}(\mu = m_e)$ does not run in this energy range because the current $\overline{\psi}\, \gamma^\alpha\, \psi$ is conserved, and so does not get renormalized.

7.4.1 Surfaces between Media

To study surface polarization near boundaries, consider again the EFT for electromagnetism defined for energies below m_e. The main new feature introduced in this section is the presence of an interface marking the transition between, say, the region exterior to a conductor or a dielectric and the description appropriate for its interior, with action

$$S_{\rm full} = \int_{\mathcal{M}_{\rm ext}} {\rm d}^4x\, \mathfrak{L}_{\rm ext} + \int_{S} {\rm d}^3x\, \mathfrak{L}_s + \int_{\mathcal{M}_{\rm int}} {\rm d}^4x\, \mathfrak{L}_{\rm int}.$$

In particular, the interest in this subsection is on instances where it is the surface action $\mathfrak{L}_s$ that influences the electromagnetic response outside of a macroscopic body, as an illustration of some of the issues described in §5.

Classical Electromagnetism

Consider first the case where all interactions are dropped if they are suppressed by powers of microscopic scales (like $1/m_e$); a limit found in §7.2.1 to correspond to Maxwell electromagnetism. In this case, both $\mathfrak{L}_{\rm int}$ and $\mathfrak{L}_{\rm ext}$ involve terms only quadratic in electromagnetic fields, which for rotation- and translation-invariant dielectric materials (that need not be Lorentz-invariant) has the form

$$\mathfrak{L}_{\rm diel} \simeq \frac{1}{2}\left(\varepsilon\, \mathbf{E}^2 - \frac{\mathbf{B}^2}{\mu}\right), \tag{7.69}$$

where ε and μ are the electric and magnetic dielectric constants. The Lorentz-invariant vacuum is the specific dielectric for which $\mu = \varepsilon = 1$ and so for objects sitting immersed in the vacuum one might expect $\mathfrak{L}_{\rm ext} \supset -\frac{1}{4}\, F_{\mu\nu}F^{\mu\nu}$.

Classical macroscopic currents can also appear in this effective theory, both in the interior and exterior of any materials and on the interface, S, where they represent surface charge or current distributions,

$$S_{\rm ext} = \int_{\mathcal{M}} {\rm d}^4x \left(-\frac{1}{4}F^{\mu\nu}F_{\mu\nu} + J^\mu A_\mu\right) \quad \text{and} \quad S_s = -\int_S {\rm d}^3x\, j^\mu A_\mu, \tag{7.70}$$

and something similar for $\mathfrak{L}_{\rm int}$. Here, $J^\mu = e\mathfrak{J}^\mu = \{\rho, \mathbf{J}\}$ denotes any classical test charges and currents external to the material and $j^\mu = \{\sigma, \mathbf{j}\}$ denotes the surface charge and current densities. As shown in Exercise 5.4, regarding S as the boundary of both $\mathcal{M}_{\rm int}$ and $\mathcal{M}_{\rm ext}$ (and using $\mathfrak{L}_{\rm int} = \mathfrak{L}_{\rm diel}$ from (7.69)) the boundary condition found by varying A_μ freely at the interface itself gives the standard relations that relate $\sigma = j^0$ and $\mathbf{j}$ to the jump in boundary values of the normal component of the electromagnetic displacement, $D_n := \mathbf{n} \cdot (\varepsilon\, \mathbf{E})$, and the tangential components of $\mathbf{H} = \mathbf{B}/\mu$. (The other components are dictated by continuity of $\mathbf{E}$ and $\mathbf{B}$.)

The conditions at the surface of a perfect conductor are similarly simple to phrase, to the extent that very good conductors do not support electromagnetic fields within their interiors. In this case, the vanishing of $\mathbf{E}$ and $\mathbf{B}$ within the interior makes $\mathfrak{L}_{\rm int}$ drop out of the variational condition on δA_μ on the interface S, making it behave instead like a bona-fide boundary for the exterior region. In this case, the condition obtained by varying Eq. (7.70) with respect to A_μ on the boundary $\partial\mathcal{M}_{\rm ext} = S$ becomes

$$n_\mu F^{\mu\nu} + j^\nu = 0 \qquad \text{on } \partial\mathcal{M}_{\rm ext}\,, \tag{7.71}$$

where n^μ is the outward-pointing normal from the point of view of $\mathcal{M}_{\rm ext}$, and so points into the material. These relate the surface charges, $\sigma = j^0$, to the normal part of the near-boundary electric field, E_n, and the surface current $\mathbf{j}$ to the tangent part of the exterior magnetic fields at the interface. Continuity implies that the other field components vanish at the interface, a condition that is expressed compactly as

$$\epsilon^{\mu\nu\lambda\rho} n_\nu F_{\lambda\rho} = 0 \qquad \text{on } \partial\mathcal{M}_{\rm ext}, \tag{7.72}$$

since this captures the vanishing of both $B_n = \mathbf{n}\cdot\mathbf{B}$ and $\mathbf{n}\times\mathbf{E}$. In practice, it is (7.72) that is the useful boundary condition for fields external to perfect conductors, with (7.71) instead to be read as determining the surface charge and currents induced by the external fields.

Effective Surface Interactions

So far, so good. But real conductors and dielectrics are not ideal: they need not be perfectly flat and any large-scale curvature they have may carry energy. Their surfaces might also be wrinkled rather than smooth over shorter distances and these fluctuations could influence nearby fields. Similarly, the bulk electromagnetic properties of the underlying materials do not turn on infinitely sharply at their surfaces; instead, there is a transition region that must be penetrated before bulk behaviour emerges as a good approximation.

What is important for the present purposes is the length scale, λ, for many of these effects is generally much smaller than the distances of interest when studying surrounding electromagnetic fields. Although in everyday examples these scales cannot be smaller than atomic dimensions, $\lambda \gtrsim r_{\rm atom} \simeq (\alpha\, m_e)^{-1}$, they are usually much larger; perhaps set by grain sizes in the material or other mesocopic scales.

Effective field theory is a natural language for their description to the extent that λ can be regarded as a microscopic scale whose detailed physics has been integrated out. In general, the localization of such effects at the surface means that they appear in S_s and not the parts describing the bulk (exterior or interior) physics. As usual, the dominant surface effects for long-distance applications should correspond to the lowest-dimension effective interactions, and the effects of these interactions can be worked out without studying the underlying microphysics in detail.

For instance, geometrical effects associated with the long-distance curvature of the surface may be parameterized in terms of the geometrical quantities that measure this curvature. These can generally be expressed in terms of the surface's intrinsic induced metric, $h_{ab} = \partial_a y^\mu \partial_b y^\nu \eta_{\mu\nu}$, and its derivatives for a surface located at $x^\mu = y^\mu(\sigma)$. Here, σ^a are parameters along the surface's world-volume. For instance, derivatives of the normal to the surface are encoded in the surface's extrinsic curvature tensor, K_{ab} (whose precise definition is not needed below), while derivatives of the intrinsic geometry appear through the Riemann curvature $R^a{}_{bcd}$ built from h_{ab} (see §A.2.1 for its definition).

Sample low-dimension terms in a surface's action might include

$$S_s \supset S_{\rm geom} = -\int_{\partial\mathcal{M}} {\rm d}^3x\, \sqrt{-h}\Big(s_0 + s_1 K + s_2 R + \cdots\Big), \tag{7.73}$$

where $K := h^{bd}K_{bd}$ has dimensions of (length)$^{-1}$, $h := \det(h_{bd})$ is dimensionless, while $R := h^{bd}R^a{}_{bad}$ has dimension (length)$^{-2}$. The tensor h^{ab} is the inverse matrix for h_{ab} so that $h^{ab}h_{bc} = \delta^a_c$. On dimensional grounds the effective couplings, s_k, should be proportional to successively higher powers of the relevant microscopic length scale, λ, with $s_0 \sim \lambda^{-3}$ an energy-per-unit-area, $s_1 \sim \lambda^{-2}$ a bending energy density and so on. (For a review with references of this kind of description for the statistical mechanics of two-dimensional membranes, see [146].)

Similarly, effective interactions describing the largest corrections to electromagnetic effects (but subdominant to the surface charges and currents) take the form,

$$S_s \supset S_{EM} = -\frac{1}{2}\int_{\partial\mathcal{M}} \mathrm{d}^3x\ \sqrt{-h}\left[c_1 E_n^2 + c_2 \mathbf{B}_{\|}^2 + c_3\ \mathbf{E}_{\|}^2 + c_4 B_n^2\right]. \tag{7.74}$$

The coefficients of these operators all should be of order $c_i \sim \lambda_{EM}$ on dimensional grounds, where λ_{EM} here is the medium's relevant electromagnetic length scale. More complicated combinations become possible if the surface breaks rotational symmetries parallel to the surface or has inhomogeneous properties.

Following the usual EFT story, the effective couplings s_i and c_i of (7.73) and (7.74) can either be regarded as phenomenological parameters (obtained by comparing with experiment) or as quantities to be matched to calculations within a calculable UV completion. It does not matter which of these is ultimately chosen when extracting the implications of terms in S_s for observables (an example involving Casimir energy is described in §7.4.2).

Sample Matching Calculation

Before exploring further how effective couplings for surface interactions contribute to observables, consider first an illustrative matching calculation to develop intuition about their size. This section uses QED as the UV theory to illustrate how integrating out electrons in the presence of boundaries can generate the electromagnetic interactions of (7.74). Although requiring only standard knowledge about the QED vacuum polarization – whose main results are briefly reviewed in §A.2.4 – this calculation is more of a pedagogical than a practical exercise in that the Compton wavelength associated with the virtual electrons being integrated out are impossibly small compared to the width of any practical boundary region.

To this end, consider the leading $1/m_e$ correction to surface interactions obtained by integrating out an electron loop starting from the UV completion of the previous sections,

$$S_{QED} = -\int_{\mathcal{M}} \mathrm{d}^4x\left[\frac{1}{4}F^{\mu\nu}F_{\mu\nu} + \overline{\psi}\,(\not{D} + m)\psi\right] + \int_{\mathcal{M}} \mathrm{d}^4x\ J^\mu A_\mu - \int_{\partial\mathcal{M}} \mathrm{d}^3x\ j^\mu A_\mu\,. \tag{7.75}$$

For simplicity (and for later comparisons with other calculations) the electrons are assumed not to acquire boundary conditions at the position of the interfaces, with just the electromagnetic field experiencing there the same perfect-conductor boundary conditions as described above.

The leading m_e-dependent contribution to the surface action, S_s, then arises from the vacuum polarization graph of Fig. 7.4, which describes how quantum fluctuations of the electron–positron field make the vacuum behave like a polarizable medium. A consequence of this polarization is that any would-be point charge gets smeared into a charge distribution over scales of order $\lambda_{EM} \sim m_e^{-1}$. An effective boundary interaction gets generated in the low-energy EFT because this is required to describe the short-range electron-cloud polarization caused by the surface charges induced on the conducting surface by the presence of any test charges nearby, as required by the boundary condition (7.71) [147].

To see this, explicitly evaluate Fig. 7.4 and compute the electron's contribution to the vacuum polarization. Because the electron field is assumed not to be directly affected by the boundary surfaces the result is the standard one from textbooks:

$$\Pi_{\mu\nu}(q^2) = \left(q^2\eta_{\mu\nu} - q_\mu q_\nu\right)\Pi(q^2) \tag{7.76}$$

with $\Pi(q^2)$ given after renormalization by (compare to Eq. (A.53))

$$\Pi(q^2) = \frac{2\alpha}{\pi}\int_0^1 du\, u(1-u)\log\left[1 + u(1-u)\frac{q^2}{m_e^2}\right]. \tag{7.77}$$

For comparisons with effective surface interactions it is more useful to have the position-space version of this result. In particular, the polarization implied for a given point-charge distribution, $Q\,\delta^3(\mathbf{x})$, is found by writing $\rho_{\text{eff}} = Q\left[\delta^3(\mathbf{x}) + \eta(\mathbf{x})\right]$, where

$$\begin{aligned}\eta(\mathbf{x}) &= \frac{1}{2}\int \frac{d^3q}{(2\pi)^3}\, e^{i\mathbf{q}\cdot\mathbf{x}}\,\Pi(\mathbf{q}^2)\\ &= \frac{\alpha}{\pi}\int_0^1 du\, u(1-u)\int\frac{d^3q}{(2\pi)^3}\, e^{i\mathbf{q}\cdot\mathbf{x}}\log\left[1 + \frac{\mathbf{q}^2}{m_e^2}u(1-u)\right]\\ &= N\,\delta^3(\mathbf{x}) - \frac{\alpha}{2\pi^2 r^3}\int_0^1 du\left(1 + \frac{m_e r}{\sqrt{u(1-u)}}\right)u(1-u)\exp\left[-\frac{m_e r}{\sqrt{u(1-u)}}\right].\end{aligned} \tag{7.78}$$

Here, $r = |\mathbf{x}|$ and N is a constant that renormalizes the bare charge Q, whose value is determined by the on-shell renormalization condition

$$\int d^3x\,\eta(\mathbf{x}) = \frac{1}{2}\Pi(\mathbf{q}^2 = 0) = 0. \tag{7.79}$$

Effective Description

For conducting surfaces what is important is that this polarization also applies to the surface charges associated with enforcing conducting boundary conditions there. The modified charge distribution caused by this polarization is obtained by integrating Eq. (7.78) over a planar sheet of charge, $\sigma\,\delta_+(z)$, where $\delta_+(z) = 2\delta(z)$ is normalized to integrate to unity on one side of the boundary: $\int_0^\infty dz\,\delta_+(z) = 1$.

The polarized charge distribution associated with a surface-charge sheet positioned at $z = 0$ then is

$$\rho(z) = \sigma\left\{(1+N)\,\delta_+(z) - \frac{2\alpha}{\pi|z|}\int_0^1 du\, u(1-u)\exp\left[-\frac{m_e|z|}{\sqrt{u(1-u)}}\right]\right\}. \tag{7.80}$$

The delta-function term here expresses how virtual electrons renormalize the bare surface charge distribution.

For matching purposes it is the second, exponential, part of (7.80) that matters because it generates detectable multipole moments around the uniform surface charge at $z = 0$. It is these moments that are replaced by effective interactions localized on the surface of the conducting plates once the electrons are integrated out. In particular, the electric dipole moment density, $\mathbf{d} = \rho(z)\, z\, \mathbf{e}_z$, implied by (7.80) is

$$\mathbf{d}(z) = -\frac{2\,\alpha\,\sigma\,\mathbf{e}_z}{\pi}\int_0^1 \mathrm{d}u\, u(1-u)\, \exp\left[-\frac{m_e|z|}{\sqrt{u(1-u)}}\right] \simeq -\frac{3\,\alpha\,\sigma\,\mathbf{e}_z}{64\,m_e}\,\delta_+(z), \tag{7.81}$$

where the approximate equality drops all but the leading power of $1/m_e$.

A matching observable can be the field energy, U, that arises when a test charge, Q, is placed in the vicinity of a conducting surface. At leading order (in the absence of vacuum polarization) any test charge placed near the conducting surface sets up an electric field whose normal component at the surface induces a nonzero charge density, σ, there, as dictated by Eq. (7.71). Virtual electrons then polarize the vacuum within a distance $1/m_e$ of both the test charge *and* this surface charge, leading to a change in the field energy.

To leading order this change is the sum of the interaction of the induced charge density σ with the polarization around the test charge, plus the interaction of the test charge Q with the induced polarization near the surface charge. Each of these effects has precisely the same size, leading to a correction of the interaction energy of size

$$\Delta U = 2 \times \left(\frac{1}{2}\right)\int \mathrm{d}^3x\; \mathbf{E}\cdot\mathbf{d} = -\frac{3\,\alpha}{64\,m_e}\int_{z=0} \mathrm{d}^2x\; \sigma\, E_n = -\frac{3\,\alpha}{64\,m_e}\int_{z=0} \mathrm{d}^2x\; E_n^2\,, \tag{7.82}$$

which uses the lowest-order result, Eq. (7.71), to write $\sigma = E_n$, where $\mathbf{E}$ is the lowest-order electric field, not including the vacuum-polarization corrections.

This is to be compared with the value for the effective coupling c_1 of Eq. (7.74) that reproduces energy shifts quadratic in E_n. The coefficient c_1 contributes to the field energy of a classical static test charge by an amount

$$\Delta U = -\frac{c_1}{2}\int_{z=0} \mathrm{d}^2x\; E_n^2 \tag{7.83}$$

and so comparing with Eq. (7.82) gives the matching result

$$c_1 = \frac{3\alpha}{32\,m_e}\,. \tag{7.84}$$

7.4.2 Casimir Energies ♠

To explore the implications of boundary interactions more concretely consider how some of the interactions in (7.74) can alter physical predictions. A simple observable that depends on the near-surface physics is the electrodynamic *Casimir energy* [148] of two parallel conducting plates separated by a distance a. From a macroscopic point of view the Casimir energy is associated with the response of quantum fluctuations to the presence of the conducting surfaces,[24] and because this response is driven by the presence of these surfaces it can be sensitive to any effective interactions that may reside there.

Casimir energies (more generally, Casimir stress-energies) are observable because of their dependence on external variables (like the distance between the plates). As seen explicitly below, they imply a force acting between the plates, and these forces have been measured [150]. Calculating how effective interactions like (7.74) alter predictions for Casimir energies is not entirely of academic interest, since modern measurements [151, 152] are becoming accurate enough to be sensitive to some subleading corrections described by such operators.

Parallel Conducting Plates

For the purposes of illustrating the method, it is simplest to specialize to perfect conductors through use of the conducting boundary conditions on $\partial \mathcal{M}_{\rm ext}$ given by (7.72). The leading part of the Casimir energy is then computed by evaluating the vacuum-expectation value of the usual Maxwell stress-energy density, $\langle 0|T_{\mu\nu}|0\rangle$, with

$$T_{\mu\nu} = F_{\mu\lambda}F_\nu{}^\lambda - \frac{1}{4}\, g_{\mu\nu}\, F_{\alpha\beta}F^{\alpha\beta}, \tag{7.85}$$

for the electromagnetic field. In particular, the time-time component of the result gives the energy density

$$\langle 0|T_{00}|0\rangle = \frac{1}{2}\langle 0|(\mathbf{E}^2 + \mathbf{B}^2)|0\rangle. \tag{7.86}$$

The matrix element is evaluated by inserting the expansion of A_μ in terms of creation and annihilation operators – see (C.39),

$$A_\mu(x) = \sum_{\lambda=\pm 1} \int \frac{\mathrm{d}^3\mathbf{k}}{\sqrt{(2\pi)^3 2\omega_k}} \left[\mathfrak{e}_\mu(\mathbf{k},\lambda;x)\,\mathfrak{a}_{k\lambda} + \mathfrak{e}^*_\mu(\mathbf{k},\lambda;x)\,\mathfrak{a}^*_{k\lambda}\right], \tag{7.87}$$

where $\omega_k = |\mathbf{k}|$ is the photon dispersion relation and $\mathfrak{e}_\mu(\mathbf{k},\lambda;x)$ are the mode functions found by solving Maxwell's equations for its two polarization states in the region between the plates, subject to the boundary conditions (7.72), and the creation- and annihilation-operators satisfy the algebra, $\left[\mathfrak{a}_{k\lambda}\,,\mathfrak{a}^*_{q\zeta}\right] = \delta_{\lambda\zeta}\delta^3(\mathbf{k}-\mathbf{q})$, together with $\mathfrak{a}_{q\zeta}|0\rangle = 0$. The answer acquires its dependence on boundary conditions through these mode-functions, $\mathfrak{e}_\mu(\mathbf{k},\lambda;x)$.

In general, the resulting expression involves an integration over mode momentum $\mathbf{k}$ which diverges for large $|\mathbf{k}|$. This UV divergence in general can be absorbed into renormalizations of effective couplings like a 'cosmological constant' (*i.e.* energy-density term in the bulk lagrangian density, $\mathfrak{L}_{\rm ext}$, as well as the effective couplings in $\mathfrak{L}_s$. This is overkill for the case of parallel flat plates, however, for which the

[24] Like any other EFT property, there is also a more microscopic – sometimes less illuminating – description directly in terms of the underlying atoms (in this case, from which the boundaries are made [149]).

divergence is independent of the separation, a, between the plates and so drops out once measurable a-dependent quantities are considered.

In what follows interest is in the difference between the energy at finite and infinite a and UV divergences cancel from this difference. Standard calculations [153] in this case (in the absence of the effective interactions of $\mathfrak{L}_s$) give the leading stress-energy for infinite parallel and perfectly conducting plates separated by a distance a as

$$\langle 0|T_{\mu\nu}|0\rangle = \frac{\pi^2}{720\, a^4}\left(\eta_{\mu\nu} - 4n_\mu n_\nu\right) = \frac{\pi^2}{720\, a^4}\begin{pmatrix} -1 & & & \\ & 1 & & \\ & & 1 & \\ & & & -3 \end{pmatrix}, \tag{7.88}$$

where the plates are chosen to be parallel to the $x - y$ plane and located at $z = 0$ and $z = a$, and n^μ is the unit vector normal to the plates (in their rest frame). The tensor structure of this result is dictated by the symmetries of the problem together with stress-energy conservation, $\partial^\mu \langle 0|T_{\mu\nu}|0\rangle = 0$ and the conformal-symmetry condition[25] $\eta^{\mu\nu}\langle 0|T_{\mu\nu}|0\rangle = 0$.

The energy-per-unit area associated with this stress tensor is

$$\varepsilon_c^{(0)} := \lim_{A\to\infty} \frac{1}{A}\int_A \mathrm{d}^3x\, \langle 0|T_{00}|0\rangle = \int_0^a \mathrm{d}z\, \langle 0|T_{00}|0\rangle = -\frac{\pi^2}{720\, a^3}, \tag{7.89}$$

where $A = \int_A \mathrm{d}^2x$ is the surface area of each plate and the plates are chosen to be parallel to the $x - y$ plane and located at $z = 0$ and $z = a$. The pressure (force-per-unit area) on either of the plates is similarly (by the definition of stress-energy) given by

$$p_c^{(0)} := \langle 0|T_{zz}|0\rangle = -\frac{\pi^2}{240\, a^4}, \tag{7.90}$$

which is negative, indicating attraction between the plates (in agreement with the result found by differentiating $\varepsilon_c^{(0)}$ and using the principle of virtual work).

Influence of Surface Operators

The boundary condition (7.72) is Lorentz-invariant in the directions parallel to the plates, and for simplicity suppose this is also true of the physics responsible for the effective boundary terms (7.74). This implies that they reduce to

$$S_s = -\frac{c_1}{2}\int_{\partial\mathcal{M}_{\rm ext}} \mathrm{d}^3x \left(E_n^2 - \mathbf{B}_\parallel^2\right). \tag{7.91}$$

In this case, the contribution of this interaction to the Casimir energy can be found in a variety of ways, such as by evaluating the vacuum-to-vacuum amplitude at very late euclidean times. A straightforward calculation using (7.91) corrects the energy-per-unit area of (7.89) to $\varepsilon_c = \varepsilon_c^{(0)} + \varepsilon_c^{(1)}$, with revised coefficient [154]

$$\varepsilon_c^{(1)} \simeq \frac{\pi^2 c_1}{240\, a^4}. \tag{7.92}$$

[25] This is one of those symmetries that has anomalies, but these vanish when evaluated for flat space in the absence of background electromagnetic fields.

Keeping in mind that $c_1 \sim O(\lambda_{EM})$ for a microscopic length-scale λ_{EM}, Eq. (7.92) represents the generic correction to ε_c due to surface-polarization effects that is of relative order λ_{EM}/a.

In the particular case where (7.84) is obtained by integrating out electrons, the matching condition (7.84) can be used, giving the leading Casimir shift due to vacuum polarization of the form

$$\varepsilon_c^{(1)} \simeq \frac{\pi^2 c_1}{240\, a^4} \simeq \frac{\pi^2 \alpha}{2560\, m_e\, a^4}. \tag{7.93}$$

This agrees with the correction found by explicitly evaluating the full one-loop correction to the Casimir energy and expanding the result to this order in $1/(m_e a)$ [155, 156].

7.5 Summary

For the most part, this section examines systems where the underlying UV theory is well-understood and calculable, making the calculation of the effective theory a convenience rather than a necessity. This shows how the general properties of effective theories arise in concrete and practical situations, and shows how low-energy behaviour is particularly transparent when viewed through an EFT lens.

The weak interactions are part of the foundations of EFT methods, with the Standard Model of electroweak unification providing an archetype for how non-renormalizable interactions (in this case the Fermi theory of weak interactions) emerge at low energies from underlying renormalizable physics. The dimensionful coupling G_F of the Fermi theory is given by powers of the mass and coupling of the W boson, $G_F \propto g^2/M_W^2$, whose renormalizable interactions are at work in the UV description. The realization that this is possible underlines that it should be possible to use non-renormalizable theories predictively.

As applied to neutrinos the weak interactions also teach other simple lessons about low-energy limits. Neutrinos illustrate how low-energy particle properties can preclude these particles from having any renormalizable interactions at all. (Gravitons furnish another real-life example of this.) Such particles are guaranteed to interact very weakly at low energies because all of their interactions are irrelevant (in the technical sense of §2.4.1).

Neutrinos and gravitons also show how hierarchies of interaction strengths can arise amongst non-renormalizable low-energy couplings. All other things being equal (as examples using QED show), it is the lightest UV particle mass that dominates in non-renormalizable couplings, but neutrinos and gravitons show how some (but not all) of the dimensions of effective couplings can be reserved for the potentially much higher scales where renormalizable interactions were last possible for the particle in question. In practice, this means that effective neutrino couplings with dimension $(\text{mass})^{-2(n+1)}$ need not be as small as G_F^{n+1}, but can instead arise with size G_F/m_e^{2n} where $1/m_e^2 \gg G_F$. Chapter §10 argues that similar statements also hold for gravitons.

Quantum electrodynamics provides the other examples explored in this chapter, with effective interactions throughout space and at simple boundaries (or interfaces between media) studied as electrons and muons are integrated out. Besides illustrating how the smallest masses dominate effective couplings at low energies, these also furnish examples where matching can be studied beyond the classical approximation. This study motivates the 'decoupling subtraction' renormalization scheme; a convenient framework for keeping decoupling explicit within renormalization-group running without losing the benefits of mass-independent schemes (like the various variations of minimal subtraction).

Exercises

Exercise 7.1 For the effective theory of QED defined below the electron mass, what is the complete list of dimension-10 interactions, $\mathfrak{L}_{10}$, that should appear in Eq. (7.14) assuming unbroken C, P and T symmetry? Remove all redundant interactions (assuming that boundaries and topology play no role). What is the lowest-dimension effective interaction in this EFT that breaks P and/or CP?

Exercise 7.2 Perform the calculation that gives the values of the effective couplings b_1 and b_2 by matching the evaluation of Fig. 7.3 in the effective theory to the low-energy limit of the box graph, Fig. 7.5, in QED (for electrons circulating in the loop). Verify in this way the validity of Eq. (7.22).

Exercise 7.3 Compute the vacuum polarization graph of Fig. 7.4 and verify using it the evolution equations (7.45) and solution (7.46) when integrating out electrons and muons in decoupling subtraction. Extend these results using also the τ lepton, u, d, s c and b quarks (don't forget that quarks come with three colours each; for the masses and charges of these particles use the Particle Data Group particle summary tables found at http://pdg.lbl.gov). What is your prediction for the value of $1/\alpha$ evaluated just below $\mu = M_W$ if you start at $\mu = m_e$ with $1/\alpha(m_e) = 137.035999074(44)$? How much of the difference between $1/\alpha(m_e)$ and $1/\alpha(M_W)$ is due to the contribution of quarks? Is it correct simply to evaluate the running of α between 100 MeV and $M_W \simeq 80$ GeV simply by summing over the contribution of virtual quarks? If not, why not and how might you do better?

Exercise 7.4 Prove there are no renormalizable interactions for the graviton field, $h_{\mu\nu}$, in Minkowski spacetime that are invariant under the symmetry transformation of Eq. (7.53) for arbitrary ζ_μ. Take renormalizable here to mean interactions for which the couplings have canonical dimension that is a non-negative power of mass.

Exercise 7.5 Expand the Einstein–Hilbert action of General Relativity – see Eqs. (C.100), (C.95) and (C.91) – about Minkowski space with $g_{\mu\nu} = \eta_{\mu\nu} + 2\kappa h_{\mu\nu}$ and $\kappa^2 = 8\pi G_N$. Verify that the term quadratic in $h_{\mu\nu}$ is given by Eq. (7.56).

Exercise 7.6 Verify that the neutrino part of the charged-current Fermi lagrangian given in Eqs. (7.9) and (7.10) can be rewritten in the form given by the terms in Eqs. (7.58), (7.59) and (7.60) that involve the matrix U_{ia}. To prove the equivalence requires using *Fierz identities* that are derived by expanding the spinor bilinear $\psi_i \bar{\psi}_j$, regarded as a 4×4 matrix in spinor-space, in terms of the basis of sixteen Dirac matrices given in (A.33).

The analog of Eq. (7.5) for the leptonic couplings of the Z boson is given by

$$\mathfrak{L}_{nc} = \frac{ig}{2\cos\theta_w} Z_\mu \left[(\bar{\nu}^a \gamma^\mu \gamma_L \nu_a) + [\bar{\ell}^j \gamma^\mu (-\gamma_L + 2\sin^2\theta_w)\ell_j] \right],$$

with an implied sum over neutrino and lepton flavour labels a and j. The Z is electrically neutral (it is its own antiparticle) and its mass is related to the W mass by $M_W = M_z \cos\theta_W$. Show that at low energies the exchange of Z bosons gives rise to the remaining 'neutral-current' terms of Eqs. (7.58), (7.59) and (7.60).

Exercise 7.7 Taking as given the low-energy light-by-light scattering lagrangian given in Eq. (7.23) and using the W-lepton couplings of (7.58), show that the Feynman graph of Fig. 7.9 leads to expression (7.67) for the leading low-energy effective coupling of neutrinos to three photons.

Exercise 7.8 Compute the contribution of the boundary couplings of Eq. (7.74) to the Casimir energy of electromagnetic fields between two parallel perfectly conducting plates, and in particular thereby verify Eq. (7.92).

8 QCD and Chiral Perturbation Theory

Much of the utility of effective theories is most fully exploited in situations where the UV theory is more poorly known, or difficult to work with (if known). The remaining chapters in Part II aim to describe several examples of this type, keeping for now to systems with relativistic kinematics.

8.1 Quantum Chromodynamics ♠

In many ways low-energy Quantum Chromodynamics (QCD) is the poster child for EFT techniques. This is because this is the area where much of the systematic treatment of Goldstone bosons was first developed. Before describing this development a brief summary of QCD and the strong interactions is in order.

8.1.1 Quarks and Hadrons

The strong interactions are the strongest interactions so far seen in nature, and were discovered when exploring what holds protons and neutrons together into atomic nuclei. According to the Standard Model (for more about which see §9), nuclei are only one piece of the strong-interaction story since the protons and neutrons themselves are understood to be composites built from point-like quarks and gluons. In the Standard Model all other strongly interacting particles – collectively called hadrons – are also postulated to be composites built from quarks, antiquarks and gluons [157–160], and QCD is the interaction believed to be responsible for binding them together. There is considerable experimental evidence that this picture is correct.

There is a strong analogy between QCD and QED, with massless spin-one gluons playing the role for QCD of photons in QED. In QCD the notion of electric charge is replaced by a new type of 3-valued charge called colour, with each species of quark coming in three variants, each carrying one of the three colours [161–163]. There are eight types of spin-one gluons that couple to colour in much the same way that photons couple to electric charge [164]. The big difference relative to electrodynamics is that the gluons themselves also carry colour (this is why there are eight of them) and so interact amongst themselves even in the absence of quarks.

More precisely, each of the six types (or flavours) of quark – for historical reasons called up, down, strange, charm, bottom and top (see Table 8.1) – comes in three

Table 8.1 Quark properties

	Up-type			Down-type		
	u	c	t	d	s	b
Elec. Charge (e)	+2/3	+2/3	+2/3	−1/3	−1/3	−1/3
Mass (GeV)	0.002	1.3	175	0.005	0.1	5

possible colours – red, yellow and blue, say. For instance, the three colours of up-type quark are represented by a Dirac spinor field, u^a, with $a = 1, 2, 3$, while the three kinds of strange quark would be described by s^a and so on. Sometimes, all six quark flavours are collectively denoted by the Dirac spinor field q^{an}, where the new index $n = 1, \ldots, 6$ labels quark types u through t. The eight types of gluon are similarly represented by eight gauge potentials G^α_μ where $\alpha = 1, \ldots, 8$ (while μ is a 4-vector index like for the electromagnetic potential, A_μ).

The (renormalizable) lagrangian for QCD is constructed from these fields and takes the deceptively simple form (see Appendix A.2 for a summary of conventions)

$$\mathfrak{L}_{QCD} = -\frac{1}{4} G^\alpha_{\mu\nu} G^{\mu\nu}_\alpha - \overline{q}(\not{D} + m)q, \tag{8.1}$$

where the gauge field-strength in this case is given by

$$G^\alpha_{\mu\nu} := \partial_\mu G^\alpha_\nu - \partial_\nu G^\alpha_\mu + g_s \, c^\alpha{}_{\beta\gamma} \, G^\beta_\mu G^\gamma_\nu = -G^\alpha_{\nu\mu}, \tag{8.2}$$

where g_s is a new dimensionless parameter called the QCD coupling constant and the 'structure constants' $c^\alpha{}_{\beta\gamma} = -c^\alpha{}_{\gamma\beta}$ are described in more detail below. The Einstein summation convention is in effect, so there is an implied sum over repeated indices like α in (8.1) and β and γ in (8.2). As before $\not{D} = \gamma^\mu D_\mu$ but now the covariant derivative of the quark field is given by

$$D_\mu q := \partial_\mu q - \mathrm{i} g_s \, G^\alpha_\mu (T_\alpha \, q), \tag{8.3}$$

where T_α are a basis of eight 3×3 traceless and hermitian matrices which matrix multiply the q's regarded as 3-component objects in colour space. Finally, the quark mass-matrix m is a 6×6 diagonal matrix in flavour space, whose diagonal elements are the quark masses, $m_u, \cdots, m_t$.

The previous formulae use a convenient and powerful matrix notation that packs a lot of information. In particular, to eliminate clutter they suppress both the Dirac spinor indices (as was also done earlier for electromagnetism) and the colour and flavour indices on both $\overline{q}_{an}$ and q^{an}. Writing flavour and colour indices in full converts (8.3) into

$$[D_\mu q]^{an} := \partial_\mu q^{an} - \mathrm{i} g_s \, G^\alpha_\mu (T_\alpha)^a{}_b \, q^{bn}, \tag{8.4}$$

with the Einstein summation convention again implying a sum over all repeated indices (even if these indices are, perversely as in (8.3), not explicitly written). Notice that the flavour index just goes along for the ride in (8.4), since the matrices T_α act only on colour.

For the present purposes what is important is that the lagrangian (8.1) has several properties. First, it is renormalizable (inasmuch as all couplings are dimensionless and all dimensionless couplings are included consistent with the field content and symmetries). Second, the absence of γ_5 in any of the interactions implies that these symmetries include parity invariance and the interchange of particles and antiparticles (charge-conjugation invariance). Third, the symmetries include a crucial one: invariance under a local (*i.e.* position-dependent) 3×3 unitary rotation of all the q's in colour space, accompanied by an appropriate gauge transformation for the G^α_μ's.

Explicitly, the couplings amongst the G^α_μ and between these and q^{an} are designed to be invariant under the infinitesimal transformations (see §C.5 for more details of how this type of nonabelian local symmetry works)

$$\delta q = \mathrm{i} g_s \zeta^\alpha T_\alpha q \quad \text{and} \quad \delta G^\alpha_\mu = \partial_\mu \zeta^\alpha + g_s c^\alpha{}_{\beta\gamma} \zeta^\beta G^\gamma_\mu, \tag{8.5}$$

where $\zeta^\alpha(x)$ are arbitrary real spacetime-dependent transformation parameters and the matrices T_α are the same ones as appear in the lagrangian, and are generators of the group of 3×3 unitary matrices with unit determinant – called $SU_c(3)$ (where the subscript denotes 'colour' to distinguish it from other instances of this group). The group structure of $SU(3)$ implies that the commutator of two T_α's gives another of the T_α's, allowing the structure constants, $c^\gamma{}_{\alpha\beta}$, to be defined through $[T_\alpha, T_\beta] = \mathrm{i}\, c^\gamma{}_{\alpha\beta} T_\gamma$ (see Appendix C.4.1 for a brief summary of useful facts about Lie groups and algebras).

Quarks and antiquarks experience an interaction mediated by the gluons that turns out to cause quarks and antiquarks to be attracted to one another when they are prepared in a colour-neutral – or $SU_c(3)$-invariant – combination. (Electromagnetism similarly causes opposite charges to attract, favouring the formation of electrically neutral bound states.) The three ways to achieve colour neutrality using quarks and antiquarks are: (a) to combine a quark with an antiquark, leading to combinations like $\mathcal{M}_n{}^m = \overline{q}_{an} q^{am}$ or, (b) to take a completely antisymmetric combination of three quarks, $\mathcal{B}^{mnp} = \epsilon_{abc} q^{am} q^{bn} q^{cp}$ or, (c) do the same for three antiquarks, $\overline{\mathcal{B}}_{mnp} = \epsilon^{abc} \overline{q}_{am} \overline{q}_{bn} \overline{q}_{cp}$.

The evidence is that all of the known hadrons can be understood as colour-neutral bound states in this way, with spin-zero and spin-one mesons being formed from quark-antiquark combinations while spin-$\frac{1}{2}$ and spin-$\frac{3}{2}$ baryons (and antibaryons) are formed from 3-quark (or 3-antiquark) combinations. For example, a proton in this picture is the combination uud while a neutron is udd and an antiproton is $\bar{u}\bar{u}\bar{d}$. The charged pion is a meson with $\pi^+ = u\bar{d}$ and its antiparticle is $\pi^- = \bar{u}\, d$. The electric charges of these bound states are consistent with the u (and c and t) quarks having electric charge $q_u = \frac{2}{3}\, e$ and the d (and s and b) quarks having charge $q_d = -\frac{1}{3}\, e$. In principle, the masses of all QCD bound states should be calculable in terms of the basic parameters, which in this case are the coupling g_s and the quark masses, m_q. At present, such *ab initio* calculations are only possible for relatively simple bound states, using numerical methods.

8.1.2 Asymptotic Freedom

Because g_s is dimensionless one might expect QCD to be scale invariant in the absence of quark masses. This turns out to be false, due mainly to the running of g_s.

Repeating the arguments leading to the running of electromagnetic couplings shows that loops of coloured particles cause the strong coupling to be scale dependent, given by

$$\frac{1}{\alpha_s(\mu)} = \frac{1}{\alpha_s(\mu_0)} - b_s \log\left(\frac{\mu^2}{\mu_0^2}\right) \tag{8.6}$$

where $\alpha_s := g_s^2/4\pi$ (by analogy with the definition of the electromagnetic fine-structure constant, $\alpha := e^2/4\pi$). Compare this result to its electromagnetic counterpart in (7.32).

An important difference from electromagnetism is that both gluons and quarks contribute to the coefficient b_s, crucially with opposite signs [165–167]. Explicit calculation using the $\overline{DS}$ scheme gives

$$b_s = \frac{1}{12\pi}\left(2n_q - 33\right), \tag{8.7}$$

where n_q represents the number of quark flavours appearing within the EFT of interest and the 33 comes from evaluating the gluon loops. Because there are only six flavours the gluon contribution in practice dominates, so b_s is negative [168, 169]. This implies that α_s gets larger for lower μ and gets weaker for larger μ – what is called 'asymptotic freedom'. For μ sufficiently small α_s eventually becomes large enough to invalidate the perturbative calculation of b_s.

Both α_s and α can be measured at energies near the mass of the Z boson, $M_Z \simeq 90$ GeV, where they are given by [170]

$$\alpha_s(\mu = M_Z) \simeq 0.12 \quad \text{and} \quad \alpha(\mu = M_Z) \simeq \frac{1}{128} = 7.8 \times 10^{-3} \tag{8.8}$$

showing that the strong coupling is the larger of the two, and both are small enough to permit perturbative calculations. But whereas the electromagnetic coupling gets smaller and smaller at lower energies (eventually reaching $\alpha \simeq 1/137$ at $\mu \simeq m_e$, as discussed in earlier chapters), the coupling α_s gets larger and larger at lower energies, until it eventually becomes too big to trust perturbation theory [171, 172].

The scale where α_s becomes nonperturbative is a fundamental scale of the strong interactions, and is the basic scale that competes with quark masses to set the dimensions of hadron masses. Precisely what this scale is depends on exactly how it is defined. A conventional definition is the *QCD scale*, Λ_{QCD}, defined as the point where an extrapolation of the perturbative running drives α_s to infinity (in the $\overline{DS}$ renormalization scheme, say). This scale is given at one-loop by

$$^{(n_q)}\Lambda_{QCD} = M_Z \exp\left[\frac{1}{2b_s\alpha_s(M_Z)}\right], \tag{8.9}$$

and so naturally depends on the value chosen for n_q when evaluating b_s. This evalutes to $^{(5)}\Lambda_{QCD} = 80$ MeV in the 5-quark EFT below $M_Z \simeq 90$ GeV, although this counting of quarks is not appropriate below $m_b \simeq 5$ GeV; the QCD scale rises to $^{(4)}\Lambda_{QCD} = 140$ MeV in the 4-quark EFT below m_b, but again this number of quarks is not appropriate below $m_c \simeq 1$ GeV. A better picture of the physical QCD scale is obtained using $^{(3)}\Lambda_{QCD} = 220$ MeV in the 3-quark regime that applies below m_c but above $m_s \simeq 100$ MeV, having matched across the thresholds at m_c and m_b when

Table 8.2 Pion properties

Particle	Charge	Parity	Spin	Mass (MeV)
$\pi^{\pm}$	$\pm e$	-1	0	140
π^0	0	-1	0	135

running down from M_Z. It makes no sense to use perturbative arguments for QCD below this scale.

Of course, the QCD coupling does not really diverge at Λ_{QCD}; it is our ability to compute its behaviour there that breaks down. Nonetheless Λ_{QCD} provides a convenient way to parameterize the characteristic physical scale of the strong interactions, lying in the ballpark of several hundred MeV. This scale is expected to typify the momentum required by the uncertainty principle for a quark within a hadron, because the growth of g_s at lower energies implies that gluon-mediated forces become stronger over longer distances. This growth is believed to confine the coloured constituents of hadrons inside a region whose size is of order $\Lambda_{QCD}{}^{-1} \sim (200\,\text{MeV})^{-1} \sim 1$ fm.

Quarks much lighter than Λ_{QCD} should therefore have zero-point momenta inside hadrons that are much larger than their masses, making their energies also of order Λ_{QCD}. Assuming interaction energies also to be roughly Λ_{QCD} one expects mesons built from quark-antiquark pairs to have masses of order several times Λ_{QCD}, while baryons built from three light quarks should be somewhat heavier than this. Indeed, hadrons built from u, d and s quarks are found at accelerators with masses roughly in this range, such as the ρ meson (mass 770 MeV), the ω meson (782 MeV), proton (938 MeV) and neutron (940 MeV).

The existence of a fundamental scale like Λ_{QCD} also provides a natural benchmark against which strong-interaction processes can be judged to be 'low-energy' or 'high-energy'. In particular, it turns out that processes at low energies – much smaller than around 1 GeV – lend themselves to a low-energy EFT description called chiral perturbation theory, whose features are the main topic of this chapter. It is mainly the lightest quarks (u and d and perhaps s) that are relevant for this effective theory, since the others are too massive to be included at such low energies.

Of course, having a low-energy regime would not be that interesting if there were no particles light enough to appear in it. If all baryons and mesons really all had similar masses then one might worry that the EFT for energies well below a GeV might be empty. But light mesons do exist: the lightest hadrons are pions: $\pi^{\pm}$ and π^0 (some of whose properties are listed in Table 8.2). Pions, whose properties are consistent with being quark-antiquark meson combinations built using only u and d quarks, are much lighter than other hadrons, making them important actors in QCD's low-energy limit.

8.1.3 Symmetries and Their Realizations

Inspection of Table 8.1 shows that for the lightest quarks it should be a good first approximation to neglect quark masses, since $m_u, m_d \ll \Lambda_{QCD} \sim 200$ MeV.

The effects of nonzero masses can then be included perturbatively as small corrections. To the extent that this perturbative treatment is an expansion in powers of m_q/Λ_{QCD} it should be much worse – though perhaps still qualitatively useful – when applied to s quarks than for u and d quarks.

The first step when analyzing the limit of massless quarks is to ask what symmetries are present in the underlying theory in this regime. To see what these are it is useful to restrict the QCD lagrangian (8.1) to up and down quarks and drop their mass terms, leaving

$$\overline{q}\not{D}q = \begin{pmatrix} \overline{u} \\ \overline{d} \end{pmatrix}^T \not{D} \begin{pmatrix} u \\ d \end{pmatrix} = \begin{pmatrix} \overline{u} \\ \overline{d} \end{pmatrix}^T \gamma_L \not{D} \begin{pmatrix} u \\ d \end{pmatrix} + \begin{pmatrix} \overline{u} \\ \overline{d} \end{pmatrix}^T \gamma_R \not{D} \begin{pmatrix} u \\ d \end{pmatrix}. \tag{8.10}$$

Here, the superscript 'T' on a two-component vector indicates taking its transpose and all colour and spinor indices are suppressed. The Dirac matrices, $\gamma_L = \frac{1}{2}(1+\gamma_5)$ and $\gamma_R = \frac{1}{2}(1-\gamma_5)$, project onto left- and right-handed Weyl spinors (see §A.2.3), and are introduced to emphasize how the left- and right-handed quarks in (8.10) interact with gluons independent of one another in the absence of quark masses.

Writing the quark part of the QCD lagrangian like (8.10) shows how the massless lagrangian enjoys a 'chiral' symmetry, meaning one for which left- and right-handed fermions can transform independently. The symmetry of (8.10) is an $SU_L(2)\times SU_R(2)$ symmetry under which left- and right-handed quarks are independently rotated by arbitrary 2×2 unitary matrices that mix up the flavours u and d:

$$\begin{pmatrix} u \\ d \end{pmatrix} \to \left[U\gamma_L + V\gamma_R\right] \begin{pmatrix} u \\ d \end{pmatrix}. \tag{8.11}$$

Here, $U \in SU_L(2)$ and $V \in SU_R(2)$ are 2×2 matrices that satisfy $U^\dagger U = I$, $V^\dagger V = I$ as well as[1] $\det U = \det V = 1$.

Having identified symmetries of the underlying action, the next question is whether or not they are spontaneously broken by the QCD ground state (or 'vacuum'), $|\Omega\rangle$. The answer to this determines whether these symmetries should be linearly or nonlinearly realized in any effective theory describing the strong interactions at low energies.

It happens that the QCD vacuum is invariant under the diagonal (or 'vector') subgroup $SU_I(2) \subset SU_L(2) \times SU_R(2)$ corresponding to the choice $V = U$ [177]. For historical reasons this particular combination is called 'isospin' symmetry [178, 179], due to the analogy with the $SU(2)$ that describes spin rotations for spin-half states in quantum mechanics.

Following the general discussion in Part I this isospin symmetry should therefore be linearly realized on low-energy fields. Since the quark pair $\binom{u}{d}$ is a doublet under these transformations, the three types of pion – $\pi^\pm$ and π^0 which, being built from u and d quark-antiquark combinations, combine two doublets – are naturally regarded as transforming as a triplet under isospin symmetry (much like two

[1] Dropping the determinant condition makes the group $U_L(2) \times U_R(2)$, which at face value is also a symmetry of (8.10) [173]. But the additional 'axial' combination (for which left- and right-handed parts of both u and d fields get multiplied by opposite phases) is anomalous (and so is not really a symmetry – see §4.3) [164, 174–176]. This leaves one symmetry beyond $SU_L(2) \times SU_R(2)$ that simply multiplies the left- and right-handed parts of both u and d quarks by a common phase. This symmetry is responsible for the conservation of baryon number, but plays no role in the remainder of this section.

spin-half objects can combine to form a spin-one state). This interpretation is supported by the comparatively small splitting in mass between $\pi^\pm$ and π^0 (see Table 8.2), since in the approximation of exact $SU_I(2)$ symmetry (*i.e.* vanishing quark masses and no electromagnetic interactions) the mass difference between states related by a linearly realized symmetry should vanish (see §C.4).

On the other hand, all the evidence[2] (experimental, numerical and circumstantial) suggests that the 'axial' part of the symmetry, defined as the subset $SU_A(2) \subset SU_L(2) \times SU_R(2)$ with $V = U^\dagger$, is spontaneously broken by the QCD vacuum, with order parameter $\langle\Omega|\,\overline{q}\,q\,|\Omega\rangle \neq 0$. If $SU_A(2)$ were an exact symmetry of QCD, Goldstone's theorem would argue there must be an exactly massless state created from the vacuum by the three conserved Noether currents for the axial symmetry, given by

$$j^\mu_\alpha = \frac{\mathrm{i}}{2}\begin{pmatrix} \overline{u} \\ \overline{d} \end{pmatrix}^T \gamma^\mu \gamma_5\, \tau_\alpha \begin{pmatrix} u \\ d \end{pmatrix}, \tag{8.12}$$

where τ_α, with $\alpha = 1, 2, 3$ represent the three Pauli matrices

$$\tau_1 = \begin{pmatrix} 0 & 1 \\ 1 & 0 \end{pmatrix}, \quad \tau_2 = \begin{pmatrix} 0 & -\mathrm{i} \\ \mathrm{i} & 0 \end{pmatrix}, \quad \tau_3 = \begin{pmatrix} 1 & 0 \\ 0 & -1 \end{pmatrix}. \tag{8.13}$$

acting in the 2×2 flavour space labelled by u and d.

Pions as Pseudo-Goldstone Bosons

The triplet of pions have precisely the right quantum numbers to be the bosons required by Goldstone's theorem, inasmuch as they can satisfy

$$\langle \pi^\beta(p)|\, j^\mu_\alpha(x)\,|\Omega\rangle = -\mathrm{i}F_\pi\, p^\mu\, e^{-ipx}\,\delta^\beta{}_\alpha \neq 0, \tag{8.14}$$

for some nonzero constant F_π (called the pion decay constant). Here the three hermitian pion-fields π^α with $\alpha = 1, 2, 3$ are related to the physical pion fields by

$$\pi^\pm = \frac{1}{\sqrt{2}}\left(\pi^1 \mp \mathrm{i}\pi^2\right) \quad \text{and} \quad \pi^0 = \pi^3, \tag{8.15}$$

while p^μ is the 4-momentum of the pion in question.

In reality, pions are not massless and quark masses are nonzero so axial transformations are not exact symmetries. But the requirement that pion masses should go to zero in the limit that the quark masses vanish ensures they are systematically lighter than other hadrons in the real world, where $m_u, m_d \ll \Lambda_{QCD}$. The spirit of chiral perturbation theory is to make this connection between m_π and m_u and m_d explicit by describing pions using the low-energy effective theory of Goldstone bosons and then incorporating the symmetry-breaking influences of quark masses perturbatively in powers of m_q/Λ_{QCD}.

[2] The first evidence for this symmetry pattern came from phenomenological successes [11, 180, 181], while the earliest theoretical argument (for three light quarks) was one of the early uses of anomaly matching [80].

8.2 Chiral Perturbation Theory

The above arguments suggest that QCD at energies much below the QCD scale is dominated by the dynamics of the lightest hadrons – pions[3] – which can be interpreted as pseudo-Goldstone bosons for the symmetry breaking pattern $SU_L(2) \times SU_R(2) \to SU_I(2)$. In this picture pions are light (and so appear in the low-energy theory) because u and d quarks are almost (but not quite) massless and pions would become honest-to-God Goldstone bosons (and so massless) in the limit $m_u, m_d \to 0$. This is the framework of *chiral perturbation theory*, whose success describing the properties of the low-energy hadrons is a success of QCD.

A similar picture of kaons and η particles as pseudo-Goldstone bosons also follows if the s-quark is regarded as being perturbatively light, though $m_s \sim 100$ MeV makes expansion in powers of m_s/Λ_{QCD} a worse approximation. The discussion below restricts just to u and d quarks.

8.2.1 Nonlinear Realization ◇

If pions are pseudo-Goldstone bosons then building their low-energy EFT involves constructing the nonlinear realization for the symmetry-breaking pattern where $G = SU_L(2) \times SU_R(2)$ breaks down to $H = SU_I(2)$, following the steps of §4.2.2. Writing the three Pauli matrices collectively as $\vec{\tau} = \{\tau_\alpha\}$, so that $\omega^\alpha \tau_\alpha = \vec{\omega} \cdot \vec{\tau}$ and so on, the elements of G can be explicitly written as 4×4 matrices with

$$g = \begin{pmatrix} U & 0 \\ 0 & V \end{pmatrix} = \begin{pmatrix} e^{i\vec{\omega}_L \cdot \vec{\tau}/2} & 0 \\ 0 & e^{i\vec{\omega}_R \cdot \vec{\tau}/2} \end{pmatrix} \in G = SU_L(2) \times SU_R(2), \tag{8.16}$$

while the unbroken group is

$$h = \begin{pmatrix} e^{i\vec{\omega}_I \cdot \vec{\tau}/2} & 0 \\ 0 & e^{i\vec{\omega}_I \cdot \vec{\tau}/2} \end{pmatrix} \in H = SU_I(2), \tag{8.17}$$

corresponding to $\vec{\omega}_L = \vec{\omega}_R =: \vec{\omega}_I$.

Transformation Rules

Following along in the footsteps of §4.2.2, the standard representation for the three Goldstone fields, $\theta^\alpha(x)$, becomes

$$\exp\left[i\,\vec{\theta} \cdot \vec{X}\right] = \begin{pmatrix} \exp\left[\frac{i}{2}\,\vec{\theta} \cdot \vec{\tau}\right] & 0 \\ 0 & \exp\left[-\frac{i}{2}\,\vec{\theta} \cdot \vec{\tau}\right] \end{pmatrix}, \tag{8.18}$$

in terms of which the standard nonlinear realization of G is obtained by multiplying (8.18) on the left by g given in (8.16) and decomposing the result into

$$g\,\exp\left[i\,\vec{\theta} \cdot \vec{X}\right] = \exp\left[i\,\tilde{\vec{\theta}} \cdot \vec{X}\right]\,\gamma, \tag{8.19}$$

[3] Later sections – §8.2.3 and §13.1 – show that particles heavier than Λ_{QCD} can also appear in the low-energy theory in some cases (such as if they are stable). These can be included either by adding new second-quantized fields as described in §8.2.3 or in the first-quantized formulation of §13.1.

where the rightmost factor appearing here is an element of H – *c.f.* (4.47) – and so

$$\gamma = \exp\left[\mathrm{i}\,\vec{u}\cdot\vec{t}\,\right] = \begin{pmatrix} \exp\left[\frac{\mathrm{i}}{2}\,\vec{u}\cdot\vec{\tau}\,\right] & 0 \\ 0 & \exp\left[\frac{\mathrm{i}}{2}\,\vec{u}\cdot\vec{\tau}\,\right] \end{pmatrix}. \tag{8.20}$$

Explicit formulae for $\vec{\tilde{\theta}}(\vec{\theta},g)$ and $\vec{u} = \vec{u}(\vec{\theta},g)$ in closed form may be found using the identity $\exp\,(\mathrm{i}\vec{\alpha}\cdot\vec{\tau}) = \cos\alpha + \mathrm{i}\hat{\alpha}\cdot\vec{\tau}\ \sin\alpha$, where $\alpha = \sqrt{\vec{\alpha}\cdot\vec{\alpha}}$, and $\hat{\alpha} = \vec{\alpha}/\alpha$. Specializing to infinitesimal transformations, this leads to

$$\begin{aligned} \delta\vec{\theta} &= \vec{\theta}\times\vec{\omega}_I + \frac{\theta}{2}\left(\tan\frac{\theta}{2} + \cot\frac{\theta}{2}\right)\left[\vec{\omega}_A - \hat{\theta}(\hat{\theta}\cdot\vec{\omega}_A)\right] + \hat{\theta}(\hat{\theta}\cdot\vec{\omega}_A), \\ &= \vec{\omega}_A + \vec{\theta}\times\omega_I + O(\theta^2);, \end{aligned} \tag{8.21}$$

and

$$\vec{u} = \vec{\omega}_I + (\hat{\theta}\times\vec{\omega}_A)\,\tan\frac{\theta}{2} = \vec{\omega}_I + \frac{1}{2}\,\vec{\theta}\times\vec{\omega}_A + O(\theta^2), \tag{8.22}$$

where (as before) $\theta := \sqrt{\vec{\theta}\cdot\vec{\theta}}$ and $\hat{\theta} := \vec{\theta}/\theta$. Notice, in particular, that the group structure of G implies that the three fields $\vec{\theta}$ shift under the broken transformations labelled by $\vec{\omega}_A$, and transform as a triplet under the unbroken subgroup, $H = SU_I(2)$, labelled by $\vec{\omega}_I$.

Invariant Action

The quantities required to build G-invariant actions for $\vec{\theta}$ and to couple $\vec{\theta}$ to other fields are computed following the steps in §4.2.2 in terms of the decomposition of $e^{-\mathrm{i}\vec{\theta}\cdot\vec{X}}\partial_\mu e^{\mathrm{i}\vec{\theta}\cdot\vec{X}}$ onto the $\vec{X}$ and $\vec{t}$ directions, leading to the 'dreibein'

$$\begin{aligned} \vec{e}_\mu &= \left(\frac{\sin\theta}{\theta}\right)\,\partial_\mu\vec{\theta} - \left(\frac{\sin\theta - \theta}{\theta^3}\right)\,(\vec{\theta}\cdot\partial_\mu\vec{\theta})\,\vec{\theta}, \\ &= \partial_\mu\vec{\theta}\left(1 - \frac{1}{6}\,\theta^2\right) + \frac{1}{6}\,(\vec{\theta}\cdot\partial_\mu\vec{\theta})\,\vec{\theta} + O(\theta^5);, \end{aligned} \tag{8.23}$$

and 'gauge connection'

$$\vec{\mathcal{A}}_\mu = -\frac{2}{\theta^2}\,\sin^2\frac{\theta}{2}\,(\vec{\theta}\times\partial_\mu\vec{\theta}) = -\frac{1}{2}\,\vec{\theta}\times\partial_\mu\vec{\theta} + O(\theta^4). \tag{8.24}$$

Notice that when $\vec{\theta}\to-\vec{\theta}$ the dreibein $\vec{e}_\mu$ changes sign while $\vec{\mathcal{A}}_\mu$ does not. This observation becomes important once one asks how the low-energy lagrangian transforms under parity (which is a symmetry of the underlying QCD dynamics), since the nonvanishing of the matrix element (8.14) implies that $\vec{\theta}$ changes sign under parity.

The most general G-invariant lagrangian with the fewest derivatives that is invariant under these transformations is constructed using the rules summarized in §4.2.2 (with details given in Appendix C.6.3). In particular, because $\vec{\theta}$ shifts under the broken $\vec{\omega}_A$ transformations, no G-invariant scalar potential is possible at all. The most general lagrangian up to overall normalization involving up to two derivatives is

$$\mathfrak{L}_{\mathrm{eff}} = -\frac{F^2}{2}\,g_{\alpha\beta}(\vec{\theta})\,\partial_\mu\theta^\alpha\partial^\mu\theta^\beta + \text{(higher-derivative terms)}, \tag{8.25}$$

where F is a real constant and the 'target-space metric,' $g_{\alpha\beta}$, on G/H is

$$g_{\alpha\beta}(\theta) = \delta_{\gamma\delta}\, e^{\gamma}{}_{\alpha}\, e^{\delta}{}_{\beta} = \delta_{\alpha\beta}\left(\frac{\sin^2\theta}{\theta^2}\right) + \theta_\alpha\theta_\beta\left(\frac{\theta^2 - \sin^2\theta}{\theta^4}\right),$$
$$= \delta_{\alpha\beta}\left(1 - \frac{1}{3}\,\theta^2\right) + \frac{1}{3}\,\theta_\alpha\theta_\beta + O(\theta^4), \tag{8.26}$$

where $e^{\gamma}{}_{\alpha}$ is obtained from $\vec{e}_\mu \cdot \vec{X}$ by writing $\vec{e}_\mu \cdot \vec{X} = X_\gamma\, e^{\gamma}{}_{\alpha}\, \partial_\mu\theta^\alpha$.

For applications it is useful to rescale the pion fields so their kinetic terms take the canonical form $-\frac{1}{2}\,\partial_\mu\vec{\pi}\cdot\partial^\mu\vec{\pi}$, in which case $\vec{\theta} = \vec{\pi}/F$. With this choice the most general two-derivative pion self-interactions are (after performing an integration by parts)

$$\mathfrak{L}_{\rm eff} = -\frac{1}{2}\,\partial_\mu\vec{\pi}\cdot\partial^\mu\vec{\pi} - \frac{1}{2F^2}\,(\vec{\pi}\cdot\partial_\mu\vec{\pi})\,(\vec{\pi}\cdot\partial^\mu\vec{\pi}) + O(\pi^6) + \cdots, \tag{8.27}$$

where all two-derivative terms out to arbitrary order in π are similarly obtained by expanding the explicit function given in Eqs. (8.25) and (8.26).

Symmetry-Breaking Terms

So far, the discussion has proceeded as if the symmetry $G = SU_L(2)\times SU_R(2)$ were an exact symmetry, and the pions were massless. To improve on this requires including the effects of the symmetry-breaking quark masses, which is done by matching the EFT to QCD perturbatively in powers of the quark masses.

In QCD the quark mass terms are

$$\mathfrak{L}_m = -\overline{q}\,Mq = -\overline{q}\,M\gamma_L q + \text{h.c.}, \tag{8.28}$$

where

$$q = \begin{pmatrix} u \\ d \end{pmatrix} \quad \text{and} \quad M = \begin{pmatrix} m_u & 0 \\ 0 & m_d \end{pmatrix}. \tag{8.29}$$

Under the $G = SU_L(2)\times SU_R(2)$ symmetry, $q \to (U\gamma_L + V\gamma_R)q$, this transforms into

$$\overline{q}\,M\gamma_L q \to \overline{q}\,V^\dagger M U\gamma_L q + \text{h.c.}. \tag{8.30}$$

Although this is not invariant, the key observation is that it *would* have been invariant if the mass matrix had been a field which had also transformed under G, according to the rule

$$M \to VMU^\dagger. \tag{8.31}$$

This is a useful observation because it suggests how to determine how M appears in the low-energy effective pion theory.

To this end, imagine the effective pion action to have an expansion in powers of the light quark masses,

$$\mathfrak{L}_{\rm eff} = \mathfrak{L}_0 + \mathfrak{L}_1 + \cdots, \tag{8.32}$$

where the subscript indicates the power of M that appears in each term. Each of the $\mathfrak{L}_i$ can also be separately expanded in powers of the fields $\vec{\pi}$ and their derivatives, as is done above for the lowest-derivative terms that appear in $\mathfrak{L}_0$. The goal now is to do the same for $\mathfrak{L}_1$, starting with those terms containing no derivatives at all. The key for doing so is to use the fact that $\mathfrak{L}_1$ must depend on the $\vec{\pi}$'s in such a way that it would be G-invariant *if* M were allowed to transform through the rule (8.31). The trick of identifying non-invariant terms by pretending that a symmetry-breaking parameter (like M) is a field is called the *spurion* trick, where 'spurion' is the name for the fictitious (or spurious) field whose expectation value is M.

The term $\mathfrak{L}_1$ is linear in M and provides the leading symmetry-breaking contribution at low energies. Its dependence on $\vec{\pi}$ is found by demanding $\mathfrak{L}_1$ be G-invariant, but only if we simultaneously transform $\vec{\pi}$ – using (8.21) – and take $M \to VMU^\dagger$. It is straightforward to construct one such a term involving only the pion fields, and the simplest method does so by building a 2×2 quantity $\Xi(\vec{\theta})$ that transforms oppositely to M: *i.e.*

$$\Xi \to \widetilde{\Xi} = U\,\Xi\,V^\dagger. \tag{8.33}$$

Once this is done, the required lagrangian density is proportional to $\mathrm{Tr}\,[M\;\Xi] + \text{c.c.}$.

The required Ξ can be built from the basic quantities $\xi_L := e^{i\vec{\theta}\cdot\vec{\tau}/2}$ and $\xi_R := e^{-i\vec{\theta}\cdot\vec{\tau}/2}$ that appear on the diagonal in (8.18), repeated again here for convenience

$$\exp\left[i\,\vec{\theta}\cdot\vec{X}\right] = \begin{pmatrix} \xi_L(\vec{\theta}) & 0 \\ 0 & \xi_R(\vec{\theta}) \end{pmatrix}, \tag{8.34}$$

since Eqs. (8.19) and (8.20) imply these transform relatively simply:

$$\xi_L \to U\,\tilde{\xi}_L\,e^{i\vec{u}\cdot\vec{\tau}/2} \quad \text{and} \quad \xi_R \to V\,\tilde{\xi}_R\,e^{i\vec{u}\cdot\vec{\tau}/2}. \tag{8.35}$$

Clearly, the transformation property (8.33) holds for the 2×2 matrix[4]

$$\Xi := \xi_L\,\xi_R^\dagger = e^{i\vec{\theta}\cdot\vec{\tau}}. \tag{8.36}$$

The leading symmetry-breaking term in the pion lagrangian therefore is

$$\begin{aligned} \mathfrak{L}_1 &= \frac{\Lambda_m^3}{2}\;\mathrm{Tr}\;[M\;(\Xi + \Xi^\dagger)] = (m_u + m_d)\,\Lambda_m^3\;\cos\theta, \\ &= m_\pi^2\left[F^2 - \frac{1}{2}\,\vec{\pi}\cdot\vec{\pi} - \frac{(\vec{\pi}\cdot\vec{\pi})^2}{4!\;F^2} + O(\pi^6)\right], \end{aligned} \tag{8.37}$$

where Λ_m is a new dimensionful parameter. The last line shows that $\mathfrak{L}_1$ generates a common mass for all three pions whose size is given by [182]

$$m_\pi^2 = (m_u + m_d)\,\frac{\Lambda_m^3}{F^2}, \tag{8.38}$$

a result of order the known pion masses given quark masses of order a few MeV, $F \sim 100$ MeV (as is shown below) and Λ_m of order 300 MeV (and so not so different from Λ_{QCD}). This confirms that small quark masses suffice to explain the size of the observed pion masses, as would be required if they are to be interpreted as pseudo-Goldstone bosons.

The pion mass term found here is proportional to $\vec{\pi}\cdot\vec{\pi}$ and so necessarily preserves the unbroken isospin symmetry, $SU_I(2)$, at least to this order in the derivative and

[4] In terms of Ξ, the two-derivative interactions of (8.25) can also be written $\mathfrak{L}_{\text{eff}} = -\frac{1}{4}F^2\,\mathrm{Tr}\,(\partial_\mu\Xi^\dagger\,\partial^\mu\Xi)$.

quark-mass expansions. This ensures, in particular, degenerate masses for all three pions despite there being an isospin-breaking quark mass difference, $m_u - m_d$. Approximate isospin invariance for the pion masses turns out to follow from the small size of both m_u and m_d relative to Λ_{QCD}, rather than any requirement that $m_d - m_u$ is small compared with $m_d + m_u$. An understanding of the mass difference between charged and neutral pions must therefore be sought elsewhere, such as the isospin-breaking electromagnetic interactions [183, 184].

Finally, notice that the mass term given in Eq. (8.37) is the unique such term that is linear in M and depends only on $\vec{\theta}$ but not its derivatives. This uniqueness follows from the impossibility of building a G-invariant scalar potential in $\mathfrak{L}_0$. To see why this is true, suppose there were two operators, $O_1(\theta)$ and $O_2(\theta)$, for which $\mathfrak{L}_{1i} = \text{Tr}\,(M\,O_i) + \text{h.c.}$ transforms in the desired way. Since both of these operators must transform the same way under G the combination $V(\theta) = \text{Tr}\,[O_1 O_2^\dagger]$ would be a G-invariant scalar potential, which does not exist. It follows that the transformation rule (8.33) determines $\Xi(\theta)$ up to normalization.

8.2.2 Soft-Pion Theorems ♠

What is important about (8.27) is that the strength of all of the two-derivative pion interactions are determined by only the one constant F. The additional interactions allowed by (8.37) introduce only one more parameter, m_π. If QCD were better understood, F could be computed from first principles (as is now becoming possible using numerical techniques). But even if this cannot be done, the lagrangian (8.25) is very predictive since any measurement of F from (say) low-energy $\pi\pi \to \pi\pi$ scattering can then be used (in principle) to compute $\pi\pi \to 4\pi$ and higher scattering as well. These relationships amongst low-energy pion properties are known as 'soft-pion' theorems.

Decays and Conserved Currents

The situation is even more predictive than just for pion scattering, as it turns out, because F also (see below) controls the decay rate of charged pions. In practice, this means that F can be determined purely using the measured $\pi^\pm$ lifetime, and once this is done its value in principle[5] completely determines the leading contribution to all low-energy pion scattering amplitudes.

This calculational windfall occurs because the current $\bar{u}\gamma^\mu\gamma_L d$ appearing in the low-energy charged-current weak interaction – *c.f.* Eqs. (7.9) and (7.10) – is also a Noether current (8.12) for the approximate $G = SU_L(2) \times SU_R(2)$ symmetry of QCD. For instance, the matrix element appearing in the π^- decay amplitude is $\langle\Omega\,|\,\bar{u}\gamma^\mu(1 + \gamma_5)d\,|\,\pi^-\rangle$, which is calculable in terms of the parameter F_π defined in (8.14) because parity invariance of QCD implies that only the axial part of the current has a nonzero matrix element.

[5] In practice, pion scattering is hard to measure and practical low-energy processes also involve protons, neutrons and other mesons, complicating the picture (but not negating the predictivity of the low-energy theory).

This observation allows the charged-pion lifetime to be computed in terms of F_π, leading to the following prediction for the lifetime for the decay $\pi \to \mu\overline{\nu}\nu$ [185, 186, 188]:

$$\frac{1}{\tau_{\rm B}} = \frac{G_F^2 |V_{ud}|^2 F_\pi^2 \, m_\mu^2 m_\pi}{4\pi}\left(1 - \frac{m_\mu^2}{m_\pi^2}\right)^2, \tag{8.39}$$

where m_μ and m_π are the masses of the muon and pion while G_F and V_{ud} are the Standard Model parameters defined in §7.1 and measured in muon and nuclear β-decays. Comparing this with the observed mean lifetime, $\tau_{\rm exp}(\pi^-) = 2.6030(24) \times 10^{-8}$ s, gives $F_\pi \simeq 92$ MeV.

To determine the value of the effective parameter F appearing in (8.25) it therefore suffices to compute the vacuum-to-pion matrix element of the conserved $SU_L(2) \times SU_R(2)$ currents within the EFT and compare the result to (8.14). This is done by computing the Noether currents directly by using the invariance of the action (8.25) under the transformations (8.21), with the result

$$\vec{j}_I^{\,\mu} = -\left(\vec{\pi} \times \partial^\mu \vec{\pi}\right) + \cdots \quad \text{and} \quad \vec{j}_A^{\,\mu} = F\,\partial^\mu \vec{\pi} + \cdots, \tag{8.40}$$

where the ellipses denote terms involving more powers of $\vec{\pi}$ and/or more derivatives. Using this to compute the matrix element $\langle \pi | \vec{j}_A^{\,\mu} | \Omega \rangle$ and comparing with (8.14) then allows the inference $F = F_\pi \simeq 92$ MeV.

Power Counting

With F determined in this way, the effective lagrangian for low-energy pion-pion scattering becomes very predictive. As usual, the first step is to identify the Feynman graphs and interactions that are required to capture the scattering to any fixed order in the low-energy expansion. This involves applying the power-counting rules of §3.2.3 to this example [2].

For a first pass it is simplest to do this using only the G-invariant terms of $\mathfrak{L}_0$, in which case the results of §3.2.3 can be carried over in whole cloth. To translate the parameters appearing here to those used in earlier sections imagine expanding $\mathfrak{L}_0$ out to include subdominant terms involving higher derivatives, with the schematic form

$$\begin{aligned}\mathfrak{L}_{\rm eff} &= -F_\pi^2\,(\partial\theta)^2\Big[\hat{c}_{2,0} + \hat{c}_{2,2}\theta^2 + \cdots\Big] - \frac{F_\pi^2}{\Lambda_\chi^2}\,(\partial\theta)^4\Big[\hat{c}_{4,0} + \hat{c}_{4,2}\theta^2 + \cdots\Big] + \cdots \\ &= -F_\pi^2\Lambda_\chi^2\left\{\frac{1}{\Lambda_\chi^2}\left(\frac{\partial\pi}{F_\pi}\right)^2\left[\hat{c}_{2,0} + \hat{c}_{2,2}\frac{\pi^2}{F_\pi^2} + \cdots\right]\right. \\ &\qquad \left. + \frac{1}{\Lambda_\chi^4}\left(\frac{\partial\pi}{F_\pi}\right)^4\left[\hat{c}_{4,0} + \hat{c}_{4,2}\frac{\pi^2}{F_\pi^2} + \cdots\right] + \cdots\right\},\end{aligned} \tag{8.41}$$

where $\Lambda_\chi \lesssim 1$ GeV is of order a typical strong-interaction scale and the dimensionless coefficients are in principle – but typically not in practice – calculable from QCD.

Comparing this form with the coefficients with (3.55) shows that the scales used there are $\mathfrak{f}^2 = F_\pi\Lambda_\chi$, $M = \Lambda_\chi$ and $v = F_\pi$. The general power-counting arguments of

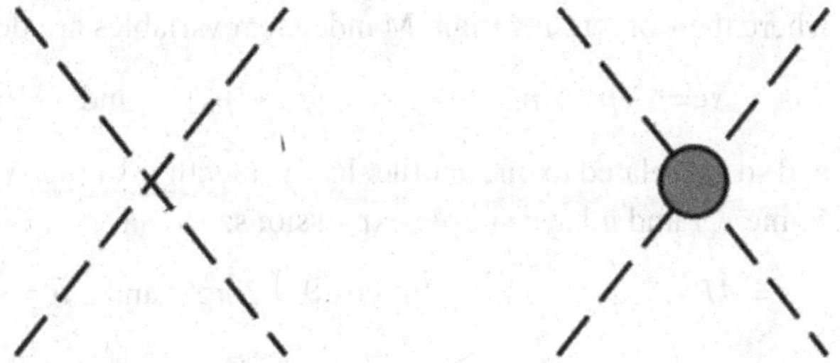

Fig. 8.1 The Feynman graphs giving the dominant contributions to pion-pion scattering in the low-energy pion EFT. The first graph uses a vertex involving two derivatives while the second involves the pion mass, but no derivatives.

§3.2.3 then show that the $\mathcal{L}$-loop contribution to a scattering amplitude for scattering involving $\mathcal{E}$ external pions at energy q is of order

$$\begin{aligned}\mathcal{A}_{\mathcal{E}}(q) &\sim q^2 F_\pi^2 \left(\frac{1}{F_\pi}\right)^{\mathcal{E}} \left(\frac{q}{4\pi F_\pi}\right)^{2\mathcal{L}} \left(\frac{q}{\Lambda_\chi}\right)^{\sum_n (d_n-2)\mathcal{V}_n} \\ &\sim q^2 F_\pi^2 \left(\frac{1}{F_\pi}\right)^{\mathcal{E}} \left(\frac{\Lambda_\chi}{4\pi F_\pi}\right)^{2\mathcal{L}} \left(\frac{q}{\Lambda_\chi}\right)^{2\mathcal{L}+\sum_n (d_n-2)\mathcal{V}_n}\end{aligned} \tag{8.42}$$

where $\mathcal{V}_n$ counts the number of vertices involving d_n derivatives. The second line of (8.42) is useful because in practice $\Lambda_\chi \sim 4\pi F_\pi \sim 1$ GeV, and because of this there is only a single quantity that controls the low-energy expansion.

These power-counting formulae are easily extended to include the symmetry-breaking interactions as well. For instance, those arising in $\mathfrak{L}_1$ involve only a single power of quark masses, and can be included by considering the corresponding dimensionless couplings to be suppressed by an additional factor of order $\hat{c} \sim m_\pi^2/\Lambda_\chi^2$ for each vertex taken from $\mathfrak{L}_1$ (as may be seen by comparing (8.37) to (3.55) using $\mathfrak{f}^2 = F_\pi \Lambda_\chi$, $M = \Lambda_\chi$ and $v = F_\pi$).

Pion-Pion Scattering

Applied to $\pi\pi \to \pi\pi$ scattering – for which $\mathcal{E} = 4$ – the above power-counting arguments show the dominant contributions come from the graphs of Fig. 8.1, involving a single 4-point vertex taken from the two-derivative terms of $\mathfrak{L}_0$ or from the no-derivative terms in $\mathfrak{L}_1$. According to (8.42), the graph built using $\mathfrak{L}_0$ contributes $\mathcal{A}_4(q) \sim (q^2/F_\pi^2)$ since $\mathcal{L} = \sum_n (d_n - 2)\mathcal{V}_n = 0$. The contribution built using $\mathfrak{L}_1$ contains an extra factor of $(\Lambda_\chi/q)^2$ because $\sum_n (d_n - 2)\mathcal{V}_n = -2$ and a factor of m_π^2/Λ_χ^2 carried by the dimensionless coupling $\hat{c}$ that is not written explicitly in (8.42). These are comparable in size for scattering at pion energies of order 100 MeV because one is of order q^2/F_π^2 while the other is of order $(q^2/F_\pi^2)(\Lambda_\chi/q)^2(m_\pi^2/\Lambda_\chi^2) \sim m_\pi^2/F_\pi^2$.

Explicitly evaluating these graphs leads to an invariant scattering amplitude for the scattering $\pi_a\pi_b \to \pi_c\pi_d$ of the form [48]

$$\mathcal{A}_4 = \frac{1}{F_\pi^2}\Big[\delta_{ab}\delta_{cd}\,(s - m_\pi^2) + \delta_{ac}\delta_{bd}\,(t - m_\pi^2) + \delta_{ad}\delta_{bc}\,(u - m_\pi^2)\Big], \tag{8.43}$$

where the Lorentz-invariant Mandelstam variables are defined by

$$s := -(p_a + p_b)^2\,, \quad t := -(p_a - p_c)^2 \quad \text{and} \quad u := -(p_a - p_d)^2 \tag{8.44}$$

and so are related to one another by the identity: $s+t+u = 4m_\pi^2$. In the centre-of-mass frame s, t and u have simple expressions:

$$s = 4E^2\,, \quad t = -2E^2 + 2p^2 \cos\vartheta + 2m_\pi^2 \quad \text{and} \quad u = -2E^2 - 2p^2 \cos\vartheta + 2m_\pi^2. \tag{8.45}$$

Here, E is the pion energy and $p = |\mathbf{p}_a| = |\mathbf{p}_b| = |\mathbf{p}_c| = |\mathbf{p}_d|$ is the magnitude of 3-momentum shared by all the pions in this frame, and ϑ is the scattering angle defined by $\mathbf{p}_a \cdot \mathbf{p}_c = \mathbf{p}_b \cdot \mathbf{p}_d = p^2 \cos\vartheta$.

Comparison with experiment is done using channels with definite angular momentum and isospin. For 2-body scattering the decomposition of $\mathcal{A}$ into combinations with definite isospin $I = 0, 1, 2$ is given by

$$\begin{aligned}\mathcal{A} = {} & \frac{\mathcal{A}^{(0)}}{3}\delta_{ab}\delta_{cd} + \frac{\mathcal{A}^{(1)}}{2}\left(\delta_{ac}\delta_{bd} - \delta_{ad}\delta_{bc}\right) \\ & + \mathcal{A}^{(2)}\left[\frac{1}{2}\left(\delta_{ac}\delta_{bd} + \delta_{ad}\delta_{bc}\right) - \frac{1}{3}\delta_{ab}\delta_{cd}\right],\end{aligned} \tag{8.46}$$

so that (8.43) gives the absolute predictions

$$\mathcal{A}^{(0)} = \frac{2s - m_\pi^2}{F_\pi^2}, \qquad \mathcal{A}^{(1)} = \frac{t-u}{F_\pi^2}, \qquad \mathcal{A}^{(2)} = -\frac{s - 2m_\pi^2}{F_\pi^2}. \tag{8.47}$$

Decomposing these into partial waves, using

$$\mathcal{A}_\ell^{(I)} \equiv \frac{1}{64\pi}\int_{-1}^{1} d\cos\vartheta\; P_\ell(\cos\vartheta)\,\mathcal{A}^{(I)}, \tag{8.48}$$

where $P_\ell(\cos\vartheta)$ are the usual Legendre polynomials, and expanding the (real part of) $\mathcal{A}_\ell^{(I)}$ in powers of the squared pion momentum: $p^2/m_\pi^2 = E^2/m_\pi^2 - 1 = (s - 4m_\pi^2)/4m_\pi^2$,

$$\mathcal{A}_\ell^{(I)} = \left(\frac{p^2}{m_\pi^2}\right)^\ell \left(a_\ell^I + b_\ell^I\,\frac{p^2}{m_\pi^2} + \cdots\right), \tag{8.49}$$

gives predictions for the pion scattering lengths, a_ℓ^I, and slopes, b_ℓ^I. Table 8.3 shows the experimental values[6] for these quantities, as well as the predictions obtained from (8.47) and from their first subdominant $O(p^2/\Lambda_\chi^2)$ correction (see §8.2.4). These support the validity of chiral perturbation theory as a systematic approximation to the low-energy strong interactions.

This example nicely illustrates the predictive power made possible by a low-energy effective lagrangian, even when the values for the effective couplings cannot be computed *ab initio* from the underlying theory. Predictive power arises because there are more observables – such as the pion scattering lengths and slopes used here – than there are effective couplings. At any order in the low-energy expansion unknown couplings can be extracted directly from experiment, much as would have been done for renormalizable interactions.

[6] It is not feasible to directly perform pion-pion scattering experiments so pion-pion scattering amplitudes at low energies are instead inferred from their influence on the final state in other processes, such as $K \to \pi\pi e\nu_e$ or pion-nucleon scattering.

Table 8.3 Theory vs experiment for low-energy pion scattering (from [189])

Parameter	Leading Order		+ Next Order	Experiment
a_0^0	$7m_\pi^2/32\pi F_\pi^2$	0.16	0.20	0.26(5)
b_0^0	$m_\pi^2/4\pi F_\pi^2$	0.18	0.26	0.25(3)
a_1^1	$m_\pi^2/24\pi F_\pi^2$	0.030	0.036	0.038(2)
a_0^2	$-m_\pi^2/16\pi F_\pi^2$	−0.044	−0.041	−0.028(12)
b_0^2	$-m_\pi^2/8\pi F_\pi^2$	−0.089	−0.070	−0.082(8)

8.2.3 Including Baryons

Chiral perturbation theory also provides an instructive example of how pseudo-Goldstone bosons couple to other degrees of freedom within the low-energy theory [187]. In this case, the natural degrees of freedom with which pions interact at low energies are nucleons – both protons and neutrons – which group together as a doublet under the unbroken isospin subgroup, $H = SU_I(2)$,

$$N := \begin{pmatrix} p \\ n \end{pmatrix}. \tag{8.50}$$

Inclusion of nucleons in the low-energy theory also raises a conceptual issue: why should nucleons be considered low-energy, given that their masses $m_N \simeq 940$ MeV are comparable to the UV scales being integrated out to obtain the low-energy effective theory?

It turns out that it is often the case that very massive particles get included within an EFT defined below their mass. As discussed at greater length in §11.1, this can be consistent provided that the energy tied up in the heavy particle's rest mass is not accessible on the time-scales of interest in the low-energy theory. In the case of the chiral effective theory, the inertness of nucleon rest masses is usually ensured by two things: (*i*) conservation of baryon number (which prevents nucleons from decaying)[7] and (*ii*) the scarcity of anti-baryons in the states of interest (which prevents nucleons from releasing their rest-mass through annihilation).

So how does the symmetry-breaking pattern $G = SU_L(2) \times SU_R(2) \to H = SU_I(2)$ constrain the low-energy couplings of pions to nucleons? This is largely dictated by their transformation properties under H, and since nucleons form an isodoublet, they transform under H according to

$$\delta N = \frac{i}{2}\, \vec{\omega}_I \cdot \vec{\tau}\, N. \tag{8.51}$$

This implies that the rule for general G tranformations is therefore

$$\delta N = \frac{i}{2}\, \vec{u}(\theta) \cdot \vec{\tau}\, N, \tag{8.52}$$

[7] Neutrons *can* decay to protons without violating B conservation, but this decay only liberates a few MeV and so poses no threat to the low-energy approximation.

with $\vec{u}(\theta)$ given explicitly by (8.22). The appropriate covariant derivative for nucleons then becomes

$$D_\mu N = \partial_\mu N - i\vec{\mathcal{A}}_\mu(\theta) \cdot \vec{\tau}\, N, \tag{8.53}$$

with $\vec{\mathcal{A}}_\mu(\theta)$ as written in (8.24).

Invariant Action

These transformation rules determine the most general couplings between nucleons and pions to lowest order in the derivative expansion. The leading terms involve only one derivative and are given by

$$\begin{aligned}\mathfrak{L}_{\pi N} &= -\overline{N}\left[\not{\partial} - \frac{i}{2}\,\vec{\not{\mathcal{A}}}(\theta)\cdot\vec{\tau} + m_N\right]N - \frac{ig_A}{2}\left(\overline{N}\gamma^\mu\gamma_5\vec{\tau}\,N\right)\cdot\vec{e}_\mu(\theta), \\ &= -\overline{N}\,(\not{\partial} + m_N)\,N - \frac{ig_A}{2F_\pi}\left(\overline{N}\gamma^\mu\gamma_5\vec{\tau}N\right)\cdot\partial_\mu\vec{\pi} \\ &\quad - \frac{i}{2F_\pi^2}\left(\overline{N}\gamma^\mu\vec{\tau}N\right)\cdot(\vec{\pi}\times\partial_\mu\vec{\pi}) + \cdots\,,\end{aligned} \tag{8.54}$$

which uses expression (8.23) for $\vec{e}_\mu(\theta)$. The ellipses in the second line represent terms that involve either three or more powers of the pion field, more than two powers of the nucleon field, or involve more than one derivative. These leading-order interactions are characterized by a single new coupling constant, g_A, whose value is related to the rate for neutron β decay.

To see why g_A appears in the rate for neutron decay, recall that the weak currents are related (as discussed for pions above) to the Noether currents for the symmetry group $G = SU_L(2) \times SU_R(2)$. Incorporating nucleons into this derivation to leading order generalizes (8.40) to

$$\begin{aligned}\vec{J}_I^{\,\mu} &= -\left(\vec{\pi}\times\partial^\mu\vec{\pi}\right) + \frac{i}{2}\,\overline{N}\gamma^\mu\vec{\tau}\,N + \cdots\,, \\ \vec{J}_A^{\,\mu} &= F_\pi\,\partial^\mu\vec{\pi} + \frac{ig_A}{2}\,\overline{N}\gamma^\mu\gamma_5\vec{\tau}\,N + \cdots\,.\end{aligned} \tag{8.55}$$

All terms not written explicitly above involve additional factors of the fields $\vec{\pi}$ or N, or more derivatives of these fields. Because the transition $n \to p$ samples the matrix elements of these currents, measurements of the neutron decay rate allow a determination of a value for g_A.

Goldberger–Treiman Relation

Historically, the trilinear pion-nucleon interaction was written as a Yukawa coupling,

$$\mathfrak{L}_{NN\pi} := ig_{NN\pi}\,(\overline{N}\,\gamma_5\vec{\tau}\,N)\cdot\vec{\pi}, \tag{8.56}$$

with the constant $g_{NN\pi}$ found from phenomenological studies to be close to 14 – which is to say, since $g_{NN\pi}/4\pi \sim O(1)$, the strong interactions are not perturbatively weak. The Yukawa coupling $g_{NN\pi}$ can be predicted in terms of the constant g_A, since the trilinear part of (8.54) differs from (8.56) by a field redefinition. To see why, perform an integration by parts in the interaction

$$\mathfrak{L}_{NN\pi} = -\frac{ig_A}{2F_\pi}\left(\overline{N}\gamma^\mu\gamma_5\vec{\tau}N\right)\cdot\partial_\mu\vec{\pi}, \tag{8.57}$$

to move the derivative to the nucleon fields, and simplifying the result using the lowest-order equations of motion for N: *i.e.* $(\not{\partial} + m_N)\,N = 0$.[8] Comparing the result with (8.56) gives

$$g_{NN\pi} = \frac{g_A m_N}{F_\pi}, \tag{8.58}$$

a result known as the *Goldberger-Treiman relation* [188].

The Goldberger–Treiman relation is a phenomenological success[9] inasmuch as neutron decay implies that $g_A \simeq 1.25$, while $m_N = 940$ MeV and $F_\pi = 92$ MeV, and so predict $g_{NN\pi} \simeq 12.8$. But chiral perturbation theory provides a much better framework for understanding pion-nucleon interactions than does the historical guess (8.56) because besides explaining why $g_{NN\pi}$ is so large, it also shows why there are circumstances under which pion-nucleon interactions are nonetheless perturbative. At energy E the expansion parameter to compare to 4π is not a strong-coupling constant like $g_{NN\pi}$, but is instead the energy ratio E/F_π.

Subdominant non-derivative pion-nucleon couplings are also allowed once G-breaking nonzero quark masses are included, suppressed by powers of m_u or m_d. Their systematic description goes beyond the scope of this book (for further reading see Appendix D).

8.2.4 Loops and Logs ◊

Another instructive feature of the chiral perturbation theory example is the need for (and structure of) next-to-leading corrections. Table 8.3 shows that these play a practical role in obtaining agreement with experiment, and their structure illustrates more precisely how EFTs can be predictive even when used at a precision where sub-dominant interactions play a role.

The general power-counting expression (8.42) shows that the next-to-leading contributions to pion scattering arise in two ways: (*i*) through one-loop graphs built using only the lagrangian's two-derivative interactions (or symmetry-breaking zero-derivative interactions), or (*ii*) tree graphs built using only two-derivative interactions plus precisely one 4-derivative interaction (or one subdominant symmetry-breaking interaction). For G-invariant interactions with momenta q the loop graph is suppressed relative to leading order by a factor of $(q/4\pi F_\pi)^2$ while the tree-level graph with a 4-derivative contribution is down by $(q/\Lambda_\chi)^2$. Graphs using the symmetry-breaking interactions replace q^2 in these estimates with m_π^2. All contributions are similar in size when $q \sim m_\pi$ because for QCD $\Lambda_\chi \sim 4\pi F_\pi \sim 1$ GeV.

[8] Since these two forms are related by use of the equations of motion for N, the general arguments of §2.5 imply they are also related by performing a field redefinition of N.

[9] This relation was derived well before the discovery of quarks. At this time the proposal that weak interactions couple to some sort of conserved current helped explain their universal strength [118, 119]. This led to a picture where the pion decays by the formation of a virtual nucleon-antinucleon pair [186] followed by a charged-current nucleon-lepton transition. This related pion decay to the nucleon-pion coupling and nucleon matrix elements arising in neutron β decay – leading to relations like (8.58). These predictions, being mainly consequences of symmetries, proved more robust than their initial derivation might have indicated [188].

There are two types of $SU_L(2) \times SU_R(2)$-invariant 4-derivative interactions, which can be written (up to total derivatives and field redefinitions) as [32]

$$\mathfrak{L}_4 = \mathfrak{c}_{41} \left[\mathrm{Tr}\left(\partial_\mu \Xi^\dagger \partial^\mu \Xi\right)\right]^2 + \mathfrak{c}_{42}\, \mathrm{Tr}\left(\partial_\mu \Xi^\dagger \partial^\mu \Xi\, \partial_\nu \Xi^\dagger \partial^\nu \Xi\right), \tag{8.59}$$

with dimensionless couplings $\mathfrak{c}_{4i}$ and Ξ as defined in (8.36). At the same order a number of symmetry-breaking terms that trade pairs of derivatives for a power of quark mass, M, or all four derivatives for two powers of M, can also be written, but for simplicity are not written here.

With these corrections, the lowest-order result, Eq. (8.43), for the scattering amplitude generalizes to [32]

$$\mathcal{A} = \delta_{ab}\delta_{cd}\, A(s,t,u) + \delta_{ac}\delta_{bd}\, A(t,s,u) + \delta_{ad}\delta_{bc}\, A(u,t,s), \tag{8.60}$$

where the function $A(s,t,u)$ has the expansion

$$A(s,t,u) \simeq A_0(s,t,u) + A_{1\ell}(s,t,u) + A_{1p}(s,t,u), \tag{8.61}$$

with the lowest-order result of (8.43) corresponding to

$$A_0(s,t,u) = \frac{s - m^2}{F^2}, \tag{8.62}$$

where F is the parameter controlling the 2-derivative couplings of (8.25) and

$$m^2 := (m_u + m_d)\,\frac{\Lambda_m^3}{F^2} \sim (m_u + m_d)\,\frac{\Lambda_\chi^3}{F^2} \tag{8.63}$$

is the parameter defined by the leading 0-derivative (symmetry breaking) term of (8.37).

At next-to-leading accuracy the parameters m^2 and F^2 need no longer precisely equal the physical quantities m_π^2 and F_π^2, and indeed are modified to expressions of the form

$$m_\pi^2 = m^2\left[1 + \frac{m^2}{32\pi^2 F^2}\,\log\left(\frac{m^2}{\mu^2}\right) + \cdots\right], \tag{8.64}$$

and

$$F_\pi = F\left[1 - \frac{m^2}{16\pi^2 F^2}\,\log\left(\frac{m^2}{\mu^2}\right) + \cdots\right], \tag{8.65}$$

where the ellipses represent $O(1)$ quantities unaccompanied by logarithms (whose precise form is not required in what follows, beyond the observation that it is their dependence on the couplings $\mathfrak{c}_{4i}(\mu)$ that ultimately cancels the explicit μ-dependence of the logarithm).

The remaining terms in (8.61) correspond to the two types of subdominant corrections, with the one-loop contributions using only lowest-order interactions giving

$$A_{1\ell}(s,t,u) = \frac{1}{96\pi^2 F^4}\Big\{3(s^2 - m^4)\, J(s) + \left[t(t-u) - 2m^2 t + 4m^2 u - 2m^4\right] J(t) + \left[u(u-t) - 2m^2 u + 4m^2 t - 2m^4\right] J(u)\Big\}, \tag{8.66}$$

with

$$J(x) := y(x) \log\left[\frac{y(x)-1}{y(x)+1}\right] + 2, \tag{8.67}$$

where $y(x) := \sqrt{1-(4m^2/x)}$.

The tree contributions using precisely one subdominant interaction similarly give

$$A_{1p}(s,t,u) = \frac{1}{96\pi^2 F^4}\left\{C_1(s-2m^2)^2 + C_2\left[s^2+(t-u)^2\right] - 12m^2 s + 15m^4\right\}, \tag{8.68}$$

where C_1 and C_2 are both UV divergent (containing poles as $\varepsilon \to 0$ in dimensional regularization, with $D = 4 - 2\varepsilon$) and linear in the constants c_{41} and c_{42}. It is the renormalization of these constants that absorbs the divergences, leaving finite parts that depend on the precise renormalization scheme used. Two observables must be used to determine the values of these two renormalized parameters, in addition to the two used earlier to fix the values of m and F.

Decomposing the scattering as before into amplitudes with definite isospin – *c.f.* (8.46) – leads to

$$\mathcal{A}^{(0)} = 3A(s,t,u) + A(t,s,u) + A(u,t,s) \tag{8.69}$$

while

$$\mathcal{A}^{(1)} = A(t,s,u) - A(u,t,s) \quad\text{and}\quad \mathcal{A}^{(2)} = A(t,s,u) + A(u,t,s). \tag{8.70}$$

Further decomposing into partial waves then leads to the order $m^4/(16\pi^2F^4)$ corrections listed in Table 8.3 for the tree-level partial waves b_0^0, a_0^2, b_0^2 and a_1^1 that are already nonzero at leading order. It also leads to order $m^4/(16\pi^2F^4)$ contributions to partial waves such as b_1^1, a_0^2 and a_2^2 that vanish at lowest order.

For the present purposes these subleading corrections illustrate two noteworthy features. First, notice that the coefficients of all terms having logarithmic singularities in the limit $s,t,u \to 0$ – which all involve the function J defined in (8.67) – depend only on the lowest-order coupling parameters m and F. In particular, they do not depend on the values of the unknown new couplings c_{4i}, which appear only premultiplying polynomials in external momenta. They are also not ultraviolet divergent, as might be expected given that UV divergences also arise at one loop as polynomials in momenta, as is required for them to be ultimately renormalized into the coefficients c_{4i}.

This is a general feature: unitarity dictates the imaginary part of $A(s,t,u)$ in the forward-scattering regime in terms of total scattering rates [38, 39], and this requires that next-to-leading contributions to $A(s,t,u)$ that are non-analytic in external momenta are absolute predictions that depend only on the lowest-order couplings.

The second noteworthy feature about the next-to-leading corrections is the presence in them of *chiral logs*; *i.e.* corrections involving factors that are not differentiable in the chiral limit $m^2 \to 0$, such as factors of $m^2 \log m^2$. These terms enter both from the explicit logarithms in $J(x)$ of Eq. (8.67) and from the revised relationships between the parameters F and m and physical quantites like F_π and m_π – *c.f.* Eqs. (8.64) and (8.65).

Chiral logs are important because the expansion in powers of small symmetry breaking parameters (like quark masses) is ultimately an expansion in powers of m^2. Although logarithms of small scales are not tracked in power-counting formulae like (8.42), they can sometimes matter in practice. For instance, the subleading correction to the soft-pion theorem for $\ell = I = 0$ turns out be

$$a_0^0 = \frac{7m_\pi^2}{32\pi^2 F_\pi^2}\left[1 - \frac{9m_\pi^2}{32\pi^2 F_\pi^2} \log \frac{m_\pi^2}{\mu^2} + \cdots \right], \tag{8.71}$$

which for $\mu \sim 1$ GeV is the 25% correction seen in Table 8.3 rather than the few percent that might have been guessed from $m_\pi^2/(4\pi F_\pi)^2$.

8.3 Summary

This chapter summarizes chiral perturbation theory; the modern understanding of the low-energy limit of Quantum Chromodynamics. This starts with a brief summary of QCD itself; in particular, its quark spectrum, asymptotic freedom and the existence of a characteristic QCD scale, $\Lambda_{QCD} \sim 200$ MeV, that sets the scale of quark and gluon momenta within hadrons. In particular, typical hadrons built from quarks lighter than Λ_{QCD} (u, d and possibly s) have masses starting slightly less than 1 GeV.

An octet of pseudo-scalar mesons is somewhat lighter than this, most notably including the lightest hadrons: pions, $\pi^\pm$ and π^0, whose masses are around 140 MeV. These are understood to be unusually light (for hadrons) because they are pseudo-Goldstone bosons for approximate symmetries that emerge in the limit of massless quarks. The symmetries in question are $SU_L(N) \times SU_R(N)$ transformations that rotate the left- and right-handed quarks, where $N = 2$ or 3 is the number of quark flavours lighter than Λ_{QCD}. Both choices for N are possible, depending on whether or not the s quark, with $m_s \sim 100$ MeV, is included.

Being pseudo-Goldstone bosons, the low-energy properties of these lightest hadrons are dictated by the assumed symmetry-breaking pattern, in which the QCD vacuum spontaneously breaks the $SU_L(N) \times SU_R(N)$ symmetry down to its diagonal $SU(N)$ subgroup. This chapter applies the tools of §4.2 to construct the EFT for pseudo-Goldstone bosons that nonlinearly realize the symmetry-breaking pattern $SU_L(2) \times SU_R(2) \to SU_I(2)$, and shows that it provides a good description of pion properties for energies well below 1 GeV. The experimental success of this picture is one reason QCD is regarded as the theory of the strong interactions.

Two new conceptual EFT issues arise in this discussion. One of is these concerns the appearance of particles in a low-energy theory whose mass is higher than the UV scale. Such particles (including baryons like nucleons in low-energy QCD) can make sense in the low-energy limit provided that their rest-energy remains locked up and inaccessible. This could be true if the heavy particle in question cannot itself decay (or does so only after sufficiently long times), or if the system involves negligible numbers of heavy antiparticles, with whom the heavy particles would otherwise annihilate.

The second conceptual issue that chiral perturbation theory illustrates is the practical value of including loop corrections. Besides being necessary for agreement with experiment, loops contribute non-analytic parts to low-energy amplitudes whose coefficients are robust functions of the lowest-order couplings that are largely insensitive to UV divergences and higher-order effective couplings.

Exercises

Exercise 8.1 The parameter Λ_m entering into the mass formula (8.64) can be related to properties of QCD as follows. Suppose the quark masses are replaced by an external local field $m_q \to \chi_q(x)$ in both QCD and the symmetry-breaking lagrangian given in (8.37). Compute the generating functional, $W[\chi]$, to leading order in χ_u and χ_d, both in QCD and in the low-energy effective theory, and by differentiating and evaluating $\delta W/\delta\chi_u$ and $\delta W/\delta\chi_d$ in the vacuum show that

$$\langle\Omega\,|\,\bar{u}u\,|\,\Omega\rangle = \langle\Omega\,|\,\bar{d}d\,|\,\Omega\rangle = -\Lambda_m^3.$$

This reveals the order parameter for the spontaneous symmetry-breaking. It also shows that the value of Λ_m (and so also the value of $m_u + m_d$ inferred from m_π) depends on the precise renormalization prescription used to define $\bar{u}u$ and $\bar{d}d$.

Exercise 8.2 Historically, the effective lagrangian first used for the G-invariant two-derivative interactions with symmetry-breaking pattern $SU_L(2) \times SU_R(2) \to SU_I(2)$ had the form [11]

$$\mathfrak{L}_{\rm eff} = -\frac{1}{2}\,\frac{\partial_\mu\vec{\pi}\cdot\partial^\mu\vec{\pi}}{(1+\vec{\pi}\cdot\vec{\pi}/F^2)^2} - \frac{m_\pi^2}{2}\,\frac{\vec{\pi}\cdot\vec{\pi}}{1+\vec{\pi}\cdot\vec{\pi}/F^2}.$$

Verify that this also leads to the $2 \to 2$ pion scattering amplitude given in Eq. (8.43).

Exercise 8.3 Show that the two-derivative term of the lagrangian in Exercise 8.2 is invariant under the linear isospin rotations, $\delta\vec{\theta} = \vec{\omega}_I \times \vec{\theta}$ (with $\vec{\theta} = \vec{\pi}/F$) as well as the nonlinearly realized broken transformations given by

$$\delta\vec{\theta} = \vec{\omega}_A(1 - \vec{\theta}\cdot\vec{\theta}) + 2\vec{\theta}\,(\vec{\omega}_A\cdot\vec{\theta}).$$

Identify the local field redefinition required to put this transformation into the standard form given in Eq. (8.19). This extends the results of Exercise 4.2 to include nonzero masses.

Exercise 8.4 Find the leading graphs that contribute within the effective theory to low-energy pion-nucleon scattering. Show that these give the leading-order predictions

$$a_0^{(1/2)} \simeq \frac{m_\pi}{\pi F_\pi^2} \quad\text{and}\quad a_0^{(3/2)} \simeq -\frac{m_\pi}{2\pi F_\pi^2},$$

for the s-wave scattering lengths for each of the two isospin channels.

Exercise 8.5 u, d and s quarks are arguably light enough to apply chiral perturbation theory, though the lightest mass for a meson containing an s quark is $m_K \simeq 500$ MeV (instead of $m_\pi \simeq 140$ MeV for u and d quarks). Using $x\log x$ with $x = [m/(4\pi F_\pi)]^2$ as the figure of merit, estimate the relative accuracy of chiral perturbation theory using s quarks compared with u and d quarks.

Compute the nonlinear realization for the symmetry-breaking pattern $G \to H$ with $G = SU_L(3) \times SU_R(3)$ and $H = SU_F(3)$, where the unbroken 'flavour' $SU_F(3)$ is the diagonal subgroup of the two factors in G. In particular, compute the quantities $\mathcal{A}^i_\mu(\theta)$ and $e^\alpha{}_\mu(\theta)$ that generalize (8.23) and (8.24) to the 3-quark case.

Exercise 8.6 Compute the invariant two-derivative action for the three-quark symmetry-breaking pattern of Exercise 8.5. (Expressing the result using the analog of the quantity Ξ defined in the main text might prove useful when doing so.) Apply Noether's theorem to this action to derive an expression for the vector and axial-vector $SU(3)$ currents relevant to the weak interactions. Identify the quartic interactions amongst the mesons predicted by this lagrangian. When doing so, it is useful to relate the fields θ^α to the physical octet of pseudoscalar mesons using

$$\lambda_\alpha \theta^\alpha = \frac{1}{F}\begin{pmatrix} \pi^0 + \frac{1}{\sqrt{3}}\eta_8 & \sqrt{2}\,\pi^+ & \sqrt{2}\,K^+ \\ \sqrt{2}\,\pi^- & -\pi^0 + \frac{1}{\sqrt{3}}\eta_8 & \sqrt{2}\,K^0 \\ \sqrt{2}\,K^- & \sqrt{2}\,\overline{K}^0 & -\frac{2}{\sqrt{3}}\eta_8 \end{pmatrix},$$

where λ_α are the eight 3×3 Gell-Mann matrices that play the same role for $SU(3)$ as the three Pauli matrices do for $SU(2)$ (and in particular satisfy $\mathrm{Tr}\,(\lambda_\alpha \lambda_\beta) = 2\delta_{\alpha\beta}$). The subscript '8' on η_8 is meant to distinguish it from the $SU(3)$-singlet state (usually denoted η_0). Because of $SU(3)$-breaking effects, the physical mass eigenstates, η and η', are linear combinations of η_8 and η_0.

Identify all of the G-invariant interactions that arise at 4-derivative level, and show that there are three independent ones:

$$\left[\mathrm{Tr}\left(\partial_\mu \Xi^\dagger \partial^\mu \Xi\right)\right]^2, \ \mathrm{Tr}\left(\partial_\mu \Xi^\dagger \partial^\mu \Xi\, \partial_\nu \Xi^\dagger \partial^\nu \Xi\right),$$
$$\mathrm{Tr}\left(\partial_\mu \Xi^\dagger \partial_\nu \Xi\right) \mathrm{Tr}\left(\partial^\mu \Xi^\dagger \partial^\nu \Xi\right).$$

Why are there more here than the two found in (8.59)?

Exercise 8.7 Prove that the electromagnetic interactions of the pseudoscalar octet of mesons explicitly break $SU_L(3) \times SU_R(3)$ but do not break an $SU_U(2) \times U(1)$ subset of each of the $SU(3)$ factors. As a consequence of this, show that the electrically neutral pseudoscalar octet mesons do not receive mass corrections due to the electromagnetic interactions.

Use this unbroken symmetry to argue that the electromagnetic corrections to the K^+ and π^+ masses must be equal for zero quark masses [193] (see also [184]).

Exercise 8.8 For the symmetry-breaking pattern of Exercise 8.5 compute the leading symmetry-breaking terms involving the quark mass matrix

$$M = \begin{pmatrix} m_u & & \\ & m_d & \\ & & m_s \end{pmatrix},$$

and show that it – once summed with the electromagnetic corrections, $\Delta_{\rm em}$, described in Exercise 8.7 – leads to the mass formulae [182, 190]

$$m^2_{\pi^0} = (m_u + m_d)\,\frac{\Lambda^3_m}{F^2_\pi}, \quad m^2_{\pi^+} = (m_u + m_d)\,\frac{\Lambda^3_m}{F^2_\pi} + \Delta_{\rm em}$$
$$m^2_{K^0} = (m_d + m_s)\,\frac{\Lambda^3_m}{F^2_\pi}, \quad m^2_{K^+} = (m_u + m_s)\,\frac{\Lambda^3_m}{F^2_\pi} + \Delta_{\rm em}$$

$$m_{\eta_8}^2 = (m_u + m_d + 4m_s)\frac{\Lambda_m^3}{3F_\pi^2},$$

as well as a π^0-η_8 mixing term

$$m_{\pi\eta_8}^2 = (m_u - m_d)\frac{\Lambda_m^3}{\sqrt{3}\,F_\pi^2}.$$

One linear combination of diagonal mass terms is independent of m_u, m_d, m_s and $\Delta_{\rm em}$, leading to a version of the Gell–Mann–Okubo mass relations [191, 192]

$$3m_{\eta_8}^2 + 2m_{\pi^+}^2 - m_{\pi^0}^2 = 2m_{K^+}^2 + 2m_{K^0}^2.$$

Derive also the following formulae expressing the quark mass ratios, m_u/m_s and m_d/m_s, as ratios of measurable pion and K meson masses,

$$\frac{m_d}{m_s} = \frac{m_{K^0}^2 + m_{\pi^+}^2 - m_{K^+}^2}{m_{K^0}^2 - m_{\pi^+}^2 + m_{K^+}^2} \quad \text{and} \quad \frac{m_u}{m_s} = \frac{2m_{\pi^0}^2 - m_{K^0}^2 - m_{\pi^+}^2 + m_{K^+}^2}{m_{K^0}^2 - m_{\pi^+}^2 + m_{K^+}^2}.$$

Because these do not depend on Λ_m the experimental ratios of quark masses, $m_d/m_s \simeq 0.053\pm0.002$ and $m_u/m_s \simeq 0.029\pm0.003$ are known more accurately than are their absolute values.

9 The Standard Model as an Effective Theory

In the applications considered up to this point the Standard Model plays the role of the underlying UV theory whose low-energy limit is to be explored using the EFT of interest. But it is also instructive to regard the Standard Model itself as an effective theory describing the low-energy limit of whatever 'new physics' ultimately replaces the Standard Model once new phenomena at much shorter distances become accessible to experiments.

There are two reasons why this point of view is useful. First, the fact that it is even possible says something about the robustness of the assumptions on which the Standard Model rests. Earlier sections show that low-energy limits are generically described by renormalizable theories: those theories involving (*i*) only couplings with non-negative mass dimension – $g_{\rm eff} \propto ({\rm mass})^p$ for $p \geq 0$ – and (*ii*) including *all* such couplings consistent with the assumed field content and symmetries.

The Standard Model ticks these two boxes, and (as argued below) does so in a nontrivial way. It does so because it is the most general theory possible that can describe the low-energy limit of the known elementary particles, together with their transformation properties under the gauge symmetries.[1] If the Standard Model fails, its failure implies either the existence of new light elementary particles in the low-energy theory or the failure of the low-energy limit (and the associated new physics at not-too-high scales that this implies).

The good news is that – as of this writing (2019) – the best evidence is that the Standard Model *does* fail, though only in a few specific ways. Its failures include providing no understanding of Dark Matter, of cosmological primordial fluctuations or of neutrino oscillations (more about which below). And this failure raises the second reason why it is useful to think of the Standard Model as an effective theory: the EFT point of view suggests how to think about what kinds of new phenomena might exist to be sought in experiments.

The examples described in the remainder of this section are meant to illustrate these points more concretely. But first we present a lightning summary of the Standard Model's field content and gauge symmetries, following the notation of [194].

[1] The requirement that there be gauge symmetries is not a separate condition once the low-energy theory includes very light spin-one particles (see the discussion leading to Eq. (4.73) as well as the arguments for massless spin-one particles in §C.3.3).

9.1 Particle Content and Symmetries♡

The Standard Model portrays the strong, weak and electromagnetic interactions as being mediated by the exchange of spin-one particles, each of which is the gauge boson of a local symmetry group. The symmetry group is chosen to be $SU_C(3) \times SU_L(2) \times U_Y(1)$, with eight gluons, G^α_μ, corresponding to the eight generators of the colour gauge group $SU_C(3)$, while the three gauge bosons, W^a_μ, for $SU_L(2)$ and the single $U_Y(1)$ boson B_μ correspond to linear combinations of the photon, the $W^\pm$ and the Z^0 boson.

The structure of the Standard Model follows from the way its field content transforms under these symmetries. It is conventional to summarize the transformation properties of any particular field, Ψ, in the format

$$\Psi \sim \Big(\mathbf{n_3}, \mathbf{n_2}, y\Big), \tag{9.1}$$

where n_3 is the dimension of the $SU_C(3)$ representation, n_2 is the dimension of the $SU_L(2)$ representation and y is the 'charge' – called *hypercharge* – of the field under the $U_Y(1)$ transformation.

Having hypercharge y means the field in question transforms under $U_Y(1)$ as $\Psi \to \Psi\, e^{iyg_1\omega_1}$, where g_1 is the $U_Y(1)$ coupling constant and ω_1 is its transformation parameter (normalized so that $B_\mu \to B_\mu + \partial_\mu\, \omega_1$). In the Standard Model the hypercharge assignments are obtained by relating the generator, Y, to the electric charge matrix, $Q_{\rm em}$, and the diagonal $SU_L(2)$ generator, T_3, by

$$Q_{\rm em} = T_3 + Y. \tag{9.2}$$

In this notation the transformation properties of the gauge bosons can be written

$$G^\alpha_\mu \sim \Big(\mathbf{8}, \mathbf{1}, 0\Big) \qquad W^a_\mu \sim \Big(\mathbf{1}, \mathbf{3}, 0\Big) \qquad B_\mu \sim \Big(\mathbf{1}, \mathbf{1}, 0\Big), \tag{9.3}$$

because $SU_C(3)$ has eight generators and $SU_L(2)$ has three.

Besides spin-one particles, the Standard Model's other boson is the scalar Higgs-doublet field, Φ, that transforms under the gauge symmetry as

$$\Phi = \begin{pmatrix} \phi^+ \\ \phi^0 \end{pmatrix} \sim \left(\mathbf{1}, \mathbf{2}, \frac{1}{2}\right). \tag{9.4}$$

Notice that because $T_3 = \frac{1}{2}\,\tau_3$ has eigenvalues $\pm\frac{1}{2}$ in the doublet representation, Eq. (9.2) gives the electric charges of the two component fields, ϕ^+ and ϕ^0, to be as indicated by their names: $q(\phi^+) = \frac{1}{2} + \frac{1}{2} = +1$ and $q(\phi^0) = -\frac{1}{2} + \frac{1}{2} = 0$.

The Standard Model's scalar potential,

$$V = \lambda \left(\Phi^\dagger \Phi - \frac{v^2}{2}\right)^2, \tag{9.5}$$

is designed to be minimized at a nonzero value $\Phi^\dagger\Phi = \frac{1}{2}\, v^2$, which spontaneously breaks the $SU_L(2) \times U_Y(1)$ subgroup down to the unbroken $U_{\rm em}(1)$ of electromagnetism. Because of this symmetry-breaking pattern, three of the four real scalars contained within the complex doublet Φ are eaten by the Higgs mechanism to give the W and Z bosons masses, leaving the massless photon and a single real physical-Higgs field, h. The resulting particle spectrum and interactions are simple to identify in *unitary gauge*, defined by the condition

$$\Phi = \frac{1}{\sqrt{2}} \begin{pmatrix} 0 \\ v + h \end{pmatrix}. \tag{9.6}$$

In particular, a simple calculation shows the W-boson mass to be $M_W = \frac{1}{2} g v$ (where $g = g_2$ is the $SU_L(2)$ gauge coupling constant), which, when used in (7.8), implies that v is related to the value of Fermi's coupling, G_F, by

$$G_F = \frac{1}{\sqrt{2}\, v^2}, \tag{9.7}$$

and so the measured value $G_F \simeq 1.166 \times 10^{-5}$ GeV^{-2} implies that $v \simeq 246$ GeV.

The remaining fields in the Standard Model are spin-half fermions, and for the present purposes what is interesting is how these fields transform under the $SU_C(3) \times SU_L(2) \times U_Y(1)$ gauge symmetry. As mentioned earlier – see, for example, Table 7.1 – these fermions come in triplicate inasmuch as there are three identical copies of a basic set, or generation, of fermions. Each generation transforms in precisely the same way as do all the other generations under the gauge group. For notational simplicity the generational index that distinguishes these copies – *e.g.* the index $n = 1, 2, 3$ of e_n that distinguishes e from μ and τ – is suppressed in much of what follows.

A key characteristic of the fermions' gauge transformations in the Standard Model is that they are *chiral*, with left- and right-handed particles transforming differently. Because of this, it is useful to adopt notation that treats left- and right-handed fermions separately. Because the antiparticle of a left-handed fermion is right-handed, and vice versa, the convention used here is to track only those fields that destroy left-handed fermions and left-handed anti-fermions (as opposed to tracking the fields that destroy the left- and right-handed parts of only fermions).[2] This way of organizing fields does not pre-judge what is a fermion and what is an anti-fermion, or whether fermions and anti-fermions are distinct from one another.

For instance, the Dirac spinor field representing the electron is often written in terms of left- and right-handed parts,

$$e = \begin{pmatrix} e_L \\ e_R \end{pmatrix} \qquad \text{(spinor space)}, \tag{9.8}$$

in Lorentz-spinor space, where $e_L(x) \sim \sum_{p\sigma} [u(p,\sigma)\, \mathfrak{a}_{p\sigma} + v(p,\sigma)\, \bar{\mathfrak{a}}^*_{p\sigma}]$ both destroys left-handed electrons and creates right-handed positrons while $e_R(x)$ destroys right-handed electrons and creates left-handed positrons. When listing all the fermions, rather than using e_L and e_R this section instead uses e_L and $e_L^c = \varepsilon\,(e_R)^*$, which destroys left-handed positrons and creates right-handed electrons. (Similarly, $e_R^c = -\varepsilon\,(e_L)^*$ destroys right-handed positrons and creates left-handed electrons. Here, $\varepsilon = i\sigma_2$ is a 2×2 antisymmetric matrix in spinor space, defined below Eq. (A.28).)

[2] These are not propagation eigenstates because the lagrangian's mass terms can mix them.

This notation is useful because neither e_L^c nor e_R transform in the same way as e_L under $SU_L(2) \times U_Y(1)$ transformations. For instance, whereas the left-handed electron and neutrino fields transform as a doublet[3] under $SU_L(2)$,

$$L_L = \begin{pmatrix} \nu_L \\ e_L \end{pmatrix} \qquad (SU_L(2) \text{ space}), \tag{9.9}$$

the right-handed electron, e_R (and so also e_L^c), is a singlet (and, in the Standard Model, the right-handed neutrino ν_R does not exist). They similarly differ in the their value of $U_Y(1)$ hypercharge, since for the doublet L we have $y_L = y(L_L) = -\frac{1}{2}$ and for the right-handed electron $y(e_R) = -1$ and so $y_E = y(e_L^c) = +1$.

With these conventions the gauge-transformation rules for the left-handed particle content of a single generation within the Standard Model may be summarized by:

$$\begin{pmatrix} \nu_L \\ e_L \end{pmatrix} \sim \left(\mathbf{1}, \mathbf{2}, -\frac{1}{2}\right) \quad \text{and} \quad \begin{pmatrix} u_L \\ d_L \end{pmatrix} \sim \left(\mathbf{3}, \mathbf{2}, +\frac{1}{6}\right), \tag{9.10}$$

for the $SU_L(2)$ doublets and

$$e_L^c \sim \Big(\mathbf{1}, \mathbf{1}, +1\Big), \quad u_L^c \sim \left(\bar{\mathbf{3}}, \mathbf{1}, -\frac{2}{3}\right) \quad \text{and} \quad d_L^c \sim \left(\bar{\mathbf{3}}, \mathbf{1}, +\frac{1}{3}\right), \tag{9.11}$$

for the rest. Here, $\bar{\mathbf{3}}$ denotes the conjugate of the $\mathbf{3}$ representation – which are inequivalent representations for the group $SU_C(3)$.

Although not present in the Standard Model, if there were a right-handed neutrino the requirement that it could combine with the left-handed neutrino and Higgs boson to give a Yukawa coupling (and so a neutrino mass once the Higgs acquires an expectation value) requires it to transform as

$$\nu_L^c \sim (\mathbf{1}, \mathbf{1}, 0)\,. \tag{9.12}$$

It would therefore not couple at all to Standard Model gauge bosons, and as a result could plausibly be very hard to detect (and so might well exist without having been detected). Hypothetical particles like this that do not transform at all under $SU_C(3) \times SU_L(2) \times U_Y(1)$ are called *sterile*, and are often called the 'dark sector' in theories that propose their existence. Many potential explanations for dark matter are founded on the idea that such a dark sector could exist and have remained undiscovered despite not being particularly heavy compared with experimentally accessible energies.

9.1.1 The Lagrangian

The Standard Model lagrangian is defined to be the most general renormalizable interactions amongst the fields discussed above that are invariant under the gauge group $SU_C(3) \times SU_L(2) \times U_Y(1)$. This consists of the terms

$$\mathfrak{L}_{SM} = \mathfrak{L}_{\text{gauge}} + \tilde{\mathfrak{L}}_{\text{gauge}} + \mathfrak{L}_{\text{Higgs}} + \mathfrak{L}_{f\text{kin}} + \mathfrak{L}_{\text{yuk}} \tag{9.13}$$

[3] As emphasized explicitly, although they look similar, Eqs. (9.8) and (9.9) are vectors in *different* spaces: the spinor space of Lorentz spinors for (9.8) and the $SU_L(2)$ gauge space for (9.9).

where $\mathfrak{L}_{\text{gauge}}$ has the form of a sum of the QCD gauge kinetic term given in (8.1), plus its counterparts for the $SU_L(2) \times U_Y(1)$ group factors:

$$\mathfrak{L}_{\text{gauge}} = -\frac{1}{4} G^{\alpha}_{\mu\nu} G^{\mu\nu}_{\alpha} - \frac{1}{4} W^{a}_{\mu\nu} W^{\mu\nu}_{a} - \frac{1}{4} B_{\mu\nu} B^{\mu\nu}. \tag{9.14}$$

The gluon field strength, $G^{\alpha}_{\mu\nu}$, is defined in Eq. (8.2), while $W^{a}_{\mu\nu} = \partial_\mu W^a_\nu - \partial_\nu W^a_\mu + g_2\, \epsilon^a{}_{bc} W^b_\mu W^c_\nu$ and $B_{\mu\nu} = \partial_\mu B_\nu - \partial_\nu B_\mu$ are its counterparts for $SU_L(2)$ and $U_Y(1)$ (see §C.5 for details). The term $\tilde{\mathfrak{L}}_{\text{gauge}}$ contains the parity and time-reversal violating counterparts to $\mathfrak{L}_{\text{gauge}}$, involving $\epsilon^{\mu\nu\lambda\rho} G^{\alpha}_{\mu\nu} G_{\alpha\lambda\rho}$, and its analogs for the other two gauge groups.

The Higgs part of the action, $\mathfrak{L}_{\text{Higgs}}$, is given by a scalar kinetic term $-D_\mu \Phi^\dagger D^\mu \Phi$ (with $D_\mu \Phi = \partial_\mu \Phi - \frac{i}{2} g_2 \tau_a W^a_\mu \Phi - \frac{i}{2} g_1 B_\mu \Phi$ the appropriate gauge covariant derivative) plus the scalar potential of Eq. (9.5); the term $\mathfrak{L}_{f\text{kin}}$ consists of a kinetic term $\overline{\psi}\, \not{D} \psi$ for each fermion field in the problem, with $D_\mu \psi$ being the covariant derivative appropriate to its transformation properties.

Most of the parameters of the Standard Model can be found in the last term of (9.13), containing its Yukawa couplings. Suppressing all gauge and Lorentz indices while re-introducing the generation labels $m, n = 1, 2, 3$, gives the most general Yukawa interactions as[4]

$$-\mathfrak{L}_{\text{yuk}} = f_{mn} (\overline{L}_m \gamma_R E_n)\, \Phi + h_{mn} (\overline{Q}_m \gamma_R D_n)\, \Phi + g_{mn} (\overline{Q}_m \gamma_R U_n)\, \widetilde{\Phi} + \text{c.c.} \tag{9.15}$$

where (suppressing indices) the $SU_L(2)$-singlet Majorana fields satisfy $\gamma_R U = u_R$, $\gamma_R D = d_R$ and $\gamma_R E = e_R$, while the $SU_L(2)$ doublets are given by (9.4) and

$$\widetilde{\Phi} = i\tau_2 \Phi^* = \begin{pmatrix} \phi^{0*} \\ -\phi^{+*} \end{pmatrix} \quad \text{while} \quad \gamma_L Q = \begin{pmatrix} u_L \\ d_L \end{pmatrix} \quad \text{and so} \quad \overline{Q}\, \gamma_R = \begin{pmatrix} u_L^\dagger \\ d_L^\dagger \end{pmatrix} i\gamma^0, \tag{9.16}$$

and similarly for $\gamma_L L$ and the lepton doublet in (9.10). The dimensionless coefficients f_{mn}, g_{mn} and h_{mn} are symmetric 3×3 matrices in generation space.[5] The interactions in $\mathfrak{L}_{\text{yuk}}$ represent fermion mass terms once Φ is replaced by its expectation value using (9.6).

Accidental Symmetries

It is worth noting some things that do *not* appear in the Standard Model lagrangian, since these contribute to its overall smell of rightness.

First, notice that all terms in the Standard Model lagrangian are by accident invariant under a common rotation for all quarks – $q_L \to e^{i\omega_b} q_L$ and $q^c_L \to e^{-i\omega_b} q^c_L$ with ω_b constant – provided all other fields are held fixed. This is 'by accident' in as much as this symmetry was not an input of the theory, so invariance (modulo anomalies – more about which below) is an automatic consequence of the assumed gauge symmetries together with the gauge quantum numbers of the known particles and the condition of renormalizability [195].

[4] This is written using Majorana spinors, both because these minimize the proliferation of sub- and superscripts and because they are more convenient when writing general higher-dimension interactions. For definitions, in general and for Standard Model fermions, see §A.2.3.

[5] One way to see they are symmetric is to work in unitary gauge for which $\phi^+ = 0$. In this gauge the couplings multiply explicitly symmetric operators like $\overline{u}_{Lm}\, u_{Rn} = (u^c_{Rm})^T \epsilon\, u_{Rn}$.

The conserved charge implied by this global symmetry is baryon number, which only quarks carry (with all quarks assigned a charge $B = 1/3$ so a proton carries $B = 1$). For the Standard Model (classical) conservation of B is a prediction since it is impossible to build renormalizable interactions using only Standard Model fields that violate B conservation. Such an economical prediction is a great success given that all observations to date are consistent with B conservation.

The same is true for lepton-number rotations $L_L \to e^{i\omega_l} L_L$ and $e_L^c \to e^{-i\omega_l} e_L^c$ with ω_l constant and all other fields held fixed. That is, all gauge-invariant renormalizable interactions for Standard Model particles are automatically also invariant (again up to anomalies – see below) under a common global phase change for all leptons. The conserved quantity associated with this 'accidental symmetry' is lepton number.

From an EFT point of view, these properties arise within the context that the Standard Model is likely just the low-energy part of a more complete theory involving much higher energies than $v \simeq 246$ GeV. Indeed, the Planck mass – $M_p = (8\pi G_N)^{-1/2} \sim 10^{18}$ GeV – inferred from Newton's constant of gravitation provides circumstantial evidence that much higher energies actually do exist in nature. At low energies relative to this renormalizable interactions should dominate. So if only Standard-Model particles survive at low energies then no baryon- or lepton-number-breaking interaction are possible at leading (zeroth) order in the low-energy expansion. Crucially, this is true regardless of whether or not the underlying UV completion itself preserves B or L at higher energies. The EFT perspective shows how B and L conservation can emerge as approximate symmetries at low energy, and so hints at a deep explanation as to why these symmetries work so well in nature.

Because of the Standard Model's assumed absence of right-handed neutrinos, leptons actually enjoy a larger set of accidental symmetries than the previous paragraphs indicate. Rather than being invariant only under a common phase rotation for all leptons, the Standard Model is invariant (again, up to potential anomalies) if each generation of lepton is rotated separately. These additional accidental symmetries imply separate conservation for lepton number for each generation: L_e, L_μ and L_τ. Though the τ had not yet been discovered [196], separate conservation for L_e and L_μ had been hypothesized [114, 197] before the Standard Model's discovery to explain properties of various radioactive decays.[6]

The Standard Model also hints at the existence of accidental approximate symmetries. The only term in the Standard Model with a dimensionful coupling is the Higgs-potential term, $\mu^2 \Phi^\dagger \Phi$, where $\mu^2 = \lambda v^2$, perhaps hinting at an underlying approximate scale invariance. In particular, there are no mass terms at all for the gauge or fermionic fields until Φ acquires its nonzero expectation value. Only one 'relevant' interaction arises because there are no $SU_c(3) \times SU_L(2) \times U_Y(1)$ invariants involving only two fermion or gauge fields with no derivatives. For the fermions this is a consequence of the gauge symmetries acting chirally (*i.e.* acting differently on the left- and right-handed fields).

Within an EFT context the absence of gauge-invariant fermion masses (without a nonzero Higgs vev) is the start of an explanation as to *why* the observed particles

---

[6] Only comparatively recent evidence [198–202] for neutrino oscillations [203, 204] shows that separate electron, muon and tau-number symmetry is not exactly conserved – more about which below (and in Chapter 16).

are light enough to be observed in the first place. In any more complete theory at higher energies the coefficient of any explicit mass term might also be expected to be large and so, if present, would give Standard Model particles large masses (perhaps at the Planck scale). But such terms are forbidden because Standard Model fermions transform in a chiral representation of the gauge group, so their masses are 'protected' from being very large provided there is a reason why the vacuum value, v, for the Higgs should be much smaller than any UV scales. Although it does not explain why v – or μ in the Higgs potential – should be small in the first place, this argument does suggest that if you find fermions in the low-energy limit of a more fundamental theory, it shouldn't be a surprise to find them transforming in chiral representations of any low-energy gauge symmetries.

A final approximate accidental symmetry starts from the fact that most fermion masses are actually quite small compared with $v \simeq 246$ GeV. In the Standard Model this happens because the corresponding fermion Yukawa couplings, $y_f = m_f/v$, are themselves small. But in the limit where these couplings vanish, and so $\mathfrak{L}_{\rm yuk}$ is completely turned off, the Standard Model fermions enjoy a very large accidental classical generation symmetry:

$$G_G = U_Q(3) \times U_L(3) \times U_E(3) \times U_U(3) \times U_D(3), \tag{9.17}$$

under which fermions in each representation of the gauge group given in (9.10) and (9.11) are rotated by separate 3×3 unitary transformations in flavour space. Although it goes beyond the scope of this book to describe it, these approximate symmetries and their breaking by the Yukawa couplings explain observed patterns of observed flavour-violating transitions (and sometimes their absence) in many reactions [205, 206].

9.1.2 Anomaly Cancellation ♣

Defining the Standard Model as the most general renormalizable theory consistent with local $SU_C(3) \times SU_L(2) \times U_Y(1)$ transformations reveals the central role played in it by gauge invariance. This is no accident. As argued in §C.3.3, gauge symmetries are required for massless spin-one particles by the interplay between the requirements of Lorentz invariance and unitarity in a quantum system. §4.2.3 extends this argument to massive spin-one particles that are sufficiently light compared with integrated-out UV scales.

Yet gauge invariance is only checked above at the classical level, by asking when the Standard Model lagrangian density is invariant, even though quantum effects are known to break classical symmetries.[7] Furthermore, quantum symmetry-breaking typically happens precisely when the transformations in question act chirally on the fields, as they do in the Standard Model. Since consistency of the Standard Model fails if such quantum effects break its gauge group, $SU_C(3) \times SU_L(2) \times U_Y(1)$ transformations must be *anomaly free* [207–209]. From an EFT perspective, this is required if the Standard Model is to emerge as a low-energy limit from some sensible

[7] As discussed in §4.3, quantum symmetry-breaking effects can be regarded as non-invariance of the path integral measure under the would-be symmetry, and when this happens the symmetry is said to be 'anomalous'.

higher-energy theory; a special case of the general principle of anomaly matching discussed in §4.3.

This section verifies that although any one species of particle actually *does* break the Standard Model gauge group at the quantum level, this breaking cancels once summed over all the particles in a single generation. The absence of gauge anomalies in the Standard Model happens in an interesting way, with gauge invariance ultimately relying on an intricate cancellation that depends on the specific hypercharge assignments of members of a fermion generation have relative to one another. These relations between hypercharge assignments turn out to explain the very precise agreement between the absolute values of the proton and electron charges [210]. The intricacy of these anomaly cancellations seems to provide a clue about how nature works at a fundamental level.

Anomaly Cancellation in the Standard Model

Recall from §4.3.1 that in four dimensions anomalies are proportional to the coefficient

$$A_{abc} = \mathrm{tr}\left(T_a\,, \left\{T_b\,T_c\right\}\right), \tag{9.18}$$

where T_a denote the (possibly reducible) generators of the symmetry group acting on a column vector built from all left-handed fermions (*i.e* all left-handed particles plus all the left-handed antiparticles of all right-handed fermions). This section evaluates these anomaly coefficients, A_{abc}, for all choices of T_a in the algebra for $SU_c(3) \times SU_L(2) \times U_Y(1)$, showing that $A_{abc} = 0$ when the trace is over all left-handed fermions in a single generation.

To this end, it is useful to adopt the notation $A(3,3,3)$ for the anomaly coefficient involving three generators all taken from within the subgroup $SU_c(3)$; $A(3,2,2)$ for the coefficients with one generator taken from $SU_c(3)$ and the other two from $SU_L(2)$; and so on, considering each possible combination in turn.

A(3, 3, 3): The $SU_c(3)$ representations are all left–right symmetric because both left- and right-handed quarks transform in the triplet **3** representation (and so left-handed antiquarks transform as anti-triplet $\bar{\mathbf{3}}$'s). The left-handed representation is therefore pseudo-real so the general arguments of §4.3.1 ensure that all of these anomaly coefficients vanish.

A(2, 2, 2): All representations of $SU_L(2)$ are pseudo-real (for the doublet representations this follows from the Pauli-matrix identity $\tau_2 \tau_i^* \tau_2 = -\tau_i$) and so these anomaly coefficients also vanish.

A(3, 3, 2) and **A(3, 2, 2)**: Because the groups $SU_c(3)$ and $SU_L(2)$ act on different spaces, the generators can be written as direct products of the form $T_\alpha = t_\alpha \otimes I$ for generators of $SU_c(3)$ and $T_i = I \otimes t_i$ for generators of $SU_L(2)$, with the first factor acting on colour indices and the second factor on $SU_L(2)$ doublet indices.

Because the generators factorize, the $A(3,3,2)$ anomaly coefficients also factorize and so are proportional to the trace over a single $SU_L(2)$ generator. Similarly, $A(3,2,2)$ coefficients are proportional to a single $SU_c(3)$ generator. But all generators of $SU(N)$ are traceless, for any N, due to the requirement that the group elements of $SU(N)$ have unit determinant (that is what the 'S' means). Consequently, these anomaly coefficients must also vanish, as must all others involving only a single $SU_c(3)$ generator or a single $SU_L(2)$ generator like $A(3, 2, 1)$, $A(3, 1, 1)$ and $A(2, 1, 1)$.

A(3, 3, 1): For the same reasons as given above this coefficient is proportional to $\text{tr}_c\, Y$, where the subscript on the trace means the sum is only over the hypercharge eigenvalues for each left-handed field that also carries colour. The anomaly coefficient (for a single generation) therefore is

$$A(3,3,1) \propto y(u_L) + y(d_L) + y(u_L^c) + y(d_L^c)$$
$$= 2y_Q + y_U + y_D = 2\left(\frac{1}{6}\right) + \left(-\frac{2}{3}\right) + \frac{1}{3} = 0. \tag{9.19}$$

A(2, 2, 1): The same argument as in the previous case shows this anomaly coefficient is proportional to the sum over the weak hypercharge eigenvalues of only the $SU_L(2)$-doublet left-handed fermions:

$$A(2,2,1) \propto \sum_{\text{LH doublets}} y = y_L + 3y_Q = \left(-\frac{1}{2}\right) + 3\left(\frac{1}{6}\right) = 0, \tag{9.20}$$

where the factor of 3 in front of y_Q arises from the sum over colours.

A(1, 1, 1): This coefficient is proportional to the sum over the cube of the hypercharge eigenvalue for all left-handed fermions:

$$A(1,1,1) \propto \sum_{\text{all LH}} y^3 = 2y_L^3 + y_E^3 + 6y_Q^3 + 3y_U^3 + 3y_D^3$$
$$= 2\left(-\frac{1}{2}\right)^3 + (+1)^3 + 6\left(\frac{1}{6}\right)^3 + 3\left(-\frac{2}{3}\right)^3 + 3\left(\frac{1}{3}\right)^3 = 0. \tag{9.21}$$

One further anomaly-cancellation condition must also be checked. This involves anomalies for the Lorentz group of spacetime symmetries [210]. The Lorentz group is also a gauge symmetry because it is a subset of the broader gauge symmetries of General Relativity, which is also part of the low-energy effective theory (see §10). Anomalies involving the Lorentz group must also cancel in order not to ruin the consistency of the description of the massless spin-two graviton (see, for example, §C.3.4).

A(J, J, 1): The only Standard Model particles not in real representations of the Lorentz group are again the fermions. Since Lorentz transformations act on spin-half fermions essentially as $SU(2)$ transformations, anomaly-cancellation arguments are similar to those for an $SU(2)$ gauge group. It follows that the only anomaly coefficient that does not trivially vanish is $A(J, J, 1)$, where J is a Lorentz generator. This anomaly coefficient is proportional to the trace of the weak hypercharge over all left-handed fermions, which for a single generation of Standard Model fermions is itself proportional to

$$\sum_{\text{all}} y = 2y_L + y_E + 6y_Q + 3y_U + 3y_D$$
$$= 2\left(-\frac{1}{2}\right) + (+1) + 6\left(\frac{1}{6}\right) + 3\left(-\frac{2}{3}\right) + 3\left(\frac{1}{3}\right) = 0. \tag{9.22}$$

It is clear that cancellation of gauge anomalies within the Standard Model requires a number of relationships amongst the fermionic weak hypercharge assignments within any given generation. Indeed, Eqs. (9.19) through (9.22) impose four conditions on the five hypercharges y_Q, y_L, y_E, y_U and y_D, generically fixing them all up to overall normalization. (The overall normalization cannot be fixed by anomaly cancellation conditions alone since these remain invariant when all y_i's are scaled by a common factor.)

One way to think about this is to imagine whether it would be possible to choose yukawa couplings to be slightly different than in the Standard Model, such as by having $y_L = -\frac{1}{2} + \epsilon$ (and similarly for other yukawa couplings) for some small ϵ. In general, such a choice would make electric charges slightly different than usual, and, in particular, could allow the positron and proton charges to differ slightly from one another.

Anomaly cancellation severely restricts the freedom to do so within the Standard Model. To see why, notice (9.20) implies that $y_L = -3y_Q$ and so the electron and quark electric charges are both given in terms of y_Q, with $q_e = -\frac{1}{2} + y_L = -\frac{1}{2} - 3y_Q$, while $q_u = \frac{1}{2} + y_Q$ and $q_d = -\frac{1}{2} + y_Q$. Since the proton and neutron charges are related to the quark charges by $q_p = 2q_u + q_d$ and $q_n = q_u + 2q_d$, the prediction for the proton/electron charge ratio is given by

$$\frac{q_p}{q_e} = \frac{2q_u + q_d}{q_e} = \frac{\frac{1}{2} + 3y_Q}{-\frac{1}{2} + y_L} = -1, \tag{9.23}$$

in agreement with very precise precision measurements of the proton/electron charge ratio.

The neutron's charge is also set by the value of y_Q through $q_n = q_u + 2q_d = -\frac{1}{2} + 3y_Q$. Similarly, the neutrino charge is $q_\nu = \frac{1}{2} + y_L = \frac{1}{2} - 3y_Q = -q_n$, once (9.20) is used. These both vanish when $y_Q = \frac{1}{6}$, although the invariance of the anomaly-cancellation conditions under a common rescaling of all hypercharges precludes them dictating a specific value for y_Q. What does fix this value is the combination of anomaly cancellation and the condition that the expectation value of the Higgs doublet does not break electromagnetic invariance, since this means that one of its component fields must be electrically neutral and so $y_H = \frac{1}{2}$ (or $y_H = -\frac{1}{2}$, in which case redefine $\Phi \to \Phi^*$). $U_Y(1)$-invariance of the lepton yukawa coupling then implies that $y_L + y_E = y_H = \frac{1}{2}$, while (9.19) and (9.22) together show $2y_L + y_E = 0$, and so $y_L = -(y_L + y_E) = -y_H = -\frac{1}{2}$. This then implies that both neutrino and neutron are neutral.

Weak-hypercharge assignments within the Standard Model are remarkably rigidly dictated by consistency issues, given the field content [210, 211].

9.2 Non-renormalizable Interactions

The Standard Model is by definition renormalizable, as would be expected at energies $E \ll M$ within a more fundamental system involving higher energies M, to the extent that all effects suppressed by powers of E/M can be ignored. One way to interpret the Standard Model's success is that this picture is basically right: the

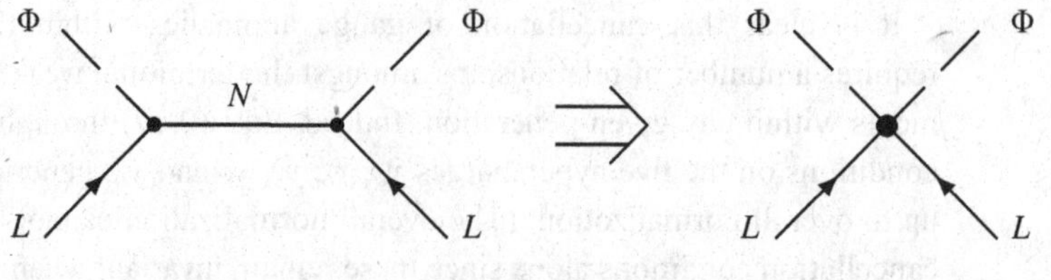

Fig. 9.1 An example of UV physics that can generate the dimension-five lepton-violating operator within SMEFT.

lightest particles yet to be discovered are very heavy, with M much higher than all energies yet experimentally accessible.[8]

In this kind of scenario the first signs that heavy new physics exists would show up as deviations from the Standard Model due to the presence of higher-dimension interactions suppressed by powers of $1/M$. Interactions with the lowest-dimension should dominate, so identifying these might usefully guide the search for new phenomena.

For this reason, it is worth listing the lowest-dimension of these higher-dimensional operators. Such a list contains all possible effective interactions that can be built using only the usual Standard Model fields that are both Lorentz-invariant and invariant under $SU_C(3) \times SU_L(2) \times U_Y(1)$. Redundant interactions should be identified and excluded as described in Part I. The result of such a construction is often called the Standard Model Effective Field Theory, or SMEFT.

9.2.1 Dimension-Five Interactions

In the Standard Model's case this list is particularly simple at linear order in $1/M$, since there is only one dimension-five interaction possible [212]. As is straightforward to show, the only dimension-5 interaction consistent with the Standard Model field content and gauge symmetries is

$$\mathfrak{L}_5 = -k_{mn}\,(\overline{L}_i^m\,\gamma_R L_j^n)\,\Phi^i\Phi^j + \text{c.c.}, \tag{9.24}$$

where, as above, L denotes the Majorana spinor representing the lepton doublet for which $i, j = 1, 2$ are $SU_L(2)$ doublet indices and $m, n = 1, 2, 3$ label the fermion generation. The matrix of effective couplings, $k_{mn} = c_{mn}/M$, is a general complex, symmetric 3×3 matrix, with dimension $(\text{mass})^{-1}$, making the coefficients c_{mn} dimensionless.

Fig. 9.1 gives an example of a UV completion that could generate an effective operator like (9.24), consisting of heavy sterile neutrinos, N^m, that mix with the light neutrinos through a yukawa coupling of the form

$$\mathfrak{L}_N = -M_{mn}\,(\overline{N}^m\,\gamma_R N^n) - y_{mn}\,(\overline{L}_i^m\,\gamma_R N^n)\,\Phi^i + \text{c.c.} \tag{9.25}$$

In this case, evaluating the graph in Fig. 9.1 gives $k_{mn} \simeq (y^T M^{-1} y)_{mn}$ to leading order in the inverse sterile-neutrino mass. Consequently, M is of order the sterile

[8] Undiscovered light particles could also exist, but only if they couple sufficiently weakly – as would happen if they are $SU_C(3)\times SU_L(2)\times U_Y(1)$ singlets (like sterile right-handed neutrinos), for example.

neutrino mass while c_{mn} is controlled by the size of the yukawa couplings that couple the sterile neutrino to the lepton and Higgs doublets. Since all other effective interactions come suppressed by at least two powers of $1/M$, they are plausibly negligible compared with (9.24) if M is very large.

The physical implications of (9.24) can be seen by going to unitary gauge, (9.6), leading to

$$\mathfrak{L}_5 = -\frac{1}{2} k_{mn} (\overline{\nu}^m \gamma_R \nu^n)(v+h)^2 + \text{c.c.} \tag{9.26}$$

where, as before, $v = 246$ GeV and h is the physical Higgs boson. Besides describing neutrino-Higgs interactions, (9.26) also contains a right-handed neutrino mass matrix of size

$$m_{mn} = k_{mn} v^2 = c_{mn} \frac{v^2}{M}. \tag{9.27}$$

The eigenvalues of the matrix $m^\dagger m$ then give the squares of the masses for the neutrino mass eigenstates.

Intriguingly, although there are only a few pieces of evidence that the Standard Model cannot be nature's whole story, the existence of neutrino masses is one of them. A variety of experiments reveal that neutrinos take part in oscillations: processes wherein the evolution of different neutrino species $|\nu_i(t)\rangle = e^{-iE_i t}|\nu_i(0)\rangle$ are seen to interfere with one another, leading to time-dependent effects in neutrino-mediated interactions that are proportional to $e^{-i(E_i - E_j)t}$. Since $E_i^2 = \mathbf{p}^2 + m_i^2$, the existence of a nonzero energy difference (at fixed momentum) shows that neutrino masses cannot all vanish.

These experiments point to neutrino mass differences of order $\Delta m^2 \sim (50 \text{ meV})^2$, corresponding to a mass scale $M \sim 10^{15}$ GeV if $c_{mn} \sim \mathcal{O}(1)$. The enormity of this scale compared with the energies currently accessible in accelerators (10^4 GeV at this writing) makes the neglect of still-higher-dimension interactions self-consistent. An EFT picture of the Standard Model links the small but nonzero size observed of neutrino masses to the potential existence of new physics at an extremely large scale: the bigger the new-physics scale M, the smaller the neutrino mass (an example of what is called a 'see-saw' mechanism for generating small masses [213, 214]).

But even if new degrees of freedom at the large scale M exist, why does the scale M dominate the effective coefficient in $\mathfrak{L}_5$ rather than a much smaller scale? After all, the lepton doublet and the Higgs doublet interact with one another even in the absence of $\mathfrak{L}_5$ because a renormalizable coupling between them already exists in the Standard Model. So why doesn't lower-energy physics contribute to k_{mn}, allowing it to be much larger?

In this regard the important property of the interaction $\mathfrak{L}_5$ is its transformation property under the Standard Model's accidental symmetries. In particular, it does not share the Standard Model's accidental conservation of total lepton number, L, because it changes lepton number by two units. Since lepton number becomes conserved in the limit, the coupling in $\mathfrak{L}_5$ vanishes, integrating out lepton-preserving lower-energy physics can only change k_{mn} in a way that vanishes as $M \to \infty$. In this sense, the selection rule of $\mathfrak{L}_5$ under the accidental lepton symmetry protects its coefficient, keeping it at most of order $1/M$.

9.2.2 Dimension-Six Interactions

There are a great many more interactions possible at the dimension-six level [195, 212, 215, 217, 218], and unlike for dimension-five there is no compelling experimental evidence for the existence of any of them. With a few exceptions (see below), most of these would not be expected to have been seen if the scale suppressing them is as large as the scale $M \sim 10^{15}$ GeV suggested for the dimension-five interactions by neutrino oscillations, since they are twice-suppressed by this large scale.

Systematic treatments of these dimension-six operators, including their enumeration (including removing redundant combinations) and one-loop evolution, was only done comparatively recently [218], with a complete description being beyond the scope of this book. But a more manageable subset (that illustrates how dimension-six operators can be usefully used more generally) are those that break the accidental baryon-number symmetry of the renormalizable Standard Model interactions [195, 215, 216]. Since dimension-six interactions are the lowest-dimension interactions that can break baryon-number invariance, they should capture very robustly the low-energy implications of *any* high-energy baryon-number violating physics (assuming it involves only the known Standard Model particles at low energies).

Six independent baryon-number violating interactions are possible, consistent with Standard Model field content and gauge symmetries. These can be written $\mathfrak{L}_B = c^I_{mnpq} O^I_{mnpq}$ + c.c., where m, n, p, q are generation indices and I labels the following basis of operators:

$$\begin{aligned}
O^1_{mnpq} &= \epsilon^{\alpha\beta\gamma}\epsilon_{ij}[\overline{Q}^i_{m\gamma}\gamma_L L^j_n][\overline{D}_{p\alpha}\gamma_R U_{q\beta}] \\
O^2_{mnpq} &= \epsilon^{\alpha\beta\gamma}\epsilon_{ij}[\overline{Q}^i_{m\alpha}\gamma_L Q^j_{n\beta}][\overline{U}_{p\gamma}\gamma_R E_q] \\
O^3_{mnpq} &= \epsilon^{\alpha\beta\gamma}\epsilon_{ij}\epsilon_{kl}[\overline{Q}^i_{m\alpha}\gamma_L Q^j_{n\beta}][\overline{Q}^k_{p\gamma}\gamma_L L^l_q] \\
O^4_{mnpq} &= \epsilon^{\alpha\beta\gamma}(\tau_a\epsilon)_{ij}(\tau_a\epsilon)_{kl}[\overline{Q}^i_{m\alpha}\gamma_L Q^j_{n\beta}][\overline{Q}^k_{p\gamma}\gamma_L L^l_q] \\
O^5_{mnpq} &= \epsilon^{\alpha\beta\gamma}[\overline{D}_{m\alpha}\gamma_R U_{n\beta}][\overline{U}_{p\gamma}\gamma_R E_q] \\
O^6_{mnpq} &= \epsilon^{\alpha\beta\gamma}[\overline{U}_{m\alpha}\gamma_R U_{n\beta}][\overline{D}_{p\gamma}\gamma_R E_q].
\end{aligned} \tag{9.28}$$

As before, $\alpha, \beta, \gamma = 1, 2, 3$ denote color-triplet indices; $i, j, k, l = 1, 2$ are $SU_L(2)$-doublet indices and $m, n, p, q = 1, 2, 3$ are generation labels. $\epsilon^{\alpha\beta\gamma}$ and ϵ_{ij} are, respectively, the invariant completely antisymmetric Levi-Civita tensors for $SU_c(3)$ and $SU_L(2)$, with $\epsilon^{123} = \epsilon_{12} = 1$. As before, τ_a, $a = 1, 2, 3$, denote the three Pauli matrices acting on $SU_L(2)$ indices.

On dimensional grounds, the effective couplings for these interactions are inversely proportional to two powers of the (presumably heavy) energy scale, M, at which baryon-number violating physics occurs: $c^I_{mnpq} = \tilde{c}^I_{mnpq}/M^2$. Once the $1/M^2$ factor is extracted, the remaining dimensionless couplings, $\tilde{c}^I_{mnpq}$, contain all the other dimensionless factors – coupling constants and such – that would also arise when generating these effective terms from some more fundamental theory.

Experimental searches for baryon-number violation – such as proton decay – are clearly of considerable interest because no fundamental principles preclude its occurrence, and if it were to occur, the accidental B conservation of the renormalizable interactions of the Standard Model would explain why it has not been observed before now. The absence of observational evidence for baryon-number violation implies that there is a smallest allowed value for M for the interactions in (9.28).

As of this writing (2019) it is known that the proton's mean decay lifetime – such as through the hypothetical reaction $p \to \pi^0 e^+$ – cannot be much shorter than of order 10^{33} years.[9] But since the dimension-6 interactions of Eq. 9.28 all satisfy $\Delta B = \pm 1$, they can produce a nonzero amplitude for proton decay at linear order in the coefficients c_{mnpq}, leading to a predicted decay rate that on dimensional grounds is of order

$$\frac{1}{\tau} \sim |c'|^2 m_p^5, \tag{9.29}$$

where phase-space integrals over the final-state momenta are estimated by the appropriate power of the proton mass, m_p. Here, c' generically represents the relevant effective coupling, for which the lower limit on τ implies an upper bound $|c'| \lesssim (10^{16}\,\text{GeV})^{-2}$. This again points to the existence of a very large mass scale, intriguingly close to the Planck scale, $M_p \sim 2 \times 10^{18}$ GeV, and similar to scales found above from estimates of the M-dependence of neutrino masses.

The list of (9.28) contains much more information than this, however [195, 212, 215]. First, since $\Delta B = \pm 1, 0$ is satisfied by *all* dimension-6 terms, $\Delta B = \pm 2$ processes – like neutron-antineutron oscillations – can only proceed suppressed by even more powers of $1/M$.

Second, all the interactions in (9.28) satisfy the selection rule $\Delta B = \Delta L$, which implies that protons must decay into anti-leptons, and so while the decays $p \to \pi^0 \ell^+$ or $n \to \pi^- \ell^+$ are allowed (with $\ell = e, \mu$), the decays $p \to \pi^+ \pi^0$, $p \to \pi^+ \pi^+ \pi^-$, or $n \to \pi^+ \ell^-$ are not. The Standard Model field content ensures $B - L$ conservation is automatically a much better approximation than B conservation alone (at low energies), regardless of whether or not this is true in detail for the underlying higher-energy particles that are ultimately responsible. In the absence of other non-standard particles at low energies, it would not be a surprise to find that proton decay (if observed) happened to satisfy $B - L$ conservation, and if so this does not say much about whether $B - L$ is actually preserved at much higher energies.

Third, all dimension-6 interactions satisfy $\text{sign}\,\Delta S = -\text{sign}\,\Delta B$, where $S = -1$ is the (only approximately conserved) charge assigned to the strange quark, s. Consequently, transitions that lower B by one unit can destroy zero, one or two strange quarks (or create the corresponding number of strange antiquarks), but cannot create any strange quarks without antiquarks. This allows decays like $p \to K^0 \ell^+$ while forbidding $p \to \overline{K}^0 \ell^+$ or $n \to K^- \ell^+$, where K^0 and K^+ are, respectively, mesons with quark content $d\overline{s}$ and $u\overline{s}$ and $\overline{K}^0$ and K^- are their antiparticles.

Similarly, all interactions in (9.28) that satisfy $\Delta S = 0$ also satisfy the isospin selection rule $\Delta I = 1/2$, where I is the 'isospin' quantum number associated with the diagonal $SU_I(2)$ symmetry (described below (8.11)) that rotates u and d quarks

[9] Lifetimes like this, much longer than the age of the universe, are measured by the absence of observed decays in very large samples of protons.

into one another in a non-chiral way. This implies relations amongst the decay rates for different channels like $\Gamma(p \to \pi^0 \ell_R^+) = \frac{1}{2}\Gamma(n \to \pi^- \ell_R^+) = \frac{1}{2}\Gamma(p \to \pi^+ \overline{\nu}) = \Gamma(n \to \pi^0 \overline{\nu})$ and $\Gamma(p \to \pi^0 \ell_L^+) = \frac{1}{2}\Gamma(n \to \pi^- \ell_L^+)$, and so on.

Of course, more detailed predictions than these are possible once the specific underlying B-violating UV theory is known [219]. But comparison with general EFT results like those given above is also instructive when trying to infer the properties of such a specific theory, since it is useful to know which predictions depend on the details and which do not.

9.3 Naturalness Issues♠

As described above, many features of the Standard Model suggest it can be usefully regarded as the low-energy limit of something more fundamental. There are also several circumstantial reasons to expect that this more fundamental physics could include scales that are quite high. This evidence includes:

- Cosmic rays bombard the Earth's upper atmosphere with energies that have been observed to be as high as 10^{10} GeV.
- Gravity exists and in fundamental units its coupling strength points to a scale $M_p = (8\pi G_N)^{-1/2} \simeq 2 \times 10^{18}$ GeV. (More about this clue in §10.)
- If interpreted in terms of an effective dimension-five operator, neutrino oscillations point to a scale associated with lepton-number breaking at or below $M \sim 10^{15}$ GeV.
- The three gauge couplings of the Standard Model, when extrapolated to higher energies using their renormalization-group running (such as in §8.6), come close to becoming equal[10] in size at a scale of order 10^{16} GeV [220].
- Cosmological measurements of primordial fluctuations, as seen in the cosmic microwave background, suggest the existence of a very early cosmological epoch during which the size of the universe, $a(t)$, had an expansion rate $H = \dot{a}/a$ of size $H/M_p \simeq 10^{-6}$ and the expansion was accelerating, $\ddot{a} > 0$. In the simplest models this occurs if the universal energy density at the time was dominated by a potential energy $V \sim M^4$ with $M \sim \sqrt{HM_p} \sim 10^{-3} M_p \sim 10^{15}$ GeV.

If these indications are to be trusted then the world revealed to us by experiments touches only the very lowest of energies compared to some of the basic scales in nature. If true, what clues does this suggest about the kinds of phenomena that might be expected to arise beyond the Standard Model but not too far beyond present experimental reach?

9.3.1 Technical and 't Hooft Naturalness ◇

One way to approach this question is to ask if there are properties of the Standard Model that do *not* resemble what would be expected as being generic for a

[10] They do not become *precisely* equal, though this need not be a big worry since the running depends on the spectrum of particles at intervening energies and plausible choices can be made for this spectrum that makes them agree more precisely [221–223]. It has been speculated that this may indicate the existence of a 'grand-unified' scale, where the three factors of $SU_c(3) \times SU_L(2) \times U_Y(1)$ all unify into one spontaneously broken simple gauge group like $SU(5)$ or $SO(10)$ [219].

low-energy theory. Features that are not generic should require some explanation, and the additional physics associated with the explanation might have observable implications. Indeed, there are a few features of the Standard Model that do seem non-generic in this way, consisting mostly of effective interactions that are 'missing' in the sense that they are much smaller than an effective interaction might be expected to be.

To clarify this expectation, recall that the examples considered earlier teach that an effective interaction of dimension[11] d_n arises in the low-energy effective theory with an effective coupling of dimension $c_i \simeq \tilde{c}_i\, M^{4-d_n}$, where the dimensionally required factor of M is made explicit so the remaining coefficient $\tilde{c}_i$ is dimensionless.

As discussed at length above, when successive scales, $M_1 \ll M_2 \ll M_3$, are integrated out in this way the effective coefficients receive contributions from each, with the smallest scale – *i.e.* M_1 – dominating for irrelevant operators (those with $d_n > 4$) and the largest scales – *i.e.* M_3 – dominating in relevant ones (with $d_n < 4$). Marginal operators (with $d_n = 4$) have dimensionless coefficients that typically depend logarithmically on the various scales, corresponding to their renormalization-group running between successive mass thresholds.[12]

Of course, this is a generic statement that can have exceptions, such as the apparent suppression of lepton-number violating dimension-five interaction by a very large scale rather than a much smaller one closer to the weak scale. But it is precisely this type of deviation from the generic expectation that calls out for explanation; in the case of the dimension-five interaction it is possible that physics at lower scales preserves lepton number and this is why smaller scales do not contribute to this particular interaction.

The interactions with $d_n \leq 4$ are particularly interesting because they should be the most sensitive to what goes on at the highest energies.[13] This is because their couplings arise *enhanced* rather than suppressed by any very large scale M associated with UV physics [79]. The Standard Model has two such operators with $d_n < 4$, and a host of interactions with $d_n = 4$, that are considered below in turn.

The two relevant interactions with $d_n < 4$ allowed by the Standard Model field content and gauge symmetries are given by

$$\mathfrak{L}_{\rm rel} = -c_4 + c_2\, \Phi^\dagger \Phi \subset \mathfrak{L}_{SM}, \tag{9.30}$$

where c_n has dimension (energy)^n, and both are contributions to the Higgs potential. The problem (to the extent there is one) with these interactions is the size of the two coefficients c_4 and c_2. Consider first c_2.

On one hand, the form of the Higgs potential (9.5) in the Standard Model dictates $c_2 = \lambda v^2$, where λ is the Higgs self-coupling and v is the size of the Higgs vacuum expectation value, as defined by $\Phi^\dagger\Phi = \frac{1}{2}\, v^2$, say. The numerical values for

[11] For strongly coupled systems an interaction operator's scaling dimension, d_n, need not be close to its naive dimension, but the argument being given goes through in this case as well, so long as it is the full scaling dimension that is used.

[12] Dimensionful couplings also depend on logarithms of scales – this is just not emphasized in what follows since the powers are usually more important.

[13] Although the rest of this discussion focuses on $d_n < 4$, a puzzle also exists for $d_n = 4$ – called the *strong CP problem* [224–226]. The problem arises because none of the parity- and time-reversal violating interactions in $\mathfrak{L}_{\rm gauge}$ defined below (9.14) are measured to be present, with dimensionless coefficients that must vanish to very good accuracy in the case of the gluon field.

these parameters follow from the relation (9.7) relating $v \simeq 246$ GeV to the Fermi constant (which is measured in muon decay), and from the measured Higgs-boson mass, $m_H^2 = 2\lambda v^2 \simeq (125 \text{ GeV})^2$, and so agreement between the Standard Model and observations dictates $c_2 = \frac{1}{2} m_H^2 \simeq (88 \text{ GeV})^2$.

On the other hand, a generic contribution to c_n obtained by integrating out particles with mass M is expected to be $\delta c_n \propto M^n$, and so can be much larger than is observed, particularly for masses as large as $M \sim 10^{15}$ GeV. To be completely concrete as to why this kind of contribution might be disturbing, it is worth examining in detail a specific model of (part of) what the UV physics might be. For this purpose a minimal example consists of a real, heavy Standard-Model-singlet scalar field, S, coupled to the Standard Model through the most general possible[14] renormalizable interactions:

$$\mathfrak{L}_S = \mathfrak{L}_{SM} - \frac{1}{2}\partial_\mu S \partial^\mu S - \frac{1}{2} M^2 S^2 - \frac{\mathfrak{g}^2}{2} S^2 \Phi^\dagger \Phi - \frac{1}{4}\lambda_s S^4, \tag{9.31}$$

where the Standard Model part includes, in particular, the offending couplings c_2 and c_4.

This extension of the Standard Model introduces a single new spinless particle and three new parameters: the two dimensionless couplings $\mathfrak{g}$, λ_s and the new-particle mass M. The existence of a renormalizable coupling like this between Standard Model fields and a singlet scalar is sometimes called the *scalar portal* (see Exercise 9.6). Sterile fields interact very weakly and so could well exist at experimentally accessible energies without having been detected. When a sterile particle can couple to Standard Model particles through renormalizable interactions, those interactions are said to be a portal to the hypothetical sterile (or dark) sector. There are only a handful of such portals, inasmuch as there are only a few kinds of sterile particles that can couple to Standard Model fields through renormalizable interactions.

With this UV extension in hand it is possible to integrate out the heavy S particle explicitly to derive the low-energy EFT appropriate for energies $E \ll M$. It is instructive to compute the size of the effective coefficient $\bar{c}_2$ in the low-energy EFT and compare it to the size of c_2 in the full theory. Following the spirit of §3.2, loops are regularized using dimensional regularization and both $\bar{c}_2$ and c_2 renormalized using the decoupling-subtraction scheme of §7.2.3.

Both $\bar{c}_2$ and c_2 are fixed by comparing to the mass of the physical Higgs boson, and because the renormalizable part of the EFT in this example *is* the Standard Model, for it the result is the standard one:

$$m_H^2 = 2\bar{c}_2 + (\text{SM loops}) = 2\bar{\lambda}\bar{v}^2 + (\text{SM loops}), \tag{9.32}$$

where $\bar{\lambda}$ and $\bar{v}$ are the corresponding Standard Model parameters within the EFT. The classical Standard Model relation $\bar{c}_2 = \bar{\lambda}\bar{v}^2$ is used in the tree-level term, and all parameters are renormalized. What is important in this expression is that loop corrections in the low-energy EFT involve only low-energy Standard Model fields and so the loop terms in (9.32) are suppressed relative to the tree-level term by small factors like $\alpha/4\pi$.

14 The existence of a symmetry $S \to -S$ is assumed for simplicity.

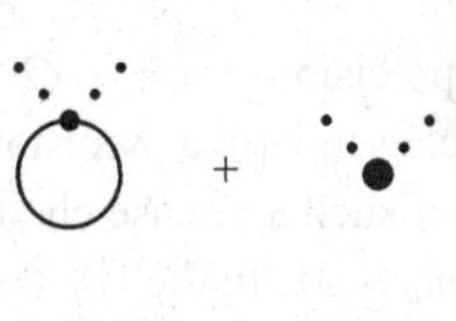

Fig. 9.2 Graphs contributing to the Higgs mass in the extended UV theory. Solid (or dotted) lines represent S (or Higgs) fields. Graph (a) is the one-loop graph through which a massive S particle contributes at the 1-loop level; Graph (b) is the direct contribution of the effective coupling c_2; the effective $\Phi^\dagger\Phi$ coupling in the low-energy Wilsonian EFT. To these are to be added all other contributions (not drawn) including one-loop Standard-Model effects. What is important is that these other effects are present in both the full theory and the low-energy EFT.

The corresponding calculation for the full UV completion is precisely the same, with the exception that the loop contributions now also include graph (a) of Fig. 9.2, in which the heavy field S also circulates.

$$\begin{aligned} m_H^2 &= 2c_2^{\text{bare}} + (\text{SM loops})_\varepsilon + \frac{g^2 M^2}{8\pi^2}\left[\frac{1}{\varepsilon} + \gamma_m - \log\left(\frac{M^2}{\mu^2}\right)\right] \\ &\simeq 2c_2 + (\text{SM loops}) - \frac{g^2 M^2}{8\pi^2}\log\left(\frac{M^2}{\mu^2}\right), \end{aligned} \tag{9.33}$$

where the g^2M^2 term comes from evaluating explicitly the contribution of graph (a) in Fig. 9.2 using dimensional regularization (with $D = 4 - 2\varepsilon$), and the precise value of the μ-independent constant γ_m is not required. The contributions of loops with SM particles are also present precisely as they were for the low-energy EFT, with a subscript ε denoting inclusion of the divergent (as $\varepsilon \to 0$) pre-renormalization result.

The relation between c_2 and $\bar{c}_2$ is inferred by comparing (9.33)–(9.32), keeping in mind that m_H is a physical observable and so is 125 GeV for both calculations. This shows that c_2 and $\bar{c}_2$ depend differently on μ, with

$$\bar{c}_2(\mu) \simeq c_2(\mu) - \frac{g^2 M^2}{16\pi^2}\log\left(\frac{M^2}{\mu^2}\right), \tag{9.34}$$

where the approximate equality neglects subdominant contributions, including the small subdominant shifts $\lambda - \bar{\lambda}$ or $v - \bar{v}$ in the values of SM couplings inferred within the UV theory and the EFT.

This comparison shows that although the parameter $\bar{c}_2 \simeq \frac{1}{2}m_H^2 \simeq (88\ \text{GeV})^2$ has a typical weak-scale size in the effective theory for a wide range of μ, the same is *not* true for the coupling c_2 in the UV theory except at the specific point $\mu = M$. For any other scale (9.34) shows c_2 must take a much larger value of order g^2M^2 that mostly cancels the equally large S-loop contribution to leave the much smaller residual weak-scale value for $\bar{c}_2$.

If there were another still heavier particle with mass $\hat{M} \gg M$ then $c_2(\mu = \hat{M})$ is order M^2 at this new threshold, and the new coupling for energies above $\hat{M}$ is of order $\hat{M}^2$ and so on. The UV theory must carefully choose to be on the RG trajectory that emerges at the threshold $\mu = M$ with a weak-scale value. What is striking about this choice is its precision: for $g \sim \mathcal{O}(1)$ and $M \sim 10^{15}$ GeV the coupling c_2 at a scale M must be chosen with a precision of over 23 decimal places.

In the absence of such a precise choice Eq. (9.34) would instead give $\bar{c}_2 \sim c_2$ and so would predict $m_H \sim M$. In the UV theory a huge hierarchy like $m_H \ll M$ between the values of two masses is not understood in terms of the small size of a coupling parameter like $c_2(\mu)$ for all μ; instead, it is controlled by an extremely precise choice for the RG trajectory that is large at all scales except at the very low energies where it is measured to be small.

This is not the way hierarchies of scale usually work. For instance, if one asks why atoms are larger than nuclei the answer can be cast in terms of the overall size of couplings within the Wilson action, and this can be done for *any* scale one chooses to define this action. Asked within the Standard Model itself, defined at hundreds of GeV, then the size of an atom is set by the Bohr radius, $1/a_0 \simeq \alpha\, m_e$, and the size of a nucleus is set by the QCD scale, $1/r_N \simeq \Lambda_{QCD}$. From the Standard Model's point of view, atoms are larger than nuclei because the fine-structure constant is small, $\alpha \simeq 10^{-2}$, and the electron is much lighter than the QCD scale, $m_e/\Lambda_{QCD} \simeq 10^{-3}$, and this is true for any μ one chooses to evaluate running couplings like α.

When asked in the effective theory below the confinement scale of QCD, the quarks and gluons of the Standard Model are replaced by the protons and neutrons (or nucleons) that are built out of them. Although the Bohr radius is still set by $\bar{a}\, \bar{m}_e$ in this new theory, the size of a nucleus is now set by the nucleon mass, $\bar{m}_N$, where (as above) 'bars' denote the corresponding renormalized parameters within this new Wilson action. The quantities $\bar{a}$, $\bar{m}_e$ and $\bar{m}_N$ can be computed in terms of the parameters α, m_e and Λ_{QCD}, of the Standard Model, and this can be done above or below the muon threshold. It is true at any scale in the low-energy theory that $\bar{a} \ll 1$ and $\bar{m}_e/\bar{m}_N \ll 1$.

There are a great many hierarchies of scale found in nature, and experience with them all teaches that there are two parts to understanding why they exist:

1. One first finds which renormalized parameters (such as c_2 or $\bar{c}_2$) must be small in the underlying microscopic theory.
2. One then asks why this parameter *remains* small as one integrates out a succession of higher-energy states to obtain the Wilson action for the effective theory appropriate to the lower energies where the parameter is measured.

When both of these questions have an answer then the small parameter is said to be *technically natural*. Our understanding of why atoms are large compared with nuclei is technically natural in this sense, as are all of the other known hierarchies in nature (with one exception, more about which below).

In the above example the hierarchy $m_H \ll M$ is not technically natural in this sense. At low-energies the small size of m_H compared to M is understood in terms of the small size of $\bar{c}_2$ compared to M^2. But it is not technically natural because it is not true that c_2 *remains* small in the EFT defined both above and below the large mass threshold at $E = M$.

't Hooft Naturalness

Approximate symmetries provide a very common reason why a parameter hierarchy can be technically natural. The reason for this is that symmetries can often forbid there being large corrections to a small quantity. They can do so because if the quantity of interest breaks a symmetry, then any corrections to it must also break the symmetry and so must contain any small factors associated with making the symmetry a good approximation in the first place.

This suggests defining a slightly stronger condition than technical naturalness, called *'t Hooft naturalness*. This states that a small parameter is ('t Hooft-) natural if additional symmetries arise once it is set to zero [80].

Small masses for pseudo-Goldstone bosons for spontaneously broken approximate symmetries fall into this category, since they must vanish in the limit that the spontaneously broken symmetry becomes exact rather than approximate. This is also why fermion masses (like the electron mass) can be small compared with other scales while still being technically natural. They can do so because additional chiral symmetries emerge once the fermion mass goes to zero, under which the fermion's left- and right-handed parts rotate separately: $\psi \to \exp(i\theta\gamma_5)\psi$.

Pretty much all of the known hierarchies are 't Hooft natural in this sense, as well as being technically natural, and the two concepts are often regarded as being synonymous. However, there are exceptions – such as within supersymmetric theories, where corrections to parameters in the super-potential can be small even when not protected by a symmetry.[15] At the end of the day, it is technical naturalness in the sense defined above that seems satisfied by well-understood effective theories, with 't Hooft naturalness providing a means to this end.

9.3.2 The Electroweak Hierarchy Problem

Returning to the Standard Model example, can the absence of large contributions to c_2 of order M^2 be understood in a technically natural way? (The puzzle of how to do so is called the *electroweak hierarchy problem*.) Three classes of proposals have been made to do so, and each has its own predictions for new phenomena to be sought at the highest energies available (none of which has been seen as of this writing – mid 2019).

Option A: Composite Higgs Particle

One way to avoid the problem of c_2 having to cancel carefully against quantum contributions of heavy particles is to have the $c_2\,\Phi^\dagger\Phi$ term not appear at all within the Wilsonian EFTs appropriate to very high energies. This might happen if the Higgs boson were a composite built out of some sort of constituents, much as atoms are built from nuclei and electrons and the proton is built from quarks and gluons.

Such substructure could have escaped detection if the compositeness scale, M_c, were high enough to put the composite nature of the Higgs beyond experimental reach. Here M_c is the lowest scale associated with compositeness, which could mean $M_c \simeq 1/a_H$ with a_H of order the radius of the Higgs bound state, or it could mean the excitation energy, $M_c \simeq E_B$, to the first excited bound state above the ground state, whichever is smaller. Although these are similar scales for protons (and also similar to the proton rest mass), they are very different from one another for atoms, for which the radius of the bound state is of order the Bohr radius $1/a_B \simeq \alpha\, m_e$

[15] Small parameters of this type that are technically natural despite not being protected by a symmetry – usually due to the structure of corrections in supersymmetric theories – are sometimes known as *supernatural*.

(where $\alpha \simeq 1/137$ is the fine-structure constant encountered earlier and m_e is the electron mass), while $E_B \sim \alpha^2 m_e \ll 1/a_B$ and both are small compared with the atomic rest mass.[16]

This kind of scenario helps with the electroweak hierarchy problem because any EFT relevant at scales above M_c would involve fields representing the new constituents rather than the Higgs field Φ. (In the same way, there are no proton or neutron fields in the Standard Model itself, despite these fields – and other composite fields, such as pions – appearing explicitly in EFTs defined below a few GeV. The Standard Model instead contains the gluon and quark fields from which these composite states are made.) Since the absence of the field Φ implies that the operator $c_2\, \Phi^\dagger \Phi$ does not appear in the Wilsonian action above M_c, there is no longer an issue as to what the size of c_2 must be.

This is the most historically conservative option, since past decades are littered with examples where particles once thought to be elementary instead prove on closer inspection to be built of smaller stuff. The hard work in this scenario is designing plausible dynamics that can bind constituents into Higgs-like objects (and nothing else that would have given away its composite nature). Part of what makes this tricky is the fact that, unlike for protons, the absence of evidence for Higgs compositeness up to energies well above 1000 GeV means M_c must satisfy $M_c \gg m_H \simeq 125$ GeV. Having a bound state with a mass much smaller than the typical binding scale is unusual, though not impossible (as the example of the pion shows). With pions in mind, this kind of scenario often arranges for the Higgs to be a pion-like pseudo-Goldstone boson for some new type of approximate internal symmetry, often by proposing a new kind of QCD-like strongly coupled 'technicolour' gauge group acting on constituent 'techni-quarks' for which the Higgs is one of the bound mesons [227, 228].

The power of using technical naturalness as a criterion to motivate such models comes from the fact that the compositeness scale, M_c, cannot be too far out of experimental reach if the compositeness is to solve the electroweak hierarchy problem. It cannot be too large because if it were all of the constituents could be integrated out to arrive at Wilsonian description of physics below M_c, and this would contain a field Φ for the composite Higgs. Once this is true, the size of the operator $c_2\, \Phi^\dagger\Phi$ again becomes relevant and we are back where we started. This illustrates how technical naturalness can suggest specific kinds of new physics – in this case a composite structure for the Higgs – that (crucially) should not be at energies too high to be accessed experimentally.

Option B: Symmetries

A second approach to solving the electroweak hierarchy problem is to suppose that the Higgs really is elementary (at least on the scales relevant to upcoming experiments), and to propose a symmetry that makes the small size of c_2 't Hooft natural (and so therefore also technically natural).

This is harder than it looks, since a term like $c_2\, \Phi^\dagger\Phi$ is automatically invariant under a large class of symmetries, since these must also preserve the scalar kinetic

[16] This particular hierarchy of scales is a consequence of the nonrelativistic kinematics of the bound electron (more about which in Part III devoted to nonrelativistic applications).

term $\partial_\mu \Phi^\dagger \partial^\mu \Phi$. These include all orthogonal linear transformations of the four real fields contained within Φ.

There are essentially three types of symmetries proposed to make the Higgs mass parameter 't Hooft natural. One of these occurs if the symmetry transformation is inhomogeneous (*i.e.* involves shifts $\Phi \to \Phi + c + \cdots$ as well as linear rotations of the fields), as do the nonlinear realizations discussed in §4 associated with a Goldstone boson. Invariance under shift symmetries of this type not only forbid interactions of the form $c_2 \Phi^\dagger \Phi$, but they also completely forbid the appearance of Φ within the scalar potential. They can therefore provide a 't Hooft-natural cover for the appearance of scalars with small coefficients in the scalar potential, and so, in particular, protect a small size for c_2. This kind of symmetry also plays a role in Option A above, but does not, in principle, require Φ to be composite.

The other two classes of symmetries that can forbid $c_2 \Phi^\dagger \Phi$ are spacetime symmetries in that they act on the spacetime coordinates (or the spacetime metric) as well as on the Higgs field (see §C.4.2 for more about the distinction between internal and spacetime symmetries). The simplest example of this type is scale invariance, under which $\Phi \to \lambda^p \Phi$ (for some nonzero p), while the spacetime metric transforms as $g_{\mu\nu} \to \lambda^2 g_{\mu\nu}$, where λ is a constant transformation parameter.[17]

For such symmetries the lagrangian density transforms as $\mathfrak{L} \to \lambda^{-4}\mathfrak{L}$, and so is *not* invariant (even though the action $S = \int d^4x\, \mathfrak{L}$ is) because the spacetime measure is also not invariant. Invariance of the kinetic term $\partial_\mu \Phi^\dagger \partial^\mu \Phi = g^{\mu\nu} \partial_\mu \Phi^\dagger \partial_\nu \Phi$ then implies that $p = -1$, in which case the only invariant form for the scalar potential is a term like $(\Phi^\dagger \Phi)^2$. In particular, $\Phi^\dagger \Phi$ is not invariant and so can be protected *provided* this symmetry itself survives quantization (and so is not anomalous). The tricky part in this case is devising couplings of Φ to heavy states that are scale invariant and not anomalous. (Unfortunately, scale invariance is one of those symmetries that usually is anomalous.)

The other kind of spacetime symmetry (and the third type of symmetry used to address the electroweak hierarchy) is the one whose implications have been explored in the most detail: *supersymmetry* [229–233]. Although a detailed explanation goes beyond the scope of this description, supersymmetry relates spinless particles to spin-half ones (and more generally relates particles whose spin differs by $\frac{1}{2}$), forcing these to have related couplings (and identical masses) in the limit that it is unbroken. Since spin-half particles can have 't Hooft-natural masses because of the new chiral symmetries that emerge in the massless limit and since supersymmetry requires spin-half and scalars to share the same mass, the combination of supersymmetry and chiral symmetries can also protect scalar masses [223, 234, 235]. Supersymmetry turns out to be a spacetime symmetry because invariance of the action again requires the lagrangian density, $\mathfrak{L}$, to transform in a specific way rather than itself be invariant.

To see more explicitly how supersymmetry helps, consider a supersymmetric extension of the simple toy UV theory given in (9.31). In the supersymmetric version

[17] This type of symmetry can sometimes be rephrased for specific metrics (like the Minkowski metric, $g_{\mu\nu} = \eta_{\mu\nu}$), as a coordinate transformation, $x^\mu \to \lambda x^\mu$, whose effect is to rescale the metric $g_{\mu\nu} \to \lambda^2 g_{\mu\nu}$. Metrics for which such a transformation $\delta x^\mu = \xi^\mu$ is possible are said to admit a 'homothetic vector field', ξ^μ.

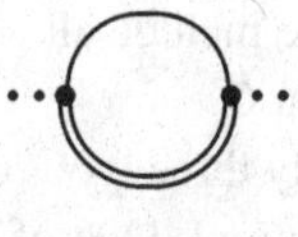

Fig. 9.3 A new graph that contributes to the shift in c_2 when the heavy fields are integrated out in the supersymmetric UV model. Dotted lines represent the scalar Φ field, while a solid line here represents its superpartner ψ (rather than the heavy scalar S). The double line represents the superpartner χ of the heavy scalar S. All order-M^2 terms in this graph precisely cancel those coming from the left-hand graph of Fig. 9.2 in the supersymmetric limit (in which the masses and couplings in this graph are related to those of Fig. 9.2).

of this theory S must be complex and both the scalar Φ and S have spin-half partners, respectively denoted ψ and χ. These spin-half fields share the same gauge transformation properties as their scalar partners, making ψ an $SU_L(2)$ doublet and χ a singlet. Φ and ψ also share the same mass so long as supersymmetry is preserved, and the same is true for S and χ.

It turns out that supersymmetry does *not* forbid the existence of the problematic $c_2\,\Phi^\dagger\Phi$ term and also allows the dangerous interaction $g^2\Phi^\dagger\Phi\,S^*S$ in the scalar potential whose use in loops gives potentially large order-M^2 contributions to c_2. But crucially, if this last term exists, supersymmetry also requires the existence of a Yukawa coupling of the form $g\,\Phi^\dagger(\overline{\chi}\,\gamma_L\psi)$ + c.c..

The existence of the new Yukawa interaction is important because it implies that there is another graph – Fig. 9.3 – that contributes to the relation between the coefficients $\bar{c}_2$ and c_2 once the heavy fields S and χ are both integrated out (as they must be, given they are both equally heavy). Because the particles in the loop are fermions, the sign of this new contribution is opposite to the left-hand graph of Fig. 9.2. In the supersymmetric limit the equality of the χ and S masses and the fact that the $g^2\Phi^\dagger\Phi\,S^*S$ interaction has coupling strength that is the square of the $g\,\Phi^\dagger(\overline{\chi}\,\gamma_L\psi)$ coupling strength then turns out to ensure all order-M^2 terms in $\partial c_2/\partial\log\mu$ cancel between the graphs of Figs. 9.2 and 9.3.

Once again, technical naturalness suggests new effects to be sought at energies not too large compared to those to which there is already access. The new effects in this case relate to the properties of the new particles required by the symmetry (such as superpartners), and these cannot be too heavy without (*e.g.*) the large difference between the boson and fermion masses ruining the cancellations required to enforce 't Hooft naturalness for the size of c_2.

Option C: Low Gravity Scale

Historically, the third option took longer to identify, and essentially involves denying the evidence for the existence of physical scales much larger than the weak scale. How is this possible, given the evidence for higher-energy physics outlined at the start of §9.3?

The strongest line of evidence[18] for higher-energy scales comes from the value of the Planck mass, $M_p = (8\pi G_N)^{-1/2} \simeq 2\times10^{18}$ GeV, reflecting the small size of Newton's gravitational constant relative to other known couplings in nature. The

[18] Note that simply observing cosmic rays with high energies does not necessarily mean that there are new types of fundamental physics at these high energies.

observation on which Option C relies is that M_p gives the strength of a coupling rather than the mass of a physical particle. As such, it likely depends both on physical masses and on the size of coupling constants, much in the same way that the Fermi constant is related to the W-boson mass by a relation of the form $G_F \propto g^2/M_W^2$ (*c.f.* Eq. (7.8)). It is because the coupling g is small that the scale $G_F^{-1/2} \simeq 300$ GeV is several times larger than the actual W mass: $M_W \simeq 80$ GeV.

Can the scale set by Newton's constant similarly be much larger than the mass of the relevant particles in the UV theory? It turns out that it can, provided more than just the usual three dimensions of space exist. Extra spatial dimensions could exist and have escaped detection if they are compact and sufficiently small, and (as explored in more detail in §10.3.2) if this is true then the effective theory at sufficiently low energies becomes four-dimensional.

One measure of extra-dimensional size is its volume, which in the case of d extra dimensions is here denoted by $\mathcal{S}_d$. For instance, if there were a single extra dimension that is a circle then $\mathcal{S}_1 = 2\pi L$ where L is the circle's radius; for two extra dimensions with the geometry of a sphere $\mathcal{S}_2 = 4\pi L^2$, and so on. In what follows, $D = 4 + d$ denotes the total number of spacetime dimensions, with d reserved for the number of additional ones besides the usual four.

It turns out that the volume of any extra dimensions enters into the relationship between the strength of Newton's constant, G_d, in the extra dimensions compared with $G_0 = G_N$ that one measures in practice in the low-energy four-dimensional effective theory.[19] As shown in more detail in §10.3.2, for a broad class of extra-dimensional geometries this relation is given by

$$G_N = \frac{G_d}{\mathcal{S}_d}. \tag{9.35}$$

There are several ways to understand this relation, one of which is to recall that for an infinite-range field like gravity, when all dimensions are infinitely large, Gauss' Law implies that the gravitational force falls off with distance like

$$|\mathbf{F}| \propto \frac{G_d}{r^{2+d}} \qquad (\text{for } r \ll L), \tag{9.36}$$

in $3+d$ spatial dimensions. This power law follows from the observation that the total gravitational flux through a sphere is independent of the size of the sphere, whose area (for all dimensions infinitely large) grows like r^{2+d} in $3 + d$ spatial dimensions. Eq. (9.36) expresses how the force falls with distance due to the spreading of force lines into the surrounding space.

When d of the extra dimensions are compact with linear size L, however, Eq. (9.36) can only be a good approximation when $r \ll L$. Distances much larger than L are only possible in the directions that are non-compact, and so for $r \gg L$ the factors of r get replaced by factors of L for each of the compact extra-dimensions, and so Eq. (9.36) becomes replaced by

$$|\mathbf{F}| \propto \frac{G_d}{L^d r^2} \qquad (\text{for } r \gg L). \tag{9.37}$$

[19] For aficianados: strictly speaking it can be the 'warped' volume of the extra dimensions that determines the size of Newton's constant in the low-energy theory [237].

This last result assumes that only the observed three spatial dimensions are much larger than L. Once the dimensionless order-unity factors are included (see below), the value of G_N inferred from measurements of the strength of gravity's inverse-square law are related to the extra-dimensional coupling, G_d, by Eq. (9.35).

Now comes the main point: it is actually possible for S_d to be large enough that the fundamental gravitational scale, G_d, can be closer to the weak scale than the 4D Planck scale, without these extra dimensions being so large they would already have been discovered [236, 237]. For instance, in order to have $8\pi G_d \simeq (10 \text{ TeV})^{-2-d}$ requires

$$S_d = \frac{8\pi G_d}{8\pi G_N} \simeq \frac{(2\times 10^{18} \text{ GeV})^2}{(10^4 \text{ GeV})^{2+d}}, \tag{9.38}$$

and so $S_2 \simeq (2\times 10^{10} \text{ GeV}^{-1})^2 \simeq (4\ \mu\text{m})^2$ if $d = 2$ while $S_6 \simeq (5 \text{ GeV}^{-1})^6 \simeq (1 \text{ fm})^6$ if $d = 6$. But (for some classes of models) the best constraint on the size of any extra dimensions comes from tests of the inverse-square law of gravity. These tests show that the inverse-square law works well down to about $r \gtrsim 45\ \mu$m, with nothing really known for distances smaller than this.

In these models the trick is to be able to calculate why S_d should happen to be this large, rather than much smaller, since in gravity the geometry is a dynamical variable. But once this is done one transitions to an extra-dimensional world at energies larger than the inverse size of the extra dimensions: providing an unusual way to explain the electroweak hierarchy. In this case, technical naturalness suggests new things to seek (the presence of the extra dimensions) at energies close to experimental reach.

To summarize: there are, broadly speaking, three classes of proposals for how new physics could arise above the electroweak scale in a way that keeps the presence of a small Higgs mass technically natural. All three classes of models have experimental signatures, and the good news is that these signatures should appear not too far above experimentally accessible energies. These signals have been sought in experiments for all three types of proposals. The bad news is that no evidence for any of these categories has been found.

Does this mean that demanding technical naturalness is a fruitless exercise? It is difficult to say at present because although naturalness motivates new physics that cannot be too far above the energies currently being accessed, it is not clear precisely how far above the electroweak scale is 'too far'. Although most would agree that $M \simeq 10^{15}$ GeV makes the electroweak hierarchy seem unnatural, is $M \simeq 10^2$ TeV acceptable? How about 10^4 TeV? Since our reach at present extends only out to 14 TeV it is hard to know for sure whether to be confident that technical naturalness will continue to be satisfied at higher energies.

9.3.3 The Cosmological Constant Problem

The same kinds of uncertainty do *not* apply to the other naturalness problem afflicting the Standard Model, mentioned in passing earlier. This problem is to do with the size of the parameter c_4 defined in Eq. (9.30). Because c_4 is only an additive constant in the scalar potential, it is not normally discussed much in particle physics. This is because it does not involve any Standard Model fields at all (apart from the spacetime metric, due to the factor of $\sqrt{-g}$, with $g = \det g_{\mu\nu}$, not made explicit in the above lagrangian).

Because this term involves only the metric, its implications are only apparent once gravity is considered. Since c_4 contributes to the stress energy by an amount

$$T^{\mu\nu} = -c_4\, g^{\mu\nu}, \tag{9.39}$$

it gravitates as does a relativistic perfect fluid (see, for example, Eq. (14.105)) whose pressure, p, and energy density, ϱ, are related by $\varrho = -p = c_4$. Because this implies that either p or ϱ must be negative, the gravitational effects of c_4 can be disentangled from those of other sorts of gravitating matter (for which these are both positive). Intriguingly, there is now observational evidence[20] [238, 239] that c_4 is nonzero and positive, with size

$$c_4 \simeq (3 \times 10^{-3}\ \text{eV})^4. \tag{9.40}$$

In cosmology (from whence the evidence comes) this inferred energy density is called *Dark Energy*.

All the naturalness issues encountered above for c_2 and the electroweak hierarchy also apply to c_4, which typically receives contributions of order M^4 when integrating out particles of mass M. As of this writing there is no understanding of how c_4 can be small at all scales within the Standard Model, and so the understanding of the small size of this vacuum-energy density is *not* technically natural. The problem of how to make it technically natural is called the (old) *cosmological constant problem.*

Modifying the Standard Model to make the observed size of c_4 technically natural proves to be much harder than the electroweak hierarchy problem. It is harder because even ordinary particles like the electron already contribute dangerously to c_4, with an initial value of order m_e^4 in the EFT above the electron mass having to cancel the contributions of electron loops. Because $m_e \simeq 511$ keV this cancellation must happen to 36 decimal places in order to make the energy density in cosmology as small as (9.40). And it only gets worse for heavier particles, with the W, Z and Higgs bosons requiring cancellations of 56 decimals.

Technical naturalness is clearly going to be harder to achieve for c_4 since it involves modifying how even the electron contribution to the vacuum energy gravitates, and this must be done at *low* energies right down to sub-eV scales. But the electron is probably the particle that we think we understand the best, so any modification must be done in such a way as not to ruin any of the many successful tests that have been made of electron properties at these energies. This seems a tall order, for which no known proposals seem to work. Although there is also no definitive no-go result showing it impossible, much effort has been invested in seeking a technically natural explanation of c_4, so far to no avail.[21]

[20] This evidence comes from two lines of argument. One measures the spatial curvature of the universe (which fixes the total energy density), assigning to Dark Energy the difference between the total energy density and the energy density present in ordinary (and dark) matter. The other evidence comes from the discovery that the universal rate of expansion is currently accelerating (as General Relativity predicts would happen if $\varrho + 3p < 0$).

[21] The leading approach so far takes the point of view that having c_4 as small as (9.40) is indeed unnatural. This is lived with by making *anthropic* arguments [240] where c_4 actually varies from place to place throughout the universe and arguing that regions where c_4 differs from (9.40) are inconsistent with the formation of life (and so our existence should correlate with finding a small size for c_4 in our

Although technical naturalness seems to hold well for all other of the many known hierarchies of scale, the failure to resolve the naturalness problem for c_4 currently undermines the proposal that all hierarchies of nature should be technically natural.

9.4 Summary

This chapter first briefly summarizes some properties of the Standard Model of particle physics, starting with its particle content and how these transform under its $SU_c(3) \times SU_L(2) \times U_Y(1)$ gauge symmetries.

The Standard Model is our most successful theory of nongravitational interactions and when combined with General Relativity provides the current framework for understanding nature. Much about the model carries the whiff of a low-energy limit: in particular, the fact that it can be defined as the most general renormalizable theory consistent with the assumed gauge symmetries and field content. This is precisely what would be expected if the Standard Model were the low-energy limit of some more fundamental theory, whose fundamental scale, M, is so high that it suffices to work at zeroth order in E/M.

Another feature that suggests the Standard Model is a low-energy limit is the fact that its fermions transform in chiral representations of the gauge group (*i.e.* left- and right-handed particles transform differently). If the fundamental scale M is large, chiral fermions help explain why a low-energy limit exists at all, because chiral fermions (similar to gauge bosons) cannot acquire masses unless the gauge symmetry spontaneously breaks. If the scale v set by this symmetry breaking proves to be low, it automatically also ensures that chiral fermions and gauge bosons are also light. Chiral fermions tend to have anomalous gauge transformations, but these turn out to cancel once summed over the particle content of a single Standard Model generation.

The Standard Model enjoys just the right accidental global symmetries, where 'accidental' means they are not assumed as part of the model's definition and instead simply emerge as consequences of gauge invariance, the Standard Model field content and renormalizability. Several successful conservation laws emerge in this way, including baryon number and a lepton number for each generation. Although evidence now exists against separate conservation of lepton number for each generation, it came quite recently and the violation found is restricted to the neutrino sector. Intriguingly, the unique dimension-five non-renormalizable interaction that would naively be the first to show up in an expansion in powers of E/M predicts precisely the symmetry-breaking pattern that is observed.

Although attractive, the Standard Model also has its flaws. The most important flaws are observational. The model gets wrong the pattern of lepton-flavour conservation seen in neutrino oscillations, and although these might just be first indications of higher-dimension interactions, they might equally well point to new species of light sterile neutrinos. The model also does not describe some cosmological observations, such as those that provide evidence for Dark Matter. This suggests our list of low-energy particles is not yet complete (something that can be possible if the undiscovered ones – the hypothetical 'dark sector' – interact too weakly with those we know). The Standard Model also becomes less and less compelling the further back it is extrapolated into the earlier universe, requiring more and more unusual initial conditions to explain what is seen now.[22]

neighborhood). My own take on the different approaches – and a definition of what the 'old' and 'new' versions of the problem are – is in [486].

22 Among these initial-condition puzzles are: the ratio of the number of baryons and antibaryons in the universe; the pattern of observed primordial density fluctuations; and so on.

The other flaw with the model is more theoretical: some effective couplings are unusually small for a low-energy effective theory. Most puzzingly, some small couplings do not remain small as heavy particles get integrated out; they are not 'technically natural'. The lower dimension the interaction involved, the more puzzling it is to find a small coupling, and on this score the cosmological constant (or vacuum energy) is the biggest such puzzle. It is not yet known how these puzzles will ultimately be resolved.

Exercises

Exercise 9.1 In the Standard Model calculate the mixed anomaly coefficients A_{abc} where two indices represent a G_{SM} generator and the other corresponds to one of the Standard Model's accidental symmetries: B, L_e, L_μ and L_τ. If these are nonzero, particularly for the strong $SU_c(3)$ interactions, then the corresponding accidental symmetry is not really a symmetry at the quantum level.

For what linear combinations of these accidental symmetries do all such anomaly coefficients vanish when summed within only a single generation? Show that additional anomaly-free combinations exist if all three generations are included in the sum.

Repeat this exercise with the Standard Model field content supplemented by a singlet sterile neutrino for each generation.

Exercise 9.2 Suppose each generation of the Standard Model is supplemented by two types of new spinless particles, transforming as a $(\mathbf{\bar{3}}, \mathbf{1}, y)$ with $y = -\frac{2}{3}$ and $y = +\frac{1}{3}$ under $G_{SM} = SU_C(3)\times SU_L(2)\times U_Y(1)$. The $y = -\frac{2}{3}$ (or $y = +\frac{1}{3}$) particle is called a 'right-handed up-squark' (or 'right-handed down-squark').[23]

What are the most general renormalizable interactions possible for such particles, both with themselves and with the rest of the Standard Model?

What are the accidental global symmetries that are allowed by these (and the usual Standard Model interactions)?

Minimize the scalar potential and identify what choices of parameters give the same symmetry-breaking pattern for G_{SM} as in the Standard Model.

Answer the previous three questions while including also left-handed squarks, transforming under G_{SM} as $(\mathbf{3}, \mathbf{2}, y)$ with $y = +\frac{1}{6}$.

Repeat the exercise including also left- and right-handed sleptons, which respectively transform under G_{SM} as $(\mathbf{1}, \mathbf{2}, -\frac{1}{2})$ and $(\mathbf{1}, \mathbf{1}, +1)$.

Exercise 9.3 Suppose the Standard Model is supplemented by new left-handed spin-half particles that transform in the same way as do the bosons in the Standard Model. That is, add left-handed 'gluinos' transforming under G_{SM} as $(\mathbf{8}, \mathbf{1}, 0)$, left-handed 'winos' transforming as $(\mathbf{1}, \mathbf{3}, 0)$, left-handed 'bino'[24] transforming as a singlet: $(\mathbf{1}, \mathbf{1}, 0)$ and left-handed 'higgsino' transforming as $(\mathbf{1}, \mathbf{2}, \frac{1}{2})$.

[23] The squarks themselves are not right- or left-handed since they are spinless, but the name comes from supersymmetric theories for which they are partners of the corresponding quarks. Despite the name 'right-handed' they share the quantum numbers of the corresponding left-handed antiquark.

[24] The wino and bino are pronounced *ween-oh* and *been-oh*, and together with the gluino are called 'gauginos'.

Compute the gauge anomalies (including the Lorentz anomaly) using these fields. Do they all vanish? If not, what happens if one also adds a new higgsino transforming as $(\mathbf{1}, \mathbf{2}, -\frac{1}{2})$?
What are the most general renormalizable interactions possible for such particles, both with themselves and with the rest of the Standard Model?
What are the accidental global symmetries that are allowed by these (and the usual Standard Model interactions)?
Identify the left-handed mass matrix for these and the Standard Model fermions. Identify the propagation eigenstates in the case where both gauginos and the two types of higgsinos described above are included.

Exercise 9.4 Prove that Eq. (9.26) is the unique dimension-five operator that can be built using only Standard Model fields that is both a Lorentz scalar (as must be the lagrangian) and is $G_{SM} = SU_C(3) \times SU_L(2) \times U_Y(1)$ invariant. Prove that this operator satisfies $\Delta L = \pm 2$ and $\Delta B = 0$.
Identify the 3×3 left-handed neutrino mass matrix in flavour space that this term predicts. Identify the mass eigenvalues and PMNS matrix elements in the special case where the coefficients k_{mn} appearing in the weak-eigenstate basis of Eq. (9.26) are given by

$$k_{mn} = \begin{pmatrix} 0 & 1 & 0 \\ 1 & 0 & 1 \\ 0 & 1 & 0 \end{pmatrix} \frac{y^2}{M}. \tag{9.41}$$

Exercise 9.5 Prove that any baryon-number violating dimension-six operator can be written as a linear combination of the ones given in Eq. (9.28), assuming they must be built using only Standard Model fields and be both a Lorentz scalar and $G_{SM} = SU_C(3) \times SU_L(2) \times U_Y(1)$ invariant. Prove that any such operator satisfies $\Delta L = \Delta B$.

Exercise 9.6 A portal to the dark sector is defined to be any renormalizable interaction between Standard Model fields and a new field that transforms as a Standard Model singlet. Any such an interaction has the general form

$$\mathfrak{L}_{\text{portal}} = \sum_I c_I \, \mathcal{O}^I_{SM} \, \mathcal{O}^I_{\text{dark}},$$

where $\mathcal{O}^I_M$ depends only on Standard Model fields while $\mathcal{O}^I_{\text{dark}}$ depend only on the new singlet fields. Being renormalizable requires the coefficients c_I to have dimension $(\text{mass})^n$ with $n \geq 0$.
Identify the three kinds of singlet fields – *i.e.* three kinds of Lorentz tensor fields – for which such a portal can exist.

Exercise 9.7 Evaluate the Feynman graph of Fig. 9.3 giving the contribution of a sterile heavy spin-half fermion to the Higgs mass and verify that it can cancel the contribution of a (complex) sterile heavy scalar particle provided its couplings and mass are chosen appropriately. What choices for couplings and masses give cancellation?
Repeat this problem for the contributions to the vacuum energy. What choices make the corrections to this vanish?

10 General Relativity as an Effective Theory

General Relativity[1] (GR) provides an excellent description of gravitating systems and in many ways is the Standard Model of gravitational physics [241, 242]. But unlike the Standard Model, General Relativity is not renormalizable [243, 244]. One way to see this is to observe that its coupling constant – *i.e.* Newton's constant of universal gravitation, G_N – has dimension $(\text{mass})^{-2}$ in fundamental units, and so (other things being equal) on dimensional grounds Feynman graphs involving more and more powers of G_N diverge more and more in the ultraviolet.

The dimensions of Newton's constant can be read in turn from the Einstein–Hilbert action [245], whose variation with respect to the spacetime metric, $g_{\mu\nu}$, reproduces the Einstein field equations. This has lagrangian density

$$\mathfrak{L}_{EH} = -\frac{1}{16\pi G_N}\sqrt{-g}\,R, \tag{10.1}$$

where the g in $\sqrt{-g}$ denotes the determinant of the metric $g_{\mu\nu}(x)$, and is required to ensure $S_{EH} = \int \mathrm{d}^4x\,\mathfrak{L}_{EH}$ is generally covariant (*i.e.* invariant under diffeomorphisms of spacetime). The quantity $R_{\mu\nu} := R^{\lambda}{}_{\mu\lambda\nu}$ is the Ricci tensor built from the curvature (Riemann) tensor, $R^{\alpha}{}_{\mu\lambda\nu}$, whose conventions and definition in terms of the metric are summarized in §A.2.1. Its trace, $R := g^{\mu\nu}R_{\mu\nu}$, is the Ricci scalar, where $g^{\mu\nu}$ is the inverse metric defined by $g^{\mu\nu}g_{\nu\lambda} = \delta^{\mu}{}_{\lambda}$. For the purposes of this chapter what is important about all of these curvatures is that they involve precisely two derivatives of the metric.

Although non-renormalizability was once regarded as poison for a serious theory, the previous sections of this book summarize why non-renormalizability in itself is no longer regarded as that remarkable. After all, other very predictive theories, like the Fermi theory of weak interactions of §7.1 or the low-energy interactions of pions encountered in §8, also share this property.

Non-renormalizability is what happens whenever couplings (like Newton's constant, G_N, or the Fermi constant, G_F) have dimension of an inverse power of mass, and the central observation that gives non-renormalizable theories predictive power is that any series in this coupling is necessarily a low-energy expansion; effective field theories are the natural language for their description.

From this point of view the Einstein–Hilbert action should not be regarded as being carved by Ancient Heroes into tablets of stone; one should instead seek the most general action built from the spacetime metric, $g_{\mu\nu}$, that is invariant under the symmetries of the problem (which in this case should include general covariance and local Lorentz invariance), organized in a derivative expansion. Since the Riemann tensor (and its traces and derivatives) is the unique covariant object built from

[1] See §C.5.2 for an extremely brief summary of the elements of GR.

derivatives of the metric [246], the effective lagrangian describing pure gravity should come as an expansion in powers of curvatures and their derivatives [2, 247].

This suggests General Relativity is the leading part of a more general Wilsonian effective action – call it GREFT – whose action is

$$-\frac{\mathfrak{L}_{\rm eff}}{\sqrt{-g}} = \lambda + \frac{M_p^2}{2} R + \left[a_{41} R^2 + a_{42} R_{\mu\nu} R^{\mu\nu} + a_{43} R_{\mu\nu\lambda\rho} R^{\mu\nu\lambda\rho} + a_{44} \Box R\right] \tag{10.2}$$

$$+ \frac{1}{M^2}\left[a_{61} R^3 + a_{62} R R_{\mu\nu} R^{\mu\nu} + a_{63} R_{\mu\nu}{}^{\lambda\rho} R_{\lambda\rho}{}^{\alpha\beta} R_{\alpha\beta}{}^{\mu\nu} + \cdots\right] + \cdots,$$

corresponding to a sum over all possible scalars built from powers of the curvature tensor and its derivatives. In this expression the first two terms represent the usual interactions of General Relativity (GR) – consisting of a cosmological constant, λ, (also called c_4 in earlier sections) and the Einstein–Hilbert action – while the first square bracket contains four-derivative terms, the second square bracket contains six-derivative terms and so on. As always, $M_p^2 = (8\pi G_N)^{-1}$ is a proxy for Newton's gravitational constant.

The effective couplings a_{di} are dimensionless (in four dimensions) and to this end an appropriate power of a UV mass scale denoted M is factored out of the curvature-cubed and higher terms, where $M \ll M_p$ is envisaged as being the lightest UV scale to have been integrated out to obtain this action.[2] Notice that the scale M is *not* written in front of the Einstein–Hilbert term, since if there it would be swamped by the M_p^2 term that is required to properly reproduce the value of G_N.

This is as would be expected from the discussion in earlier chapters – see *e.g.* §3 and §7.3 – which argue that it is generically the smallest UV scale that dominates in all denominators, while the largest UV scale dominates in all numerators. Although this expectation works fine for the large coefficient M_p^2 of the Einstein–Hilbert term, as discussed in the previous chapter it is an unsolved puzzle why the cosmological constant is not also dominated by contributions from the largest values of M. In what follows the observational information that λ is measured to be extremely small is used to justify its neglect, at least until discussing cosmological models in later sections.

Finally, it should also be noted that many of the interactions appearing in the action (10.2) are redundant, in the precise sense used in §2.5. Terms like $\sqrt{-g}\,\Box R$ are total derivatives, as is (locally, in four dimensions) the combination $\sqrt{-g}\left(R_{\mu\nu\lambda\rho}R^{\mu\nu\lambda\rho} - 4\,R_{\mu\nu}R^{\mu\nu} + R^2\right)$ (called the Euler invariant) which allows the Reimann-squared term to be traded for terms involving only the Ricci tensor [248, 249].

Furthermore – as is also argued in §2.5 – field redefinitions can be used to remove terms that vanish when using the lowest-order field equations. For just gravity (in the absence of other fields – more about the inclusion of other fields below) with vanishing cosmological constant ($\lambda = 0$) the lowest-order metric field equations imply $R_{\mu\nu} = 0$ and so any terms involving Ricci tensors or Ricci scalars can also be

[2] To the extent that M is discussed in the literature, it is often chosen to be $M \sim M_p$, although this usually dramatically underestimates the influence of these effective interactions. For instance, for applications to very large distances (such as in late-time cosmology) it could be that M is the electron mass once electrons are integrated out (see the related discussion in §7.3.2).

dropped. Under these circumstances, the first nontrivial effective interactions arise at curvature-cubed level, such as the term $(a_{63}/M^2)\, R_{\mu\nu}{}^{\lambda\rho} R_{\lambda\rho}{}^{\alpha\beta} R_{\alpha\beta}{}^{\mu\nu} \subset \mathfrak{L}_{\rm eff}$ of (10.2).

10.1 Domain of Semi-Classical Gravity ◇

The next step is to power-count using this effective theory, to identify the small expansion parameter underlying the semiclassical expansion.[3] To this end, expand the metric about a classical solution to Einstein's equations,[4]

$$g_{\mu\nu}(x) = \tilde{g}_{\mu\nu}(x) + \frac{2h_{\mu\nu}(x)}{M_p}, \tag{10.3}$$

and rewrite (10.2) as a sum of effective interactions of the form

$$\mathfrak{L}_{\rm eff} = \tilde{\mathfrak{L}}_{\rm eff} + M^2 M_p^2 \sum_n \frac{c_n}{M^{d_n}}\, \mathcal{O}_n\left(\frac{h_{\mu\nu}}{M_p}\right), \tag{10.4}$$

where $\tilde{\mathfrak{L}}_{\rm eff} = \mathfrak{L}_{\rm eff}(\tilde{g}_{\mu\nu})$ is the lagrangian density evaluated at the background configuration.

As in previous sections the sum over n runs over the labels for a complete set of non-redundant interactions, $\mathcal{O}_n$, each of which involves $f_n \geq 2$ powers of the field $h_{\mu\nu}$ (with $f_n \neq 1$ because of the background field equations satisfied by $\tilde{g}_{\mu\nu}$). The parameter d_n counts the total number of derivatives appearing in $\mathcal{O}_n$ (acting either on the background or the perturbation), and so the factor M^{-d_n} is what is required to keep the coefficients, c_n, dimensionless. For instance, an example of an interaction appearing in (10.4) (call it $n = n_0$) that has $d_{n_0} = 2$ and $f_{n_0} = 3$ is (*c.f.* Exercise 10.3)

$$M^2 M_p^2 \left(\frac{c_{n_0}\mathcal{O}_{n_0}}{M^2}\right) = \frac{c_{n_0}}{M_p} \sqrt{-\tilde{g}}\; h^{\mu\nu} h^{\lambda\rho} \tilde{\nabla}_\mu \tilde{\nabla}_\nu h_{\lambda\rho}. \tag{10.5}$$

In this interaction (and also for the others) indices are raised and covariant derivatives are built using the background metric, $\tilde{g}_{\mu\nu}$, so $h^{\mu\nu} = \tilde{g}^{\mu\lambda}\tilde{g}^{\nu\rho}h_{\lambda\rho}$ and so on. The overall prefactor, $M^2 M_p^2$, is chosen so that the kinetic terms – *i.e.* those terms in the sum for which $d_n = f_n = 2$ – are M and M_p independent. As is clear from the example, the operators $\mathcal{O}_n$ depend implicitly on the classical background, $\tilde{g}_{\mu\nu}$, about which the expansion is performed.

The coefficients c_n are calculable in terms of the a_{di} of (10.2), and if $M \ll M_p$ the c_n's cannot be order unity if the a_{di}'s are. Comparing Eqs. (10.2) and (10.4) shows that the absence of M_p in all of the curvature-squared and higher terms in (10.2) implies that the c_n for these interactions should be of order

$$c_n = \left(\frac{M^2}{M_p^2}\right) g_n \qquad (\text{if } d_n > 2), \tag{10.6}$$

where g_n is (at most) order unity and depends only logarithmically on M.

[3] This section uses the power-counting logic [2] of §3.2.1, whose application to gravity is given in [24].
[4] Beware of a convenient – though somewhat perverse – notational change in this chapter: here tildes denote the background metric (whereas they marked the classical deviation from the background in earlier chapters). The factor of 2 in (10.3) is not required in what follows, but achieves canonical normalization – see Exercise 10.3.

Perturbation theory proceeds as in earlier chapters, by separating $\mathfrak{L}_{\rm eff} - \tilde{\mathfrak{L}}_{\rm eff}$ into quadratic and higher order parts,

$$\mathfrak{L}_{\rm eff} = \left(\tilde{\mathfrak{L}}_{\rm eff} + \mathfrak{L}_0\right) + \mathfrak{L}_{\rm int}, \tag{10.7}$$

where $\mathfrak{L}_0$ denotes the 'unperturbed' lagrangian density consisting of those terms in $\mathfrak{L}_{\rm eff}$ for which $f_n = 2$ and $d_n \leq 2$. All other terms are lumped into $\mathfrak{L}_{\rm int}$. Expanding the path integral in powers of $\mathfrak{L}_{\rm int}$ allows the integral over $h_{\mu\nu}$ to be expressed as a sum of Gaussian integrals, classifiable in terms of Feynman graphs, with $\mathfrak{L}_0$ defining the propagators, $G_{\mu\nu,\lambda\rho}(x,y) = \langle h_{\mu\nu}(x)\, h_{\lambda\rho}(y)\rangle$, appearing in these graphs and $\mathfrak{L}_{\rm int}$ defining their vertices in the usual way. For the purposes of power counting what matters about these propagators is that they do not depend on M and M_p (though they do depend on scales associated with $\tilde{g}_{\mu\nu}$).

From here on the discussion follows earlier treatments, like that of the Toy Model in §1.1 or of chiral perturbation theory in §8.2. For gravitational applications it is important to recognize that the power-counting arguments of earlier sections are in essence dimensional: one counts factors of M and M_p coming from vertices and assigns the power of the low-energy scale purely on dimensional grounds (assuming all low-energy scales are similar in size). In particular, nothing in the power-counting argument requires working in momentum space, or that the background metric be flat.[5]

In position space UV divergences emerge as singularities experienced by the propagators, $G_{\mu\nu,\lambda\rho}(x,y)$, in the coincidence limit $y \to x$. As usual, regulating these using dimensional regularization – see *e.g.* Exercise 10.6 – and renormalizing using a scale-independent scheme (like modified minimal subtraction) makes the dimensional power-counting argument particularly simple. As usual, the spacetime dimension is written as $D = 4 - 2\varepsilon$ and divergences arise as $\varepsilon \to 0$.

The relevant low-energy scale appearing when power counting might be set by the energies, E, of scattered gravitons, or it might be set by the generic size of a derivative of the background configuration and denoted by H, so that $\partial^2 \tilde{g} \sim H^2$, for instance. For simplicity there is assumed to be only one low-energy scale relevant for the process of interest, so when scattering particles with energy E on a background geometry with curvature scale H it is assumed $E \sim H$.

With these choices, the arguments used in §3.2.1 imply that a graph involving $\mathcal{E}$ (amputated) external graviton lines, $\mathcal{L}$ loops and $\mathcal{V}_n$ vertices (each involving d_n derivatives) depends on the scales H, M and M_p as

$$\mathcal{A}_{\mathcal{E}}(H) \simeq H^2 M_p^2 \left(\frac{1}{M_p}\right)^{\mathcal{E}} \left(\frac{H}{4\pi\, M_p}\right)^{2\mathcal{L}} \prod_n \left[c_n \left(\frac{H}{M}\right)^{d_n - 2}\right]^{\mathcal{V}_n}, \tag{10.8}$$

with factors of 4π also included as in earlier sections. Keeping in mind the factors of M and M_p hidden in the c_n's for $d_n > 2$ – *c.f.* Eq. (10.6) – the above expression is more usefully rewritten as

[5] This is not to say that curved space introduces no new issues; one such is the existence sometimes of horizons and the way this makes an external observer's description more similar to the open systems described in §16 than like traditional Wilsonian EFTs [250, 251].

$$\mathcal{A}_{\mathcal{E}}(H) \simeq H^2 M_p^2 \left(\frac{1}{M_p}\right)^{\mathcal{E}} \left(\frac{H}{4\pi M_p}\right)^{2\mathcal{L}} \left[\prod_{d_n=2} c_n^{\mathcal{V}_n}\right] \prod_{d_n \geq 4} \left[g_n \left(\frac{H}{M_p}\right)^2 \left(\frac{H}{M}\right)^{d_n-4}\right]^{\mathcal{V}_n}. \tag{10.9}$$

Here, the condition $d_n > 2$ is traded for $d_n \geq 4$ because d_n must be even. General covariance requires d_n to be even because each derivative contributes a spacetime index, ∇_μ, and all indices must be contracted with a rank-two metric, $\tilde{g}_{\mu\nu}$, or a rank-four Levi-Civita symbol, $\tilde{\epsilon}_{\mu\nu\lambda\rho}$ (whose definition is given in §A.2.1).

Because (as in earlier sections) external lines are amputated in (10.9), the amplitude $\mathcal{A}_{\mathcal{E}}$ has dimension (mass)$^{4-\mathcal{E}}$, as would be expected for the coefficient of $\mathcal{E}$ powers of $h_{\mu\nu}$ in an expansion of the 1PI action. This makes the dependence of $\mathcal{A}_{\mathcal{E}}$ on scales useful when estimating the size of effective couplings in this action.

Eq. (10.9) shows that the validity of the semiclassical expansion in General Relativity (or, more precisely, the loop expansion in GREFT) is controlled by the small parameter

$$\left(\frac{H}{4\pi M_p}\right)^2 \ll 1, \tag{10.10}$$

since this is what allows for graphs with fewer loops to dominate those with more loops (for a fixed number of external lines). Similarly, the suppression of interactions involving higher-derivatives additionally requires

$$g_n \left(\frac{H}{M_p}\right)^2 \left(\frac{H}{M}\right)^{d_n-4} \ll 1 \quad (\text{for } d_n \geq 4). \tag{10.11}$$

By contrast, repeated insertions of two-derivative interactions (*i.e.* those coming from the Einstein–Hilbert action) are not generically suppressed unless the graph in question also involves higher loops. Because the Einstein–Hilbert term generically predicts[6]

$$c_n \simeq 1 \quad (\text{for } d_n = 2), \tag{10.12}$$

the low-energy semiclassical expansion in itself therefore does not also require the neglect of the nonlinearity of General Relativity. The lack of suppression of these interactions resembles what was found earlier when power counting using the EFT for non-abelian Goldstone bosons, and has its roots in the dominance of derivatively coupled interactions.

Why Classical Methods Work in GR

Conceptually, Eq. (10.9) (and its analog once matter is included – see *e.g.* §10.2) is the foundation for *all* applications of General Relativity to observations, since it identifies systematically which interactions are important in any given physical process and quantifies the theoretical error implicit in any classical calculation. Because H is always in the numerator in this expression, it shows that control over

[6] Exercise 10.3 calculates c_n for cubic two-derivative interactions in an expansion about flat space. Later examples explore cosmological situations with two low-energy scales, for which hierarchies can arise amongst the dimensionless couplings c_n.

semiclassical methods is essentially a low-energy approximation (as expected due to the presence of the dimensionful non-renormalizable coupling G_N).

In particular, the vast majority of applications use General Relativity as a classical field theory and Eq. (10.9) shows why this is usually valid: for a fixed number of external lines the least-suppressed contributions come if two conditions are satisfied:

C1a: $\mathcal{L} = 0$ (that is, no loops – corresponding to the classical limit);
C1b: $\mathcal{V}_n = 0$ for all interactions for which $d_n > 2$.

Since d_n is even for all interactions (and the neglect of the cosmological constant, λ, implies that $d_n \geq 2$); this means that the dominant contributions to observable processes arise from tree graphs using only interactions having precisely $d_n = 2$ derivatives. That is, from classical processes computed using only interactions coming from the Einstein–Hilbert action.

But Eq. (10.9) says much more than this: it also identifies the contributions that enter at subleading, and sub-subleading order, and so on. The dominant subleading corrections arise in one of two ways: either

C2a: use only $d_n = 2$ interactions but with $\mathcal{L} = 1$, or
C2b: use only $\mathcal{L} = 0$ with $\mathcal{V}_n = 0$ for $d_n > 4$ interactions and $\sum_n \mathcal{V}_n = 1$ for the $d_n = 4$ interactions.

These say that the leading subdominant corrections to classical GR arise suppressed by $(H/4\pi M_p)^2$ and come either from one-loop General Relativity or from tree graphs containing exactly one curvature-squared interaction. It is renormalizations of the coefficients of the curvature-squared interactions appearing in tree graphs that cancel the UV divergences that arise in the one-loop graphs. One can proceed similarly to any order in H/M and H/M_p.

To make this more concrete, consider quantum corrections in the Schwarzschild geometry that describes the gravitation field of the Sun in our solar system. These corrections are part of the theoretical error for any classical solar-system predictions obtained using General Relativity (which are routinely compared with observations of how objects move – see, for example, [252, 253]). In this case, derivatives of the classical background geometry at a distance r from the Sun are characterized by the curvature scale, $H^2 \simeq r_s/r^3$, where

$$r_s = 2\,G_N\mathcal{M} = \frac{\mathcal{M}}{4\pi M_p^2} \tag{10.13}$$

is the geometry's Schwarzschild radius, with $\mathcal{M}$ the mass of the gravitating source. For the Sun $\mathcal{M} = \mathcal{M}_\odot \simeq 1.99 \times 10^{30}$ kg, and so (10.13) evaluates to $r_s \simeq 3$ km. This makes the estimate (10.10) of the fractional size of quantum corrections to local physics at radius r from a source of order

$$\left(\frac{H}{4\pi M_p}\right)^2 \simeq \frac{r_s}{(4\pi M_p)^2 r^3} \simeq 6 \times 10^{-94} \left(\frac{\mathcal{M}}{\mathcal{M}_\odot}\right)\left(\frac{R_\odot}{r}\right)^3, \tag{10.14}$$

which gets smaller the larger r is. The numerical estimate uses the solar mass and the solar radius, $R_\odot \simeq 6.96 \times 10^5$ km, as would be appropriate at the solar surface.

This estimate shows that quantum effects in the solar system are *extremely* small because the Planck length, $\ell_p := 1/M_p \simeq 5 \times 10^{-19}\ \text{GeV}^{-1} \simeq 1 \times 10^{-34}$ m, is so tiny compared with the macroscopic scales of practical interest. Even for radii as small as $r \simeq r_s$ the size of loop corrections is of order

$$\left(\frac{\ell_p}{4\pi r_s}\right)^2 \simeq \left(\frac{M_p}{\mathcal{M}}\right)^2, \tag{10.15}$$

showing that quantum effects remain under control at the Schwarzschild radius provided the gravitating source is much more massive than the (reduced) Planck mass, which is $M_p = (8\pi G_N)^{-1/2} \simeq 4\ \mu$g in macroscopic units.

It is the small size of $M_p/\mathcal{M}$ that ultimately justifies the validity of semiclassical calculations near the event horizon of a black hole, such as those leading to the prediction of Hawking radiation[7] [254]. Although (10.15) is extremely small for astrophysical objects, it is enormous for elementary particles (like electrons or protons, for example). This is why classical black hole physics is trusted (and small) in astrophysical settings, but need not give a good description of spacetime at distances $r \sim r_s$ from an electron or proton.

10.2 Time-Dependence and Cosmology ♠

Although quantum effects involving gravity are incredibly small in the solar system, and even just outside relativistic astrophysical objects, there is a situation where they might actually be large enough to be observable while still believing semiclassical methods (and indeed may have already been observed).

The observations that might be sensitive to a practical quantum/gravity interplay study the distribution of matter within the universe. This can be measured using a variety of astrophysical methods, including surveys of the large-scale distribution of galaxies and by measurements of temperature fluctuations seen in the Cosmic Microwave Background (CMB) radiation. Although a complete description of these observations is well beyond the scope of this book, suffice it to say that a good quantitative understanding is emerging for the overall distribution of matter and its evolution in time [255]. The large-scale distribution of matter in the Universe seems consistent with what would be expected from gravitational amplification of initially small density fluctuations due to the *Jeans instability*, wherein initially slight over-densities get amplified as they gravitationally accrete more material onto themselves[8] [256].

7 Although control over semiclassical methods inevitably means quantum processes are small, the saving grace of phenomena like Hawking radiation is that – in ideal systems – there is no classical black-hole radiation with which it must compete.

8 This picture is only successful in the presence of Dark Matter and Dark Energy – two poorly understood types of matter – and indeed part of the evidence for their existence comes from the details of how large-scale structures form.

More precisely, gravitational amplification of small density fluctuations explains current observations *provided* the relatively recent Universe inherits a specific pattern of primordial density fluctuations from much-earlier epochs. To produce the observed pattern of large-scale structure these primordial fluctuations must be small in amplitude – with $\delta\rho/\rho \sim 10^{-5}$ – and have a spectrum that is close to scale invariant. Intriguingly, the much earlier epochs that produce these fluctuations plausibly have a Hubble scale, H, much closer to (though still smaller than) the Planck scale, making quantum effects more important (but still under control). Best of all, semiclassical calculations in simple models [257] reveal that the required primordial fluctuations would be well-described (both in amplitude and in spectrum) by vacuum fluctuations, if these were stretched across the sky by an intervening epoch during which the expansion of the Universe accelerates.

A precondition for comparing observations to the predictions of such models is a controlled framework within which semiclassical calculations can be reliably performed. Without this, no meaningful assessment of theoretical error can be made. The purpose of this section is to explore how the above picture of quantum fluctuations in GREFT can be extended to provide this framework.

In cosmology universal expansion is described by an explicitly time-dependent background metric, $\tilde{g}_{\mu\nu}$, of the form

$$d\tilde{s}^2 = \tilde{g}_{\mu\nu}\, dx^\mu dx^\nu = -dt^2 + a^2(t)\, \gamma_{ij}\, dx^i dx^j, \tag{10.16}$$

where $a(t)$ describes the evolution of length scales with cosmic time, t, and γ_{ij} is a 3-dimensional metric describing the geometry of spatial slices at fixed t. The observed homogeneity and isotropy of the Universe on the largest scales suggests γ_{ij} is maximally symmetric (*i.e.* describes a 3-sphere, 3-plane or 3-hyperbola). Measurements of the CMB further indicate that these slices are consistent with being flat (so $\gamma_{ij} = \delta_{ij}$) [258]. The generic size of background-metric derivatives for this class of metrics is set by the Hubble scale, $H(t) = \dot{a}/a$, where over-dots denote differentiation with respect to t.

Simple cosmological models achieve the required accelerated expansion, $\ddot{a} > 0$, by supposing the universe's energy density is dominated in the past by a collection of $N \geq 1$ scalar fields, θ^a (with the simplest models restricting to a single field and so $N = 1$). In many of these models[9] – called *inflationary models* [259] – the scalars are designed to evolve very slowly in time because when this is so their energy density is dominated by their potential energy, V, and the Einstein equations imply that the Hubble scale is approximately constant in time, $H(t) \simeq H_I$ where $H_I^2 \simeq V/(3M_p^2)$. Under these circumstances, the Universe expands exponentially:

$$a(t) \simeq a_0 \exp[H_I(t - t_0)]. \tag{10.17}$$

The controlled study of primordial fluctuations requires an effective description of the early universe (such as a scalar-gravity system) as a systematic semiclassical expansion.

[9] Inflationary models comprise that subset of accelerating cosmologies (*i.e.* those for which $\ddot{a} > 0$) that take an initially expanding universe and increase its expansion (as opposed, say, to changing an initially contracting universe into an expanding one).

The effective lagrangian relevant to a scalar-metric system can be expressed as a derivative expansion, including both scalar fields and the metric, with the leading terms being

$$-\frac{\mathfrak{L}_{\rm eff}}{\sqrt{-g}} = v^4 U(\theta) + \frac{M_p^2}{2}\, g^{\mu\nu}\left[W(\theta)\, R_{\mu\nu} + G_{ab}(\theta)\, \partial_\mu\theta^a \partial_\nu\theta^b\right] \tag{10.18}$$
$$+ \left[A(\theta)(\partial\theta)^4 + B(\theta)\, R^2 + C(\theta)\, R\, (\partial\theta)^2 + \cdots\right]$$
$$+ \left[\frac{E(\theta)}{M^2}(\partial\theta)^6 + \frac{F(\theta)}{M^2} R^3 + \cdots\right],$$

with terms involving up to two derivatives written explicitly in the first line, those involving four derivatives in the square bracket of the second line, those with six derivatives in the square bracket of the third line and so on. The second and third lines are schematic inasmuch as R^2 and R^3 collectively respectively represent all possible independent curvature invariants involving four and six derivatives, each with a separate coefficient function.

Redundant interactions are eliminated from these interactions as usual, leaving an independent basis whose precise details are not important for the discussion to follow. The detailed form of the dimensionless functions $W(\theta), A(\theta), \ldots$ is also not very important in the power-counting arguments made below. Successful cosmology assumes some broad properties for the scalar potential, $V(\theta) = v^4 U(\theta)$, as is elaborated in more detail below.

As for GREFT, the scale M appearing in all denominators denotes the mass of the lightest states that are imagined to have been integrated out to obtain the Wilson action (10.18). More care is needed for scales appearing in numerators, however, and v and M_p are extracted so that the accompanying functions $U(\theta)$, $W(\theta)$ and $G_{ab}(\theta)$ are also dimensionless. The cosmologies of interest normalize the scalar fields so that their kinetic term has Planck mass coefficient, as above when $G_{ab}(\theta)$ is order unity, and (as usual) take $M \ll M_p$. The scale v is pulled out of the scalar potential and $U(\theta) \sim \mathcal{O}(1)$ is assumed when $\theta \simeq \mathcal{O}(1)$, so that $V \simeq v^4$. Of particular interest are models for which[10] $v \ll M$.

10.2.1 Semiclassical Perturbation Theory

Since the phenomenologically successful description of primordial fluctuations relates them to vacuum fluctuations, the first step is to identify the domain of validity of such a semiclassical calculation. This section shows that power counting broadly goes through as it did for GREFT (following the discussion in [260]), with loops being suppressed by powers of H^2/M_p^2. But the presence of the scalar potential also introduces an important new complication.

Semiclassical calculations are performed in the usual way, by expanding about a classical solution

$$\theta^a(x) = \vartheta^a(x) + \frac{\phi^a(x)}{M_p} \quad \text{and} \quad g_{\mu\nu}(x) = \tilde{g}_{\mu\nu}(x) + \frac{h_{\mu\nu}(x)}{M_p}, \tag{10.19}$$

[10] As discussed in §9.3, naturalness issues can often arise within this class because the assumption $v \ll M$ is not generic when v appears in the numerator.

where in practice $\vartheta^a = \vartheta^a(t)$ describes a homogeneous but time-dependent scalar evolution and $\tilde{g}_{\mu\nu}$ has the form given in (10.16). Because the background is time-dependent, all of the considerations of §6 apply, including the proviso that the low-energy EFT only aspires to capture slowly varying backgrounds for which *all* background time-rates-of-change – such as $\dot{\vartheta}^a$ and $H = \dot{a}/a$ – are much smaller than all UV scales (such as M).

Expanding the effective lagrangian of (10.18) in powers of ϕ^a and $h_{\mu\nu}$ allows it to be written in a form resembling (10.4):

$$\mathfrak{L}_{\rm eff} = \tilde{\mathfrak{L}}_{\rm eff} + M^2 M_p^2 \sum_n \frac{c_n}{M^{d_n}}\, O_n\left(\frac{\phi}{M_p}, \frac{h_{\mu\nu}}{M_p}\right). \tag{10.20}$$

As before, $\tilde{\mathfrak{L}}_{\rm eff} = \mathfrak{L}_{\rm eff}(\vartheta, \tilde{g}_{\mu\nu})$ and the interactions, O_n, involve $f_n = f_n^{(\phi)} + f_n^{(h)} \geq 2$ powers of the fields ϕ^a and $h_{\mu\nu}$. Also as before, the parameter d_n counts the number of derivatives appearing in O_n, the coefficients c_n are dimensionless and the prefactor, $M^2 M_p^2$, ensures the kinetic terms (and so also the propagators) are M and M_p independent.

Power counting is performed for this system following the now-familiar steps of previous sections. In particular, requiring Eq. (10.20) to capture the same dependence on M and M_p as does Eq. (10.18) requires the coefficients c_n (for $d_n > 2$) to satisfy (10.6), repeated again here for convenience of access:

$$c_n = \left(\frac{M^2}{M_p^2}\right) g_n \qquad (\text{if } d_n > 2), \tag{10.21}$$

with g_n at most order-unity, depending only logarithmically on M.

Similarly, properly reproducing the scales coming from the scalar potential, $V(\theta)$, requires

$$c_n = \left(\frac{v^4}{M^2 M_p^2}\right) \lambda_n \qquad (\text{if } d_n = 0), \tag{10.22}$$

where the dimensionless couplings λ_n are also largely independent of M_p and M. This choice amounts to assuming the scalar potential has the schematic form

$$V(\phi) = v^4 \left[\lambda_0 + \lambda_2 \left(\frac{\phi}{M_p}\right)^2 + \lambda_4 \left(\frac{\phi}{M_p}\right)^4 + \cdots\right], \tag{10.23}$$

which shows that V ranges through values of order v^4 as the ϕ^a range through values of order M_p (and so the θ^a range through values that are order unity).

Following the same steps as for earlier examples, a perturbative calculation writes $\mathfrak{L}_{\rm eff} = \left(\tilde{\mathfrak{L}}_{\rm eff} + \mathfrak{L}_0\right) + \mathfrak{L}_{\rm int}$ and expands $\exp\left[i \int d^4x\, \mathfrak{L}_{\rm int}\right]$ in powers of $\mathfrak{L}_{\rm int}$ within the path integral. As usual, the goal is to quantify how Feynman graphs with $\mathcal{E}$ external lines, $\mathcal{L}$ loops and $\mathcal{V}_n$ vertices (involving d_n derivatives) depend on the various scales v, M, M_p and the assumed single low-energy scale,[11] H.

[11] For cosmology with flat spatial slices correlation functions typically depend on both the Hubble scale H and mode momentum k/a, but these variables are both similar in size when evaluated for momenta comparable to H, as is in particular true for the epoch of 'horizon exit' of interest for primordial fluctuations.

In particular, the focus for cosmological applications is on primordial fluctuations, and these are related in cosmological models to correlation functions of the scalar and metric fluctuation fields, like $\langle \theta^a \cdots \theta^b \rangle$ or $\langle h_{\mu\nu} \cdots h_{\lambda\rho} \rangle$ or $\langle \theta^a \cdots h_{\mu\nu} \rangle$. Unlike the amputated amplitudes $\mathcal{A}_{\mathcal{E}}$ studied earlier for the pure-gravity case, these correlation functions – denoted $\mathcal{B}_{\mathcal{E}}$ – are not amputated and so include a propagator for each external line relative to an amputated graph. Since power counting associates factors of H on dimensional grounds, this means the Feynman amplitude for an unamputated $\mathcal{E}$-point amplitude, $\mathcal{B}_{\mathcal{E}}$, scales relative to an amputated amplitude, $\mathcal{A}_{\mathcal{E}}$, according to $\mathcal{B}_{\mathcal{E}} \simeq \mathcal{A}_{\mathcal{E}} H^{2\mathcal{E}-4}$.

Combining everything leads to the result

$$\mathcal{B}_{\mathcal{E}}(H) \simeq M_p \left(\frac{H^2}{M_p}\right)^{\mathcal{E}-1} \left(\frac{H}{4\pi M_p}\right)^{2\mathcal{L}} \left[\prod_{d_n=2} c_n^{\mathcal{V}_n}\right]$$
$$\times \prod_{d_n \geq 4} \left[g_n \left(\frac{H}{M_p}\right)^2 \left(\frac{H}{M}\right)^{d_n-4} \right]^{\mathcal{V}_n} \prod_{d_n=0} \left[\lambda_n \left(\frac{v^4}{H^2 M_p^2}\right) \right]^{\mathcal{V}_n}, \tag{10.24}$$

where all but the very last product are much as found in the previous section for pure gravity. What is new is the product over vertices with $d_n = 0$ containing the contribution of interactions coming from the scalar potential. What is dangerous about these scalar-potential contributions is the appearance within them of the low-energy scale H in the *denominator*, rather than numerator. These contributions are dangerous because at face value this enhancement for small H undermines the validity of the entire low-energy expansion.

In practice, however, cosmological models remain under control, provided the coefficients λ_n of the scalar potential are $O(1)$ (or smaller). This is because in the models of most interest the parameters v and H are not independent of one another. Indeed, H is determined by the classical Einstein equations for the background fields, and (as discussed above) in these models the potential $V(\theta)$ is assumed to dominate the scalar stress energy, leading to the relation $H \sim v^2/M_p$. Using this allows the potentially dangerous $d_n = 0$ terms of (10.24) to be rewritten as

$$\prod_{d_n=0} \left[\lambda_n \left(\frac{v^4}{H^2 M_p^2}\right) \right]^{\mathcal{V}_n} \simeq \prod_{d_n=0} \lambda_n^{\mathcal{V}_n}. \tag{10.25}$$

leading to the more transparent power-counting result

$$\mathcal{B}_{\mathcal{E}}(H) \simeq M_p \left(\frac{H^2}{M_p}\right)^{\mathcal{E}-1} \left(\frac{H}{4\pi M_p}\right)^{2L} \left[\prod_{d_n=2} c_n^{\mathcal{V}_n}\right] \left[\prod_{d_n=0} \lambda_n^{\mathcal{V}_n}\right] \prod_{d_n \geq 4} \left[g_n \left(\frac{H}{M_p}\right)^2 \left(\frac{H}{M}\right)^{d_n-4} \right]^{\mathcal{V}_n} \tag{10.26}$$

Although insertions of scalar interactions can in principle undermine the underlying expansion in powers of H/M_p, this does not happen for potentials of the form assumed in the cosmological models of most interest.[12]

[12] This power-counting argument shows that *if* the potential has the form given in (10.23) then the potential does not ruin the low-energy expansion. It does *not* show why a potential of the form (10.23) should emerge in the first place when integrating out heavier fields. For this there are two kinds of potentially dangerous interactions: the usual ones that get flagged by naturalness problems – the super-renormalizable (or relevant) interactions like ϕ^2 whose coefficients involve positive powers

Eq. (10.26) shows that the validity of the semiclassical expansion in the scalar-tensor models of cosmological interest again relies heavily on the low-energy approximation given in (10.10): $H^2 \ll (4\pi M_p)^2$. The leading contribution again comes from no-loop – *i.e.* classical – physics, built using just the zero- and two-derivative parts of the action. This is what justifies standard classical treatments of cosmological models.

For theories where primordial fluctuations have a quantum origin it is important that no observable primordial fluctuations arise from the leading classical contributions. Otherwise, small quantum effects are easily swamped by the classical results. Happily enough, classical erasure of pre-existing classical fluctuations often occurs automatically in expanding cosmologies. For instance, the contribution of spatial gradients like

$$\tilde{g}^{ij}\,\partial_i\vartheta^a\partial_j\vartheta^b = \frac{1}{a^2(t)}\,\gamma^{ij}\,\partial_i\vartheta^a\partial_j\vartheta^b, \tag{10.27}$$

in energy densities tends to drop rapidly at late times due to the growth of $a(t)$. This ironing out of initial fluctuations is particularly ruthless in inflationary models, for which $a(t) \propto e^{H_I t}$ grows exponentially.

In the absence of dominant classical perturbations the leading contributions to primordial fluctuations arise at subdominant order. In inflationary models classical fluctuations get inflated away by exponential expansion, but near-scale-invariant quantum fluctuations of fields in their vacuum state persist indefinitely and so eventually can dominate even though they are small. The power-counting formula (10.26) shows that for fixed $\mathcal{E}$ the size of subdominant contributions is suppressed by the small loop factor $(H_I/4\pi M_p)^2$ and comes from one-loop quantum contributions built using the zero- and two-derivative interactions (together with classical contributions using precisely one insertion of the appropriate counterterms involving up to four derivatives). Although standard treatments often do not go through the power-counting exercise, the dominant ingredients to which it leads are indeed the ones used in practice in the simplest models.

10.2.2 Slow-Roll Suppression

In principle, since the spectrum of primordial fluctuations is actually measured, one might hope to be able to use observations to infer the value of H_I/M_p during the primordial epoch when the fluctuations were generated. This turns out not to be possible at present because H_I/M_p is not the only parameter that is important for these models.[13]

This complication arises because successful cosmological models actually usually involve more than one low-energy scale, and so violate the single-scale assumption made when deriving (10.26). The underlying reason for this (at least within simple inflationary models) is that phenomenological success relies on there being near-exponential expansion, $a(t) \simeq e^{H_I t}$, for a sufficiently long time. The precise amount

of mass – but also non-renormalizable scalar interactions like ϕ^n that (10.23) says have coefficients v^4/M_p^n rather than M^{4-n}.

[13] The ratio H_I/M_p could be inferred if primordial gravitational waves were observed, but as of 2020 these had as yet escaped detection.

of time required depends on some of the details, but in many models the criterion is that a should expand exponentially for $H_I t \gtrsim 50$. Because of this, $H(t) = \dot{a}/a$ must be approximately constant for a long period of time, and this means that $\dot{H}/H$ is a new low-energy scale that is hierarchically different from H_I itself. This appearance of two scales is usually quantified in terms of their ratio: a dimensionless parameter $\epsilon = -\dot{H}/H^2$ that is generally positive for the cosmologies of interest. In terms of ϵ the regime of phenomenological success is $0 < \epsilon \ll 1$; a regime that fits nicely with the assumptions about slow time-variation underlying the use of EFT methods in time-dependent situations, as discussed in §6.

In the simplest models there is just this one new low-energy scale, rather than there being an independent new scale associated with each new derivative, H, $\dot{H}$, $\ddot{H}$ and so on [261]. For these models power-counting predictions can be fairly simply generalized to include the effects of the second low-energy scale. This is most simply done by quantifying how the dimensionless coefficients – like g_n, λ_n in (10.26) – depend on the new slow-roll parameter $\epsilon \ll 1$, rather than being order unity.

To see how this works, consider what is required for the leading classical solutions to have a long period of near exponential expansion. Since the leading contributions come from classical evolution (*i.e.* zero-loops) using those parts of the action involving two or fewer derivatives, the classical solutions must satisfy the scalar field equations

$$M_p^2 G_{ab}\left[\frac{D^2\vartheta^b}{dt^2} + 3H\,\dot{\vartheta}^b\right] + \frac{\partial V}{\partial \vartheta^a} = 0 \tag{10.28}$$

where $\dot{\vartheta}^a := \partial_t \vartheta^a$ and $D^2\vartheta^a/dt^2 = \ddot{\vartheta}^a + \Gamma^a_{bc}\,\dot{\vartheta}^b\dot{\vartheta}^c$ where (as usual) $H = \dot{a}/a$ and

$$\Gamma^a_{bc} := \frac{1}{2}G^{ad}\Big(\partial_b G_{cd} + \partial_c G_{bd} - \partial_d G_{bc}\Big), \tag{10.29}$$

is the target-space Christoffel symbol (of the 2nd kind) built from the target-space metric, $G_{ab}(\vartheta)$, that appears in the two-derivative terms of the EFT of (10.18). These are supplemented by the Einstein equations, which in this case boil down to the Friedmann equation that determines H in terms of ϑ^a:

$$3H^2 = \frac{1}{2}G_{ab}\,\dot{\vartheta}^a\dot{\vartheta}^b + \frac{V(\vartheta)}{M_p^2}. \tag{10.30}$$

Exponential expansion occurs when $H \simeq H_I$ is approximately constant, and (10.30) reveals this to be assured if ϑ^a is itself approximately constant (that is to say, if $\vartheta^a(t)$ evolves slowly enough that its kinetic energy is negligible compared with its potential energy). When this is true, then $H^2 \simeq V(\vartheta)/3M_p^2$, and so in order of magnitude the Hubble scale of interest is given by $H_I^2 \sim v^4/M_p^2$, while the slow-roll parameter obtained by taking its derivative is

$$\epsilon = -\frac{\dot{H}}{H^2} \simeq -\frac{\dot{\vartheta}^a}{6H^3M_p^2}\frac{\partial V}{\partial \vartheta^a}. \tag{10.31}$$

But if time-derivatives of ϑ^a are small, then second derivatives in (10.28) can be dropped, leading to an expression for how $\dot{\vartheta}^a$ is related to the choices made in the lagrangian:

$$\dot{\vartheta}^a \simeq -\frac{1}{3HM_p^2}G^{ab}\frac{\partial V}{\partial \vartheta^b}, \tag{10.32}$$

where G^{ab} is (as usual) the inverse matrix of G_{ab}. Using (10.32) in (10.31) then gives

$$\epsilon \simeq \frac{1}{18H^4M_p^4}\, G^{ab}\, \frac{\partial V}{\partial \vartheta^a}\, \frac{\partial V}{\partial \vartheta^b} \simeq \frac{1}{2V^2}\, G^{ab}\, \frac{\partial V}{\partial \vartheta^a}\, \frac{\partial V}{\partial \vartheta^b}. \tag{10.33}$$

These arguments show what is in any case intuitive: a sufficiently slowly rolling classical background can be ensured by choosing a sufficiently shallow scalar potential. But the above arguments quantify precisely how shallow: in order to have only a single new low-energy scale (with each background time-derivative suppressed by a factor of $\sqrt{\epsilon}\, H$), we ask each additional derivative of V to be suppressed by an additional power of $\sqrt{\epsilon}$:

$$\frac{\partial^s V}{\partial \vartheta^{a_1} \cdots \partial \vartheta^{a_s}} \sim \epsilon^{s/2} V \sim \epsilon^{s/2} v^4. \tag{10.34}$$

In terms of the quantities appearing in the power-counting result (10.26), this implies that the previously order-unity quantities λ_n should now be regarded as suppressed by

$$\lambda_n \simeq \epsilon^{f_{sn}/2}\, \hat{\lambda}_n, \tag{10.35}$$

where $\hat{\lambda}_n$ is now order unity and f_{sn} counts the number of scalar lines that meet at the $d_n = 0$ vertex in question (and so differs from f_n, which also counts metric lines). Using this in Eq. (10.26) shows how insertions of interactions from the scalar potential cost powers of the slow-roll parameter $\sqrt{\epsilon}$ (but no powers of H/M_p because $V \sim v^4$ ensures $H \sim v^2/M_p$).

Notice that no slow-roll suppression by powers of ϵ need also be assumed in $G_{ij}(\vartheta)$ or $W(\vartheta)$ or other places where the scalar field appears undifferentiated in the action, besides the scalar potential.[14]

The other way slow-roll parameters enter into Eq. (10.26) is through scalar background-field derivatives, which we assume satisfy Eq. (10.32) and its higher slow-roll extensions, so

$$\frac{\mathrm{d}^s \vartheta^a}{\mathrm{d}t^s} \sim \left(\sqrt{\epsilon}\, H\right)^s. \tag{10.36}$$

This kind of suppression is distinct from the factors of $\sqrt{\epsilon}$ in the λ_n, and arises in the effective lagrangian once differentiated scalar-fields are expanded about their background (assuming all slow-roll parameters are similar in size), as in

$$\partial_\mu \left(\vartheta^a + \frac{\phi^a}{M_p} \right) = \dot{\vartheta}^a\, \delta^0_\mu + \frac{\partial_\mu \phi^a}{M_p}, \tag{10.37}$$

and so on. Whenever a factor $\mathrm{d}^s \vartheta^a/\mathrm{d}t^s$ arises in the EFT in this way it therefore counts as a suppression by a factor of $\epsilon^{s/2}$ in addition to the factors of H that had been tracked earlier.

For example, expanding a kinetic term like $M_p^2 \sqrt{-g}\; g^{\mu\nu} \partial_\mu \theta\, \partial_\nu \theta$ in this way includes an order $\sqrt{\epsilon}\, H$ contribution to a bilinear ϕ^a-$h_{\mu\nu}$ mixing term coming from keeping a term linear in fluctuations from each of $\sqrt{-g}\; g^{\mu\nu}$ and $\partial_\mu \theta^a$ while evaluating $\partial_\nu \theta^a$ as $\dot{\vartheta}^a\, \delta^0_\nu$. This kind of term is crucial for cosmological

[14] Ignoring ϵ suppression in functions like $W(\vartheta)$ likely over-estimates its size in models where the small size of $V(\vartheta)$ is understood because ϑ is a pseudo-Goldstone boson.

applications, as it turns out, because the metric-scalar mixing that accompanies time-dependent backgrounds allows quantum fluctuations of the scalar field to induce related fluctuations in the Newtonian gravitational potential. It is ultimately these fluctuations in the Newtonian potential that are observable using galaxy distributions (regarded as proxies for density fluctuations) and the properties of the CMB.

To track these factors of ϵ in a power-counting formula, replace the two labels d_n and f_n with a new set that more finely resolves the properties of a Feynman graph. Instead of counting just the total number of derivatives, d_n, in a graph, it is useful also to count the number of derivatives acting only on background scalar fields, d_{sn}, since these come accompanied by powers of $\sqrt{\epsilon}$. It is similarly useful to track the number of background scalar fields, f_{sn}, meeting at a vertex in addition to the total number of fields (background plus fluctuation). Slow-roll suppression due to background evolution is then included by requiring any vertex with these labels to be suppressed by

$$c_n \simeq \epsilon^{d_{sn}/2}\hat{c}_n \quad (\text{for } 2 = d_n \geq d_{sn}) \qquad \text{and} \qquad g_n \simeq \epsilon^{d_{sn}/2}\hat{g}_n \quad (\text{for } d_n > 2), \tag{10.38}$$

where now it is $\hat{c}_n$ and $\hat{g}_n$ that are order-unity constants.

Using these choices in Eq. (10.26) leads to the very useful power-counting estimate for an unamputated Feynman graph with $\mathcal{E}$ external lines, $\mathcal{L}$ loops and $\mathcal{V}_n$ vertices of type 'n',

$$\mathcal{B}_{\mathcal{E}}(H) \simeq M_p \left(\frac{H^2}{M_p}\right)^{\mathcal{E}-1} \left(\frac{H}{4\pi M_p}\right)^{2\mathcal{L}} \prod_{d_n=2}\left(\epsilon^{d_{sn}/2}\,\hat{c}_n\right)^{\mathcal{V}_n}$$
$$\times \prod_{d_n=0}\left(\epsilon^{f_{sn}/2}\,\hat{\lambda}_n\right)^{\mathcal{V}_n} \prod_{d_n\geq 4}\left[\epsilon^{d_{sn}/2}\hat{g}_n\left(\frac{H}{M_p}\right)^2\left(\frac{H}{M}\right)^{d_n-4}\right]^{\mathcal{V}_n}. \tag{10.39}$$

This expression summarizes the dependence of a general Feynman graph on the two small parameters ϵ and H/M_p appearing in the simplest inflationary models.

Of most practical interest when comparing to observations are single-field models, which involve only one scalar field. For these, comparisons with observations potentially involve the two- and three-point correlation functions of scalar and tensor perturbations,[15] $\langle\phi\phi\rangle \simeq \mathcal{B}_{\phi\phi}$, $\langle h\phi\rangle \simeq \mathcal{B}_{h\phi}$ and $\langle hh\rangle \simeq \mathcal{B}_{hh}$. Here, '$h$' generically denotes both the transverse-traceless tensor fluctuations, h_{ij}, corresponding to gravitational waves, and the scalar metric components that mix with the scalar field. For two-point functions the above power counting estimates show that the leading contributions come from Feynman graphs with $\mathcal{E} = 2$ external lines, built using $\mathcal{L} = 0$ graphs using vertices taken only from the 2-derivative interactions, predicting them to arise at leading order with size

$$\mathcal{B}_{hh} \sim \mathcal{B}_{\phi\phi} \sim H_I^2 \quad \text{while} \quad \mathcal{B}_{\phi h} \sim \sqrt{\epsilon}\, H_I^2. \tag{10.40}$$

This specializes to $H = H_I$, since this is the epoch of observational interest.

[15] Although it goes beyond the scope of this presentation, these correlation functions are taken with the fields prepared in a specific vacuum state, called the Bunch–Davies vacuum state.

The leading contribution for 3-point correlations instead comes from graphs with $\mathcal{E} = 3$, though still using $\mathcal{L} = 0$ and taking only 2-derivative interactions, which for the quantities $\langle hhh \rangle$ and $\langle h\phi\phi \rangle$ are (for single-field models) ϵ-unsuppressed

$$\mathcal{B}_{hhh} \sim \mathcal{B}_{h\phi\phi} \sim \frac{H_I^4}{M_p}. \tag{10.41}$$

For these contributions the ϵ-unsuppressed cubic vertex comes from either the Einstein–Hilbert action or the inflaton kinetic term. No similarly unsuppressed contributions arise for $\langle hh\phi \rangle$ or $\langle \phi\phi\phi \rangle$, however, since no cubic interactions of these types arise unsuppressed by ϵ in the $d_n \leq 2$ lagrangian. (This need no longer be true for multiple-scalar models.) Since the cubic interaction of the scalar potential is order $\epsilon^{3/2}$, it is subdominant to the interactions obtained by inserting a single $O(\sqrt{\epsilon})$ h-ϕ kinetic mixing into $\langle hhh \rangle$ or $\langle hh\phi \rangle$, giving the slow-roll suppressed size

$$\mathcal{B}_{hh\phi} \sim \mathcal{B}_{\phi\phi\phi} \sim \frac{\sqrt{\epsilon}\, H_I^4}{M_p}. \tag{10.42}$$

For tensor fluctuations the above expressions basically tell the whole story. Since tensor fluctuations arise purely in the gravitational sector there are no issues about mixing with the scalar sector. The result for fluctuations in the normalized strain, $t_{ij} = h_{ij}/M_p$, can therefore be directly read off from $\mathcal{B}_{hh}$ and $\mathcal{B}_{hhh}$ above, to give

$$\mathcal{B}_{tt} \sim \frac{H_I^2}{M_p^2} \quad \text{and} \quad \mathcal{B}_{ttt} \sim \frac{H_I^4}{M_p^4} \quad \text{for} \quad t_{ij} = \frac{h_{ij}}{M_p}. \tag{10.43}$$

Scalar fluctuations are a bit more subtle, since for these issues of scalar-metric mixing and of gauge-choice are relevant. Although a proper discussion goes beyond the scope of this presentation, some simple statements can be made for completeness' sake. In particular, only one combination of the scalar fluctuation ϕ and the scalar part of the metric fluctuation h is physical, because the other combination can be modified merely by changing coordinates – *i.e.* changing gauge. (The discussion to this point essentially works in a non-unitary gauge for which the scalar field and scalar-metric fluctuations are tracked separately, even though only one combination of these survives in physical quantities.)

As it turns out, the physical gauge-invariant combination of these two fields can be found by moving to unitary gauge, which for power-counting purposes amounts to rescaling

$$\zeta \simeq \frac{\phi}{\dot{\varphi}/H} \sim \frac{\phi}{\sqrt{\epsilon}\, M_p}, \tag{10.44}$$

where $\varphi = \vartheta M_p$. Combining Eq. (10.44) with previous estimates leads to the expectations

$$\mathcal{B}_{\zeta\zeta} \sim \frac{H_I^2}{\epsilon\, M_p^2}, \quad \mathcal{B}_{tt\zeta} \sim \frac{H_I^4}{M_p^4} \quad \text{and} \quad \mathcal{B}_{\zeta\zeta\zeta} \sim \frac{H_I^4}{\epsilon\, M_p^4}. \tag{10.45}$$

and so on, to any order desired.

It is the prediction of (10.45) for $\mathcal{B}_{\zeta\zeta} = \langle \zeta\zeta \rangle$ that agrees well with observations, both in its overall size and in the dependence implicitly made for how its spectrum depends on scales (it is close to, but not exactly, scale invariant largely because H_I

and ϵ are approximately constant since both only evolve very slowly with time). Current observations allow all correlation functions except $\mathcal{B}_{\zeta\zeta}$ to vanish, with agreement between observations and more detailed calculations of $\mathcal{B}_{\zeta\zeta}$ requiring $H_I^2/\epsilon \simeq (1 \times 10^{15}\ \text{GeV})^2$ [262].

More can be learned once any of the other correlation functions are also detected as being nonzero, including redundant tests (in principle) of the entire single-field slow-roll inflationary framework. Redundant tests become possible – for instance, once $\mathcal{B}_{tt}$ is measured – because more observables are then available than there are parameters in the predictions. At present, the best one can do is use the non-observation of primordial gravitational waves in cosmology to put an upper limit on $\mathcal{B}_{tt}/\mathcal{B}_{\zeta\zeta}$, which requires $\epsilon \lesssim 0.064$ [262].

Encouragingly, both of these values are consistent with small ϵ and H_I/M_p, as is required for the validity of EFT methods. Although it is not yet clear whether they provide the ultimate explanation of primordial fluctuations, inflationary models set the current standard for observational success and calculational control against which all other theoretical proposals are compared.

10.3 Turtles All the Way Down? *

So far, it seems that there is always another, deeper, effective theory that plays the role of UV completion at smaller distances for long-distance (or low-energy) effective theories. Does this nesting of effective theories continue forever? Is it effective theories all the way down?

Although nobody knows for sure how nature ultimately answers this question, some conceptual progress has been made by identifying what a theory can look like in which endlessly new microscopic scales might not be necessary.

10.3.1 String Theory

At present, the main clues about what goes on at the smallest distances come from gravity. That is because, with very few exceptions, nature seems to be described very well by the Standard Model of particle physics combined with General Relativity (GR). But while the Standard Model is renormalizable – and therefore relatively insensitive to physics at much higher energies – GR is not. This fact that it is not renormalizable makes GR potentially more sensitive to UV physics, perhaps providing a clue as to how things work at the highest energies, at and above the Planck mass (which, after all, is the largest fundamental energy scale known in physics).

To put it slightly differently, because it is not renormalizable the only known control of gravitational predictions are as a low-energy expansion, using the GREFT framework described above. Some sort of UV completion for General Relativity must therefore intervene at sufficiently short distances (at the Planck length, or possibly at some other, longer, length scale) to allow more complete gravitational predictions using quantum gravity at these short distances. What might a candidate

UV completion of General Relativity look like? Using as a guide the Standard Model's completion of the Fermi theory of the weak interactions suggests it should include both the massless graviton and new massive particles (with masses at or below M_p) whose exchange generates the effective non-renormalizable couplings of GR once they are integrated out. Ideally, these new interactions might themselves be renormalizable, so they can be insensitive to physics at still-higher energies.

The Good News is that this suggests a method for guessing what the UV completion might be: one just collects all renormalizable systems that include the gravitational field, $g_{\mu\nu}$, and seeks within it the subset that generates General Relativity once the heavier particles are integrated out. The Bad News – at least before the early 1980s – is that the list of renormalizable field theories including a dynamical metric, $g_{\mu\nu}$, appeared to be empty.

This bleak picture changed in the early 1980s when a consensus developed that a candidate renormalizable theory including $g_{\mu\nu}$ might actually exist.[16] Better yet, the candidate theory in question appears not only to be renormalizable (that is, any dependence on still-higher energies can be absorbed into a small number of couplings), but it seems to be *ultraviolet finite* (*i.e.* seems not to depend on still-higher energies at all).

If renormalizability is motivated by a desire that the physics in question be insensitive to unknown physics at still-higher energies (because the contributions of such scales can be absorbed into unknown couplings), then a UV-finite theory is what one might expect to find if there were no unknown higher-energy physics at all. It is not that short-distance physics is present but can be absorbed into a few couplings; rather the calculations of the low-energy theory are complete in themselves, and make no reference to any unknown high-energy scales.

The rest of this section describes (in very broad brush-strokes) some of the features of this candidate UV completion. This is done not so much because this must be the theory of everything. It is instead done because it is the only known example that *could* be the theory of everything. As such, it provides a useful illustration of how it can be that there might not be just more EFTs on and on forever at higher energies. It also shows concretely how the energy scale where the UV completion kicks in can be much smaller than the Planck mass, and how and why this can occur.

String Theory: The Basic Idea

The theory that does all this is called string theory (or superstring theory or M-theory, as various variants are known [264, 265]), and the remainder of this section gives a brief cartoon of a few of its properties. As its name suggests, its main change of perspective relative to an ordinary field theory is its proposal that the fundamental constituents of nature are one-dimensional objects in space having zero thickness (*i.e.* strings, that sweep out two-dimensional world-sheets in spacetime), rather than being zero-dimensional objects in space having zero thickness (*i.e.* point particles, that sweep out one-dimensional world-lines in spacetime).

[16] This line of argument takes the point of view motivated by experience with EFTs elsewhere – that non-renormalizability is a central clue. But this is not the only point of view, and at this writing (2019) there are a variety of other approaches to quantum gravity being vigorously pursued for which renormalizability is not the main motivation [263].

How could this possibly be a viable proposal, given that all known elementary particles behave as (quantum-mechanical) point particles? The idea is this: fundamental strings are envisioned to be *very* short; possibly almost as short as the Planck size. This makes them too short to have their nonzero length be experimentally resolved, even with the best present-day efforts. More quantitatively, relativistic strings are characterized by their tension (or energy-per-unit-length), T, which for historical reasons is often also represented as a 'slope parameter': $\alpha' := 1/(2\pi T)$. For fundamental strings the absence of a string thickness makes the 'string scale' $M_s^2 \sim 1/\alpha'$ the only characteristic scale in the problem, and so having strings be too short to have measured their lengths in practice means taking M_s to be much higher than the highest energies accessible experimentally.

The ability to confuse a short string for a point particle also carries with it a potentially enormous economy of description: in string theory all known particles (and all particles not yet discovered) are imagined to be different modes of vibration of a single type (or a very few types) of string. How does this work? The idea is this: a string has infinitely many normal modes of vibration, each with a specific oscillation energy roughly determined by having the wavelength of oscillation be commensurate with the length of the string. But if the string is too short to be resolved and so is mistaken as a particle, then this energy of oscillation appears to be an energy that is independent of the particle's motion. That is to say: the oscillation energy would be interpreted as the particle's rest mass. Crucially for what follows, because there are an infinite number of string oscillation modes, a single short string can be mistaken as an infinite number of different types of particles having a specific pattern of masses.

As it turns out, fundamental strings come in several types: those with ends, called 'open' strings, and those that are loops (without ends), called 'closed' strings. The basic interaction that fundamental strings experience is interconnection: when two strings cross, they can reconnect differently as they pass through one another. The quantum amplitude for this process to happen is proportional to a dimensionless quantity called the string coupling constant, g_s.

String theory naturally involves supersymmetry in its formulation (for reasons [266] not explained here for brevity's sake), and usually involves more than one type of supersymmetry at that. Its implications are often easiest to understand when background fields for the various particle states are chosen so as to leave one or more of these supersymmetries unbroken. The spectrum of string oscillations is simplest for such vacua, particularly when the geometry of the spacetime through which the strings move is flat.

String Theory: Spectrum at Weak Coupling

Weakly coupled strings are those for which g_s is small, and for these the spectrum of energies for oscillating string states can be computed perturbatively in g_s, with the free-string spectrum providing a good first approximation.

This spectrum, for strings moving in flat space, includes a collection of massless states of various spins. The massless states found in the spectrum of open strings include massless spin-one states (whose low-energy interactions prove to be well-described by gauge-boson interactions), while for closed strings they include a massless spin-two excitation (which turns out to be the graviton). In general each of these massless states has an associated field, with the massless spin-two particle associated with the spacetime metric, the massless spin-one particles with various gauge potentials, A^a_μ, spinless particles associated with scalar fields, and so on. All of these fields (not just the metric) in principle can take nontrivial background values, with different values corresponding to different string vacua.

The string spectrum also contains many massive excitations with masses quantized in units of $1/\alpha'$. Broadly speaking, the spectrum of free strings about a supersymmetric and flat field configuration has the general form

$$M_N^2 = N\, M_s^2, \tag{10.46}$$

where N is a non-negative integer[17] and $M_s \sim (\alpha')^{-1/2}$ up to a dimensionless order-unity coefficient (whose precise value depends on the particular string involved). There is more than one state for each N and these generally fill out multiplets for all of the unbroken symmetries, be these spacetime symmetries like Poincaré invariance or supersymmetry, or internal transformations for the gauge symmetries associated with the massless spin-one gauge bosons. For large N the degeneracy of states with mass M_N grows very quickly, asymptotically going exponentially with N for large N.

In the simplest vacua the only nonzero background field that is turned on is the flat (Minkowski) metric, and such vacua only exist for specific numbers of dimensions. The weak-coupling examples involving the maximal spacetime symmetries occur in 10 spacetime dimensions (nine space plus one time),[18] and for these the states fall into multiplets of 10D supergravity. There is more than one type of 10D supergravity – with variants called Type I, Type IIA, Type IIB and Heterotic [268] – and the modern picture is that each of them describes the low-energy limit of string excitations about a particular supersymmetric 10D string vacuum (rather than each being a separate kind of string theory).

There are also solutions with fewer maximally symmetric dimensions, which is possible if other background fields are turned on. Among these are background geometries that are more complicated solutions to the same 10-dimensional field equations (more about these equations below) that describe alternatives to the maximally symmetric 10D Minkowski background configurations. Examples include 'compactified' spacetime geometries like $\mathfrak{M}_d \times X_n$, say, where $\mathfrak{M}_d$ is d-dimensional Minkowski (or anti-de Sitter) space and X_n is an $(n = 10 - d)$-dimensional compact space, plus possibly other nonzero background fields. When the background curvatures and fields of these new solutions are small each of the levels of (10.46) gets split by a small amount (as illustrated in Fig. 10.1), into multiplets that represent the new background's smaller number of symmetries.

EFTs and String Interactions

The spectrum (10.46) brings several important lessons about string theory and EFTs.

[17] I have superstrings in mind when writing this, since for bosonic strings N can be negative.
[18] There is also an 11-dimensional vacuum, though not a weakly coupled one [267].

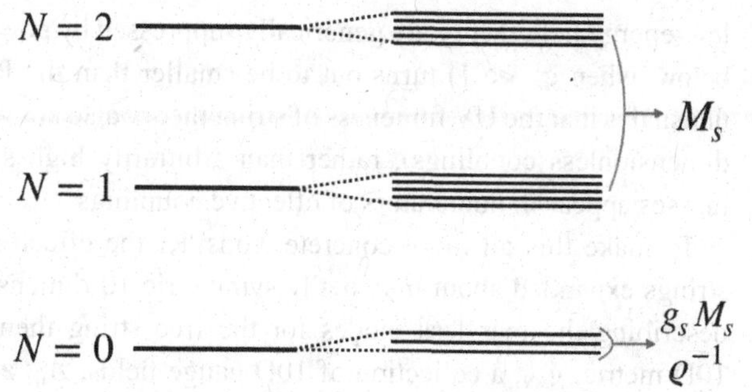

Fig. 10.1 Cartoon of how free string levels (labelled by $N = 0, 1, 2, \ldots$ and spaced by order M_s) are split at weak coupling into a 'fine structure' whose size is either suppressed by a power of string coupling, $g_s M_s$, or a Kaluza–Klein compactification scale.

The first lesson builds on the observation that (10.46) says M^2 comes equally spaced in steps of M_s^2. This implies that there is never a regime $\Lambda > M_s$ for which the heavier states with $M > \Lambda$ are all hierarchically heavier than the lighter states with $M < \Lambda$. For instance, if $M_N < \Lambda < M_{N+1}$ the ratio of masses for the heaviest light state over the lightest heavy state is

$$\frac{M_N}{M_{N+1}} = \sqrt{\frac{N}{N+1}} \qquad \text{and so} \qquad \frac{1}{\sqrt{2}} \simeq 0.7 \leq \frac{M_N}{M_{N+1}} \leq 1 \quad \text{for } N \geq 1. \tag{10.47}$$

Consequently, although there are only a finite number of states with mass less than any fixed Λ, once Λ is larger than M_s it is never a very good approximation to expand in inverse powers of the masses of the states that have been integrated out. Intuition based on Wilsonian EFTs involving a finite number of fields is likely to be misleading about the high-energy behaviour of string theory, opening up the possibility that this behaviour might be interestingly different.

The exception to the above assertion is the case $N = 0$; for the massless sector energies can be chosen to be hierarchically low compared with M_s, and in this regime standard EFT arguments should apply. Consequently, one expects ordinary field theories to furnish Wilsonian descriptions for the 'massless' sectors[19] of string vacua to the extent that effective interactions suppressed by inverse powers of M_s are hierarchically small. For maximally symmetric 10D vacua the terms in the low-energy EFT involving two and fewer derivatives are precisely the 10D supergravities mentioned above.

The leading, two-derivative, terms are, in general, also supplemented by higher-derivative interactions, as usual, that capture the exchange of the massive string states that are integrated out. As usual for EFTs the masses of the lightest states dominate in the denominators of effective couplings having dimensions of inverse powers of mass. For string theory this means that higher-dimension interactions in the

[19] The word 'massless' is in quotes here because, strictly speaking, (10.46) only holds in the limit $g_s \to 0$ (and for maximally symmetric flat space and so on). More generally, the free-string massless state splits into states with a variety of masses when g_s and background curvatures are nonzero (see Fig. 10.1). Since the masses of any of these massive $N = 0$ states can be parametrically small relative to M_s, they can be described by an EFT, and in this case the low-energy EFT expansion for physical quantities reproduces the g_s (or small-curvature) expansion of the full theory.

low-energy field theory are generically suppressed by powers of M_s, which (as shown below, when $g_s \ll 1$) turns out to be smaller than the Planck mass. What is more unusual is that the UV finiteness of string theory also allows it to be M_s (together with dimensionless couplings), rather than arbitrarily high scales, that dominate when masses appear in numerators of effective couplings.

To make this all more concrete, consider the effective description for heterotic strings expanded about maximally symmetric 10-dimensional flat space. The fields describing the massless modes for the free string then consist of the symmetric 10D metric, g_{MN}, a collection of 10D gauge fields, A^a_M, a 10D scalar, D, (called the 'dilaton') and an anti-symmetric 'Kalb-Ramond' potential, $B_{MN} = -B_{NM}$, subject to a gauge symmetry of the form $B_{MN} \to B_{MN} + \partial_M \Omega_N - \partial_N \Omega_M$, for an arbitrary 10D vector field Ω_M. Plus fermionic superpartners for each of these bosonic fields.

The leading part of the low-energy action for this string sector turns out to be governed by the following effective field theory [248, 269]

$$S_{\rm het} = -\int d^{10}x\,\sqrt{-g}\left[\frac{1}{2\kappa_{10}^2}\,g^{MN}\Big(\mathcal{R}_{MN} + \partial_M D\,\partial_N D\Big) + \frac{1}{6}\,e^{-\sqrt{2}D} H_{MNP}H^{MNP}\right.$$
$$\left. + \frac{1}{4}\,e^{-D/\sqrt{2}} F^a_{MN}F_a^{MN} + \cdots\right], \tag{10.48}$$

where dots represent both fermionic and other types of terms involving up to two-derivatives, as well as a myriad of higher-derivative interactions. Here, $\kappa_{10}^2 = 8\pi G_{10}$ represents the reduced 10D gravitational constant, $\mathcal{R}_{MN}$ is the Ricci tensor for the 10D metric, g_{MN}, and the gauge field-strength tensor, F^a_{MN}, for the nonabelian gauge group is defined as usual by

$$F^a_{MN} = \partial_M A^a_N - \partial_N A^a_M + g_{10}\, c^a{}_{bc} A^b_M A^c_N. \tag{10.49}$$

Here, g_{10} is the gauge-field coupling constant with the structure constants $c^a{}_{bc}$ defined by the gauge commutation relations $[T_b, T_c] = i c^a{}_{bc} T_a$ with the representation matrices T_a assumed normalized by the condition $\text{tr}\,(T_a T_b) = 2\,\delta_{ab}$. Finally, with these conventions the field strength H_{MNP} is defined by

$$H_{MNP} = \left[\partial_M B_{NP} - \frac{\kappa_{10}}{4}\,\text{tr}\left(A_M F_{MP} - \frac{1}{3}\,g_{10}\,A_M A_N A_P\right) + \cdots\right] + (\text{cyclic}), \tag{10.50}$$

where the gauge-field quantity is a 'Chern-Simons form' and the ellipses denote a similar quantity built from the metric's spin connection rather than the gauge potential.

The action (10.48) captures the leading low-energy behaviour of perturbative string scattering amplitudes, and agreement with these amplitudes fixes the value of the coupling parameters to be [248, 269]

$$\frac{\kappa_{10}^2}{g_{10}^2} = \frac{\alpha'}{2} \quad \text{and} \quad \kappa_{10} = 2\,g_s(\alpha')^2. \tag{10.51}$$

Notice that the second of these implies that weak coupling, $g_s \ll 1$, makes the string scale parametrically small compared with the 10D Planck scale: writing $\kappa_{10} = M_{10}^{-4}$ and defining $\alpha' = M_s^{-2}$ gives $M_s^4 = 2g_s M_{10}^4 \ll M_{10}^4$.

Another important property of (10.48) is the way it depends on the dimensionless string coupling, g_s. This turns out to appear everywhere in the lagrangian together with the dilaton, as is most easily seen by rescaling fields $A^a_M = \hat{A}^a_M/g_{10}$ and $B_{MN} = \kappa_{10}\,\hat{B}_{MN}$ and defining the 'string-frame' metric, $\hat{g}_{MN}$, through[20]

$$g_{MN} = e^{-\phi/2}\,\hat{g}_{MN}, \tag{10.52}$$

where $\phi = \sqrt{2}\,D$. With these redefinitions the action (10.48) becomes

$$S_{\rm het} = -\int {\rm d}^{10}x\,\sqrt{-\hat{g}}\;e^{-2\phi}\left[\frac{1}{2\kappa_{10}^2}\hat{g}^{MN}\left(\hat{\mathcal{R}}_{MN} - 4\,\partial_M\phi\,\partial_N\phi\right)\right.$$
$$\left. + \frac{\kappa_{10}^2}{6\,g_{10}^4}\,\hat{g}^{MQ}\hat{g}^{NR}\hat{g}^{PT}\,\hat{H}_{MNP}\hat{H}_{QRT} + \frac{1}{4g_{10}^2}\,\hat{g}^{MP}\hat{g}^{NQ}\hat{F}^a_{MN}\hat{F}_{aPQ} + \cdots\right], \tag{10.53}$$

where $F^a_{MN} = \hat{F}^a_{MN}/g_{10}$ and $H_{MNP} = \kappa_{10}\hat{H}_{MNP}/g_{10}^2$ with $\hat{F}^a_{MN}$ and $\hat{H}_{MNP}$ given in terms of $\hat{A}^a_M$ and $\hat{B}_{MN}$ by (10.49) and (10.50), but without the dependence on κ_{10} and g_{10}, which have been scaled out.

What is important about (10.53) is that (for the two-derivative terms) ϕ appears undifferentiated only through the factor $e^{-2\phi}$ in front of the whole action. But this also ensures that it only appears in the combination $g_s\,e^{\phi}$, since κ_{10} and g_{10} are both proportional to g_s. As a consequence, g_s can be completely absorbed into a shift of ϕ. This shows that g_s is not a free parameter of string theory; it is better regarded as the expectation value of the dilaton field. The replacement of g_s by e^{ϕ} is the rule in string theory: there are no fundamental dimensionless parameters at all. The one fundamental parameter is α', which simply sets the overall units of the problem. The role normally played by parameters in other theories is in string theory instead played by the values of background fields [270].

From this point of view, the string loop expansion – which in the full theory is a series in powers of g_s – can be regarded as an expansion in powers of $e^{2\phi}$. This can be seen independently by power counting in the low-energy theory because in the action (10.48) ϕ appears undifferentiated only through the overall factor $e^{-2\phi}$, and so systematically appears together with the factors of $1/\hbar$ in $e^{iS/\hbar}$. Semiclassical methods therefore provide a good approximation only for that part of field space for which e^{ϕ} is small.

The two expansions that control the use of (10.48) or (10.53) as the low-energy limit of the perturbative string are: powers of e^{ϕ} (for the string loop expansion) as well as powers of $\alpha' E^2$ that control the low-energy approximation (the α' expansion). Couplings like the effective 10D gravitational coupling are, in practice, combinations of these:

$$\kappa := \kappa_{10}\,e^{\phi} \propto (\alpha')^2 g_s\,e^{\phi}. \tag{10.54}$$

Having string loops counted by the value of a field is useful, in some cases leading to what are called 'non-renormalization' theorems. For instance, in the above example supersymmetry alone dictates the dilaton-dependence of the two-derivative terms of the action (10.48). Because this gives a dilaton-dependence proportional to $e^{-2\phi}$, appropriate to a tree-level contribution, supersymmetry implies that there can be no higher-loop contributions to the action's two-derivative terms. After all,

[20] The metric g_{MN} appearing in (10.48) is called the 'Einstein-frame' metric, and is defined by the property that there is no scalar-field dependence premultiplying the Einstein–Hilbert action.

if there were such corrections they would involve a different power of $e^{2\phi}$, which supersymmetry forbids. (This argument is, in practice, most commonly applied to compactifications to 4D maximally symmetric vacua [271] rather than to the 10D action.)

Some of the subleading corrections to (10.48) have also been computed by working at string tree level but to subdominant order in powers of α'. This is possible because the full string tree-level 4-point graviton amplitude due to string exchange has been calculated explicitly as a function of energy. The dependence often involves products of Euler Gamma-functions, sometimes called a Veneziano form based on the earliest discovered examples [272].

For instance, for Type II strings the tree-level amplitude for $2 \to 2$ scattering of massless gravitons in 10D flat space turns out [273] to be proportional to

$$\mathcal{A} \propto \kappa^2(\alpha')^3 \left[\frac{\Gamma(-\alpha' s/4)\Gamma(-\alpha' t/4)\Gamma(-\alpha' u/4)}{\Gamma(1+\alpha' s/4)\Gamma(1+\alpha' t/4)\Gamma(1+\alpha' u/4)}\right], \tag{10.55}$$

where $s = -(p_1 + p_2)^2$, $t = -(p_1 - p_2)^2$ and $u = -(p_1 - p_3)^2$ are the usual three Mandelstam invariants built from the momenta of the scattering states. $\Gamma(z)$ is Euler's gamma function – which has poles at non-positive integer arguments – showing that (10.55) has multiple poles corresponding to the exchange of all possible string excitations as intermediate states.

The low-energy EFT applies at energies well below the string scale, in which case (10.55) reduces to[21]

$$\mathcal{A} \simeq \kappa^2(\alpha')^3 \left[-\frac{64}{(\alpha')^3 stu} - 2\zeta_R(3) + \mathcal{O}(\alpha')\right], \tag{10.56}$$

where the Riemann zeta-function is defined for sufficiently large real argument by $\zeta_R(r) := \sum_{n=1}^{\infty} n^{-r}$, and by analytic continuation for other complex values. The first term of (10.56) depends only on κ^2 (and not also on α') and is captured by the corresponding two-derivative supergravity contribution. The second term gives the first subleading correction, and implies the existence of a quartic-curvature term in the Wilsonian EFT for the metric. Recalling that each factor of the canonically normalized metric perturbation comes with a factor of κ, the effective coupling required for this quartic-curvature term by matching to (10.56) is of order $(\alpha')^3/\kappa^2 \propto e^{-2\phi}/\alpha'$ [274]. The proportionality to $e^{-2\phi}$ is the right dependence for a contribution at string tree level and the power of α' is dimensionally appropriate in 10D for a term with four powers of the background curvature tensor.

10.3.2 Extra Dimensions

The above discussion makes the connection between the string scale and the Planck scale explicit, but so far only does so in 10 dimensions. To make contact with the measured 4D Planck mass requires identifying how a four-dimensional world might emerge from a 10-dimensional one.

21 Strictly speaking, taking $s \sim t \sim u$ all small but similar in size is a low-energy, large-angle approximation because the variables t and u can be $\ll s$ for small-angle scattering.

Dimensional Reduction

A comparatively simple way to make the connection from 10 to 4 dimensions explicit is by seeking more general background solutions to the field equations of actions like (10.48), such as by seeking geometries with background metrics, $\tilde{g}_{MN}$, of the form,[22]

$$\tilde{g}_{MN}(x,y)\,dx^M dx^N = \tilde{g}_{\mu\nu}(x)\,dx^\mu dx^\nu + \tilde{g}_{mn}(y)\,dy^m dy^n, \tag{10.57}$$

with possibly nonzero backgrounds allowed for other fields as well. Here, $x^M = \{x^\mu, y^m\}$ with $M = 0, \ldots, 9$ with $\mu = 0, 1, 2, 3$ corresponding to the 4D geometry and $m = 4, \cdots, 9$ to the rest. As discussed in §6, solutions found using the 10D EFT should be a good approximation to the full UV theory if all background curvatures and derivatives of background fields are small compared with the string scale.

Many explicit solutions of this type are known with a maximally symmetric 4D metric, $\tilde{g}_{\mu\nu}$. Much of the focus is on solutions that break the 10D supersymmetry down to a single 4D supersymmetry, in which case $\tilde{g}_{\mu\nu}$ is either flat space or anti-de Sitter space [275]. Unbroken 4D supersymmetry gets a lot of attention, partly for phenomenological reasons (to help solve the electroweak hierarchy problem discussed in §9.3), and partly because it turns out to help control the size of the g_s and α' corrections to such solutions.

In the semiclassical limit the spectrum of fluctuations about a background configuration like (10.57) is found in the usual way by expanding fields,

$$g_{MN}(x,y) = \tilde{g}_{MN}(x,y) + h_{MN}(x,y), \tag{10.58}$$

and expanding the action in powers of the fluctuations, treating all cubic and higher terms perturbatively. In this kind of expansion $h_{\mu\nu}(x,y)$ transforms as a 4D symmetric tensor, $h_{m\nu}(x,y) = h_{\nu m}(x,y)$ is a 4D 4-vector and $h_{mn}(x,y)$ transform as a collection of 4D scalars. A similar decomposition goes through for other 10D fields, like $A^a_\mu(x,y)$ and $A^a_m(x,y)$.

In addition to this, the dependence on the extra-dimensional coordinate y^m can often be traded for a discrete label, corresponding to what is called a 'Kaluza–Klein' tower of 4D fields that depend on 4D position, x^μ, only [276]. This kind of discrete expansion applies, in particular, when the extra-dimensional geometries are compact.

As a simple example of how this works, consider a massless 10D scalar field, $\Phi(x,y)$, whose equation of motion is linearized about this background. The resulting (linearized) 10D field equation might be

$$\tilde{g}^{MN}\tilde{\nabla}_M\tilde{\nabla}_N\Phi = \tilde{g}^{\mu\nu}\tilde{\nabla}_\mu\tilde{\nabla}_\nu\Phi + \tilde{g}^{mn}\tilde{\nabla}_m\tilde{\nabla}_n\Phi = 0. \tag{10.59}$$

The idea is to expand

$$\Phi(x,y) = \sum_b \phi_b(x)\,u_b(y), \tag{10.60}$$

in a complete set of eigenfunctions satisfying

$$\tilde{g}^{mn}\tilde{\nabla}_m\tilde{\nabla}_n u_b(y) = -\lambda_b u_b(y), \tag{10.61}$$

[22] A product space is chosen here for simplicity. 4D maximal symmetry allows more general geometries, such as if the 4D part of the metric is pre-multiplied by a 'warp factor' $W^2(y)$.

in which case the 10D field equation, (10.59), becomes a Kaluza–Klein tower of 4D equations for the $\phi_b(x)$:

$$\tilde{g}^{\mu\nu}\tilde{\nabla}_\mu\tilde{\nabla}_\nu\phi_b - \lambda_b\phi_b = 0, \tag{10.62}$$

showing that each mode $\phi_b(x)$ behaves as a 4D field and satisfies a 4D Klein–Gordon equation with mass $m_b^2 = \lambda_b$. Similar expansions in terms of eigenfunctions also go through for higher-spin fields.

For stable background configurations the eigenvalue spectrum satisfies $\lambda_b \geq 0$. For example, if there were two extra dimensions with the geometry of a 2-sphere of radius ϱ, then the desired eigenfunctions are spherical harmonics, $Y_{\ell m}(\theta, \phi)$, and $\lambda_{\ell m} = \ell(\ell+1)/\varrho^2 \geq 0$ is non-negative because $\ell = 0, 1, 2, \cdots$.

This example shows that the generic size of the Kaluza–Klein masses, m_b, is set by a geometrical scale appearing in the internal metric, $\tilde{g}_{mn}$, with $m \propto 1/\varrho$ in the case of the 2-sphere. These masses are parametrically small compared with the string mass scale, $m_b \ll M_s$, in the regime where curvatures are small enough to justify calculating them using the 10D effective theory rather than using string theory in its full glory. From that point of view the Kaluza–Klein mass spectrum can be regarded as the fine-structure of string energy levels: 10D string states with masses of order M_s become split into many 4D states separated by masses of order m_b (as illustrated in Fig. 10.1).

Modulus Stabilization

Because the internal metric is a dynamical quantity, ideally ϱ (and consequently also the m_b^2) should be calculable using the 10D field equations, though this is often – but not always (see for example [277]) – a challenge in practice. One reason this can be a challenge arises because higher-dimensional supergravities generically enjoy a classical scaling symmetry under which a transformation like $g_{MN} \to c\, g_{MN}$ (plus possibly transformations of other fields) causes the two-derivative parts of the 10D lagrangian to scale: $\mathfrak{L}_W \to c^p\mathfrak{L}_W$, for some p. Here, c is the constant symmetry parameter.

An example of this can be seen in (10.53), which transforms as $S_W \to c^2 S_W$ when $e^{-\phi} \to c\, e^{-\phi}$ and all other fields are fixed. Although this does not transform the string-frame metric $\hat{g}_{MN}$, it does transform the Einstein-frame metric (10.52), with $g_{MN} \to c^{1/2} g_{MN}$. In particular, it rescales the overall volume of the extra-dimensional Einstein-frame metric (such as its radius, ϱ, if this metric were that of a sphere).

Because this kind of transformation rescales the action, it is not a symmetry in the usual sense (which involves being an invariance of the action). It does preserve the equations of motion, however, since any classical solution (or saddle point, $\delta S_W = 0$, of the action) gets mapped to another such a solution. Consequently, if a classical solution exists for any nonzero parameter like ϱ and if this parameter transforms nontrivially under such a classical rescaling symmetry, then other classical solutions must also exist for all values of ϱ. These must all exist because they are all just rescalings of one another under a symmetry of the classical equations of motion.

Variables like ϱ on which solutions can depend parametrically like this are called 'moduli'. Another simple example of a modulus is also visible in (10.59), because $\Box\phi = 0$ is satisfied by a one-parameter family of solutions $\phi =$ constant. Notice that constant ϕ is also a zero-mode, $\tilde{g}^{mn}\tilde{\nabla}_m\tilde{\nabla}_n u_0 = 0$, corresponding to the massless state, $\varphi_0(x)$, found in the example above. This connection between moduli and massless 4D Kaluza-Klein modes is one of the reasons why moduli are of interest: their identification is central to reliably identifying the field content of the low-energy 4D theory.

Because the above scaling argument is explicitly classical, it need no longer apply once corrections are included – such as quantum corrections – that violate the invariance of the field equations. Incorporating such corrections is therefore very important, since predictions for the stabilized values of modulus fields (and for the mass of the associated particle) cannot be made until they are included.

4D EFTs

Because Kaluza–Klein states come in towers, there is usually not a sufficiently clean hierarchy of scales to allow a low-energy EFT to be defined that keeps any of the nonzero-mass states, $\varphi_b(x)$, in the low-energy sector while integrating out the others – for reasons similar to those given surrounding Eq. (10.47) for the string spectrum.

Intuition based on 4D Wilsonian EFTs can therefore easily be misleading about the high-energy behaviour of higher-dimensional field theories, as is indeed borne out by experience.[23] For instance, earlier sections argue that integrating out heavy 4D states of mass M contribute to 4D Wilsonian action terms like $\sqrt{-g}\left(c_0M^4+c_2M^2R+\cdots\right)$, where R is the 4D Ricci scalar and the c_i are calculable dimensionless constants. Integrating out a particle of mass M in D dimensions instead contributes to a Wilsonian action an amount $\sqrt{-g}\left(\mathfrak{c}_0M^D + \mathfrak{c}_2M^{D-2}\mathcal{R} + \cdots\right)$ where $\mathcal{R}$ denotes the D-dimensional Ricci scalar and $\mathfrak{c}_i$ are a different set of calculable dimensionless constants. Higher-dimensional EFTs depend differently on M and are subject to all of the constraints of higher-dimensional symmetries (like higher-dimensional diffeomorphism or Lorentz invariance): symmetries whose implications are more difficult to see when written as a tower of 4D Kaluza–Klein fields.

What 4D theories *can* capture is the physics of the massless (or near-massless) modes, for which $\lambda_b = 0$. These are the states present at energies that are hierarchically small compared with the lowest-lying Kaluza–Klein mass, which in the rest of this section is denoted M_c (for 'compactification scale'). Because the corresponding zero-mode fields, like $\varphi_0(x)$, depend only on x^μ, in practice they live in four dimensions, and so their low-energy EFT is four-dimensional (with the leading higher-dimensional interactions having effective couplings suppressed by powers of[24] $M_c \sim 1/\varrho \ll M_s \ll M_p$, with the last hierarchy applying in the perturbative limit $g_s \ll 1$).

[23] Although the form for divergences computed in higher dimensions [278] generally differs from what is found by performing a sum of 4D KK divergent contributions, these can agree for logarithmic divergences [279] (in what is a variant of the general argument for the magic of logarithms given around Eq. (3.30)).

[24] Although simple geometries – *e.g.* spheres – only involve one scale, more complicated ones can involve a wide variety of scales and M_c need not be simply related to the mean geometrical curvature or volume [280].

A zero mode that is generically present whenever the background 4D metric, $\tilde{g}_{\mu\nu}(x)$, is maximally symmetric is the massless 4D graviton. To see why its presence is generic, recall that when a gravitational wave is written as an expansion about the background geometry, $g_{MN} = \tilde{g}_{MN} + h_{MN}$, the graviton part of the fluctuation can be chosen to satisfy $\tilde{\nabla}^M h_{MN} = \tilde{g}^{MN} h_{MN} = 0$ [281]. But linearizing the extra-dimensional vacuum Einstein equations, $\mathcal{R}_{MN} = 0$, shows that the linearized graviton fluctuation satisfies the Lichnerowicz equation, $\tilde{\Delta}_{MN}^{PQ} h_{PQ} = 0$, where [282]

$$\tilde{\Delta}_{MN}^{PQ} h_{PQ} = \frac{1}{2}\tilde{\nabla}^2 h_{MN} + \frac{1}{2}\left(\tilde{\mathcal{R}}^P{}_M h_{NP} + \tilde{\mathcal{R}}^P{}_N h_{MP}\right) - \tilde{\mathcal{R}}_{PMRN} h^{PR} \tag{10.63}$$

defines the 'Lichnerowicz operator': $\tilde{\Delta}_{MN}^{PQ}$. Here, tildes indicate that the corresponding derivative or curvature is built from the background metric, $\tilde{g}_{MN}$.

The 4D spin-2 graviton is contained in the part, $h_{\mu\nu}$, where both indices take values in 4D and $\tilde{\nabla}^\mu h_{\mu\nu} = \tilde{g}^{\mu\nu} h_{\mu\nu} = 0$. Assuming the background metric is maximally symmetric in 4D – which implies that $\tilde{\mathcal{R}}^m{}_\mu = 0 = \tilde{\mathcal{R}}_{\mu m n \nu}$, with the proof of the second of these requiring use of the Bianchi identity (C.96) – and evaluating (10.63) with $M = \mu$ and $N = \nu$ then gives

$$\begin{aligned}\tilde{\Delta}_{\mu\nu}^{PQ} h_{PQ} &= \frac{1}{2}\left(\tilde{g}^{\lambda\rho}\tilde{\nabla}_\lambda\tilde{\nabla}_\rho + \tilde{g}^{mn}\tilde{\nabla}_m\tilde{\nabla}_n\right)h_{\mu\nu} + \frac{1}{2}\left(\tilde{\mathcal{R}}^\rho{}_\mu h_{\nu\rho} + \tilde{\mathcal{R}}^\rho{}_\nu h_{\mu\rho}\right) - \tilde{\mathcal{R}}_{\rho\mu\sigma\nu} h^{\rho\sigma} \\ &= \tilde{\Delta}_{\mu\nu}^{\lambda\rho} h_{\lambda\rho} + \tilde{g}^{mn}\tilde{\nabla}_m\tilde{\nabla}_n h_{\mu\nu},\end{aligned} \tag{10.64}$$

where $\tilde{\Delta}_{\mu\nu}^{\lambda\rho}$ is the 4D Lichnerowicz operator (which describes graviton fluctuations in the 4D Einstein–Hilbert action). Consequently, the Kaluza–Klein mode decomposition for the spin-two part of $h_{\mu\nu}(x,y) = \sum_s \mathfrak{h}^s_{\mu\nu}(x) u_s(y)$ involves modes $u_s(y)$ that are eigenvalues of the Laplacian, $\tilde{g}^{mn}\tilde{\nabla}_m\tilde{\nabla}_n$, acting on extra-dimensional scalars. The zero mode for this operator is simply given by $u_0(y) =$ constant, and so a single massless 4D graviton exists on very general grounds for compact extra-dimensional geometries.

On equally general grounds the low-energy action governing the two-derivative self-interactions of this massless spin-two 4D graviton are given by the 4D Einstein–Hilbert lagrangian (to which one can add possible couplings to other low-energy 4D fields). The 4D Planck mass appearing in this lagrangian is most easily seen by restricting the higher-dimensional Einstein–Hilbert action to the quadratic part in the zero-mode $\mathfrak{h}_{\mu\nu}(x)$, and so – assuming the product-space background geometry (10.57),

$$\begin{aligned}S_{EH(10D)} &= -\frac{1}{2\kappa_{10}^2}\int d^{10}x\ \sqrt{-\tilde{g}_{10}}\ h^{MN}\tilde{\Delta}_{MN}^{PQ} h_{PQ} + \cdots \\ &= -\frac{1}{2\kappa_{10}^2}\int d^6y\,\sqrt{\tilde{g}_6}\int d^4x\ \sqrt{-\tilde{g}_4}\ \mathfrak{h}^{\mu\nu}\tilde{\Delta}_{\mu\nu}^{\lambda\rho}\,\mathfrak{h}_{\lambda\rho} + \cdots,\end{aligned} \tag{10.65}$$

where $\tilde{g}_{10} = \det \tilde{g}_{MN}$, $\tilde{g}_4 = \det \tilde{g}_{\mu\nu}$ and $\tilde{g}_6 = \det \tilde{g}_{mn}$ and (as above) $\kappa_{10}^2 = 8\pi G_{10}$, where G_{10} is the higher-dimensional Newton's constant of universal gravitation.

Eq. (10.65) agrees with the corresponding expansion of the 4D Einstein–Hilbert action,

$$S_{EH(4D)} = -\frac{1}{2\kappa_4^2}\int d^4x\ \sqrt{-\tilde{g}_4}\ \mathfrak{h}^{\mu\nu}\Delta_{\mu\nu}^{\lambda\rho}\,\mathfrak{h}_{\lambda\rho} + \cdots, \tag{10.66}$$

provided the 4D gravitational coupling (defined by $\kappa_4^2 = 8\pi G_N$) is given by $\kappa_4^{-2} = \kappa_{10}^{-2} S_6$, where $S_6 = \int d^6y \sqrt{-\tilde{g}_6}$ is the volume of the extra dimensions. Using the Planck-mass definitions, $\kappa_{10}^{-1} = M_{10}^4$ and $\kappa_4^{-1} = M_p$ then shows

$$M_p^2 = M_{10}^8 S_6 \qquad \text{or} \qquad G_N = \frac{G_{10}}{S_6}, \tag{10.67}$$

a formula used earlier, such as in Eq. (9.35).

Recall that the neglect of higher-curvature terms when performing dimensional reduction is only valid if Kaluza–Klein scales are much smaller than the string scale, $M_c \sim 1/\varrho \ll M_s$. Furthermore, the string scale itself is much smaller than M_{10} for weakly coupled strings – *c.f.* the discussion below (10.51) for $g_s \ll 1$. Provided the length scale defined by the volume is comparable to the Kaluza–Klein scale of the background geometry (or is larger), $S_6 \geq \varrho^6$, this also means that $M_{10}^6 S_6 \geq M_{10}^6 \varrho^6 \gg M_s^6 \varrho^6 \gg 1$ and this in turn implies that $M_p \gg M_{10}$. We are left with the generic hierarchy

$$M_p \gg M_{10} \gg M_s \gg M_c, \tag{10.68}$$

for weakly coupled string compactifications performed within the higher-dimensional EFT.

In this regime the new physics describing high-energy gravity can kick in at energies well below the Planck mass. The first big change occurs at the compactification scale, $M_c \ll M_s$, above which the 4D Wilsonian EFT gets replaced by a higher-dimensional field theory. This higher-dimensional field theory itself then fails at scales of order M_s, above which a description in terms of a finite number of fields becomes inadequate and the proper treatment involves the full spectrum of string theory.[25]

String vacua indeed seem to explore a wide range of this parameter space [275, 283], with (at this writing) the only completely model-independent constraint being the requirement that the string scale be larger than observable energies, $M_s \gtrsim 1$ TeV [284]. This is a surprisingly weak bound because it happens that the stronger contraint $M_c \gtrsim 1$ TeV need *not* be true, since it can happen that ordinary particles can interact too weakly with extra-dimensional physics to rule out having M_c at surprisingly low energies [285]. At present, the strongest robust upper bound on the size of extra dimensions comes from tests of the gravitational inverse-square law [286], which as of this writing (2019) allow a few extra dimensions to be as large as 40 μm.

10.4 Summary

This chapter brings some gravity to the more general themes about effective theories that run through this book.

[25] Notice that the new physics at each of these scales can be – though of course need not be – weakly coupled, in the sense that it is well described by perturbing in the dimensionless couplings. This is one of many examples arguing against use of the term 'strong coupling' to describe the breakdown of the low-energy approximation at higher energies.

The most important message of this chapter is that the non-renormalizability of General Relativity need not preclude being able to make precise quantum predictions. Reliable predictions are possible because the same power-counting tools used for other non-renormalizable theories (like the Fermi theory of weak interactions or chiral perturbation theory) also apply to gravity, once it is recognized that semiclassical predictions intrinsically involve a low-energy expansion.

Indeed, in the end it is power-counting formulae like (10.9) that define the boundaries of validity for the classical (and semiclassical) tools that are universally used in all practical applications of General Relativity. Because they quantify the theoretical error associated with *any* classical GR calculation, low-energy power counting rules are implicit in any meaningful comparisons between theory and observations when testing classical theories of gravity.

Although laughably small for applications within the solar-system, quantum corrections can be important in other gravitational situations. Most notable of these are black-hole and cosmological spacetimes, for which quantum predictions – such as those leading to Hawking radiation or a quantum origin for primordial fluctuations – can introduce novel phenomena not present at all within a purely classical regime.[26]

Continued exploration of both black-hole and cosmological examples has proven very instructive, helping identify conceptual challenges to the current understanding that arise within a controlled context [287]. A quantum origin for primordial fluctuations would also bring quantum-gravitating effects into the concrete realm of observations. Conceptual puzzles and practical applications both motivate quantifying theoretical error by power counting the precise size of quantum effects for more involved systems, such as the cosmological models used to derive (10.24) or (10.26) (which could yet be tested by precision observations).

Gravitational systems also provide insight in other, more conceptual ways. The fact that GR is not renormalizable suggests seeking more complete descriptions of quantum-gravity effects at the much higher energies where low-energy techniques inevitably fail. Although it is not yet known what this UV completion might be in nature, theories now exist that, in principle, *could* play this role. This chapter provides a whirlwind summary of string theory to illustrate some of the things study of UV complete systems might teach us about how effective theories may work at much shorter length-scales. Theories such as this provide the first potentially realistic examples of what UV finite physics could look like, and so provide insight into what might be possible at the shortest of distances. If the successive effective theories describing nature are regarded as successive layers of a cosmic onion, UV finite theories provide a first glimpse into its core.

Exercises

Exercise 10.1 Fill in the steps and derive the power-counting rule (10.9) starting from the GREFT effective action (10.2). Repeat the exercise for relativistic scalar-tensor theories and derive (10.24) starting from the effective action (10.18).

26 More interesting novel quantum predictions may also be possible, perhaps resolving the curvature singularities that plague these geometries. Unfortunately, at the present writing no proposals along these lines have yet shown complete control over the semiclassical methods used.

Exercise 10.2 Eliminate the redundant curvature-squared and curvature-cubed interactions for pure gravity in an expansion about flat space in D spacetime dimensions, dropping total derivatives and performing field redefinitions as necessary. Identify in this way a basis of non-redundant 4- and 6-derivative interactions.

Specialize the result to $D = 4$ and $D = 6$, keeping in mind that some combinations of curvatures might be total derivatives in specific numbers of dimensions.

Repeat this exercise for an expansion about de Sitter space rather than flat space.

Exercise 10.3 Expand the Einstein–Hilbert action (10.1) using $g_{\mu\nu} = \eta_{\mu\nu} + 2\kappa\, h_{\mu\nu}$, with $\kappa^2 = 8\pi G_N = 1/M_p^2$ and show using the quadratic term that $h_{\mu\nu}$ is the canonically normalized fluctuation. Show also that the cubic term in $h_{\mu\nu}$ includes the two-derivative interaction

$$\mathfrak{L}^{(3)}_{EH} = -\kappa\left(h^{\mu\nu}h^{\lambda\rho}\partial_\mu\partial_\nu h_{\lambda\rho} + 2\partial^\rho h_{\mu\nu}\partial^\mu h^{\nu\lambda}h_{\lambda\rho}\right).$$

Exercise 10.4 For two-body graviton-graviton scattering show that the power-counting result (10.9) implies that the leading-order contribution has amplitude $\mathcal{A} \propto (E/M_p)^2$. Draw all of the Feynman graphs that contribute to this order. Evaluate these graphs and show that the amplitude for tree-level unpolarized graviton-graviton scattering in flat space is given by

$$\mathcal{A} = 8\pi i\, G_N\, \frac{s^3}{tu},$$

where s, t and u are the Mandelstam invariants defined in (8.44).

Exercise 10.5 For single-field cosmological models – *i.e.* the system of §10.2 with a single scalar field – explicitly expand the action built from (10.18) to quadratic order about a spatially flat FRW background geometry (10.16) with a homogeneous time-dependent scalar, $\vartheta(t)$, and compute the dimensionless coefficients c_n appearing in (10.20) for the $f_n = d_n = 2$ terms as functions of the effective couplings in (10.18).

Exercise 10.6 To see how to work with dimensional regularization in curved position space, calculate the Feynman propagator, $G(x, x')$, for a free massive scalar field in n-dimensional de Sitter space, where the field equation is $(-\Box + m^2)\phi = 0$.

Define n-dimensional de Sitter space as the surface $\eta_{MN}\xi^M\xi^N = \kappa^{-1}$ within $(n+1)$-dimensional Minkowski space, with flat metric η_{MN}. Show that the surface defined in this way has Riemann curvature $R_{\mu\nu\lambda\rho} = \kappa(g_{\mu\lambda}g_{\nu\rho} - g_{\mu\rho}g_{\nu\lambda})$. Show that a de Sitter invariant measure of the separation of two points can be defined in terms of the embedding space by $\sigma(x, x') = \frac{1}{2}\,\eta_{MN}(\xi - \xi')^M (\xi - \xi')^N$, and that this satisfies the identities $\nabla_\mu\sigma\nabla^\mu\sigma = \sigma(2 - \kappa\sigma)$ and $\nabla_\mu\nabla_\nu\sigma = g_{\mu\nu}(1 - \kappa\sigma)$.

de Sitter invariance implies that $G(x, x') = G[\sigma(x, x')]$ is a function of x and x' only through the variable $\sigma(x, x')$. Use this to show that the Klein–Gordon equation becomes the ordinary differential equation

$$(-\Box + m^2)G(x, x') = \sigma(\kappa\sigma - 2)G'' + n(\kappa\sigma - 1)G' + m^2 G = 0,$$

where $G' := \mathrm{d}G/\mathrm{d}\sigma$. Use this to show that the only propagator that is singular only at $\sigma = 0$ and that is analytic in the upper-half σ-plane and the lower-half m^2-plane is given by the Hypergeometric function

$$G(x,x') = \frac{i\kappa^{(n-2)/2}}{(4\pi)^{n/2}} \left\{ \frac{\Gamma\left[\frac{1}{2}(n-1)+i\alpha\right]\Gamma\left[\frac{1}{2}(n-1)-i\alpha\right]}{\Gamma[n/2]} \right\} \times {}_2F_1\left[\frac{1}{2}(n-1)+i\alpha, \frac{1}{2}(n-1)-i\alpha; \frac{n}{2}; 1-\frac{1}{2}\kappa(\sigma+i\epsilon)\right],$$

where $\epsilon \to 0^+$ in the end and $\alpha^2 = (m^2/\kappa) - \frac{1}{4}(n-1)^2$. Use this result to evaluate the dimensionally regularized coincident limit $G(x,x)$ obtained by taking $x' \to x$. Expand this result in powers of $n-4$ to identify its divergence in the $n \to 4$ limit. Position-space dimensional regularization can also be adapted to more general geometries using *heat-kernel* techniques [84, 288–291].

Exercise 10.7 Evaluate the displayed terms of the low-energy Einstein-frame effective action (10.48) using the string-frame metric defined by (10.52) and thereby verify (10.53).

Exercise 10.8 Identify all of the poles in the string scattering amplitude given in (10.55). Given that these poles occur once s, t or u is evaluated at the masses of a string state, identify the string spectrum and compare it with (10.46). What is the residue of the amplitude at the poles in the s-channel? What should this residue be for resonant s-wave scattering?

Part III

Nonrelativistic Applications

About Part III

This part of the book makes the switch to nonrelativistic applications. By restricting to systems involving a relatively small number of particles, nonrelativistic issues are addressed in a framework where many-body issues are not also needed. This is meant to isolate the new complications associated with nonrelativistic kinematics from the other independent complications associated with having a large number of degrees of freedom.

The core applications in Part III are to the electrodynamics of slowly moving particles (such as electrons in bound states) and to the electrodynamics of neutral particles (like atoms) built from charged constituents. As such, they are the natural framework for understanding atoms and their properties and lay the foundations for many of the more complicated systems studied in later sections.

The final chapter of this part then broadens the focus again to more complicated types of nonrelativistic objects, in situations where their substructure is much smaller than the size of the physics of interest. Examples considered range from classical motion of 'lumps' whose substructure is explicitly known – such as solitons or large-scale defects (monopoles, vortex lines, domain walls, *etc.*) in field theories – through to potentially complicated composite systems – such as nuclei within atoms or the influence of the Earth itself (or other planets or stars) in applications to orbital motion. For these systems EFT techniques generalize the logic of multipole expansions to a broader class of settings than electromagnetism.

In these last applications the interest is usually in specific numbers of nonrelativistic objects (several specific planets, say, or an atom's individual nucleus), so the description is more efficiently made using first-quantized methods rather than second-quantized techniques. In this way one comes full circle: conceptually retrieving single-particle Schrödinger quantum mechanics as the EFT governing a specific type of low-energy limit of a full quantum field theory.

11 Conceptual Issues (Nonrelativistic Systems)

Up to this point in this book, all of the applications of EFT methods deal with relativistic systems. This is done partly on historical grounds but also partly because of simplicity: the comparatively large number of symmetries helps reduce the number of independent interactions in relativistic applications. But restricting just to relativistic applications also gives a distorted picture of the real power of EFT methods, which apply essentially anywhere in physics where a hierarchy of scales exists.

To illustrate the extent of this Wilsonian reach, the rest of the book relaxes the relativistic assumption and concentrates instead on the new features that emerge when effective theories are applied to intrinsically nonrelativistic systems. That starts here in Part III, with a discussion of some of the new scaling issues that arise for nonrelativistic kinematics, followed by a selection of instructive examples. (Part IV continues the story with a discussion of a slightly different kind of EFT generalization: to many-body and open systems.)

A second point of the following chapters is to show how second-quantized, nonrelativistic Schrödinger field theory – and, indeed, even ordinary single-particle Schrödinger quantum mechanics – systematically emerges as the low-energy limit of relativistic systems, as do many of the other effective theories used at much lower energies in more complicated many-body settings. Making this connection explicit allows the influence of small relativistic effects – like QED radiative corrections or parity violation from the weak interactions – to be tracked systematically within slowly moving bound systems like atoms, to which nonrelativistic Schrödinger methods normally are best suited.

11.1 Integrating Out Antiparticles ◇

Nonrelativistic particles arise in effective theories where the UV scale, Λ, of the low-energy theory is smaller than the mass, M, of the particle in question: $\Lambda \lesssim M$. Such a particle's kinetic energy is necessarily much smaller than M, making its kinematics nonrelativistic, but its *total* energy satisfies $E \geq M$ since it cannot be smaller than its rest mass. The very first question to ask is why it makes sense for particles with $E \geq M \gtrsim \Lambda$ to be present in the low-energy theory – whose energies are by assumption much smaller than Λ – in the first place.

The presence in an EFT of such massive particles is only consistent if later time evolution does not involve any integrated-out high-energy states. This requires the energy locked away in the rest-mass M to be inert, in the sense that it cannot be released to produce these other high-energy states. If that were to happen, their exclusion (as is done by assumption in the EFT below Λ) would be inconsistent.

Among other things, having the heavy objects be inert implies that they must be relatively stable (*i.e.* not able to liberate their rest mass by decaying into much lighter fields). If they are unstable, release of too much rest-energy might be precluded if one of the decay daughters is also nonrelativistic, with $M - M' < \Lambda$ so that enough of the rest mass remains locked up. Alternatively, the decay might be very slow relative to the time-scales under study, so that although EFT methods eventually do break down this takes longer than the timeframe of interest.[1]

Being in a low-energy theory also implies that nonrelativistic particles must encounter their antiparticles very infrequently, since once they do, the resulting annihilation releases order M in energy; too much to allow a purely low-energy description. Most often the absence of annihilation happens because antiparticles are completely absent, but it might also just be that antiparticles are only rarely encountered. In the presence of antiparticles EFT methods are valid only until particle and antiparticle meet.

Should annihilations (or energetic decays) occur, the effective theory can sometimes be used so long as no attempt is made to follow any high-energy states. The price paid for doing so is that the low-energy theory becomes non-unitary: EFT evolution is not unitary because probability is being lost to higher-energy sectors whose evolution is not being tracked. Examples along these lines are examined below for the EFT relevant to precision calculations with positronium (an e^+e^- bound state).

Nonrelativistic Field Theories

In relativistic systems the relationship between fields and creation/annihilation operators always involves both particles and antiparticles, and as a result it is impossible to build a local interaction that does not change the number of particles in a relativistic theory. For instance, a relativistic complex scalar field $\phi(x)$ consists of the combination

$$\phi(x) = \int \frac{d^3\mathbf{p}}{\sqrt{(2\pi)^3 2E_p}} \left(\mathfrak{a}_p \, e^{ip\cdot x} + \bar{\mathfrak{a}}_p^* \, e^{-ip\cdot x} \right), \tag{11.1}$$

where $\mathfrak{a}_p$ destroys particles with 4-momentum p^μ while $\bar{\mathfrak{a}}_p^*$ creates antiparticles with 4-momentum p^μ. Both are required in order to ensure the fields (and so also the interaction Hamiltonian densities) commute at space-like separations.[2]

By contrast, nonrelativistic field theories involve separate fields for each particle type (with antiparticles, if they appear, getting their own separate field). For instance,[3]

[1] The treatment of baryons in chiral perturbation theory provides a concrete example of these alternatives, such as in the discussion of §8.2.3.

[2] As argued in Appendix C.3, having Hamiltonian densities commute at spacelike separation is compulsory in relativistic theories because otherwise time-ordering – such as arises when computing the S matrix – is ill-defined for spacelike-separated events (on whose ordering in time different observers can disagree).

[3] Notice the conventional change of normalization between relativistic fields (11.1) and nonrelativistic fields (11.2), with $\phi = (\Psi + \overline{\Psi})/\sqrt{2M}$ when $E_p \simeq M$ (see Appendices B.1, C.2 and C.3).

$$\Psi(x) = \int \frac{d^3\mathbf{p}}{(2\pi)^{3/2}}\, \mathfrak{a}_p\, e^{ip\cdot x} \quad \text{and} \quad \overline{\Psi}(x) = \int \frac{d^3\mathbf{p}}{(2\pi)^{3/2}}\, \bar{\mathfrak{a}}_p\, e^{ip\cdot x}, \tag{11.2}$$

with nothing forcing Ψ and $\overline{\Psi}$ to appear in the specific combination $\Psi + \overline{\Psi}^*$. From this point of view nonrelativistic effective field theories are obtained from relativistic ones by integrating out the antiparticle part of a field while keeping the particle (or vice versa).

At first sight this seems an odd thing to do because a particle and an antiparticle have precisely the same mass. Since neither is lighter than the other, why does it make sense to integrate one out and not both? Indeed, in a world with equal numbers of particles and antiparticles all particles and antiparticles can mutually annihilate, and so the low-energy sector is usually devoid of both.[4] In what types of situations are nonrelativistic particles *ever* relevant to an EFT below their mass?

Conserved Charges and Selection Rules

An important instance where nonrelativistic particles naturally appear in low-energy EFTs arises for states containing more stable particles than antiparticles (or vice versa). In this case, the excess particles can survive at the lowest-available energies just for want of a way to disappear. Unequal numbers of particles and antiparticles can be consistent with being at low energies when the particles in question carry a conserved charge, $Q = Q_0$ (like baryon number or electric charge). When this is true then the lowest-energy state in a sector of the theory with nonzero net charge, say $Q = NQ_0$, necessarily contains $|N|$ particles without their antiparticles (or antiparticles without the particles, depending on the sign of N).

Conserved charges can also keep the heavy particles from decaying too quickly (or at all), keeping them sufficiently stable to survive on the relatively slow time-scales appropriate at low energies. Both annihilations and decays can then be sufficiently rare (or absent) on the long timeframes of interest. The existence of such a charge is assumed throughout most of the remainder of this part of the book (§12.2.4 is an exception), with the first few sections specializing to situations where N is relatively small, and later sections considering systems for which N can be extremely large.

The remainder of this chapter sets out some of the main ways in which the formalism of Part I differs when applied to nonrelativistic systems. One such a difference comes because nonrelativistic systems treat space and time differently. This changes the basic scaling properties of fields, and so also changes the assessments of which types of interactions are relevant, irrelevant or marginal. The precise way this scaling changes can differ for different systems (depending, for example, on how energies and momenta are related by dispersion relations) and so nonrelativistic systems can exhibit a broad class of novel types of scaling regimes in the low-energy and/or low-momentum limit. Indeed, for nonrelativistic systems low energy need not mean low momentum, and so it is important to keep these two limits conceptually separate.

[4] As §12.2.4 emphasizes, even with equal number of particles and antiparticles EFTs can remain useful for specific kinds of questions.

A second important new feature generic to nonrelativistic systems is the inevitable appearance of multiple scales in the low-energy theory. More scales become inevitable simply for kinematic reasons: because the nonrelativistic assumption ensures that speeds in the low-energy theory satisfy $v \ll 1$ (or $v \ll c$, in units with $c \neq 1$). For any particular low-energy scale m there is always a number of other low-energy scales also present – $m \gg mv \gg mv^2 \gg \cdots$ and so on. The presence of so many scales in the low-energy sector complicates the power-counting story for these theories, independent of the presence of any other scales associated with interactions.

The next few sections address each of these new features in turn.

11.2 Nonrelativistic Scaling ◇

The first step is to revisit the discussion of scaling given for perturbative systems in §2.4 to see how it changes in a nonrelativistic regime. To this end, the next few paragraphs recall how scaling works for relativistic systems, following [19].

In particular, the key argument in the (perturbative) discussion of §2.4.2 assumes the path integral to be dominated by kinetic terms in the action, such as[5] $\partial_\mu \phi^* \partial^\mu \phi$ for a relativistic scalar field. The demand that this part of field space be preserved under rescalings of coordinates, $x^\mu \to x'^\mu := s\, x^\mu$, dictates a scalar field scales as $\phi(x) \to \phi_s(x) := s^{-1}\phi(x)$. The same analysis for a spin-half fermion starts from the invariance of the kinetic term $\overline{\psi}\partial\!\!\!/\psi$ to conclude that $\psi(x) \to \psi_s(x) = s^{-3/2}\psi(x)$.

In both cases, this leads to the conclusion that an effective interaction of the form $\mathfrak{L}_{\rm int} = c_n O_n$, involving an operator of dimension $[O_n] = \Delta_n$, has an effective coupling that scales as $c_n \to s^{\Delta_n - 4} c_n$, implying irrelevance for $\Delta_n > 4$ and relevance for $\Delta_n < 4$, in agreement with the expectations of naive dimensional analysis. In particular, mass terms like $m^2\phi^*\phi$ or $m\overline{\psi}\psi$ are relevant, implying they become more important at sufficiently low energies – making their neglect eventually a bad approximation when setting scaling dimensions.

11.2.1 Spinless Fields

To see how the mass term changes things consider (for concreteness' sake) a complex scalar field, and recall that fields representing a particle state with energy E_p depend on time like $\phi \propto e^{-iE_p t}$. Because $E_p \simeq m$ in the nonrelativistic regime, the kinetic term $\partial_t\phi^*\partial_t\phi \simeq m^2\phi^*\phi$, implying a big cancellation occurs between it and the mass term in quantities like the lagrangian density or the equations of motion.

This near-cancellation of two large contributions obscures the scaling properties of physical quantities, so it is useful to adopt a more convenient set of variables

$$\phi(x) = \frac{1}{\sqrt{2m}}\,\Phi(x)\, e^{-imt}\,. \tag{11.3}$$

[5] As discussed above, a complex field is used with the idea that it carries a conserved charge to ensure the nonrelativistic particle is stable and/or has an excess of particles over antiparticles.

The extraction of the phase e^{-imt} can be regarded as a change of the overall zero of energy so that particle energy in the low-energy theory really measures kinetic energy, relative to the rest mass.[6]

Making this change of variables in the free action gives

$$-\left[\partial_\mu \phi^* \partial^\mu \phi + m^2 \phi^* \phi\right] = \frac{i}{2}(\Phi^* \partial_t \Phi - \Phi \partial_t \Phi^*) - \frac{1}{2m} \nabla \Phi^* \cdot \nabla \Phi + \frac{1}{2m} |\partial_t \Phi|^2. \tag{11.4}$$

For free nonrelativistic particles with kinetic energy $E_p \sim |\mathbf{p}|^2/2m$ – and so which satisfy $(i\partial_t + \nabla^2/2m)\Phi = 0$ (on shell) – the first two terms in the last line are similar in size in the dominant part of the path integral while the last one is order $E_p^2/m = |\mathbf{p}|^4/4m^3$ and so is much smaller. The field Φ can then be split into particle and antiparticle parts, as in (11.2), and the antiparticles integrated out. This shows how (second-quantized) Schrödinger field theory emerges as the low-energy limit of a relativistic Klein–Gordon field.

Scaling

The goal now is to study how interactions scale under the assumption that the path integral over Φ is dominated by the Schrödinger lagrangian – *i.e.* all but the last term on the right-hand side of (11.4). In particular, the nonrelativistic scaling of the field Φ is chosen to be whatever is required to preserve the relative size of the terms in the Schrödinger action.

That is, consider rescaling time and space coordinates according to $\mathbf{x} \to \mathbf{x}' := s\,\mathbf{x}$ and $t \to t' := s^2\, t$, as required to have $\Phi^* \partial_t \Phi$ and $\nabla\Phi^* \cdot \nabla\Phi$ scale in the same way. This implies that the unperturbed action, $S_{\rm unp}$, transforms under rescalings as

$$S_{\rm unp}[\Phi(s\,\mathbf{x}, s^2 t)] = \int \frac{dt'}{s^2} \frac{d^3x'}{s^3}\, s^2 \left[\frac{i}{2}(\Phi^* \partial_{t'} \Phi - \Phi \partial_{t'} \Phi^*) - \frac{1}{2m} \nabla' \Phi^* \cdot \nabla' \Phi\right], \tag{11.5}$$

where the spacetime integration variable is changed from t and $\mathbf{x}$ to t' and $\mathbf{x}'$. This shows that the property $S_{\rm unp}[\Phi(s\,\mathbf{x}, s^2 t)] = S_{\rm unp}[\Phi_s(\mathbf{x}, t)]$ requires the field rescaling $\Phi(x) \to \Phi_s(x) := s^{-3/2}\Phi(x)$.

With this definition, various interaction terms can be classified as relevant, irrelevant and marginal at lower energies (as $s \to 0$). For instance, for interaction terms like

$$S_{\rm int}[\Phi(x); c_1, c_2, c_3, \cdots] = \int dt\, d^3x \left[c_1 \partial_t \Phi^* \partial_t \Phi + c_2 (\Phi^* \Phi)^2 + \frac{c_3}{|\mathbf{x}|^n} \Phi^* \Phi + \cdots\right], \tag{11.6}$$

the scaling of couplings is given by

$$\begin{aligned} &S_{\rm int}[\Phi(s\,x); c_1, c_2, c_3, \cdots] \\ &= \int \frac{dt'}{s^2} \frac{d^3x'}{s^3} \left[s^4 c_1 \partial_{t'} \Phi^* \partial_{t'} \Phi + c_2 (\Phi^* \Phi)^2 + \frac{s^n c_3}{|\mathbf{x}'|^n} \Phi^* \Phi + \cdots\right] \\ &= S_{\rm int}[\Phi_s(x); s^2 c_1, s\, c_2, s^{n-2} c_3, \cdots], \end{aligned} \tag{11.7}$$

[6] At least this is what happens for particles. Because the antiparticle part of the field varies with time like e^{+iEt}, antiparticle energies instead get shifted up to $E \geq 2m$, ensuring that particle-antiparticle pairs still cost total energy $2m$.

showing, as before, that most interactions are irrelevant, although there are a few that are not (such as the c_3 interaction with $n \le 2$).

Notice, in particular, that relevance or irrelevance of operators can differ when explored using relativistic or nonrelativistic scaling. For example, the interaction $\lambda(\phi^*\phi)^2$ is marginal for relativistic scalars but becomes irrelevant once written in terms of Φ, since

$$\lambda(\phi^*\phi)^2 = \frac{\lambda}{4m^2}\,(\Phi^*\Phi)^2, \tag{11.8}$$

and nonrelativistic scaling of Φ implies that $\lambda \to s\,\lambda$ — *c.f.* (11.7) — as $s \to 0$.

11.2.2 Spin-Half Fields

In practice, many of the nonrelativistic particles in everyday life, like electrons and nucleons, are spin-half fermions. Although the arguments leading to nonrelativistic scaling in this case are similar to those used above for spinless fields, spin introduces a few complications that later applications make worth working through explicitly.

For spin-half particles the basic argument for relativistic scaling goes through as for scalars: for relativistic scaling the mass term $m\,\overline{\psi}\psi$ is relevant and so eventually competes with the kinetic term $\overline{\psi}\not{\partial}\psi$. To see how scaling works past this point it again helps to remove the large m-dependence hidden in the time derivatives by scaling out a factor of e^{-imt}. But in the spin-half case integrating out the anti-particle spin-states plays a more prominent part in arriving at the low-energy nonrelativistic theory [292].

To see how this works, recall that in the particle/antiparticle expansion of a fermion field

$$\psi(x) = \sum_{\sigma=\pm\frac{1}{2}} \int \frac{d^3\mathbf{p}}{\sqrt{(2\pi)^{3}2E_p}} \Big[\mathfrak{a}_{p\sigma}\, u_{p\sigma}\, e^{ip\cdot x} + \bar{\mathfrak{a}}^*_{p\sigma}\, v_{p\sigma}\, e^{-ip\cdot x}\Big] \tag{11.9}$$

the Dirac equation $(\not{\partial} + m)\psi = 0$ implies that the particle and antiparticle spinors, respectively, satisfy

$$\Big(\mathrm{i}\not{p} + m\Big)u_{p\sigma} = 0 \quad \text{and} \quad \Big(-\mathrm{i}\not{p} + m\Big)v_{p\sigma} = 0. \tag{11.10}$$

Because of this, it is convenient to write the fermion 4-momentum as $p^\mu = m\,u^\mu + k^\mu$, where k^μ is of the order of momentum transfers found in the low-energy theory (and so is small) while u^μ is the 4-velocity of its rest frame (and so is order unity). In the conventions used here u^μ satisfies $u^\mu u_\mu = \not{u}^2 = -1$ and in the rest frame gives $\not{u} = -\gamma^0 = \gamma_0$ and $u \cdot x = -t$, where t is rest-frame time.

With these definitions the $O(m)$ part of Eqs. (11.10) become $(1 + \mathrm{i}\not{u})u_{p\sigma} = (1 - \mathrm{i}\not{u})v_{p\sigma} = 0$, so the particle and antiparticle parts of a spinor field ψ can be, respectively, written $\psi = \psi_- + \psi_+$ where

$$\psi_\pm := \left(\frac{1 \pm \mathrm{i}\not{u}}{2}\right)\psi, \tag{11.11}$$

satisfy $\mathrm{i}\not{u}\Psi_\pm = \pm\Psi_\pm$. In terms of these fields the free Dirac action (in the particle rest frame) becomes

$$-\overline{\psi}(\not{\partial} + m)\psi = \psi_-^\dagger(i\partial_t - m)\psi_- + \psi_+^\dagger(i\partial_t + m)\psi_+ - \psi_-^\dagger(\gamma \cdot \nabla)\psi_+ + \psi_+^\dagger(\gamma \cdot \nabla)\psi_- \tag{11.12}$$

which uses the rest-frame result $\not{u} = -\gamma^0 = \gamma_0$ as well as $\overline{\psi} = i\psi^\dagger\gamma^0$.

As before, the dependence on the large mass, m (for particles), is removed by extracting a power of e^{-imt}, writing

$$\psi(x) = \Psi(x)\, e^{-imt}, \tag{11.13}$$

so that the free Dirac action becomes

$$-\overline{\psi}(\not{\partial} + m)\psi = i\Psi_-^\dagger\partial_t\Psi_- + \Psi_+^\dagger(2m + i\partial_t)\Psi_+ - \Psi_-^\dagger(\gamma \cdot \nabla)\Psi_+ + \Psi_+^\dagger(\gamma \cdot \nabla)\Psi_-. \tag{11.14}$$

Because the spatial-derivative terms mix Ψ_+ with Ψ_-, the next step – integrating out Ψ_+ – is *not* the same as simply truncating it to zero. Instead, evaluating the gaussian integral over Ψ_+ works out to be equivalent to replacing it in the action using its equations of motion $(2m + i\partial_t)\Psi_+ = -\gamma \cdot \nabla\Psi_-$, and so

$$\Psi_+ = (2m + i\partial_t)^{-1}(-\gamma \cdot \nabla)\Psi_- \simeq -\frac{1}{2m}(\gamma \cdot \nabla)\Psi_- + \frac{i\partial_t}{4m^2}(\gamma \cdot \nabla)\Psi_- + \cdots, \tag{11.15}$$

leading to the following low-energy lagrangian for Ψ_+ (after a spatial integration by parts):

$$-\overline{\psi}(\not{\partial} + m)\psi = i\Psi_-^\dagger\partial_t\Psi_- + \frac{1}{2m}\Psi_-^\dagger\nabla^2\Psi_- + \cdots. \tag{11.16}$$

Here, the ellipses include all terms suppressed by more than two powers of $1/m$, which encode the higher-order terms of the nonrelativistic expansion of kinetic energy,

$$\sqrt{|\mathbf{p}|^2 + m^2} - m = \frac{|\mathbf{p}|^2}{2m} - \frac{|\mathbf{p}|^4}{8m^3} + \cdots, \tag{11.17}$$

and the hermiticity of the spatial gamma matrices, $\gamma_i^\dagger = \gamma_i$ is used, as well as $(\gamma \cdot \nabla)^2 = \nabla^2$.

For later use notice that the projection (11.11) implies that the spinor Ψ_- is essentially the two-dimensional Pauli spinor [293] describing rotations. It is convenient to regard Ψ this way, dropping the subscript '–' when doing so. Spatial Dirac matrices, γ_i, are then converted to Pauli matrices, σ_i, using the rest-frame representation

$$\left(\frac{1 \pm i\not{u}}{2}\right)\gamma_i\left(\frac{1 \mp i\not{u}}{2}\right) \to \pm i\,\sigma_i. \tag{11.18}$$

Scaling

From the point of view of scaling, what is important is that the leading terms in (11.16) are again those of the Schrödinger lagrangian density, just as was true for the spinless case. Consequently, nonrelativistic scaling is defined precisely as was done in the previous section:

$$S_{\text{unp}}[\Psi(s\,\mathbf{x}, s^2 t; c_1, c_2, \cdots] = S_{\text{unp}}[\Psi_s(\mathbf{x}, t); c_1', c_2', \cdots], \tag{11.19}$$

where $\Psi_s(x) = s^{-3/2}\Psi(x)$ and c'_a denote the rescaled effective couplings. Interactions then scale precisely as they do for spinless particles, leading to scaling transformations as given in (11.7).

As the following sections show, scaling becomes more interesting once nonrelativistic and relativistic degrees of freedom couple to one another.

11.3 Coupling to Electromagnetic Fields ♠

Many practical uses of nonrelativistic EFTs involve slowly moving particles that interact through electromagnetic, gravitational, or strong interactions, so it is useful to extend the above nonrelativistic scaling arguments to gauge interactions [294]. This section does so using electromagnetism as the example, highlighting the two reasons why it introduces several new complications into low-energy scaling arguments.

One complication arises because photons are massless and so are never themselves nonrelativistic. Because of this, photons with momenta similar to those of the nonrelativistic particles, $|\mathbf{k}| \sim |\mathbf{p}| \ll m$, necessarily have energies that are much larger than the nonrelativistic kinetic energies, $\omega_k = |\mathbf{k}| \gg E_p = |\mathbf{p}|^2/2m$. Similarly, photons with energies similar to nonrelativistic kinetic energies, $\omega_k \sim E_p$, necessarily have momenta that are much smaller than their slowly moving counterparts, $|\mathbf{k}| \ll |\mathbf{p}|$. This means there are several different regimes to consider when studying how photons interact with slowly moving matter, each of which has its own scaling properties.

The second complication arises because electromagnetic interactions involve more than just photons. In addition to photons, the electromagnetic field also includes 'constrained', non-propagating, physics like the electrostatic Coulomb interaction. For nonrelativistic systems these contributions can also scale differently than do photon interactions.

Gauge-Invariant Interactions

The coupling between electromagnetism and an electrically charged particle (with charge[7] e_q) is obtained for nonrelativistic systems in precisely the same way as for relativistic ones (see *e.g.* Appendix C.5): the requirement of gauge invariance dictates that all spacetime derivatives are replaced by gauge-covariant derivatives,

$$\partial_t\Phi \to D_t\Phi = (\partial_t - ie_q A_0)\Phi \qquad \text{and} \qquad \nabla\Phi \to \mathbf{D}\Phi = (\nabla - ie_q\mathbf{A})\Phi, \quad (11.20)$$

and similarly for Ψ. To this must be added all possible interactions built directly using the field strengths $\mathbf{E}$ and $\mathbf{B}$.

At lowest order in $1/m$ this modifies (for spinless particles with charge e_q) the free Schrödinger lagrangian density to become

[7] Particle charge is denoted e_q rather than q to avoid confusion with momentum q. e without a subscript is reserved to be the proton charge.

$$\mathfrak{L}_{\rm sch} = \frac{i}{2}(\Phi^* D_0 \Phi - \Phi D_0 \Phi^*) - \frac{1}{2m}\, \mathbf{D}\Phi^* \cdot \mathbf{D}\Phi + \cdots , \tag{11.21}$$
$$= \mathfrak{L}_{\rm unp} + e_q A_0\, \Phi^*\Phi + \frac{ie_q}{2m}\mathbf{A}\cdot(\Phi\nabla\Phi^* - \Phi^*\nabla\Phi) - \frac{e_q^2}{2m}\mathbf{A}^2(\Phi^*\Phi) + \cdots .$$

A similar story goes through for spin-half particles, but with one important difference. At linear order in $1/m$ it is important to remember that the $\mathbf{D}^2$ term initially arises as $(\gamma \cdot \mathbf{D})^2$, and so has a spin-dependence that contributes a new term in the presence of an electromagnetic field. Keeping in mind that $[D_i, D_j] = -ie_q F_{ij}$ when acting on Ψ (which assumes Ψ destroys particles with charge e_q), after integrating by parts the $D_i D_j$ term as derived above becomes

$$\frac{1}{2m}\Psi_-^\dagger(\gamma\cdot\mathbf{D})^2\Psi_- = \frac{1}{2m}\Psi_-^\dagger\left(\mathbf{D}^2 - \frac{ie_q}{4}F_{ij}[\gamma^i,\gamma^j]\right)\Psi_-$$
$$= \frac{1}{2m}\left(\Psi^\dagger\mathbf{D}^2\Psi + e_q\,\mathbf{B}\cdot\Psi^\dagger\sigma\Psi\right). \tag{11.22}$$

The second term on the right-hand side does not arise at all for scalars, and it vanishes for spin-half fermions in the absence of an electromagnetic field. The lowest-order spin-half electromagnetic interactions – up to and including $O(1/m)$ – therefore contain a magnetic coupling to spin, as in

$$\mathfrak{L}_{\rm pauli} = i\Psi^\dagger\partial_t\Psi + \frac{1}{2m}\Psi^\dagger\nabla^2\Psi + e_q A_0\,\Psi^\dagger\Psi \tag{11.23}$$
$$+ \frac{ie_q}{2m}\mathbf{A}\cdot\left(\nabla\Psi^\dagger\Psi - \Psi^\dagger\nabla\Psi\right) - \frac{e_q^2}{2m}\mathbf{A}^2(\Psi^\dagger\Psi) + \frac{e_q}{2m}\mathbf{B}\cdot(\Psi^\dagger\sigma\Psi) + \cdots ,$$

for spin-half particles with charge e_q.

11.3.1 Scaling

In order to assess the relative importance of these interactions at low energy, the next step is to identify how electromagnetic interactions affect the scaling encountered in previous sections.

Classical Electro- and Magneto-Static Interactions

For later comparisons, consider first interactions between a nonrelativistic charged particle and an applied classical electromagnetic field (*i.e.* the field generated by a macroscopic distribution of electric charge, $\varrho = J^0$, and electric current, $\mathbf{J}$). As in earlier sections, these would be implemented by modifying the lagrangian density couplings according to

$$\mathfrak{L} \to \mathfrak{L}_J := \mathfrak{L}_{\rm sch} + A_\mu J^\mu, \tag{11.24}$$

with J^μ regarded as fixed distributions. The new interaction preserves gauge invariance provided the macroscopic distribution is conserved, $\partial_\mu J^\mu = \partial_t\varrho + \nabla\cdot\mathbf{J} = 0$, which includes the static case $\partial_t\varrho = \nabla\cdot\mathbf{J} = 0$. (A concrete choice explored in some detail below is the case of a massive static point charge: $\varrho(\mathbf{x}) = Q\,\delta^3(\mathbf{x})$ and $\mathbf{J} = 0$.)

Electromagnetic fields satisfy Maxwell's equations, which in this case are

$$\nabla \cdot \mathbf{E} = \varrho + e_q \, \Phi^* \Phi + \cdots$$
$$\nabla \times \mathbf{B} = \mathbf{J} + \frac{ie_q}{2m}(\Phi \nabla \Phi^* - \Phi^* \nabla \Phi) - \frac{e_q^2}{m}\mathbf{A}(\Phi^* \Phi) + \cdots , \tag{11.25}$$

where ellipses denote terms suppressed by more than one power of $1/m$. For sufficiently large ϱ and $\mathbf{J}$ the response to the quantum field Φ should become negligible, with the electromagnetic field well-described by a classical configuration that satisfies (11.25) with only ϱ and $\mathbf{J}$ on the right-hand side. Much of the discussion to follow is aimed at quantifying the size of deviations from this classical picture.

When the classical Maxwell equations dominate they imply $\mathbf{E}$ and $\mathbf{B}$ inherit a scaling behaviour from the scaling properties of their sources. Since $\int d^3\mathbf{x}\, \varrho = Q$ is fixed, it follows that $\varrho \to \varrho_s = s^{-3}\varrho$ when $\mathbf{x} \to \mathbf{x}' := s\,\mathbf{x}$. Notice that this scaling is also shared by the contribution, $e\,\Phi^*\Phi$, to the electric charge density of non relativistic quantum matter to (11.25), since previous sections argue $\Phi_s = s^{-3/2}\,\Phi$.

Assuming $\mathbf{J}$ also scales similarly to $\Phi^*\nabla\Phi - \Phi\nabla\Phi^*$ (or to $\varrho\,\mathbf{v}$) then invariance of (11.25) implies that $\mathbf{E}$ and $\mathbf{B}$ must scale as

$$\mathbf{E} \to \mathbf{E}_s = s^{-2}\mathbf{E} \quad \text{and} \quad \mathbf{B} \to \mathbf{B}_s = s^{-3}\mathbf{B}. \tag{11.26}$$

Although this electric scaling agrees with what would have been obtained if t and $\mathbf{x}$ were to scale in the same way, the magnetic scaling does not. This reflects the assumption that $\mathbf{J}$ scales the same way as does the contribution from nonrelativistic matter, and so is suppressed by factors of charged-particle speed, v.

In principle, the implications of (11.26) for the vector and scalar potentials are found from the usual relations

$$\mathbf{E} = \nabla A_0 - \partial_t \mathbf{A} \quad \text{and} \quad \mathbf{B} = \nabla \times \mathbf{A}, \tag{11.27}$$

and for the electrostatic potential this does imply

$$A^0 \to A^0_s = s^{-1}A^0, \tag{11.28}$$

consistent with what would be found for a Coulomb potential, $A^0 \propto 1/|\mathbf{x}|$.

Using the scaling of electric and magnetic fields in (11.26) to infer the scaling of $\mathbf{A}$ from (11.27) gives inconsistent implications, however, depending on whether or not one makes the inference using $\partial_t\mathbf{A} \subset \mathbf{E}$ or $\nabla \times \mathbf{A} = \mathbf{B}$. The problem has its root in the mismatch between the relativistic kinematics of photons and the nonrelativistic kinematics assumed for the sources (and in particular for the different scaling that this implies for t and $\mathbf{x}$). This inconsistency plays an important role when understanding nonrelativistic charged systems since it forces different kinds of scalings in situations where a photon's momentum or energy are matched to the corresponding nonrelativistic quantity.

For quasistatic sources $\partial_t\mathbf{A}$ can be dropped in $\mathbf{E}$, in which case (11.26) and (11.27) imply that $\mathbf{A}$ scales according to

$$\mathbf{A} \to \mathbf{A}_s = s^{-2}\mathbf{A}. \tag{11.29}$$

With these choices the corresponding effective interactions

$$S_{\rm int}[\Phi(\mathbf{x},t), A_0(\mathbf{x},t), \mathbf{A}(\mathbf{x},t); c_1, c_2, c_3, \cdots] = \int \mathrm{d}t\, \mathrm{d}^3\mathbf{x} \Big[c_1 A_0\, \Phi^*\Phi + \mathrm{i} c_2\, \mathbf{A}\cdot(\Phi\nabla\Phi^* - \Phi^*\nabla\Phi) - c_3 \mathbf{A}^2 (\Phi^*\Phi) + \cdots \Big], \tag{11.30}$$

scale to lower energies according to

$$S_{\rm int}[\Phi(s\,\mathbf{x}, s^2 t), A_0(s\,\mathbf{x}, s^2 t), \mathbf{A}(s\,\mathbf{x}, s^2 t); c_1, c_2, c_3, \cdots] = S_{\rm int}[\Phi_s(\mathbf{x},t), A_{0s}(\mathbf{x},t), \mathbf{A}_s(\mathbf{x},t); s^{-1} c_1, s\, c_2, s^2 c_3, \cdots]. \tag{11.31}$$

This scaling starts with couplings taking the values $c_1 = e_q$, $c_2 = e_q/2m$ and $c_3 = e_q^2/2m$ at lowest order when matched at scales $\mu \sim m$.

The scaling of c_1 in Eq. (11.31) shows, in particular, that the Coulomb interaction is relevant and so grows in importance at lower energies for nonrelativistic systems. This growth reflects the fact that low-energy Coulomb interactions of slowly moving charges are enhanced by positive powers of $1/v$, as later sections show in detail. Magnetostatic interactions, on the other hand, are irrelevant in the nonrelativistic regime and so become less important (being suppressed at low energies by positive powers of v).

Electromagnetic Fluctuations I: Soft Regime

For the study of how electromagnetic waves – *i.e.* photons – interact with nonrelativistic systems it is useful to scale the electromagnetic field differently than above, in such a way as to preserve the form of its kinetic term. Since photon propagation is controlled by the Maxwell action,

$$S_{\rm Maxwell} = \frac{1}{2}\int \mathrm{d}t\, \mathrm{d}^3\mathbf{x}\left(\mathbf{E}^2 - \mathbf{B}^2\right), \tag{11.32}$$

it is this action that dominates the path integral in these applications, and so controls the perturbative scaling of electromagnetic fluctuations. As $\mathbf{x} \to \mathbf{x}' = s\,\mathbf{x}$ and $t \to t' = s^2 t$, this implies that

$$\mathbf{E} \to \mathbf{E}_s = s^{-5/2}\mathbf{E} \quad \text{and} \quad \mathbf{B} \to \mathbf{B}_s = s^{-5/2}\mathbf{B}, \tag{11.33}$$

which differs from the scaling found for the classical quasi-static systems in (11.26).

So far, so simple. But as mentioned earlier, having time and space transform differently complicates the inference of how the vector potential, $\mathbf{A}$, transforms. Physically, this reflects the clash between photon's relativistic kinematics, $\omega_k = |\mathbf{k}|$, and the nonrelativistic kinematics of the charged particles, $E_p \simeq |\mathbf{p}|^2/2m$, which for $|\mathbf{p}| \ll m$ precludes their agreeing on both the energy and momentum scales involved in any particular process. If $|\mathbf{k}| \sim |\mathbf{p}|$ then $\omega_k \gg E_p$ and when $\omega_k \sim E_p$ then $|\mathbf{k}| \ll |\mathbf{p}|$.

Since the higher-energy of these two alternatives corresponds to $|\mathbf{k}| \sim |\mathbf{p}|$, this section extrapolates to lower energies by keeping spatial derivatives and dropping $\partial_t \mathbf{A}$ in $\mathbf{E}$, deferring the opposite choice to the next section. In this regime the scaling (11.33) implies that

$$A_0 \to A_{0s} = s^{-3/2} A_0 \quad \text{and} \quad \mathbf{A} \to \mathbf{A}_s = s^{-3/2}\mathbf{A}, \tag{11.34}$$

causing the interactions of (11.30) to scale according to

$$S_{\rm int}[\Phi(s\,\mathbf{x}, s^2t), A_0(s\,\mathbf{x}, s^2t), \mathbf{A}(s\,\mathbf{x}, s^2t); c_1, c_2, c_3, \cdots] \\ = S_{\rm int}[\Phi_s(\mathbf{x}, t), A_{0s}(\mathbf{x}, t), \mathbf{A}_s(\mathbf{x}, t); s^{-1/2}c_1, s^{1/2}c_2, s\,c_3, \cdots]. \tag{11.35}$$

This shows that the Coulomb interaction remains relevant, though grows more slowly than in the regime considered earlier. Magnetic interactions remain irrelevant, though they also evolve more slowly than previously.

The growth of the Coulomb interaction at low energies is saying something physical. If $c_1(\mu) \sim e_q$ starts perturbatively small at $\mu \sim m$ then $c_1(\mu_\star) \sim 1$ for $\mu_\star \sim e_q^2\, m$, indicating that perturbation theory in c_1 is likely to break down at length scales of order $(\alpha_q\, m)^{-1}$. It is no accident that this is of order the Bohr radius when $\alpha_q = \alpha$ and m is the electron mass, since bound states signal that the Coulomb interaction is no longer a small perturbation.

The neglect of time derivatives when inferring the scaling of **A** amounts to lumping the $(\partial_t\mathbf{A})^2$ terms from the Maxwell action in with the perturbative interactions rather than including them in the unperturbed action. This implies that unperturbed electromagnetic propagators appearing in Feynman rules are instantaneous: $G(t, \mathbf{x}; t', \mathbf{x}') = \mathcal{G}(\mathbf{x}, \mathbf{x}')\,\delta(t - t')$, as appropriate for the effects of photons whose momenta are similar to the momenta of the nonrelativistic particles. Such photons have much higher energies than the kinetic energies of the nonrelativistic particles, and so indeed act effectively instantaneously over typical orbital times for the slowly moving massive particles.

Electromagnetic Fluctuations II: Ultra-Soft Regime

How does scaling work at very low energies where time-derivatives of **A** cannot be neglected, and how does the effective theory for this regime differ from what has been found to this point? This is not a purely academic question because this is the regime relevant for low-energy photon exchange, for instance, given that for photons the dispersion relation $\omega_k = |\mathbf{k}|$ ensures equally rapid variations for **A** in space and time. (Exercise 11.3 explores how interactions scale if one naively scales the electromagnetic field to ignore spatial derivatives of **A**.)

Photons in this energy regime capture non-instantaneous effects such as time-retardation, amongst other things [295–297]. In this regime because photon energies are order $\omega_k \sim mv^2$ their momenta are much lower than typical nonrelativistic momenta, $|\mathbf{k}| \sim mv^2 \ll mv$. This makes the influence of such photons effectively nonlocal in space, suggesting a framework in which their influence is captured through an ultra-soft effective theory involving a series of effective potentials [298].

The different scaling for soft and ultra-soft photons shows that they influence slowly moving charges very differently at different scales. Relative impact on different observables can be identified by matching in the effective theory all dependence on small hierarchical parameters like mass ratios and/or powers of small velocities (as may be determined from the full theory using techniques such as the method of regions discussed in §3.2.1). Some instances of this type of reasoning are given in the examples of §12. Although it goes beyond the scope of this book to describe, more specialized EFT techniques have been developed to systematize the study of these types of regimes, such as for ultra-soft [295–298] and collinear gauge physics [299], or their analogs for nonrelativistic gravitating objects [247, 359].

11.3.2 Power Counting

To further quantify the implications of various effective interactions, a power-counting estimate is useful (with the soft and ultra-soft cases treated separately). Although precisely the same Wilsonian action,[8] $S_W(\Phi, A_0, \mathbf{A})$, is of interest for both soft and ultrasoft regimes, for each the division of the action into unperturbed and perturbing parts, $S_W = S_{\rm unp} + S_{\rm int}$, is done differently.

Power Counting in the Soft Regime

Consider first the simplest situation, for which $\omega_k \sim |\mathbf{k}| \sim |\mathbf{p}| \gg E_p \sim |\mathbf{p}|^2/2m$, where photon momenta are comparable to the momentum of the heavy charged particles, but recoil energies for the heavy particle are negligible compared with the photon energy. In this regime the term $(2m)^{-1}\Phi^*\nabla^2\Phi$ is subdominant to $\mathrm{i}\Phi^*\partial_t\Phi$ in the action, suggesting perturbing around

$$\mathfrak{L}_{\rm unp} = \mathrm{i}\Phi^*\partial_t\Phi + \frac{1}{2}\left(\mathbf{E}^2 - \mathbf{B}^2\right). \tag{11.36}$$

All other terms of S_W in this case are in $S_{\rm int}(\Phi, A_0, \mathbf{A}) = \int \mathrm{d}^4x\, \mathfrak{L}_{\rm int}$, which consists of all possible local operators built from powers of the fields and their derivatives,[9]

$$\mathfrak{L}_{\rm int} = m^4 \sum_n \frac{1}{m^{d_n}}\, O_n\left(\frac{\Phi}{m^{3/2}}, \frac{A_0}{m}, \frac{\mathbf{A}}{m}\right). \tag{11.37}$$

Here, appropriate powers of the non relativistic particle mass, m, are included on dimensional grounds, assuming this is the only scale – such as is true, in particular, for the nonrelativistic limits, (11.4) and (11.23), of the Klein–Gordon and Dirac lagrangians. For these lagrangians all terms in $S_{\rm int}$ come suppressed by at least one power of $1/m$, except the (relevant) Coulomb interaction, $eA_0\,\Phi^*\Phi$, whose coefficient is dimensionless.

The Feynman rules are computed as usual, with the propagator for Φ given in Fourier space by

$$G(p) \propto \frac{\mathrm{i}}{p^0 + \mathrm{i}\epsilon}, \tag{11.38}$$

(where ϵ is the positive infinitesimal chosen to implement the time-ordered boundary conditions) and so falls off only linearly for large p_0. The electromagnetic propagators can be evaluated in a gauge where its propagator falls off like $1/k^2$ for large $k^0 \sim |\mathbf{k}| = k$. (When required to choose, electromagnetic propagation is

[8] For simplicity the spinless case is described explicitly in this section, but the dimensional power-counting arguments used apply equally well to the spin-half case.

[9] It is assumed for simplicity that the particle mass m provides the only important UV scale, such as would be true (for different reasons) for comparatively simple particles like electrons or protons. In later sections this assumption is re-examined for particles like nuclei whose size, R, and inverse mass, $1/m$, are not comparable.

evaluated using Coulomb gauge, for which $\nabla \cdot \mathbf{A} = 0$, since this removes any mixing between A_0 and $\mathbf{A}$.)

This lagrangian has the form used earlier in (3.61), with the choices $\mathfrak{f} = M = v_B = v_F = m$, and electromagnetic fields counting as 'B' fields and Φ counting as 'F' fields (Φ counts as an 'F' field even though it could be a boson rather than a fermion, since what counts is that its propagator falls off only like $1/k$ for large $k^0 \sim |\mathbf{k}| = k$). This means the power-counting arguments culminating in Eq. (3.64) are appropriate, provided that all external momenta and energies are similar in size $k^0 \sim |\mathbf{k}| = k$.

This leads to the following power-counting result for amputated amplitudes with $\mathcal{E}_A$ external electromagnetic field lines and $\mathcal{E}_\Phi$ external Φ lines:

$$\mathcal{A}_{\mathcal{E}_A \mathcal{E}_\Phi}(k) \sim m^2 k^2 \left(\frac{1}{m}\right)^{\mathcal{E}_A} \left(\frac{1}{m^2 k}\right)^{\mathcal{E}_\Phi/2} \left(\frac{k}{4\pi m}\right)^{2\mathcal{L}} \left(\frac{k}{m}\right)^{\widehat{\mathcal{P}}}. \tag{11.39}$$

Here, $\mathcal{L}$ counts the number of loops and $\widehat{\mathcal{P}} = \mathcal{P} - 2 + \frac{1}{2}\mathcal{E}_\Phi$ where $\mathcal{P}$, is given by (3.65). Consequently, $\widehat{\mathcal{P}} = \sum_n \mathcal{P}_n$ with:

$$\mathcal{P}_n := \left(d_n + \frac{1}{2} f_n - 2\right) \mathcal{V}_n. \tag{11.40}$$

As before, $\mathcal{V}_n$ counts the number of vertices taken from interaction O_n, having d_n derivatives, f_n powers of Φ or Φ^* and b_n powers of A_0 or $\mathbf{A}$.

Notice that most interactions – that is, interactions with four or more Φ fields or interactions with two Φ fields and at least one derivative – contribute non-negative amounts to $\widehat{\mathcal{P}}$. Exceptions for which $\mathcal{P}_n < 0$ include the Coulomb interaction, $A_0\, \Phi^* \Phi$ (denoted here by $n = c$), or the 'seagull' interaction, $\mathbf{A}^2 \Phi^* \Phi$ (denoted $n = s$), since these satisfy $d_c = d_s = 0$ and $f_c = f_s = 2$, and therefore $\mathcal{P}_c = -\mathcal{V}_c$ and $\mathcal{P}_s = -\mathcal{V}_s$. Each insertion of this kind of interaction costs a potentially dangerous power of m/k.

To see how this works, consider the Coulomb scattering process described by the Feynman graphs of Fig. 11.1, obtained by multiply iterating the basic two-body Coulomb interaction. Applying (11.39) to the contribution of the graph involving n such Coulomb interactions gives (using $\mathcal{E}_A = 0$, $\mathcal{E}_\Phi = 4$ and $\mathcal{L} = n - 1$ while $\widehat{\mathcal{P}} = \mathcal{P}_c = -\mathcal{V}_c = -2n$):

$$\mathcal{A}_{04}^{nc}(k) \sim \frac{e_q^{2n}}{m^2} \left(\frac{k}{4\pi m}\right)^{2(n-1)} \left(\frac{k}{m}\right)^{-2n} \sim \frac{e_q^2}{k^2} \left(\frac{e_q}{4\pi}\right)^{2(n-1)}. \tag{11.41}$$

This expression includes also powers of the dimensionless charge, e_q, of the Φ particle while k denotes the generic size of the external momentum/energy scale.

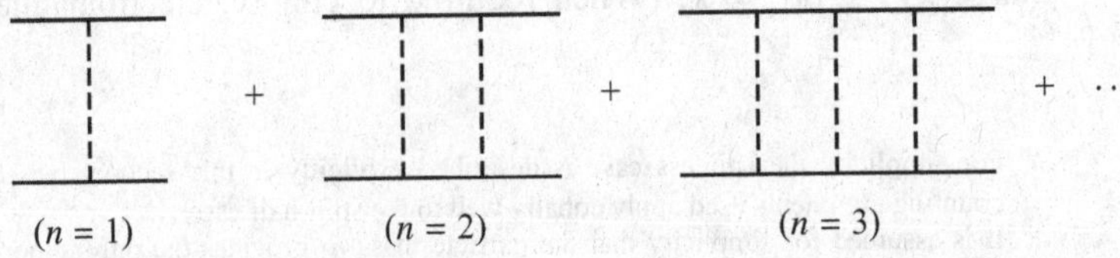

Fig. 11.1 'Ladder' graphs describing multiple Coulomb interactions that are unsuppressed at low energies. Solid (dashed) lines represent Φ (A_0), propagators.

Eq. (11.41) shows what might have been obvious from the very beginning: because both $\mathfrak{L}_{\rm unp}$ and the Coulomb interaction $e_q A_0 \Phi^* \Phi$ involve no powers of m, any graph built only with these components cannot be suppressed by powers of k/m. This comes about in (11.41) by having the $(m/k)^{\mathcal{V}_c}$ enhancement cancel the $(k/m)^{2\mathcal{L}}$ suppression coming from loops. It follows that the Coulomb interaction is suppressed only by the small value of its dimensionless coupling; in this case, the electromagnetic coupling. In particular, for $e_q = \pm e$ each loop is suppressed by a factor of $(e/4\pi)^2 = \alpha/4\pi$, where $\alpha := e^2/4\pi \simeq 1/137$ is the usual fine-structure constant.

The overall factor of $1/k^2$ is also easily understood in the $n = 1$ special case, where it arises simply as the contribution, $1/|\mathbf{k}|^2$, of the Coulomb interaction (written in momentum space), where $\mathbf{k} = \mathbf{p} - \mathbf{p}'$ is the momentum transfer of the scattering.

Contributions to $\mathcal{A}_{04}(k)$ involving all other (non-Coulomb) interactions necessarily involve suppressions by powers of m, and (as usual) (11.39) identifies which interactions and which graphs arise at next-to-leading order in k/m. Next-to-leading order arises in one of two ways:

1. use interactions (like the seagull interaction, $e_q^2 \mathbf{A}^2 \Phi^* \Phi$), for which $\widehat{\mathcal{P}} = -2$ is the same as for the leading ($n = 1$) Coulomb contribution, but with $\mathcal{L} = 1$; or
2. remain at tree level ($\mathcal{L} = 0$) but use interactions (like $e_q \mathbf{A} \cdot \Phi^* \nabla \Phi$), for which $\widehat{\mathcal{P}} = 0$.

The first of these – a one-loop graph built using two seagull interactions – contributes $\mathcal{A}_{04} \sim (e_q^2/4\pi m)^2$, while the second – a tree graph built by single-photon (or $\mathbf{A}$) exchange – gives $\mathcal{A}_{04} \sim e_q^2/m^2$. Both are suppressed relative to the leading Coulomb result by $(k/m)^2$, with the seagull loop also suppressed by the dimensionless loop factor $(e_q/4\pi)^2$.

Power Counting in the Ultra-Soft Regime

Experience with the Coulomb potential suggests dramatic things should happen for fluctuations with energies and momenta related by $E \sim |\mathbf{p}|^2/m$ since this is the regime appropriate for Coulomb bound states.[10] Because the size of such states is set by the Bohr radius, $a_B \simeq (m\alpha_q)^{-1}$, they are characterized by the scales $|\mathbf{p}| \sim m\alpha_q$ in momentum space while their typical energy is set by $E \sim m\alpha_q^2$ [301]. Here $\alpha_q = e_q^2/4\pi$ is the relevant fine-structure constant. The existence of bound states suggests that power counting should indicate that perturbation theory in dimensionless couplings like α_q should eventually break down (as is indeed found, below).

The presence of two scales – both $|\mathbf{p}| \sim mv$ and $E \sim mv^2$ for $v \sim \alpha_q \ll 1$ – precludes directly applying earlier power-counting results, such as (3.64), to this regime. A different estimate is required that can expose the presence of the small dimensionless factors of v. It is, after all, these factors of v that are responsible for the different scaling between $\mathbf{x}$ and t and so also for the novelty of the low-energy nonrelativistic scaling.

[10] This assumes all particles have roughly the same mass, and is revisited in §12.2, which considers bound states built from particles with different masses (such as atoms).

To this end, it is helpful to rescale the coordinates and fields of the problem in a way suggested by the scaling behaviour of ultra-soft photons found in §11.3 [296]. In particular, change variables to dimensionless coordinates, $\hat{\mathbf{x}} = \mathbf{x}\,mv$ and $\hat{t} = tmv^2$, so that $\nabla = mv\widehat{\nabla}$ and $\partial_t = mv^2\hat{\partial}_t$ since this ensures virtual momenta are automatically order mv and energies are order mv^2. Then, in order to keep the kinetic terms in

$$S_{\rm unp} = \int {\rm d}t\, {\rm d}^3\mathbf{x}\, \left[\Phi^* \left({\rm i}\partial_t + \frac{\nabla^2}{2m} \right) \Phi + \frac{1}{2}\left[(\nabla A_0)^2 - (\nabla \times \mathbf{A})^2 \right] \right] \tag{11.42}$$

invariant, also rescale to $\Phi = (mv)^{3/2}\,\widehat{\Phi}$ as well as $A_0 = mv^{3/2}\,\hat{A}_0$ and $\mathbf{A} = mv^{3/2}\,\hat{\mathbf{A}}$.

With these redefinitions, the action of (11.21) becomes

$$\begin{aligned} S_{\rm sch} = \int {\rm d}\hat{t}\, {\rm d}^3\hat{\mathbf{x}} \Bigg\{ & \widehat{\Phi}^* \left({\rm i}\hat{\partial}_t + \frac{\widehat{\nabla}^2}{2} \right) \widehat{\Phi} + \frac{e_q}{\sqrt{v}}\, \hat{A}_0\, \widehat{\Phi}^*\widehat{\Phi} + \frac{{\rm i}e_q}{2}\, \sqrt{v}\, \hat{\mathbf{A}} \cdot (\widehat{\Phi}\widehat{\nabla}\widehat{\Phi}^* - \widehat{\Phi}^*\widehat{\nabla}\widehat{\Phi}) \\ & - \frac{e_q^2 v}{2} \hat{\mathbf{A}}^2 (\widehat{\Phi}^*\widehat{\Phi}) + \frac{1}{2}\left[(\widehat{\nabla}\hat{A}_0 - v\,\hat{\partial}_t \hat{\mathbf{A}})^2 - (\widehat{\nabla} \times \hat{\mathbf{A}})^2 \right] + \cdots \Bigg\}, \end{aligned} \tag{11.43}$$

showing how $v \ll 1$ acts to enhance the Coulomb interaction while couplings to $\hat{\mathbf{A}}$ are suppressed by powers of v. With these variables all factors of v are explicit in the couplings and in the propagators, so repeating the power-counting arguments of earlier sections allows Feynman graphs to be ordered according to the number of powers of v that appear within them. It is useful to use Coulomb gauge when doing so because this allows A_0 and $\mathbf{A}$ exchange to be treated separately.

In this kind of framework the graphs that give the leading contributions in powers of v are those of Fig. 11.1 involving multiple insertions of the Coulomb interaction. The relative strength of each successive exchange is in this case controlled by powers of e_q^2/v (rather than just e_q^2, as in (11.41)). The enhancement of the scattering cross section for two-body Coulomb scattering in the regime $\alpha_q = e_q^2/4\pi \ll v \ll 1$ by powers of $1/v$ is a well-known result (sometimes called Sommerfeld enhancement), with the leading contribution being order e_q^4/v^2. Perturbation theory in powers of e_q eventually fails once $e_q^2/v \simeq 4\pi$, or $v \simeq \alpha_q \ll 1$, precisely in the regime appropriate to Coulomb bound states.

Using the Coulomb interaction and propagators,

$$G^{(\Phi)}(\mathbf{p}, p^0) = \frac{{\rm i}}{p^0 - \frac{1}{2}\mathbf{p}^2 + {\rm i}\epsilon}, \qquad G^{(A_0)}(\hat{\mathbf{p}}) = \frac{{\rm i}}{\mathbf{p}^2}, \tag{11.44}$$

obtained from (11.43) ensures that it is the ladder graphs of Fig. 11.1 that dominate for small v since crossed graphs vanish due to the absence of energy-dependence in the A_0 propagator. Summing these graphs is equivalent to solving for the two-body Schrödinger–Coulomb propagator (as is made more explicit below for the interactions of two charged particles with differing masses).

Subdominant contributions in powers of v are found in the usual way using the propagators and vertices read off from (11.43). Since the couplings of $\mathbf{A}$ are of order $e_q\sqrt{v}$ while those of A_0 are order $e_q/\sqrt{v}$, naively the $\mathbf{A}$ couplings are suppressed relative to those of the Coulomb sector by replacing $e_q^2 \to e_q^2 v^2$. This conclusion is modified, however, by the appearance of v in the Coulomb-gauge photon propagator, which reads

$$G^{(A)}_{jk} = \frac{i}{v^2(p^0)^2 - \mathbf{p}^2 + i\epsilon}\left(\delta_{jk} - \frac{p_j p_k}{|\mathbf{p}|^2}\right). \tag{11.45}$$

This introduces factors of $1/v$ (from the residues of the poles at $p^0 = |\mathbf{p}|/v$) that can reduce the suppression of $\mathbf{A}$ exchange to a single power of v relative to the Coulomb contribution. The factors of $1/v$ arising from these poles reflect the fact that on-shell photons have energies and momenta both order mv and so, in particular, their energies are larger than mv^2 by order $1/v$.

The propagator (11.45) illustrates a fundamental complication that can arise when power counting in situations where different regimes of integration compete in their dependence on small parameters. For instance, for graphs built using (11.45) in a regime where $p^0 \sim |\mathbf{p}|$ the $v^2(p^0)^2$ term is small and so can be regarded as a perturbative two-point interaction rather than part of the propagator. Doing so leads to a relatively straightforward expansion in powers of v. But for graphs dominated by on-shell photon exchange the pole at $p^0 = |\mathbf{p}|/v \gg |\mathbf{p}|$ controls the integration over p^0, leading to contributions that scale differently with v, carrying as they do the non-instantaneous effects of photon exchange. In this regime the path integral is dominated by configurations where $v^2(\partial_t \mathbf{A})^2$ competes in size with $\mathbf{A} \cdot \nabla^2 \mathbf{A}$, making it wrong to treat the $v^2(\partial_t \mathbf{A})^2$ as a perturbation. As later examples show, in any particular graph both regimes must be examined (such as through the method of regions described in §3.2.1) when matching to low-energy properties.

11.4 Summary

Effective theories by definition integrate out states with energies E larger than some scale Λ to focus exclusively on low-energy processes. This section elaborates an observation already made in §8.2.3, that it can be useful to include particles within a low-energy EFT, even if the heavy particle's mass satisfies $m \gg \Lambda$. Doing so can make sense if the particle's rest mass remains locked up, and so cannot be released to produce other high-energy states whose absence the effective theory assumes.

There are two necessary pre-conditions for this type of rest-mass sequestration. First, the particle should be exactly (or approximately) stable, since this precludes (or delays) liberation of energy through radioactive disintegration. Second, the particle's antiparticles should be absent (or rare) in the environment of interest, since this precludes (or makes infrequent) energy release through annihilation.

When these conditions are satisfied it is useful to integrate out the heavy particle's antiparticle, leading to a nonrelativistic Schrödinger EFT describing the low-energy interactions of the slowly moving heavy states. For any particle that survives down to low-enough energies a nonrelativistic EFT must eventually dominate below the rest mass. The inevitability of this transition from relativistic to nonrelativistic kinematics is reflected by the particle mass being generically a relevant operator when perturbative scaling is assessed using relativistic kinematics.

The scaling of interactions differs in the nonrelativistic regime, and, in particular, the Coulomb interaction switches from being marginal (in a relativistic counting) to relevant. In particular, the small ratio $v \sim E_{\rm kin}/|\mathbf{p}|$ can compete with other small dimensionless couplings, like $\alpha_q = e_q^2/4\pi$, to radically change the relative importance of these interactions at energies well below the mass. The relative growth of newly relevant interactions at lower energies can allow initially weak interactions eventually to become strong, sometimes leading to the appearance of new phenomena like bound states at these lower energies.

Exercises

Exercise 11.1 Evaluate the position-space form, $G(t, \mathbf{x}; t', \mathbf{x}')$, of the nonrelativistic propagator given in (11.38).

Exercise 11.2 Evaluate the position-space propagator, $D(t, \mathbf{x}; t', \mathbf{x}')$, for the Coulomb part of the electromagnetic field, given that its action is given by the Maxwell term

$$S_0 = -\frac{1}{2}\int d^4x\, (\nabla A_0)^2.$$

Use this to argue why 'crossed' graphs (obtained by swapping the top ends of any two lines) are not as large as the graphs shown explicitly in Fig. 11.1. (Imagine time points to the right in this figure.)

Exercise 11.3 For ultra-soft photons imagine scaling the electromagnetic potentials, A_0 and $\mathbf{A}$, in such a way as to ensure the scaling (11.33) is satisfied with the scaling of $\mathbf{A}$ determined from $\mathbf{E} = \nabla A_0 - \partial_t \mathbf{A}$ (and ignoring the connection between $\mathbf{B}$ and $\mathbf{A}$). Show that in this case one would have found

$$A_0 \to A_{0s} = s^{-3/2} A_0 \qquad \text{and} \qquad \mathbf{A} \to \mathbf{A}_s = s^{-1/2}\mathbf{A},$$

instead of (11.34). Show that these imply that the interactions of (11.30) would now scale according to

$$\begin{aligned} &S_{\rm int}[\Phi(s\,\mathbf{x}, s^2t), A_0(s\,\mathbf{x}, s^2t), \mathbf{A}(s\,\mathbf{x}, s^2t); c_1, c_2, c_3, \cdots] \\ &\quad = S_{\rm int}[\Phi_s(\mathbf{x}, t), A_{0s}(\mathbf{x}, t), \mathbf{A}_s(\mathbf{x}, t); s^{-1/2}c_1, s^{-1/2}c_2, s^{-2}c_3, \cdots], \end{aligned}$$

and in particular that all three interactions would grow strongly in the infrared in a regime dominated by perturbation theory around this type of unperturbed action.

Exercise 11.4 This exercise provides practice using the method of regions (from §3.2.1) to identify 'Sudakov' double logarithms [302] (following [30, 303]). Consider the following integral, coming from a vertex-correction diagram involving effectively massless particles

$$I(p, \ell) := \frac{i}{\mu^{D-4}} \int d^Dk\, \frac{1}{[(p+k)^2 - i\epsilon][(\ell+k)^2 - i\epsilon][k^2 - i\epsilon]},$$

for which our interest is in large logarithms that arise in the limit $|p^2| \sim |\ell^2| \ll Q^2 := |(p-\ell)^2|$. To explore this limit, define the small parameter $\lambda^2 := |p^2/Q^2| \sim |\ell^2/Q^2|$ and seek terms that are large as λ gets small. To this end, decompose p^μ, ℓ^μ and k^μ in terms of a basis that includes null vectors $n^\mu_\pm$ that satisfy $n^2_\pm = 0$ and $n_+ \cdot n_- = -2$, so that $p^\mu = p^\mu_+ + p^\mu_- + p^\mu_\perp$ (and similarly for ℓ^μ and k^μ), where $p^\mu_+ = -\frac{1}{2}(p \cdot n_-)\, n^\mu_+$, $p^\mu_- = -\frac{1}{2}(p \cdot n_+)\, n^\mu_-$ and $n_\pm \cdot p_\perp = 0$.

Then choosing n_+^μ to lie along the direction of p^μ and n_-^μ to be parallel to ℓ^μ means the components of p^μ and ℓ^μ are of order

$$[p \cdot n_+, p \cdot n_-, p_\perp] = [O(\lambda^2), O(1), O(\lambda)]\, Q$$
$$[\ell \cdot n_+, \ell \cdot n_-, \ell_\perp] = [O(1), O(\lambda^2), O(\lambda)]\, Q,$$

since this ensures $p^2 \sim \ell^2 \sim O(\lambda^2)Q^2$.

By Taylor expanding the integrand in each of these regimes and integrating the result in dimensional regularization, show that the leading behaviour of $\mathcal{I}(p,\ell)$ for small λ receives contributions from each of the regions corresponding to k^μ of size

$$\begin{aligned}
[k \cdot n_+, k \cdot n_-, k_\perp] &= [O(1), O(1), O(1)]\, Q && \text{(hard)}\\
[k \cdot n_+, k \cdot n_-, k_\perp] &= [O(\lambda^2), O(1), O(\lambda)]\, Q && \text{(collinear-1)}\\
[k \cdot n_+, k \cdot n_-, k_\perp] &= [O(1), O(\lambda^2), O(\lambda)]\, Q && \text{(collinear-2)}\\
[k \cdot n_+, k \cdot n_-, k_\perp] &= [O(\lambda^2), O(\lambda^2), O(\lambda^2)]\, Q && \text{(ultrasoft).}
\end{aligned}$$

Show that all other scaling of k^μ in powers of λ contribute zero in dimensional regularization, but that the above regions contribute the leading behaviour

$$\mathcal{I}_{\rm h}(p,\ell) = \frac{\pi^{D/2}\Gamma(1+\epsilon)}{Q^2}\left[\frac{1}{\epsilon^2} + \frac{1}{\epsilon}\ln\left(\frac{\mu^2}{Q^2}\right) + \frac{1}{2}\ln^2\left(\frac{\mu^2}{Q^2}\right) - \frac{\pi^2}{6}\right]$$

$$\mathcal{I}_{\rm c1}(p,\ell) = \frac{\pi^{D/2}\Gamma(1+\epsilon)}{Q^2}\left[-\frac{1}{\epsilon^2} - \frac{1}{\epsilon}\ln\left(\frac{\mu^2}{|p^2|}\right) - \frac{1}{2}\ln^2\left(\frac{\mu^2}{|p^2|}\right) + \frac{\pi^2}{6}\right]$$

$$\mathcal{I}_{\rm c2}(p,\ell) = \frac{\pi^{D/2}\Gamma(1+\epsilon)}{Q^2}\left[-\frac{1}{\epsilon^2} - \frac{1}{\epsilon}\ln\left(\frac{\mu^2}{|\ell^2|}\right) - \frac{1}{2}\ln^2\left(\frac{\mu^2}{|\ell^2|}\right) + \frac{\pi^2}{6}\right]$$

$$\mathcal{I}_{\rm us}(p,\ell) = \frac{\pi^{D/2}\Gamma(1+\epsilon)}{Q^2}\left[\frac{1}{\epsilon^2} + \frac{1}{\epsilon}\ln\left(\frac{\mu^2 Q^2}{|p^2\ell^2|}\right) + \frac{1}{2}\ln^2\left(\frac{\mu^2 Q^2}{|p^2\ell^2|}\right) + \frac{\pi^2}{6}\right],$$

where $D = 4 - 2\epsilon$ in dimensional regularization. These sum to give a result that is finite as $\epsilon \to 0$, giving:

$$\mathcal{I}(p,\ell) = \frac{\pi^2}{Q^2}\left[\ln\left(\frac{Q^2}{|p^2|}\right)\ln\left(\frac{Q^2}{|\ell^2|}\right) + \frac{\pi^2}{3} + O(\lambda)\right].$$

12 Electrodynamics of Nonrelativistic Particles

The world around us is filled with slowly moving charged particles, and this chapter exploits this fact to identify practical examples that illustrate the utility of the previous chapter's power-counting procedure. These include applications to the properties of both atomic constituents as well as to the interactions of atoms (and larger objects) as a whole.

The presentation starts by explicitly matching to determine the leading effective interactions of the nonrelativistic theory for particles like electrons with simple underlying relativistic UV completions. For these the utility of EFT methods is illustrated using examples for which the matching can be performed beyond leading order in the electromagnetic coupling, showing how precision radiative corrections can be systematically incorporated into Schrödinger bound-state calculations.

The discussion then widens to include more complicated nonrelativistic systems, including more complicated charged particles (like nucleons and nuclei) that are built from smaller constituents. A particular focus is systems (like real atoms) that involve more than one species of slowly moving particles with very different masses. The discussion then closes by treating the universal low-energy electromagnetic response of composite systems that are electrically neutral but built from smaller charged constituents. Although the results obtained are not surprising, their derivation within an EFT framework shows how standard treatments can be embedded from first principles into a systematic field-theoretic framework using controlled approximations.

12.1 Schrödinger from Wilson ◇

This section starts by enumerating the possible lowest-dimension effective interactions for a non-relativistic charged particle, a theory called Nonrelativistic Quantum Electrodynamics (or NRQED) [294]. Because these are the most general, consistent with general symmetries, they equally well describe simple elementary particles and more complicated composite charged objects. A matching calculation then follows to determine the values taken by the leading effective couplings in a situation where the relativistic UV completion is well-understood: the quantum electrodynamics of a point-like lepton.

12.1.1 Leading Electromagnetic Interactions

Consider the low-energy EFT governing the dynamics of a nonrelativistic spin-half particle described by a two-component Pauli–Schrödinger field, Ψ, interacting with the electromagnetic potential, $A_\mu = \{A_0, \mathbf{A}\}$. To start, the particle mass m and charge e_q are left general, but when applied to electrons these will take values $m = m_e$ and $e_q = -e$.

As usual, there are an infinite number of possible interactions, with the most important interactions having the smallest operator dimension. Since electromagnetic interactions are weak, this operator dimension is computed perturbatively assuming dominance in $\mathfrak{L}_{\rm eff}$ of terms like $\mathrm{i}\,\Psi^\dagger \partial_t \Psi$ as well as $(\nabla A_0)^2$ and $(\nabla\times\mathbf{A})^2$. With applications to electrons or nucleons in mind it is conventional to extract powers of m and factors of e_q so that the dimensionless effective couplings would satisfy $c_i = 1$ in lowest-order expressions like (11.23).

With applications to QED and QCD in mind it suffices to restrict to interactions that are rotationally invariant and preserve both parity and time-reversal (see §C.4.3 for a summary of parity and time-reversal transformation properties in electrodynamics). The resulting Wilsonian action has the form $S_{\rm eff} = \int \mathrm{d}t\,\mathrm{d}^3\mathbf{x}\;\mathfrak{L}_{\rm eff}$ with $\mathfrak{L}_{\rm eff} = \mathfrak{L}_0 + \mathfrak{L}_1 + \mathfrak{L}_2 + \mathfrak{L}_3 + \cdots$ where $\mathfrak{L}_n$ is proportional to m^{-n}. The terms $\mathfrak{L}_0$ and $\mathfrak{L}_1$ are given by interactions familiar from (11.23), though here assigned arbitrary coefficients [294]

$$\mathfrak{L}_0 = \mathrm{i}\,\Psi^\dagger \partial_t \Psi + e_q A_0 (\Psi^\dagger\Psi) + \frac{\epsilon_E}{2}\,\mathbf{E}^2 - \frac{1}{2\mu_B}\,\mathbf{B}^2, \tag{12.1}$$

of which the first two terms are equivalent to $\frac{\mathrm{i}}{2}[\Psi^\dagger(D_t\Psi) - (D_t\Psi^\dagger)\Psi]$. Similarly,

$$\begin{aligned}\mathfrak{L}_1 &= \frac{1}{2m}\Psi^\dagger \mathbf{D}^2\Psi + \frac{e_q}{2m}\,c_F\,\mathbf{B}\cdot(\Psi^\dagger\boldsymbol{\sigma}\Psi) \qquad (12.2)\\ &= \frac{1}{2m}\Psi^\dagger\nabla^2\Psi + \frac{\mathrm{i}e_q}{2m}\mathbf{A}\cdot\Big[(\nabla\Psi^\dagger)\Psi - \Psi^\dagger\nabla\Psi\Big]\\ &\quad - \frac{e_q^2}{2m}\mathbf{A}^2(\Psi^\dagger\Psi) + \frac{e_q}{2m}\,c_F\,\mathbf{B}\cdot(\Psi^\dagger\boldsymbol{\sigma}\Psi)\cdots,\end{aligned}$$

where (as usual) $\mathbf{E} = \nabla A_0 - \partial_t\mathbf{A}$ and $\mathbf{B} = \nabla\times\mathbf{A}$ while $\boldsymbol{\sigma} = \{\sigma_1,\sigma_2,\sigma_3\}$ represents a vector of two-by-two Pauli matrices acting on the (unwritten) spinor indices of Ψ. Covariant derivatives of Ψ are given by $D_t\Psi = (\partial_t - \mathrm{i}e_q A_0)\Psi$ and $\mathbf{D}\Psi = (\nabla - \mathrm{i}e_q\mathbf{A})\Psi$.

In these expressions Ψ is rescaled[1] to remove any parameter in front of $\mathrm{i}\Psi^\dagger D_t\Psi$ and the coefficient of $\Psi^\dagger\mathbf{D}^2\Psi$ is taken to *define* the charged-particle mass.[2] At this order in $1/m$ the effective couplings are ϵ_E, μ_B and c_F, which §11 shows are unity at lowest order in e_q for slowly moving particles moving within a Lorentz-invariant environment (like the vacuum). Notice also that no term $\mathfrak{L}_{-1} = m\Psi^\dagger\Psi$ is written in $\mathfrak{L}_{\rm eff}$ because this can be removed using the field redefinition $\Psi \to \Psi e^{-imt}$ (as indeed is done explicitly in §11). The removal of any such a term is a prerequisite for cleanly identifying the low-energy $1/m$ expansion.

[1] In principle A_0 and $\mathbf{A}$ can be similarly rescaled to remove ϵ_E and μ_B (if these are constants), but this is not done here to allow for later applications where they vary in space.

[2] That is, m is defined by the coefficient of $E_p = (\mathbf{p}^2/2m) + \cdots$ in the particle dispersion relation. For nonrelativistic systems this need not, in general, agree with other definitions of mass, such as the size of the particle gap $E_p(\mathbf{p} = 0)$, and the relation between different definitions must be computed on a case-by-case basis.

The terms involving higher powers of $1/m$ are constructed in a similar way, with care taken at each order to remove redundant interactions as described in §2.5. In particular, the freedom to perform field redefinitions to remove terms in $\mathfrak{L}_n$ that vanish on use of the lowest-order field equations allows the elimination of any terms involving time derivatives like $D_t\Psi$ or $D_t\Psi^*$.

At order m^{-2} a basis of independent interactions can be written

$$\mathfrak{L}_2 = \frac{e_q}{8m^2}\, c_D(\Psi^\dagger\Psi)(\nabla\cdot\mathbf{E}) - \frac{ie_q}{8m^2}\, c_S\,\Psi^\dagger\boldsymbol{\sigma}\cdot\Big(\mathbf{D}\times\mathbf{E}-\mathbf{E}\times\mathbf{D}\Big)\Psi + \frac{d_1}{m^2}(\Psi^\dagger\boldsymbol{\sigma}\Psi)\cdot(\Psi^\dagger\boldsymbol{\sigma}\Psi) + \frac{d_2}{m^2}(\Psi^\dagger\Psi)(\Psi^\dagger\Psi), \qquad (12.3)$$

with undetermined coefficients c_D ('Darwin' term), c_S ('Spin-orbit' term) and the 'two-body contact' couplings d_1 and d_2. Integrating out the antiparticle to lowest order in e_q – either as in §11 or through the matching calculation below – gives $c_D = c_S = 1$ and $d_1 = d_2 = 0$, so deviations from these values indicate contributions beyond leading order.

Strictly speaking, when ϵ_E is constant the coefficients c_D and d_2 are not independent because the Darwin term can be removed by performing the field redefinition

$$\delta A_0 = \frac{e_q}{8m^2\epsilon_E}\, c_D\,\Psi^\dagger\Psi, \qquad (12.4)$$

at the expense of causing the shift $d_2 \to d_2 + e_q^2 c_D/(8\epsilon_E)$. They are nonetheless kept here separately, partly for historical reasons and partly because the argument relating them becomes more subtle once more than one species of nonrelativistic particle is included.

Not all of the effective couplings appearing in these (and higher-order) interactions are independent if the underlying physics being described is Lorentz invariant (such as for slow particles moving in the vacuum). When this is true then $\mathfrak{L}_{\rm eff}$ must be invariant under Lorentz boosts, and although the absence of antiparticles makes the transformation rules for these more complicated than in the high-energy theory (and beyond the scope of this book to describe in detail) they relate the values of some of the effective couplings. A simple and intuitive example of this is the operator $\Psi^\dagger\mathbf{D}^4\Psi$, whose coefficient becomes dictated by Lorentz invariance to be $-1/(8m^3)$ because it represents the order $\mathbf{p}^4$ correction to the underlying relativistic dispersion relation $E = \sqrt{m^2+\mathbf{p}^2} = m + \mathbf{p}^2/(2m) - \mathbf{p}^4/(8m^3) + \cdots$. For the effective couplings listed out to order m^{-2} in (12.1) through (12.3), Lorentz invariance turns out also to imply both that $\epsilon_E\,\mu_B = 1$ and the constants c_S and c_F are related by [304–306]

$$c_S = 2\,c_F - 1, \qquad (12.5)$$

as is satisfied in particular by the lowest-order results $c_S = c_F = 1$.

12.1.2 Matching

In later applications to precision calculations it becomes necessary to know the values predicted by QED for the effective couplings to subdominant order in $\alpha_q = e_q^2/4\pi$ so that $c_i = 1 + c_i^{(1)} + c_i^{(2)} + \cdots$ and $d_i = d_i^{(1)} + d_i^{(2)} + \cdots$ with $c_i^{(n)},\ d_i^{(n)} \propto \alpha_q^n$. The hard way to obtain these is to explicitly integrate out the antiparticle within the path integral along the lines used at leading order in §11. Much easier is to determine these couplings by demanding EFT and QED calculations agree for a few simple observable quantities. It is important when doing this matching to focus on quantities that are insensitive to field redefinitions, since these are often used (as above) to arrive at a basis of independent effective interactions. For this reason it is useful to use either in-principle observable quantities or their close proxies, because observables do not change when field redefinitions are performed.

To see how this works in detail, consider extending the leading-order calculation [294] of the couplings c_i to subdominant order [305, 307, 308]. A convenient matching calculation for fixing these couplings compares the matrix element of the conserved electromagnetic current, $\langle p'\sigma'|J^\mu(x)|p\,\sigma\rangle$, calculated both within QED and in the nonrelativistic effective theory between single-fermion states. This matrix element is a good proxy for an observable because it controls the emission/absorption amplitude for real photons as well as the coupling to applied electromagnetic fields.

Within a relativistic theory, parity- and Lorentz-invariance together with current conservation dictate that the matrix element must have the form [309]

$$\langle p'\sigma'|J^\mu(x)|p\,\sigma\rangle = \mathrm{i}e\,\overline{u}(p',\sigma')\left[\gamma^\mu F_1(k^2) + \frac{\mathrm{i}}{2m}F_2(k^2)\,\gamma^{\mu\nu}k_\nu\right]u(p,\sigma)\,e^{ik\cdot x}, \tag{12.6}$$

where e is the electromagnetic coupling constant – conventionally, the charge of a proton – while $u(p,\sigma)$ is the relativistic spinor representation of a spin-half particle with 4-momentum p^μ and third component of spin $\sigma = \pm\frac{1}{2}$ (see §A.2.3 and §C.3.2 for details) while $k^\mu := (p'-p)^\mu$ and $\gamma^{\mu\nu} := \frac{1}{2}[\gamma^\mu,\gamma^\nu]$. The matrix element of J^μ for *any* spin-half fermion therefore depends only on the two dimensionless and Lorentz-invariant 'form factors', $F_1(k^2)$ and $F_2(k^2)$, which are functions of the invariant 4-momentum transfer.

Eq. (12.6) follows purely as a consequence of symmetries and quantum numbers and does *not* rely on the spin-half particle being weakly interacting or elementary, and so applies equally well to strongly interacting and composite spin-half particles, like protons and neutrons, as it does to electrons. Indeed, it is the shape of the form factors, $F_1(k^2)$ and $F_2(k^2)$ as functions of k^2, that provides an operational diagnostic for whether or not the particle is elementary. Form factors are useful discriminants of particle structure because their shape can be inferred experimentally by measuring the cross section for electromagnetic Ψ-scattering as a function of energy and scattering angle. Part of the evidence that protons and neutrons have substructure (they are built from quarks and gluons) while electrons do not comes from the fact that such measurements reveal that for electrons $F_1(k^2)$ and $F_2(k^2)$ agree with the expectations (see below) for weakly coupled elementary fermions while the same is not true for nucleons [310].

Physically, the functions $F_i(0)$ capture the fermion's static electromagnetic properties, with the response to applied electric fields showing that its electric charge is given by $e_q = eF_1(0)$. Similarly, $eF_2(0)$ parameterizes the fermion's 'anomalous' magnetic moment, since using (12.6) to compute particle energy in a magnetic field reveals the (z-component of the) magnetic moment to be

$$\mu_M = \frac{e}{2m}\Big[F_1(0) + F_2(0)\Big] = \frac{e_q}{2m}\left[1 + \frac{eF_2(0)}{e_q}\right]. \tag{12.7}$$

To compare with the nonrelativistic EFT it is useful to specialize (12.6) to slowly moving fermions, expanding the spinors $u(p,\sigma)$ to linear order in the momenta $\mathbf{p}$ and $\mathbf{p}'$ and expressing the result in terms of the 2-component spinor, $\chi(\sigma)$. This gives

$$\langle p'\sigma'|J^0(0)|p\,\sigma\rangle = \mathrm{i}e\,\overline{u}(p',\sigma')\left[\gamma^0 F_1(k^2) + \frac{\mathrm{i}}{2m}\,F_2(k^2)\,\gamma^{0n}k_n\right]u(p,\sigma)$$
$$= eF_1(0)\,(\chi'^{\dagger}\chi) + (\text{quadratic in momenta}), \tag{12.8}$$

and

$$\langle p'\sigma'|J^\ell(0)|p\,\sigma\rangle$$
$$= \mathrm{i}e\,\overline{u}(p',\sigma')\left[\gamma^\ell F_1(k^2) + \frac{\mathrm{i}}{2m}\,F_2(k^2)\,\gamma^{\ell n}k_n\right]u(p,\sigma) \tag{12.9}$$
$$= \frac{e}{2m}\left\{F_1(0)(\chi'^{\dagger}\chi)(\mathbf{p}'+\mathbf{p}) + [F_1(0)+F_2(0)](\chi'^{\dagger}\boldsymbol{\sigma}\chi)\times(\mathbf{p}'-\mathbf{p})\right\}^\ell,$$

which again drops terms quadratic or higher in momenta.

Matching proceeds by computing $F_1(k^2)$ and $F_2(k^2)$ in the UV theory and then comparing (12.8) and (12.9) to the same matrix element in the low-energy EFT. The conserved current for the low-energy theory is again found by differentiating the matter part of the action with respect to the electromagnetic potential, and as applied to the lagrangian $\mathfrak{L}_{\rm eff} \simeq \mathfrak{L}_0 + \mathfrak{L}_1$ given by (12.1) and (12.2) this gives

$$J^0 = \frac{\delta S_{\rm eff}}{\delta A_0} = e_q\,\Psi^\dagger\Psi \tag{12.10}$$

and

$$\mathbf{J} = \frac{\delta S_{\rm eff}}{\delta \mathbf{A}} = \frac{e_q}{2m}\left\{\mathrm{i}\left[(\nabla\Psi^\dagger)\Psi - \Psi^\dagger\nabla\Psi\right] + c_F\,\nabla\times(\Psi^\dagger\boldsymbol{\sigma}\Psi)\right\}, \tag{12.11}$$

at least up to order $1/m$. The matrix element for this current must be computed within the low-energy EFT to the same order in small quantities as is done in the UV theory, with the constants c_i chosen to ensure the results agree.

Evaluating the single-fermion matrix element of (12.10) to lowest order and equating to (12.8) then gives the matching condition $e_q = eF_1(0)$, as expected. The same conclusion follows from the comparison of (12.11) and the $\mathbf{p}'+\mathbf{p}$ term of (12.9). The $\mathbf{p}'-\mathbf{p}$ term, on the other hand, gives

$$e_q\,c_F = e\left[F_1(0) + F_2(0)\right]. \tag{12.12}$$

Repeating this exercise at quadratic order in momenta similarly implies that

$$e_q\,c_D = e\left[F_1(0) + 2F_2(0) + 8F_1'(0)\right] \quad\text{and}\quad e_q\,c_S = e\left[F_1(0) + 2F_2(0)\right], \tag{12.13}$$

where $F_1' := m^2\mathrm{d}F_1/\mathrm{d}k^2$.

Elementary Fermions: Lowest-Order Matching

Further progress requires choosing a specific UV theory. To this end, specialize now to the case of an elementary fermion (such as an electron) for which the UV completion is simply QED with spin-half part of the UV action given by

$$\mathfrak{L}_{UV} = -\overline{\psi}(\not{D} + m)\psi \tag{12.14}$$

with $\not{D} = \gamma^\mu(\partial_\mu - ie_q A_\mu)$. The electromagnetic current obtained using Noether's theorem for this action is

$$J^\mu = \frac{\delta S_{UV}}{\delta A_\mu} = ie_q\,\overline{\psi}\gamma^\mu\psi, \tag{12.15}$$

and it is the matrix elements of this operator between single-fermion states that defines the functions $F_1(k^2)$ and $F_2(k^2)$ through (12.6) and so also (12.8) and (12.9).

Evaluating the matrix element $\langle p'\sigma'|\overline{\psi}\gamma^\mu\psi|p\,\sigma\rangle \simeq \overline{u}(p',\sigma')\gamma^\mu u(p,\sigma)\,e^{ik\cdot x}$ (to lowest order in e_q) predicts [294]

$$eF_1(k^2) = e_q \quad \text{and} \quad eF_2(k^2) = 0, \tag{12.16}$$

for all k^2, which includes, in particular – *c.f.* Eq. (12.7) – the standard prediction $\mu_M = e_q/2m$. Using (12.16) in (12.12) and (12.13) implies that point fermions satisfy

$$c_F \simeq c_D \simeq c_S \simeq 1 \qquad \text{(zeroeth order in } \alpha_q) \tag{12.17}$$

reproducing the leading results obtained in §11 by integrating out the antiparticle.

Elementary Fermions: Matching to Order α_q/m^2

Similar logic applies if the matching is performed for elementary fermions at higher order in the fine-structure constant, $\alpha_q = e_q^2/4\pi$, again assuming the UV theory is described by the QED lagrangian (12.14). On the UV side the current matrix element must be computed to one-loop order, which involves evaluating the vertex-correction Feynman graphs of Fig. 12.1. These graphs are evaluated using dimensional regularization, since this keeps gauge invariance explicit and so ensures automatically that $eF_1(0) = e_q$ remains true even once loop corrections are included. Since evaluation of the graphs is straightforward, the result is simply quoted below, together with a few general observations [305, 307].

It turns out that graph (d) of Fig. 12.1 does not contribute at all when matching the coefficients c_i, and this is because it contributes in the same way in both the full and low-energy theories and so cancels in the difference once these are compared to read off the c_i. That is, to order $1/m^2$ the vacuum polarization graph of Fig. 7.4 can be regarded as an $O(\alpha_q)$ contribution to the effective operator $F^{\mu\nu}\Box F_{\mu\nu}$, rather than to the vertex correction, and graph (d) of Fig. 12.1 simply captures the influence of this operator to fermion-photon scattering, rather than a contribution to the c_is.

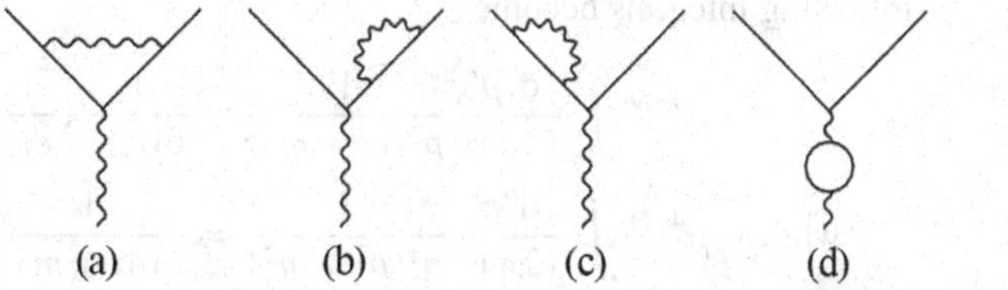

Fig. 12.1 The graphs used when matching the fermion-fermion-photon vertex at one-loop order. Not shown explicitly are the counter-term graphs. Graphs (b), (c) and (d) contribute wave-function renormalization contributions, though gauge invariance ensures graph (d) need not be evaluated explicitly in a matching calculation, since fermion charge e_q does not get renormalized. Graphs (b) and (c) do contribute nontrivially through the fermion wave-function renormalization, δZ, with graph (a) contributing the rest.

The only graph in Fig. 12.1 with the right Dirac-matrix structure to contribute to $F_2(k^2)$ is graph (a), which when Taylor expanded in powers of k^2 evaluates to give the finite result

$$\frac{e}{e_q}\, F_2^{(a)}(k^2) = \frac{\alpha_q}{2\pi}\left[1 + \frac{k^2}{6m^2} + O(k^4)\right]. \tag{12.18}$$

The remaining graphs all contribute to $F_1(k^2)$, and do so in a way that diverges in both the ultraviolet and infrared. For this reason a few words are in order about how these divergences are regularized and (for UV divergences) renormalized. Since the external fermion lines of Fig. 12.1 are on shell – *i.e.* their momenta satisfy $p^2 + m^2 = 0$ – divergences are cancelled using an on-shell renormalization scheme, rather than a mass-independent scheme like modified minimal subtraction. In practice, this means counterterms are chosen to make the fermion propagator have residue unity at the position of the pole at $p^2 + m^2 = 0$, ensuring the fields remain canonically normalized at one-loop order. Such a scheme keeps the intermediate fermion propagators in graphs (b) and (c) from being a problem since it ensures that the sum of the self-energy loop with the corresponding counter-term (whose graph is not explicitly drawn) vanishes on shell, and so cancels the propagator's on-shell pole.

The use of an on-shell renormalization scheme leads to infrared divergences in places in which they do not otherwise arise, such as in the self-energy graphs (b) and (c) of Fig. 12.1. IR divergences normally arise (in four dimensions) when external lines in an on-shell scattering process are connected by a massless-particle propagator — such as in graph (a) of Fig. 12.1. In this graph the on-shell condition for the external fermion implies that the two internal fermion lines each depend on small virtual photon momentum like $1/k$, while the photon propagator goes like $1/k^2$, leading to an IR divergence of the form $\int d^4k/k^4$ for small k. But IR divergences can also arise in other places – like self-energy graphs – if on-shell subtractions are made when renormalizing ultraviolet divergences, since these improve the large-momentum behaviour at the expense of worsening the asymptotic form for small momenta (as is nicely described in [40], for instance).

In what follows both ultraviolet and infrared divergences are regularized using dimensional regularization, making them arise as poles in the limit $\varepsilon = (D-4)/2 \to 0$. IR and UV divergences are distinguished by artificially labelling the regularization parameter ε during intermediate steps, with poles of the form $1/\varepsilon_{IR}$ representing IR divergences and those coming from UV divergences labelled $1/\varepsilon_{UV}$. Thus the following integrals become

$$\mu^{4-D}\int \frac{d^Dp}{(2\pi)^D}\,\frac{1}{p^2(p^2+m^2)} = \frac{1}{(4\pi)^2}\left[\frac{1}{\varepsilon_{UV}} - \ln\frac{m^2}{\mu^2} - \gamma + 1\right]$$

$$\text{while}\quad \mu^{4-D}\int \frac{d^Dp}{(2\pi)^D}\,\frac{1}{p^4(p^2+m^2)} = -\frac{1}{(4\pi)^2m^2}\left[\frac{1}{\varepsilon_{IR}} - \ln\frac{m^2}{\mu^2} - \gamma + 1\right], \tag{12.19}$$

for $D = 4 - 2\varepsilon$ (see §A.2.4) with, at the end of the day, $\varepsilon_{UV} = \varepsilon_{IR} = \varepsilon$. Keeping these divergences separate during intermediate steps is useful because they eventually disappear from physical quantities for entirely different reasons: $1/\varepsilon_{UV}$ poles are removed by renormalization counterterms in both the UV and effective theories while $1/\varepsilon_{IR}$ poles cancel automatically amongst themselves once a physical observable is computed.

Distinguishing UV from IR poles also illuminates an otherwise puzzling feature of dimensional regularization (that turns out to play an important role in what follows). Consider, for example, the integral $I \propto \int d^D p/p^4$, which for general D has mass-dimension $[I] = D - 4$. Dimensional regularization evaluates such integrals as zero because there is no dimensionful parameter on which I can depend for $D \neq 4$ that can carry the right dimensions. This vanishing is a bit less obscure if the integral is regarded as a sum of IR- and UV-divergent parts, such as by writing

$$\int \frac{d^D p}{(2\pi)^D} \frac{1}{p^4} = \int \frac{d^D p}{(2\pi)^D} \frac{1}{p^2(p^2+m^2)} + \int \frac{d^D p}{(2\pi)^D} \frac{m^2}{p^4(p^2+m^2)} = \frac{1}{(4\pi)^2}\left(\frac{1}{\varepsilon_{UV}} - \frac{1}{\varepsilon_{IR}}\right), \tag{12.20}$$

which evaluates the integrals using (12.19). Vanishing occurs once $\varepsilon_{UV} = \varepsilon_{IR} = \varepsilon$ is used.

With these comments in mind a straightforward evaluation of Fig. 12.1 gives a prediction for $F_1(k^2)$ at order α_q, of which only the Taylor expansion out to quadratic powers of k^2 is relevant for the matching calculation to order $1/m^2$. For graph (a) this leads to

$$\frac{e}{e_q} F_1^{(a)}(k^2) = \frac{\alpha_q}{\pi}\left[\left(\frac{1}{2\varepsilon_{UV}} + \frac{1}{\varepsilon_{IR}} + 1 - \frac{3}{4}\ln\frac{m^2}{\mu^2}\right) + \frac{k^2}{m^2}\left(-\frac{1}{3\varepsilon_{IR}} - \frac{1}{8} + \frac{1}{6}\ln\frac{m^2}{\mu^2}\right) + O(k^4)\right] \tag{12.21}$$

while graphs (b) and (c) sum to

$$\frac{e}{e_q} F_1^{(b,c)} = -\frac{\alpha_q}{\pi}\left(\frac{1}{2\varepsilon_{UV}} + \frac{1}{\varepsilon_{IR}} + 1 - \frac{3}{4}\ln\frac{m^2}{\mu^2}\right), \tag{12.22}$$

where μ is the usual arbitrary dimensional regularization scale. Notice that the UV divergent factors $1/\varepsilon_{UV}$ cancel once all three graphs are summed, indicating that no renormalization is necessary.[3]

Combining all contributions gives the two form factors out to order α_q:

$$\frac{e}{e_q} F_1(k^2) = 1 + \frac{\alpha_q}{\pi}\left(\frac{k^2}{m^2}\right)\left(-\frac{1}{3\varepsilon_{IR}} - \frac{1}{8} + \frac{1}{6}\ln\frac{m^2}{\mu^2}\right) + O(k^4) \tag{12.23}$$

and

$$\frac{e}{e_q} F_2(k^2) = \frac{\alpha_q}{2\pi}\left(1 + \frac{k^2}{6m^2}\right) + O(k^4). \tag{12.24}$$

The remaining infrared divergence in (12.23) is the usual one that ultimately cancels against the rate for soft-photon emission in any physical scattering process [135].

Eqs. (12.23) and (12.24) are to be compared to the result computed using the same approximations within the low-energy EFT, which amounts to again evaluating

[3] Not needing to renormalize e_q to cancel divergences in the vertex correction is a famous consequence of gauge invariance. Because $e_q A_\mu$ always appears together in the covariant derivative $D_\mu = \partial_\mu - ie_q A_\mu$, renormalizations of e_q and A_μ always cancel. This ensures that any UV divergences in graph (a) can be absorbed into the fermion wave-function renormalization – *i.e.* must cancel those of graphs (b) and (c).

the graphs of Fig. 12.1, but (at order k^2/m^2) with vertices and propagators found using the lagrangian $\mathfrak{L}_{\rm eff} \simeq \mathfrak{L}_0 + \mathfrak{L}_1 + \mathfrak{L}_2$ given in (12.1) through (12.3). Working in Coulomb gauge allows the electromagnetic propagators to be split into separate A_0 and $\mathbf{A}$ exchange. Furthermore, only the term $i\Psi^\dagger \partial_t \Psi$ is needed when defining the fermion propagator, leading to the result given in (11.38).

Using these Feynman rules the loop corrections to $\langle p'\sigma'|J^\mu|p\,\sigma\rangle$ computed in the effective theory are then particularly simple: they all vanish. They do so because none of the propagators carry any dimensionful parameters besides the virtual momenta and energies over which loop integrals get performed. Consequently, the loop graphs in the low-energy theory all evaluate to zero in dimensional regularization, along the lines of (12.20).

When performing the matching it is useful to keep in mind how this vanishing of loop graphs can be regarded as a cancellation between $1/\varepsilon_{UV}$ and $1/\varepsilon_{IR}$. This is useful because the $1/\varepsilon_{UV}$ part of this cancellation should really instead be absorbed into renormalizations of the effective couplings,

$$c_i^{\rm bare} = c_i^{\rm ren} + \frac{c_i^\infty}{\varepsilon_{UV}}, \tag{12.25}$$

within the effective theory. But once this is done the factors of $1/\varepsilon_{UV}$ are no longer available to cancel with the $1/\varepsilon_{IR}$ terms of the loops, which in the low-energy theory actually then remain uncancelled. This is just as well since the IR divergences also do not cancel in the UV result (12.23), and should have precisely the same form when calculated in the UV or effective theories (which after all do not differ in their low-energy IR physics).

The upshot is that $F_i(k^2)$ is given at this order in terms of the bare couplings by precisely the same formula as at lowest-order, with

$$\begin{aligned}\frac{e}{e_q}\,F_1(k^2) &= (2c_F^{\rm bare} - c_S^{\rm bare}) + \frac{c_D^{\rm bare} - c_S^{\rm bare}}{8}\left(\frac{k^2}{m^2}\right) + \frac{\alpha_q}{3\pi}\left(\frac{1}{\varepsilon_{UV}} - \frac{1}{\varepsilon_{IR}}\right)\left(\frac{k^2}{m^2}\right) \\ &= (2c_F^{\rm ren} - c_S^{\rm ren}) + \left(\frac{c_D^{\rm ren} - c_S^{\rm ren}}{8} - \frac{\alpha_q}{3\pi\varepsilon_{IR}}\right)\left(\frac{k^2}{m^2}\right),\end{aligned} \tag{12.26}$$

where the IR divergence is precisely the same as for the UV theory. As a result, it cancels once UV and EFT results are compared to infer the values of $c_i^{\rm ren}$. Notice the general condition $eF_1(0) = e_q$ is enforced in this expression by the relation (12.5) between c_S and c_F. The second form factor similarly becomes

$$\frac{e}{e_q}\,F_2(0) = c_S^{\rm bare} - c_F^{\rm bare} = c_S^{\rm ren} - c_F^{\rm ren}. \tag{12.27}$$

Including order α_q corrections into matching of the renormalized quantities $c_F^{\rm ren}$, $c_D^{\rm ren}$ and $c_S^{\rm ren}$ is therefore very simple: the result is simply Eqs. (12.12) and (12.13), but evaluated using the loop-corrected expressions, (12.23) and (12.24), for F_1 and F_2, with all UV- and IR-divergent poles simply thrown away. Dropping the superscript 'ren', this leads to

$$c_F = 1 + \frac{\alpha_q}{2\pi} + O(\alpha_q^2), \quad c_D = 1 + \frac{4\alpha_q}{3\pi}\ln\frac{m^2}{\mu^2} + O(\alpha_q^2) \quad \text{and}$$
$$c_S = 1 + \frac{\alpha_q}{\pi} + O(\alpha_q^2), \tag{12.28}$$

and the same procedure can be adapted to any desired accuracy in α_q and/or k^2/m^2.

As usual, the great power of EFT methods is that these coefficients can now be used to compute *any* other low-energy observable. A particularly simple example is given by the magnetic moment, given by (12.7) as

$$\mu_M = \frac{e_q c_F}{2m} = \frac{e_q}{2m}\left[1 + \frac{\alpha_q}{2\pi} + O(\alpha_q^2)\right], \tag{12.29}$$

capturing the leading QED contribution to the fermion's anomalous magnetic moment. Later sections use Eqs. (12.28) to illustrate further how high-energy radiative corrections propagate through to contribute to precision calculations of low-energy observables.

Composite Particles

The effective lagrangian $\mathfrak{L}_{\rm eff}$ of Eqs. (12.1)–(12.3) applies equally well to composite particles (such as protons, neutrons, nuclei or atoms and ions) as for elementary fermions (like electrons or muons), since it contains the most general interactions consistent with the assumed spacetime symmetries (parity, rotation invariance and so on). What does *not* hold for such particles is the matching results of Eqs. (12.28).

For applications to composite particles it is less useful to normalize all effective couplings by powers of the mass, m. Although this makes sense if the mass is the only relevant scale in the problem (as is true both for elementary particles and for composite particles like protons and neutrons), it need not for nonrelativistic composite particles which may exhibit many different low-energy scales.

Suppose, for example, Ψ represents a $^4\mathrm{He}^+$ ion, which is a spin-half ion consisting of a spinless ^{4}He nucleus orbited by a single spin-half electron. In this case, the ion mass is approximately its nuclear mass, which is of order the sum of four nucleon masses, $M \simeq 4m_N \simeq 3.8$ GeV. This is much larger than the energy scale associated with ion's radius, R_i, say, since this is of order the appropriate Bohr radius, $1/R_i \sim Z\alpha\, m_e = 2\alpha\, m_e \simeq 7$ keV. And $1/R_i$ is bigger still than typical electronic binding energies, $|E_B| \simeq \frac{1}{2}(Z\alpha)^2 m_e \simeq 55$ eV.

Mass can also be much larger than inverse size for nuclei, for which M is often the largest of many scales and effective nuclear properties instead often involve the nuclear size, R_N. Although $1/R_N$ and M are similar for nucleons, for larger nuclei usually $M \gg 1/R_N$. These scales differ because the nuclear mass is largely dominated by the rest mass of its constituent nucleons, and the density of nuclear matter turns out to be approximately independent of the number of nucleons within a nucleus. Consequently, the nuclear mass is roughly proportional to both the nuclear volume, R_N^3, and to the total number, A, of nucleons in a nucleus, $M \sim m_N A$. It follows that its radius scales with A like $R_N \sim A^{1/3}/m_N$ (where m_N is of order the nucleon mass) and so $M \sim A^{4/3}/R_N \gg 1/R_N$ when $A \gg 1$.

As discussed at length in Part I, if all other things are equal it is usually the smallest energy scale (largest distance) that suppresses the size of higher-dimension effective coefficients within an EFT. Since this is often not the mass it is less useful to scale factors of M out of effective couplings when working with composite particles, such as by writing[4]

$$\mathfrak{L}_1 = \frac{1}{2M}\Phi^\dagger \mathbf{D}^2\Phi + \mathfrak{m}\,\mathbf{B}\cdot(\Phi^\dagger\sigma\Phi), \tag{12.30}$$
$$= \frac{1}{2M}\Phi^\dagger\nabla^2\Phi + \frac{ie_q}{2M}\mathbf{A}\cdot\left[(\nabla\Phi^\dagger)\Phi - \Phi^\dagger\nabla\Phi\right] - \frac{e_q^2}{2M}\mathbf{A}^2(\Phi^\dagger\Phi) + \mathfrak{m}\,\mathbf{B}\cdot(\Phi^\dagger\sigma\Phi)\cdots,$$

where (as usual) e_q denotes the particle's electric charge while $\mathfrak{m}$ denotes its magnetic moment. For typical nuclei or ions $\mathfrak{m}$ is of order the radius, $e_q R$, of the bound-state's charge-distribution rather than its inverse mass, e_q/M.

12.1.3 Thomson Scattering

As a first, fairly trivial, example of how $\mathfrak{L}_{\rm eff}$ can simplify low-energy calculations, consider computing the low-energy limit of Compton scattering between a (possibly composite) charged particle and a photon: $\gamma + \Psi \to \gamma + \Psi$.

Since the energy and momentum transfers are comparable in such a reaction, the contributions of interactions in $\mathfrak{L}_{\rm eff}$ can be estimated using the $E_p \sim |\mathbf{p}|$ power counting of §11.3.2. In this case, no factors of m arise in the unperturbed lagrangian and so power counting in powers of $1/m$ is relatively simple. In particular, since real photons only appear in $\mathbf{A}$ and not in A_0, and since all $\mathbf{A}$ interactions come with a factor of e_q and are suppressed by at least one power of $1/m$, the low-energy scattering amplitude must satisfy $\mathcal{A}_{2,2}(k) \lesssim O(\alpha_q/m)$. More precisely, (11.39) implies that

$$\mathcal{A}_{2,2}(k) \sim \frac{k}{m^2}\left(\frac{k}{4\pi m}\right)^{2\mathcal{L}}\left(\frac{k}{m}\right)^{\widehat{\mathcal{P}}}, \tag{12.31}$$

with $\mathcal{L}$ counting the number of loops and $\widehat{\mathcal{P}}$ given by (11.40). The leading contribution at low energies therefore comes from $\mathcal{L}=0$ and $\widehat{\mathcal{P}}=-1$, corresponding to a single insertion of one $\mathbf{A}^2\Psi^\dagger\Psi$ interaction.[5] Including also the factors of electric charge e_q this gives $\mathcal{A}_{2,2}(k) \sim e_q^2/m$ and so $d\sigma/d\Omega \propto |\mathcal{A}_{2,2}|^2 \lesssim O(\alpha_q^2/m^2)$.

Putting in the order-unity factors, taking the matrix element of the $\mathbf{A}^2\Psi^\dagger\Psi$ term in $\mathfrak{L}_1$, as given in (12.2), leads to the invariant amplitude (see §B.2 for a refresher on scattering)

$$\mathcal{A}_{2,2}[\Psi(p),\gamma(k) \to \Psi(\tilde{p}),\gamma(\tilde{k})] = -2i\left(\frac{e_q^2}{2m}\right)(\tilde{\chi}^\dagger\chi)\,\tilde{\epsilon}\cdot\epsilon, \tag{12.32}$$

where $\epsilon = \epsilon(k,\lambda)$ and $\tilde{\epsilon} = \epsilon(\tilde{k},\tilde{\lambda})$ are the polarization vectors for the initial and final photons (with helicity $\lambda, \tilde{\lambda} = \pm 1$), while $\chi = \chi(p,\sigma)$ and $\tilde{\chi} = \chi(\tilde{p},\tilde{\sigma})$ are the 2-component spinors for the initial and final fermions (with spin components $\sigma, \tilde{\sigma} = \pm\frac{1}{2}$). The initial factor of 2 comes from the two ways $\mathbf{A}^2$ can destroy the initial particle and create the final one.

[4] A factor of $1/M$ remains in front of the $\Phi^\dagger\mathbf{D}^2\Phi$ term because the coefficient of this term is taken to *define* the object's inertial mass, through the dispersion relation $E(\mathbf{p}) = \mathbf{p}^2/2M$.

[5] Multiple Coulomb exchanges do not complicate this power counting due to the presence of only a single fermion.

Averaging over initial spins and summing over final spins gives the unpolarized squared matrix element

$$\langle |\mathcal{A}_{2,2}|^2 \rangle := \frac{1}{4} \sum_{\sigma\tilde{\sigma}\lambda\tilde{\lambda}} |\mathcal{A}_{2,2}|^2 = 2\left(\frac{e_q^2}{2m}\right)^2 \sum_{\lambda\tilde{\lambda}} |\tilde{\epsilon} \cdot \epsilon|^2 = \frac{e_q^4}{2m^2}(1 + \cos^2\theta), \quad (12.33)$$

which uses $\sum_\lambda \epsilon_j^*(k,\lambda)\epsilon_l(k,\lambda) = \delta_{jl} - \hat{k}_j\hat{k}_l$, where $\hat{\mathbf{k}} = \mathbf{k}/|\mathbf{k}|$, and ditto for $\sum_{\tilde{\lambda}} \tilde{\epsilon}_j^*(\tilde{k},\tilde{\lambda})\,\tilde{\epsilon}_l(\tilde{k},\tilde{\lambda})$. Here θ is the scattering angle between the momenta of the incoming photon, $\mathbf{k}$, and the outgoing one, $\tilde{\mathbf{k}}$, in the charged particle's rest frame. The differential cross section then is

$$\frac{d\sigma}{d\Omega} = \frac{1}{(4\pi)^2}\langle |\mathcal{A}_{2,2}|^2 \rangle = \frac{\alpha_q^2}{2m^2}(1 + \cos^2\theta) \quad (12.34)$$

leading to the standard expression for the total unpolarized Thomson cross section,

$$\sigma = \frac{8\pi}{3}\left(\frac{\alpha_q}{m}\right)^2, \quad (12.35)$$

where $\alpha_q = e_q^2/4\pi = Z^2\alpha$ for particles with electric charge $e_q = Ze$.

Eqs. (12.34) and (12.35) agree with the low-energy limit of the lowest-order QED result [311] for photon scattering from an elementary fermion, but is here computed with much less effort and with a much broader domain of validity. For instance, as derived here the result depends only on the total charge and mass, equally for spin-half and spinless particles (and this is true regardless of whether the particle in question is elementary or composite). Furthermore, knowing that the $1/m^2$ contribution to $d\sigma/d\Omega$ is completely controlled by the coefficient of $O_{\rm sg} := \mathbf{A}^2\Psi^\dagger\Psi$ in $\mathfrak{L}_{\rm eff}$ also shows that this $1/m^2$ contribution to the cross section does not receive radiative corrections to any order in α_q. It cannot receive any corrections because the only way they could contribute is by modifying the coefficient of $O_{\rm sg}$. But the coefficient of this interaction is tied by gauge invariance to the term $\Psi^\dagger\nabla^2\Psi$, whose coefficient is determined by the energy-momentum dispersion relation to be precisely $(2m)^{-1}$.

12.2 Multiple Particle Species ♠

With applications to atoms (and single-electron ions) in mind it is useful to extend the low-energy lagrangian $\mathfrak{L}_{\rm eff}$ to include two different species of nonrelativistic charged particles (such as electrons and nuclei). To this end, the same arguments as above are repeated to construct the low-energy couplings of A_0 and $\mathbf{A}$ to a pair of nonrelativistic particles: an electron, Ψ with charge $e_q = -e$ and mass m, and a nucleus, Φ with charge $Q = Ze$ and mass M. For definiteness both Ψ and Φ are chosen here to be spin-half particles (such as for the ^{1}H atom), though in other applications Φ could equally well have different spin (such as spin zero for even-even nuclei like ^{4}He). Of particular interest in this section is the practical situation where the two species have very different masses, $M \gg m$, since this allows a systematic exploration of the approximations involved when replacing the heavy particle by a Coulomb potential [312].

The low-energy action appropriate to such systems has a similar form as given in (12.1) through (12.3), but written as a double series in $1/m$ and the relevant heavy-particle scale (either a scale of order its size – *e.g.* the nuclear radius, R_N – or its inverse mass, $1/M$):

$$\mathfrak{L}_{\rm eff} = \mathfrak{L}_{0,0} + \mathfrak{L}_{1,0} + \mathfrak{L}_{0,1} + \cdots \tag{12.36}$$

where $\mathfrak{L}_{k,l}$ involves k powers of $1/m$ and l powers of the relevant nuclear scale. For example, the first few terms are[6]

$$\mathfrak{L}_{0,0} = \mathrm{i}\,\Psi^\dagger \partial_t \Psi + \mathrm{i}\,\Phi^\dagger \partial_t \Phi + eA_0\Big(Z\,\Phi^\dagger\Phi - \Psi^\dagger\Psi\Big) + \frac{1}{2}\,\mathbf{E}^2 - \frac{1}{2}\,\mathbf{B}^2, \tag{12.37}$$

and

$$\begin{aligned}\mathfrak{L}_{1,0} = \;& \frac{1}{2m}\Psi^\dagger\nabla^2\Psi - \frac{ie}{2m}\mathbf{A}\cdot\Big[(\nabla\Psi^\dagger)\Psi - \Psi^\dagger\nabla\Psi\Big] \\ & - \frac{e^2}{2m}\mathbf{A}^2(\Psi^\dagger\Psi) - \frac{e}{2m}\,c_F\,\mathbf{B}\cdot(\Psi^\dagger\sigma\Psi) + \cdots ,\end{aligned} \tag{12.38}$$

while

$$\begin{aligned}\mathfrak{L}_{0,1} = \;& \frac{1}{2M}\Phi^\dagger\nabla^2\Phi + \frac{iZe}{2M}\mathbf{A}\cdot\Big[(\nabla\Phi^\dagger)\Phi - \Phi^\dagger\nabla\Phi\Big] \\ & - \frac{Z^2e^2}{2M}\mathbf{A}^2(\Phi^\dagger\Phi) + \mathfrak{m}_N\,\mathbf{B}\cdot(\Phi^\dagger\sigma\,\Phi) + \cdots .\end{aligned} \tag{12.39}$$

These expressions use $D_\mu\Psi = (\partial_\mu + ieA_\mu)\Psi$ and $D_\mu\Phi = (\partial_\mu - \mathrm{i}ZeA_\mu)\Phi$, and adopt the notation $\mathfrak{m}_N$ for the nuclear magnetic moment for a spin-half nucleus[7] (which, if Φ represents a proton, is often instead written $\mathfrak{m}_N = e\mathfrak{c}_F/2M$, with $\mathfrak{c}_F$ order unity).

More terms arise suppressed by two powers of the microscopic UV energy scale, such as terms similar to those in (12.3) including contact interactions like

$$\mathfrak{L}_{\rm contact} = \mathfrak{e}_1(\Psi^\dagger\sigma\Psi)\cdot(\Phi^\dagger\sigma\Phi) + \mathfrak{e}_2(\Psi^\dagger\Psi)(\Phi^\dagger\Phi). \tag{12.40}$$

No terms of the form $\Phi^\dagger\Psi$ arise even for spin-half nuclei because these are forbidden by the symmetries responsible for making each particle type stable (which allow them to be in the low-energy action in the first place). For atoms these include global symmetries encoding conservation of lepton number, $\Psi \to e^{\mathrm{i}\theta_L}\Psi$, and baryon number, $\Phi \to e^{\mathrm{i}\theta_B}\Phi$.

Power Counting

The same power-counting issues arise with two species of fields as do when only one species is present, although any hierarchy in masses, $m \ll M$, introduces similar hierarchies in small power-counting parameters like $|\mathbf{p}|/M \ll |\mathbf{p}|/m$. For both fields the Coulomb interaction is relevant using nonrelativistic power counting, eventually leading to a breakdown of perturbation theory for fluctuations with energies $E \ll |\mathbf{p}|$.

Intuition for where this breakdown occurs can be found from the known properties of two-body Coulomb bound states, whose centre-of-mass wave-functions involve

6 This expression assumes a Lorentz-invariant UV theory and so sets $\epsilon_E = \mu_B = 1$, together with further relations – such as (12.5) – amongst higher-order couplings.

7 Rotation invariance forbids the term $\Phi^\dagger\sigma\,\Phi$ for spinless nuclei, and so $\mathfrak{m}_N = 0$ in this particular case.

equal momenta for both fields, of size $|\mathbf{p}| \sim \tilde{m}\, Z\alpha$, where $\tilde{m} := mM/(m+M) \simeq m + O(m^2/M)$ denotes the system's reduced mass and $\alpha = e^2/4\pi$ is (as usual) the fine-structure constant. The relevant bound-state energies are similarly of order $E \sim \tilde{m}\,(Z\alpha)^2$. Since these scales satisfy $E \sim |\mathbf{p}|^2/m \gg |\mathbf{p}|^2/M$ when $m \ll M$, the regime of interest when applying the power-counting arguments of §11.3.2 to bound-state properties can treat interactions like $(\Phi^\dagger \nabla^2 \Phi)/2M$ as a perturbation, but cannot equally neglect $(\Psi^\dagger \nabla^2 \Psi)/2m$.

12.2.1 Atoms and the Coulomb Potential

Because the large mass appears only in powers of $1/M$ within bound-state energies, the leading features of atomic systems can be explored in the $M \to \infty$ limit, with $O(m/M)$ corrections added later perturbatively. In this regime the dominant contributions to two-particle propagation unsuppressed by any powers of $1/m$ or $1/M$ come from multiple Coulomb exchange, as in the graphs of Fig. (12.2).

Crucially, in this figure the nucleus and Coulomb propagators are

$$G^{(\Phi)}(\mathbf{p}, p^0) = \frac{i}{p^0 + i\epsilon} \quad \text{and} \quad G^{(A_0)}(\mathbf{p}, p^0) = \frac{i}{\mathbf{p}^2}, \tag{12.41}$$

and so because $G^{(\Phi)}$ depends only on energy and $G^{(A_0)}$ depends only on 3-momentum, the argument of the Φ propagator remains unchanged before and after the emission of an A_0 line. These propagators are more informatively written in position space by Fourier transforming, leading to

$$G^{(\Phi)}(\mathbf{x}-\mathbf{x}', t-t') = \int \frac{d^4p}{(2\pi)^4} \left(\frac{i}{p^0 + i\epsilon} \right) e^{-ip^0(t-t') + i\mathbf{p}\cdot(\mathbf{x}-\mathbf{x}')} = \delta^3(\mathbf{x}-\mathbf{x}')\,\Theta(t-t'), \tag{12.42}$$

where the p^0 integration is performed by contours and $\Theta(x) = \{0 \text{ if } x < 0;\ 1 \text{ if } x > 0\}$ is the Heaviside step function. Similarly,

$$\begin{aligned} G^{(A_0)}(\mathbf{x}-\mathbf{x}', t-t') &= \lim_{m\to 0} \int \frac{d^4p}{(2\pi)^4} \left(\frac{i}{\mathbf{p}^2 + m^2} \right) e^{-ip^0(t-t') + i\mathbf{p}\cdot(\mathbf{x}-\mathbf{x}')} \\ &= \frac{i}{4\pi|\mathbf{x}-\mathbf{x}'|}\,\delta(t-t'). \end{aligned} \tag{12.43}$$

Notice that the two propagators depend on time through a step-function and a delta-function, and this is what prevents crossed versions of Fig. 12.2 from contributing a nonzero result.

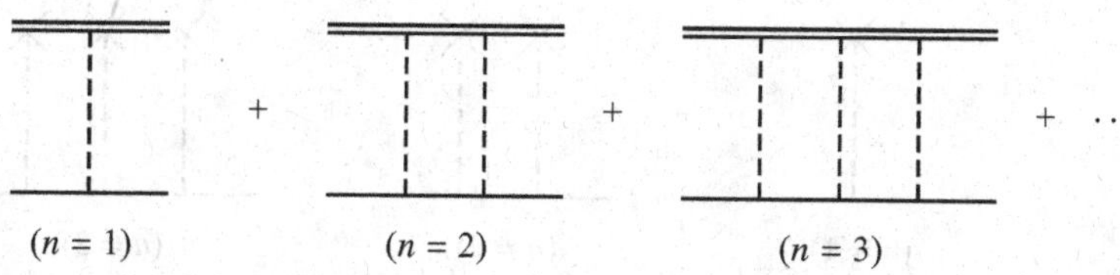

Fig. 12.2 'Ladder' graphs describing multiple Coulomb interactions that are unsuppressed at low energies. Solid lines represent electrons (Ψ propagators), double lines represent nuclei (Φ propagators) and dashed lines represent A_0 propagators.

Using these propagators to evaluate the top half of the ladder graphs of Fig. 12.2 (in position space) gives

$$(Ze)^n \int \mathrm{d}^4x_1\mathrm{d}^4x_2\cdots\mathrm{d}^4x_n\,\phi_f^*(x_n)\,G^{(\Phi)}(x_n-x_{n-1})\,G^{(A_0)}(x_n-y_n)\cdots$$
$$\times\, G^{(\Phi)}(x_2-x_1)\,G^{(A_0)}(x_1-y_1)\,\phi_i(x_1) \qquad (12.44)$$
$$= \int \mathrm{d}^3\mathbf{x}\,\phi_f^*(\mathbf{x},y_n^0)\left(\frac{\mathrm{i}Ze}{4\pi|\mathbf{x}-\mathbf{y}_1|}\right)\left(\frac{\mathrm{i}Ze}{4\pi|\mathbf{x}-\mathbf{y}_2|}\right)\cdots\left(\frac{\mathrm{i}Ze}{4\pi|\mathbf{x}-\mathbf{y}_n|}\right)\phi_i(\mathbf{x},y_1^0),$$

where the unwritten step functions enforce the inequalities $y_1^0 < y_2^0 < \cdots < y_n^0$ and the factors $\phi_f^*(\mathbf{x},y_n^0)$ and $\phi_i(\mathbf{x},y_1^0)$ represent the initial and final nuclear wave-function, as represented by the external double lines. This is to be multiplied by the Feynman rule for the electron field describing the bottom solid line of Fig. 12.2, and integrated over $\mathrm{d}^4y_1\cdots\mathrm{d}^4y_n$.

Notice that assuming initial and final nuclear states to have vanishing energy does not constrain their **x**-dependence because all momentum states contribute to the energy suppressed by $1/M$, which is negligible to the order of interest. With this in mind, it is convenient to work in the rest-frame of the nucleus, with the nucleus in an approximate position eigenstate chosen to be the origin of coordinates, so $\phi_f^*(\mathbf{x})\,\phi_i(\mathbf{x}) = \delta^3(\mathbf{x})$.

Now comes the key observation. So far as the electron sector is concerned, the ladder graphs of Fig. 12.2 are completely equivalent to interacting with a classical background field, $A_0 = \mathcal{A}_0$, with

$$\mathcal{A}_0(\mathbf{y}) := \frac{Ze}{4\pi|\mathbf{y}|}. \qquad (12.45)$$

Interactions with this background potential are described by summing the Feynman graphs of Fig. 12.3, where the dashed lines ending in crosses represent the field of (12.45).

What is important about this is that it allows the ladders to be resummed by reorganzing which terms in $\mathfrak{L}_{\rm eff}$ lie in the unperturbed lagrangian and which are perturbations. In particular, rather than perturbing around the lagrangian

$$\mathfrak{L}_{\rm unp} = \mathrm{i}\,\Phi^\dagger\partial_t\Phi + \mathrm{i}\,\Psi^\dagger\partial_t\Psi - \frac{1}{2m}\nabla\Psi^\dagger\cdot\nabla\Psi + \frac{1}{2}\mathbf{E}^2, \qquad (12.46)$$

the ladder graphs are resummed by instead expanding in graphs describing perturbations around the alternative lagrangian

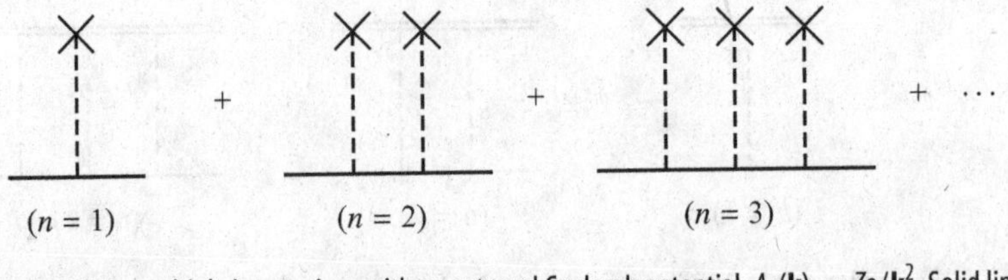

Fig. 12.3 Graphs describing multiple interactions with an external Coulomb potential, $A_0(\mathbf{k}) = Ze/\mathbf{k}^2$. Solid lines represent Ψ propagators while dashed lines capped by an 'X' represent insertions of the external Coulomb potential.

$$\mathfrak{L}'_{\rm unp} = {\rm i}\,\Phi^\dagger \partial_t \Phi + {\rm i}\,\Psi^\dagger \partial_t \Psi - e\mathcal{A}_0 \Psi^\dagger \Psi - \frac{1}{2m}\nabla\Psi^\dagger \cdot \nabla\Psi + \frac{1}{2}\mathbf{E}^2, \tag{12.47}$$

with the Coulomb part of the interaction terms rewritten to become

$$\mathfrak{L}_{\rm int} \ni ZeA_0\Phi^\dagger\Phi - e(A_0 - \mathcal{A}_0)\Psi^\dagger\Psi. \tag{12.48}$$

Alternative Feynman rules are then obtained by replacing the free electron propagator with the result obtained for an unperturbed electron field interacting with the static Coulomb potential of (12.45), satisfying

$${\rm i}\partial_t\Psi = -\frac{1}{2m}\nabla^2\Psi + \frac{Ze}{4\pi|\mathbf{x}|}\Psi, \tag{12.49}$$

with solutions

$$\Psi(\mathbf{x}, t) = \sum_{\sigma=\pm\frac{1}{2}}\sum_n \chi_{n\sigma}(\mathbf{x})\, \mathfrak{a}_{n\sigma}\, e^{-{\rm i}E_n t}. \tag{12.50}$$

Here, $\chi_{n\sigma}(\mathbf{x}, t)$ represent the usual spinor basis of Coulomb energy eigenstates of elementary quantum mechanics courses (with spin), satisfying

$$-\frac{1}{2m}\nabla^2\chi_{n\sigma} + \frac{Ze}{4\pi|\mathbf{x}|}\chi_{n\sigma} = E_n\chi_{n\sigma} \quad \text{and} \quad S_z\chi_{n\sigma} = \sigma\,\chi_{n\sigma}, \tag{12.51}$$

and $\mathfrak{a}_{n\sigma}$ represents the annihilation operator for this state, satisfying the fermionic algebra $\left\{\mathfrak{a}_{n\sigma}, \mathfrak{a}^*_{m\lambda}\right\} = \delta_{nm}\delta_{\sigma\lambda}$. In these expressions n and m denote the complete set of labels for these Coulomb states, and the resulting fermionic propagator becomes

$$\begin{aligned}\langle 0|T\left[\Psi(\mathbf{x}, t)\Psi^\dagger(\mathbf{x}', t')\right]|0\rangle &= \Theta(t'-t)\sum_{n\sigma}\chi_{n\sigma}(\mathbf{x})\chi^\dagger_{n\sigma}(\mathbf{x}')\, e^{{\rm i}E_n(t-t')} \\ &= {\rm i}\sum_{n\sigma}\int \frac{{\rm d}\omega}{2\pi}\left[\frac{\chi_{n\sigma}(\mathbf{x})\chi^\dagger_{n\sigma}(\mathbf{x}')}{\omega - E_n + {\rm i}\epsilon}\right] e^{{\rm i}\omega(t-t')}.\end{aligned} \tag{12.52}$$

Power counting with this revised perturbation theory no longer implies that the infinite sequence of ladder graphs of Fig. 12.2 are all leading order since it is only $A_0 - \mathcal{A}_0$ that now couples to $\Psi^\dagger\Psi$. This is the way the low-energy EFT connects with standard single-particle atomic Rayleigh–Schrödinger perturbation theory. The advantage of making this connection is the ability to systematically integrate these atomic calculations with higher-order quantum-field effects (such as radiative corrections that enter through higher-order matching conditions like (12.1.2), or effects of the weak interactions) as perturbations. Although it goes beyond the scope of this book to explore further, considerable effort has been made to systematize this kind of perturbative expansion [313].

12.2.2 Dipole Approximation

A simple example of the practical use of $\mathfrak{L}_{\rm eff}$ in atomic systems computes the low-energy limit of photon absorption by a charged particle, $\gamma + \Psi \to \Psi'$ (or emission $\Psi \to \Psi' + \gamma$), for an electron bound in an atom. Besides illustrating the utility of EFT methods this example also introduces the dipole approximation that proves useful for later sections.

The leading graphs for this process involve attaching an external photon line in all possible ways to the bottom (electron) line of the graphs of Fig. 12.2 (or to the graphs of Fig. 12.3). Since real photons only appear in **A**, the photon-electron vertex that is relevant to leading order in $1/m$ comes from either the 2nd or 4th terms of the lagrangian of (12.38), reproduced here for convenience of reference

$$\mathfrak{L}_{\rm int} = -\frac{ie}{2m}\,\mathbf{A}\cdot\left[(\nabla\Psi^\dagger)\Psi - \Psi^\dagger\nabla\Psi\right] - \frac{e}{2m}\,c_F\,\mathbf{B}\cdot(\Psi^\dagger\sigma\Psi). \tag{12.53}$$

It is the first of these – the (electric) 'dipole' term – that dominates in atomic absorption and emission processes. This is because for these the photon energy is given by the difference in electron energy levels, and so is of order mv^2, where $v \sim Z\alpha \ll 1$. This implies that the photon momentum is also of this order, making it much smaller than the typical electron momentum: $|\mathbf{k}| \sim mv^2 \ll |\mathbf{p}_e| \sim mv$. Since the derivative in the first term of (12.53) involves electron momenta, it dominates the second term, whose derivative (inside $\mathbf{B} = \nabla\times\mathbf{A}$) samples only the photon's momentum.

As a consequence of this, only the dipole term contributes to leading order in v, and the photon momentum can be dropped in this term. So initial and final electron momenta are approximately equal, $\mathbf{p}_f = \mathbf{p}_i + \mathbf{k} \simeq \mathbf{p}_i$, and the matrix element relevant for photon absorption become

$$\langle 0|\mathbf{A}(\mathbf{x},t)|\mathbf{k},\lambda\rangle = A_c\,\boldsymbol{\epsilon}(\mathbf{k},\lambda)\,e^{i(\mathbf{k}\cdot\mathbf{x}-k^0t)} \simeq A_c\,\boldsymbol{\epsilon}(\mathbf{k},\lambda)\,e^{-ik^0t}, \tag{12.54}$$

where λ denotes the photon helicity and A_c can be regarded as the amplitude of an incoming classical electromagnetic wave, or can be evaluated as the matrix element of the photon destruction operator in an initial configuration with multiply occupied photon states (see §B.1 and §C.3.3, respectively, for conventions about momentum-state and field normalization). Writing the density of initial photons as $n = N/\mathcal{V} = n_{\mathbf{k}}\mathrm{d}^3\mathbf{k}/(2\pi)^3$, where $\mathcal{V}$ is the volume of space, then[8]

$$|A_c|^2 = \frac{n_{\mathbf{k}}}{(2\pi)^3 2k^0}. \tag{12.55}$$

Keeping only the electric-dipole interaction of (12.53) and dropping the dependence on photon momentum is called the 'dipole approximation'.

The summation over arbitrary numbers of Coulomb interactions between nucleus and electron makes evaluation of the graphs daunting at first sight; however, these are elegantly resummed – as discussed above – by expanding the electron field Ψ as in (12.50), using the single-particle Schrödinger–Coulomb wave-functions for the electron, $\chi_{n\sigma}(x)$, found as solutions to (12.51). In practice, what is required is the matrix element of $\mathfrak{L}_{\rm int}$ between the Coulomb wave-functions for the initial and final states.

Denoting the initial and final electronic spins by σ and $\tilde\sigma$, these observations lead to a transition matrix element

$$\begin{aligned}\mathcal{A}_{\rm abs}[\Psi(i,\sigma)+\gamma(\mathbf{k},\lambda)\to\Psi(f,\tilde\sigma)] &= -\int \mathrm{d}^3x\langle\Psi(f,\tilde\sigma)|\mathfrak{L}_{\rm int}|\Psi(i,\sigma),\gamma(\mathbf{k},\lambda)\rangle\\ &\simeq \frac{ie}{m}\,A_c\,\boldsymbol{\epsilon}(\mathbf{k},\lambda)\cdot\int \mathrm{d}^3x\,\chi^\dagger_{f\tilde\sigma}\nabla\chi_{i\sigma},\end{aligned} \tag{12.56}$$

8 These formulae use nonrelativistic normalization for momentum states: $\langle\mathbf{k}|\mathbf{k}'\rangle = \delta^3(\mathbf{k}-\mathbf{k}')$.

where $\chi_{n\sigma}$ is the Schrödinger–Coulomb wave-function of (12.50). An example of the utility of making contact with Coulomb eigenstates is the ability to use single-particle quantum reasoning to rewrite this matrix element in terms of the dipole moment:

$$\begin{aligned}\mathcal{A}_{\rm abs}[\Psi(i,\sigma)+\gamma(\mathbf{k},\lambda)\to\Psi(f,\tilde{\sigma})] &= -\frac{e}{m}\,A_c\,\epsilon(\mathbf{k},\lambda)\cdot\langle f|\hat{\mathbf{P}}|i\rangle\,\delta_{\sigma\tilde{\sigma}}\\ &= -ie\,A_c\,\epsilon(\mathbf{k},\lambda)\cdot\langle f|[\hat{H},\hat{\mathbf{x}}]|i\rangle\,\delta_{\sigma\tilde{\sigma}} \\ &= i\omega_{fi}\,A_c\,\epsilon(\mathbf{k},\lambda)\cdot\mathbf{d}_{fi}\,\delta_{\sigma\tilde{\sigma}},\end{aligned} \tag{12.57}$$

where $\omega_{fi} := E_f - E_i$ is the energy change of the electron, and the first-quantized atomic dipole-moment matrix element is defined by

$$\mathbf{d}_{fi}(t) := -e\langle f|\hat{\mathbf{x}}|i\rangle = -e\int \mathrm{d}^3\mathrm{x}\;\mathbf{x}\,\chi_f^*(\mathbf{x},t)\chi_i(\mathbf{x},t), \tag{12.58}$$

where $\chi_f(\mathbf{x},t)$ without an index σ indicates the spin-independent Schrödinger wave-function. The first line of (12.57) follows from the single-particle Schrödinger representation of the momentum operator, $\hat{\mathbf{P}} = -i\nabla$, and the second and third lines use $\langle f|\hat{\mathbf{P}}|i\rangle = m(\mathrm{d}/\mathrm{d}t)\langle f|\hat{\mathbf{x}}|i\rangle = im\langle f|[\hat{H},\hat{\mathbf{x}}]|i\rangle$ and the fact that both initial and final states diagonalize the Schrödinger–Coulomb Hamiltonian $\hat{H}$.

For later reference, notice that the matrix element (12.57) has the form expected from an interaction Hamiltonian, $H \propto \mathbf{E}\cdot\mathbf{d}$, since $\mathbf{E} = -\partial_t\mathbf{A} + \nabla A_0$ implies that $\langle f|\mathbf{E}|i\rangle = i\omega_{fi}\langle f|\mathbf{A}|i\rangle$.

The differential lab-frame absorption rate for an atomic electron exposed to polarized photons with 3-momentum within a volume element $\mathrm{d}^3\mathbf{k}$ of $\mathbf{k}$ can then be obtained from Fermi's Golden rule (*c.f.* §B.2.2)

$$\mathrm{d}\Gamma_{\rm abs}[\Psi(i,\sigma)+\gamma(\mathbf{k},\lambda)\to\Psi(f,\tilde{\sigma})] = 2\pi|\mathcal{A}_{\rm abs}|^2\delta(k-\omega_{fi})\,\mathrm{d}^3\mathbf{k}. \tag{12.59}$$

As shown in Exercise 12.1, using (12.55) and (12.57) in this expression gives the following unpolarized lab-frame absorption rate for an atomic electron exposed to a bath of electromagnetic radiation:

$$\Gamma_{\rm abs} = \frac{\omega_{fi}^3 n_k}{6\pi}\,|\mathbf{d}_{fi}|^2 = \frac{\pi}{3}\,\rho(\omega_{fi})\,|\mathbf{d}_{fi}|^2, \tag{12.60}$$

where the final equality writes the rate in terms of the energy density of the initial radiation per unit frequency

$$\rho(\omega_{fi}) := \left(k\,\frac{\mathrm{d}n_\gamma}{\mathrm{d}k}\right)_{k=\omega_{fi}} = \frac{n_k\,\omega_{fi}^3}{2\pi^2}. \tag{12.61}$$

Here, the density of initial photons with wave-number $\mathbf{k}$ is assumed to be independent of photon direction, so $\mathrm{d}n_\gamma = n_k\,\mathrm{d}^3\mathbf{k}/(2\pi)^3$ and so the initial density of photons in a frequency range $\mathrm{d}k$ is $\mathrm{d}n_\gamma/\mathrm{d}k = n_k\,k^2/(2\pi^2)$ once summed over photon direction.

Although (12.60) is a standard result, this derivation makes clear that it relies only on working to leading order in the $1/m$ expansion and on the energy difference $\omega_{fi} \sim mv^2$ being much smaller than the inverse size of the bound state: $1/a \sim mv$. For situations where v and α are independent of one another there would be no radiative corrections (in powers of α) that are not also accompanied by additional powers of v. These cannot arise because the normalization of the dipole term is completely dictated by the charged particle's mass and charge, being tied as it is by gauge invariance to the term $\Psi^\dagger \nabla^2 \Psi$, whose coefficient is determined by the energy-momentum dispersion relation $E = \mathbf{p}^2/2m$. For systems (like many atoms) for which $v \sim \alpha$ corrections in powers of α can arise but are at least partly due to the presence of subdominant powers of v. The EFT derivation shows that the above rate also holds equally well for other, possibly composite, charged particles inside bound states, provided these satisfy the basic hierarchy $k = \omega_{fi} \sim mv^2 \ll p \sim mv$, since nothing in the EFT relies on the charged-particle involved being fundamental.

12.2.3 HQET

Previous sections risk leaving the impression that the utility of nonrelativistic field theories is limited to electromagnetic systems like atoms. A practical non-electromagnetic application for the above framework is to the properties of mesons in QCD for which one of the constituent quarks is heavy enough to be treated nonrelativistically. This application is known as *heavy-quark effective theory*, or HQET for short [314].

As described in §8, mesons are colour-neutral bound states involving one valence quark and one valence antiquark, and a framework using nonrelativistic particles is relevant when the mass of one (or both) of the quark/antiquarks involved is much larger than the typical quark momentum, which is of order the strong-interaction energy scale: $m_Q \gg \Lambda_{QCD} \sim 200$ MeV. Table 8.1 shows this is appropriate for mesons involving c, b and t quarks (although the t quark decays so quickly that its mesons are usually not of practical interest).

Casting the interactions of heavy quarks in terms of nonrelativistic EFTs reveals how the limit of infinite quark mass enjoys symmetries that relate different heavy quark species and spins. Because these symmetries are not broken by the QCD couplings to gluons, they should be good approximations for describing mesons containing these quarks, up to corrections in powers of heavy-quark velocity, $v \sim \Lambda_{QCD}/m_Q$. This leads to a systematic approximation scheme for describing many features of mesons built from heavy quarks.

To display these symmetries explicitly start from the QCD lagrangian, (8.1), and integrate out the antiparticles[9] of the three heavy-quark species. Dropping all terms suppressed by heavy mass scales (including m_Q for the heavy quarks), this leads to an EFT of the form

$$\mathfrak{L}_{HQET} = \mathfrak{L}_{NR} + \mathfrak{L}_{QCD}^{\rm light}, \qquad (12.62)$$

where the second term on the right-hand side describes the relativistic degrees of freedom with lagrangian

$$\mathfrak{L}_{QCD} = -\frac{1}{4} G^\alpha_{\mu\nu} G^{\mu\nu}_\alpha - \sum_{q=u,d,s} \overline{q}(\not{D} + m_q)q, \qquad (12.63)$$

with gauge field-strength given by (8.2) and $\not{D} = \gamma^\mu D_\mu$ with covariant derivative of the quark fields given by (8.3).

[9] For mesons involving antiparticles, one instead integrates out the particles.

By contrast, the terms involving the heavy-quark fields start off with

$$\mathfrak{L}_{NR} = \mathrm{i} \sum_{Q} \Psi_Q^\dagger u_Q^\mu D_\mu \Psi_Q \tag{12.64}$$

where $\Psi_Q = \{\Psi_{Qi}^a\}$ represents the heavy-quark fields, with the sum on the 'flavour' index Q running over whichever of the heavy-quark species, $Q = c, b, t$ are relevant to the process of interest. The index $i = 1, 2$ similarly represents the 2-component nonrelativistic spinor index, while $a = 1, 2, 3$ represents their three colours. The covariant derivative relevant in this case acts only on the colour index and is the same as for light quarks:

$$(D_\mu \Psi_{Qi})^a := \partial_\mu \Psi_{Qi}^a - \mathrm{i}g\, G_\mu^\alpha (T_\alpha)^a_b\, \Psi_{Qi}^b, \tag{12.65}$$

with T_α an $SU_c(3)$ generator and Einstein summation convention assumed (as usual) for the repeated indices $\alpha = 1, \cdots, 8$ and $b = 1, 2, 3$.

The 4-vector u^μ represents the heavy quark's 4-velocity, and it comes with a subscript Q to emphasize that this 4-velocity, in general, depends on which heavy quark is of interest. In particular, for a decay $Q \to Q'$ the rest frame of the initial and final heavy quarks differ due to the recoil against any other relativistic degrees of freedom participating in the decay. Notice that

$$\mathrm{i}\Psi_Q^\dagger u_Q \cdot D\Psi_Q = \mathrm{i}\gamma_Q \left[\Psi_Q^\dagger D_t \Psi_Q + \Psi_Q^\dagger \mathbf{v}_Q \cdot \mathbf{D}\Psi_Q\right], \tag{12.66}$$

with $\gamma_Q = (1 - v_Q^2)^{-1/2}$, and momentum conservation implies that the magnitude, v_Q, of the 3-velocity of the daughter heavy quark, Q', is of order $v_{Q'} \sim (m_Q - m_{Q'})/m_{Q'}$ in the rest frame of the decaying heavy quark. Consequently, v_Q need not be small in the heavy-quark limit because $m_Q - m_{Q'}$ need not be much smaller than $m_{Q'}$.

For mesons involving one heavy and one light quark interactions like[10] $(\Psi^\dagger \mathbf{D}^2 \Psi)/2m_Q$ are small relative to those shown in (12.62). This is because the energy and momentum transfer within such mesons between the heavy quark and the relativistic quarks and gluons is order Λ_{QCD}. As a consequence, the heavy quark recoil kinetic energy is order Λ_{QCD}^2/m_Q and therefore is negligible when $m_Q \gg \Lambda_{QCD}$.

This neglect of $1/m_Q$ interactions is important from the point of view of symmetries. This is particularly clear in the kinematic regime where $v_Q \to 0$, since in this limit the heavy-quark term becomes

$$\mathrm{i}\Psi^\dagger D_t \Psi, \tag{12.67}$$

where now all indices on Ψ involving spin i or flavour Q are suppressed. For N species of heavy quark this has a large symmetry group, $SU(2N)$, corresponding to unitary rotations acting on both the 'flavor' index $n = 1, \cdots, N$ and the 2-valued spin index i. The mixing of spin and flavor symmetries in this way is possible when $m_Q \to \infty$ because in this limit spin-orbit couplings disappear and spin essentially acts as just another internal symmetry.[11] For systems involving only c and b quarks the

[10] Here $\mathbf{D}$ represents the projection of D_μ in the direction perpendicular to u_Q^μ, and so is, in principle, also Q-dependent.

[11] For relativistic systems the *Coleman–Mandula* theorem [315] excludes similar symmetries that mix spin with internal symmetries. It states that (for systems with a nontrivial S matrix at least) under broad assumptions the most general symmetry algebra consists of the product of spacetime symmetries (*i.e.* ordinary Poincaré or conformal transformations) and internal symmetries.

appropriate value is $N = 2$, and so this suggests the heavy sector has an approximate $SU(4)$ symmetry that mixes the four flavour and spin states – $b_\uparrow$, $b_\downarrow$, $c_\uparrow$ and $c_\downarrow$ – into one another.

To display these symmetry implications more explicitly consider adding the low-energy weak interactions perturbatively, $\mathfrak{L}_{HQET} \to \mathfrak{L}_{HQET} + \mathfrak{L}_{\rm weak}$, and computing the rates for spinless B mesons (with quark content $B^- = b\,\overline{u}$, $\overline{B}^0_d = b\,\overline{d}$ or $\overline{B}^0_s = b\,\overline{s}$) to decay into spinless D mesons (with quark content $D^+ = c\,\overline{d}$ and $D^0 = c\,\overline{u}$). These decays arise due to the underlying heavy-quark decay through virtual W-boson emission, $b \to cW$ via the charged-current weak interactions in $\mathfrak{L}_{\rm weak}$.

To see what such a symmetry can imply, following [316] imagine computing the rate for the semi-leptonic decay, $\overline{B} \to D\ell^-\overline{\nu}$, where $\overline{B}$ and D are mesons containing heavy b and c quarks, respectively. Assuming the decay occurs due to the weak interaction responsible for $b \to cW$, evaluation of this rate requires computing the matrix element

$$\langle D(p')|\overline{c}\gamma^\mu b|\overline{B}(p)\rangle = f_+(q^2)\,(p+p')^\mu + f_-(q^2)\,(p-p')^\mu \\ \to \Big[f_+(q^2_{\rm mx})(m_B + m_D) + f_-(q^2_{\rm mx})(m_B - m_D)\Big]u^\mu, \tag{12.68}$$

where the first line uses only lorentz-invariance, the spinless character of the B and D mesons and the parity-invariance of the strong interactions. The 'form factors' $f_\pm = f_\pm(q^2)$ are functions of the Lorentz-invariant combination, $q^2 = -(p - p')^2$, and the axial current does not contribute to the left-hand side because of the parity invariance of the strong interactions. The second line, however, specializes to the kinematic regime where initial and final heavy quarks do not move relative to one another, and u^μ denotes the 4-velocity of their common rest frame. In this limit the leptons carry off as much momentum as they can, so $q^2 \to q^2_{\rm mx} = (m_B - m_D)^2$.

HQET makes a number of predictions for the otherwise unknown form factors, f_+ and f_-, and because HQET is a consistent EFT for a low-energy regime of QCD its predictions are also robust consequences of QCD itself up to small quantities like Λ_{QCD}/m_Q or α_s that are neglected in its derivation (but whose implications can, in principle, be included to any desired order).

The first prediction is a relation that must be satisfied by the functions $f_\pm$ when evaluated at $q^2 = q^2_{\rm mx}$ (for which both heavy quarks share the same rest frame – *i.e.* $v_Q \to 0$). The reasoning goes as follows. Charged-current weak decays are governed in the UV theory (above m_Q) at leading order by matrix elements like (12.68) of the operator $i\overline{q}'\gamma^\mu q$. After transitioning to the nonrelativistic EFT below m_Q, this operator matches at leading order to the operator $i\overline{q}'\gamma^\mu\Psi_Q$, and this can be run (using perturbative QCD in the leading-log approximation) down to $m_{Q'}$ to give the operator [317]

$$i\Psi^\dagger_{Q'}\gamma^\mu\Psi_Q = C_{QQ'}(\Psi^\dagger_{Q'}\Psi_Q)u^\mu + \cdots, \tag{12.69}$$

where

$$C_{QQ'} = \left[\frac{\alpha_s(m_Q)}{\alpha_s(m_{Q'})}\right]^{-6/25}, \tag{12.70}$$

with the power of 6/25 coming from the QCD beta function $6/(33 - 2n_q)$ evaluated with $n_q = 4$ quark flavours. Ellipses in Eq. (12.69) represent contributions that are suppressed relative to those shown by powers of Λ_{QCD}/m_Q and/or small couplings like α_s or α.

The factor $C_{QQ'}$ captures the renormalization of the quark current by gluon loops, and is simply multiplicative (rather than involving mixing with other effective operators, say) because the quantity of interest is a conserved Noether current for the $SU(2) \subset SU(4)$ symmetry of (12.62) that rotates the Q and Q' quarks (*i.e.* the b and c quarks) into one another without touching their spins. The Noether current for this symmetry computed from the action (12.62) is

$$j_\alpha^\mu = (\Psi^\dagger t_\alpha \Psi)u^\mu + \cdots , \qquad (12.71)$$

where t_α is an $SU(2)$ generator and (again) the ellipses denote terms suppressed by powers of Λ_{QCD}/m_Q or small couplings.

Evaluating (12.68) by computing the matrix element of the current (12.69) gives – keeping in mind (see §B.1) that the normalization of a relativistic state, $|\mathbf{p}\rangle$, differs from the normalization of nonrelativistic states by a factor of $\sqrt{2E_p}$ – then implies

$$\left[f_+(q_{mx}^2)(m_B + m_D) + f_-(q_{mx}^2)(m_B - m_D)\right]u^\mu = \sqrt{4m_B m_D}\, C_{BD}, \qquad (12.72)$$

up to order Λ_{QCD}/m_Q and α_s corrections. As above, $q_{mx}^2 = (m_B - m_D)^2$ denotes the maximum momentum transfer possible in the decay, and C_{BD} is given by (12.70), with $Q \to B$ and $Q' \to D$.

But there is also much more symmetry information in the low-energy HQET EFT, and this also constrains the form-factors for nonzero recoil 3-velocity v_Q (and so for $u^\mu \neq u^{\mu'}$ and $q^2 \neq q_{mx}^2$). In this case, the relevant information relates the form factors $f_\pm(q^2)$ to the form factors arising in the matrix elements measured in other reactions, such as

$$\langle D(p')|\bar{c}\gamma^\mu c|D(p)\rangle = f_D\,(p + p')^\mu \quad \text{and} \quad \langle \bar{B}(p')|\bar{b}\gamma^\mu b|\bar{B}(p)\rangle = f_B\,(p + p')^\mu, \qquad (12.73)$$

where f_D and f_B are again functions of $q^2 = -(p - p')^2$. All of these form factors can equivalently be regarded as functions of $u \cdot u'$ since

$$q_D^2 = -(p_D - p_D')^2 = 2m_D^2(1 + u \cdot u'), \quad q_B^2 = -(p_B - p_B')^2 = 2m_B^2(1 + u \cdot u')$$
$$\text{and} \quad q_{BD}^2 = -(p_B - p_D)^2 = (m_B - m_D)^2 + 2m_B m_D(1 + u \cdot u'). \qquad (12.74)$$

Now comes the main point. Because these currents are all related to combinations of the form $(\Psi^\dagger \Psi')u^\mu$, for Ψ and Ψ' representing either Ψ_B or Ψ_D it follows that all four form factors (at low velocities) are determined by one unknown function of $u \cdot u'$. That is,

$$f_B(q^2) = \xi(q^2), \quad f_D(q^2) = \xi(q^2 m_B^2/m_D^2) \qquad (12.75)$$

$$\text{and} \quad f_\pm(q^2) = \pm C_{BD}\left[\frac{m_B \pm m_D}{\sqrt{4m_B m_D}}\right]\xi[(q^2 - q_{mx}^2)(m_B/m_D)],$$

for a single function $\xi(q^2)$, at least to leading order in Λ_{QCD}/m_c and Λ_{QCD}/m_b and dropping powers of $\alpha_s(m_c)$ that are not pre-multiplied by $\ln(m_b/m_c)$. Furthermore, direct evaluation shows that $\xi(0) = 1$, and the robustness of this result can be understood as a general consequence of the fact that the operators involved are conserved currents in the infinite-mass limit. Notice, in particular, that using $\xi(0) = 1$ in (12.75) implies (12.72), as it must.

What is important here is that these results are *exact* in the non-perturbative strong interactions at or below Λ_{QCD} since it is a consequence of general symmetry properties in the heavy-quark limit.

12.2.4 Particle-Antiparticle Systems

The picture given above treats nonrelativistic EFTs as being obtained by integrating out antiparticles (or particles), with the idea that heavy particles could still appear in the low-energy effective theory provided they are stable and so cannot release the energy tied up in their rest mass (and thereby ruin the low-energy approximation). Since particle-antiparticle annihilation also liberates the energy locked up in the rest mass, the examples considered up to this point involve only heavy particles or heavy antiparticles, but not both.

This section steps beyond this framework by examining systems involving *both* heavy particles and heavy antiparticles. This can be consistent with the existence of a sensible low-energy limit so long as rates for heavy-particle decay or annihilation are sufficiently slow. In this case, EFT methods can be appropriate so long as questions are only asked about physics before any decays or annihilations occur, but not afterwards.

NRQED and Positronium

Positronium provides a simple example of this type, since it is the electromagnetic bound state of an electron and its antiparticle the positron. Once captured into ordinary matter, positrons produced in experiments often pair off with electrons into bound states after which the electron–positron pair eventually annihilates. But the lifetime for annihilation proves to be long enough that hydrogen-like energy levels form whose properties can be accurately measured. Because these systems are so theoretically simple, considerable effort is made computing their properties, in order to perform high-precision comparisons between experiments and calculations.

The effective theory useful for studying these energy levels is again NRQED [294], but specialized to two kinds of nonrelativistic fields, with one (Ψ) destroying the nonrelativistic electron and another (Φ) that destroys the nonrelativistic positron [318, 319]. NRQED facilitates computing radiative corrections (such as the Lamb shift) in positronium by efficiently separating the high-energy radiative corrections – that enter in the low-energy theory through higher-order matching determinations of effective couplings, as in (12.1.2) – from low-energy bound-state physics – described in the low-energy theory using well-tested Schrödinger–Coulomb techniques.

The EFT is then constructed along the lines given in Eqs. (12.37)–(12.39) (and higher orders), specialized to the case $Ze = e$ and $M = m$ (and the resulting simplification that higher dimensions are suppressed only by a single suppression scale: m^{-1}). The leading terms are given by

$$\mathfrak{L}_0 = i\,\Psi^\dagger \partial_t \Psi + i\,\Phi^\dagger \partial_t \Phi + eA_0(\Phi^\dagger\Phi - \Psi^\dagger\Psi) + \frac{1}{2}\,\mathbf{E}^2 - \frac{1}{2}\,\mathbf{B}^2, \tag{12.76}$$

while the leading subdominant interactions become

$$\mathfrak{L}_1 = \frac{1}{2m}\Psi^\dagger\nabla^2\Psi + \frac{1}{2m}\Phi^\dagger\nabla^2\Phi - \frac{e^2}{2m}\mathbf{A}^2(\Phi^\dagger\Phi + \Psi^\dagger\Psi) + \frac{e}{2m}\,c_F\,\mathbf{B}\cdot(\Phi^\dagger\boldsymbol{\sigma}\Phi - \Psi^\dagger\boldsymbol{\sigma}\Psi)$$
$$+ \frac{ie}{2m}\mathbf{A}\cdot\left[(\nabla\Phi^\dagger)\Phi - \Phi^\dagger\nabla\Phi - (\nabla\Psi^\dagger)\Psi + \Psi^\dagger\nabla\Psi\right] + \cdots. \tag{12.77}$$

For later purposes the $1/m^2$ interactions are also required, and generalize those of (12.3) to include terms involving both Ψ and Φ, specialized to $e_q = \pm e$.

$$\mathfrak{L}_2 = \frac{e}{8m^2}\,c_D(\Phi^\dagger\Phi - \Psi^\dagger\Psi)(\nabla\cdot\mathbf{E}) + \frac{c_U}{m^2}\,F^{\mu\nu}\Box F_{\mu\nu}$$
$$- \frac{ie}{8m^2}\,c_S\left[\Phi^\dagger\boldsymbol{\sigma}\cdot\left(\mathbf{D}\times\mathbf{E} - \mathbf{E}\times\mathbf{D}\right)\Phi - \Psi^\dagger\boldsymbol{\sigma}\cdot\left(\mathbf{D}\times\mathbf{E} - \mathbf{E}\times\mathbf{D}\right)\Psi\right] \tag{12.78}$$
$$+ \frac{d_v}{m^2}(\Psi^\dagger\boldsymbol{\sigma}\Psi)\cdot(\Phi^\dagger\boldsymbol{\sigma}\Phi) + \frac{d_s}{m^2}(\Psi^\dagger\Psi)(\Phi^\dagger\Phi) + \mathfrak{L}_{\text{contact}},$$

where $\mathfrak{L}_{\text{contact}}$ involves four-fermion terms that involve either four powers of Ψ or four powers of Φ. These interactions are not written explicitly because they are not used in what follows. Other interactions like $(\Psi^\dagger\Phi)(\Phi^\dagger\Psi)$ are not independent of the ones shown, as can be seen using a 'Fierz' spinor identity [320].[12]

The purely electromagnetic interaction of (12.78) (with coefficient c_U, called the Uehling term) is often not written since – as described surrounding Eq. (7.15) – it is redundant inasmuch as it can be traded for a particular combination of 4-fermion couplings using a field redefinition. This is not done here because it proves simpler to keep it rather than track the additional 4-fermion terms in $\mathfrak{L}_{\text{contact}}$.

So far as power counting is concerned, for bound e^+e^- states the energies and momenta of interest are of order $E \sim |\mathbf{p}|^2/m \sim mv^2$ with $v \sim \alpha \ll 1$. This makes it impossible to neglect the interactions $(\Psi^\dagger\nabla^2\Psi)/2m$ and $(\Phi^\dagger\nabla^2\Phi)/2m$ relative to $i\Psi^\dagger\partial_t\Psi$ and $i\Phi^\dagger\partial_t\Phi$.

Because both electron and positron are elementary, their various effective couplings are obtained by matching along the lines of §12.1.2. For vertex corrections this leads to the results of (12.1.2), which are repeated for convenience here (specialized to $e_q = \pm e$)

$$c_F = 1 + \frac{\alpha}{2\pi} + O(\alpha^2), \quad c_D = 1 + \frac{8\alpha}{6\pi}\ln\frac{m^2}{\mu^2} + O(\alpha^2) \quad c_S = 1 + \frac{\alpha}{\pi} + O(\alpha^2), \tag{12.79}$$

together with the contribution of the QED vacuum polarization graph – Fig. 7.4 – to c_U

$$c_U = \frac{\alpha}{60\pi} + O(\alpha^2). \tag{12.80}$$

For positronium it is also necessary to perform a matching calculation for the 4-fermion couplings d_v and d_s. These are obtained by matching tree- and one-loop-4-fermion graphs in QED to the graphs in NRQED at the same order, also computed in dimensional regularization with minimal subtraction. The leading contributions arise by demanding NRQED reproduce the tree graph of Fig. 12.4.

[12] A Fierz identity is obtained by decomposing the dyadic matrix $\Phi\Phi^\dagger$ in terms of the unit and Pauli matrices as follows: $\Phi\Phi^\dagger = -\frac{1}{2}(\Phi^\dagger\Phi)I - \frac{1}{2}(\Phi^\dagger\boldsymbol{\sigma}\Phi)\cdot\boldsymbol{\sigma}$.

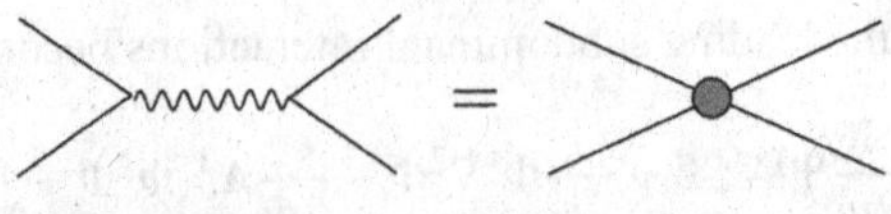

Fig. 12.4 The tree graphs whose matching determine d_s and d_v to $\mathcal{O}(\alpha)$. All graphs are evaluated for scattering nearly at threshold, with the ones on the left evaluated in QED and the ones on the right in NRQED.

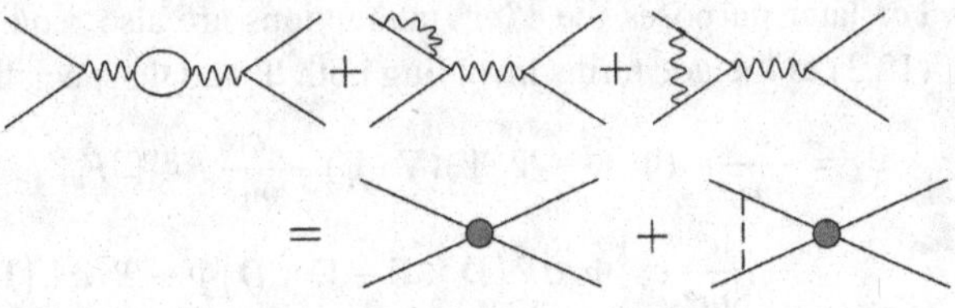

Fig. 12.5 Loop corrections to one-photon exchange graphs whose matching contributes to d_s and d_v at $\mathcal{O}(\alpha^2)$. Dashed lines on the NRQED (*i.e.* right-hand) side represent 'Coulomb' A_0 exchange.

Fig. 12.6 Diagrams whose matching contributes the two-photon annihilation contributions (and imaginary parts) to d_s and d_v.

The effects of the tree-level s-channel annihilation graph in QED must be reproduced by an effective interaction in NRQED because the exchanged virtual photon necessarily has four-momenta of order m, and so does not appear in the low-energy theory. Although the t-channel photon-exchange graph also contributes to electron–positron collisions at tree level in QED, the energy of the exchanged photon is well below the electron mass and so is described by the same t-channel graph within NRQED. Consequently, tree-level t-channel photon exchange cancels once low and high-energies are compared, and does not contribute to d_s or d_v in the matching process. The leading-order matching results obtained by evaluating this graph give [294]

$$d_s = \frac{3\pi\alpha}{2} \quad \text{and} \quad d_v = -\frac{\pi\alpha}{2}. \tag{12.81}$$

A qualitatively new feature emerges once this matching is performed at subdominant order in α. This is achieved by demanding NRQED reproduce the same scattering amplitude as do the graphs of Figs. 12.5–12.7 computed for energies at threshold (*i.e.* with external particles essentially at rest). Comparing results implies that QED is reproduced properly only if Eq. (12.81) is generalized to [318, 321, 322]:

$$\begin{aligned} d_s &= \frac{3\pi\alpha}{2} - \alpha^2\left[\ln\frac{m^2}{\mu^2} + \frac{23}{3} - \ln 2 + \frac{i\pi}{2}\right] + O(\alpha^3) \\ d_v &= -\frac{\pi\alpha}{2} + \alpha^2\left[\frac{22}{9} + \ln 2 - \frac{i\pi}{2}\right] + O(\alpha^3). \end{aligned} \tag{12.82}$$

The qualitatively new feature here is the imaginary part that d_s and d_v acquire at $O(\alpha^2)$, whose origin traces back to matching with the graphs of Fig. 12.6. On the QED side these particular graphs develop imaginary parts due to unitarity [39], because they incorporate the physics of electron–positron annihilation.

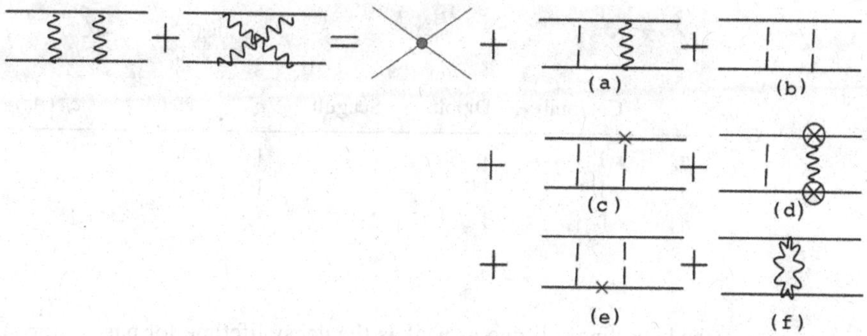

Fig. 12.7 One-loop t-channel matching diagrams that contribute to d_s and d_v to $\mathcal{O}(\alpha^2)$. Vertices and self-energy insertions marked with crosses represent terms in NRQED that are subdominant in $1/m$. Dashed and wavy lines on the right-hand (NRQED) side are, respectively, Coulomb gauge A_0 and $\mathbf{A}$ propagators. For brevity's sake not all of the time-orderings of the $\mathbf{A}$ propagator are explicitly drawn.

This annihilation contributes in the effective theory to local 4-Fermi contact interactions because annihilation occurs only when electron and positron are within of order a Compton wavelength, m^{-1}, of one another.

Because these annihilation contributions to d_s and d_v are not real, the lagrangian density is also not real, $\mathfrak{L}_{\rm eff} \neq \mathfrak{L}^*_{\rm eff}$, and so the corresponding Hamiltonian is not hermitian. Having $H_{\rm eff} \neq H^*_{\rm eff}$ implies that $U(t) = \exp[-iH_{\rm eff}t]$ is not unitary within the effective theory, reflecting the loss of probability from the low-energy electron-positron sector as an electron and positron occasionally annihilate one another. From the point of view of the low-energy theory annihilation looks like a loss of probability because the photons produced by annihilation are not low-energy states and so are not present in the EFT.

In particular, the non-hermitian operators describing annihilation contribute imaginary parts into positronium energy levels, $E_n = E^R_n - iE^I_n$, and so the time-evolution $\psi(t) \propto e^{-iE_nt}$ of an energy eigenmode implies that the probability density falls exponentially, $\psi^*\psi \propto e^{-\Gamma_nt}$, showing that the energy eigenvalue and decay rate, Γ_n, are related by $E^I_n = \Gamma_n/2 \geq 0$.

As usual, the good news is that the matching calculation is the only place where the full complexity of QED enters, and matching can be done using the simplest possible scattering process, and in particular need not involve any bound states at all. But once known, the effective couplings of the NRQED lagrangian can be used to calculate *any* observable in the low-energy theory, including properties of bound states. This sequestering of issues lies at the heart of NRQED's simplicity, for instance allowing different gauges to be used in the relativistic and nonrelativistic parts of the calculation if convenient (assuming one matches using gauge-invariant quantities). This permits the convenience of using a covariant gauge, like Feynman gauge, in the QED part of the calculation while keeping Coulomb gauge for bound-state calculations.

Positronium Decay and Hyperfine Structure

This section follows [318] and uses the previous matching arguments in an illustrative calculation: computing both the leading and next-to-leading contributions to

Table 12.1 Powers of $e^{n_e}v^{n_v}$ (and of α^{n_α} when $e^2 \sim v \sim \alpha$) appearing in leading effective couplings

	Coulomb	Dipole	Seagull	c_F	$c_S\,(A_0)$	$c_D\,(A_0)$	d_s	d_v
n_e	1	1	2	1	1	1	2	2
n_v	−1/2	1/2	1	1/2	3/2	3/2	1	1
n_α	0	1	2	1	2	2	2	2

the hyperfine splitting as well as the decay lifetime for para-positronium. Along the way this example also provides a concrete illustration of how to use the ultra-soft power-counting rule.

Hyperfine splitting is the energy difference between positronium levels for states that differ only in the combined spin of the electron–positron system, which can be either $S = 0$ or $S = 1$. It provides a convenient example partly because it is complicated enough to show the merits of the method, but also because it is relatively simple: only spin-dependent energy shifts need be considered.

The first step (as always) is to use power counting to identify which interactions within the EFT contribute to observables at any particular order. Because the application of interest involves the positronium bound state, the relevant power-counting analysis from §11.3 is the ultra-soft regime for which $p \sim mv$ and $E \sim mv^2$ with $v \sim \alpha$. In the present instance this requires working non-perturbatively in the Coulomb interaction, which – as described above – amounts to perturbing about the Schrödinger–Coulomb system (*i.e.* using Coulomb wave-functions).

Consider computing a graph with two external electron and two external positron lines, such as would arise when computing a time-ordered correlation function of the form $\langle T\,[\Psi^*_{i_1}(x_1)\,\Phi^*_{i_2}(x_2)\,\Psi_{i_3}(x_3)\,\Phi_{i_4}(x_4)]\rangle$, where i_k are two-component spinor indices. On general grounds [323] positronium bound-state energies can be extracted by taking $x^\mu_1 = x^\mu_2 = 0$ and $x^0_3 \to x^0_4 = x^0$, say, and Fourier transforming the variables x^0 and $\mathbf{x} = \mathbf{x}_3 - \mathbf{x}_4$ (see §C.7 for details). Positronium energies appear as the positions of poles in Fourier space for such an object.

The Feynman rules used to evaluate the graphical representation of such a correlation function are taken from the NRQED lagrangian given in Eqs. (12.76), (12.77) and (12.78) (and so on, to higher orders). According to the ultra-soft power-counting rules built using the rescaled lagrangian (11.43), each vertex in the graph comes with a factor $e^{n_e}v^{n_v}$, with values for n_e and n_v given in Table 12.1 for the lowest-dimension interactions.[13]

Once evaluated at $v \sim \alpha$ (as appropriate for bound states) any graph built from these vertices contributes a total power of α to bound state energies that are (modulo logarithms of α) $\delta E \sim m\,\alpha^{p_E}$, where

$$p_E = 2 + \kappa_v + \frac{\kappa_e}{2} - L. \tag{12.83}$$

[13] In this table 'dipole' denotes the $\mathbf{A}\cdot\Psi^*\nabla\Psi$ interaction, while 'seagull' denotes the $\mathbf{A}^2\Psi^*\Psi$ term and the powers are quoted for the Coulomb field $\nabla A_0 \subset \mathbf{E}$ for the spin-orbit and Darwin terms. For d_s and d_v the quoted power of e uses the result of (12.81) that the leading coupling arises at order α.

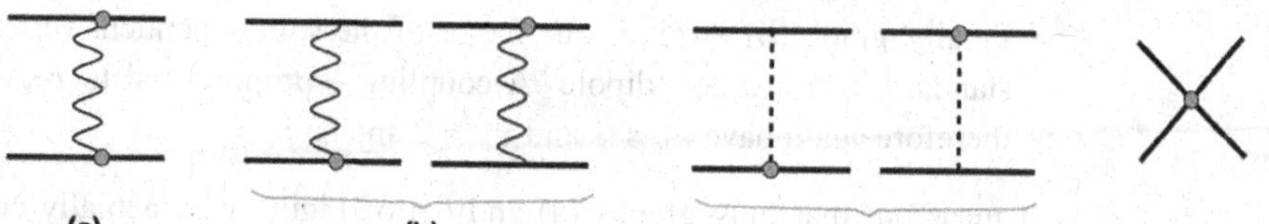

Fig. 12.8 The NRQED graphs contributing to the hyperfine structure at order $m\alpha^4$ (and order $m\alpha^5$). The fat vertex in graphs (a) and (b) represents c_F, and is c_S in graphs (c). The contact interaction in (d) involves d_s and d_v.

Here, the initial 2 comes because all energies are order $mv^2 \propto m\alpha^2$ once one converts to ordinary units from the rescaled time, $\hat{t}$, of (11.43). $\kappa_v = \sum n_v$ and $\kappa_e = \sum n_e$ represent the total power of e and v contributed by all the graph's vertices and L counts the number of loops containing **A** particle lines, each of whose energy integrations contributes a factor of $1/v$ because of the pole at $k^0 = |\mathbf{k}|/v$ that arises due to the v-dependence of the vector-potential's propagator (as given in (11.45)). Although the quantity L enters this expression with a negative sign, increasing L by inserting additional interactions also involves adding sufficiently many new vertices to ensure that the net contribution to p increases, giving a net suppression by positive powers of v.

Eq. (12.83) confirms that repeated insertions of the Coulomb interaction – each of which increases κ_e by +1 and decreases κ_v by $\frac{1}{2}$ while leaving L unchanged – are not suppressed in the ultrasoft regime, forcing its non-perturbative resummation (as described above) using Coulomb–Schrödinger wave functions. Adding any other interaction, however, increases the value of p_E.

For applications to hyperfine splitting it suffices to restrict to graphs for which at least one of the vertices involves a spin-dependent coupling; *i.e.* at least one involvement of c_F, c_S or a combination of the contact couplings d_s and d_v. Since the leading contributions have $L = 0$ they involve only a single instantaneous interaction. The graphs with $L = 0$ contributing the smallest value for p_E, for which at least one vertex is spin-dependent, are displayed in Fig. 12.8. Consider each of these in turn.

- Simplest to understand are the graphs in figure (d), involving the four-fermi couplings d_s and d_v. Since $d_{s,v} \sim \alpha v$ this graph has $\kappa_v = 1$ and $\kappa_e = 2$, which using (12.83) implies that $p_E = 4$ and so $\delta E \sim m\alpha^4$. Indeed, an energy shift this large is intuitively easy to understand: because these are contact interactions, at leading order they shift the energy by an amount proportional to the wave-function at the origin,

$$\delta E \sim d_{s,v}|\psi(0)|^2 \propto \frac{\alpha}{m^2}(\alpha m)^3 \sim m\,\alpha^4. \tag{12.84}$$

- Similarly, the graphs in (c) contain one Coulomb interaction proportional to $e/\sqrt{v}$ and one spin-dependent spin-orbit interaction proportional to $c_S e v^{3/2}$. Since matching implies that $c_S \sim O(1)$ this graph contributes with $\kappa_v = 1$ and $\kappa_e = 2$, and so again $p_E = 4$.
- The graph labelled (a) is quadratic in the spin-dependent vertex c_F, and so is proportional $(e\sqrt{v}\,c_F)^2$. Since matching gives $c_F \sim O(1)$ this graph also gives $\kappa_v = 1$ and $\kappa_e = 2$, leading again to $p_E = 4$.

- Finally, graph (b) involves one factor of the spin-dependent vertex $e\sqrt{v}\,c_F$ and a standard lowest-order 'dipole' **A** coupling – proportional to $e\sqrt{v}$. These graphs therefore again have $\kappa_v = 1$ and $\kappa_e = 2$ and so $p_E = 4$.

It turns out that only graphs (a) and (d) of Figure 12.8 actually contribute to the hyperfine splitting of the ground state, since the others vanish when evaluated in an s-wave configuration. This is because the other two graphs always contain only one vector, σ, which has no other vector to contract with once evaluated in a state with $\ell = 0$.

To illustrate the power of EFT methods next ask: what provides the leading subdominant contribution to hyperfine splitting? Additional exchange of **A** inevitably causes $L > 0$; how large are these contributions? Having $L = 1$ requires at least one interaction more than the diagrams shown in Figure 12.8. As is easily checked, all such additional interactions (except the Coulomb interaction) increase the combination $\kappa_v + \kappa_e/2$ by at least 2, and so first contribute to δE at order $m\,\alpha^6$. For example, adding a transverse photon coupled to two dipole vertices introduces the square of the coefficient $e\sqrt{v}$ and so increases κ_v by 1 and κ_e by 2. Similarly, adding another Coulomb photon, with a Coulomb interaction of order $e/\sqrt{v}$ at one end and a Darwin vertex, $ec_D v^{3/2}$, at the other, again increases κ_v by 1 and κ_e by 2. Alternatively, adding a relativistic kinetic vertex $\sim \Psi^\dagger \nabla^4 \Psi$ doesn't change κ_e at all but increases κ_v by 2 (because of the two extra spatial derivatives). The upshot is this: all graphs other than those given in Fig. 12.8 contribute at best $\delta E \sim m\,\alpha^6$.

This doesn't mean there are no $O(m\,\alpha^5)$ contributions to the hyperfine structure. What it means is that all $O(m\,\alpha^5)$ contributions come from the *same* graphs – those of Fig. 12.8 – but using the effective couplings that are matched to next-to-leading order (*i.e.* using the matching to the precision given explicitly in Eqs. (12.79) and (12.82).

To compute the hyperfine splitting in detail one evaluates the graphs of Fig. 12.8 explicitly (see §C.7 for details of how to extract δE from these graphs). Because $L = 0$ there is no integration over photon energy and all of the graphs involve instantaneous interactions. Their contribution to the energy shift amounts to stripping off the external fermion lines and treating the rest of the graph as if it were a perturbing interaction in old-fashioned Rayleigh–Schrödinger perturbation theory, using Schrödinger–Coulomb wave-functions, $\psi(\mathbf{p})$, to resum multiple insertions of the Coulomb interaction.

For simplicity only the result for s-wave states is given explicitly here, for which only graphs (a) and (d) need be evaluated. This gives

$$\begin{aligned}
\delta E_n(\mathrm{a}) &= c_F^2 \int \frac{\mathrm{d}^3\mathbf{p}\,\mathrm{d}^3\mathbf{k}}{(2\pi)^3}\,\psi_n^*(\mathbf{p}) \left[\frac{-ie(\mathbf{p}-\mathbf{k})\times\sigma_1}{2m}\right]_i \left[\frac{-ie((\mathbf{p}-\mathbf{k})\times\sigma_2}{2m}\right]_j \\
&\quad \times \left[-\frac{1}{(\mathbf{p}-\mathbf{k})^2}\right]\left[\delta_{ij} - \frac{(\mathbf{p}-\mathbf{k})_i(\mathbf{p}-\mathbf{k})_j}{(\mathbf{p}-\mathbf{k})^2}\right]\psi_n(\mathbf{k}) \\
&= \frac{2\pi\alpha c_F^2}{3m^2}\langle \sigma_1\cdot\sigma_2\rangle |\psi_n(0)|^2 = \frac{m\alpha^4 c_F^2}{6n^3}\left[S(S+1) - \frac{3}{2}\right] \quad (s\text{-wave}),
\end{aligned} \tag{12.85}$$

where n is the principal quantum number and $S = 0, 1$ denotes the total intrinsic spin of the electron–positron system. Similarly, evaluating graph (d) gives

$$\delta E_n(\mathrm{d}) = \frac{m\,\alpha^3}{8\pi n^3}\Big[d_s + d_v\big(3 - 2S(S+1)\big)\Big] \quad (s\text{-wave}). \tag{12.86}$$

Summing these results gives the following expression

$$\delta E_n(\ell = 0) = \frac{m\,\alpha^3}{2\pi n^3}\left[\frac{1}{4}\left(d_s + 3d_v - 2\pi\alpha c_F^2\right) + \left(\frac{\pi\alpha c_F^2}{3} - \frac{d_v}{2}\right)S(S+1)\right], \tag{12.87}$$

for the spin-dependent part of the positronium s-wave energy shifts.

Using $\mathrm{Im}\ c_F = 0$ and $\mathrm{Im}\ d_s = \mathrm{Im}\ d_v = -\frac{1}{2}\pi\alpha^2$ – *c.f.* (12.82) – together with the formula $\Gamma_n = -2\ \mathrm{Im}\ E_n$ – these expressions give the well-known result for the $O(m\,\alpha^5)$ decay rate of the s-wave state of para-positronium (for which $S = 0$) [324]:

$$\Gamma_n(\ell = S = 0) = -\frac{m\,\alpha^3}{4\pi n^3}\,\mathrm{Im}\left(d_s + 3d_v\right) = \frac{m\,\alpha^5}{2n^3}. \tag{12.88}$$

The corresponding result for ortho-positronium ($S = 1$) vanishes to this order because it is proportional to $\mathrm{Im}\ (d_v - d_s) = 0$. This is also standard: charge-conjugation (C) invariance forbids an $\ell = 0$ and $S = 1$ state from annihilating into fewer than three-photons [325].

For the hyperfine splitting itself the imaginary part of δE_n can be dropped. Subtracting the result for $S = 0$ from that for $S = 1$ in (12.87) and using the matching conditions of Eqs. (12.79) and (12.82) in the result gives:

$$\begin{aligned}\Delta E_{\rm hfs}(\ell = 0) &:= \delta E_n(S = 1) - \delta E_n(S = 0) = \frac{m\,\alpha^3}{n^3}\left(\frac{\alpha c_F^2}{3} - \frac{d_v}{2\pi}\right) \\ &= \frac{m\alpha^4}{2n^3}\left[\frac{7}{6} - \frac{\alpha}{\pi}\left(\ln 2 + \frac{16}{9}\right)\right],\end{aligned} \tag{12.89}$$

reproducing $O(m\alpha^4)$ and $O(m\alpha^5)$ QED calculations [326] (though with much less effort). The hyperfine splitting for arbitrary ℓ is similarly computed to this order by evaluating the graphs of Fig. 12.8 without the restriction to s-wave Schrödinger–Coulomb states.

To summarize, the calculation of hyperfine splitting to next-to-leading order is hardly more difficult to obtain within NRQED than is the leading-order result, and both are much more simply obtained than is the case when a fully relativistic treatment of positronium is performed in QED without first separating scales. The only real effort required is to obtain the higher-order matching of all spin-dependent effective couplings, but this can be done for scattering in a convenient kinematic regime without any bound-state complications.

NRQCD and Quarkonia

QCD contains a close counterpart to the previous positron example, consisting of 'quarkonium' mesons involving a heavy-quark and its antiparticle: $\overline{Q}Q$. For such systems the Coulombic energy levels are of order $m_Q\alpha_s^2$ where $\alpha_s = \alpha_s(\mu = m_Q)$ is the QCD coupling evaluated at the relevant quark mass. For $m_Q \gg \Lambda_{QCD} \sim 200$ MeV the QCD coupling is weak and so the Coulomb energy satisfies the hierarchy $\Lambda_{QCD} \ll m_Q\alpha_s^2 \ll m_Q$, and so lends itself to an analysis within an EFT for scales below m_Q, involving a nonrelativistic quark-antiquark pair interacting with relativistic gluons and light quarks. The effective theory that results is called Nonrelativistic Quantum Chromodynamics, or NRQCD for short [294, 327].

This treatment resembles NRQED, with the heavy quark described by a nonrelativistic colour-triplet field, Ψ_Q^a, and a nonrelativistic colour-antitriplet field, $\Phi_{\overline{Q}a}$, representing the heavy antiquark. To these are coupled the various relativistic degrees of freedom, including the gluons and light quarks of QCD. The purpose of writing this EFT lies in its ability to isolate the Coulomb interaction of QCD as being the dominant contribution to the heavy meson binding energy, with the strong low-energy QCD interactions being systematically suppressed by powers of Λ_{QCD}/m_Q.

There are two key differences between the NRQCD lagrangian and the HQET effective theory described earlier. A fairly trivial difference is the inclusion of fields for both the heavy quark and heavy antiquark, similar to what is encountered in NRQED above. But a more important difference lies in the parts of the lagrangian that are chosen as the 'unperturbed' part, about which the perturbative expansion is ultimately performed. For HQET the heavy-quark (or antiquark) kinetic term, $(\Psi_Q^\dagger \mathbf{D}^2\Psi)/2m_Q$, is negligible because the bound state energies and momenta are both of order Λ_{QCD}, making heavy-quark recoil effects of order Λ_{QCD}^2/m_Q and so perturbatively small. But for NRQCD these same terms are *not* perturbative because the momenta of interest are order $\alpha_s m_Q$, while bound-state energies are order $\alpha_s^2 m_Q$. The physics of NRQCD is very rich, but a full description lies beyond the scope of this book.

12.3 Neutral Systems

The previous sections develop the EFT for slowly moving electrically charged objects interacting with electromagnetic fields. This section turns to the low-energy description of how electromagnetic fields interact with electrically neutral objects. These interactions need not be trivial since the neutral objects might themselves be constructed from smaller charged particles (as indeed are neutral atoms and most other macroscopic neutral objects).

An EFT description aims to identify the kinds of interactions that dominate when electromagnetic fields vary only over length scales much larger than the size of the composite neutral bodies. In this long-distance, low-energy regime interactions might be expected to be captured by a multipole expansion, and this is indeed part of the story. But it is not the whole story, and part of this section's purpose is to identify how the multipole expansion fits into the broader EFT context.

12.3.1 Polarizability and Rayleigh Scattering

Consider first the simplest case where the neutral particle is spinless. In this case, the only dynamical degree of freedom appearing in the low-energy theory is the particle's centre-of-mass motion, and the field $\Phi(\mathbf{x}, t)$ is a rotational scalar. Since the weak interactions usually play a negligible role in a particle's substructure, the Wilson action describing its low-energy properties can be required to be invariant under rotations, translations, gauge invariance, discrete parity and time-reversal transformations. To these should also be added a global rephasing symmetry like $\Phi \to e^{i\theta}\Phi$ (in practical examples perhaps the symmetry responsible for conservation of baryon number) that is responsible for the nonrelativistic particle's stability (or approximate stability) and so also its presence at low energies.

The lowest-dimension lagrangian density describing interactions between the Φ and electromagnetic fields satisfying all of these properties has the form $\mathfrak{L}_{\rm neut} = \mathfrak{L}_{n2} + \mathfrak{L}_{n3} + \cdots$, with $\mathfrak{L}_{nk}$ having effective couplings with dimensions (length)^k. For neutral objects like atoms, whose size is much larger than their inverse mass – *i.e.* $R \gg 1/m$ – these couplings would naturally be of order R^k in size.

The first interactions arise at order (length)^2, and are given by[14]

$$\mathfrak{L}_{n2} = \mathfrak{g}\,\Phi^*\Phi\,\nabla\cdot\mathbf{E} + \mathfrak{h}\,(\Phi^*\Phi)^2. \tag{12.90}$$

These parallel the Darwin term and spinless two-body interactions of (12.3) for charged particles, but different notation is used here for the couplings $\mathfrak{g}$ and $\mathfrak{h}$ to emphasize that their natural scale is R^2 rather than e_q/m^2.

As usual – see the discussion below (12.3) – in the absence of other particles, one combination of the two couplings in (12.90) is redundant, and so either one can be dropped in favour of the other. Both are kept here for later comparison with §13.3.4, in which other fields are also present. In these later applications the couplings of (12.90) turn out to describe the mean-square electromagnetic charge radius of the neutral object in question: the radius as measured using electromagnetic probes.

At order (length)^3 a number of possible couplings arise, of which the ones having precisely two powers of electromagnetic fields are of particular interest:

$$\mathfrak{L}_{n3} = \frac{1}{2}\Phi^*\Phi\left(\mathfrak{p}_E\,\mathbf{E}^2 - \mathfrak{p}_B\,\mathbf{B}^2\right) + \text{(linear in EM fields)}. \tag{12.91}$$

The couplings $\mathfrak{p}_E$ and $\mathfrak{p}_B$ are called electric and magnetic 'polarizabilities' and are later shown – *c.f.* section §16.3.3 – to arise at second order from microscopic dipole couplings (of the form also described in §12.2.2) to the constituent charged particles inside the neutral object of interest. The interactions of Eq. (12.91) are particularly interesting because they often control how neutral particles scatter long-wavelength electromagnetic waves.

Rayleigh Scattering

As an application of the EFT for neutral particles interacting with electromagnetism, consider the cross section for scattering long-wavelength electromagnetic waves (*i.e.* those with wavelengths much larger than the particle size, R) from spinless neutral particles. Low-energy scattering from small neutral polarizable objects is called 'Rayleigh scattering'. It is this kind of scattering, applied to small

[14] Terms of the form $\mathbf{j}\cdot\mathbf{E}$ or $\mathbf{j}\cdot\mathbf{B}$, with $\mathbf{j} = i(\Phi^*\nabla\Phi - \nabla\Phi^*\Phi)$, are not included in $\mathfrak{L}_{n2}$ because they, respectively, violate time-reversal or parity invariance (see §C.4.3 for the relevant transformation properties).

particles suspended in the atmosphere, that ultimately explains why the sky is both blue and polarized.

To lowest order in powers of kR, where k is the photon frequency, this scattering arises from the polarizability terms of (12.91). These dominate because the linear 'Darwin' coupling of (12.90) vanishes for radiation – because $\mathbf{k} \cdot \boldsymbol{\epsilon}(\mathbf{k}) = 0$ – and because scattering requires two factors of electromagnetic fields, so the terms linear in electromagnetic fields in (12.91) only contribute at second order. The six powers of the microscopic scale R that this implies are therefore subdominant to the three powers appearing in the couplings $\mathfrak{p}_E$ and $\mathfrak{p}_B$.

For composite objects built from charged constituents (as opposed to magnetic materials) it is the electric polarizability that is more important for photon scattering. Putting in the order-unity factors, taking the matrix element of $\mathfrak{L}_{n3}$, and restricting to electric polarizability, leads to the invariant amplitude (see §B.2)

$$\mathcal{A}_{2,2}[\Phi(p) + \gamma(k) \to \Phi(\tilde{p}) + \gamma(\tilde{k})] = i\,\mathfrak{p}_E\, k\, \tilde{k}\, \tilde{\boldsymbol{\epsilon}} \cdot \boldsymbol{\epsilon}, \tag{12.92}$$

where $\boldsymbol{\epsilon}=\boldsymbol{\epsilon}(\mathbf{k}, \lambda)$ and $\tilde{\boldsymbol{\epsilon}}=\boldsymbol{\epsilon}(\tilde{\mathbf{k}}, \tilde{\lambda})$ are the polarization vectors for the initial and final photons (with helicity $\lambda, \tilde{\lambda} = \pm 1$), while $k = |\mathbf{k}|$ and $\tilde{k} = |\tilde{\mathbf{k}}|$.

Averaging over the initial polarization, λ, and summing over the final spin, $\tilde{\lambda}$, gives the unpolarized squared matrix element (see Exercise 12.1)

$$\langle|\mathcal{A}_{2,2}|^2\rangle := \frac{1}{2}\sum_{\lambda\tilde{\lambda}} |\mathcal{A}_{2,2}|^2 = \frac{k^2\tilde{k}^2\mathfrak{p}_E^2}{2}\sum_{\lambda\tilde{\lambda}} |\tilde{\boldsymbol{\epsilon}} \cdot \boldsymbol{\epsilon}|^2 = \frac{k^2\tilde{k}^2\mathfrak{p}_E^2}{2}(1 + \cos^2\theta), \tag{12.93}$$

where θ is the scattering angle between the momenta of the incoming photon, $\mathbf{k}$, and the outgoing one, $\tilde{\mathbf{k}}$, as measured in the (lab) rest-frame of the initial heavy particle. The lab-frame differential cross section then is

$$\frac{d\sigma_R}{d\Omega} = \frac{1}{(4\pi)^2}\langle|\mathcal{A}_{2,2}|^2\rangle = \frac{\mathfrak{p}_E^2 k^4}{32\pi^2}(1 + \cos^2\theta), \tag{12.94}$$

leading to the following expression for the total unpolarized Raleigh cross section,

$$\sigma_R = \frac{\mathfrak{p}_E^2 k^4}{6\pi}, \tag{12.95}$$

for light with wavelength much larger than the scattered object's size.

As shown in more detail in §16.3.3, for scattering from neutral atoms the natural scale for the polarizabilities is given by $|\mathbf{d}_{ij}|^2/\Delta E$, where $|\mathbf{d}_{ij}| \sim e a_B$ is a typical transition dipole moment (where a_B is the Bohr radius) and $\Delta E \sim e^2/a_B$ is a typical energy denominator for the difference between two electronic energy levels. This leads to an estimate wherein the polarizability is given by the atomic size, $\mathfrak{p}_E \sim a_B^3$, with no additional factors of e.

Rayleigh scattering accounts for much of the systematics of how light is scattered in transparent media. The proportionality of the cross section to k^4 is what favours the scattering of blue relative to more reddish colours when visible sunlight passes through transparent materials like the atmosphere. The dependence on six powers of the size of the scattering object helps explain why a medium's transparency can vary so strongly depending on the kinds of objects that are suspended within it.

The cross section (12.95) predicts that light passing through a medium containing a density, $n =: 1/\ell^3$, of polarizable particles of size a is likely to scatter after travelling a distance of order the scattering length,

Table 12.2 Scattering lengths for $\lambda_{\text{blue}} = 400$ nm and $\lambda_{\text{red}} = 600$ nm

	a in μm	n in cm^{-3}	D_{red} in m	D_{blue} in m
gas atoms	10^{-4}	10^{19}	2×10^8	3×10^7
liquid atoms	10^{-4}	10^{24}	2000	300
haze aerosol	1	10	200	30
fog droplets	1	100	20	3

$$D = \frac{1}{n\,\sigma_R} = 3\lambda \left(\frac{\lambda}{2\pi a}\right)^3 \left(\frac{\ell}{a}\right)^3, \tag{12.96}$$

where $\mathfrak{p}_E = a^3$ and $\lambda = 2\pi/k$ is the light's wavelength. Defining 'blue' and 'red' light to have wavelength $\lambda_{\text{red}} = 600$ nm and $\lambda_{\text{blue}} = 400$ nm (putting them on opposite sides of the visible range), then (12.96) predicts the values given in Table 12.2 given a few representative numbers for the size and density of scatterers. Notice that the approximation $\lambda \gg 2\pi a$ is not very good for the last two rows of this table.

The polarization of initially unpolarized light due to its scattering from neutral objects is also predicted by the Rayleigh cross section. Measuring the final-state photon polarization amounts to not summing over its possible values, so using

$$\langle|\hat{\mathcal{A}}_{2,2}|^2\rangle := \frac{1}{2}\sum_{\lambda} |\mathcal{A}_{2,2}|^2 = \frac{k^2\tilde{k}^2\mathfrak{p}_E^2}{2}\sum_{\lambda} |\tilde{\epsilon}\cdot\epsilon|^2 = \frac{k^2\tilde{k}^2\mathfrak{p}_E^2}{2}\left[1 - (\hat{\mathbf{k}}\cdot\tilde{\epsilon})^2\right], \tag{12.97}$$

shows that the polarized cross section is maximized when the final polarization, $\tilde{\epsilon}$, is perpendicular to the initial photon direction, $\hat{\mathbf{k}} := \mathbf{k}/|\mathbf{k}|$, and vanishes when they are parallel or antiparallel.

Scattering formulae are often derived using specific, often simplistic, models of electrons within an atom (such as where they are modelled as charges in a simple-harmonic potential), and then taking the long-wavelength limit. The derivation given here shows that the Rayleigh scattering formula, (12.94), is much more robust than any particular model-dependent derivation of this type. Of course, the flip side of this generality is the inability of EFT methods to capture any of the detailed microscopic effects (such as resonance) that arise once the photon wavelength becomes comparable to the size of the scatterers.

The EFT derivation shows that the validity of the Rayleigh scattering formula ultimately rests only on the long-wavelength approximation, $ka \ll 1$, together with the dominance of the effective action (12.91) in this regime. For spherically symmetric neutral particles the above discussion shows that this dominance is guaranteed by the symmetries of the general low-energy action, but it can also happen that (12.91) dominates in this regime even for non-spherical objects (like most practical molecules or dust particles). Whether this is so or not depends on whether a lower-dimension effective interaction is also possible that can be larger than (12.91).

An example of an interaction that can successfully compete with (12.91) is an interaction of the dipole form

$$\mathfrak{L}_{\text{dipole}} = \mathfrak{D}\cdot\mathbf{E}, \tag{12.98}$$

where $\mathfrak{D}(\Phi^*, \Phi)$ represents the density of any permanent electric dipole moment arising from the microscopic distribution of the charged constituents within the neutral object. On dimensional grounds $\mathfrak{D} \sim n\,\mathbf{d}$ where n is the particle density and $|\mathbf{d}| \sim e_q R$ scales only linearly in the object's size. Because photon scattering arises at second order in $\mathfrak{L}_{\rm dipole}$ the scattering amplitude produced by a dipole interaction contributes at order R^2 (and so at long wavelengths potentially beats the $O(R^3)$ Rayleigh result). This is why dipole scattering (rather than Rayleigh scattering) dominates the low-energy interactions of photons with some non-spherical objects, such as water molecules.

12.3.2 Multipole Moments

This section follows up on several questions suggested by the above discussion. How do multipole moments fit into an EFT description? How many other moments (besides dipole moments) can compete with Rayleigh scattering at low energies, and what does this say about when Rayleigh scattering should control the low-energy behaviour of non-spherical particles? Many scatterers of real interest, be they neutrons or molecules, are not spherically symmetric; what governs how these scatter electromagnetic waves at low energies?

The constraints of gauge-invariance, locality and conservation of heavy-particle number imply that any effective interaction with lower dimension than (12.91) should be linear in either **E** or **B** or their derivatives, and this class of interactions is closely related to the EFT representation of multipole moments. Since the heavy particle of interest is not rotationally invariant (by assumption), assume it is represented by a collection of non-scalar fields Φ_a, on which rotations act nontrivially. Calling an interaction 'n-body' if it involves n powers of bilinears, $\Phi_a^*\Phi_b$, then electromagnetic multipole interactions have the general form

$$\mathfrak{L}_{\rm mp} = \sum_{k=1}^{\infty} \Big[\Phi^\dagger N_E^{i_1\cdots i_k} \Phi\, \partial_{i_k} \cdots \partial_{i_2} E_{i_1} + \Phi^\dagger N_B^{i_1\cdots i_k} \Phi\, \partial_{i_k} \cdots \partial_{i_2} B_{i_1}\Big], \qquad (12.99)$$

where, without loss, the coupling coefficients $N_E{}^{i_1\cdots i_k}$ and $N_B{}^{i_1\cdots i_k}$ (more about which below) can be taken to be completely symmetric in the vector indices[15] $i_2, \ldots, i_k = 1, 2, 3$. No terms involving A_0 or **A** undifferentiated are written here because these are forbidden by electromagnetic gauge invariance when Φ_a describes an electrically neutral particle.

The couplings in (12.99) can also be chosen to be traceless on any pair of indices involving i_1 – *e.g.* $\delta_{ij} N_E^{i_1\cdots i\cdots j\cdots i_k} = 0$ and so on – because any effective operators built from the trace involve one of the quantities $\nabla \cdot \mathbf{E}$ and $\nabla \cdot \mathbf{B}$, and so can be simplified using the Maxwell equations (which state $\nabla \cdot \mathbf{B} = 0$ and trade $\nabla \cdot \mathbf{E}$ for the density of electrical charge). As such, they are either completely redundant or can be rewritten, using the arguments of §2.5, in terms of two-body interactions (or interactions with other, non-electromagnetic fields). The Maxwell equations similarly allow omitting time derivatives $\partial_t \mathbf{E}$ and $\partial_t \mathbf{B}$.

[15] Notice this symmetry excludes the index i_1 of the electromagnetic field itself.

The completely symmetric couplings in (12.99) correspond to multipole interactions in the limit of zero momentum, inasmuch as they contain interaction energies that are proportional to the local electromagnetic field, or its symmetric derivatives, at the position of the particle in question. For instance, the first terms in the expansion (12.99) are given by

$$\mathfrak{L}_{\rm mp} = -\Phi^\dagger N_E^i \Phi\, E_i - \Phi^\dagger N_B^i \Phi\, B_i - \Phi^\dagger N_E^{ij} \Phi\, \partial_i E_j - \Phi^\dagger N_B^{ij} \Phi\, \partial_i B_j + \cdots, \quad (12.100)$$

and if these survive in the static limit they define electric and magnetic multipole moments. For instance, the first two give static energies proportional to electric and magnetic fields of the dipole form, $\mathfrak{D}_E \cdot \mathbf{E}$ or $\mathfrak{D}_B \cdot \mathbf{B}$, suggesting the quantities $\mathfrak{D}_E = \Phi^\dagger \mathbf{N}_E \Phi$ and $\mathfrak{D}_B = \Phi^\dagger \mathbf{N}_B \Phi$ are the electric and magnetic dipole-moment density operators. The static and symmetric parts of the next two terms are similarly related to quadrupole moments, and so on.

But what are the coefficient matrices $N_E^{ab;i_1\cdots i_k}$ and $N_B^{ab;i_1\cdots i_k}$ in concrete examples? In practice, these quantities are independent couplings whose form could be relatively complicated combinations of microscopic degrees of freedom (such as moments of the microscopic charge distributions for a molecule like water). Things are simpler in the special case where the underlying particle only has particle spin and centre-of-mass position as independent low-energy dynamical degrees of freedom. In this case, the $N^{i_1\cdots i_n}$ must be built from these two quantities, in a way consistent with the transformation properties under rotations and discrete symmetries (like parity or time-reversal). In all cases, the values of the coupling coefficients are determined by matching to the UV theory (or experiments).

To make this concrete, consider the case discussed in some detail above where Φ represents a neutral spin-half particle, like a ^{3}He atom or a neutron, and so indices a, b are two-valued. In this case, invariance under translations, rotations and fermion-number transformations requires $N^{i_1\cdots i_n}$ to be built from Pauli matrices and spatial derivatives, such as $\Phi^\dagger \sigma\, \Phi$ or $i(\Phi^\dagger \nabla\Phi - \nabla\Phi^\dagger \Phi)$, and so on. Parity and time-reversal invariance impose further conditions leading at the zero-derivative level to

$$N_B^i = \mathfrak{m}\, \sigma^i, \quad N_E^i = N_E^{ij} = N_B^{ij} = 0, \quad (12.101)$$

where $\mathfrak{m}$ denotes the magnetic dipole moment. Once parity breaking is included (such as arises in the weak interactions) then an 'anapole' moment, $N_B^{ij} = a\, \epsilon^{ijk} \sigma_k$ is also possible [328], as is an electric dipole moment N_E^i once both P and T invariance are broken.[16]

As discussed above, persistent dipole moments play particularly important roles for low-energy scattering because they can successfully compete with the general polarizability interaction of (12.91). They can do so because dipole moments can scale linearly with a neutral object's size and so contribute to $2 \to 2$ electromagnetic scattering amplitudes at order $\mathcal{A}_{22} \propto \mathfrak{m}^2$ (and so give scattering cross sections of

[16] P- and T-preserving electric dipoles are possible for particles for which more than just spin breaks rotation invariance, such as for water molecules whose microstructure picks out a plane (of the $H-O-H$ bonds) plus a direction within the plane, along which its electric dipole moment points.

order $\sigma \propto k^2 R^4$, as compared with the $k^4 R^6$ predicted by Rayleigh scattering). But higher multipoles do not similarly compete with Rayleigh scattering since a quadrupole moment of size R^2 would contribute to scattering amplitudes at order $\mathcal{A}_{22} \propto R^4$, which is already smaller than the R^3 dependence found for Rayleigh scattering.

These arguments show why the Rayleigh scattering result applies so robustly. Even if an object is not spherically symmetric, all that is required is an absence of sizeable dipole moments to ensure the effective interaction (12.91) can prevail at low energies.

12.4 Summary

This chapter applies the tools of §11 to a variety of practical applications taken from the electromagnetic and strong interactions. The electromagnetic examples show how nonrelativistic EFT methods robustly capture the low-energy electromagnetic interactions of slowly moving particles, both when they are electrically charged (Thomson scattering) and when they are neutral but built from charged constituents (dipole and Rayleigh scattering). In all cases, the implications of the predictions are revealed to have a much broader domain of validity than any particular microscopic derivation.

The strong-interaction example (HQET) illustrates a different virtue of nonrelativistic methods: the power of symmetry arguments in the limit that recoil kinetic energies can be neglected. This is particularly valuable for applications to QCD, where an absence of small expansion parameters makes explicit calculation of kinematic decay distributions relatively rare. Although the HQET action might not solve this problem, the large spin-flavour symmetry of the leading term instead systematically allows such distributions to be related for different decays.

The examples described here also illustrate how the coupling of relativistic photons (or gluons) to nonrelativistic charged particles complicates power counting at low energies, as described in §11. While Thomson and Rayleigh scattering fall into the soft-regime (for which photon energies and momenta are both order mv), absorption and bound-state spectra are sensitive to the ultra-soft regime (for which photon energies and momenta are order mv^2). As expected, the small energy denominators associated with loops of non-static ultra-soft photons contribute less suppressed by powers of v than is true for harder photons.

The inclusion of both particles and antiparticles within the framework of a nonrelativistic EFT is explored in some detail, using positronium (an electron–positron bound state) as the practical example for doing so. Positronium also provides a vehicle for displaying the utility of EFT techniques for precision calculations more generally, via the next-to-leading hyperfine splitting and the annihilation rate. This utility relies on the separation between the relativistic higher-energies relevant to the radiative corrections and the low-energy complications to do with the bound-state structure. From the point of view of the low-energy theory, radiative corrections enter when determining the values of the low-energy effective couplings, by matching beyond leading order in the fine-structure constant. The nonrelativistic character of the low-energy theory then simplifies bound-state calculations by allowing use of well-developed Schrödinger methods. It is the derivation of these Schrödinger methods from first principles that allows all effects to be treated within a systematic approximation scheme.

Exercises

Exercise 12.1 Use the dipole-approximation for the photon-absorption amplitude given in Eqs. (12.55) and (12.57) to show that the absoprtion rate of (12.59) can be written

$$d\Gamma_{\rm abs}[\Psi(i,\sigma)+\gamma(\mathbf{k},\lambda)\to\Psi(f,\tilde{\sigma})] = \frac{\omega_{fi}^3}{8\pi^2}\, n_{\mathbf{k}}\, |\boldsymbol{\epsilon}\cdot\mathbf{d}_{fi}|^2 \delta_{\sigma\tilde{\sigma}} d^2\Omega.$$

Use the polarization completeness identity $\sum_\lambda \epsilon_j^*(k,\lambda)\epsilon_l(k,\lambda) = \delta_{jl} - \hat{k}_j\hat{k}_l$, where $\hat{\mathbf{k}} = \mathbf{k}/|\mathbf{k}|$, to average over initial polarizations and perform the integral over photon directions to show

$$\int d^2\Omega\, \langle|\boldsymbol{\epsilon}\cdot\mathbf{d}_{fi}|^2\rangle = \frac{4\pi}{3}\,|\mathbf{d}_{fi}|^2,$$

and thereby find that the unpolarized lab-frame absorption rate for an atomic electron in a bath of electromagnetic radiation is

$$\Gamma_{\rm abs} = \frac{\omega_{fi}^3 n_k}{6\pi}\,|\mathbf{d}_{fi}|^2,$$

which is Eq. (12.60) of the main text.

Exercise 12.2 Explicitly evaluate the matching of NRQED parameters at leading and next-to-leading order and thereby derive Eqs. (12.79), (12.80) and (12.82). This involves evaluating the form factors $F_1(q^2)$ and $F_2(q^2)$, the vacuum polarization and fermion-fermion scattering at threshold in both QED and NRQED and fixing NRQED couplings to ensure they agree to the required order. Be careful when performing the matching to keep track of the conventional difference between normalization for relativistic and nonrelativistic momentum eigenstates (about which, see §B.1 for details).

Exercise 12.3 Identify which effective interactions in the low-energy nonrelativistic effective theory contribute to electronic energy shifts in single-electron ions to order $m_e(Z\alpha)^4$, where Z is the nuclear charge. Use the effective couplings as matched in the main text to compute the fine-structure contribution to the positronium energy levels.

Exercise 12.4 Imagine a world where the muon mass is $M \simeq 10$ MeV (rather than its real-world value of 105 MeV), so its Bohr radius is much larger than the electron Compton wavelength: $\alpha M \ll m_e$. In this regime the vacuum polarization graph, Fig. 7.4, should be more important for muonic hydrogen-like atoms – *i.e.* a muon-nucleus electromagnetic bound state – than other one-loop corrections because $c_U \propto \alpha/m_e^2$ is suppressed by the electron mass (see, *e.g.* (12.80), where the Uehling coupling is defined in Eq. (12.78)), while other one-loop corrections might be expected to be suppressed instead by $1/M$. (For real-world muonic Hydrogen m_e is not so different from αM and so it is not such a good approximation to expand the vacuum polarization, $\pi(k^2)$, in powers of k^2/m_e^2 and thereby use just the Uehling interaction of (12.78) [329].)

For the Lamb shift ($2S_{1/2} - 2P_{1/2}$ energy splitting) power count the size of one-loop contributions to muonic energy levels and quantify the amount by which the vacuum polarization dominates other one-loop effects. Evaluate the vacuum polarization contribution to the energy level of an S-wave state ($\ell = 0$) with principal quantum number n and show that it is given by

$$\delta E_n(\ell = 0) \simeq -\frac{4(Z\alpha)^4 \alpha M^3}{15\pi n^3 m_e^2},$$

where Z is the nuclear charge.

Exercise 12.5 Power count the contributions to Hydrogen's electronic atomic energy levels and identify all of the graphs in the low-energy non relativistic EFT that can contribute at order $\delta E \simeq m_e \alpha^5$. Identify in this way which effective couplings must be determined by matching (as well as the order in α with which the matching must be performed) in order to compute the full Lamb shift (between the $2S_{1/2} - 2P_{1/2}$ levels). Which graph involves a loop with ultra-soft photons? (For this graph be sure to use the Coulomb propagator of Eq. (12.52) obtained by resumming multiple Coulomb interactions.) Identify how the infrared divergences that arise in the parameter matchings cancel to give an IR-finite energy shift. (For hints see [307, 330].)

Exercise 12.6 Show that the lowest-dimension effective electromagnetic interactions for the nucleus (besides those describing the total mass and charge) in the non-relativistic EFT for the $^4He^+$ ion (*i.e.* for a spinless nucleus) has an effective coupling with dimension (length)^2. Use this observation to show that the leading contribution to atomic energy levels from nuclear substructure arises at order $\delta E \sim (Z\alpha)^4 m_e^3 R^2$ where $R \sim 1$ fm is a typical nuclear size. What does your argument imply for the dependence of δE on principal quantum number n and angular quantum number ℓ?

Exercise 12.7 Use the dipole approximation to compute the rate for spontaneous emission of a photon as an atomic electron transits from an initial excited state $|i\rangle$ to a another state $|f\rangle$ with lower energy, as a function of the energy difference $E_i - E_f$ and the dipole transition matrix element, $\mathbf{d}_{fi}$, as given in (12.58).

13 First-Quantized Methods

In the discussion of the previous two chapters heavy particles are treated nonrelativistically, but still within second-quantized field theory. However, for some problems keeping the entire superstructure of second-quantized field theory unnecessarily complicates matters. This is particularly true for systems where there is only a single heavy particle, whose presence may be largely passive, since in this case all of the action occurs purely within the single-particle sector of the Hilbert space for which first-quantized methods suffice and can be more convenient.

Atomic energy levels provide an important example of this type, as alluded to in §12.2. In this case, the heavy particle is often the nucleus, and the EFT exploits both the small electron/nucleus mass ratio, $m_e/M \simeq 5 \times 10^{-4}/A$ (for A the atomic number), and the small ratio between nuclear and atomic sizes, $R/a_B \simeq 2 \times 10^{-5} A^{1/3}$, to determine how atomic properties depend on those of the nucleus. Keeping the nucleus as a second-quantized field is largely a distraction in this type of problem, which is more fruitfully cast in terms of light particles interacting with a specific localized source (to which many tools built[1] for 1st-quantized Schrödinger–Coulomb quantum mechanics can be applied).

For such systems the formalism comes full circle, with ordinary nonrelativistic quantum mechanics emerging as the systematic low-energy EFT in some sectors.[2] This chapter develops the description of EFT techniques using a first-quantized description for the heavy particle (or particles, if more than one is present). There turn out to be two parts to the story, of which the first simply identifies how the low-energy dynamical quantities – typically the centre-of-mass coordinate and (possibly) spin – of the heavy particle emerge as quantum variables from the full field theory. This is explored in §13.1, using solitons as a concrete and well-understood example of how an emergent low energy first-quantized description arises within field theory.

The second part of the story then asks how these heavy-particle variables interact with the quantum fields describing any other low-energy degrees of freedom that may be present. The upshot of this second part is that the matching conditions that dictate the influence of the single heavy particle on other low-energy fields boils down to a set of boundary conditions these fields must satisfy in the region near the heavy particle. It turns out that the precise form of these boundary conditions can be derived directly from the effective action for the massive classical particle. How these

[1] Among these tools is the Born–Oppenheimer approximation [312], which exploits the small electron-nucleus mass ratio, m_e/M, to organize calculations of molecular energies and can be regarded as one of the earlier examples of EFT reasoning (see §16.1 for still earlier examples).

[2] Not *all* sectors need be well-described by single-particle methods, so formulating them as EFTs provides a foundation for coupling 1st- and 2nd-quantized systems to one another (as was historically done first [120]).

boundary conditions emerge is described in §13.2, with illustrative applications to nuclear effects in atoms given in §13.3.4 and §13.3.3 then shows that a spin-off of this framework is the systematic understanding of how to handle singular interactions (like the inverse-square potential) in quantum mechanics. Together these two parts provide the key to efficiently extracting how individual heavy particles interact with their surroundings.

13.1 Effective Theories for Lumps ◇

A common idealization in physics describes how point particles interact with one another and with applied fields (perhaps electromagnetic or gravitational). Although called 'point particles', in practical applications the objects of interest usually are not; they are often macroscopic things like projectiles or planets. They are point-like only inasmuch as their size is smaller than the spatial resolution of interest, and to the extent that their centre-of-mass motion (and perhaps rigid rotation) are the only relevant degrees of freedom.

This section sets up how this kind of description arises systematically when small massive and relatively inert objects – collectively called[3] 'lumps' – are described using EFT methods. In particular, the focus here is on how centre-of-mass position emerges as a collective variable when an object's physical radius, a, is much smaller than the shortest wavelengths, ℓ, available as probes in the low-energy theory: $\ell \gg a$. It is perhaps worth emphasizing that this regime need not automatically follow from the conditions $1/\ell \lesssim \Lambda$ and $M \gg \Lambda$ expected for measurements on a heavy particle in an EFT with a UV scale, Λ. For example, for Λ at everyday energies like those associated with room temperatures, $\Lambda \sim 100$ K $\sim 10^{-2}$ eV, we have $\Lambda^{-1} \sim 100\ \mu$m and so atoms and molecules are both point-like and heavy while the Earth or the Sun would be heavy, but not necessarily point-like.

The discussion given here is broad enough to include objects that span more dimensions than point particles, such as one-dimensional strings or vortices (lumps that span one spatial dimension as well as time, but whose transverse size is neglected) or two-dimensional domain walls (lumps that span two spatial dimensions – plus time – whose thickness in the third dimension is negligible). For the present purposes it doesn't really matter if these lumps arise as complicated objects (like the Earth, say) or as relatively simple classical soliton solutions within a specific field theory.

The description can be so general because the limit $\ell \gg a$ ensures that the object is captured by only a few properties, such as its centre-of-mass coordinate, y^μ, and perhaps its spin. The leading part of the effective field theory where only such quantities are relevant becomes the ordinary quantum mechanics of these degrees of freedom. Furthermore, these degrees of freedom have a symmetry origin inasmuch as they quantify how the lump breaks an underlying invariance of the problem, with

[3] Inspired by Coleman [331], the word 'lump' is adapted here to describe a generic small and heavy object, as opposed to specific examples such as a topological soliton or a bound system (like an atom or the Earth).

position and spin describing, in particular, the response of the lump to translations and rotations.

The effective theory to which one is led therefore realizes Poincaré invariance nonlinearly, and as such represents an example of spontaneously broken spacetime symmetries.[4] As argued below, the Goldstone mode in this case is the centre-of-mass coordinate itself, since this shifts under symmetries like spacetime translations. In this sense, the ordinary 'first-quantized' quantum mechanics of the position of a massive object emerges as the EFT appropriate to its low-energy motion, whose generality follows from the general robustness of the description of Goldstone modes [230, 332–334].

13.1.1 Collective Coordinates ♡

This section starts the discussion with a simple and relatively explicit example of a heavy object, imagined to be a classical soliton solution for a microscopic scalar field theory. This is done to see explicitly how the centre-of-mass coordinate, y^μ, emerges as a low-energy degree of freedom within a concrete framework. The next section then looks at what the spacetime symmetries imply for the effective field theory of the centre-of-mass variable.

Domain-Wall Example

As ever, it is useful to have a concrete example in mind when describing the issues. This could be done using the toy model of Part I involving a complex scalar field, which contains vortex solitons in its low-energy limit. However, it is simpler instead to restrict to a real field ϕ with action otherwise similar to the toy model one

$$S = -\int d^4x \left[\frac{1}{2}(\partial\phi)^2 + \frac{\lambda}{4}(\phi^2 - v^2)^2\right], \tag{13.1}$$

for which the classical equations of motion are

$$\Box\phi = \lambda(\phi^2 - v^2)\phi. \tag{13.2}$$

Much like for the toy model, the classical energy

$$E = \int d^3x \left[\frac{1}{2}(\partial_t\phi)^2 + \frac{1}{2}(\nabla\phi)^2 + \frac{\lambda}{4}(\phi^2 - v^2)^2\right], \tag{13.3}$$

is minimized by more than one solution: in this case, constant configurations with $\phi = \pm v$.

The classical soliton solution playing the role of the lump in this model is found by seeking solutions that depend only on one cartesian coordinate, $\phi = \phi(z)$. The lump is selected (rather than the vacuum) by minimizing the energy subject to the boundary condition $\phi(z) \to \pm v$ as $z \to \pm\infty$, since this excludes a constant solution. The minimum-energy configuration satisfies

[4] §14.3 describes a regime for which spacetime symmetries are broken in homogeneous many-body systems without breaking translation invariance (unlike the situation here, where it is the heavy-particle's position that breaks the spacetime symmetries).

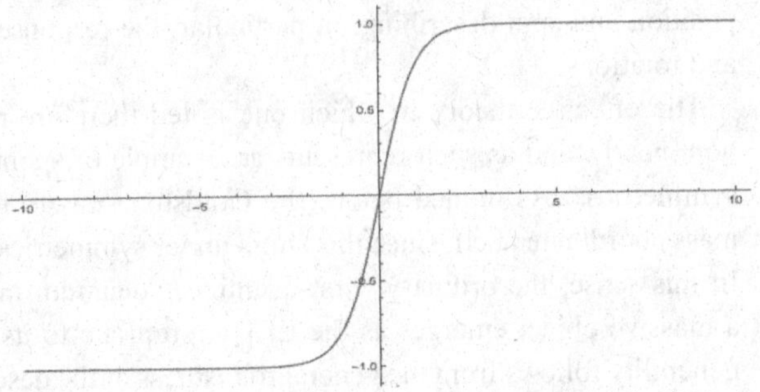

Fig. 13.1 Plot of the kink solution, $\varphi(z)/v$, as a function of $\kappa(z - z_0)$.

$$-\phi'' + \lambda(\phi^2 - v^2)\phi = 0, \tag{13.4}$$

where primes denote differentiation with respect to z.

The resulting solution – called the 'kink' (or domain wall) – is $\phi = \varphi(z)$ with

$$\varphi(z) = v \tanh\left[\kappa(z - z_0)\right], \tag{13.5}$$

where z_0 is an arbitrary constant, and the equations of motion require

$$\kappa^2 = \frac{\lambda v^2}{2}. \tag{13.6}$$

The field configuration $\phi = -\varphi(z)$ – the 'antikink' – is similarly the minimum-energy solution satisfying $\phi \to \mp v$ as $z \to \pm\infty$. As shown in Fig. 13.1, for $\kappa|z - z_0| \gg 1$ the profile $\varphi(z)$ approaches the vacuum configuration, $\varphi \to \pm v$, exponentially quickly. Its energy density is therefore concentrated within a region of width $\Delta z \sim \kappa^{-1}$ centred about $z = z_0$.

Moduli

For the present purposes what is important is that the kink/antikink solutions exist for any value of the parameter z_0. Parameters such as these carried by background solutions are called *moduli* (see also §10.3.2), and this particular modulus arises on general symmetry grounds: because the kink is not translation invariant in the z direction, whereas the equations of motion are. Translation-invariance of the field equations guarantees that the translation of any solution is another solution. Only translations in the z direction generate a parameter because the kink is invariant under translations in the x–y plane.

A similar story holds for objects with different dimensions. For instance, translation invariance ensures that a straight one-dimensional string-like object localized parallel to the x-axis has two moduli, y_0 and z_0, describing the transverse position where the localization takes place. A particle-like lump localized in all three dimensions instead has three translation-related moduli, x_0, y_0 and z_0. These moduli represent the transverse centre-of-mass position for each of these types of objects.

Other moduli can also arise for other spacetime symmetries $\delta x^\mu = \xi^\mu(x)$ – like rotations – not respected by the lump solution. A general kink is described not just by a single position, z_0, but also by two angles that describe the direction normal to the plane of the kink. But not all parameters obtained by acting with spacetime symmetries necessarily lead to independent labels that can be used to specify the state of a quantum lump.[5] For example, a Lorentz boost of a kink explicitly changes its energy (unlike time-independent translations or rotations) and so the new parameters associated with boosting the background φ are not strictly moduli. They instead describe the rate of change of parameters like z_0 with time and do not label a new degeneracy of vacua sharing the same minimum energy. Parameters in φ associated with the action of time-dependent transformations (like boosts) instead show up as contributions to the canonical momenta for honest-to-God moduli like z_0 (and x_0 and y_0 if these are present).

Moduli are important because they correspond to zero-energy excitations of the basic lump solution and as a result generically appear in the low-energy effective theory describing lump interactions. To see how this arises in more detail consider the quantum theory of small fluctuations about the classical lump (not restricted to be a kink, so possibly involving more than a single translational modulus). In a semiclassical expansion this involves expanding about the classical solution, $\phi(x) = \varphi(x - \chi) + \hat{\phi}(x)$, and quantizing all of the modes appearing in $\hat{\phi}(x)$. The notation here emphasizes the dependence of the background solution on both the spacetime position, $x^\mu = \{t, x, y, z\}$, and on the moduli describing the world-line of the centre-of-mass, $\chi^\mu(t) = \{t, x_0(t), y_0(t), z_0(t)\}$, of the lump in question (if it is localized in all three spatial directions, or with fewer components if localization happens only in a few directions such as for a kink). If the lump is assumed to be inertial but not in its rest frame then the $x_0^i(t) = \chi_0^i + v^i t$ are at most linear in t.

Expanding the action in this way leads to $S[\varphi+\hat{\phi}] = S[\varphi]+S_2[\varphi,\hat{\phi}]+S_{\rm int}[\varphi,\hat{\phi}]$, where

$$S_{\rm int}[\varphi + \hat{\phi}] = \sum_{n=3}^{\infty} S_n[\varphi, \hat{\phi}], \tag{13.7}$$

and S_n denotes those terms involving n powers of the fluctuation, $\hat{\phi}$. (There is no term linear in $\hat{\phi}$ because φ solves the classical equations of motion.) The leading quantum effects correspond to truncating this expansion at quadratic order, and so dropping $S_{\rm int}$. In this limit the action governing how the modes of $\hat{\phi}$ are to be quantized is given by the quadratic part

$$S_2(\varphi, \hat{\phi}) = -\int {\rm d}^4x\, \hat{\phi}\Delta\hat{\phi} \tag{13.8}$$

for some φ-dependent differential operator Δ. [For instance, for the kink example this operator is $\Delta = -\Box + \lambda(3\varphi^2 - v^2)$.]

A key point is that some of the modes of the fluctuation $\hat{\phi}$ being quantized can be traded for the variables $\chi^\mu(t)$ because the fluctuations, $\hat{\phi}$, include as a specific case the translations of the background configuration, $\delta\varphi = \chi^\mu\partial_\mu\varphi$. The integration over $\mathcal{D}\hat{\phi}$ can therefore be broken up into a part over the independent components of χ^μ and a part, $\mathcal{D}\hat{\phi}'$, over those modes of $\hat{\phi}$ orthogonal to $\delta\varphi$.

This is a useful decomposition of $\hat{\phi}$ because the mode $u_0 = \delta\varphi = \chi^\mu\partial_\mu\varphi$ is a zero-eigenvector: $\Delta u_0 = 0$. This is a consequence of the statement that $(\delta S/\delta\phi)_{\phi=\varphi} = 0$ holds as an identity for all χ^μ, which is what it means to say χ^μ are moduli of

[5] A broader counting of Goldstone modes for broken spacetime symmetries is deferred to §14.3.

the classical solution. [For example, with the kink solution it is easily verified that $u_0 \propto \varphi' \propto \cosh^{-2}(\kappa z)$ satisfies $-u_0'' + \lambda(3\varphi^2 - v^2)u_0 = 0$ when $\varphi = v\tanh(\kappa z)$.]

In a path integral formulation the integration over e^{iS} in the semiclassical approximation writes $e^{iS_{\rm int}} = \sum_{k=0}^{\infty}(iS_{\rm int})^k/k!$, leaving a series of gaussian integrals to be evaluated weighted by

$$\int \mathcal{D}\hat{\phi}\ \exp[iS_2(\varphi,\hat{\phi})](1+\cdots) = \left(\det\Delta\right)^{-1/2}\left[1+\cdots\right], \tag{13.9}$$

where $\det\Delta = \prod_n \lambda_n$ is formally the product over all of the eigenvalues of Δ. This can be evaluated in a basis that diagonalizes Δ, by expanding $\hat{\phi}(x) = \sum_n a_n u_n(x)$ in terms of the eigenfunctions, $u_n(x)$, that satisfy $\Delta u_n = \lambda_n u_n$, and writing the path integral measure as $\mathcal{D}\hat{\phi} \propto \prod_n da_n$.

The appearance in (13.9) of factors of $\lambda_n^{-1/2}$ (from the determinant) shows that zero eigenvalues of Δ are a problem, and so the integral over these – and in particular over the χ^μ – requires more care. Generically, the problem arises because the description of the zero modes as small gaussian fluctuations breaks down. In particular, the zero modes count as low energy when dividing $\int \mathcal{D}\hat{\phi}$ up into low- and high-energy parts, so their integral should be reserved to the low-energy theory (leaving only the product over nonzero eigenvalues to be performed when integrating out UV modes).

These arguments show how the ordinary quantum mechanics of lump motion – *i.e.* the first-quantized path integral over $\chi^\mu(t)$ – can be regarded as emerging as the low-energy limit of the more complete quantum field theory describing the UV sector.

13.1.2 Nonlinearly Realized Poincaré Symmetry *

Rather than continuing down the road of explicitly integrating out the UV modes in a specific model, what is more useful for what follows is a discussion of the symmetry constraints that govern the general ways that the centre-of-mass modes, χ^μ, can enter the low-energy action [334]. Because these symmetry constraints arise from the breaking of spacetime symmetries, they are generic for any low-energy description of interacting lumps.

The upshot of the previous section is that translation invariance of the UV theory implies that the centre-of-mass coordinates, $\chi^\mu(t)$, are generic in the low-energy effective theory of large slowly moving lumps. For point-like particles these coordinates describe the world-line of the lump's trajectory through spacetime, $\chi^\mu(t) = \{t, x(t), y(t), z(t)\}$ while for higher-dimensional objects they instead describe the world-volume,

$$\chi^\mu(\sigma^a) = \{t(\sigma^a), x(\sigma^a), y(\sigma^a), z(\sigma^a)\}, \tag{13.10}$$

swept out as time evolves. Here, $\sigma^a = \{\tau, \sigma\}$ might be two parameters describing the two-dimensional world-sheet swept out by a string (see Fig. 13.2), or $\sigma^a = \{\tau, \sigma^1, \sigma^2\}$ could be the three parameters decribing the world-volume of a domain wall.

The dependence of the low-energy action on these variables is strongly constrained by many symmetries. First off, if the 'target' 4D spacetime is invariant under motions

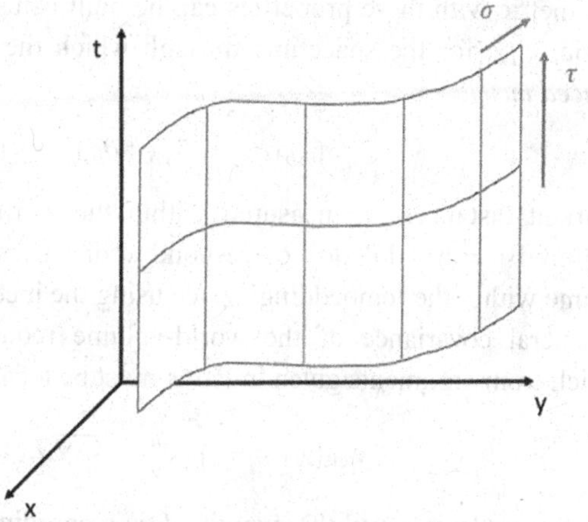

Fig. 13.2 Sketch of the world-sheet swept out in spacetime by a one-dimensional lump (*i.e.* a string) as time evolves. The world-sheet coordinates $\sigma^a = \{\tau, \sigma\}$ label points on the world-sheet while $\chi^\mu(\sigma^a)$ describes the embedding of the world-sheet into spacetime.

$\delta x^\mu = \xi^\mu(x)$ then the action must not change under the shifts $\delta\chi^\mu = \xi^\mu(\chi)$. For instance, for a flat target space, for which $g_{\mu\nu} = \eta_{\mu\nu}$, this can include spacetime translations for which $\xi^\mu = a^\mu$ is constant, as well as Lorentz transformations

$$\chi^\mu(\sigma^a) \to \Lambda^\mu{}_\nu \chi^\nu(\sigma^a). \tag{13.11}$$

By definition, the matrix $\Lambda^\mu{}_\nu$ satisfies $\Lambda^\mu{}_\alpha \Lambda^\nu{}_\beta \eta_{\mu\nu} = \eta_{\alpha\beta}$, which implies that its infinitesimal form is $\xi^\mu = \omega^\mu{}_\nu x^\nu$, where $\omega_{\mu\nu} := \eta_{\mu\lambda}\omega^\lambda{}_\nu$ is completely antisymmetric under $\mu \leftrightarrow \nu$. These symmetries act inhomogeneously on χ^μ because they are spontaneously broken by the lump solution in the underlying theory, much as happened for the $U(1)$ symmetry of the toy model developed in Part I.

Furthermore, the parameterization, σ^a, of the lump's world-volume is arbitrary, so the action also must be invariant under nonsingular reparameterizations,

$$\sigma^a \to \tilde{\sigma}^a(\sigma). \tag{13.12}$$

This requires it to be built in a generally covariant way (see, for example, the discussion of general covariance in the review of gravity in Appendix C) from the point of view of world-volume reparameterizations. This typically requires[6] a metric γ_{ab} to be defined within the lump's world-volume. Any such a metric should transform under reparameterizations, (13.12), as a covariant second-rank tensor,

$$\gamma_{ab}(\sigma) = \tilde{\gamma}_{cd}(\tilde{\sigma}) \frac{\partial \tilde{\sigma}^c}{\partial \sigma^a} \frac{\partial \tilde{\sigma}^d}{\partial \sigma^b}, \tag{13.13}$$

and should be symmetric under $a \leftrightarrow b$, and have Lorentzian signature (*i.e.* have one negative eigenvalue corresponding to the time direction, with all the others positive).

[6] The requirement of a metric can be avoided in a few cases with specific field contents [335, 336].

A metric with these properties can be built using only χ^μ and the 'target-space' metric, $g_{\mu\nu}$ (for the spacetime through which the lump moves), by defining[7] the *induced metric*

$$\mathfrak{h}_{ab}(\sigma) := \partial_a \chi^\mu \, \partial_b \chi^\nu \, g_{\mu\nu}[\chi(\sigma)]. \tag{13.14}$$

Invariant distances, s, measured within the world-volume using the differential relation $\mathrm{d}s^2 = \mathfrak{h}_{ab}\, \mathrm{d}\sigma^a\, \mathrm{d}\sigma^b$ correspond to distances measured tangent to the world-volume within the 'embedding' space using the metric, $\mathrm{d}s^2 = g_{\mu\nu}\, \mathrm{d}\chi^\mu\, \mathrm{d}\chi^\nu$.

General covariance of the world-volume requires the low-energy lagrangian (which from arguments given in Part I must be local) to have the form

$$S_{\rm eff}[\chi] = -\int \mathrm{d}^n\sigma \, \sqrt{-\gamma}\, L(\chi, \partial\chi, \cdots), \tag{13.15}$$

where $\gamma = \det \gamma_{ab}$ and the quantity L is a coordinate scalar – *i.e.* built from fields like $\chi^\mu(\sigma)$ in such a way that $L(\sigma) = L(\tilde{\sigma})$. The factor $\sqrt{-\gamma}$ is required to cancel the transformation properties of the measure $\mathrm{d}^n\sigma$, where n is the dimension of the world-volume.

The low-energy couplings of χ^μ are therefore dictated by the choice of L, but further information about L requires more information about the kinds of fields that appear in the low-energy theory of the underlying system of interest. In general, L is built as a derivative expansion involving all of the relevant fields, such as γ_{ab} and χ^μ and any other low-energy fields that may be present localized on the lump (such as those associated with spin or other internal degrees of freedom, some of which are discussed below). All terms must also be invariant under any assumed symmetries, so (for instance) if $\delta\chi^\mu = a^\mu$ is a symmetry then χ^μ must always appear differentiated, through the combination $\partial_a \chi^\mu$, and so on.

For example, the simplest situation is if the low-energy lump dynamics is translation-invariant and involves no other fields on the lump's world-volume beyond $\chi^\mu(\sigma)$. In this case, L is built only using $\partial_a \chi^\mu$ and the metric is the induced metric $\gamma_{ab} = \mathfrak{h}_{ab}$. In this case, L can be written

$$L = T_0 + \cdots \tag{13.16}$$

where T_0 is a constant (with the interpretation of the energy-per-unit-volume of the lump) and the ellipses all involve more derivatives. These derivatives include powers of the curvature tensor built from h_{ab} and its first and second derivatives in the usual way (see Appendix C for details) as well as its (covariant) derivatives. The detailed form of this expansion is not needed below.

Point Particles

For later purposes (and to be concrete) it is useful to write out what the above expressions become in the special case where the lump moves through Minkowski space (with metric $g_{\mu\nu} = \eta_{\mu\nu}$) and is localized in all three dimensions (and so at low energies resembles a point particle). Since $n = 1$ for a world-line, the $n \times n$ 'metric'

[7] The induced metric is denoted $\mathfrak{h}_{ab}$ rather than γ_{ab} because the metric in the lump action can, but need not, be given by $\mathfrak{h}_{ab}$.

$\gamma_{ab}(\sigma)$ degenerates into the single function $\gamma(\sigma)$. The induced metric similarly becomes $\mathfrak{h} = \eta_{\mu\nu}\,\dot{\chi}^\mu\,\dot{\chi}^\nu$, where dots denote differentiation with respect to the single parameter[8] σ.

To write down explicit lagrangians, first assume that $\gamma(\sigma)$ is an independent field in addition to $\chi^\mu(\sigma)$, and is *not* assumed to be the induced metric $\mathfrak{h}$. The leading contributions to the action then are

$$S_{\rm eff}[\chi,\gamma] = -\frac{1}{2}\int_W d\sigma\,\sqrt{-\gamma}\,\Big[M\left(1+\gamma^{-1}\eta_{\mu\nu}\dot{\chi}^\mu\dot{\chi}^\nu\right)+\cdots\Big]. \tag{13.17}$$

Here, W denotes the range traced by the arbitrary parameter, σ, along the world-line, $\chi^\mu(\sigma)$, along which the particle in question moves. The overall factor of $\frac{1}{2}$ is conventional and the freedom to rescale χ^μ is used to remove any independent parameter in front of the $\dot{\chi}^2$ term. Once this is done, the same cannot also be done for the terms involving more derivatives (indicated by the ellipses). The factor of $\gamma^{-1}(\sigma)$ in the second term is required for reparameterization invariance, as may be seen by comparing it to its higher-dimensional analog: $\gamma^{ab}\partial_a\chi^\mu\partial_b\chi^\nu\eta_{\mu\nu} = \gamma^{ab}\mathfrak{h}_{ab}$.

Including both γ_{ab} and χ^μ as independent fields is called the 'Polyakov' formulation of the action [337–339], as opposed to the 'Nambu' formulation [264], which uses only the induced metric, $\mathfrak{h}$, from the get-go. (The relation between these formulations is described below.) The interpretation of (13.17) is found at the classical level by using it to find the equations of motion for both $\gamma(\sigma)$ and $\chi^\mu(\sigma)$, which (neglecting the ellipses) leads to the equations

$$\frac{\delta S_{\rm eff}}{\delta\chi^\mu} = -\frac{M}{2\sqrt{-\gamma}}\,\eta_{\mu\nu}\left[\ddot{\chi}^\nu - \frac{\dot{\gamma}\,\dot{\chi}^\nu}{2\gamma}\right] = 0,$$
$$\text{and}\quad \frac{S_{\rm eff}}{\delta\gamma} = \frac{M}{4\sqrt{-\gamma}}\left[1-\frac{\eta_{\mu\nu}\dot{\chi}^\mu\,\dot{\chi}^\nu}{\gamma}\right] = 0, \tag{13.18}$$

the second of which has as solution $\gamma = \eta_{\mu\nu}\dot{\chi}^\mu\dot{\chi}^\nu$ and so $\gamma = \mathfrak{h}$. Evaluating $S_{\rm eff}$ at this solution for γ leads to

$$S_{\rm eff}[\chi,\gamma=\mathfrak{h}] = -\int_W d\sigma\,\sqrt{-\mathfrak{h}}\,M\left(1+\cdots\right), \tag{13.19}$$

which is the leading form for the action in the Nambu formulation. This shows that these two formulations have the same classical predictions.

For point particles, carrying around a world-line metric like γ or $\mathfrak{h}$ is overkill, since any such a metric can be made into a constant by performing an appropriate reparameterization of σ. To this end, it is conventional to trade σ for the proper time, τ (or 'arc-length' as measured along the world-line):

$$d\tau^2 := -\mathfrak{h}\,d\sigma^2 = -\eta_{\mu\nu}\,\dot{\chi}^\mu\dot{\chi}^\nu\,d\sigma^2 = -\eta_{\mu\nu}\,d\chi^\mu d\chi^\nu, \tag{13.20}$$

with the negative sign required because $\mathfrak{h} < 0$ along any time-like world-line. Changing variables from σ to τ allows $\mathfrak{h}$ to be written

$$\mathfrak{h} = \eta_{\mu\nu}\frac{d\chi^\mu}{d\tau}\frac{d\chi^\nu}{d\tau} = -1, \tag{13.21}$$

where the last equality follows from (13.20).

[8] If σ is target-space time, t, then $\mathfrak{h} = -1 + \mathbf{v}^2(t)$, where $\mathbf{v} = d\mathbf{y}/dt$ is the lump's centre-of-mass velocity.

The action (13.19) then becomes

$$S_{\rm eff}[\chi] = -\int_W d\tau\, M\Big(1+\cdots\Big), \tag{13.22}$$

and the equation (13.18) for χ^μ reduces to

$$\frac{d^2\chi^\mu}{d\tau^2} = 0. \tag{13.23}$$

That is to say: within the approximations made the lump described by $S_{\rm eff}$ moves in straight lines at constant speed.[9] The interpretation of the parameter M can be found by computing the canonical momentum for χ^μ, found in the usual way from

$$p_\mu := \frac{\delta S_{\rm eff}}{\delta \dot\chi^\mu} = \frac{M}{\sqrt{-h}}\,\eta_{\mu\nu}\dot\chi^\nu = M\,\eta_{\mu\nu}\,\frac{d\chi^\nu}{d\tau}. \tag{13.24}$$

Combined with the condition (13.21) this implies that $\eta^{\mu\nu}p_\mu p_\nu = -M^2$, verifying that the lump's energy and 3-momenta are related by $E^2 = \mathbf{p}^2 + M^2$ and so M is its rest-mass.

13.1.3 Other Localized Degrees of Freedom

Any lagrangian (such as $L = T_0$ in particular) built using only $\partial_a\chi^\mu$ contracted using only target-space fields (like the metric $g_{\mu\nu}$) describes a relativistic lump in the sense that its action is invariant under any symmetries of the target space fields – including, in particular, the Lorentz transformations of (13.11) when $g_{\mu\nu} = \eta_{\mu\nu}$. For these systems preferred-frame effects arise through interactions with ambient fields, like $\dot\chi^\mu A_\mu(\chi)$ or $\dot\chi^\mu\dot\chi^\nu R_{\mu\nu}(\chi)$, once $\dot\chi^\mu$ is evaluated at its classical background value, $d\chi^\mu/d\tau = U^\mu$, which is the 4-velocity of the lump's centre-of-mass.

The low-energy dynamics can be much richer if there are also other fields besides χ^μ localized on the world-line in the low-energy limit, as can arise if the underlying physics of the lump gives it other physical properties. These properties could include effects like an intrinsic spin or more complicated multipole moments, some of which might point along directions perpendicular to the lump's world-volume (unlike $\partial_a\chi^\mu$). If these new properties are free to evolve in the low-energy regime they become dynamical degrees of freedom localized at the lump's position, and so are described by new fields defined on the lump's world-volume. Among these can be vectors, $N_\mu(\sigma)$, normal to the world-volume, and their derivatives, $\partial_a N_\mu$ (which are covariantly expressed in terms of the world-volume's extrinsic curvatures), which can express an energy cost for bending the lump in question.

Intrinsic spin provides perhaps the simplest example of this type. For a spinning particle the position field $\chi^\mu(\sigma)$ is supplemented by adding a new target-space vector field [337, 338] of the form $\zeta^\mu(\sigma)$, which at the classical level is a fermionic (or Grassman [340]) variable (which means it *anticommutes* with itself $\{\zeta^\mu, \zeta^\nu\} = 0$, where (as usual) curly brackets denote the anticommutator: $\{A, B\} := AB + BA$). The field $\zeta^\mu(\sigma)$ transforms under reparameterizations of the world-line like a scalar (just as does χ^μ).

9 For particles moving in a curved spacetime Eq. (13.23) generalizes to imply that the particle moves along a target-space geodesic (see Exercise 13.1).

The lowest-derivative kinetic action for such a field is

$$S_{\rm kin} = \frac{i}{2}\int d\sigma\,(\bar{\zeta}\dot{\zeta} - \dot{\bar{\zeta}}\zeta), \tag{13.25}$$

which (because of the σ derivative) is reparameterization invariant without the need for any factors of $\mathfrak{h}(\sigma)$. Quantizing this system leads to the quantum variable $\hat{\zeta}^\mu$ whose equation of motion implies is τ-independent, and for which the canonical commutation relations imply the algebra

$$\left\{\hat{\zeta}^\mu, \hat{\zeta}^\nu\right\} = 2\,\eta^{\mu\nu}, \tag{13.26}$$

generalizing the classical result $\{\zeta^\mu, \zeta^\nu\} = 0$.

The quantum dynamics of $\hat{\zeta}^\mu$ therefore takes place in the Hilbert space that represents this algebra, whose finite-dimensional representations fix the object's spin. For instance, spin-half particles correspond to choosing a 2-dimensional Hilbert space for these degrees of freedom, with $\hat{\zeta}^\mu$ represented by Pauli matrices (and the unit matrix). If both heavy particle and antiparticle are present a 4-dimensional representation [337, 338] using the Dirac matrices, γ^μ (as defined in §A.18, for example), is preferable.

13.2 Point-Particle EFTs

The explicit point-particle effective field theories (PPEFTs) studied to this point involve lumps in isolation, but things really only get interesting once they interact with the quantum fields that make up their environments. This section mostly explores the interactions of first-quantized heavy particles with bulk electromagnetic fields, with a short aside exploring also their gravitational interactions. Here, the adjective 'bulk' is meant to convey that the field in question (in this case electromagnetic) permeates all of space and is not localized at the position of the lump. The spirit of this discussion is similar to the theory of 'quantum defects' [341], wherein a potential localized near the nucleus, is introduced, for outer valence electrons, to capture incomplete screening due to the inner electrons. One way to think about this chapter is that it embeds these techniques into the broader EFT canon.

To couple a first-quantized heavy particle to a bulk field only requires including this field in $S_{\rm eff}$, evaluated at the lump position $x^\mu = \chi^\mu(\sigma)$. For instance, coupling to electromagnetic fields requires allowing $S_{\rm eff}$ to depend on $A_\mu[\chi(\sigma)]$ in addition to fields discussed above, like $\gamma_{ab}(\sigma)$ or $\chi^\mu(\sigma)$ or $\zeta^\mu(\sigma)$, that describe degrees of freedom localized at the lump's position.

The main result of this section comes in §13.2.3, which reveals a virtue of treating the heavy particle within a first-quantized framework. The first-quantized treatment allows all of the information in the lump's effective action to be translated into a boundary condition for bulk fields in the vicinity of the heavy object. This boundary condition succinctly summarizes the matching conditions that determine the lump's effective couplings, since it ultimately transmits all of the information about the heavy object to the bulk fields, and thereby to any observables (like scattering amplitudes or bound-state energies) built from them [342]. In §13.2.3 the boundary condition is first derived for the special case of electric multipole moments, where the argument is a familiar one. It is then extended in §13.3 to the less familiar situation where the heavy object couples nonlinearly to the bulk field, using bulk Schrödinger, Klein–Gordon and Dirac fields as examples.

13.2.1 Electromagnetic Couplings

Electromagnetic couplings of a lump are built by asking how $S_{\rm eff}$ can be generalized to include a dependence on the electromagnetic gauge potential, $A_\mu[\chi(\sigma)]$. As always for low-energy effective theories this is done by writing down the most general possible couplings consistent with the required symmetries, to which is now added the requirement of invariance under gauge transformations,

$$A_\mu \to A_\mu + \partial_\mu \zeta. \tag{13.27}$$

The lowest-dimension interactions found in this way are no longer as model-independent as is the mass term considered above, and this is a reflection that different types of microscopic UV physics can give rise to different kinds of low-energy electromagnetic response. In particular, the kinds of allowed effective interactions depend on whether or not the lump in question is rotationally invariant or has nonzero intrinsic angular momentum.

If not rotationally invariant, the EFT depends on precisely how rotational invariance is broken. How many independent order parameters are there? In the simplest cases the only non-symmetric quantity is the spin degree of freedom, and in this case the EFT is required to be rotationally invariant but couplings involve both $\chi^\mu(\sigma)$ and $\zeta^\mu(\sigma)$ (and so break rotational invariance once the spin is chosen to point in a specific direction). Examples of this type could include electrons or (to good approximation) nucleons, or a spinning but spherical planet or neutron star. Alternatively, the underlying object might break rotational invariance through more than just spin, in which case more kinds of interactions can be entertained. Examples of this type could include molecules, or a moving aspherical macroscopic object like a spinning chair.

For simplicity of presentation suppose the physics governing the lump is rotationally invariant, and that the lump is spinless and so does not carry any dipole or higher-multipole moments. The lowest-dimension interactions in $S_{\rm eff}$ involving the electromagnetic field and consistent with these symmetries then are (in the Nambu formulation)[10]

$$S_{\rm eff} = Q \int_W d\sigma\, A_\mu \dot{\chi}^\mu + \int_W d\sigma\, C_D\, \dot{\chi}^\mu \partial^\nu F_{\nu\mu} + \frac{1}{2}\int_W d\sigma \sqrt{-\mathfrak{h}} \left[(C_E + C_B)\mathfrak{h}^{-1} \dot{\chi}^\mu \dot{\chi}^\nu F_{\mu\nu} F_\nu{}^\lambda + \frac{1}{2} C_B\, F_{\mu\nu} F^{\mu\nu} + \cdots \right]$$

[10] The same issues of redundant interactions described in §2.5 arise here as for the second-quantized case, in the sense that terms in the world-line action that vanish using lowest-order (bulk) equations of motion (like Maxwell equations) can be removed using a field redefinition whose parameter now is localized at the position of the lump.

$$= Q \int_W \mathrm{d}\tau\, A_\mu U^\mu + \int_W \mathrm{d}\tau\, C_D U^\mu \partial^\nu F_{\nu\mu} \tag{13.28}$$
$$+ \frac{1}{2}\int_W \mathrm{d}\tau \left[-(C_E + C_B) U^\mu U^\nu F_{\mu\lambda} F_\nu{}^\lambda + \frac{1}{2} C_B F_{\mu\nu} F^{\mu\nu} + \cdots \right]$$

where the fields A_μ and $F_{\mu\nu}$ are evaluated at the lump's position: $x^\mu = \chi^\mu$, and the second equality specializes to proper time parameter, with lump 4-velocity given by

$$U^\mu = \gamma \begin{pmatrix} 1 \\ \mathbf{v} \end{pmatrix} \quad \text{with} \quad \gamma := \frac{1}{\sqrt{1 - \mathbf{v}^2}}. \tag{13.29}$$

This is particularly simple in the lump's rest frame, since then $U^\mu = \mathrm{d}\chi^\mu/\mathrm{d}\tau = \delta^\mu_0$ and so

$$S_{\text{eff}} \to Q \int_W \mathrm{d}\tau\, A_0 + \int_W \mathrm{d}\tau \left[C_D \nabla \cdot \mathbf{E} + \frac{1}{2}(C_E \mathbf{E}^2 + C_B \mathbf{B}^2) + \cdots \right] \quad \text{(rest frame).} \tag{13.30}$$

These interactions are to be added to the terms in (13.17) discussed earlier.

Notice that the first interaction does not involve the world-line metric, $\mathfrak{h}$, at all but is nonetheless generally covariant because the factor $\dot{\chi}^\mu$ transforms under world-line reparameterizations in precisely the way $\sqrt{-\mathfrak{h}}$ does. Notice also that this first term is gauge invariant despite it depending explicitly on the gauge potential, A_μ, because under the replacement $\delta A_\mu = \partial_\mu \zeta$ this term transforms as

$$\delta \int_W \mathrm{d}\sigma\, A_\mu \dot{\chi}^\mu = \int_W \mathrm{d}\sigma\, \dot{\chi}^\mu \partial_\mu \zeta = \int_W \mathrm{d}\sigma\, \frac{\mathrm{d}\zeta}{\mathrm{d}\sigma}, \tag{13.31}$$

and so because the integrand is a total derivative the result depends only on the values taken by ζ at the boundaries of the integration range, ∂W. (This is a special case of similar higher-dimensional 'Chern-Simons' terms [335] – see, for example, the discussion in Eq. (15.56) – that are also reparameterization and gauge invariant despite being built with A_μ not appearing only through $F_{\mu\nu}$ and without reference to the world-sheet metric.)

As usual, the interpretation of the parameters in this lagrangian can be inferred by matching, and this time the matching is to the spinless interactions in the second-quantized version, Eqs. (12.1)–(12.3), of the electromagnetic couplings of a heavy particle. One way to make this comparison is to evaluate the expectation value $\langle \Psi^*(\mathbf{x}, t)\Psi(\mathbf{x}, t)\rangle = \delta^3[\mathbf{x} - \chi]$ in a static centre-of-mass position eigenstate. This reveals the dimensionless parameter Q to be the lump's net electric charge, while the parameter C_D is its (redundant) Darwin coupling. If the first-quantized action is obtained by matching to a second-quantized action like (12.3) then

$$Q = e_q \qquad C_D = \frac{e_q c_D}{8m^2}. \tag{13.32}$$

For composite particles like nuclei the size of C_D is instead expected to be set by its radius, $C_D \sim QR^2$, rather than Q/m^2 (see, for example, Eq. (13.57) below). A similar comparison using couplings with dimension $(\text{length})^3$ – *c.f.* Eq. (12.91), for example – shows the parameters $C_E = \mathfrak{p}_E$ and $C_B = \mathfrak{p}_B$ represent the particle's electric and magnetic polarizabilities.

Alternatively, matching can be done in the first-quantized theory by reproducing motion of the heavy particle in the classical limit. For instance, including the Q term in the field equation for χ^μ modifies (13.23) to become

$$M\,\eta_{\mu\nu}\frac{d^2\chi^\nu}{d\tau^2}+Q\,F_{\mu\nu}\frac{d\chi^\nu}{d\tau}=0, \tag{13.33}$$

which is recognizable as the Lorentz force law for a particle of mass M and charge Q.

13.2.2 Gravitational Couplings

The low-energy couplings of lumps to gravitational fields are explored in much the same way. To first approximation this is done by everywhere promoting the flat-space Minkowski metric $\eta_{\mu\nu}$ to the curved metric $g_{\mu\nu}(x)$ expressing the spacetime gravitational field. In particular, the induced metric becomes

$$\mathfrak{h}(\sigma)=g_{\mu\nu}[\chi(\sigma)]\,\dot\chi^\mu\dot\chi^\nu. \tag{13.34}$$

Once this is done, keeping only the M and Q terms in the action and varying χ modifies the Lorentz-force equation (13.33) to

$$M\,g_{\mu\nu}\left(\frac{d^2\chi^\mu}{d\tau^2}+\Gamma^\mu_{\nu\lambda}\frac{d\chi^\nu}{d\tau}\frac{d\chi^\lambda}{d\tau}\right)+Q\,F_{\mu\nu}\frac{d\chi^\nu}{d\tau}=0, \tag{13.35}$$

where $\Gamma^\mu_{\nu\lambda}$ is the usual Christoffel symbol constructed from the metric $g_{\mu\nu}$ (see Appendix C). This expression predicts, in particular, that the lump moves along geodesics of the target-space metric, $g_{\mu\nu}$, in the absence of an electric charge.

Similarly varying $S_{\rm eff}$ with respect to $g_{\mu\nu}$ gives the lump's contribution to the stress energy to be

$$T^{\mu\nu}=\frac{2}{\sqrt{-g}}\frac{\delta S_{\rm eff}}{\delta g_{\mu\nu}}=M\int_W d\tau\,\frac{d\chi^\mu}{d\tau}\frac{d\chi^\nu}{d\tau}\,\delta^4[x-\chi(\tau)], \tag{13.36}$$

appropriate for a point source of mass M localized on the world-line $x^\mu=\chi^\mu(\tau)$.

These expressions show how the point-particle lump actions, regarded as functionals of χ^μ, reproduce standard expressions for how the heavy lump moves in the presence of bulk electromagnetic and gravitational fields. This includes a framework for computing the corrections to the point-particle picture by including in the action higher-dimension interactions, order by order in powers of the object's size (relative to any other, larger, length scales). The next section asks the reverse question: how does the presence of the lump back-react onto these bulk fields?

13.2.3 Boundary Conditions I

To ask how the lump acts as a source for electromagnetic fields one instead varies $A_\mu(x)$ in the action formed by combining $S_{\rm eff}$ with the bulk Maxwell action,

$$S=-\int d^4x\left\{\frac14 F_{\mu\nu}F^{\mu\nu}+\int_W d\sigma\left[M\sqrt{-\mathfrak{h}(\sigma)}-QA_\mu(x)\dot\chi^\mu(\sigma)+\cdots\right]\delta^4\left[x-\chi(\sigma)\right]\right\}. \tag{13.37}$$

Here, the 4D delta-function appearing with the $S_{\rm eff}$ terms enforces the evaluation of fields like A_μ at $x^\mu=\chi^\mu(\sigma)$.

In the rest-frame of the lump the temporal delta-function $\delta[t-\chi^0(\tau)]$ can be used to evaluate the proper-time integration over τ, leaving a 3D spatial $\delta^3(\mathbf{x}-\chi)$

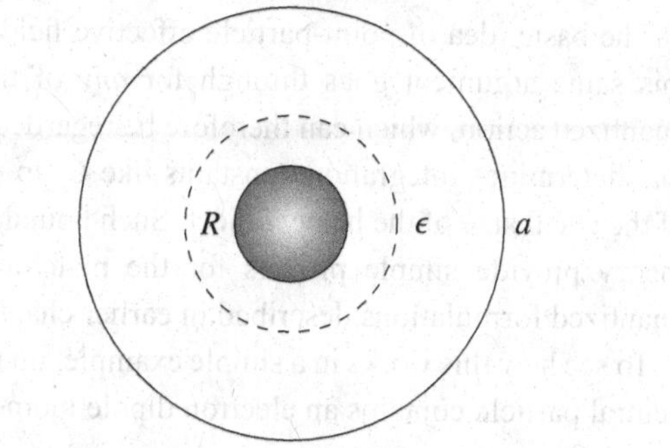

Fig. 13.3 The relative size of scales arising when setting near-source boundary conditions to the source action: R represents an actual UV scale characterizing the size of the source; a is a (much longer) size of the external physical system; ϵ is the radius between these two where boundary conditions are imposed. The precise value of ϵ is arbitrary, subject to the condition $R \ll \epsilon \ll a$.

expressing the localization of the source. The electrostatic Maxwell equation then becomes

$$\nabla \cdot \mathbf{E} = \nabla^2 A_0 = \left(Q + \cdots\right) \delta^3[\mathbf{x} - \chi], \tag{13.38}$$

showing that the lump's charge density is $Q\,\delta^3[\mathbf{x} - \chi]$ at leading order in the low-energy expansion. Before discussing the implications of subdominant terms in the PPEFT it is useful first to dispense with the delta-function by trading the localized source in (13.38) for a statement about boundary conditions for the electromagnetic field near the heavy charged particle.

For the point-charge interaction with coupling Q this boundary condition is found using a standard argument: at any given time t integrate (13.38) over a small spherical region of radius ϵ, S_ϵ (see Fig. 13.3), surrounding the heavy particle's instantaneous position, leading to the standard Gauss' law result

$$\oint_{\partial S_\epsilon} \mathrm{d}^2x\; \hat{\mathbf{x}} \cdot \mathbf{E} = \int_{S_\epsilon} \mathrm{d}^3x\; \nabla \cdot \mathbf{E} = Q, \tag{13.39}$$

with ∂S_ϵ denoting the bounding surface of S_ϵ. For a static spherically symmetric source this is a boundary condition on the electrostatic potential, implying

$$4\pi\epsilon^2 \left(\frac{\partial A_0}{\partial r}\right)_{r=\epsilon} = Q, \tag{13.40}$$

where $r = |\mathbf{x} - \chi|$. The utility of this boundary condition is its determination of an integration constant in the solution to $\nabla^2 A_0 = 0$ everywhere in the bulk (away from the heavy particle).

For later purposes notice that despite appearances the Gaussian boundary condition (13.40) does not really depend on the arbitrary radius, ϵ, of the integration region S_ϵ. This is because the general spherically symmetric solution to $\nabla^2 A_0 = 0$ is

$$A_0 = C_1 + \frac{C_2}{r}, \tag{13.41}$$

with integration constants C_i, and so the combination $\epsilon^2(\partial A_0/\partial r)_{r=\epsilon} = -C_2$ is independent of ϵ. The sole content of Eq. (13.40) is to determine the value of the integration constant: $C_2 = -Q/4\pi$.

The basic idea of point-particle effective field theory (PPEFT) is that essentially this same argument goes through for *any* of the effective couplings in the first-quantized action, which can therefore be regarded as providing a boundary condition that determines integration constants like C_2 to ever-increasing accuracy in powers of the small size of the heavy object. Such boundary conditions in the first-quantized theory provide simple proxies for the matching arguments used in the second-quantized formulations described in earlier chapters.

To see how this works in a simple example, imagine that the EFT for an electrically neutral particle contains an electron dipole moment interaction of the form

$$S_d = \int_W d\tau \, \mathbf{d}_E \cdot \mathbf{E}, \tag{13.42}$$

which when combined with the Maxwell action in the heavy-particle rest-frame becomes

$$S_{\rm tot} = \int d^4x \, \left\{ \frac{1}{2}\left(\mathbf{E}^2 - \mathbf{B}^2\right) + \mathbf{d}_E \cdot \mathbf{E}\, \delta^3(\mathbf{x}) \right\}, \tag{13.43}$$

where for simplicity coordinates are chosen here so that $\chi = 0$. This modifies the Maxwell equation obtained by varying A_0 to

$$\nabla \cdot \mathbf{E} = -(\mathbf{d}_E \cdot \nabla)\delta^3(\mathbf{x}). \tag{13.44}$$

What is important about (13.44) is the boundary condition it predicts for the external electric field, in precisely the same way that (13.38) leads to the boundary condition (13.39) for a point charge. Away from the position, $\mathbf{x} = 0$, of the dipole itself the solution constructed using only $\mathbf{d}_E$ and $\mathbf{x}$ that satisfies $\nabla \cdot \mathbf{E} = \nabla \times \mathbf{E} = 0$ is $\mathbf{E} = C_3\, \mathbf{E}_d$ where

$$\mathbf{E}_d := -\nabla \left[\frac{\mathbf{x} \cdot \mathbf{d}_E}{4\pi |\mathbf{x}|^3} \right] = \frac{1}{4\pi |\mathbf{x}|^3} \left[-\mathbf{d}_E + 3\, \frac{\mathbf{x}\,(\mathbf{x} \cdot \mathbf{d}_E)}{|\mathbf{x}|^2} \right], \tag{13.45}$$

and C_3 is an undetermined constant. But do $\nabla \cdot \mathbf{E}_d$ and $\nabla \times \mathbf{E}_d$ also vanish at $\mathbf{x} = 0$? If not (and it turns out not) then the solution to (13.44) should be sought in the form

$$\mathbf{E} = C_3\, \mathbf{E}_d + C_4\, \mathbf{d}_E\, \delta^3(\mathbf{x}), \tag{13.46}$$

since this differs from $\mathbf{E}_d$ only at $\mathbf{x} = 0$ and is also built only from $\mathbf{x}$ and $\mathbf{d}_E$ (with no derivatives). How are the constants C_3 and C_4 determined given the singularities of the solutions at $\mathbf{x} \to 0$?

It is here that boundary conditions play the decisive role. To see how, consider the identity

$$\begin{aligned} \mathbf{d}_E \cdot [\mathbf{x} \times (\nabla \times \mathbf{E})] &= \mathbf{d}_E \cdot \Big[\nabla(\mathbf{x} \cdot \mathbf{E}) - \mathbf{E} - (\mathbf{x} \cdot \nabla)\mathbf{E}\Big] \\ &= \nabla \cdot \Big[\mathbf{d}_E(\mathbf{x} \cdot \mathbf{E}) - \mathbf{x}(\mathbf{d}_E \cdot \mathbf{E})\Big] + 2\, \mathbf{d}_E \cdot \mathbf{E}. \end{aligned} \tag{13.47}$$

The left-hand side ensures this is a quantity that vanishes everywhere provided only that $\nabla \times \mathbf{E} = 0$ everywhere, *including* at $\mathbf{x} = 0$. Although one might not know the precise electric fields deep within the heavy particle's interior, Eq. (13.47) relies only on this configuration being smooth and curl-free. Furthermore, the right-hand side contains a total divergence that does not depend at all on the interior solution because it can be rewritten (using Gauss' theorem) as a surface integral once both sides are integrated over the small sphere, $\mathcal{S}_\epsilon$, centred on the heavy-particle's position. Finally, (13.47) is chosen bilinear in $\mathbf{x}$ and $\mathbf{d}_E$ in order for this surface integral to pick off the $\ell = 1$ spherical harmonic (unlike the $\ell = 0$ harmonic seen by (13.39), for example).

It follows that *any* electric field that is curl-free everywhere throughout $\mathcal{S}_\epsilon$ must satisfy

$$\oint_{\partial\mathcal{S}_\epsilon} d^2x\, \hat{\mathbf{x}} \cdot \Big[\mathbf{d}_E(\mathbf{x}\cdot\mathbf{E}) - \mathbf{x}(\mathbf{d}_E\cdot\mathbf{E})\Big] = -2\int_{\mathcal{S}_\epsilon} d^3x\, \mathbf{d}_E \cdot \mathbf{E}, \tag{13.48}$$

which when applied to (13.46) implies that

$$C_3 \oint_{\partial\mathcal{S}_\epsilon} d^2x\, \hat{\mathbf{x}} \cdot \Big[\mathbf{d}_E(\mathbf{x}\cdot\mathbf{E}_d) - \mathbf{x}(\mathbf{d}_E\cdot\mathbf{E}_d)\Big] = -2\,C_4\,\mathbf{d}_E^2, \tag{13.49}$$

because the delta-function cannot contribute to the surface integral at $|\mathbf{x}| = \epsilon$ and because the angular integrations ensure that $\int d^2\Omega\; \mathbf{d}_E\cdot\mathbf{E}_d \propto \int d^2\Omega\; [3(\mathbf{d}_E\cdot\hat{\mathbf{x}})^2 - \mathbf{d}_E^2] = 0$. The integral on the left-hand side of (13.49) evaluates to $\frac{2}{3}\,\mathbf{d}_E^2$ and so (13.49) implies that $C_4 = -\frac{1}{3}\,C_3$, and so

$$\mathbf{E} = C_3\left[\mathbf{E}_d - \frac{1}{3}\,\mathbf{d}_E\,\delta^3(\mathbf{x})\right]. \tag{13.50}$$

Now comes the main point: to determine the constant C_3 requires the boundary condition implied by (13.44), which implies that $\mathbf{E} = \mathbf{E}_{\rm hom} - \mathbf{d}_E\,\delta^3(\mathbf{x})$, where $\mathbf{E}_{\rm hom}$ satisfies $\nabla\cdot\mathbf{E}_{\rm hom} = 0$ everywhere within $\mathcal{S}_\epsilon$, including at $\mathbf{x} = 0$. An identical argument to the one culminating in (13.50), but using the identity

$$(\mathbf{x}\cdot\mathbf{d}_E)\nabla\cdot\mathbf{E} = \nabla\cdot\Big[(\mathbf{x}\cdot\mathbf{d}_E)\mathbf{E}\Big] - \mathbf{d}_E\cdot\mathbf{E} \tag{13.51}$$

shows that if $\nabla\cdot\mathbf{E}_{\rm hom} = 0$ everywhere within $\mathcal{S}_\epsilon$ then

$$\oint_{\partial\mathcal{S}_\epsilon} d^2x\, \hat{\mathbf{x}} \cdot \Big[(\mathbf{x}\cdot\mathbf{d}_E)\mathbf{E}_{\rm hom}\Big] = \int_{\mathcal{S}_\epsilon} d^3x\, \mathbf{d}_E\cdot\mathbf{E}_{\rm hom}, \tag{13.52}$$

and so

$$\mathbf{E} = \mathbf{E}_{\rm hom} - \mathbf{d}_E\,\delta^3(\mathbf{x}) = C_3'\left[\mathbf{E}_d + \frac{2}{3}\,\mathbf{d}_E\,\delta^3(\mathbf{x})\right] - \mathbf{d}_E\,\delta^3(\mathbf{x}). \tag{13.53}$$

Combining this with (13.50) then implies that $C_3 = C_3' = 1$, reproducing the standard electric dipole expression.

A similar story goes through for the spherically symmetric effective interactions of (13.30). Focussing on the Darwin-like term, the A_0 field equation then alters (13.38) to become

$$\nabla^2 A_0 = Q\,\delta^3(\mathbf{x}) + C_D\nabla^2\delta^3(\mathbf{x}), \tag{13.54}$$

whose formal solution now is

$$A_0 = A_0^{\rm c} + C_D\,\delta^3(\mathbf{x}). \tag{13.55}$$

In this case, because the solution remains spherically symmetric the boundary condition does not change from (13.40), because the C_D term does not contribute to the surface term at $r = \epsilon$. This leads to a standard Coulomb homogeneous solution, $A_0^{\rm c} = -Q/(4\pi r)$, simply supplemented by the delta-function potential as in (13.55).

The physical interpretation of C_D is most easily seen by computing the effective electric charge density of the point particle within the effective theory. This is identified (as always) by differentiating its effective action with respect to A_0:

$$\rho(\mathbf{x}) = \frac{\delta S_{\text{eff}}}{\delta A_0(\mathbf{x})} = Q\,\delta^3(\mathbf{x}) + C_D\,\nabla^2\,\delta^3(\mathbf{x}), \tag{13.56}$$

which both confirms $Q = \int d^3\mathbf{x}\ \rho(\mathbf{x})$ to be the total charge and reveals C_D couples to classical electromagnetic fields as does the mean-square charge radius, r_C, defined by

$$r_C^2 := \frac{1}{Q}\int d^3\mathrm{x}\ \mathbf{x}^2 \rho(\mathbf{x}) = \frac{C_D}{Q}\int d^3\mathrm{x}\ \mathbf{x}^2 \nabla^2 \delta^3(\mathbf{x}) = \frac{6C_D}{Q}. \tag{13.57}$$

The main lesson is this: the point-particle EFT completely determines both the response of the heavy particle to external fields and the response of the fields to the presence of the heavy particle. In particular, the response of the field is completely contained within the integration constants of the solution to the bulk field equations, and is solely fixed through the boundary condition that is dictated by the heavy-particle action.

Although these are standard results for electromagnetic multipole moments when restricted to terms in the action linear in electromagnetic fields, subsequent sections in this chapter show the same statements also holds for other types of fields, including for terms in the action that are not linear in these fields.

13.2.4 Thomson Scattering Revisited

But first a reality check on the above arguments. The point-particle action (13.30) appears to differ fundamentally from the second-quantized expression (12.2), in particular by not including a coupling involving $\mathbf{A}^2$. Since §12.1.3 shows that the $\mathbf{A}^2$ interaction is responsible for successfully reproducing the Thomson scattering cross section, this cross section is now recalculated within the PPEFT framework to verify that it obtains the same results.

The lowest-order contribution to low-energy Thomson scattering uses just the net charge Q and so comes purely from the $A_\mu\,\dot{\chi}^\mu$ coupling of (13.37). In particular, writing the particle position as $\chi^\mu(\tau) = \mathcal{X}^\mu(\tau) + y^\mu(\tau)$, where the unperturbed position represents a particle at rest and so satisfies $\dot{\mathcal{X}}^\mu = \delta^\mu_0$, the leading interaction lagrangian describing the coupling of $\mathbf{y}$ to the vector potential, $\mathbf{A}$, is given by

$$\mathfrak{L}_{\text{int}} = Q\,\mathbf{A}\cdot\dot{\mathbf{y}}\,\delta^3(\mathbf{x}-\mathcal{X}). \tag{13.58}$$

Similarly, the part of the lagrangian (13.37) describing free propagation of the variable $\mathbf{y}$ is given by expanding the Nambu action using $\mathfrak{h} = \eta_{\mu\nu}\dot{\chi}^\mu\dot{\chi}^\nu = -1 + \dot{\mathbf{y}}\cdot\dot{\mathbf{y}} + \cdots$ so

$$\mathfrak{L}_0 = -M\sqrt{-\mathfrak{h}} = -M\left[1 - \frac{1}{2}\,\dot{\mathbf{y}}\cdot\dot{\mathbf{y}} + \cdots\right], \tag{13.59}$$

where over-dots denote differentiation with respect to proper time for the background trajectory $\mathcal{X}^\mu$.

To compute the Thomson scattering amplitude requires evaluating the Feynman graph of Fig. 13.4, wherein the vertex represents the coupling given in (13.58) and

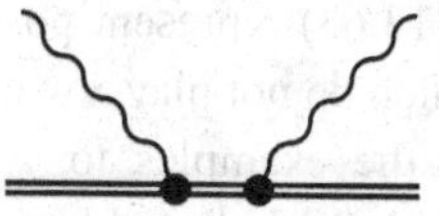

Fig. 13.4 Graph giving the leading Thomson scattering amplitude for photon scattering by a heavy charged particle in the first-quantized formulation.

the internal double line represents the propagator, $G^{ij}(\tau - \tau')$, for the 'field' $y^i(\tau)$. Inspection of (13.59) shows that the propagator satisfies

$$M \frac{d^2}{d\tau^2} G^{ij}(\tau - \tau') = i\delta^{ij}\,\delta(\tau - \tau'). \tag{13.60}$$

The graph in Fig. 13.4 then evaluates to

$$\begin{aligned} Q^2 \int d\tau d\tau' d^3\mathbf{x}\, d^3\mathbf{x}'\; \epsilon_i(\mathbf{k}, \lambda)\tilde{\epsilon}^*_j(\tilde{\mathbf{k}}, \tilde{\lambda})\; e^{i\mathbf{k}\cdot\mathbf{x} - ik\tau - i\tilde{\mathbf{k}}\cdot\mathbf{x}' + i\tilde{k}\tau'} \\ \times \frac{d}{d\tau}\frac{d}{d\tau'} G^{ij}(\tau - \tau')\delta^3(\mathbf{x} - \mathcal{X})\delta^3(\mathbf{x}' - \mathcal{X}) \\ = -\frac{Q^2}{M}\,\epsilon \cdot \tilde{\epsilon}^*\, e^{i(\mathbf{k} - \tilde{\mathbf{k}})\cdot\mathcal{X}}\; 2\pi i\delta(k - \tilde{k}), \end{aligned} \tag{13.61}$$

which should be equated to $2\pi i\delta(k - \tilde{k})\,\mathcal{A}_{22}$. (The second way of writing this result uses (13.60) to simplify the differentiated propagator.) Since the heavy particle being considered here is spinless, the spin-averaging only involves photon polarizations, giving

$$\langle |\mathcal{A}_{22}|^2 \rangle = \frac{1}{2} \sum_{\lambda\tilde{\lambda}} |\mathcal{A}_{22}|^2 = \frac{Q^4}{2M^2} \sum_{\lambda\tilde{\lambda}} |\epsilon \cdot \tilde{\epsilon}|^2, \tag{13.62}$$

in agreement with (12.33).

13.3 PPEFT and Central Forces ♠

This section extends the couplings of first-quantized effective actions to other bulk fields besides electromagnetic ones. The sections to follow then extend the discussion to atomic problems for which both electromagnetic and other matter fields appear.

To start, consider problems for which the bulk degrees of freedom are described by a Schrödinger field with bulk action

$$S_B = \int d^D x \left\{ \frac{i}{2}\left(\Psi^* \partial_t \Psi - \partial_t \Psi^* \Psi\right) - \frac{1}{2m} \nabla\Psi^* \cdot \nabla\Psi - V(\mathbf{x})\,\Psi^*\Psi + \cdots \right\}, \tag{13.63}$$

where $D = d + 1$ and d counts the number of spatial dimensions. Although $d = 3$ is obviously the case of interest for a central source localized in all three dimensions, such as a massive compact particle, keeping d general is nonetheless useful in order to include also the interactions of line ($d = 2$) or domain-wall ($d = 1$) defects.

The ellipses in (13.63) represent possible subdominant terms in the bulk effective theory for Ψ, which do not play any role in what follows.

The point of the examples to come is to determine how the first-quantized lump interacts with the bulk field, and in particular identify precisely how different interactions in the low-energy PPEFT influence Ψ observables order-by-order in the lump's small size. These effects get passed to the bulk degrees of freedom purely through a set of boundary conditions, whose detailed form is governed completely by the first-quantized action of the compact source, similar to the discussion of electromagnetic multipoles given above.

The new feature that arises in the Schrödinger example is that the point-particle action in general depends nonlinearly on the bulk field Ψ, and as a result the boundary condition obtained involves evaluating the bulk field itself at the position of the compact source. But it is generic that the extrapolation of bulk fields to the source position diverges (as does, for example, the Coulomb potential $A_0 \propto 1/r$ when extrapolated to $r = 0$) so the boundary condition obtained is also ill-defined in this limit. (Of course, the actual fields in nature do not really diverge, because the extrapolation using only external fields breaks down once one enters the interior of the compact central object.) Because of this it is important to define the boundary condition at a distance $\epsilon > R$ a short distance outside of the actual compact source, with ϵ regarded as a cutoff that regulates the divergent near-source behaviour of the bulk fields (as in Fig. 13.3).

What is crucial is that the radius $|\mathbf{x}| = \epsilon$ is essentially arbitrary, provided only that it is much smaller than the scales of physical interest in the bulk. This is because ϵ is really just a hypothetical scale associated with deriving the boundary condition, and is not intrinsic at all to the structure of the physical central source. As a result, nothing physical actually depends on the precise value of ϵ. This gets expressed by the first-quantized EFT in a familiar way: any apparent dependence on ϵ that observables appear to have gets cancelled by the implicit dependence on ϵ that is carried by all the first-quantized theory's effective couplings. That is, the couplings of the first-quantized EFT acquire an ϵ-dependence that renormalizes the divergences that otherwise would have appeared in the limit $\epsilon \to 0$. As the examples below show in detail, a renormalization-group equation expresses how the couplings must depend on ϵ in order to ensure that physical observables are ϵ-independent.

13.3.1 Boundary Conditions II

To make all of this concrete, suppose the bulk degrees of freedom are the low-energy nonrelativistic particles described by Ψ, whose bulk action is (13.63) with no potential: $V(\mathbf{x}) = 0$. Then

$$S_B = \int \mathrm{d}^D x \left\{ \frac{\mathrm{i}}{2} \left(\Psi^* \partial_t \Psi - \partial_t \Psi^* \Psi \right) - \frac{1}{2m} \nabla \Psi^* \cdot \nabla \Psi + \cdots \right\}. \tag{13.64}$$

What are the leading interactions such a field can have with a first-quantized PPEFT describing a compact massive particle? As usual, these are found by writing down all possible local interactions consistent with the symmetries, built from $\Psi(x)$ and its complex conjugate, along the compact particle's world line. Among the symmetries to be imposed is the one responsible for conservation of Ψ particle number: $\Psi(x) \to e^{i\eta}\,\Psi(x)$, where η is a spatially constant real parameter.

There is a unique nontrivial interaction term that arises at lowest dimension, given by

$$S_b = -\int_W \mathrm{d}\tau\; h\,\Psi^*\Psi = -\int \mathrm{d}^D x \int_W \mathrm{d}\tau\; h\,\Psi^*\Psi\,\delta^D[x - \chi(\tau)], \qquad (13.65)$$

where h is the corresponding effective coupling (not to be confused with the induced metric $\mathfrak{h}$). As mentioned earlier, the main new feature relative to the examples discussed above is the nonlinearity of S_b in the bulk field (in this case, Ψ), and this complicates the story because of the divergences bulk fields often experience when evaluated at the position of the source. The symmetry $\Psi(x) \to e^{i\eta}\,\Psi(x)$ clearly forbids including terms linear in Ψ or Ψ^*.

Since Ψ is canonically normalized, its dimension is $(\text{mass})^{d/2}$ and so h has dimension $(\text{length})^{d-1}$. If obtained from an EFT in which the heavy compact source is also represented by a second-quantized field, Φ, then (13.65) corresponds to a two-body interaction of the form $\mathfrak{L}_{2\text{-body}} = -h\,(\Phi^*\Phi)(\Psi^*\Psi)$.

Choosing the heavy compact particle to be at rest, with coordinates chosen so that $\chi(\tau) = 0$, the field equations for Ψ obtained by varying the action $S_B + S_b$ become

$$i\partial_t\Psi + \frac{\nabla^2\Psi}{2m} = h\,\delta^d(\mathbf{x})\,\Psi, \qquad (13.66)$$

showing that (13.65) contributes to the bulk Schrödinger equation in the same way as does a delta-function potential. As usual, the implications of such a potential for physics in the bulk is obtained by integrating over the standard small sphere S_ϵ, chosen to surround the source out to radius $|\mathbf{x}| = \epsilon$.

In the usual presentation integrating (13.66) over S_ϵ and dropping terms that are subdominant in the limit $\epsilon \to 0$ leads to

$$\frac{1}{2m}\oint_{\partial S_\epsilon} \mathrm{d}^{d-1}x\; \hat{\mathbf{x}}\cdot\nabla\Psi = h\,\Psi(0), \qquad (13.67)$$

where only the ∇^2 term in (13.66) is taken to contribute on the left-hand side. If $\Psi(0)$ is identified with $\Psi(|\mathbf{x}| = \epsilon)$ – in the spirit that distances of size ϵ are too small to be distinguished in the effective theory – then (13.67) leads to the following boundary condition for Ψ at $r = |\mathbf{x}| = \epsilon$:

$$\left(\Omega_{d-1} r^{d-1}\,\frac{\partial\Psi}{\partial r}\right)_{r=\epsilon} = 2mh\,\Psi(r = \epsilon), \qquad (13.68)$$

at least for the spherically symmetric modes with no angular momentum. Here, $\Omega_{d-1} = 2\pi^{d/2}/\Gamma[\tfrac{1}{2}d]$ – where $\Gamma(x)$ is Euler's gamma-function – is the volume of the unit $(d-1)$-dimensional sphere (or area of a d-dimensional ball with unit radius), with

$$\Omega_1 = 2\pi\,,\quad \Omega_2 = 4\pi\,,\quad \Omega_3 = 2\pi^2\,,\quad \Omega_4 = \frac{8\pi^2}{3}\,,\quad \Omega_5 = \pi^3 \quad \text{and so on.} \qquad (13.69)$$

To justify the neglect of the $\partial_t\Psi$ term in (13.67) imagine expanding

$$\Psi(x) = \sum_N u_N(x)\,\mathfrak{a}_N \quad \text{with} \quad -\frac{\nabla^2 u_N}{2m} = \omega_N u_N, \qquad (13.70)$$

in terms of modes, $u_N(x)$, that satisfy $i\partial_t u_N = \omega_N u_N$, where N represents a complete set of single-particle state labels. (For example, when $d = 3$ we have $N = \{n, \ell, \ell_z\}$, where $\ell = 0, 1, 2, \cdots$ and $\ell_z = -\ell, -\ell+1, \ldots, \ell-1, \ell$ are the integers that label the state's angular momentum. For $d = 2$ instead $N = \{n, \ell\}$, where $\ell = 0, \pm 1, \cdots$.) The modes u_N – which can be computed, say, by separation of variables in spherical polar coordinates – come as two linearly independent solutions, whose small-r asymptotic behaviour is $u_N \propto r^p$ where $p = \ell$ or $p = -\ell - d + 2$ for the two solutions, respectively (for $d \geq 2$). But if $\Psi \sim r^p$ for small r, then $\nabla^2\Psi \sim r^{p-2}$ and so for sufficiently small ϵ the integral over $\mathcal{S}_\epsilon$ of $\partial_t\Psi$ is order ϵ^{d-1+p} while that of $\nabla^2\Psi$ is order ϵ^{d-3+p}, showing that the relative error of dropping the $\partial_t\Psi$ term when deriving (13.68) is order ϵ^2.

The Boundary Action

Since these boundary conditions are being argued to be very general, it is worth looking more closely at the assumptions underlying this derivation. For instance, what if the Schrödinger equation contains an interaction potential, $V(\mathbf{x})$, for which $V \geq \mathcal{O}(1/r^2)$ for small r? Then the integration of $V(\mathbf{x})\,\Psi$ over $\mathcal{S}_\epsilon$ need not be subdominant to the integral over $\nabla^2\Psi$ when $\epsilon \to 0$, undermining faith in the validity of (13.68).

To arrive at a better argument refer again to Fig. 13.3, which shows how the radius where boundary conditions are inferred relates to the problem's underlying hierarchy of scales. In what follows it is again important to recognize that the scale ϵ arises purely as a calculational crutch, dividing the calculation into the following two steps:

- Part I: starting (in principle) from the microscopic properties of the compact source (typically as specified by its action, S_b), one imagines computing the values Ψ and its derivatives at the surface of $\mathcal{S}_\epsilon$.
- Part II: Calculate the behaviour of observables well outside of $\mathcal{S}_\epsilon$, referring only to the boundary data derived in Part I on the surface of $\mathcal{S}$ (as opposed to using detailed microscopic properties of the compact source).

These two steps reveal the utility of choosing $\epsilon \gg R$, since when this is true Part I should only depend on a few properties of the source, such as the lowest few multipole moments (and perhaps their generalizations), since successive terms are suppressed by higher powers of R/ϵ. The utility of choosing $a \gg \epsilon$ comes in Part II, since it ensures that observables are not inordinately sensitive to boundary effects at $\mathcal{S}_\epsilon$, which are suppressed by powers of ϵ/a.

The key question is: how is the boundary condition at $\mathcal{S}_\epsilon$ determined in practice? When answering this it is important conceptually that Part I above does not literally specify Ψ or its derivatives at the boundary, $\mathcal{B} := \partial\mathcal{S}_\epsilon$ at $r = \epsilon$. In general, specifying Ψ (or perhaps its derivative) would overdetermine the problem exterior to $\mathcal{S}_\epsilon$ because, in practice, the actual values taken by fields on $\partial\mathcal{S}_\epsilon$ also depend somewhat on the positions of other possible heavy compact particles elsewhere in the problem, outside and far from $\mathcal{S}_\epsilon$. This dependence on external compact particles gets weaker the further away they are from $\mathcal{S}_\epsilon$ (hence the condition $\epsilon \ll a$), but it is there in principle and so the boundary information at $\partial\mathcal{S}_\epsilon$ must be encoded in a way that leaves the fields free to adjust as required in response to their distant motions.

A simple and efficient way to do so is to specify the boundary data at $\mathcal{B}$ in terms of the variation of a *boundary action*, $I_{\mathcal{B}}$, along the lines considered in §5 of Part I. This action is related to, but not the same as, the original point-particle effective action, S_b, discussed above – such as found in (13.65). Whereas S_b is an integral over the source's world-tube, $I_{\mathcal{B}}$ is always a d-dimensional integral over the codimension-1 world-volume that $\mathcal{B} = \partial S_\epsilon$ sweeps out as time evolves. For example, for a point particle in 3 spatial dimensions S_b comes as a one-dimensional integral over the particle world-line while $I_{\mathcal{B}}$ is a 3-dimensional integral over time plus the two angular directions of a 2-sphere surrounding the particle.

In principle, $I_{\mathcal{B}}$ is constructed given S_b through a matching calculation. For N effective couplings in S_b one computes N convenient observables exterior to S_ϵ from which the couplings can be determined. Computing these same N observables using the most relevant interactions on $I_{\mathcal{B}}$ and equating results gives the couplings of $I_{\mathcal{B}}$ in terms of those of S_b.

In practice, it is often simpler than this. For $\ell = 0$ modes about spherically symmetric sources the connection between S_b and $I_{\mathcal{B}}$ at lowest order is fairly direct: $I_{\mathcal{B}}$ is simply S_b multiplied by $\Omega_{d-1}\epsilon^{d-1}$, which is the surface area of S_ϵ (or the volume of its boundary, ∂S_ϵ). For example, in the rest-frame of the compact source, with S_b given by (13.65), the surface action is simply

$$I_{\mathcal{B}} = -\int \mathrm{d}^d x\, \tilde{h}\, \Psi^*\Psi = -\int \mathrm{d}\tau \int \mathrm{d}^{d-1}\Omega\, \epsilon^{d-1}\, \tilde{h}\, \Psi^*\Psi, \tag{13.71}$$

and so

$$h = \Omega_{d-1}\epsilon^{d-1}\, \tilde{h}. \tag{13.72}$$

Whereas h has dimension (length)$^{d-1}$ the coupling $\tilde{h}$ is dimensionless. It is similar for non-spherically symmetric sources and higher multipoles, such as the dipole-moment system considered in §13.2.3, with the difference that the angular integration is weighted by the appropriate spherical harmonic.

Once $I_{\mathcal{B}}$ is specified, the surface $\mathcal{B}$ can be regarded as a boundary of the exterior region, with its influence on physics exterior to S_ϵ obtained along the lines described in §5: by requiring that the total action, $S_{\mathcal{B}} + I_{\mathcal{B}}$, is stationary with respect to variations of the fields on $\mathcal{B}$ in addition to the bulk:

$$\left[\frac{\delta S_{\mathcal{B}}}{\delta \Psi^*} + \frac{\delta I_{\mathcal{B}}}{\delta \Psi^*}\right]_{r=\epsilon} = 0. \tag{13.73}$$

For the example where $S_{\mathcal{B}}$ is given by (13.64) and $I_{\mathcal{B}}$ by (13.71), this takes the form found in (13.68):

$$\frac{1}{2m}\left(\Omega_{d-1} r^{d-1}\, \frac{\partial \Psi}{\partial r}\right)_{r=\epsilon} = -\frac{\delta I_{\mathcal{B}}}{\delta \Psi^*} = \Omega_{d-1}\epsilon^{d-1}\, \tilde{h}\, \Psi(\epsilon) = h\, \Psi(\epsilon). \tag{13.74}$$

This can be regarded as a regularization of the formal boundary condition.

$$\frac{1}{2m}\left(\Omega_{d-1} r^{d-1}\, \frac{\partial \Psi}{\partial r}\right)_{r=\epsilon} = -\frac{\delta S_b}{\delta \Psi^*}. \tag{13.75}$$

This last equation is formal in the sense that its right-hand side involves fields evaluated at the position of the source, like $\Psi(\mathbf{x}, t)\, \delta^d(\mathbf{x})$, where Ψ generically diverges at $\mathbf{x} = 0$ and so needs regularization. These kinds of terms first arise when

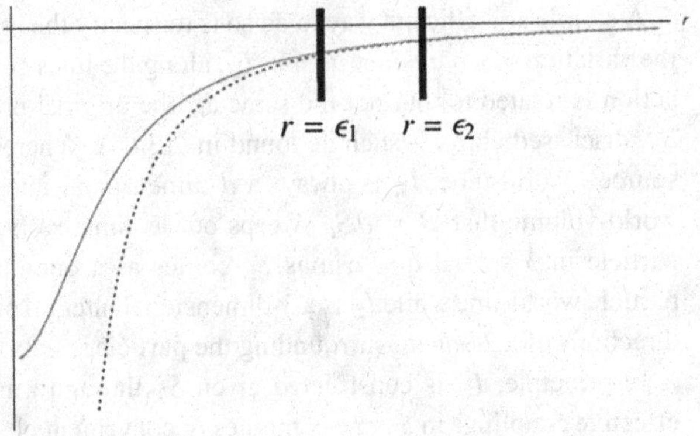

Fig. 13.5 Sketch of a real bulk-field profile produced by a localized source in the UV theory (solid line) superimposed on the diverging profile obtained by extrapolating towards the source from outside within the external PPEFT (dotted line). Two radii, $r = \epsilon_1$ and $r = \epsilon_2$, are shown where boundary conditions are applied using the boundary action $\mathcal{I}_{\mathcal{B}}(\epsilon)$ in the external EFT. The ϵ-dependence of $\mathcal{I}_{\mathcal{B}}(\epsilon)$ is defined to ensure that the external profile approximates the fixed real profile, no matter what particular value of ϵ is chosen. This shows how the ϵ-dependence of the effective boundary couplings is designed to reproduce the r-dependence of the real field profile as predicted by the bulk field equations.

the action is nonlinear in the field, and this is why the divergence issue does not arise for linear multipole couplings discussed in §13.2.3. The advantage of the boundary construction is that it goes through equally well in the presence of nonlinear terms, singular bulk potentials and other potential complications to the naive delta-function treatment.

RG As Field Equations

The boundary action also provides a simple geometrical interpretation of the renormalization group alluded to at the beginning of §13.3, and described in more detail in the examples to follow. As always, the RG expresses how physical quantities cannot depend on arbitrary scales used to regularize short-distance physics, since any apparent dependence cancels – *i.e.* is renormalized into – dependence that is implicit in the theory's coupling constants.

In the particular case of the boundary action described above, the renormalization group expresses the fact that physical quantities do not depend on the precise radius of $\mathcal{S}_\epsilon$, since this can be chosen fairly arbitrarily. This ϵ-independence arises in detail because any explicit dependence cancels an ϵ-dependence that is implicit in effective couplings like $\tilde{h}$.

The required ϵ-dependence of couplings is most easily found simply by differentiating observables with respect to ϵ while holding fixed all physical quantities. This corresponds to adjusting the couplings in $I_{\mathcal{B}}$ in such a way as to not change the physical bulk-field profile as the radius of $\mathcal{S}_\epsilon$ is varied (see Fig. 13.5). Because the bulk-field profile is determined by the bulk field equations, this Callan–Symanzik type of condition relates the RG evolution of couplings in $I_{\mathcal{B}}$ to the classical bulk evolution.[11] The couplings of S_b, such as h, then also inherit an ϵ-dependence because of relations like (13.72).

13.3.2 Contact Interaction

To see in detail how this renormalization story implies boundary conditions can be imposed without introducing ϵ-dependence into physical predictions, return to the example where S_B is given by (13.64) and S_b by (13.65) (or $\mathcal{I}_B$ by (13.71)). Working in the rest-frame of a static source the field equation for this system then is (13.66) and the boundary condition is (13.68). Expanding the field $\Psi(x) = \sum_N u_N(x)\,\mathfrak{a}_N$ as in (13.70) then implies that the modes satisfy

$$-\frac{\nabla^2 u_N}{2m} = \omega_N u_N, \tag{13.76}$$

for all $r = |\mathbf{x}| \neq 0$, with the near-source boundary condition

$$\left(r^{d-1}\frac{\partial}{\partial r}\ln u_N\right)_{r=\epsilon} = \frac{2mh}{\Omega_{d-1}}. \tag{13.77}$$

Expand now in spherical harmonics, $Y_L(\Omega)$, where $\Omega = \{\theta_1, \cdots, \theta_{d-2}\}$ denote the angular variables and the mode label N becomes $\{n, L\}$. Then $u_N(r,\Omega) = \mathcal{R}_{nL}(r)Y_L(\Omega)$, where (13.76) implies that the radial mode-function satisfies

$$r^2\frac{d^2\mathcal{R}}{dr^2} + (d-1)r\frac{d\mathcal{R}}{dr} + \left[-\varpi_L + k^2r^2\right]\mathcal{R} = 0, \tag{13.78}$$

where $k^2 = 2m\omega_{nL}$ and ϖ_L is the total angular momentum eigenvalue.

For later reference, notice that for scalar functions on a $(d-1)$-sphere for general $d \geq 2$ it is a standard result that the label L always includes a non-negative integer ℓ for which the angular-momentum spectrum is

$$\varpi_\ell = \ell(\ell + d - 2), \tag{13.79}$$

with each level arising with degeneracy $D_\ell(d)$ given by

$$D_\ell(d) = \binom{d-1+\ell}{d-1} - \binom{d-3-\ell}{d-1} = \frac{(d-3+\ell)!\,(d-2+2\ell)}{\ell!\,(d-2)!}. \tag{13.80}$$

The above formula for $D_\ell(d)$ breaks down in the special case $d = 2$ and $\ell = 0$, in which case direct calculation shows $D_0(2) = 1$.

These expressions reproduce the familiar simplest cases. For $d = 2$ angular modes are labelled by $L = \pm\ell = 0, \pm 1, \cdots$ and $\varpi_\ell = \ell^2$, and so $D_0(2) = 1$, while $D_\ell(2) = 2$ for all $\ell \neq 0$. Similarly, for $d = 3$ the modes are labelled by $L = \{\ell, \ell_z\}$, with $\ell = 0, 1, 2, \cdots$ and $\ell_z = -\ell, -\ell+1, \cdots, \ell-1, \ell$, while $\varpi_\ell = \ell(\ell+1)$ and $D_\ell(3) = 2\ell + 1$.

Returning to the radial equation, Eq. (13.78) is solved by

$$\mathcal{R}_{nL}(r) = C_+\mathcal{R}^+_{nL}(r) + C_-\mathcal{R}^-_{nL}(r), \tag{13.81}$$

[11] This argument qualitatively resembles similar arguments [343] relating field equations and RG-evolution in holographic models based on the AdS/CFT correspondence [344].

where $C_\pm$ are integration constants and the $\mathcal{R}^\pm_{nL}(r)$ are linear combinations of Bessel functions. For the present purposes it is useful to choose these functions so that they differ in their asymptotic form for small r, behaving there as

$$\mathcal{R}^\pm_{nL}(r) = (2kr)^{s_\pm}\left[1 + \mathcal{O}(kr)\right], \tag{13.82}$$

where the factor of 2 is for later convenience and

$$s_+ = \ell \quad \text{while} \quad s_- = -\ell - (d-2). \tag{13.83}$$

This power-law form breaks down in the special case $\ell = 0$ and $d = 2$, for which one of the solutions instead varies logarithmically for small r.

Now comes the main point. The boundary condition, (13.77), in this instance becomes

$$\frac{2mh}{\Omega_{d-1}\epsilon^{d-2}} = \left(r\,\frac{\partial}{\partial r}\ln \mathcal{R}_{n\ell}\right)_{r=\epsilon} = \left[\frac{r\,\partial_r \ln \mathcal{R}^+_{n\ell} + \Xi\, r\,\partial_r \ln \mathcal{R}^-_{n\ell}}{1+\Xi}\right]_{r=\epsilon}, \tag{13.84}$$

where

$$\Xi := \left(\frac{C_-}{C_+}\right)\frac{\mathcal{R}^-_{n\ell}(\epsilon)}{\mathcal{R}^+_{n\ell}(\epsilon)}. \tag{13.85}$$

There are two complementary ways to read Eqs. (13.84) and (13.85).

The naive way to read (13.84) and (13.85) is to regard them as determining C_-/C_+ once values are specified for both h and ϵ. Read this way Eq. (13.84) shows how the boundary condition coming from S_b dictates the ratio of integration constants, C_-/C_+, and thereby affects physical observables like scattering amplitudes or energy levels for the Ψ particles. At face value, C_-/C_+ obtained in this way depends explicitly on ϵ, and so also must all observables that can be expressed in terms of this ratio.

A more sophisticated reading starts from the observation that if physical observables are to be ϵ-independent then so must also be C_-/C_+. This requirement is consistent with (13.84) if $h = h(\epsilon)$ is not held fixed as ϵ is varied. In this point of view Eqs. (13.84) and (13.85) are instead read as dictating the functional form of $h(\epsilon)$ given that the physical requirement that C_-/C_+ cannot depend on ϵ. In this language Eqs. (13.84) and (13.85) give $h(\epsilon)$ as a one-parameter family of formulae, where the parameter is C_-/C_+. This formula defines the RG flow of $h(\epsilon)$. The value of C_-/C_+ then determines which particular RG trajectory of this one-parameter family of flows describes the system of interest.

These alternatives can be made more explicit by rewriting (13.84) as

$$\frac{2mh}{\Omega_{d-1}\epsilon^{d-2}} - \frac{1}{2}\left[r\,\partial_r \ln\left(\mathcal{R}^+_{n\ell}\mathcal{R}^-_{n\ell}\right)\right]_{r=\epsilon} = \frac{1}{2}\left(r\,\partial_r \ln \frac{\mathcal{R}^+_{n\ell}}{\mathcal{R}^-_{n\ell}}\right)_{r=\epsilon}\left(\frac{1-\Xi}{1+\Xi}\right), \tag{13.86}$$

with Ξ defined by (13.85). As above, the naive way to read this solves for C_-/C_+ (which appears only in Ξ), to find

$$\frac{C_-}{C_+} = -\frac{\mathcal{R}^+_{n\ell}(\epsilon)}{\mathcal{R}^-_{n\ell}(\epsilon)}\left[\frac{2mh - \Omega_{d-1}\epsilon^{d-1}\partial_r \ln \mathcal{R}^+_{n\ell}}{2mh - \Omega_{d-1}\epsilon^{d-1}\partial_r \ln \mathcal{R}^-_{n\ell}}\right]_{r=\epsilon} \tag{13.87}$$

which explicitly gives C_-/C_+ for any specified pair (ϵ, h).

The second, renormalization-group, way to view (13.84) instead defines the variable

$$\lambda(\epsilon) := \frac{1-\Xi}{1+\Xi}, \tag{13.88}$$

since this satisfies

$$\begin{aligned} \epsilon\,\partial_\epsilon \lambda &= \frac{2\,\Xi}{(1+\Xi)^2}\Big(r\,\partial_r \ln \mathcal{R}^+_{n\ell} - r\,\partial_r \ln \mathcal{R}^-_{n\ell}\Big)_{r=\epsilon} \\ &= \frac{1}{2}\Big(1-\lambda^2\Big)\Big(r\,\partial_r \ln \mathcal{R}^+_{n\ell} - r\,\partial_r \ln \mathcal{R}^-_{n\ell}\Big)_{r=\epsilon}, \end{aligned} \tag{13.89}$$

when ϵ is varied with C_-/C_+ held fixed.

The main difference between Eqs. (13.88) and (13.89) is that the differential version can be regarded as an equation that relates the running with ϵ of λ to $\mathcal{R}^\pm$ in a way that does not depend explicitly on C_-/C_+. The appearance of the ratio C_-/C_+ in (13.88) (through expression (13.85) giving Ξ in terms of $\mathcal{R}^\pm$) shows that it can be regarded as the integration constant found when integrating (13.89). Once $\lambda(\epsilon)$ is known then the corresponding RG evolution for $h(\epsilon)$ is found by writing (13.86) as

$$\frac{2mh}{\Omega_{d-1}\epsilon^{d-2}} - \frac{1}{2}\Big[r\,\partial_r \ln\Big(\mathcal{R}^+_{n\ell}\mathcal{R}^-_{n\ell}\Big)\Big]_{r=\epsilon} = \frac{\lambda(\epsilon)}{2}\left(r\,\partial_r \ln \frac{\mathcal{R}^+_{n\ell}}{\mathcal{R}^-_{n\ell}}\right)_{r=\epsilon}. \tag{13.90}$$

RG Evolution at Leading Order in $2k\epsilon$

It is worth pausing at this point to explore more fully the properties of the RG evolution defined by (13.89) – in particular, how $h(\epsilon)$ and $\lambda(\epsilon)$ evolve – in the low-energy regime of most practical interest: $2k\epsilon \ll 1$. Evolution formulae for this regime are collected here since they are also useful in later sections.

In this regime simplification occurs because the mode functions satisfy $\mathcal{R}^\pm_{n\ell}(\epsilon) \simeq (2k\epsilon)^{s_\pm}$ (*c.f.* Eq. (13.82)). With this asymptotic form equation (13.87) reduces to

$$\frac{C_-}{C_+} = \Xi\left[\frac{\mathcal{R}^+_{n\ell}(\epsilon)}{\mathcal{R}^-_{n\ell}(\epsilon)}\right] \simeq \frac{1-\lambda(\epsilon)}{1+\lambda(\epsilon)}\,(2k\epsilon)^{s_+-s_-}. \tag{13.91}$$

This inverts (13.88) to eliminate Ξ in terms of $\lambda(\epsilon)$. Eq. (13.90) giving h in terms of λ simplifies similarly,

$$\frac{2mh}{\Omega_{d-1}\epsilon^{d-2}} \simeq \frac{1}{2}(s_+ + s_-) + \frac{\lambda(\epsilon)}{2}(s_+ - s_-), \tag{13.92}$$

while Eq. (13.89), giving the differential running of λ, becomes [357]

$$\epsilon\,\partial_\epsilon \lambda \simeq \frac{1}{2}(s_+ - s_-)\big(1-\lambda^2\big), \tag{13.93}$$

when ϵ is varied with C_-/C_+ held fixed.

The approximate equality in these last three expressions indicates how the right-hand sides drop a factor of the form $[1 + O(2k\epsilon)]$. These last expressions do *not* also use the explicit formulae (13.83) for $s_\pm$ because they apply equally to the more general situations encountered in later examples.

The evolution defined by (13.93) has fixed points at $\lambda_{\star\pm} = \pm 1$. Eq. (13.92) shows that if λ sits at the fixed point $\lambda_{\star\pm} = \pm 1$ while ϵ varies, then the original coupling $2mh_{\star\pm}(\epsilon)$ must scale with its naive scaling dimension,

$$2mh_{\star\pm}(\epsilon) \simeq \Omega_{d-1} s_\pm \, \epsilon^{d-2}. \tag{13.94}$$

From this point of view, nontrivial running for $\lambda(\epsilon)$ corresponds to $2mh$ acquiring an 'anomalous' scaling dimension.

A perhaps surprising observation hides within Eq. (13.94): the fixed-point condition need *not* be consistent with the absence of a first-quantized coupling: $h = 0$. When $h = 0$ is not a fixed point it is inconsistent to ignore this particular bulk-source interaction, since even if it is set to zero at some specific scale ϵ_0, RG evolution implies that it cannot remain zero at other values for ϵ.

A second potential surprise lies in the observation that effective couplings like $2mh$ run in a way that depends on bulk-particle quantum numbers – such as angular momentum, ℓ, which appears in $s_\pm$ through formulae like Eqs. (13.83). On one hand, this seems natural since different choices for ℓ lead to radial mode-functions with different near-source asymptotic behaviour. On the other hand, having effective couplings depend on bulk quantum numbers seems counter-intuitive since normally physical properties of a compact source (like multipole moments) are intrinsic properties of the source alone. As is argued below, however, physical properties are characterized by RG-invariants, so it is only these that should be expected to be intrinsic to the source alone.

To see how this works, it helps to identify convenient RG-invariant descriptions of the coupling flow. To do so notice that the general solution to the flow equation (13.93) is

$$\lambda(\epsilon) = \frac{(1+\lambda_0)(\epsilon/\epsilon_0)^{s_+-s_-} - (1-\lambda_0)}{(1+\lambda_0)(\epsilon/\epsilon_0)^{s_+-s_-} + (1-\lambda_0)}, \tag{13.95}$$

where the integration constant is chosen so that $\lambda(\epsilon_0) = \lambda_0$. But this expression has (by construction) precisely the same ϵ-dependence as does (13.88), and so comparing them shows that the integration constant λ_0 is related to C_-/C_+ by

$$\frac{C_-}{C_+} = \frac{1-\lambda_0}{1+\lambda_0}\,(2k\epsilon_0)^{s_+-s_-} = \frac{1-\lambda(\epsilon)}{1+\lambda(\epsilon)}\,(2k\epsilon)^{s_+-s_-}. \tag{13.96}$$

Here the second equality uses (13.95) to eliminate (λ_0, ϵ_0) in favour of (λ, ϵ). This verifies how the physical quantity C_-/C_+ is ϵ-independent and instead depends only on a choice for a particular RG trajectory, $\lambda(\epsilon)$.

Eq. (13.95) also reveals that the general RG evolution runs from the fixed point at $\lambda_{\star-} = -1$ to the fixed point at $\lambda_{\star+} = +1$ as ϵ runs from zero to infinity, as illustrated in Fig. 13.6. The figure reveals two distinct categories of flow, distinguished by $\eta_\star = \text{sign}(\lambda^2 - 1)$, which is itself an RG-invariant quantity. Any flow line is uniquely characterized in an RG-invariant way by specifying both $\eta_\star$ and $\epsilon_\star$ defined by the condition $|\lambda(\epsilon_\star)| = \infty$ if $\eta_\star = +1$ or $\lambda(\epsilon_\star) = 0$ if $\eta_\star = -1$. Inspection of (13.95) shows that these are defined in terms of an initial-condition pair, (λ_0, ϵ_0), by $\eta_\star = \text{sign}(\lambda_0^2 - 1)$ and

$$\frac{\epsilon_\star}{\epsilon_0} = \left|\frac{\lambda_0 - 1}{\lambda_0 + 1}\right|^{1/(s_+-s_-)}. \tag{13.97}$$

Finally, (13.95) and (13.96) give the flow and C_-/C_+ in terms of the RG-invariant parameters $\eta_\star$ and $\epsilon_\star$, as follows

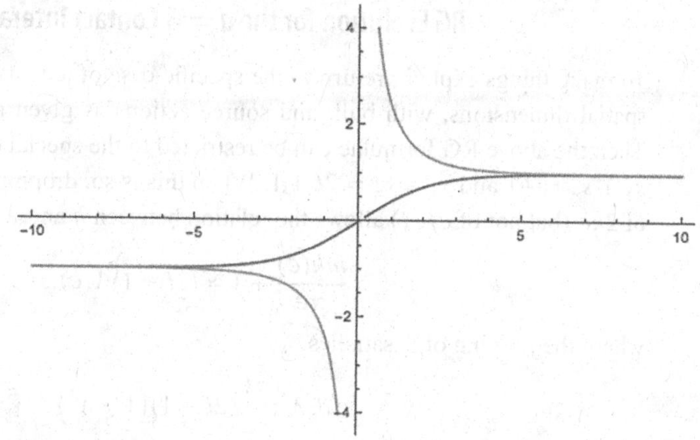

Fig. 13.6 Plot of the RG flow predicted by Eq. (13.95) for λ vs $\ln(\epsilon/\epsilon_\star)$ where the RG-invariant scale $\epsilon_\star$ is chosen to be the unique value of ϵ for which λ either vanishes or diverges, depending on the RG-invariant sign $\eta_\star = \text{sign}(\lambda^2 - 1)$.

$$\lambda(\epsilon) = \frac{(\epsilon/\epsilon_\star)^{s_+-s_-} + \eta_\star}{(\epsilon/\epsilon_\star)^{s_+-s_-} - \eta_\star} \quad \text{and} \quad \frac{C_-}{C_+} = -\eta_\star\,(2k\epsilon_\star)^{s_+-s_-}. \tag{13.98}$$

Since physical observables (like the scattering cross sections or energy-level shifts – see Exercises 13.2–13.6 and 13.7 for example) are computed in terms of C_-/C_+, this last equation shows that these only depend on the RG-invariant parameters $\epsilon_\star$ and $\eta_\star$ describing the coupling's RG trajectory.

Eq. (13.98) also reveals the physical meaning of the scale $\epsilon_\star$: it is the radius where the solution transitions between being dominated by each type of small-r asymptotic behaviour: $\mathcal{R}^+ \simeq (2kr)^{s_+}$ and $\mathcal{R}^- \simeq (2kr)^{s_-}$. From this point of view the RG evolution merely traces the crossover between these two asymptotic forms as one moves into the bulk away from the compact source. The fixed points then correspond to the special cases where no such transition takes place: *i.e.* either C_- or C_+ vanishes, as is most easily seen by setting $\lambda_0 = \pm 1$ in (13.96).

Notice that because $\epsilon_\star$ is a derived scale it need not be similar to the physical underlying size, R, of the source. A hierarchy between $\epsilon_\star$ and R occurs if initial conditions for very small scales, $\epsilon_0 \sim \mathcal{O}(R)$, give values for λ_0 very near one of the fixed points $\lambda_{\star\pm}$. Whenever $\lambda_0 \simeq \lambda_{\star-}$ inspection of Fig. 13.6 shows that $\epsilon_\star \gg \epsilon_0$, and so physical quantities (like scattering cross sections – see Exercise 13.2) can be much larger than the geometrical size of the compact central object. Practical examples along these lines include some light nuclei [345], or trapped atoms tuned near a Feshbach resonance [347], for which scattering lengths can be significantly larger than the object's size.[12]

Conversely, $\lambda_0 \simeq \lambda_{\star+}$ implies that $\epsilon_\star \ll \epsilon_0$. The extreme case of this is when there is no central source at all. This corresponds to the RG-invariant condition $\epsilon_\star \to 0$, in which case (13.98) implies that the coupling sits at the IR fixed point, $\lambda = \lambda_{\star+}$ and so (13.98) implies that $C_- = 0$. This implies that the radial solution remains smooth at the origin (as is usually assumed in the absence of a central object).

[12] See [346] for a description of systems with large scattering lengths within 2nd-quantized EFTs.

RG Evolution for the $d = 3$ Contact Interaction

To make things explicit, return to the specific case of a contact interaction in $d = 3$ spatial dimensions, with bulk and source actions as given in (13.64) and (13.65). Then the above RG formulae can be restricted to the special case (13.83), for which $s_+ + s_- = -1$ and $s_+ - s_- = 2\ell + 1$. When this is so, dropping subdominant powers of $2k\epsilon$ (but not of $\epsilon/\epsilon_\star$) allows the relation between h and λ to be written

$$\frac{mh(\epsilon)}{\pi\epsilon} + 1 \simeq (2\ell + 1)\lambda(\epsilon), \tag{13.99}$$

where the running of λ satisfies

$$\epsilon\, \partial_\epsilon \lambda \simeq \frac{1}{2}(2\ell + 1)\big(1 - \lambda^2\big), \tag{13.100}$$

with solution

$$\lambda(\epsilon) = \frac{(1 + \lambda_0)(\epsilon/\epsilon_0)^{2\ell+1} - (1 - \lambda_0)}{(1 + \lambda_0)(\epsilon/\epsilon_0)^{2\ell+1} + (1 - \lambda_0)} = \frac{(\epsilon/\epsilon_\star)^{2\ell+1} + \eta_\star}{(\epsilon/\epsilon_\star)^{2\ell+1} - \eta_\star}. \tag{13.101}$$

In terms of these the physical integration constant C_-/C_+ is given by

$$\frac{C_-}{C_+} = \frac{1 - \lambda_0}{1 + \lambda_0}\, (2k\epsilon_0)^{2\ell+1} = -\eta_\star\, (2k\epsilon_\star)^{2\ell+1}\,. \tag{13.102}$$

Once C_-/C_+ is known, any bulk observable (such as cross sections and energy levels) can be computed, such as is done explicitly in Exercises 13.2–13.6 at the end of this chapter. Using (13.102) in these expressions then shows how these observables depend only on the parameters $\epsilon_\star$ and $\eta_\star$, and so depend on $h(\epsilon)$ and ϵ only in an RG-invariant way. For instance, the scattering phase δ for elastic Ψ scattering from the heavy central object predicted by the actions (13.64) and (13.65) is given by (see Exercise 13.2)

$$e^{2i\delta} = \frac{1 - i\eta_\star k\epsilon_\star}{1 + i\eta_\star k\epsilon_\star}, \tag{13.103}$$

where $k = \sqrt{2mE}$ is the momentum of the scattered Ψ particle (whose kinetic energy is E).

The *absence* of the central source similarly corresponds to the choice $C_- = 0$ (since this makes the radial solution smooth at $r = 0$), or equivalently to $\epsilon_\star = 0$, in which case $\lambda(\epsilon) = \lambda_{\star+} = +1$ for all ϵ. As Eq. (13.99) shows, the coupling h then is

$$h = h_{\star+} = \frac{2\pi\ell\,\epsilon}{m}, \tag{13.104}$$

which vanishes for nonzero ϵ when $\ell = 0$, and not otherwise. This nonvanishing expression for $h(\epsilon)$ when $\ell \neq 0$ is precisely what is required to ensure $\mathcal{R}_{n\ell}(r) \propto (2kr)^\ell$ once used in the original boundary condition (13.84):

$$\frac{mh}{2\pi} = \ell\,\epsilon = \left(r^2\, \frac{\partial}{\partial r} \ln \mathcal{R}_{n\ell}\right)_{r=\epsilon}. \tag{13.105}$$

The presence of a nontrivial central compact source is signalled by any deviation from the above specific asymptotic form, which necessarily involves a nonzero overlap with $\mathcal{R}^-_{n\ell}(r)$. This occurs if there should be any radius ϵ_0 for which $\lambda_0 = \lambda(\epsilon_0) \neq +1$. If so, the physical content of this source lies in the RG-invariant quantities $\eta_\star = \text{sign}(\lambda_0^2 - 1)$ and $\epsilon_\star$ found using the second equality of (13.102). Although h depends on bulk quantum numbers like ℓ, the physical RG-invariant quantities $\eta_\star$ and $\epsilon_\star$ do not.

RG methods are usually useful inasmuch as they can be used to resum corrections, such as large logarithms of scale. The above formulae also provide some insight into what is being resummed in this case. To see why, consider the weak-coupling limit where $\epsilon_\star$ is much smaller than the central object's intrinsic size, $\epsilon_\star \ll \epsilon_0$, and specialize to the case where $\ell = 0$. With these choices λ is close to $+1$ for the entire region $\epsilon > \epsilon_0$ outside the source, and so Eq. (13.101) can be expanded in powers of $\epsilon_\star/\epsilon$ to give

$$\lambda(\epsilon) \simeq 1 + \frac{2\eta_\star\epsilon_\star}{\epsilon} + O\left[\left(\frac{\epsilon_\star}{\epsilon}\right)^2\right]. \tag{13.106}$$

Comparing this with Eq. (13.99) gives $mh/(\pi\epsilon) = \lambda - 1 \simeq 2\eta_\star\epsilon_\star/\epsilon + \cdots$, and so to leading order in $\epsilon_\star/\epsilon$ the coupling h does not evolve with scale, with

$$mh \simeq 2\pi\eta_\star\epsilon_\star + O\left(\frac{\epsilon_\star^2}{\epsilon}\right) \qquad (\text{for } \ell = 0). \tag{13.107}$$

Using this to trade $\eta_\star\epsilon_\star$ for h in the scattering phase given in (13.103) then gives

$$e^{2i\delta} = \frac{1 - i\eta_\star k\epsilon_\star}{1 + i\eta_\star k\epsilon_\star} \simeq \frac{1 - i(mh\,k/2\pi)}{1 + i(mh\,k/2\pi)}, \tag{13.108}$$

reproducing standard formulae [348] for scattering from a delta-function potential $h\,\delta^3(\mathbf{x})$. When $\epsilon_\star$ is not quite so small, however, the first equality of (13.108) remains true but the connection between $\epsilon_\star$ and $h(\epsilon_0)$ becomes more complicated, as found by combining Eqs. (13.99) and (13.101). This more complicated RG-improved expression resums corrections in the dimensionless variable $mh(\epsilon_0)/\epsilon_0$ as this combination becomes larger, as it does once $\epsilon_\star/\epsilon_0$ becomes order unity.

13.3.3 Inverse-Square Potentials: Fall to the Centre

This section generalizes the above discussion by extending the PPEFT formalism to the case where the bulk field experiences a long-range inverse-square attraction towards the heavy central object in addition to their contact interaction. It is shown that the presence of such an inverse-square potential makes the presence of the contact interaction compulsory, in the sense that the inverse-square potential modifies the running of the effective coupling h in such a way that makes it *inconsistent* to set $h = 0$ for all scales.

Attractive inverse-square potentials (and those more singular than this for small r) have long been studied as quantum systems [349] because the competition between the potential and the angular momentum barrier necessarily modifies the asymptotic shape of the wave-function near the origin. In particular, these systems provide concrete examples where the wave-function cannot remain bounded at the origin, which makes predictions depend sensitively on precisely how boundary conditions at the origin are chosen.

From the PPEFT point of view, this sensitivity to boundary conditions simply reflects the fact that the attractive potential concentrates the bulk-particle probability to lie near the compact source, making its properties sensitive to the point-particle action, S_b. Because the properties of S_b dictate the required boundary condition, its form removes all guess-work from the choice of boundary conditions to be made at the origin. A concrete example explored in Exercise 13.3 also illustrates how the boundary conditions chosen at the source need not always be chosen to be self-adjoint, depending on whether the UV physics allows probability to be lost at the source. Just as in §12.2.4, such probability loss shows up here through the appearance of complex couplings within the low-energy EFT [350].

Consider, therefore, the complex Schrödinger field, Ψ, governed by the bulk action of Eq. (13.63) (repeated for convenience here)

$$S_B = \int d^D x \left\{ \frac{i}{2} (\Psi^* \partial_t \Psi - \partial_t \Psi^* \Psi) - \frac{1}{2m} \nabla \Psi^* \cdot \nabla \Psi - V(\mathbf{x}) \Psi^* \Psi + \cdots \right\}, \tag{13.109}$$

where the potential is now chosen to be

$$V(\mathbf{x}) = -\frac{g}{r^2}, \tag{13.110}$$

with $g > 0$ and $r = |\mathbf{x}|$ is the distance to the compact, massive central source, which is assumed to be approximately localized at $\mathbf{x} = 0$. Spacetime dimension $D = d + 1$ is again kept open, though the cases of practical interest are $d = 2$ (a line source) and $d = 3$ (point-particle). See Exercise 13.3 for a concrete atomic system described by this action.

The Schrödinger modes in the presence of such a potential satisfy

$$-\frac{\nabla^2 u_N}{2m} - \frac{g}{r^2} u_N = \omega_N u_N, \tag{13.111}$$

for all $r \neq 0$, and once decomposed in terms of spherical harmonics, $u_N(r, \Omega) = \mathcal{R}_{nL}(r) Y_L(\Omega)$, the radial function $\mathcal{R}_{nL}$ satisfies

$$r^2 \frac{d^2 \mathcal{R}}{dr^2} + (d-1) r \frac{d\mathcal{R}}{dr} + \left[2mg - \varpi_L + k^2 r^2 \right] \mathcal{R} = 0, \tag{13.112}$$

instead of (13.78), where as before $k^2 = 2m\omega_{nL}$ and $\varpi_L = \ell(\ell + d - 2)$ is the total angular momentum eigenvalue in d spatial dimensions.

Because the inverse-square potential competes with the angular-momentum barrier, the independent radial solutions

$$\mathcal{R}_{nL}(r) = C_+ \mathcal{R}^+_{nL}(r) + C_- \mathcal{R}^-_{nL}(r), \tag{13.113}$$

have small-r asymptotic forms that depend on g, with $\mathcal{R}^\pm_{nL}(r) = (2kr)^{s_\pm} [1 + \mathcal{O}(kr)]$ where $s_\pm^2 + (d-2)s_\pm + 2mg - \varpi_L = 0$ and so

$$s_\pm = \frac{1}{2} [2 - d \pm \zeta] \quad \text{with} \quad \zeta := \sqrt{(2\ell + d - 2)^2 - 8mg}. \tag{13.114}$$

The convention is adopted that $\zeta > 0$ when it is real and nonzero. Notice that ζ initially decreases as g increases from zero, but eventually becomes imaginary (so both powers $s_\pm$ become complex) when $g > g_c$ where

$$g_c := \frac{1}{8m} (2\ell + d - 2)^2. \tag{13.115}$$

The interesting thing about this asymptotic behaviour is that for some quantum numbers *both* solutions are singular at the origin, underlining that boundedness at the origin cannot in general be the right criterion for choosing the boundary condition for small r. The singularity of the wavefunction reflects the accumulation of probability near the origin due to the suppression of the centrifugal barrier by the attractive potential, making energy levels and scattering amplitudes depend sensitively on the boundary conditions chosen at the origin.

What is important is that these near-origin boundary conditions are not arbitrary; they are dictated by the PPEFT action, S_b, describing the source that resides there. In the present instance, the lowest-dimension interaction for a Schrödinger field is the same as in the previous example, Eq. (13.65). Choosing coordinates so that the central source sits at $\mathbf{x} = 0$ in its rest frame, the leading term in the first-quantized source action becomes

$$S_b = -\int \mathrm{d}^D x \; h \, \Psi^* \Psi \, \delta^d(\mathbf{x}). \tag{13.116}$$

Because the derivation of the boundary conditions that follow from this action so closely resembles the treatment of §13.3.1, in what follows only the main steps are highlighted, with an emphasis on places where the inverse-square potential modifies the conclusions.

RG Evolution for Real ζ

The RG evolution of the interaction of (13.116) is as given in the previous sections, with the presence of the bulk inverse-square potential implying $s_+ + s_- = -(d-2)$ and $s_+ - s_- = \zeta$, with ζ as given in (13.114).

The connection between h and λ in this case therefore becomes

$$\frac{4mh}{\Omega_{d-1}\epsilon^{d-2}} + (d-2) \simeq \hat{\lambda}(\epsilon) := \zeta\, \lambda(\epsilon), \tag{13.117}$$

in which the last equality defines the useful variable $\hat{\lambda}$. The running of $\hat{\lambda}$ inherited from the evolution equation (13.93) is

$$\epsilon\, \partial_\epsilon \hat{\lambda} \simeq \frac{1}{2}\left(\zeta^2 - \hat{\lambda}^2\right), \tag{13.118}$$

with solutions

$$\frac{\hat{\lambda}}{\zeta}(\epsilon) = \frac{(\zeta + \hat{\lambda}_0)(\epsilon/\epsilon_0)^\zeta - (\zeta - \hat{\lambda}_0)}{(\zeta + \hat{\lambda}_0)(\epsilon/\epsilon_0)^\zeta + (\zeta - \hat{\lambda}_0)} = \frac{(\epsilon/\epsilon_\star)^\zeta + \eta_\star}{(\epsilon/\epsilon_\star)^\zeta - \eta_\star}, \tag{13.119}$$

and RG-invariant quantities

$$\eta_\star = \mathrm{sign}(\hat{\lambda}^2 - \zeta^2) \quad \text{and} \quad \frac{\epsilon_\star}{\epsilon_0} = \left|\frac{\hat{\lambda}_0 - \zeta}{\hat{\lambda}_0 + \zeta}\right|^{1/\zeta}. \tag{13.120}$$

In terms of these the physical mode-function integration constants are

$$\frac{C_-}{C_+} = -\eta_\star\, (2k\epsilon_\star)^\zeta. \tag{13.121}$$

In the absence of the inverse-square potential the special case $C_- = 0$ corresponded to choosing bounded mode-functions at the origin, and made $\lambda = \lambda_{\star +} = +1$ sit at the

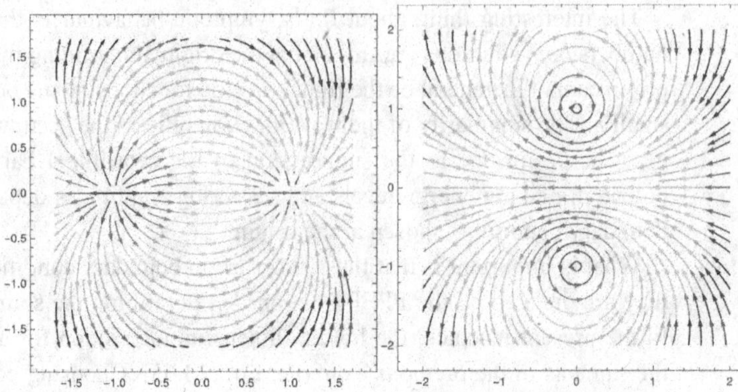

Fig. 13.7 The RG evolution predicted by Eq. (13.119) in the complex $\hat{\lambda}/|\zeta|$ plane. The left (right) panel uses a real (imaginary) value for ζ. Arrows (shading) show direction (speed) of flow as ϵ increases. Figure taken from [350]

IR fixed point for all ϵ. For the $\ell = 0$ mode this also implied $h(\epsilon) = 0$ for all ϵ. These statements change in the presence of the inverse-square potential, however, since in this case Eqs. (13.114) and (13.117) together show that $g \neq 0$ is an obstruction to the coupling h vanishing for all ϵ, even when $\hat{\lambda} = \hat{\lambda}_{\star+} = +\zeta$ and $\ell = 0$. This shows how the presence of an inverse-square potential makes it *inconsistent* to set to zero the effective coupling h, even when $\epsilon_\star = \ell = 0$. Although $h(\epsilon_0)$ might vanish at some scale ϵ_0, if $g \neq 0$ then (for all ℓ) $h(\epsilon)$ *cannot* vanish for $\epsilon \neq \epsilon_0$. This is the RG version of the message that inverse-square potentials are intrinsically sensitive to the details of the boundary conditions near the origin, and so also require information about the properties of S_b.

RG Evolution for Complex $\hat{\lambda}$

For sufficiently strong inverse-square potentials – *i.e.* when $g > g_c$ as defined in (13.115) – the parameter ζ is no longer real even if the original coupling h is. When ζ becomes imaginary the flow qualitatively changes, as might be expected given that both solutions $\mathcal{R}^\pm$ then share the same value for Re $s_\pm$, which implies that the behaviour of $|\mathcal{R}|$ is the same for both at small r.

Fig. 13.7 draws the resulting flow lines within the complex $\hat{\lambda}$ plane, with the left panel showing the flow for real ζ and the right panel illustrating the case where ζ is imaginary. As the figure makes clear, the condition $\hat{\lambda} = \hat{\lambda}^*$ is RG invariant, so once $\hat{\lambda}$ is chosen real at any scale it remains so for all scales, regardless of the sign of ζ^2.

What changes when ζ is imaginary is the ability to hit a fixed point starting from a real initial value $\hat{\lambda}(\epsilon_0) = \hat{\lambda}_0$. The flow of an initially real $\hat{\lambda}$ instead displays a limit-cycle behaviour which reflects the emergence of a discrete scale-invariance. Further exploration of this kind of flow goes beyond the scope of this section, though it has practical applications, such as to the Efimov effect [351]: a universal discrete scale-invariance that emerges in the low-energy limit of many-body scattering.

Why entertain complex values for $\hat{\lambda}$ in the first place if RG flow preserves the reality of $\hat{\lambda}$, even when ζ is imaginary? This can be worth doing because for some systems complex values for $\hat{\lambda}$ (and so also for h) are appropriate, corresponding

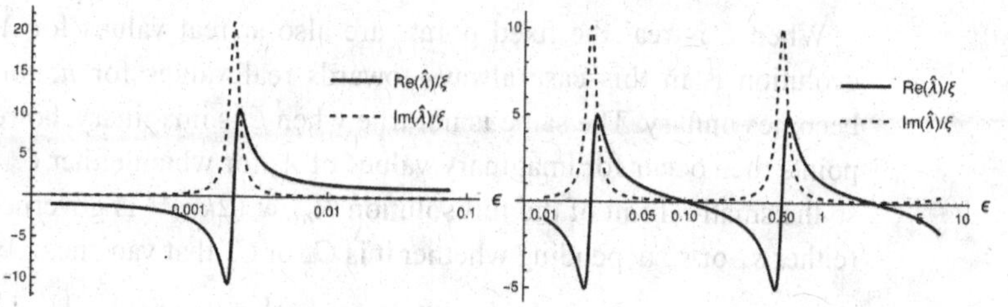

Fig. 13.8 RG flows predicted by (13.121) for Re $\hat{\lambda}/\xi$ and Im $\hat{\lambda}/\xi$ (where $\xi = |\zeta|$) for ζ real (left panel) and ζ imaginary (right panel). Each flow defines an RG-invariant scale $\epsilon_\star$ defined by Re $\hat{\lambda}(\epsilon_\star) = 0$, at which point Im $\hat{\lambda}(\epsilon_\star) = iy_\star$ is a second RG-invariant label. $\epsilon_\star$ is multiply defined when ζ is imaginary. Figure taken from [350]

to the occurrence of probability loss at the position of the central object (in much the same way that annihilation gave an imaginary part to couplings in the EFT for positronium described in §12.2.4). A concrete example of this is provided by the gas of polarizable atoms around a charged wire introduced in Exercise 13.3, for which the atoms in question could be in an atomic trap from which they are ejected on scattering from the charged wire (with a consequent loss of probability localized at the position of the wire).

To quantify the relation between Im h and probability loss at the source compute the radial probability flux operator at $r = \epsilon$:

$$J_r(\epsilon) = \frac{i}{2m}(\Psi\partial_r\Psi^* - \Psi^*\partial_r\Psi)_{r=\epsilon} = i\,(h^* - h)\,\frac{|\Psi(\epsilon)|^2}{\Omega_{d-1}\epsilon^{d-1}}, \tag{13.122}$$

where as before Ω_{d-1} is the surface area of the unit $d-1$ sphere and the final equality uses the boundary condition (13.74). The operator controling the net rate of probability flow out of a sphere of radius $r = \epsilon$ then is

$$\mathcal{P} := \oint_{r=\epsilon} J_r\,\epsilon^{d-1}\mathrm{d}^{d-1}\Omega = 2\,|\Psi(\epsilon)|^2\,\mathrm{Im}\,h, \tag{13.123}$$

and so positive (negative) Im h corresponds to the compact object at $\mathbf{x} = 0$ being a net probability source (sink).

Once $\hat{\lambda}$ is complex then the RG evolution is richer, even when ζ is real, as is illustrated by the plots of Fig. 13.7. The projections of these plots giving Re $\hat{\lambda}$ and Im $\hat{\lambda}$ as functions of ϵ are seen in Fig. 13.8. These flows can be characterized using two RG-invariant real numbers, $\epsilon_\star$ and $y_\star$, where $\epsilon_\star$ is defined as the scale where the flow crosses the imaginary axis – *i.e.* where Re $\hat{\lambda}(\epsilon_\star) = 0$ – and $y_\star$ defined by the imaginary value taken by $\hat{\lambda}$ at this crossing: Im $\hat{\lambda}(\epsilon_\star) =: iy_\star$.

When the flow crosses the imaginary axis more than once (as happens when ζ is imaginary) $y_\star$ is chosen as the larger of the absolute values taken by Im $\hat{\lambda} = y_\star$ on the trajectory of interest. (Equivalently, $y_\star$ is defined as the value at the axis crossing for which the real part of $\hat{\lambda}$ satisfies $\mathrm{d}\hat{\lambda}_R/\mathrm{d}\epsilon > 0$.) The parameter $y_\star$ generalizes the RG-invariant sign $\eta_\star$ defined above for real $\hat{\lambda}$, inasmuch as when specialized to real $\hat{\lambda}$ the new parameter $y_\star$ takes one of two values: $y_\star = 0$ (corresponding to $\eta_\star = -1$ in earlier sections) or $y_\star \to \pm\infty$ (corresponding to $\eta_\star = +1$ in earlier sections).

When ζ is real the fixed points are also at real values for $\hat{\lambda}$, showing that RG evolution is in this case always towards real values for h, for which the system becomes unitary. The same is not true when ζ is imaginary, however, since the fixed points then occur for imaginary values of $\hat{\lambda}$, for which either C_+ or C_- vanishes and so the small-r limit of the full solution $\mathcal{R}_{n\ell} \propto (2kr)^{s_\pm}$ is governed by a single power (either s_+ or s_- depending whether it is C_+ or C_- that vanishes). Writing $\zeta = \pm i\xi$ and

$$(kr)^\zeta e^{-i\omega t} = \exp\left\{-i\left[\omega t \mp \xi \ln(kr)\right]\right\} \tag{13.124}$$

shows that these fixed-point solutions correspond to completely in-falling or completely out-going boundary conditions (with no admixture of the other type) in the small-r limit.[13] Exercises 13.5 and 13.6 give explicit expressions for bound-state energies and scattering cross sections (both elastic and inelastic) for Ψ particles as functions of the RG-invariant quantities $y_\star$ and $\epsilon_\star$.

13.3.4 Nuclear Effects in Atoms

Atoms provide a very natural place to apply first-quantized tools, with the nucleus playing the role of the compact central source. Atoms are also of considerable practical interest because high-precision measurements of their energy levels call for equally accurate theoretical calculations. Reliable determinations of how atomic energy levels respond to nuclear structure are valuable, since nuclear uncertainties often dominate theoretical errors.

This section explores several issues that arise when studying how electrons couple to nuclei, representing the electron as a Schrödinger field (with a brief discussion of how to include relativistic effects at the end). For simplicity only individual electrons orbiting spinless nuclei are considered here, such as for $^4He^+$ ions.

Using a 2-component Pauli spinor field, Ψ, to represent the bulk nonrelativistic electrons leads to a bulk action similar to the one used in earlier sections, see *e.g.* Eq. (13.63). To this is added the Maxwell action for electromagnetism. Altogether $S_B = \int d^4x\, \mathfrak{L}_B$ with

$$\mathfrak{L}_B = \frac{1}{2}\left(\mathbf{E}^2 - \mathbf{B}^2\right) + \frac{i}{2}\left(\Psi^\dagger D_t\Psi - D_t\Psi^\dagger\,\Psi\right) - \frac{1}{2m}\mathbf{D}\Psi^\dagger\cdot\mathbf{D}\Psi - \frac{ec_F}{2m}\mathbf{B}\cdot(\Psi^\dagger\sigma\,\Psi) + \cdots, \tag{13.125}$$

where $D_t\Psi = (\partial_t + ieA_0)\Psi$ and $\mathbf{D}\Psi = (\nabla + ie\mathbf{A})\Psi$ while $c_F = 1 + \mathcal{O}(\alpha)$ is the effective magnetic-moment coupling whose value is obtained by matching to QED, as in (12.1.2). For atoms the ellipses in (13.125) can also be important, in principle containing the subdominant single-body terms discussed in more detail in §12.2, such as kinematic corrections to the dispersion relation, $E = m + \mathbf{p}^2/(2m) + \mathbf{p}^4/(8m^3) + \cdots$, spin-orbit couplings, and so on.

The leading couplings of electron and electromagnetic fields to the first-quantized nucleus are given by (13.30) and (13.65) as

[13] These fixed-point boundary conditions are called perfect absorbers or perfect emitters. See [352] for an application of these boundary conditions to black-hole physics, for which having a perfect absorber as a fixed point corresponds physically to the horizon not being a particularly special place.

$$S_N = \int d^4x \left\{ M + QA_0 + C_D \nabla \cdot \mathbf{E} - h\Psi^\dagger\Psi + \cdots \right\} \delta^3(\mathbf{x}), \tag{13.126}$$

where $Q = Ze$ is the nuclear charge and we neglect (for simplicity) nuclear recoil effects, and so specialize to the nuclear rest-frame, choosing coordinates that place it at $\mathbf{x} = 0$. Here, the ellipses contain only terms suppressed by more than two powers of the nuclear size, R.

Using this action to infer the near-nucleus boundary conditions for the electromagnetic field, and perturbing in the electron-photon couplings, leads – see (13.55) and the discussion that follows it – to A_μ acquiring a nontrivial background configuration, $A_\mu = \mathcal{A}_\mu + \delta A_\mu$, with

$$\mathcal{A}_0(\mathbf{x}) = -\frac{Ze}{4\pi r} + C_D\, \delta^3(\mathbf{x}). \tag{13.127}$$

Furthermore, for bound-state calculations the power-counting arguments of §12.2.1 argue that this background should be included in the unperturbed action, corresponding in the second-quantized language to resumming multiple Coulomb exchange.

The lowest-order electron-field expansion then takes the form $\Psi(x) = \sum_N u_N(x)\, \mathfrak{a}_N$, where $u_N(x)$ satisfies

$$\begin{aligned} 0 &= i\partial_t u_N + \frac{\nabla^2 u_N}{2m} - e\mathcal{A}_0(\mathbf{x})u_N(\mathbf{x}) - h\,\delta^3(\mathbf{x})\,u_N \\ &= i\partial_t u_N + \frac{\nabla^2 u_N}{2m} + \frac{Z\alpha}{r}u_N(\mathbf{x}) - \left(h + eC_D\right)\delta^3(\mathbf{x})\,u_N, \end{aligned} \tag{13.128}$$

with $\alpha = e^2/4\pi$ being the usual electromagnetic fine-structure constant. Energy eigenstates in the presence of such a potential in particular satisfy the usual time-independent Schrödinger–Coulomb equation

$$-\frac{\nabla^2 u_N}{2m} - \frac{Z\alpha}{r}u_N = \omega_N u_N, \tag{13.129}$$

for all $r \neq 0$, subject to the near-nucleus boundary condition – *c.f.* Eqs. (13.68) and (13.77):

$$\left(\frac{\partial}{\partial r}\ln u_N\right)_{r=\epsilon} = \frac{2mh_{\rm eff}}{\Omega_2} = \frac{mh_{\rm eff}}{2\pi}, \tag{13.130}$$

where

$$h_{\rm eff} := h + eC_D. \tag{13.131}$$

Having h and C_D only appear together through the one combination $h_{\rm eff}$ reflects the fact that these two couplings are effectively redundant. This redundancy can be seen using the arguments of §2.5 because the Darwin term can be simplified using the Maxwell equation $\nabla \cdot \mathbf{E} = -e\Psi^\dagger\Psi + Ze\,\delta^3(\mathbf{x})$.

The main conclusion to which these expressions lead is that the leading effects of the nucleus' finite size for electronic energy levels is communicated through the near-nucleus boundary condition (13.130). At leading order there is only a single parameter, $h_{\rm eff}$, describing the nuclear shape that can contribute to atomic energy-level shifts. Furthermore, this nuclear parameter is related to its mean-square charge radius – *c.f.* Eq. (13.57) – in a way that is established in more detail below.

Atomic Energy Levels

To compute nucleus-dependent atomic energy shifts explicitly, decompose u_N in terms of spherical harmonics, $u_N(r,\theta,\phi) = \mathcal{R}_{n\ell}(r) Y_{\ell\ell_z}(\theta,\phi)$, with the radial function $\mathcal{R}_{n\ell}$ satisfying

$$r^2 \frac{\mathrm{d}^2\mathcal{R}_{n\ell}}{\mathrm{d}r^2} + (d-1)r\frac{\mathrm{d}\mathcal{R}_{n\ell}}{\mathrm{d}r} + \left[wr - \ell(\ell+1) - \kappa^2 r^2\right]\mathcal{R}_{n\ell} = 0, \tag{13.132}$$

where $w = 2mZ\alpha$ and $\kappa^2 = -2m\,\omega_{NL} > 0$ (because for bound states $\omega_{NL} < 0$).

A choice for the two independent solutions with the desired asymptotic behaviour at small r are

$$\mathcal{R}^{\pm}_{n\ell}(r) = (2\kappa r)^{\frac{1}{2}(-1\pm\zeta)}\, e^{-\kappa r}\mathcal{M}\left[\frac{1}{2}\left(-\frac{w}{\kappa}+1\pm\zeta\right), 1\pm\zeta; 2\kappa r\right], \tag{13.133}$$

where

$$\mathcal{M}(a,b;z) = 1 + \left(\frac{a}{b}\right)z + \frac{1}{2}\left[\frac{a(a+1)}{b(b+1)}\right]z^2 + \cdots, \tag{13.134}$$

is the confluent hypergeometric function that is regular at $z = 0$ (but with a series that breaks down when b is a non-positive integer). For the Coulomb solution $\zeta = 2\ell + 1$ and because $1 - \zeta = -2\ell$ is a negative integer the series for $\mathcal{R}^-$ becomes problematic (and so is instead obtained by a limiting procedure).

The properties of hypergeometric functions imply the linear combination that is normalizable at $r \to \infty$ is[14]

$$\mathcal{R}_{n\ell}(r) = C\left[\frac{\Gamma(-\zeta)}{\Gamma\left[\frac{1}{2}\left(-\frac{w}{\kappa}+1-\zeta\right)\right]}\mathcal{R}^+_{n\ell}(r) + \frac{\Gamma(\zeta)}{\Gamma\left[\frac{1}{2}\left(-\frac{w}{\kappa}+1+\zeta\right)\right]}\mathcal{R}^-_{n\ell}(r)\right] \tag{13.135}$$

for arbitrary normalization constant C. Clearly, this fixes C_-/C_+ to be

$$\frac{C_-}{C_+} = \frac{\Gamma(\zeta)\Gamma\left[\frac{1}{2}\left(-\frac{w}{\kappa}+1-\zeta\right)\right]}{\Gamma(-\zeta)\Gamma\left[\frac{1}{2}\left(-\frac{w}{\kappa}+1+\zeta\right)\right]} = -\frac{\Gamma(1+\zeta)\Gamma\left[\frac{1}{2}\left(-\frac{w}{\kappa}+1-\zeta\right)\right]}{\Gamma(1-\zeta)\Gamma\left[\frac{1}{2}\left(-\frac{w}{\kappa}+1+\zeta\right)\right]}, \tag{13.136}$$

and it is the demand that this be consistent with the near-nucleus boundary condition (13.130) that gives the quantization conditions for κ (and so for the bound-state energy).

Consider first a point-like nucleus, for which standard treatments assume the mode functions to be bounded at the origin, implying $C_- = 0$. In this case the energy levels are found by choosing κ such that (13.136) vanishes, leading to the condition

$$\frac{\Gamma\left[\frac{1}{2}\left(-\frac{w}{\kappa}+1-\zeta\right)\right]}{\Gamma\left[\frac{1}{2}\left(-\frac{w}{\kappa}+1+\zeta\right)\right]} = 0, \tag{13.137}$$

which implies that $\kappa = \overline{\kappa}_N$ where[15]

$$\frac{1}{2}\left(-\frac{w}{\overline{\kappa}_N}+1+\zeta\right) = -N, \tag{13.138}$$

[14] The singular factor $\Gamma(-\zeta)$ for $\zeta = 2\ell+1$ is an artefact of the degeneracy of the solution $\mathcal{R}^-$ in (13.133), and again can be finessed using a limiting procedure.

[15] These solutions use the fact that $\Gamma(z)$ has no real zero and has a pole for non-positive integers $z = -N$.

for a non-negative integer $N = 0, 1, 2, \cdots$. Using $w = 2mZ\alpha$ and defining $n = N + \ell + 1$ (so $n = 1, 2, \cdots$ and $\ell = 0, 1, \cdots, n-1$) this implies that

$$\overline{\kappa}_n = \frac{mZ\alpha}{n} \quad \text{and so} \quad E_n = -\frac{\overline{\kappa}_n^2}{2m} = -\frac{m(Z\alpha)^2}{2n^2}, \tag{13.139}$$

reproducing the standard lowest-order hydrogen-like energy levels.

Nuclear Charge Radius

To compute the leading nuclear-size shift to atomic energy levels requires repeating the above exercise and tracing the dependence that C_-/C_+ acquires on the effective coupling $h_{\rm eff}$ due to the near-nucleus boundary condition. This then alters the resulting energy eigenvalues when its nonzero value is equated to (13.136).

To verify the ϵ-independence of the result, recall that the RG evolution of $h_{\rm eff}(\epsilon)$ appropriate here is identical to the running used in the example considered above of a contact interaction in $d = 3$ spatial dimensions. In particular, the RG formulae given earlier can be restricted to the special case (13.83), for which $s_+ + s_- = -1$ and $s_+ - s_- = 2\ell + 1$, leading to Eqs. (13.99) and (13.100), reproduced here for convenience:

$$\frac{mh_{\rm eff}(\epsilon)}{\pi\epsilon} + 1 \simeq (2\ell+1)\lambda(\epsilon) \quad \text{with} \quad \epsilon\,\partial_\epsilon\lambda \simeq \frac{1}{2}(2\ell+1)\left(1-\lambda^2\right). \tag{13.140}$$

These have (13.101) as solution:

$$\lambda(\epsilon) = \frac{(1+\lambda_0)(\epsilon/\epsilon_0)^{2\ell+1} - (1-\lambda_0)}{(1+\lambda_0)(\epsilon/\epsilon_0)^{2\ell+1} + (1-\lambda_0)} = \frac{(\epsilon/\epsilon_\star)^{2\ell+1} + \eta_\star}{(\epsilon/\epsilon_\star)^{2\ell+1} - \eta_\star}. \tag{13.141}$$

These expressions work to lowest nontrivial order in $2\kappa\epsilon$ but do not restrict the size of $\epsilon/\epsilon_\star$.

The physical integration constant C_-/C_+ to be compared with (13.136) is then given by (13.102), which becomes

$$\frac{C_-}{C_+} = \frac{1-\lambda_0}{1+\lambda_0}(2\kappa\epsilon_0)^{2\ell+1} = -\eta_\star\,(2\kappa\epsilon_\star)^{2\ell+1}. \tag{13.142}$$

It is clear that any energy shift inferred by equating Eqs. (13.136) and (13.142) can depend on the pair (λ_0, ϵ_0) only through the RG-invariant combinations $\eta_\star$ and $\epsilon_\star$. Furthermore, because $2\kappa\epsilon_\star \simeq Zm\alpha\epsilon_\star \simeq \epsilon_\star/a_B \ll 1$ (where $a_B = (mZ\alpha)^{-1}$ is the appropriate Bohr radius) the ℓ-dependence of (13.142) shows that at leading order only the $\ell = 0$ (S-wave) states are shifted in this way. For $\ell = 0$ Eq. (13.142) then shows that $\epsilon_\star$ and $\eta_\star$ only appear as the product $\eta_\star\epsilon_\star$.

Since it is the same parameter $\eta_\star\epsilon_\star$ – or the pair $h_{\rm eff}(\epsilon_0)$ and ϵ_0 – that also controls deviations from the point-nucleus Rutherford cross section when electrons scatter from the nucleus (see Exercises 13.2 and 13.5), it is possible to trade $\eta_\star\epsilon_\star$ for the charge radius as measured through low-energy electromagnetic scattering from nuclei. The result of doing so is most easily seen after performing a field redefinition $\delta A_0 \propto (h/e)\,\delta^3(\mathbf{x})$ to eliminate the coupling h so that $h_{\rm eff} = eC_D$, and then repeat the argument leading to (13.57) to conclude

$$mh_{\rm eff} \simeq \frac{2\pi}{3}\,Z\alpha\,mr_c^2. \tag{13.143}$$

In the limit $\epsilon_\star/\epsilon_0 \ll 1$ the RG evolution of $h_{\rm eff}$ is superfluous because (13.107) shows for $\ell = 0$ that

$$m h_{\rm eff} \simeq 2\pi\, \eta_\star \epsilon_\star + O\left(\frac{\epsilon_\star^2}{\epsilon}\right), \tag{13.144}$$

and so in this limit

$$\eta_\star \epsilon_\star \simeq \frac{1}{3} Z\alpha\, m r_c^2. \tag{13.145}$$

Since r_c is measured to be similar to typical nuclear sizes, this justifies *ex post facto* the assumption $\epsilon_\star \ll \epsilon_0$ for any ϵ_0 of order nuclear sizes (or larger).

When RG evolution is not negligible (such as when moving beyond leading-order accuracy, such as when including relativistic corrections, as described below) it is important to realize that it is (13.145) and not (13.143) that remains valid as the relationship between $\eta_\star\epsilon_\star$ and $h_{\rm eff}$ becomes more complicated than (13.144). This is because r_c, being a measureable quantity, is primarily tied to RG-invariant combinations like $\eta_\star\epsilon_\star$ rather than to ϵ-dependent couplings like $h_{\rm eff}(\epsilon)$.

Leading Nuclear Energy Shift

With these tools the atomic energy shift as a function of r_c can be computed. In this case, the energy found by equating Eqs. (13.136) and (13.142) changes from its point-nucleus value, and this is the leading contribution to the energy due to finite nuclear size. In this case, $\kappa = \kappa_n$ is determined by solving the equation

$$\frac{C_-}{C_+} = -\eta_\star (2\kappa\epsilon_\star)^\zeta = \frac{\Gamma(\zeta)\Gamma\left[\frac{1}{2}\left(-\frac{w}{\kappa} + 1 - \zeta\right)\right]}{\Gamma(-\zeta)\Gamma\left[\frac{1}{2}\left(-\frac{w}{\kappa} + 1 + \zeta\right)\right]}, \tag{13.146}$$

with $w = 2mZ\alpha$ and the limit $\zeta \to 2\ell + 1$ taken at the end.

In particular, solutions to (13.146) are sought perturbatively close to (13.139): $\kappa = \overline{\kappa}_n + \delta\kappa$, where the small perturbation parameter is

$$2\overline{\kappa}_n \epsilon_\star = \frac{2Z\alpha m \epsilon_\star}{n} = \frac{2\epsilon_\star}{n a_B}. \tag{13.147}$$

The perturbative solution is found by using

$$\frac{1}{2}\left(-\frac{w}{\kappa} + 1 + \zeta\right) = -N + \left(\frac{w}{2\overline{\kappa}_n^2}\right)\delta\kappa \tag{13.148}$$

as well as $\frac{1}{2}\left(-\frac{w}{\overline{\kappa}_n} + 1 - \zeta\right) = -N - \zeta$ to rewrite (13.146) as

$$-\eta_\star (2\overline{\kappa}_n \epsilon_\star)^\zeta \simeq \frac{\Gamma(\zeta)\Gamma(-N-\zeta)}{\Gamma(-\zeta)\Gamma\left[-N + \left(w/2\overline{\kappa}^2\right)\delta\kappa\right]} \simeq \frac{\Gamma(\zeta)\Gamma(\zeta+1)\, N!}{\Gamma(\zeta+N+1)}\left(\frac{w}{2\overline{\kappa}_n^2}\right)\delta\kappa, \tag{13.149}$$

where the second equality uses the near-pole expansion

$$\Gamma(z - N) \simeq \frac{(-)^N}{N!\, z}\Big[1 + O(z)\Big], \tag{13.150}$$

as well as $\Gamma(-N-\zeta)/\Gamma(-\zeta) = (-)^N \Gamma(\zeta+1)/\Gamma(\zeta+N+1)$.

The limit $\zeta \to 2\ell + 1$ is now safe to take, leading to

$$\frac{\delta\kappa_{n\ell}}{\overline{\kappa}_n} = -\eta_\star \frac{(n+\ell)!}{n(2\ell)!\,(2\ell+1)!\,(n-\ell-1)!}\left(\frac{2Z\alpha m\epsilon_\star}{n}\right)^{2\ell+1}, \tag{13.151}$$

and so using $\delta E_n = -\overline{\kappa}_n \delta\kappa/m$ gives

$$\delta E_{n\ell} = \eta_\star \frac{(n+\ell)!}{(2\ell)!\,(2\ell+1)!\,(n-\ell-1)!}\left(\frac{2Z\alpha m\epsilon_\star}{n}\right)^{2\ell+1}\frac{m(Z\alpha)^2}{n^3}. \tag{13.152}$$

The factor $(2\overline{\kappa}_n\epsilon_\star)^{2\ell+1}$ on the right-hand side shows that $\ell = 0$ states get the largest shift, so specializing to S-wave states gives

$$\delta E_{n0} = \eta_\star \frac{2(Z\alpha)^3 m^2 \epsilon_\star}{n^3}. \tag{13.153}$$

Using (13.145) in (13.153) gives a standard formula [353]

$$\delta E_{n0} = \left(\frac{Z\alpha m}{n}\right)^3 \frac{2}{3}(Z\alpha) r_c^2, \tag{13.154}$$

relating the nuclear charge radius to the leading S-wave atomic energy shift. As might be expected, this has the delta-function form

$$\delta E_{n0} = |\psi_{n0}(0)|^2 h_{\text{eff}}, \tag{13.155}$$

with h_{eff} given by (13.143) for hydrogen-like S-wave states.

Mesonic Atoms

Mesonic atoms provide a similar example of calculable nuclear shifts in atomic energy levels, but one for which $\epsilon_\star$ is not as small as above, and so for which the RG evolution is comparatively more important. In mesonic atoms a negatively charged pion or kaon (pions are considered here for concreteness) is electromagnetically bound to a nucleus. Such states form when mesons are stopped in materials, and the formation of the atomic levels can be detected by seeking the photons that are emitted as the meson cascades through a series of excited states down to its ground state [354].

The two main kinematic differences between mesonic states and electronic ones are the absence of meson spin and the much larger meson mass. For pions, because $m_\pi/m_e \sim 280$ the corresponding Bohr radius,

$$a_{B\pi} = \frac{1}{Z\alpha\, m_\pi}, \tag{13.156}$$

is much smaller than for electrons, ensuring that the meson nestles deep down in a hydrogen-like orbit well inside any clouds of screening electrons. Furthermore, the larger binding energy, $E_n = -(Z\alpha)^2 m_\pi/(2n^2)$, of these hydrogen-like mesonic states is of order keV, so the photons emitted during the meson cascade are X-rays.

Unlike electrons, mesons participate in the strong interactions (being made, as they are, of quarks and gluons) and this is the main dynamical difference between them. The strong interaction between mesons and nuclei is short-ranged, however, acting only over distances of order $r_n \sim m_\pi^{-1} \sim 1$ fm. Because this is about 100 times smaller than the size, $a_{B\pi}$, of the electromagnetic mesonic orbit, it lends itself well to being treated using the PPEFT methods described above.

Indeed, there are practical reasons for doing so. This is because observations of X-rays from the meson cascade provide precise measurements of the mesonic energy levels, and so inferences can be drawn from them about how badly these energies are perturbed by the short-range strong interactions. This provides an experimental window on meson-nucleus interactions at lower energies than those sampled with other methods.

The calculation proceeds much as in the previous section, with only a few changes to highlight. To start with, the leading-order action simply replaces the electron field Ψ with the spinless meson field, Φ, thereby changing Eqs. (13.125) and (13.126) to

$$S_B = \int d^4x \left\{ \frac{1}{2}\left(\mathbf{E}^2 - \mathbf{B}^2\right) + \frac{i}{2}\left(\Phi^* D_t \Phi - D_t \Phi^* \Phi\right) - \frac{1}{2m_\pi} \mathbf{D}\Phi^* \cdot \mathbf{D}\Phi + \cdots \right\}, \tag{13.157}$$

and

$$S_N = \int d^4x \left\{ M + ZeA_0 - h_\pi \Phi^* \Phi + \cdots \right\} \delta^3(\mathbf{x}), \tag{13.158}$$

where, as before, $D_\mu \Phi = (\partial_\mu + ieA_\mu)\Phi$ and a field redefinition is used to eliminate a redundant interaction $\nabla \cdot \mathbf{E}\, \delta^3(\mathbf{x})$, thereby absorbing its coefficient, $C_{D\pi}$, into h_π, along the lines described below Eq. (13.131).

The resulting effective coupling h_π now does double duty: it contains the leading contributions from the pion-nucleus strong force in addition to any contributions to do with the nuclear charge radius. It can do both because both are short-distance effects, and h_π is the unique lowest-dimension effective coupling (and as a result captures the leading long-wavelength implications of *any* short-distance effects localized near the nucleus).

Given the action of (13.157) and (13.158), the calculation of the leading energy shift goes through word-for-word as in the previous section, culminating, for S-wave states, in the main result, Eq. (13.153):

$$\delta E_{n0} = 2 \left(\frac{Z\alpha}{n} \right)^3 m_\pi^2 \, \eta_{\star\pi} \epsilon_{\star\pi}. \tag{13.159}$$

What is no longer true for pions is formula (13.145) relating $\eta_\star \epsilon_\star$ to the nuclear charge radius, since h_π now contains (larger) contributions from pion-nucleus strong interactions.

The difference can be made more explicit by trading $\epsilon_\star$ for λ_π using the second equality of (13.102), or for h_π using Eq. (13.99) with $\ell = 0$. This gives

$$\delta E_{n0} = 2 \left(\frac{Z\alpha}{n} \right)^3 \left[\frac{\lambda_\pi(\epsilon_0) - 1}{\lambda_\pi(\epsilon_0) + 1} \right] m_\pi^2 \epsilon_0 = 2 \left(\frac{Z\alpha}{n} \right)^3 \left[\frac{m_\pi h_\pi(\epsilon_0)}{m_\pi h_\pi(\epsilon_0) + 2\pi\epsilon_0} \right] m_\pi^2 \epsilon_0, \tag{13.160}$$

where ϵ_0 might be a typical near-nuclear scale at which the interaction's strength is determined. This last expression reduces to (13.155) only if $m_\pi h_{\pi 0} \ll 2\pi\epsilon_0$, which is true for nuclear charge radii but typically not so for pion-nuclear interactions. In this sense, the RG resums all orders in the dimensionless combination $m_\pi h_{\pi 0}/(\pi\epsilon_0)$.

A more useful expression trades the combination $\eta_\star\epsilon_\star$ for something else physical, so that (13.159) becomes a prediction relating two observables. For pions a useful observable to use is its strong-interaction elastic scattering length with the nucleus, defined to be the scattering amplitude obtained for the residual short-range nuclear forces once Rutherford scattering from the Coulomb potential is subtracted out.

This amounts to fixing $\eta_\star\epsilon_\star$ by comparing with the S-wave part of the result found in Exercise 13.2:

$$e^{2i\delta_0} = \frac{1 - ik\,\eta_\star\epsilon_\star}{1 + ik\,\eta_\star\epsilon_\star}. \tag{13.161}$$

Since the wave-numbers associated with atomic states are extremely low compared with nuclear scales, it suffices when comparing to use only the scattering length, a_s, given by phase-shift's low-energy limit

$$k \cot \delta_0 \simeq -1/a_s + O(k^2), \tag{13.162}$$

which is read from the low-energy cross section: $\sigma_{\rm le} \simeq 4\pi a_s^2$. Eq. (13.161) then shows

$$a_s = \eta_{\star\pi}\epsilon_{\star\pi}. \tag{13.163}$$

The experimental fact that this scattering length is not that different from nuclear scales verifies that the strong pion-nucleus interaction is not in the small-$\epsilon_\star$ limit.

Using this in (13.159) gives a direct connection between the fractional nucleus-induced atomic energy shift and the scattering length for low-energy pion-nucleus scattering:

$$\frac{\delta E_{n0}}{|E_{n0}|} \simeq 4\left(\frac{Z\alpha\, m_\pi}{n}\right)\eta_{\star\pi}\epsilon_{\star\pi} = \frac{4}{n}\left(\frac{a_s}{a_{B\pi}}\right), \tag{13.164}$$

which uses the definition (13.156) of the pionic Bohr radius. For the ground state $n = 1$ this reproduces what is known as the *Deser formula* for mesonic atoms [355]. As is usual for an EFT analysis, corrections to this expression should arise from higher-dimension interactions localized at the source, and because of their higher dimension would be expected to be suppressed by further powers of a_s/a_B (and *not* by some not-small strong-interaction coupling, like $g_{N\pi\pi}$ of §8.2.3).

Relativistic Near-Nucleus Effects

Although the compact central source described by the first-quantized action is necessarily heavy, nothing forces the various bulk fields discussed above to be non-relativistic. In fact, one might worry that a relativistic treatment eventually becomes mandatory, if the near-source boundary condition is imposed sufficiently near to the compact object. In particular, one might imagine relativistic bulk kinematics to be important if the evolution of couplings like $h(\epsilon)$ are to be followed down to distances small enough that relativistic kinematics become relevant where the boundary conditions are to be applied.

For instance, if R is a nuclear size then once $mR \lesssim Z\alpha$ near-nucleus Coulomb potential energies are of order $Z\alpha/R \gtrsim m$, raising the possibility that a non-relativistic treatment might not get the nuclear boundary conditions quite right because the behaviour of fields near $r = R$ are not treated sufficiently accurately. This regime is also not hypothetical, since for $R \sim 1$ fm the electron mass satisfies $m_e R \sim 1/400$ and so can be smaller than $Z\alpha \sim Z/137$.

On the other hand, any error due to this mistreatment of the boundary condition lies only in a short-distance component of the states relative to atomic scales, and so should be captureable by the usual expansion in powers of v/c, as is done using standard methods (like NRQED, as described in §12). This section shows both of these statements are true: on one hand, a relativistic treatment of the boundary condition does change the relation between effective couplings and RG-invariants; on the other hand, it does not change the relationship between observables and these RG invariants in a way not captured by the usual nonrelativistic expansions.

Relativistic Crossover

The crossover to relativistic running of $h(\epsilon)$ is most simply described for relativistic bosons, and is discussed here since it closely resembles the Schrödinger treatment given above (see [356] for the treatment of Dirac fermions). It turns out that the RG evolution of $h(\epsilon)$ changes in a relativistic regime because the bulk potential seen by the relativistic boson transitions from being dominated by $1/r$ behaviour to $1/r^2$ for small enough r. To see why, recall that for relativistic spinless bosons the bulk action replaces (13.157) with the Klein–Gordon form

$$S_B = \int d^4x \left\{ -\frac{1}{4} F_{\mu\nu}F^{\mu\nu} - D_\mu\phi^* D^\mu\phi - m^2\phi^*\phi + \cdots \right\},$$

while the leading coupling to the first-quantized system remains as in (13.165),

$$S_N = \int d^4x \, \left\{ M + ZeA_0 - h_\phi \, \phi^*\phi + \cdots \right\} \delta^3(\mathbf{x}), \tag{13.165}$$

and as before, $D_\mu\phi = (\partial_\mu + ieA_\mu)\phi$.

The bulk Klein–Gordon field equation for ϕ in a Coulomb potential, $\mathcal{A}_0 = -Ze/r$, is given by $(D_\mu D^\mu - m^2)\phi = 0$, and so for a stationary mode function, $U_N(\mathbf{x}, t) = u_N(\mathbf{x})\, e^{-i\omega t}$, the spatial part satisfies

$$0 = \left[-(\partial_t + ie\mathcal{A}_0)^2 + \nabla^2 - m^2 \right] U_N = \left[\nabla^2 - 2\omega e\mathcal{A}_0 + (e\mathcal{A}_0)^2 - \kappa^2 \right] u_N, \tag{13.166}$$

where $\kappa^2 = m^2 - \omega^2$ for bound state solutions (for which $\omega < m$).

Notice that (13.166) has the same form as for the Schrödinger problems studied earlier – *i.e.* the form $\nabla^2\phi - U\,\phi = \kappa^2\phi$ – with 'potential' $U(\mathbf{x})$ defined by

$$U(r) := 2\omega e\mathcal{A}_0 - (e\mathcal{A}_0)^2 = -\frac{2\omega Z\alpha}{r} - \frac{(Z\alpha)^2}{r^2}, \tag{13.167}$$

containing both $1/r$ and $1/r^2$ components. As a result, the explicit radial mode functions now solve the bulk radial equation (using $d = 3$)

$$r^2 \frac{d^2 \mathcal{R}_{n\ell}}{dr^2} + 2r\frac{d\mathcal{R}_{n\ell}}{dr} + \left[wr + v - \ell(\ell+1) - \kappa^2 r^2 \right] \mathcal{R}_{n\ell} = 0, \tag{13.168}$$

where

$$w = 2\omega Z\alpha\,, \quad \text{and} \quad v = 2mg = (Z\alpha)^2. \tag{13.169}$$

Comparing this with the Schrödinger–Coulomb radial equation, (13.132), shows that this is again solved by confluent hypergeometric functions as in (13.133),

$$\mathcal{R}^{\pm}_{n\ell}(r) = (2\kappa r)^{\frac{1}{2}(-1\pm\zeta)}\, e^{-\kappa r}\, \mathcal{M}\left[\frac{1}{2}\left(-\frac{w}{\kappa}+1\pm\zeta\right), 1\pm\zeta; 2\kappa r\right], \tag{13.170}$$

with $\mathcal{M}(a,b;z)$ as given in (13.134) but with the inverse-square potential implying

$$\zeta = \sqrt{1+4\ell(\ell+1)-4v} = \sqrt{(2\ell+1)^2-4(Z\alpha)^2}. \tag{13.171}$$

The presence of an inverse-square potential means that a nonzero effective coupling, h_ϕ, arises even for S-wave states, since $h_\phi = 0$ never solves the fixed-point equations. To see this in detail recall that (13.165) implies that the near-nucleus boundary condition for the Klein–Gordon field is the same as for the Schrödinger field, but with the replacement $2mh \to h_\phi$, and so

$$4\pi\left(r^2\frac{\partial}{\partial r}\ln u_N\right)_{r=\epsilon} = h_\phi. \tag{13.172}$$

The RG flow implied by this is cast in terms of λ_ϕ defined by (13.117),

$$\frac{h_\phi}{2\pi\epsilon}+1 \simeq \hat{\lambda}_\phi(\epsilon) := \zeta\,\lambda_\phi(\epsilon), \tag{13.173}$$

with $\hat{\lambda}_\phi$ naively satisfying the evolution equation (13.118), which is solved (as usual) by

$$\lambda_\phi(\epsilon) = \frac{\hat{\lambda}_\phi}{\zeta}(\epsilon) = \frac{(\epsilon/\epsilon_\star)^\zeta+\eta_\star}{(\epsilon/\epsilon_\star)^\zeta-\eta_\star} \simeq 1+2\eta_\star\left(\frac{\epsilon_\star}{\epsilon}\right)^\zeta+\cdots, \tag{13.174}$$

where the last equality takes the limit $\epsilon_\star \ll \epsilon$. Specializing to $\ell = 0$ gives

$$\zeta_s = \zeta(\ell=0) = \sqrt{1-4(Z\alpha)^2} \simeq 1-2(Z\alpha)^2+\cdots \tag{13.175}$$

and because the right-hand side is not unity the fixed points of this flow are inconsistent with $h_\phi = 0$, as is seen by using (13.174) in (13.173), leading to

$$\begin{aligned} h_\phi(\epsilon) &\simeq 2\pi\epsilon\left[(\zeta_s-1)+2\zeta_s\eta_\star\left(\frac{\epsilon_\star}{\epsilon}\right)^{\zeta_s}+\cdots\right] \\ &\simeq 4\pi\eta_\star\epsilon_\star - 4\pi(Z\alpha)^2\left[\epsilon+2\eta_\star\epsilon_\star\ln\left(\frac{\epsilon_\star}{\epsilon}\right)\right]+\cdots. \end{aligned} \tag{13.176}$$

It is the first term in the second line of (13.176) that dominates in the Schrödinger case and leads to the leading ϵ-independent identification $h_\phi \simeq \frac{4\pi}{3}m(Z\alpha)r_c^2$ used when discussing the nuclear charge radius. But for situations where $\epsilon_\star/\epsilon \lesssim (Z\alpha)^2$ it is instead the second term that dominates, leading to the alternate RG evolution $h_\phi \simeq -4\pi(Z\alpha)^2\epsilon$. This new regime would be the appropriate one when $\epsilon \sim r_c$ is of order a nuclear radius and $mr_c \lesssim Z\alpha$ (making the near-nucleus kinematics relativistic).

Although this shows that the relation between $\eta_\star\epsilon_\star$ and $h(\epsilon)$ changes once relativistic kinematics dominate in the boundary-condition regime, what does *not* change at leading order is the relation between RG invariants and charge radius, (13.145), or energy-shift formulae like (13.153) or (13.154). It is for this reason that the leading energy-shift formula is not changed, and standard nonrelativistic expansions remain unchanged.

Although qualitatively correct, the above RG discussion must be done with more care once higher orders in α are sought. This is because once the order $(Z\alpha)^2$ term in ζ is included, so must subdominant $O(2kr)$ corrections in the small-r expansion of mode functions like $\mathcal{R}_{nL}(kr)$, since these are of relative order $2\kappa\epsilon \simeq 2Z\alpha\, m\epsilon/n$ and it is the regime $m\epsilon \sim Z\alpha$ that is of interest. For the Klein–Gordon equation the small-r expansion of the S-wave radial mode functions takes the form

$$\mathcal{R}^{+}_{n0}(\epsilon) \simeq (2\kappa\epsilon)^{\frac{1}{2}(-1+\zeta_s)}\left[1 - Z\alpha m\epsilon + \frac{2n^2+1}{6n^2}(Z\alpha m\epsilon)^2 + \cdots\right] \tag{13.177}$$

$$\mathcal{R}^{-}_{n0}(\epsilon) \simeq (2\kappa\epsilon)^{\frac{1}{2}(-1-\zeta_s)}\left[1 - \frac{m\epsilon}{Z\alpha} + (m\epsilon)^2 - \frac{2n^2+1}{6n^2}Z\alpha(m\epsilon)^3 + \cdots\right],$$

using $1-\zeta_s \simeq 2(Z\alpha)^2$ and $\kappa = \sqrt{(m-\omega)(m+\omega)} \simeq Z\alpha m/n$, where n is the principal quantum number. In principle, the evolution of h_ϕ in this regime is captured by returning to the full boundary conditions, (13.88) and (13.89), rather than the simpler version found using only the leading power of $2\kappa\epsilon$. Energy shifts and scattering amplitudes are then computed as before by finding C_-/C_+ using (13.172) and comparing with the conflicting demands coming from $r \to \infty$.

13.4 Summary

First-quantized methods bring the EFT discussion full circle: successively integrating out high-energy modes can lead to effective theories at lower energies involving very massive but approximately stable and slowly moving nonrelativistic particles. In many situations these massive particles only appear individually and are too small to have their structure resolved, making their centre-of-mass and overall rigid-body orientation the important low-energy degrees of freedom. In these circumstances second-quantized EFT methods can be overkill: it is simpler to directly focus on the centre-of-mass motion using first-quantized techniques, leading to a good old-fashioned single-particle quantum description of these heavy states. This represents a full circle inasmuch as single-particle quantum mechanics – which is usually one's first experience with quantum systems – here emerges systematically as the low-energy EFT of slowly moving and compact particles. It turns out that we were all using effective field theories right from day one.

Having a systematic framework for having first-quantized techniques emerge from second-quantized EFTs offers additional advantages beyond the usual ones associated with efficiently isolating any hierarchy of scales. In particular, it shows how to systematically couple first- and second-quantized systems to one another with corrections included order-by-order in any small parameters and scale ratios. It also reinforces how exploiting separations of scales is equally useful to classical systems, where using the hierarchy between fast and slow degrees of freedom has also been the subject of much study [358], recently including computing gravitational radiation from compact orbiting black holes [359].

A useful feature that the first-quantized formulation highlights is the role played by boundary conditions for the interaction between fields and first-quantized effective systems. Standard matching procedures for the effective couplings in a first-quantized formulation of small compact sources are efficiently expressed in terms of near-source boundary conditions for the bulk fields with which it interacts, that can be derived directly from the source's first-quantized effective action. More details of the source's structure get included into the boundary condition by including more effective interactions in this action, thereby generalizing standard 'multipole' techniques from electromagnetism to arbitrary bulk fields. Furthermore, this grounding of boundary conditions into the choice of the source's low-energy effective action explains why some boundary conditions (*e.g.* linear Neumann, Dirichlet and Robin boundary conditions) arise so frequently in physical problems.

The ability to turn an effective first-quantized particle action into boundary conditions for bulk fields also has other uses, such as allowing EFT methods to be employed in ordinary quantum mechanical settings. In particular, they remove the guesswork associated with singular potentials, which have long been known to be very sensitive to choices of boundary conditions near the potential's singularity [349]. The EFT picture described in this chapter directly relates the freedom to choose boundary conditions to the choices available in the action describing the heavy compact object responsible for the singular potential in the first place.

Exercises

Exercise 13.1 Generalize the point-particle action of (13.19) to a curved target space with metric $g_{\mu\nu}(x)$ by using $\mathfrak{h} = g_{\mu\nu}(\chi)\dot{\chi}^\mu\dot{\chi}^\nu$ as the induced metric on the particle world-line. By repeating the steps leading to Eq. (13.23) show that the classical equations of motion expressed using proper time $d\tau^2 = -g_{\mu\nu}\, dx^\mu\, dx^\nu$ become the equations for an affinely parameterized geodesic

$$\frac{d^2\chi^\mu}{d\tau^2} + \Gamma^\mu_{\nu\lambda}\frac{dx^\nu}{d\tau}\frac{dx^\lambda}{d\tau} = 0,$$

where $\Gamma^\mu_{\nu\lambda}$ represents the Christoffel symbol built from $g_{\mu\nu}$, defined by (C.91).

Exercise 13.2 Consider a Schrödinger field coupled through a contact interaction to a first-quantized source – with action given in (13.64) and (13.65) – and use the boundary condition (13.68) to compute the cross section for S-wave ($\ell = 0$) scattering of Ψ particles from a stationary compact centralized source when $d = 3$. Express your result in terms of the RG-invariant quantities $\epsilon_\star$ and $\eta_\star$ that govern the evolution of $h(\epsilon)$, and show that the scattering phase shift satisfies

$$e^{2i\delta} = \frac{1 - i\eta_\star k\epsilon_\star}{1 + i\eta_\star k\epsilon_\star},$$

where $k^2 = 2mE$ for E the incident energy in the source's rest frame. For low-energy scattering the scattering length, a_s, is defined by $k \cot\delta \simeq -1/a_s + O(k^2)$. Show that a_s is in this case given by

$$a_s = \eta_\star\epsilon_\star.$$

Exercise 13.3 Consider a charged wire with charge/length σ, whose electric field points radially (perpendicular to the wire) and is given by

$$\mathbf{E} = \frac{\sigma}{2\pi r}\,\hat{\mathbf{x}},$$

where $r = |\mathbf{x}|$ is the perpendicular distance from the wire and $\hat{\mathbf{x}}$ denotes the unit vector pointing radially outward in the plane perpendicular to the wire.

Imagine placing a polarizable atom in this electric field, recalling that – as discussed for neutral particles in §12.3.1 – the electromagnetic interactions of a polarizable atom are proportional to $\mathfrak{p}_E \mathbf{E}^2$ (with $\mathfrak{p}_E$ denoting the atomic polarizability). Show that an atom a distance $r = |\mathbf{x}|$ from the charged wire experiences an interaction energy

$$V(\mathbf{x}) = -\frac{1}{2}\,\mathfrak{p}_E \mathbf{E}^2 = -\frac{\mathfrak{p}_E \sigma^2}{8\pi^2 r^2},$$

of inverse-square form, $V = -g/r^2$, with $g = \mathfrak{p}_E \sigma^2/(8\pi^2)$. The negative sign arises because the atom's polarizability gives it a dipole moment $\mathbf{d}_E \propto \mathbf{E}$ in the presence of an applied field, and this dipole lowers its energy by aligning with the field. In this case, cylindrical symmetry (in three spatial dimensions) allows one to use the analysis in the main text for inverse-square potentials with $d = 2$. For any angular momentum, ℓ what is the critical charge density, corresponding to the value $g = g_C$ above which ζ becomes imaginary? What about the special case $\ell = 0$?

For a gas of polarizable atoms in a trap surrounding such a charged wire assume that any atoms that hit the wire get ejected from the trap (so the wire is a perfect absorber – see the discussion surrounding Eq. (13.124)). What is the value of the effective coupling λ that corresponds to this boundary condition? What is the corresponding value of the contact interaction $h(\epsilon)$?

Exercise 13.4 For $d = 3$ use boundary condition (13.105) together with the condition that the u_N is normalizable as $r \to \infty$ to find the bound states for a Schrödinger field interacting with a compact point source through both a long-range inverse-square potential and a delta-function contact interaction. (General solutions to the radial equation satisfied by $\mathcal{R}(r)$ with an inverse-square potential can be found in terms of confluent hypergeometric functions, as described in the main text.) Show that the energy of these bound states is given by

$$E = -\frac{2}{m\epsilon_\star^2}\left[\frac{\Gamma\left(-\frac{1}{2}\zeta\right)}{\Gamma\left(+\frac{1}{2}\zeta\right)}\right]^{-2/\zeta}.$$

What is ζ in this formula? Under what conditions on the parameters can these bound-state energies be trusted? (Self-consistency requires them to lie within the EFT regime.)

For ζ real show that these bound states of the joint inverse-square/delta-potential system amount to those that are supported by the compulsory delta-function interaction alone. When ζ is imaginary (so $g > g_C$) but $\hat{\lambda}$ is real show that there is instead a tower of bound states with energies related by a discrete scale transformation that are supported by the inverse-square potential itself.

Exercise 13.5 When $\hat{\lambda}$ is complex show that the bound-state energies of the joint inverse-square/delta-potential system are complex, $\mathcal{E} = E - \frac{1}{2}\mathrm{i}\hat{\gamma}$, and given by

$$\mathcal{E} = -\frac{2}{m\epsilon_\star^2}\left[\left(\frac{\zeta - \mathrm{i}y_\star}{\zeta + \mathrm{i}y_\star}\right)\frac{\Gamma\left(-\frac{1}{2}\zeta\right)}{\Gamma\left(+\frac{1}{2}\zeta\right)}\right]^{-2/\zeta}.$$

The imaginary part of this energy gives the bound state's decay rate, or inverse mean life $\hat{\gamma} = 1/\tau$, due to the loss of probability at the source. For $d = 3$

compute $\hat{\gamma}$ as a function of the real and imaginary parts of $h(\epsilon_0)$, defined at a microscopic scale ϵ_0. How does the result depend on the RG invariants $y_\star$ and $\epsilon_\star$?

Exercise 13.6 For complex h scattering phase shifts need no longer be real and so are written $e^{i\gamma}$ with $\gamma_\ell = \delta_\ell + i\eta_\ell$. Unitary scattering corresponds to $\eta_\ell = 0$. For $d = 3$ the elastic and absorptive cross sections are

$$\sigma_{\rm el}^{(\ell)} = f_\ell(k^2)\left[1 + e^{-4\eta_\ell} - 2e^{-2\eta_\ell}\cos 2\delta_\ell\right] \quad \text{and} \quad \sigma_{\rm abs}^{(\ell)} = f_\ell(k^2)\left(1 - e^{-4\eta_\ell}\right),$$

where $f_\ell(k^2) = \pi(2\ell+1)/k^2$ and $k^2 = 2mE$ for incident particles of energy E. For each partial wave show that scattering from an inverse-square potential in $d = 3$ dimensions implies that

$$e^{2i\gamma} = \left[\frac{1 - \mathcal{A}\, e^{i\pi\zeta/2}}{1 - \mathcal{A}\, e^{-i\pi\zeta/2}}\right] e^{i\pi(\ell - l)} \quad \text{with} \quad \mathcal{A} = \left(\frac{iy_\star - \zeta}{iy_\star + \zeta}\right)\left(\frac{k\epsilon_\star}{2}\right)^{\zeta} \frac{\Gamma\left(1 - \frac{1}{2}\zeta\right)}{\Gamma\left(1 + \frac{1}{2}\zeta\right)},$$

where l is defined by $2l + 1 := \zeta = \sqrt{(2\ell+1)^2 - 8mg}$.

Use these expressions to prove that there is no absorption – *i.e.* $\eta_\ell = 0$ – for real $\hat{\lambda}$ (for which $y_\star = 0$ or $y_\star \to \pm\infty$), both when ζ is real and when ζ is pure imaginary. Show that for small incident momentum the elastic cross section approaches $\epsilon_\star^2$ while the absorptive cross section is proportional to $\epsilon_\star/k$. (Historically, this small-k dependence was what convinced people that ordinary quantum mechanics could account for measured rates for low-energy neutron absorption reactions [360].)

Exercise 13.7 Use the small-r boundary condition of the main text to show that for scattering from the combined Coulomb/inverse-square/delta-function potential,

$$V(r) = -\frac{v}{r} - \frac{g}{r^2} + h\,\delta^3(\mathbf{x}),$$

in $d = 3$, with real v, h and g, the scattering phase shift is

$$e^{2i\delta_\ell} = \frac{N_{++} - \eta_\star(2ik\epsilon_\star)^\zeta N_{+-}}{N_{-+} - \eta_\star(-2ik\epsilon_\star)^\zeta N_{--}}\, e^{i\pi(\ell - l)},$$

where

$$N_{ab} := \frac{\Gamma(1 + b\zeta)}{\Gamma\left[\frac{1}{2}\left(a\frac{iw}{k} + 1 + b\zeta\right)\right]},$$

with $a, b = \pm$ and $w = 2mv$. Assume $g < g_c$ so that $\zeta = \sqrt{(2\ell+1)^2 - 8mg}$ is real, and define l using $\zeta =: 2l + 1$.

Show that this reproduces of the results of Exercise 13.6 when the Coulomb contribution is turned off ($w \to 0$), and those of Exercise 13.2 for S-wave scattering from a delta-function potential. Show that the above expression also reproduces the usual Rutherford expression

$$e^{2i\delta_\ell} = \frac{\Gamma(\ell + 1 - iw/2k)}{\Gamma(\ell + 1 + iw/2k)} \qquad \text{(Rutherford limit)},$$

when $g \to 0$ and in the absence of a delta-function potential ($\epsilon_\star \to 0$).

Exercise 13.8 In Type IIB string theory D7 branes (objects with seven spatial dimensions that sweep out an 8-dimensional world-surface) in 10 spacetime dimensions provide completely different examples of localized sources to which PPEFT methods can be applied. Since D7 branes are 'fundamental', this example also provides insight in how the RG evolution might behave for such objects (the brane couplings do not run).

For the present purposes the fields sourced by D7 branes are the 10-dimensional spacetime metric, g_{MN}, and a complex scalar field

$$\tau = C_0 + \mathrm{i}\, e^{-\phi} \quad \text{and} \quad \overline{\tau} = C_0 - \mathrm{i}\, e^{-\phi},$$

where C_0 is called a 'Ramond-Ramond' scalar and ϕ is the 10D dilaton, related to the string coupling by $g_s = e^{\phi}$ along the lines described in §10.3. The bulk action for these fields can be written

$$S_B = -\frac{1}{2\kappa^2}\int \mathrm{d}^{10}x\,\sqrt{-g}\; g^{MN}\left[\mathcal{R}_{MN} + \frac{\partial_M\overline{\tau}\,\partial_N\tau}{2\,(\mathrm{Im}\,\tau)^2}\right],$$

where $\mathcal{R}_{MN}$ is the Ricci tensor for the 10D metric (see Appendix A.2.1 for definitions). Assuming the world-volume of the D7 brane spans the $\{x^0, x^1, \cdots, x^7\} = \{x^a\}$ dimensions, use complex coordinates $z = x^8 + \mathrm{i}x^9$ and $\overline{z} = x^8 - \mathrm{i}x^9$ in the two transverse directions and show that the 10D field equations for τ and the metric are solved if the 10D metric is written

$$g_{MN}\,\mathrm{d}x^M\mathrm{d}x^N = \eta_{ab}\,\mathrm{d}x^a\mathrm{d}x^b + e^{2L(z,\overline{z})}\,\mathrm{d}z\mathrm{d}\overline{z},$$

with τ and L satisfying the equations

$$\partial\overline{\partial}\tau + \frac{2\partial\tau\overline{\partial}\tau}{\overline{\tau}-\tau} = 0 \quad \text{and} \quad 2\partial\overline{\partial}L = \frac{\partial\tau\overline{\partial}\overline{\tau}}{(\tau-\overline{\tau})^2}.$$

The first of these is solved by any holomorphic function, $\tau = \tau(z)$, for which $\overline{\partial}\tau = 0$.

It happens that the field C_0 is multi-valued very close to a D7 brane situated at $z = z_i$, requiring a monodromy $\tau \to \tau + 1$ as $z - z_i \to (z - z_i)\,e^{2\pi \mathrm{i}}$. This, together with holomorphy – *i.e.* $\tau = \tau(z)$ – requires that solutions to the field equations take the following asymptotic form [361]

$$\tau(z) \simeq \frac{1}{2\pi\mathrm{i}}\ln(z - z_i) + \cdots \quad \text{and} \quad e^{2L(z,\overline{z})} \simeq k\;\mathrm{Im}\,\tau,$$

for constant k. Notice this implies a profile for ϕ, with $e^{\phi} \to 0$ as $|z - z_i| \to 0$.

The first-quantized action for a D7 brane can be taken as

$$S_b = -\int \mathrm{d}^8x\,\sqrt{-\gamma}\;T_b(\tau,\overline{\tau}),$$

where $\gamma_{ab} = \eta_{ab}$ is the usual induced metric and $T_b = T_*/\mathrm{Im}\,\tau$, for constant T_*. Use the techniques of this chapter to show that this brane action implies that the bulk scalar must satisfy the near-brane boundary condition

$$\frac{2\pi}{\kappa^2}\left[\frac{r\,\partial_r\tau}{4\,(\mathrm{Im}\,\tau)^2}\right]_{z_i} = \frac{\partial T_b}{\partial\overline{\tau}},$$

where $r = |z|$. Similar arguments applied to the metric [362] imply that the metric function L also must satisfy

$$-\frac{2\pi}{\kappa^2}\Big[r\,\partial_r L\Big]_{z_i} = T_b(\tau,\overline{\tau}).$$

Show that the near-brane profiles of τ and L are such that both sides of these conditions share the same dependence on $|z|$ and so can be evaluated without requiring a renormalization of the parameter T_*. Show that both boundary conditions boil down to the single condition $2\kappa^2 T_* = 1$ that relates T_* to the gravitational coupling κ. Notice that this relation is satisfied by the string theory predictions for D7 branes [363]

$$T_* = \frac{1}{(2\pi)^7(\alpha')^4} \quad \text{and} \quad \kappa^2 = \frac{1}{2}(2\pi)^7(\alpha')^4,$$

with α' as defined in §10.3.1.

Part IV

Many-Body Applications

About Part IV

The systems whose low-energy limit has been analyzed to this point all involve small numbers of particles, with both relativistic and nonrelativistic kinematics. Effective field theories are also useful tools for more complicated many-body systems, however, and some representative examples of these are explored in the remaining parts of this book.

Several conceptual issues arise when treating many-body systems using these tools. A relatively trivial one of these remarks that from the fundamental point of view many-body systems are not really at low energies. Because in everyday life they can involve enormous numbers of atoms, the energy tied up in their rest mass can be huge. This is similar to the issue encountered with baryons in chiral perturbation theory, at the end of §8, and its resolution is the same: one works at low energies within a subsector of the Hilbert space carrying a specified amount of a conserved charge. Rather than choosing the ground state of the $B = 0$ or $B = 1$ sector, as in chiral perturbation theory, one might work within the $B = 10^{23}$ sector when thinking about a macroscopic number of atoms.

An important change relative to the nonrelativistic systems described in Part III is the loss in many-body applications of conclusions like those surrounding Eq. (12.5) that are implications of non-linearly realized Poincaré invariance. Formally, these relations amongst the low-energy couplings arose in Part III as a consequence of matching the nonrelativistic theory to a relativistic underlying theory, for which the *only* breaking of Lorentz invariance comes from integrating out the antiparticle. For many-body systems the UV physics usually breaks Lorentz invariance in many ways due to the preferred frame effects provided by the system's many background atoms, preventing the inference of relations like (12.5). Indeed, other spacetime symmetries are often broken by the background matrix of atoms as well (such as rotation invariance or translation invariance).

A final issue concerns dissipation and the very existence of an action. This issue arises because for macroscopic systems one rarely follows *all* of the low-energy degrees of freedom and as a result one necessarily ignores some degrees of freedom with which the system can entangle or exchange energy and information. In short, most many-body applications really are open systems, rather than proper Wilsonian systems.

The final chapter of the book aims at systems for which this open nature of many-body problems plays a more central role. Although methods for handling quantum open systems have been known for some time, the interpretation of these methods in terms of effective field theories in open systems is comparatively less well-developed.

14 Goldstone Bosons Again

Goldstone bosons feature prominently in the low-energy limit of many systems, and because they interact weakly at low energies they can often be studied in relative isolation. As discussed in §4.1.2, both their appearance at low energies and their weak low-energy interactions have their origin in Goldstone's theorem, which guarantees on very general grounds that their energy and interactions must vanish at long wavelengths.

This chapter examines several examples of Goldstone modes in many-body systems, with examples involving both gauged and global internal symmetries, as well as broken spacetime symmetries. These examples are presented with several goals in mind. A first goal is simply to show that §4.2 applies equally well within nonrelativistic contexts. Second, some of the ideas encountered in these examples also prove useful in later sections. But most importantly, effective field theories and spontaneous symmetry breaking in nonrelativistic many-body systems also involve features that are interesting in their own right: Goldstone bosons can have unusual dispersion relations, and their number can differ from the number of broken symmetry generators; effective theories can be nonlocal; low energies can coincide with large momenta; and so on.

14.1 Magnons ◇

The first many-body example to be considered aims at the low-energy behaviour of spin systems, both for ferromagnets and antiferromagnets. Both of these systems consist of atoms for whom electron spin plays a significant role in how the atoms order once brought together in bulk. What is important about the relevant spins is that they are typically free to rotate with comparatively low energy cost, subject to spin-dependent inter-atomic interactions. When this is so, the interplay of how each atom responds to the spins on its various neighbours can determine how atomic spins align relative to one another.

This spin-spin interaction can arise microscopically in a variety of ways – often due to exchange effects for the Coulomb interactions of the electrons most relevant for inter-atomic interactions [364] – whose details are largely not important for the purposes of this section. The interaction is often modelled in terms of nearest-neighbour spin-spin couplings with Hamiltonians of the form [365, 366]

$$H_{\rm spin} = \sum_{\langle ij \rangle} J_{ij}\, \mathbf{s}_i \cdot \mathbf{s}_j, \qquad (14.1)$$

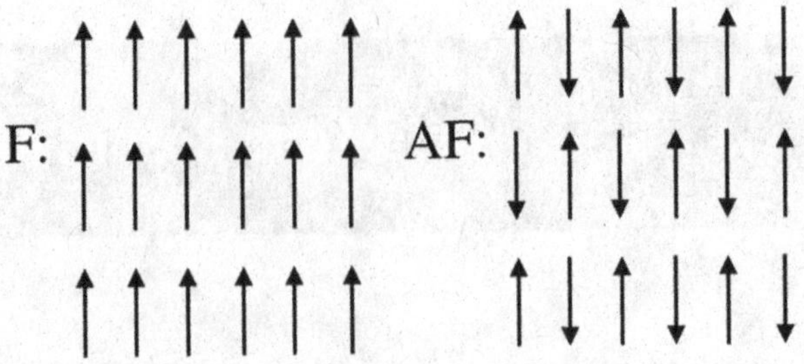

Fig. 14.1 A cartoon illustrating how any given spin is parallel (antiparallel) to its four nearest neighbours for ferromagnetic (antiferromagnetic) order in two dimensions.

where $\mathbf{s}_i$ is the spin on the atom labelled by 'i' and the sum is over atomic positions (typically sites on some lattice) with '$\langle ij \rangle$' indicating that the sum runs over only nearest neighbours rather than all possible atomic pairs. Here, J_{ij} represents an energy of alignment that, when negative, favours spins that point in the same direction and, when positive, favours spins that instead *anti-align* so that their spins point in opposite directions.

For models like (14.1) macroscopic numbers of spins tend either to align or anti-align within their ground state, as sketched in the two panels of Fig. 14.1, according to which of these takes less energy. A system for which neighbouring spins tend to align is called a ferromagnet, and it acquires a macroscopic magnetization in which the magnetic moment of each atom adds coherently to give a large sum.

If neighbouring spins prefer to anti-align, however, then the ground state consists of spins that point in alternating directions, for which no net macroscopic magnetization need be present. Such an arrangement is called antiferromagnetic, and is still ordered even though the net magnetization in the ordered state vanishes. This type of ordering is called Néel ordering [367].

Both ferromagnets and antiferromagnets exhibit spontaneous symmetry breaking, because a preferred direction is picked out once the spins all align or anti-align, and this means the ground state is not invariant under the symmetry of $H_{\rm spin}$ that rotates all spins through a common angle. The remainder of this chapter aims to describe the Goldstone bosons associated with this ordering, and the effective action that robustly describes their low-energy interactions.

The next two subsections examine antiferromagnetism and ferromagnetism in turn, following the treatment in [107]; treating them separately because the results obtained turn out to differ considerably. From the low-energy perspective, what makes ferromagnets and antiferromagnets differ is the way they realize time-reversal symmetry.

14.1.1 Antiferromagnetism

Consider first antiferromagnets, since these resemble the spontaneous symmetry breaking encountered earlier for relativistic systems more closely than do ferromagnets. For an antiferromagnet, the order parameter, $\mathbf{N}$, can be taken to be the staggered sum of the spins, $\mathbf{s}_i$, for each lattice site, 'i':

$$\mathbf{N} = \sum_i (-)^i \, \mathbf{s}_i, \tag{14.2}$$

where the sign, $(-)^i$, alternates for adjacent atomic sites in such a way that it is $+1$ for the sublattice of sites on which the spin is up (say) and is -1 for sublattice of down spins. This ensures these two lattices contribute in the same way to the average of $\mathbf{N}$ in the system's ground state, which allows $\langle \mathbf{N} \rangle$ to be nonzero.

Time-reversal invariance, T, imposes important constraints on the low-energy action of an antiferromagnet. This might seem surprising given that T reverses the sign of each individual spin, $\mathbf{s}_i \to -\mathbf{s}_i$, and so also reverses the order parameter, $\mathbf{N} \to -\mathbf{N}$. At first sight, one might therefore expect the existence of a nonzero $\langle \mathbf{N} \rangle$ therefore to break time-reversal invariance.

This turns out not to be true, however, because T can be combined with another broken symmetry, S, to obtain a transformation, $\tilde{T} = TS$, that is *not* broken when $\langle \mathbf{N} \rangle \neq 0$. This other symmetry, S, consists of a translation that maps the whole sublattice of sites pointing in one direction onto the sublattice aligned in the opposite direction. For instance, in the simple configuration illustrated in Fig. 14.1 S could correspond to a translation to the right (say) by a single lattice site. Since both S and T act to reverse the direction of $\mathbf{N}$, when they are performed together $\mathbf{N}$ remains unchanged.

The basic ordering, $\langle \mathbf{N} \rangle \neq 0$, breaks the underlying rotation symmetry of the atomic spins, $\mathbf{s}_i \to O\,\mathbf{s}_i$, for $O^T O = I$, so the general arguments of §4.1.2 imply the spectrum of low-energy fluctuations should contain Goldstone particles – called *magnons* – whose low-energy EFT is the main focus in what follows. As usual, the quantum numbers and dominant interactions at low energies for Goldstone modes are dictated purely by the symmetry-breaking pattern and so can be identified very robustly using the general tools described in §4.2.2 (and Appendix C.6).

The first step is to identify the symmetry-breaking pattern, $G \to H$, relevant to antiferromagnets. In this case, the symmetry of the action (or Hamiltonian) consists of the group of spin rotations in three dimensions: $G = SO(3)$, the group of 3×3 orthogonal matrices,[1] $\{O\}$, describing 3-dimensional spin rotations. [Alternatively, this can also be described as $G = SU(2)$, the group of 2×2 unitary matrices, $\{U\}$, acting on the underlying electron spins, where (schematically) Pauli matrices satisfy $U^\dagger \sigma^i U = \sigma^j O^i{}_j$ for $U \in SU(2)$ and $O \in SO(3)$. These are interchangeable at the level of their Lie algebra since $SU(2)$ shares the same Lie algebra as $G = SO(3)$, and so the distinction between these two groups doesn't play an important role in this section's discussion.]

It is tempting to conclude that because G is at heart a rotation, it is therefore a spacetime symmetry that acts on vectors like ∇ as well as spins (as opposed to being an 'internal' symmetry that acts only on field labels and not also positions). This would be important if so, because the tools developed in §4.2.2 assume the symmetry to be internal.

There are several reasons to see why G really is an internal symmetry, at least to a very good approximation. One such starts from the specific Hamiltonian given in (14.1), which is invariant under rotations $\mathbf{s}_i \to O\,\mathbf{s}_i$ without requiring any transformation on the spin's position (or label 'i').

Furthermore, for spins sitting on a lattice of atoms the lattice itself already breaks rotational and translational symmetries. So these are not available to be rebroken by

[1] The 'S' in $SO(3)$ and $SU(2)$ stands for 'special' and means that these matrices must also have unit determinant.

the antiferromagnetic order. (The Goldstone bosons for the spontaneous breaking of translational and/or rotational symmetry by the lattice itself also exist in the low-energy spectrum in the form of phonons, but these are not the focus of the present section.)

Microscopically, atomic spins ultimately have their origins in the spins of slowly moving particles like electrons or nucleons, whose electromagnetic interactions are argued in §12.1.1 to be dominated at low energies by the interactions of $\mathfrak{L}_0$ of Eq. (12.1). The point is that $\mathfrak{L}_0$ is invariant under *independent* rotations of spins and spatial vectors (like derivatives or electromagnetic fields), because it never involves a dot product between spins and other vectors. This is why spin rotations effectively behave as if they are internal symmetries in many systems.

The same is no longer true once $\mathfrak{L}_1$ and $\mathfrak{L}_2$ of Eqs. (12.2) and (12.3) are included, due to the appearance of magnetic moment (c_F) and spin-orbit (c_S) interactions. These are invariant only if both spins and spatial vectors rotate by a common amount, and so in real systems invariance under separate spin and spatial rotations is ultimately broken by effects suppressed by at least one power of e/m.

What is the unbroken group, H? Since the order parameter is the vector, $\langle \mathbf{N} \rangle$, the group of transformations in G left invariant by $\langle \mathbf{N} \rangle$ consists of rotations about the axis defined by $\langle \mathbf{N} \rangle$ itself. This corresponds to the subgroup $H = SO(2) \subset SO(3)$ (or $H = U(1) \subset SU(2)$), and so the coset space parameterized by the Goldstone bosons is $G/H = SO(3)/SO(2)$ (or $G/H = SU(2)/U(1)$). Geometrically, this makes G/H the space swept out as a unit vector,

$$\vec{n}(\vartheta, \varphi) := \frac{\mathbf{N}}{|\mathbf{N}|}, \tag{14.3}$$

of fixed length is rotated in all directions.[2] This two-dimensional space is a definition of the two-sphere, S_2, and so can be parameterized by two angles, (ϑ, φ), that define the direction of $\vec{n}$. Because $\mathbf{N}(\mathbf{x}, t)$ is a field, the same is true for $\vec{n}(\mathbf{x}, t)$, and so also are the two component Goldstone fields $\vartheta(\mathbf{x}, t)$ and $\varphi(\mathbf{x}, t)$.

At this point, the formalism of §4.2.2 could be launched in all its glory, allowing a systematic construction of the low-energy effective interactions of the Goldstone fields. However, in this instance this formalism is not necessary since the variables $\vartheta(x)$ and $\varphi(x)$ also provide an equally valid representation of these fields. The formalism of §4.2.2 uses the freedom to perform field redefinitions to put the nonlinear transformations of the Goldstone fields into a standard form, for which the problem of building invariant actions is solved once and for all. But nothing requires the use of these variables if the action can be derived in other ways. Because field redefinitions cannot change anything observable, the predictions of Eq. (14.8) below are completely equivalent to what would have been obtained if instead the formalism of §4.2.2 had been deployed (see Exercise 14.1).

The action of the nonlinear realization of G transformations on the variables ϑ and φ is straightforward to work out, starting with the standard $SO(3)$ vector transformation rule for $\vec{n}$:

$$\delta\vec{n} = \vec{\omega} \times \vec{n}, \tag{14.4}$$

[2] The vector $\vec{n}$ is not denoted using a bold-face font in order to emphasize that it is regarded as rotating independent of spacetime vectors like ∇ and $\mathbf{B}$.

with the vector $\vec{\omega}$ representing the three $SO(3)$ group transformation parameters. Writing the components of the unit vector $\vec{n}$ out explicitly

$$n_x = \sin\vartheta\cos\varphi\,, \quad n_y = \sin\vartheta\sin\varphi \quad \text{and} \quad n_z = \cos\vartheta, \tag{14.5}$$

shows that $\vec{n}\cdot\vec{n} = 1$ for any ϑ and φ, and (for $\sin\vartheta \neq 0$) implies that the transformation rule is

$$\begin{aligned}\delta\vartheta &= \omega_y\cos\varphi - \omega_x\sin\varphi \\ \delta\varphi &= \omega_z - \omega_x\cot\vartheta\cos\varphi - \omega_y\cot\vartheta\sin\varphi.\end{aligned} \tag{14.6}$$

Notice the characteristic inhomogeneous shifting of a Goldstone field.

The effective lagrangian involving the fewest derivatives that governs the low-energy Goldstone dynamics is then given (assuming the modified time-reversal invariance, $\tilde{T}$) by

$$\mathfrak{L}_{AF} = \frac{F_t^2}{2}\,\dot{\vec{n}}\cdot\dot{\vec{n}} - \frac{1}{2}\,Z^{ab}\;\partial_a\vec{n}\cdot\partial_b\vec{n} + (\text{higher derivatives}), \tag{14.7}$$

in which the matrix Z^{ab} must be invariant under any residual rotation symmetries of the underlying spacetime lattice. For some lattice symmetries this imposes as powerful constraints as would full-blown rotational invariance in space, in which case $Z^{ab} = F_s^2\,\delta^{ab}$, and this is assumed in what follows.[3] At the very long wavelengths relevant at low energies invariance under lattice translations implies ordinary translation invariance, so that both F_t^2 and F_s^2 are constants.

The lagrangian (14.7) makes clear the physical interpretation of the Goldstone fields: they describe long-wavelength variations in the direction of the order parameter $\vec{n} \propto \langle \mathbf{N}\rangle$ evaluated for low-energy states that are near, but not precisely in, the ordered vacuum. The quanta of these waves are called magnons. The lagrangian (14.7) obscures, however, that there are self-interactions amongst the Goldstone fields despite its being purely quadratic in $\vec{n}$. These self-interactions are hidden in the constraint $\vec{n}\cdot\vec{n} = 1$, but are made explicit by rewriting (14.7) in terms of the unconstrained fields ϑ and φ, leading to

$$\mathfrak{L}_{AF} = \frac{F_t^2}{2}\left(\dot{\vartheta}^2 + \sin^2\vartheta\,\dot{\varphi}^2\right) - \frac{F_s^2}{2}\left(\nabla\vartheta\cdot\nabla\vartheta + \sin^2\vartheta\,\nabla\varphi\cdot\nabla\varphi\right) + (\text{higher derivatives}). \tag{14.8}$$

What role did time-reversal invariance play in this story? What is important is $\tilde{T}$-invariance forbids terms in $\mathfrak{L}_{AF}$ that are linear in time derivatives, implying that the lowest-derivative combinations have the general form

$$\mathfrak{L}_{AF} = \frac{F_t^2}{2}\,g_{\alpha\beta}(\vartheta)\,\dot{\vartheta}^\alpha\dot{\vartheta}^\beta - Z^{ab}\;g_{\alpha\beta}(\vartheta)\,\partial_a\vartheta^\alpha\partial_b\vartheta^\beta + (\text{higher derivatives}) \tag{14.9}$$

where $\{\vartheta^\alpha\} = \{\vartheta,\varphi\}$ and $g_{\alpha\beta}(\vartheta)$ is the standard $SO(3)$-invariant metric on the two-sphere.[4]

[3] The agreement between the implications of lattice rotations and full spatial rotations is an accident for terms in $\mathfrak{L}$ with the fewest derivatives, which does not also hold for generic effective interactions.

[4] Comparison to (14.8) reveals the G-invariant metric to be the standard 2-sphere metric, $g_{\alpha\beta}\mathrm{d}\vartheta^\alpha\mathrm{d}\vartheta^\beta = \mathrm{d}\vartheta^2 + \sin^2\vartheta\,\mathrm{d}\varphi^2$, in the coordinates used.

In order to interpret the constants F_s and F_t in (14.8) consider the interactions of small spacetime-dependent fluctuations in the order parameter, corresponding to perturbations $\delta\vec{n}$ (or, equivalently, $\delta\vartheta$ and $\delta\varphi$) about the vacuum configuration, $\vec{n}_0 = \langle\mathbf{N}\rangle$. The direction of $\vec{n}_0$ can be chosen to point in any convenient direction by performing an appropriate $SO(3)$ transformation, which is used to ensure $\vec{n}_0$ points along the positive x-axis (in field space). With this choice the vacuum values for ϑ and φ become $\vartheta_0 = \frac{\pi}{2}$ and $\varphi_0 = 0$.

Writing the canonically normalized fluctuation fields as $\chi := (\vartheta - \frac{\pi}{2})F_t$ and $\psi = \varphi F_t$, the lagrangian for small values of χ and ψ becomes

$$\begin{aligned}\mathfrak{L}_{AF} &= \frac{1}{2}\left(\dot{\chi}^2 - v^2\,\nabla\chi\cdot\nabla\chi\right) + \frac{1}{2}\cos^2\left(\frac{\chi}{F_t}\right)\left(\dot{\psi}^2 - v^2\,\nabla\psi\cdot\nabla\psi\right) + \text{(higher derivatives)},\\ &= \frac{1}{2}\left(\dot{\chi}^2 - v^2\,\nabla\chi\cdot\nabla\chi + \dot{\psi}^2 - v^2\,\nabla\psi\cdot\nabla\psi\right) - \frac{\chi^2}{2F_t^2}\left(\dot{\psi}^2 - v^2\,\nabla\psi\cdot\nabla\psi\right) + \cdots,\end{aligned} \tag{14.10}$$

where $v^2 = F_s^2/F_t^2$ and terms not written in the second line involve at least six powers of the fields, or involve more than two derivatives with respect to either position or time.

The quadratic piece of (14.10) describes two real modes that propagate at low energies according to the dispersion law:

$$E^2(\mathbf{p}) \simeq v^2\mathbf{p}^2. \tag{14.11}$$

These modes physically represent spin waves: small, long-wavelength precessions of the vector $\vec{n}$ about its vacuum value, $\vec{n}_0$. Their quanta, magnons, carry ± 1 unit of the conserved $SO(2)$ spin in the direction parallel to $\langle\mathbf{N}\rangle$. The parameter v is the propagation speed for these modes, and the condition that it must be smaller than the speed of light – *i.e.* $v \leq 1$ in fundamental units – implies that $F_s \leq F_t$. The lagrangian (14.10) reveals how the spin waves interact at low energies, with a strength governed by $1/F_t^2$.

Understanding magnon phenomenology also requires knowing their couplings to electromagnetic fields. To construct these, notice that when the microscopic physics is described by (12.1) through (12.3) magnetic fields couple to electron spins dominantly through the effective interaction $(e_q c_F/2m)\,\mathbf{B}\cdot(\Psi^\dagger\sigma\Psi) \subset \mathfrak{L}_1$ of Eq. (12.2) – or more generally to $\mathfrak{m}\,\mathbf{B}\cdot(\Psi^\dagger\sigma\Psi) \subset \mathfrak{L}_1$ of Eq. (12.39). But for the microscopic lagrangian the combination

$$\vec{\rho} := \vec{\jmath}^{\,0} = \frac{1}{2}\Psi^\dagger\vec{\sigma}\Psi, \tag{14.12}$$

is also the charge density of the Noether current, $\vec{\jmath}^{\,\mu}$, for the $SO(3)$ spin-rotation symmetry.

The interaction between magnons and electromagnetic fields is therefore also given by a term of the form:

$$\mathfrak{L}_{\rm em} = \mu\,\vec{\rho}\cdot\mathbf{B}, \tag{14.13}$$

where $\vec{\rho}$ is the Noether charge (or spin) density, computed as a function of $\vec{n}$ using Noether's theorem for the $SO(3)$ symmetry of the low-energy action $\mathfrak{L}_{AF}$. Matching the couplings to the underlying theory gives at leading order $\mu = e_q c_F/m$.

Applying Noether's theorem to the effective magnon action (14.7) gives the conserved charge and current,

$$\vec{\rho} = F_t^2\,(\vec{n}\times\dot{\vec{n}}) + \cdots \qquad \text{and} \qquad \vec{\mathbf{j}} = -F_s^2\,(\nabla\vec{n}\times\vec{n}) + \cdots, \tag{14.14}$$

where the dots are a reminder of the unwritten higher-derivative contributions.

Writing this interaction in terms of unconstrained fields then gives the lowest-dimension effective interaction between magnons and magnetic fields to be

$$\begin{aligned}\mathfrak{L}_{\rm em} &= \mu F_t^2\,\mathbf{B}\cdot(\vec{n}\times\dot{\vec{n}}) + \cdots\\ &= \mu F_t^2\Big[B_x\,(\dot{\vartheta}\,\sin\varphi + \dot{\varphi}\,\sin\vartheta\cos\vartheta\cos\varphi)\\ &\qquad + B_y\,(-\dot{\vartheta}\,\cos\varphi + \dot{\varphi}\,\sin\vartheta\cos\vartheta\sin\varphi) - B_z\,\dot{\varphi}\,\sin^2\vartheta)\Big],\end{aligned} \tag{14.15}$$

which at lowest order in the canonical fluctuation fields involves mixing with the components of **B** that are perpendicular to the ground-state configuration $\vec{n}_0$:

$$\mathfrak{L}_{\rm em} = -\mu F_t\left(B_y\,\dot{\chi} + B_z\,\dot{\psi}\right) + \cdots. \tag{14.16}$$

Notice that the time derivative in this interaction ensures invariance with respect to $\tilde{T}$ transformations, under which $\mathbf{B}\to-\mathbf{B}$. Some implications of this magnon EFT are explored together with a similar treatment for Ferromagnets, in §14.1.3 below.

14.1.2 Ferromagnetism

For ferromagnets, the order parameter is simply the total spin,

$$\mathbf{S} = \sum_i \mathbf{s}_i. \tag{14.17}$$

whose nonzero expectation $\langle\mathbf{S}\rangle \neq 0$ again spontaneously breaks the spin symmetry, $G = SO(3)$, down to $H = SO(2)$, precisely as for an antiferromagnet. What is different in this case is the absence of any time-reversal symmetry like $\tilde{T}$ that was present for antiferromagnets. The low-energy effective theory can therefore contain T-violating terms, and this changes the properties of its Goldstone bosons in an important way.

Since the symmetry-breaking pattern for both ferromagnets and antiferromagnets is $SO(3)\to SO(2)$, the nonlinear realization of this symmetry on the Goldstone boson fields can be carried over from the antiferromagnetic case in whole cloth. Again defining the unit vector $\vec{s} = \mathbf{S}/|\mathbf{S}|$ and using polar coordinates, $\vartheta(\mathbf{x},t)$ and $\varphi(\mathbf{x},t)$, to describe its direction,

$$s_x = \sin\vartheta\cos\varphi \qquad s_y = \sin\vartheta\sin\varphi \qquad \text{and} \qquad s_z = \cos\vartheta\,, \tag{14.18}$$

so that $\vec{s}\cdot\vec{s} = 1$, the field, $\vec{s}(\mathbf{r},t)$, describes long-wavelength variations in the direction of **S**.

The action of $SO(3)$ on these variables is again given by Eq. (14.6), and the term in the effective lagrangian with two spatial derivatives is the same as for an antiferromagnet,

$$\mathfrak{L}_{F,s} = -\frac{F_s^2}{2}\left(\nabla\theta\cdot\nabla\theta + \sin^2\theta\,\nabla\phi\cdot\nabla\phi\right), \tag{14.19}$$

where invariance under spatial rotations is again assumed for simplicity.

The new features appear once the leading term involving time derivatives is constructed, because this time a term is possible having only a single time derivative (possible because of the absence of time-reversal symmetry). It has the following general form:

$$\mathfrak{L}_{F,t} = -A_\alpha(\vartheta)\, \dot{\vartheta}^\alpha, \tag{14.20}$$

where the coefficient function, $A_\alpha(\theta)$, is to be determined. At low energies this type of term always dominates the two-derivative term encountered for antiferromagnets.

To determine $A_\alpha(\vartheta)$ requires knowing how it transforms under G transformations. Performing a general field redefinition of the form $\delta\vartheta^\alpha = \xi^\alpha(\vartheta)$ shows that the coefficient A_α transforms like a vector field on the target space, G/H,

$$\delta A_\alpha = \pounds_\xi A_\alpha := \xi^\beta \partial_\beta A_\alpha + A_\beta \partial_\alpha \xi^\beta. \tag{14.21}$$

What is puzzling at first sight is that there is no nonzero choice for $A_\alpha(\vartheta)$ that remains invariant under all G transformations. This is intuitively clear because if any such quantity existed the vector $A_\alpha(\vartheta)$ would be a rotationally invariant vector field tangent to a 2-sphere, which are known not to exist [368]. (To see why intuitively, recall that all rotationally invariant vector fields in three dimensions point radially and so cannot lie tangent to a 2-sphere centred at the origin.)

But now comes the main point: it is only the action that must remain G-invariant, and not the lagrangian density. So if a choice for $A_\alpha(\vartheta)$ could be found that satisfies

$$\pounds_\xi A_\alpha = \xi^\beta \partial_\beta A_\alpha + A_\beta \partial_\alpha \xi^\beta = \partial_\alpha \Omega_\xi, \tag{14.22}$$

for some choice of functions $\Omega_\xi(\vartheta)$, then the action would be invariant. It suffices that the right-hand side is the gradient of a scalar because if $A_\alpha = \partial_\alpha \Omega$ then $A_\alpha \dot{\vartheta}^\alpha = \dot{\Omega}$ is a total derivative, and so drops out of the action (up to temporal boundary terms).

Geometrically, this says that A_α may be considered to be proportional to a G-invariant gauge field (rather than a vector field) defined on the target space G/H, because the condition that the action be G invariant is that A_α must only be G-invariant up to a gauge transformation: $A_\alpha \to A_\alpha + \partial_\alpha \Omega$. This last condition is equivalent to the invariance of the field strength for A_α:

$$\pounds_\xi F_{\alpha\beta} := \xi^\gamma \partial_\gamma F_{\alpha\beta} + F_{\gamma\beta} \partial_\alpha \xi^\gamma + F_{\alpha\gamma} \partial_\beta \xi^\gamma = 0, \tag{14.23}$$

where $F_{\alpha\beta} := \partial_\alpha A_\beta - \partial_\beta A_\alpha$. A G-invariant action exists if there is a G-invariant choice for $F_{\alpha\beta}$ tangent to $G/H = SO(3)/SO(2) \equiv S_2$.

The construction of such an invariant field strength is actually quite simple. Since the coset space is two dimensional, it is always possible to write the field strength in terms of a scalar field, $F_{\alpha\beta} = \mathcal{F}(\vartheta)\, \epsilon_{\alpha\beta}$, where $\epsilon_{\alpha\beta}$ is the antisymmetric Levi-Civita tensor – see *e.g.* Appendix A.2.1 – constructed using the coset's G-invariant metric. The condition that $F_{\alpha\beta}$ must be G invariant is then equivalent to the invariance of $\mathcal{F}$:

$$\pounds_\xi \mathcal{F} \equiv \xi^\alpha \partial_\alpha \mathcal{F} = 0, \tag{14.24}$$

which is only possible for all G transformations on a 2-sphere if $\mathcal{F}$ is a constant.

All that is required is a gauge potential for which $F_{\alpha\beta} = \mathcal{F}\, \epsilon_{\alpha\beta}$ on the two-sphere, $S_2 = SO(3)/SO(2)$. But this is the field strength of a magnetic monopole positioned at the centre of the two-sphere, so what is required is the gauge potential for a magnetic monopole configuration. Since for a unit 2-sphere the nonzero component of the Levi-Civita tensor is $\epsilon_{\vartheta\varphi} = \sin\vartheta$, the solution to $F_{\vartheta\varphi} = \mathcal{F}\epsilon_{\vartheta\varphi}$ may be written (locally) as:

$$A^{\pm}_{\alpha}\,\mathrm{d}\vartheta^{\alpha} = \mathcal{F}\,(\pm 1 - \cos\vartheta)\,\mathrm{d}\varphi, \tag{14.25}$$

where $A^{+}_{\alpha} \neq 0$ throughout the region $\cos\vartheta \neq +1$, while $A^{-}_{\alpha} \neq 0$ for the region $\cos\vartheta \neq -1$.

Although neither A^{+}_{α} nor A^{-}_{α} is everywhere nonzero on G/H, the field strength $F_{\alpha\beta}$ obtained from them can nonetheless be globally defined provided its magnitude satisfies a quantization condition [369]. To see why, require $A^{\pm}_{\alpha}$ to differ by a single-valued gauge transformation in the region $-1 < \cos\vartheta < 1$ — *i.e.* $A^{+}_{\alpha} - A^{-}_{\alpha} = \partial_{\alpha}\Omega = -ig^{-1}\partial_{\alpha}g$ for some group element $g = \exp(i\Omega)$ satisfying $g(\varphi + 2\pi) = g(\varphi)$, and so $\Omega(\varphi + 2\pi) = \Omega(\varphi) + 2\pi l$ for integer l. Writing $\Omega(\varphi) = l\varphi + \hat{\Omega}(\varphi)$ where $\hat{\Omega}(\varphi + 2\pi) = \hat{\Omega}(\varphi)$ then shows that the only φ-independent possibility is $A^{+}_{\varphi} - A^{-}_{\varphi} = l$, which leads to a quantization condition for the coefficient $\mathcal{F}$ of the form[5]

$$\mathcal{F} = \frac{l}{2}, \tag{14.26}$$

where l is a nonzero integer.

The corresponding lagrangian is then given by

$$\mathfrak{L}_{F,t} = \frac{lF_t^3}{2}(\pm 1 - \cos\vartheta)\,\dot{\varphi}, \tag{14.27}$$

for an arbitrary mass scale F_t. Either sign is equally good for most purposes because the difference is a total derivative in the action. Combining the contributions of Eqs. (14.19) and (14.27) then gives the lowest-derivative terms for the magnon lagrangian,

$$\mathfrak{L}_F = \frac{lF_t^3}{2}(\pm 1 - \cos\vartheta)\,\dot{\varphi} - \frac{F_s^2}{2}\left(\nabla\vartheta\cdot\nabla\vartheta + \sin^2\vartheta\,\nabla\varphi\cdot\nabla\varphi\right). \tag{14.28}$$

As is easily verified, the classical equations of motion for the lagrangian, (14.28), are equivalent to

$$\dot{\vec{s}} + k\left(\vec{s}\times\nabla^2\vec{s}\right) = 0, \tag{14.29}$$

an equation – known as the *Landau–Lifshitz–Gilbert* equation [370] – long known to describe long-wavelength spin waves in ferromagnets. The constant, k, in this equation is given by

$$k = \frac{2F_s^2}{lF_t^3}. \tag{14.30}$$

For small fluctuations, $\vec{s}(\mathbf{x},t) = \vec{s}_0 + \delta\vec{s}(\mathbf{x},t)$, the linearization of (14.29) predicts that $\delta\vec{s}$ satisfies

$$\delta\dot{\vec{s}} + k\left(\vec{s}_0\times\nabla^2\delta\vec{s}\right) = 0, \tag{14.31}$$

[5] For gauge transformations of the form $g(\theta) = e^{ie\Omega}$ this quantization condition becomes $\mathcal{F} = n/(2e)$. It is the freedom to define what one means by 'charge' on the target space that allows the introduction of an arbitrary scale F_t^3 in (14.27).

which describes spin waves transverse to $\vec{s}_0$ that at low energies have a dispersion relation

$$E(\mathbf{p}) = k\,\mathbf{p}^2, \tag{14.32}$$

that rises much more steeply with $|\mathbf{p}|$ than does its antiferromagnetic counterpart (14.11).

For the purposes of coupling to magnetic fields (and also as a reality check), it is instructive to compute the Noether currents for the $SO(3)$ symmetry that are implied by this lagrangian density. Since the terms involving spatial derivatives are the same as for antiferromagnets, the conserved current density is the same as was found in the previous section

$$\vec{\mathbf{j}} = F_s^2\,(\vec{s} \times \nabla\vec{s}) + \text{(higher derivatives)}. \tag{14.33}$$

When computing the charge density, it is important to keep in mind that $\mathfrak{L}_F$ is not invariant under G-transformations, but instead transforms into a total derivative:

$$\delta\mathfrak{L}_F = \frac{\mathrm{d}}{\mathrm{d}t}\left\{\frac{lF_t^3}{2}\left[\frac{\omega_x\,\cos\varphi + \omega_y\,\sin\varphi}{\sin\vartheta}\right]\right\}. \tag{14.34}$$

Using this in the general expression, Eq. (4.7), for the Noether current gives the conserved charge density:

$$\vec{\rho} = \frac{lF_t^3}{2}\,\vec{s} + \text{(higher derivatives)}. \tag{14.35}$$

This is the reality check alluded to above: (14.35) confirms that $\vec{\rho}$ indeed does give the spin density – just as does its microscopic counterpart, (14.12) – inasmuch as its direction is completely set by the direction of $\vec{s}$. This reveals the total ground-state spin density to be

$$\langle\mathbf{S}\rangle = \frac{lF_t^3}{2}\,\vec{s}_0, \tag{14.36}$$

thereby providing a phenomenological value for lF_t^3.

As is easy to check, using (14.33) and (14.35) in the equation for current conservation, $\partial_t\vec{\rho} + \nabla\cdot\vec{\mathbf{j}} = 0$ reproduces the Landau–Lifshitz equation (14.29). Eq. (14.35) also shows that it is the conserved charge density, $\vec{\rho}$, itself that is the order parameter whose expectation value breaks the symmetry:

$$\left\langle\vec{\rho}\right\rangle = \frac{lF_t^3}{2}\langle\vec{s}\rangle \neq 0, \tag{14.37}$$

something that would have been forbidden for a Lorentz-invariant system.

From here on, the determination of the lowest-dimension effective coupling between $\vec{s}$ and magnetic fields proceeds as for antiferromagnetism, with the basic coupling having the form $\mu\,\mathbf{B}\cdot\vec{\rho}$. The only difference for ferromagnets is the use of (14.35) instead of (14.14) when expressing $\vec{\rho}$ in terms of ϑ and φ. In the ferromagnetic case this gives

$$\begin{aligned}\mathfrak{L}_{\text{em}} = \mu\,\vec{\rho}\cdot\mathbf{B} &= \frac{\mu lF_t^3}{2}\Big(B_x\,\sin\vartheta\cos\varphi + B_y\,\sin\vartheta\sin\varphi + B_z\,\cos\vartheta\Big),\\ &= \frac{\mu lF_t^3}{2}\,B_x + \frac{\mu lF_t^3}{2}\Big(B_y\,\delta\varphi - B_z\,\delta\vartheta\Big) + \cdots,\end{aligned} \tag{14.38}$$

where μ is the effective coupling parameter, expected to match onto $e_q c_F/m$ at UV scales, and the expansion in powers of fluctuations chooses variables to ensure $\vec{s}_0 \propto \langle \mathbf{S} \rangle$ points in the positive x direction, corresponding to $\vartheta = \frac{\pi}{2} + \delta\vartheta$ and $\varphi = \delta\phi$. Eq. (14.38) does not involve derivatives of $\vec{s}$ because invariance of the interaction $\mathbf{B} \cdot \vec{s}$ requires spin rotations to be accompanied by rotations in real space, thereby making them spacetime rather than internal symmetries. (See §14.3.1 for more detail about counting Goldstone modes for spacetime symmetries.)

The first term in the final line of (14.38) gives the $\mu \mathbf{B} \cdot \langle \vec{\rho} \rangle$ interaction energy between the magnetic field and the spins in the ordered vacuum and this is why it survives when evaluated at $\vec{s} = \vec{s}_0$. This also reveals the effective coupling μ as the material's magnetic moment per particle, as is also clear from the matching condition $\mu \simeq e_q c_F/m$.

14.1.3 Physical Applications

For both ferromagnets and antiferromagnets it is useful to couple the magnons to other degrees of freedom with which the system might be probed. Of particular interest are those involving magnetic fields, since these couple microscopically to the underlying spin degrees of freedom that are of interest.

An example of such a probe is the scattering of slow neutrons from magnetically ordered materials [371]. A nonrelativistic neutron dominantly couples to the magnetic field through the magnetic-moment interaction coming from $\mathfrak{L}_1$ of Eq. (12.39),

$$\mathfrak{L}_{\rm int} = \mathfrak{m}_n \, \mathbf{B} \cdot \Psi_n^\dagger \, \sigma \Psi_n, \tag{14.39}$$

where $\Psi_n(x)$ denotes the neutron field and $\mathfrak{m}_n$ is the neutron magnetic moment (which is negative and of order e/m_n in size). Because of this coupling, neutrons undergo magnetic-moment scattering from a sample's atomic electrons.[6] The very different nuclear and electron masses also mean that the recoil distribution of magnetically scattered neutrons can be distinguished from their much stronger (but short-ranged) nuclear interactions with a sample's nuclei.

Integrating out the instantaneous photon exchange using (14.13) and (14.39) leads to a dipole–dipole interaction, which when used in Fermi's Golden Rule leads to a standard expression for the differential cross section for inelastic magnetic neutron scattering [372] (see Exercise 14.2)

$$\frac{d\sigma}{dE' d\Omega} = \frac{p'}{p} (2m_n \mathfrak{m}_n)^2 \left(\delta_{ij} - \frac{q_i q_j}{\mathbf{q}^2} \right) S_{ij}(\omega, \mathbf{q}), \tag{14.40}$$

where $p = |\mathbf{p}|$ (and $p' = |\mathbf{p}'|$) is the magnitude of the initial (final) neutron momentum and $E' = (p')^2/2m_n$ is the final neutron energy. $S_{ij}(\omega, \mathbf{q})$ is the magnetization dynamical structure function, defined by[7]

$$S_{ij}(\omega, \mathbf{q}) = \frac{1}{2\pi} \int dt \, \Big\langle M_i(-\mathbf{q}, t) \, M_j(\mathbf{q}, 0) \Big\rangle \, e^{-i\omega t}, \tag{14.41}$$

6 A medium's electrons dominate over its nuclei in magnetic scattering for the same reason they dominate in spin ordering: the electron has a much larger magnetic moment due to its much smaller mass.

7 Unlike common practice this definition does not divide S_{ij} by the total number of scattering sites. Furthermore, the cross section as derived neglects the contribution of electronic spin-orbit coupling to the magnetization (which is a good approximation for many – but not all – materials).

with

$$\vec{M}(\mathbf{q},t) := \mu \int d^3\mathbf{r}\, \vec{\rho}\,(\mathbf{r},t)\, e^{i\mathbf{q}\cdot\mathbf{r}}, \tag{14.42}$$

and the expectation-value is taken in whatever state the material is initially in. Here $\omega = E - E'$ and $\mathbf{q} = \mathbf{p} - \mathbf{p}'$ are the energy and momentum transferred by the neutron to the medium. Expressions like these show that measurements of scattered neutron momenta and energies can provide information about two-point functions for the spin density $\vec{\rho}$.

Although this response function can receive contributions from many sources, the specific contribution from magnon excitations is easily computed using the effective theory, within which the magnons are weakly coupled states. The result differs for antiferromagnets and ferromagnets. For antiferromagnets, Eq. (14.14) relates $\vec{\rho}$ to $\vec{n}$ and so using

$$\vec{n} = \sin\vartheta\cos\varphi\,\mathbf{e}_x + \sin\vartheta\sin\varphi\,\mathbf{e}_y + \cos\vartheta\,\mathbf{e}_z \simeq \mathbf{e}_x + \frac{1}{F_t}\big(\psi\,\mathbf{e}_y - \chi\,\mathbf{e}_z\big) + \cdots, \tag{14.43}$$

leads to the lowest-order result $\vec{\rho} \simeq F_t\,(\dot{\chi}\,\mathbf{e}_y + \dot{\psi}\,\mathbf{e}_z) + \cdots$ and so $S_{ij}(\omega,\mathbf{q})$ can be evaluated using the properties of χ and ψ read from the lagrangian (14.10) together with knowledge of the state in which the magnon field is prepared (such as in a thermal state).

The simplest case takes the medium initially to be in the no-magnon ground state, with energies low enough to neglect magnon self-interactions. In this case, $S_{ij}(\omega,\mathbf{q})$ is related to the Wightman function [373] for what are essentially Klein–Gordon fields (see Appendix C.3.1), $\langle\chi(x)\chi(x')\rangle$ and $\big\langle\psi(x)\psi(x')\big\rangle$, leading to

$$S_{ij}(\omega,\mathbf{q}) \simeq \frac{\Omega\mu^2F_t^2\omega}{2}\,\Big[\delta_{ij} - (n_0)_i(n_0)_j\Big]\,\delta(\omega - v|\mathbf{q}|) \qquad \text{(AF magnons)}. \tag{14.44}$$

Here $v = F_s/F_t$ is the speed of magnon propagation and $\vec{n}_0 = \mathbf{e}_x$ is the direction of the ground-state alignment. This shows the appearance of a narrow peak centred on the magnon dispersion relation that provides a way to measure this relation. (Once interactions are included, the infinitely narrow peak typically widens to a line-shape characterized by a magnon decay width Γ.) Ω is the system volume, which arises because the ground state is assumed translation invariant.

A similar calculation goes through for ferromagnets, though with a number of instructive differences. For ferromagnets the relation between $\vec{\rho}$ and the magnon fields is instead found from (14.35), which states

$$\vec{\rho} = \frac{lF_t^3}{2}\Big[\sin\vartheta\cos\varphi\,\mathbf{e}_x + \sin\vartheta\sin\varphi\,\mathbf{e}_y + \cos\vartheta\,\mathbf{e}_z\Big] \simeq \mathbf{S}_0 + \frac{1}{F_t}\big(\psi\,\mathbf{e}_y - \chi\,\mathbf{e}_z\big) + \cdots, \tag{14.45}$$

where in this case the canonical fields are defined so that the quadratic lagrangian in the expansion of (14.28) is

$$\mathfrak{L}_{\rm quad} = -\frac{lF_t^3}{2}\cos\vartheta\,\dot{\varphi} - \frac{F_s^2}{2}\big(\nabla\vartheta\cdot\nabla\vartheta + \nabla\varphi\cdot\nabla\varphi\big) = \chi\dot{\psi} - k\big(\nabla\chi\cdot\nabla\chi + \nabla\psi\cdot\nabla\psi\big), \tag{14.46}$$

with k as given in (14.30).

Eq. (14.46) shows that the pair $\{\chi, \psi\}$ includes fewer degrees of freedom for ferromagnets than for antiferromagnets. This can be seen because χ enters (14.46) with no time derivatives, more like a canonical momentum for ψ than a separate field. Alternatively, for ferromagnets it suffices to specify $\chi(\mathbf{x}, t_0)$ and $\psi(\mathbf{x}, t_0)$ at an initial time to have a well-posed initial-value problem for the linearized field equations, whereas $\dot{\chi}(\mathbf{x}, t_0)$ and $\dot{\psi}(\mathbf{x}, t_0)$ would also have to be given separately in the antiferromagnetic case. As a consequence, the pair $\{\chi, \psi\}$ get quantized more as the real and imaginary parts of a Schrödinger field than like a Klein–Gordon field, making their Feynman rules follow as in the nonrelativistic examples of Part III (see Appendix C.2) rather than a relativistic field [107, 108].

With this in mind a similar calculation as above gives the magnon contribution to the spin response function, $S_{ij}(\omega, \mathbf{q})$, of Eq. (14.41). Just as was true for antiferromagnets this shows peaks where the neutron scatters to produce a magnon, although this time satisfying the ferromagnetic dispersion relation $\omega = k\,\mathbf{q}^2$.

The momentum-dependence of the magnon dispersion relation also has implications for the temperature dependence of the magnitude of the magnetization, $M = |\langle \mathbf{M} \rangle|$, of a ferromagnet at very low temperatures. Since it is the magnon field itself that describes the long-wavelength deviations of the magnetization, deviations at very low temperatures are controlled by the average magnon occupation number:

$$M(0) - M(T) \propto M(0) \int \mathrm{d}^3\mathbf{p}\, f\left(E_p/T\right), \tag{14.47}$$

where $f(x) = (e^x - 1)^{-1}$ is the Bose–Einstein distribution function and $E(p) \propto |\mathbf{p}|^z$ is the long-wavelength magnon dispersion relation.

The temperature dependence of (14.47) is sensitive to z, as is made explicit by changing integration variables from $p = |\mathbf{p}|$ to the quantity $x = E_p/T$. If $E(p) \propto p^z$ then:

$$p^2\,\mathrm{d}p = p^2 \frac{\mathrm{d}p}{\mathrm{d}E}\,\mathrm{d}E \propto E^{2/z}\, E^{-(z-1)/z}\,\mathrm{d}E \propto T^{3/z} x^{(3-z)/z}\mathrm{d}x. \tag{14.48}$$

For $z = 2$ this predicts $[M(0) - M(T)]/M(0) \propto T^{3/2}$ – a result known as Bloch's Law [374] – which agrees well with low-temperature observations with ferromagnets [375].

14.2 Low-Energy Superconductors ♠

Spontaneous gauge symmetry breaking and the Higgs mechanism[8] also arises for nonrelativistic systems, with superconductors providing the best-known practical example.

Once cooled below a critical temperature, T_c, superconductors display a number of remarkable electromagnetic properties, including: the absence of resistance to electrical currents (with super-currents known to persist undiminished for years [376]),

[8] More aptly: the Anderson–Higgs mechanism in this case, since the understanding of the mechanism arguably started with superconductors [60].

the Meissner effect (in which magnetic fields are expelled from the interior of a superconductor) [377], flux quantization (in which flux threading a superconducting ring is quantized with apparently arbitrary precision) [378, 379] and more.

These properties are consequences of the spontaneous breakdown of electromagnetic gauge invariance by the superconducting material. This spontaneous breaking is understood to arise because of a tendency for pairs of electrons to bind together and condense in the ground state in response to comparatively weak attractive forces that in specific circumstances can dominate their naive Coulomb repulsion. For many 'traditional' superconductors the attractive force responsible arises as a consequence of interactions between electrons and the vibrations of the material's lattice of ions [380]. As of this writing the underlying reason for pairing is not equally well understood for the more recently discovered [381] class of high-T_c superconductors.

Effective field theories contribute in several ways to the understanding of how electron pairing works. As discussed in more detail in §15.2, EFT methods are useful for understanding which interactions amongst electrons dominate at low energies. For traditional superconductors such low-energy methods are relevant because for them T_c is typically much smaller than typical interaction energies of atomic electrons.

However, this section deploys EFT arguments to superconductors with a different goal. The purpose here – closely following the presentation in [382] – is not to understand why electrons pair (and so why electromagnetic gauge invariance spontaneously breaks). Instead, it aims to show how the striking electromagnetic response of superconductors mentioned above depends only on the fact of electromagnetic symmetry breaking, and not on the details of *why* this symmetry breaks. This helps understand why the predictions for these effects are so much more accurate than the understanding of the details of electron pairing; the predictions are general low-energy consequences of the symmetry-breaking pattern itself for which the microscopic details are (literally) irrelevant.

14.2.1 Implications of the Goldstone Mode

The most robust consequence of spontaneous symmetry-breaking is the existence of a Goldstone boson [3]. This boson is typically not gapless when the symmetry is gauged (due to its mixing to become a component of – and thereby generating a mass for – the relevant gauge boson [60]). It nonetheless mediates physical effects, and (as usual) its low-energy properties are largely governed purely by the symmetry-breaking pattern itself.

For superconductors the symmetry-breaking pattern is $G \to H$, where $G = U(1)$ is the gauge symmetry of electromagnetism and $H = Z_2$ is the two-element group $\{h_0, h_1\}$ with h_0 the unit element and multiplication rule $h_1^2 = h_0$. The residual Z_2 symmetry here arises because the microscopic order parameter for real superconductors comes from the pairing of electrons, and so has electric charge $e_q = -2e$.

To see how this works, recall that the gauge group G acts on charged fields through the rule

$$\psi(x) \to g\,\psi(x) = e^{ie_q\zeta(x)}\,\psi(x), \tag{14.49}$$

which defines the $U(1)$ group element $g = e^{ie_q\zeta}$, where e_q is the field's electric charge. At the same time, the $U(1)$ gauge potential transforms as

$$A_\mu(x) \to A_\mu(x) + \partial_\mu\zeta(x) = A_\mu - \frac{i}{e_q} g^{-1}\partial_\mu g. \tag{14.50}$$

To the extent that all free charges are quantized[9] in units of e – *i.e.* $e_q = N_q\, e$ for N_q an integer, and $N_q = \pm 1$ for at least one particle type – the single-valuedness of $g[\zeta]$ implies that the gauge parameters ζ and $\zeta + 2\pi/e$ should be identified. The unbroken group H consists of those transformations for which $\zeta = \pi/e$, since these act nontrivially on fields with $N_q = \pm 1$ but leave unchanged fields (like those for paired electrons) with $N_q = \pm 2$.

Like for any internal symmetry, the Goldstone boson field, $\phi(x)$, parameterizes the coset G/H, which for real superconductors is $U(1)/Z_2$. Normalizing ϕ so that its transformation rule under G is

$$\phi(x) \to \phi(x) + \zeta(x), \tag{14.51}$$

a typical element of G can then be written $g[\phi(x)] = e^{ie_q\phi}$. The fact that the discrete group Z_2 remains unbroken means that $\phi(x)$ lives on a circle with ϕ and $\phi + \pi/e$ identified.

But having a semiclassical Goldstone field means more than simply having a field that shifts under the gauge symmetry. After all, defining $\chi = \ln\psi$, where ψ transforms as in (14.49), generates a field that shifts under the symmetry without the need for spontaneous symmetry-breaking. The change of variables from ψ to χ breaks down, however, inasmuch as a semiclassical expansion for χ around any finite vacuum χ_0 provides a poor description of the expansion of ψ about $\psi_0 = 0$. So, besides having the field ϕ shift under gauge transformations it should also be required to have a sensible semiclassical expansion. In what follows this is built in by demanding that the effective lagrangian for ϕ deep within a superconductor have an equilibrium ground state for which $\phi = \phi_0$ is constant (and can be chosen to vanish), about which the effective lagrangian can be expanded at low energies in powers of ϕ and its derivatives. The key assumption is that this expansion does not start only with higher derivatives, but also includes terms with two or fewer derivatives.

Now imagine writing down the low-energy effective lagrangian, L_s, for a superconductor, well below its symmetry-breaking scale. Given that a field exists that shifts under a symmetry, it is always possible to take any other charged field and make it gauge invariant by absorbing the appropriate power of $e^{i\phi}$. That is, suppose $\psi(x)$ is a local operator that transforms like (14.49) for some nonzero value of e_q. Then the new operator $\tilde\psi(x) := \psi(x)\, e^{-ie_q\phi(x)}$ is gauge invariant. It is useful to use the gauge-invariant variable $\tilde\psi_i(x)$ to describe all charged degrees of freedom because when this is done gauge invariance completely ties ϕ to the electromagnetic field:

$$L_s\big(\mathbf{A}, A_0, \phi, \psi_i\big) = L_s\big(\mathbf{A} - \nabla\phi, A_0 - \partial_t\phi, \tilde\psi_i\big). \tag{14.52}$$

In previous sections one would proceed by writing $L_s = \int d^3x\, \mathfrak{L}_s$ as a local expression, and expand it in powers of the fields and their derivatives. However, this

[9] We know quark charges come as fractions of e, but because these are confined within colour-neutral hadrons, in practice all present evidence is consistent with free charges being quantized in units of e.

is one of the places where many-body applications can differ from the discussion of previous sections. They can differ because the underlying material can contain the effects of long-range forces, such as (but not restricted to) the Coulomb force.

In many-body applications this can happen in a variety of ways. First, because a material always has a preferred rest frame, special relativity is (spontaneously) broken and high-energy modes need not also have short wavelengths. If a high-energy mode involves large spatial correlations it can produce nonlocal effects once integrated out.

Second, it can also be true that not all of a body's low-energy modes are of equal interest for any given application. Because of this it can be useful to integrate out some modes that actually belong in the low-energy theory, simply because no measurements on them are ever intended. This also can give nonlocal effects (in addition to other issues, that are the topic of §16) since such modes often have long wavelengths. When this is done the action of interest is no longer, strictly speaking, a Wilsonian action (since Wilsonian actions by definition remove states only based on their energy – plus perhaps also their charges for some other conserved quantities).

For traditional superconductors it happens that the pairing of electrons takes place over a correlation length, ξ, that can be quite large (compared, say, to the lattice spacing, a, of the underlying atoms). The effective low-energy description therefore need not be local if it includes distances, x, satisfying $a \ll x \lesssim \xi$.

What should the effective lagrangian on these scales look like? Since we assume ground states for which ϕ is constant, the lagrangian describing fluctuations near this ground state can be built using powers of ϕ and its derivatives. Gauge invariance then implies that the expansion is in powers of $\mathbf{A} - \nabla\phi$ and $A_0 - \partial_t\phi$, and the expansion starts quadratically in these fields because the ground-state configuration solves the field equations. Specializing to time-independent fluctuations – for which $\partial_t\phi - A_0 = 0$ – this leads to the form

$$L_s = L_{s0} - \frac{1}{2}\int \mathrm{d}^3\mathbf{x}\,\mathrm{d}^3\mathbf{y}\,\mathfrak{C}^{ij}(\mathbf{x},\mathbf{y})\big[A_i(\mathbf{x}) - \partial_i\phi(\mathbf{x})\big] \times \big[A_j(\mathbf{y}) - \partial_j\phi(\mathbf{y})\big] + (\text{higher orders}), \tag{14.53}$$

with a positive-definite symmetric kernel, $\mathfrak{C}^{ij}(\mathbf{x},\mathbf{y}) = \mathfrak{C}^{ji}(\mathbf{y},\mathbf{x})$, that can be nonlocal up to distances of order ξ. For non-superconducting materials this kernel would vanish, leaving the energy to be governed only by derivatives of $\mathbf{A}$ (*i.e.* field strengths like $\mathbf{B} = \nabla\times\mathbf{A}$) rather than $\mathbf{A}$ itself.

For simplicity of presentation, in what follows it is assumed that $\mathfrak{C}^{ij}(\mathbf{x},\mathbf{y})$ is invariant under translations, rotations and parity (though invariance under more than the rotational symmetries of the underlying lattice, in particular, need not be true for real systems). When this is so the kernel simplifies to

$$\mathfrak{C}^{ij}(\mathbf{x},\mathbf{y}) = \delta^{ij}\,\mathfrak{c}_1\left(|\mathbf{x}-\mathbf{y}|^2\right) + \partial^i\partial^j\,\mathfrak{c}_2\left(|\mathbf{x}-\mathbf{y}|^2\right), \tag{14.54}$$

where $\mathfrak{c}_{1,2}(x)$ are positive functions of a single scalar argument.

The electromagnetic charge and current densities predicted by the superconducting effective lagrangian are $J^\mu = \delta L_s/\delta A_\mu$, which (14.52) links to variations with respect to ϕ,

$$\sigma = J^0 = \frac{\delta L_s}{\delta A_0} = -\frac{\delta L_s}{\delta(\partial_t\phi)} \quad\text{and}\quad \mathbf{J} = \frac{\delta L_s}{\delta\mathbf{A}} = -\frac{\delta L_s}{\delta(\nabla\phi)}. \tag{14.55}$$

The field equations for ϕ (with $\tilde{\psi}_i$ and A_μ fixed) boil down to $\partial_t \sigma + \nabla \cdot \mathbf{J} = 0$, which simply expresses local conservation of electromagnetic charge.

The first of Eqs. (14.55) implies that $-\sigma$ is the canonical momentum for ϕ, and so once written in a Hamiltonian framework one trades $\partial_t \phi$ for σ so that, $H_s = H_s[\sigma, \phi]$. In terms of H_s the gauge-invariant ϕ evolution equation is

$$\partial_t \phi - A_0 = \frac{\delta H_s}{\delta(-\sigma)} =: -\mathfrak{v}(\phi, \sigma), \tag{14.56}$$

where the voltage, $\mathfrak{v}(\phi, \sigma) = \delta H_s / \delta \sigma$, is defined as the rate of change of energy with respect to changes in charge density.

The rest of this section is devoted to showing that the above properties suffice to capture many of the characteristic low-energy properties of superconductors.

Meissner Effect

The assumption that the equilibrium condition deep inside a superconductor occurs for $\mathbf{A} - \nabla\phi = 0$ locally implies that $\mathbf{B} = \nabla \times \mathbf{A} = 0$ there, immediately implying that magnetic fields must vanish deep within a superconductor. The Meissner effect [377] is thereby seen to emerge as a low-energy theorem in the effective description.

For static configurations – those satisfying, in particular, $\partial_t \phi - A_0 = 0$ – the approach to zero field within superconductors with boundaries is found by combining Eqs. (14.55) and (14.53), to give

$$\begin{aligned} J^i(\mathbf{x}) &= -\int \mathrm{d}^3\mathbf{y}\, \mathfrak{C}^{ij}(\mathbf{x}, \mathbf{y}) \left[A_j(\mathbf{y}) - \partial_j \phi(\mathbf{y})\right] \\ &= -\int \mathrm{d}^3\mathbf{y}\, \mathfrak{c}_1(|\mathbf{x} - \mathbf{y}|^2) \left[A^i(\mathbf{y}) - \partial^i \phi(\mathbf{y})\right] \\ &\quad - \partial^i \partial^j \int \mathrm{d}^3\mathbf{y}\, \mathfrak{c}_2(|\mathbf{x} - \mathbf{y}|^2) \left[A_j(\mathbf{y}) - \partial_j \phi(\mathbf{y})\right]. \end{aligned} \tag{14.57}$$

Putting this into the static Maxwell equation $\nabla \times \mathbf{B} = \mathbf{J}$ and taking the curl of both sides (using $\nabla \cdot \mathbf{B} = 0$) then implies that

$$\nabla^2 \mathbf{B} = -\nabla \times \mathbf{J} = \int \mathrm{d}^3\mathbf{y}\, \mathfrak{c}_1(|\mathbf{x} - \mathbf{y}|^2)\, \mathbf{B}(\mathbf{y}). \tag{14.58}$$

For a superconductor filling the half-space $x \geq 0$ a solution to (14.58) and $\nabla \cdot \mathbf{B} = 0$ is $B_x = B_y = 0$ and $B_z = B_0 e^{-x/\lambda}$ where the parameter λ gives the magnetic penetration depth [383], and satisfies the implicit equation

$$\frac{1}{\lambda^2} = \int_0^\infty \mathrm{d}x \int_{-\infty}^\infty \mathrm{d}y\, \mathrm{d}z\, \mathfrak{c}_1(x^2 + y^2 + z^2)\, e^{-x/\lambda}. \tag{14.59}$$

This has a unique real solution because $\mathfrak{c}_1(r^2)$ is a positive function that is bounded at $r = 0$ and falls to zero for $r \gg \xi$ (ensuring the integral converges for any non-negative λ).

Flux Quantization

If the deep interior of a superconductor is topologically trivial then because $\mathbf{A} - \nabla\phi = 0$ a gauge can be chosen for which $\phi = 0$ everywhere. The effects of

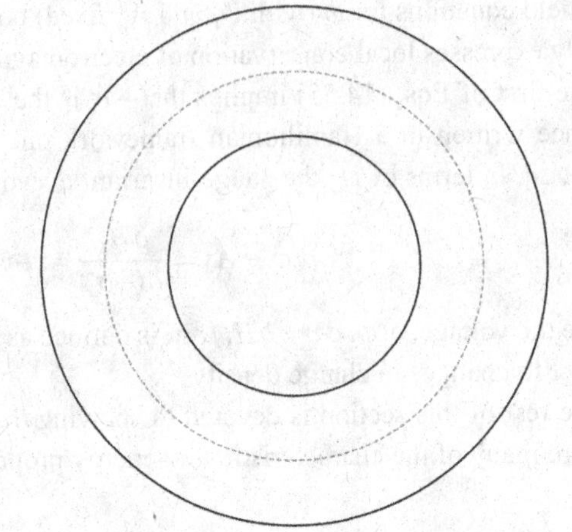

Fig. 14.2 Diagram of a superconducting annulus. Dashed line marks a path deep within the annulus along which $\mathbf{A} - \nabla\phi = 0$.

topology can be studied by considering instead a superconducting annulus that is thick enough to ensure $\mathbf{A} - \nabla\phi = 0$ deep within its interior (such as along the dashed line of Fig. 14.2).

Because ϕ is equivalent to $\phi + \pi/e$, the single-valuedness of physics as one circles the midline of the annulus implies that $\phi(\mathbf{r})$ must satisfy

$$\oint_C \mathrm{d}\mathbf{r} \cdot \nabla\phi = \phi(\theta + 2\pi) - \phi(\theta) = \frac{\pi n}{e}, \tag{14.60}$$

for some integer n, where C is the circle around the midline of the annulus. The vanishing of $\mathbf{A} - \nabla\phi$ along this midline then implies that the net flux through the disc bounded by C is

$$\Phi_B := \oint_C \mathrm{d}\mathbf{r} \cdot \mathbf{A} = \oint_C \mathrm{d}\mathbf{r} \cdot \nabla\phi = \frac{\pi n}{e}, \tag{14.61}$$

showing how the flux through any closed curve deep within a superconductor must be quantized. This is not so informative for curves entirely contained within a simply connected superconductor because the Meissner effect implies that Eq. (14.61) is just satisfied with $n = 0$. But for a superconducting annulus like in Fig. 14.2 the integer n can in principle be – and is measured to be [379] – nonzero.

Infinite Conductivity

Deep within the superconductor the conditions $\mathbf{A} - \nabla\phi = A_0 - \partial_t\phi = 0$ imply vanishing fields, $\mathbf{B} = \mathbf{E} = 0$, as well as (from Eq. (14.57)) vanishing currents, $\mathbf{J} = 0$. Imagine, however, perturbing the system away from its ground state in such a way that a nonzero current flows, $\mathbf{J} \neq 0$.

If this is done in such a way that voltage varies with position within the sample, $\nabla v \neq 0$, then (14.56) implies that $\partial_t \nabla\phi - \nabla A_0 = -\nabla v \neq 0$, and so the configuration cannot be static inasmuch as $\partial_t\phi - A_0$ cannot vanish everywhere. Using $\mathbf{E} = -\partial_t(\mathbf{A} - \nabla\phi) + \nabla v$, relations like (14.57) imply that the electric current is also time-dependent

$$\partial_t \mathbf{J}(\mathbf{x}, t) = \int d^3\mathbf{y}\, c_1(|\mathbf{x}-\mathbf{y})|^2)\left[\mathbf{E} - \nabla \upsilon\right] + \nabla\left[\int d^3\mathbf{y}\, c_2(|\mathbf{x}-\mathbf{y}|^2)\left(\sigma - \nabla^2 \upsilon\right)\right] + \cdots, \tag{14.62}$$

which uses the Maxwell equation $\nabla \cdot \mathbf{E} = \sigma$ and ellipses denote the contribution of other terms in the expansion in powers of $\mathbf{A} - \nabla\phi$ and $A_0 - \partial_t\phi$.

Since voltage gradients lead generically to time-dependent currents,[10] then a nonzero and time-independent current, $\partial_t \mathbf{J} = 0$, between two points deep within a superconductor requires the voltage difference between these points to vanish. Constant currents flowing for vanishing voltage difference is what it means to have infinite conductivity.

Josephson Effect

The final example to be explored here is the AC version of the Josephson effect [384, 385], which predicts a periodic current to flow between two superconductors that are brought into proximity, separated by a gap of non-superconducting material. For the present purposes the significance of this example lies in the fantastic precision of the prediction and measurements [386] for the frequency of the oscillating current.

Consider therefore two semi-infinite superconductors separated by a gap of width d oriented along the $x - y$ plane, perpendicular to the z axis. Suppose first that all external electromagnetic potentials are turned off. Then the energy per unit area associated with the interface between the superconductors depends on the difference, $\Delta\phi = \phi_2 - \phi_1$, between the value taken by ϕ in the two superconductors: $L_s/\mathcal{A} = F(\Delta\phi)$. But because ϕ and $\phi + \pi/e$ are two coordinates for the same point on G/H it must be true that the function $F(x)$ is periodic:

$$F(x + \pi/e) = F(x). \tag{14.63}$$

Suppose now that a time-independent voltage difference, $\Delta\upsilon$, is applied across the gap between the two superconductors. Because of (14.56) it follows that $\Delta\phi$ must be time-dependent:

$$\Delta\phi = -\Delta\upsilon\, t + \text{constant}. \tag{14.64}$$

But in the presence of electromagnetic fields the energy can only depend on the gauge invariant quantity, $F = F(\Delta\Phi)$, where

$$\Delta\Phi := \int_1^2 dz\, (\partial_z\phi - A_z) = \Delta\phi - \int_1^2 dz\, A_z. \tag{14.65}$$

Because this introduces a dependence of the energy on A_z it follows that a current must also flow between the two superconductors,

$$\frac{J_z}{\mathcal{A}} = \frac{\delta F}{\delta A_z} = -F'(\Delta\Phi). \tag{14.66}$$

Together with (14.63) and (14.64) this implies an oscillatory current with period

$$\tau = \frac{\pi}{e|\Delta\upsilon|}. \tag{14.67}$$

This prediction of the AC Josephson period is thereby seen to be very robust; not depending on microscopic details at all.

[10] *E.g.* the analog of Eq. (14.57) for σ does not allow $\sigma = \nabla^2\upsilon$.

14.2.2 Landau–Ginzburg Theory

It is instructive to contrast the above effective theory of a superconductor's Goldstone mode with the Ginzburg and Landau effective theory for superconductors [387]. Unlike the effective theory just described, the Landau–Ginzburg theory is *not* meant to apply to energies very small in comparison to the superconducting scale, $E \ll \epsilon_{sc} \sim T_c$ (where T_c denotes the superconductor's critical temperature). The Landau-Ginzburg theory instead describes the superconductor's behaviour very close to this critical temperature, *i.e.* $|T - T_c| \ll T_c$.

In particular, there are more degrees of freedom relevant for the Landau–Ginzburg theory than just the Goldstone mode. The lagrangian (or free energy) $\mathfrak{L}_{GL}$ of the Landau–Ginzburg theory still depends on the electromagnetic field and the Goldstone mode, ϕ, since both of these degrees of freedom are necessarily in the low-energy spectrum in the superconducting phase. The new feature is that this theory also contains another scalar degree of freedom, χ, in addition to the Goldstone mode. This field describes the mode which must combine with the Goldstone field, ϕ, in order to fill out a linear representation – *i.e.* a complex field, $\psi = \chi\, e^{2ie\phi}$, of the full symmetry group, $G = U(1)$, *above* the transition point.

The appearance of χ in the effective description near the critical point, T_c, is guaranteed so long as the particle spectrum remains continuous as T varies across the transition temperature. Continuity across T_c requires the existence of an additional degree of freedom because of the interplay of the properties of the normal and superconducting phases. In particular, we know that ϕ and χ must have identical dispersion relations *above* T_c, because they are related in the normal phase by the unbroken $U(1)$ transformations. On the other hand, ϕ must also be gapless *below* T_c, since it is a Goldstone mode. These two conditions, together with continuity at T_c, then require χ to also become gapless precisely at T_c. The gapless condition for χ at T_c implies that it must be included in any effective description of the system response near the critical point.

It should be remarked at this point that such a continuity requirement does *not* always hold for a phase transition. In particular, it can fail if phase transition is of first order. As a result, a Landau–Ginzburg description need not be appropriate for an arbitrary phase transition, even near the transition temperature.

As usual, it is the lowest-dimension interactions that dominate the long-wavelength, low-energy response near the critical temperature. Assuming the relevant physics to be local, the most general form for these is given in terms of the complex field, ψ, by the well-known expression:

$$\mathfrak{L}_{GL} = \mathfrak{L}_{EM} + i\,\psi^*(\dot{\psi} + 2ie\,A_0\,\psi) - w_s^2(\nabla\psi + 2ie\,\mathbf{A}\,\psi)^* \cdot (\nabla\psi + 2ie\,\mathbf{A}\,\psi) - V(\psi^*\psi) + \cdots, \tag{14.68}$$

where ψ is given charge $e_q = -2e$ and some freedom has been used to give the time-derivative terms a standard normalization. Once expanded in powers of the fields the potential energy has the form

$$V(\psi^*\psi) = V_0 - \mu^2\,\psi^*\psi + \frac{g}{2}\,(\psi^*\psi)^2 + \cdots, \tag{14.69}$$

where $g > 0$ and μ is a new parameter (which, unlike in the previous section, is unrelated to magnetic moments).

In this language the normal and superconducting phases differ in the sign they have for μ^2, since the classical energy[11]

$$H = \int d^3x \left[\frac{1}{2}\left(\mathbf{E}^2 + \mathbf{B}^2\right) + w_s^2(\nabla\psi + 2ie\mathbf{A}\,\psi)^*(\nabla\psi + 2ie\mathbf{A}\,\psi) + V(\psi^*\psi)\right] + \cdots, \tag{14.70}$$

corresponding to (14.68) is minimized at $\psi^*\psi = 0$ when $\mu^2 < 0$ (normal phase) and at $\psi^*\psi = \mu^2/g$ when $\mu^2 > 0$ (superconducting phase). The transition region between the two therefore lies in the regime where μ^2 is close to zero. Eq. (14.69) predicts that the classical difference in energy density between these phases is

$$V_n - V_{sc} \simeq \frac{\mu^4}{2g}. \tag{14.71}$$

For $\mu^2 < 0$ the $U(1)$ symmetry $\psi \to e^{-2ie\zeta}\psi$ is unbroken by the ground state and small fluctuations of the field ψ about the ground state describe degenerate quasiparticles with energy gap $|\mu|$. For $\mu^2 > 0$ the ground state breaks the $U(1)$ symmetry, and the field $\chi = |\psi|$ acquires a gap of order $|\mu|$ while the phase of ψ is the Goldstone boson (and so would be gapless in the absence of mixing with the electromagnetic field, but acquires an energy gap of order $e|\mu|$ once this mixing turns on).

These semiclassical arguments show how the non-Goldstone mode χ becomes light enough to be in the low-energy theory near the transition, where μ is close to zero. Qualitatively, the Landau–Ginzburg dynamics goes over to the dynamics of the Goldstone mode when $\mu^2 > 0$ is large, because then the modulus, χ, acquires a large gap and can be integrated out (much as is done for the toy model in Part I) leaving only the phase ϕ.

Vortices and Type II Superconductors

One piece of physics well-suited to a Landau–Ginzburg description is the development of vortices [387]. In this context vortices are line-like regions of non-superconducting material that can thread a superconducting bulk, and because a vortex's centre does not superconduct, it also does not exclude magnetic fields. Such vortices can arise – at least in some superconductors, called Type II superconductors [388] – in sufficiently strong magnetic fields, and provide a way for these fields to penetrate into a superconducting sample; a region that would otherwise have been forbidden to them by the Meissner effect.

A Landau–Ginzburg description of vortices is appropriate because their physics is about the energetic trade-off between having regions of the sample superconduct or not. Within a Landau–Ginzburg description vortices arise as 'lumps', in the sense used in Chapter 13. That is, they arise as soliton solutions to the classical field equations, found by minimizing the classical energy (14.70) subject to the boundary condition

$$\psi(r \to \infty, z, \theta) \to \sqrt{\frac{\mu^2}{g}}\, e^{i\theta}, \tag{14.72}$$

[11] Beware: in low dimensions (and at nonzero temperatures in 3 dimensions) IR divergences (due to large fluctuations tied to μ being zero) typically cause semiclassical perturbation theory to break down very near T_c.

for all z, in cylindrical coordinates (r, z, θ). Notice that this boundary condition assures that the field lies in its ground state at infinity, since $\psi^*\psi \to \mu^2/g$. But because this boundary condition implies that the phase of ψ satisfies $\phi(\theta + 2\pi) = \phi(\theta) + \pi/e$ at $r \to \infty$, the profile of χ must vanish somewhere at smaller r. Otherwise the profile of ψ would have to have a discontinuity somewhere. It is useful to choose coordinates so that χ vanishes at $r = 0$.

It is straightforward to verify that a solution exists that minimizes the energy subject to these boundary conditions, subject to the cylindrically symmetric ansatz [387, 389]

$$A_\mu dx^\mu = A_\theta(r)\, d\theta \quad \text{and} \quad \psi = \chi(r)\, e^{i\theta}, \tag{14.73}$$

with $\chi \to v := \sqrt{\mu^2/g}$ and so $A_\theta \to 1/(2e)$ as $r \to \infty$ in order to ensure that $(\partial_\theta + 2ieA_\theta)\psi$ vanishes at infinity. Notice that this implies that the vortex carries precisely one quantum of magnetic flux, since the line integral $\oint \mathbf{A} \cdot d\mathbf{x}$ at $r \to \infty$ evaluates to

$$\Phi = \lim_{r\to\infty} \oint A_\theta\, d\theta = \frac{\pi}{e} =: \Phi_0. \tag{14.74}$$

Examining the asymptotic form of the solutions to the field equations at large r reveals the approach to the values at infinity is exponential, with

$$A_\theta - \frac{1}{2e} \propto e^{-r/\lambda} \quad \text{and} \quad \chi - v \propto e^{-r/\xi}, \tag{14.75}$$

where

$$\lambda \simeq \frac{1}{ev} \quad \text{and} \quad \xi \propto \frac{1}{|\mu|} \simeq \frac{1}{\sqrt{g}\, v}, \tag{14.76}$$

are, respectively, the magnetic penetration depth described above and the superconducting correlation length of the field ψ.

These scales provide a criterion for understanding the energy trade-off that controls whether vortices will form, and so whether or not a superconductor should be Type I (vortices do not form) or Type II (vortices do form). For a Type-I superconductor, turning on a large enough magnetic field ruins the sample's superconductivity because eventually the energy density, B^2, paid to exclude the magnetic field (because of the Meissner effect) is not worth the energy density, $\mu^4/g \sim gv^4$, gained – see (14.71) – by becoming a superconductor. This shows that for Type-I superconductors the critical field (above which superconductivity is ruined) is of order

$$B_c \sim \sqrt{g}\, v^2. \tag{14.77}$$

For a Type II superconductor vortices can form once the field is large enough to squeeze a single flux quantum, $\Phi_0 = \pi/e$, into an area set by the magnetic length, $\pi\lambda^2$. The minimum field required therefore is

$$B_{c1} \sim \frac{\Phi_0}{\pi\lambda^2} \sim ev^2. \tag{14.78}$$

For smaller fields than this a magnetic field cannot penetrate into the sample, but for fields $B > B_{c1}$ fields penetrate by forming more vortices, each of which drives an area $\pi\xi^2$ normal in its immediate vicinity (because this is the region over which χ

is appreciably smaller than its vacuum value). The entire sample is therefore driven normal once the total number of flux quanta, $N = B\mathcal{A}/\Phi_0$, is equal to the total area, $\mathcal{A}$, divided by $\pi\xi^2$. This happens at a field value B_{c2} given by

$$B_{c2} \sim \frac{\Phi_0}{\pi\xi^2} \sim \frac{gv^2}{e}. \tag{14.79}$$

Notice B_c is the geometric mean of B_{c1} and B_{c2}. But having vortices penetrate the sample without ruining superconductivity only makes sense if $B_{c2} > B_{c1}$, which requires $g > e^2$ (and so $\lambda > \xi$). It follows that a superconductor will be Type I if $\lambda < \xi$ and Type II if $\lambda > \xi$. This criterion proves to be useful in later sections (*e.g.* §15.3).

14.3 Phonons ♣

Real materials are not Poincaré invariant since, at the very least, their ground state picks out a preferred reference frame: the rest-frame of the material. For such systems Poincaré invariance itself can be regarded as a spontaneously broken symmetry, and it is natural to ask about the properties of the corresponding Goldstone modes [230, 390].

As might be expected, the properties of these bosons depend to some extent on the symmetry-breaking pattern: if G is the Poincaré group, what is H in the breaking $G \to H$? The answer depends on the system of interest. The Poincaré group has ten generators: four translations, three rotations and three boosts.[12] For solids the ground state breaks nine of these, leaving only time-translation invariance plus whichever discrete group of translations and rotations preserves the underlying lattice. For some systems (like gelatin) boosts are broken but the ground state is translation- and rotation-invariant. Fluids support even more symmetries, since the relative positions of different fluid elements can be changed without energy cost provided the fluid is not compressed when doing so.

14.3.1 Goldstone Counting Revisited

This section argues that phonons are the Goldstone modes for all of the above examples, and sketches the effective field theory of Goldstone states to which the usual symmetry arguments lead. As already discussed in §6.3.2 there is not a one-to-one relation between the number of Goldstone states and the number of broken generators in G/H, contrary to what one might naively expect from the discussion in §4, and this is ultimately because Poincaré symmetries are spacetime symmetries – *i.e.* they act on spacetime coordinates, x^μ, as well as on the fields.

To see why the counting of Goldstone particles changes, one must return to the proof of Goldstone's theorem given in §4.1.2. The key equation there is (4.19), which states that spontaneous symmetry-breaking guarantees the existence of a Goldstone state, $|G\rangle$, for which the following matrix element is nonzero,

[12] Reminder: boosts are those Lorentz transformations that relate inertial observers with different velocities.

$$\langle G|J^0(\mathbf{x},t)|\Omega\rangle \neq 0, \tag{14.80}$$

where $|\Omega\rangle$ is the system's ground state and $J^\mu(\mathbf{x},t)$ is the Noether current for the broken symmetry of interest. All of the properties of $|G\rangle$ follow as consequences of (14.80). What changes for spacetime symmetries is the number of independent Noether currents that exist: the current associated with *all* spacetime symmetries is built from the stress-energy tensor, $T^\mu{}_\nu(\mathbf{x},t)$.

To see why this is so, it is worth digressing to recall how spacetime symmetries work.[13] In general, spacetime symmetries like Poincaré transformations are defined as symmetries (isometries) of the background spacetime metric, $g_{\mu\nu}(x)$. A metric has an isometry if it is left unchanged by a transformation (diffeomorphism) of the form $\delta x^\mu = \xi^\mu(x)$. Like any rank-two tensor, under this type of transformation $g_{\mu\nu}$ transforms as

$$\begin{aligned}\delta g_{\mu\nu} = \pounds_\xi g_{\mu\nu} &:= \partial_\mu \xi^\lambda g_{\lambda\nu} + \partial_\nu \xi^\lambda g_{\mu\lambda} + \xi^\lambda \partial_\lambda g_{\mu\nu} \\ &= \nabla_\mu \xi_\nu + \nabla_\nu \xi_\mu,\end{aligned} \tag{14.81}$$

where the first line defines the Lie derivative acting on the metric, and the second line is a consequence of the definitions, where $\xi_\mu := g_{\mu\nu}\xi^\nu$ and $\nabla_\mu \xi_\nu := \partial_\mu \xi_\nu + \Gamma^\lambda_{\mu\nu}\xi_\lambda$ is the usual covariant derivative, with $\Gamma^\lambda_{\mu\nu}$ the Christoffel symbol – see §C.5.2 for the precise definition – built from $g_{\mu\nu}$.

So for any given metric, $g_{\mu\nu}$, there is a symmetry for each independent solution to the *Killing equation*,

$$\delta g_{\mu\nu} = \nabla_\mu \xi_\nu + \nabla_\nu \xi_\mu = 0. \tag{14.82}$$

There are ten such solutions for the Minkowski metric, $g_{\mu\nu} = \eta_{\mu\nu}$, for which $\Gamma^\lambda_{\mu\nu} = 0$. These are given by $\xi^\mu = a^\mu + \omega^\mu{}_\nu x^\nu$, for constant a^μ and $\omega_{\mu\nu} = -\omega_{\nu\mu}$, corresponding to the 10-parameter Poincaré group.

Solutions to (14.82) are called Killing vector fields and whenever one exists the corresponding Noether current

$$J^\mu := T^\mu{}_\nu\, \xi^\nu, \tag{14.83}$$

is (covariantly) conserved, $\nabla_\mu J^\mu = 0$, as a consequence of the symmetry of the stress-energy tensor, $T^{\mu\nu} = T^{\nu\mu}$, stress-energy conservation, $\nabla_\mu T^\mu{}_\nu = 0$, and Eq. (14.82).

Now comes the main point. Clearly, the way a Goldstone state satisfies (14.80) is by having a nonzero matrix element

$$\langle G|J^0(\mathbf{x},t)|\Omega\rangle = \langle G|T^0{}_\nu(\mathbf{x},t)|\Omega\rangle \xi^\nu \neq 0, \tag{14.84}$$

so what matters for counting Goldstone states is the number of independent currents of this type that exist, not the number of independent Killing vectors, ξ^μ. Since there are at most four independent components of the form $T^0{}_\mu(x)$ for the stress-energy tensor, there are at most four independent Goldstone modes that can be produced by spontaneously broken space time symmetries.

For systems that do not break time-translation invariance the symmetry associated with the energy density, $T^0{}_0$, is not broken, leaving a total of three independent

[13] The treatment here slightly generalizes the discussion of §6.3.2 away from Minkowski space to a broader class of background fields (extensions to a much broader class can be found in [391]).

Goldstone modes for which $\langle G|T^0{}_i(\mathbf{x},t)|\Omega\rangle \neq 0$ for $i = x, y, z$. These three modes are the three phonon modes – one longitudinal and two transverse – corresponding to the gapless sound waves that are generic to many-body systems.

14.3.2 Effective Action

Once sound waves are seen in this way to be generic to the low-energy spectrum of many-body systems, the next step is to identify the effective lagrangian that governs their properties. This section specializes for simplicity to the lagrangian governing their propagation and self-interactions. Although the results obtained for real phonons are old ones (for a modern textbook treatment see [392]), their reformulation in terms of a modern EFT description has continued until quite recently [108, 230, 333]. The description here closely follows [393–395].

The procedure is similar to the procedure used in §4.2.2 when identifying the low-energy Goldstone boson action for internal symmetries: parameterize the Goldstone state as a slowly varying symmetry transformation of the ground state, and ask how the energy changes when doing so. For spacetime symmetries this amounts to asking how the energy of a medium responds when its component parts are moved relative to one another.

To track this, imagine dividing a medium up into many small volume elements, perhaps labelled by painting co-moving coordinates onto the medium itself and watching how these move relative to a fixed set of spatial coordinates x^μ as the medium deforms. In three spatial dimensions denote the co-moving labels by ϕ^I with $I = 1, 2, 3$, and the spatial coordinates by x^i, with every point (volume element) of the medium corresponding *at all times* to a particular value for the three labels ϕ^I.

Although one might initially choose

$$\phi^I(t = 0) = x^I, \tag{14.85}$$

in general this need not remain true at later times, after which their new spatial positions might be $x^i(\phi^I, t)$. An exception to this might be the ground state, which would usually be static and so (14.85) could hold for all times.

Provided the medium doesn't do anything singular the relation between ϕ^I and x^i at any given time is invertible, so one might equally well think about $\phi^I(x^i, t)$ instead of $x^i(\phi^I, t)$. This is convenient because the motion can then be simply regarded as the evolution of three scalar fields. In this case, the ground state can be characterized as the configuration for which (14.85) holds for all times, which is to say that the ϕ^I have ground-state expectation values

$$\langle \phi^I(x, t)\rangle = x^I. \tag{14.86}$$

In this language any residual spacetime symmetries of the medium (*i.e* the group H) can be regarded as internal symmetries acting on the index I of ϕ^I, independent of spacetime. This is how the lagrangian knows whether it is describing a solid, a gelatin or a fluid. If the medium's ground state is translation-invariant then the action must be invariant under the shifts

$$\phi^I \to \phi^I + a^I, \tag{14.87}$$

for constant a^I. If the medium allows a residual rotational symmetry then the action is invariant with respect to transformations of the form

$$\phi^I \to O^I{}_J \phi^J, \tag{14.88}$$

for some orthogonal 3×3 matrix O. (For instance, for a solid $O^I{}_J$ and a^I might live within the invariance group of lattice rotations and translations, while for gelatin $O^I{}_J$ might consist of the full group of $O(3)$ transformations and a^I might consist of arbitrary translations.)

What is important is that these internal symmetries are *a-priori* independent of the Poincaré symmetry of spacetime (assuming a flat metric $\eta_{\mu\nu}$), so the action must be separately invariant under both. It is the expectation value (14.86) that breaks the product of these independent symmetries down to a diagonal subgroup acting on both the internal label 'I' and the spatial label 'i'.[14]

Consider next constructing the lowest-derivative terms for the low-energy EFT for ϕ^I consistent with Poincaré invariance in spacetime and invariance under the internal transformations (14.87) and (14.88). All of the issues from the previous sections also apply here. For instance, the action for L could be nonlocal if the distances of interest are not much larger than the medium's underlying correlation length ξ. The action might contain multiple order parameters breaking Poincaré symmetry, and so on.

The most general shift-invariant and Lorentz-invariant quantity can be built from the matrix-valued scalar field

$$B^{IJ}(\mathbf{x}, t) := \eta^{\mu\nu}\, \partial_\mu \phi^I\, \partial_\nu \phi^J, \tag{14.89}$$

which transforms only under the internal rotation symmetries of (14.88) according to $B \to OBO^T$. If it is assumed – as was also done in §14.1 – that the intended applications are to distances long enough to allow the lagrangian (or free energy) to be given by a local expression, $L_{\rm ph} = \int d^3x\, \mathfrak{L}_{\rm ph}$, then the equations of motion become

$$\partial^\mu \left[\partial_\mu \phi^J \frac{\partial \mathfrak{L}_{\rm ph}}{\partial B^{IJ}}\right] = 0. \tag{14.90}$$

These are satisfied by any configuration for which ϕ^I is linear in x^i (such as (14.86) in particular). Stability of the ground state then imposes positivity conditions on the second derivatives of the action around this particular solution, as usual.

More detailed calculations of the two-derivative low-energy lagrangian depend on how many spacetime symmetries remain unbroken. Consider for simplicity the case of gelatin, where the lagrangian must be invariant under both translations and the full $O(3)$ group of rotations. In this case, the part of $L_{\rm ph}$ involving only single derivatives of ϕ^I must be built out of rotationally invariant combinations of B^{IJ}, which can be taken to be its three eigenvalues. Equivalently, one might use tr B, tr B^2 and tr B^3, or trade any one of these for

[14] This construction closely parallels the treatment of spin in curved spacetime. The action is invariant under independent local Lorentz transformations, such as $\delta\psi(x) = \frac{1}{2}\,\omega^{ab}(x)\Gamma_{ab}\psi(x)$ for a spinor field, and local coordinate transformations (diffeomorphisms), $\delta\psi(x) = \xi^\mu(x)\partial_\mu\psi(x)$, and these are broken down to an unbroken subset acting in both spin-space and real space by the 'expectation value' of the vierbein fields, $e^a{}_\mu(x)$, that is related to the metric by $g_{\mu\nu} = \eta_{ab}\, e^a{}_\mu e^b{}_\nu$ (see e.g. [396–398]).

$$\det B = \frac{1}{6}\left[(\text{tr}\, B)^3 - 3(\text{tr}\, B)(\text{tr}\, B^2) + 2\,\text{tr}\, B^3\right]. \tag{14.91}$$

A conventional basis of invariants is

$$X := \frac{1}{3}\,\text{tr}\, B\,, \quad Y := \frac{3\;\text{tr}\, B^2}{(\text{tr}\, B)^2} \quad \text{and} \quad Z := \frac{9\;\text{tr}\, B^3}{(\text{tr}\, B)^3}, \tag{14.92}$$

since these all approach unity in the ground state, for which (14.86) holds. In terms of these, the part of the lagrangian density involving only single derivatives can be written

$$\mathfrak{L}_{\rm ph}[\partial\phi] = -F(X,Y,Z) \qquad (SO(3)\ \text{invariant case}), \tag{14.93}$$

for some function F (with the overall sign in $L_{\rm ph}$ chosen for later convenience).

Fluctuation Spectrum

Continuing with the case of gelatin, (14.93), to examine the spectrum of phonons, expand about the ground-state background,

$$\phi^I(x^i,t) = x^i + \pi^I(x^i,t), \tag{14.94}$$

and keep that part of $\mathfrak{L}_{\rm ph}$ that is quadratic in the fluctuation fields $\pi^I(\mathbf{x},t)$. Using

$$B^{IJ} = \delta^{IJ} + \partial^I\pi^J + \partial^J\pi^I + \partial_\mu\pi^I\partial^\mu\pi^J, \tag{14.95}$$

then gives

$$X = 1 + \frac{2}{3}(\partial_i\pi^i) - \frac{1}{3}(\partial_t\vec{\pi})^2 + \frac{1}{3}(\partial_i\pi_j\partial^i\pi^j) + \cdots$$

$$Y = 1 + \frac{2}{3}(\partial_i\pi_j\partial^i\pi^j) + \frac{2}{9}(\partial_i\pi^i)^2 + \cdots \tag{14.96}$$

$$Z = 1 + 2(\partial_i\pi_j\partial^i\pi^j) + \frac{2}{3}(\partial_i\pi^i)^2 + \cdots \tag{14.97}$$

and so, after integrating by parts and dropping boundary terms, the quadratic action is

$$\begin{aligned}\mathfrak{L}_2 = &\left(\frac{1}{3}\,F_X\right)_0 \delta_{ij}\,\partial_t\pi^i\partial_t\pi^j - \left(\frac{1}{3}\,F_X + \frac{2}{3}\,F_Y + 2\,F_Z\right)_0 \partial^i\pi^j\,\partial_i\pi_j \\ &- \left(\frac{2}{9}\,F_{XX} + \frac{2}{9}\,F_Y + \frac{2}{3}\,F_Z\right)_0 (\partial_i\pi^i)^2,\end{aligned} \tag{14.98}$$

in which $F_X := \partial F/\partial X$ and so on. The subscript '0' here denotes evaluation at the background configuration $\phi^I = x^I$, for which B^{IJ} is the unit matrix and $X_0 = Y_0 = Z_0 = 1$.

Dividing the fluctuation field into longitudinal and transverse components, $\vec{\pi} = \vec{\pi}_T + \vec{\pi}_L$, with $\nabla \times \vec{\pi}_L = \nabla\cdot\vec{\pi}_T = 0$ then reveals that the linear field equations obtained from (14.98) become wave equations, so admit wave solutions (*i.e.* sound) that propagate with a linear dispersion relation at low energy. That is, if $\pi^i(\mathbf{r},t) = \epsilon^i(\mathbf{p})\exp[-iEt + i\mathbf{p}\cdot\mathbf{r}]$ then solutions require

$$E(\mathbf{p}) = c\,|\mathbf{p}|, \tag{14.99}$$

with differing phase velocities for transverse and longitudinal polarizations: $c = c_T$ if $\mathbf{p} \cdot \epsilon(\mathbf{p}) = 0$ and $c = c_L$ if $\mathbf{p}$ and ϵ are parallel. Long-wavelength modes indeed lie in the low-energy theory, as required by Goldstone's theorem, and for sufficiently long wavelengths the locality assumption made for the action is justified.

The propagation speeds are calculable in terms of the parameters in the action, with

$$c_T^2 = 1 + 2\left(\frac{F_Y + 3F_Z}{F_X}\right)_0$$
$$c_L^2 = 1 + \frac{2}{3}\left(\frac{F_{XX}}{F_X}\right)_0 + \frac{8}{3}\left(\frac{F_Y + 3F_Z}{F_X}\right)_0. \tag{14.100}$$

Positive kinetic energy in (14.98) evidently requires the parameters to satisfy $F_{X0} > 0$ and having $0 < c_{L,T}^2 < 1$ requires $-F_{X0} < 2(F_Y + 3F_Z)_0 < 0$ (with a similar argument in the longitudinal sector constraining F_{XX0}/F_{X0}).

Self-interactions amongst the phonons are straightforwardly (and tediously) identified by continuing the expansion of the action to higher orders in $\pi^I(x)$.

14.3.3 Perfect Fluids

Perfect fluids provide a special case of the previous discussion, for which the internal symmetry is very large since energies are unchanged by arbitrary volume-preserving transformations. In this case, the internal symmetry group acting on B^{IJ} is larger than $O(3)$, corresponding to the group of transformations

$$\phi^I \to \xi^I(\phi) \qquad \text{with} \quad \det\left(\frac{\partial \xi^I}{\partial \phi^J}\right) = 1. \tag{14.101}$$

In this case, the above construction still goes through, with the lagrangian more restricted because of the larger set of symmetry conditions. In particular, requiring invariance under (14.101) implies that any local term involving only first derivatives of ϕ^I has the general form

$$L_{\rm fl} = -\int \mathrm{d}^3x\, G(\mathcal{B}), \tag{14.102}$$

where $\mathcal{B} := \det B$.

To verify that this describes a fluid, check its stress-energy. This is most simply found by making the replacements $\eta^{\mu\nu} \to g^{\mu\nu}$ in (14.89) and $\mathrm{d}^3x \to \mathrm{d}^3x\sqrt{-g}$ in (14.102), where $g = \det g_{\mu\nu}$, and varying the metric using $T^{\mu\nu} = (2/\sqrt{-g})(\delta S_{\rm fl}/\delta g_{\mu\nu})$, with $S_{\rm fl} = \int \mathrm{d}t\, L_{\rm fl}$. The result is

$$T_{\mu\nu} = 2\mathcal{B}\left(\frac{\partial G}{\partial \mathcal{B}}\right) B_{IJ}\partial_\mu\phi^I\partial_\nu\phi^J - G\,\eta_{\mu\nu}, \tag{14.103}$$

where $g_{\mu\nu} \to \eta_{\mu\nu}$ is used at the end and the determinant identity $\epsilon_{IJK}B^{IL}B^{JM}B^{KN} = \mathcal{B}\,\epsilon^{LMN}$ (for completely antisymmetric Levi-Civita tensor ϵ_{IJK}) shows that

$$B_{IJ} = \frac{1}{2\mathcal{B}}\,\epsilon_{IKR}\epsilon_{JMN}B^{KM}B^{RN} \tag{14.104}$$

satisfies $B_{IJ}B^{JK} = \delta_I^K$ and so is the inverse matrix to B^{IJ}.

This stress-energy tensor has the perfect-fluid form

$$T_{\mu\nu} = (\rho + p)U_\mu U_\nu + p\,\eta_{\mu\nu}, \tag{14.105}$$

where the fluid pressure, p, energy density, ρ, and local rest-frame 4-velocity, U_μ (with $U_\mu U^\mu = -1$), are given by

$$\rho := G(\mathcal{B}), \qquad p := 2\mathcal{B}\left(\frac{\partial G}{\partial \mathcal{B}}\right) - G(\mathcal{B}), \tag{14.106}$$

and

$$U^\mu := \frac{1}{6\sqrt{\mathcal{B}}}\,\epsilon^{\mu\nu\lambda\rho}\epsilon_{IJK}\partial_\nu\phi^I\,\partial_\lambda\phi^J\,\partial_\rho\phi^K. \tag{14.107}$$

Both the pressure and the energy density are determined purely by $\mathcal{B}$, and so once G is known the one can use it to infer p given ρ, say. The functional form of $G(\mathcal{B})$ therefore builds in an equation of state $p = p(\rho)$. Furthermore, as shown in more detail in §16.1.2, conservation of stress-energy, $\partial_\mu T^{\mu\nu} = 0$, for a stress-energy tensor of the form (14.105) implies that the fluid satisfies the proper Navier–Stokes equations for a perfect fluid. §16.1.2 also shows why finding a perfect fluid makes sense in the present instance, since at face value the use of lagrangian methods at all (without introducing additional degrees of freedom as in [399], for instance) typically requires dissipation to be negligible.

Fluctuation Spectrum

The fluctuation spectrum for sound in a fluid is found by following precisely the same steps as for solids, by writing $\phi^i(x^i,t) = x^i + \pi^i(x^i,t)$ in the field equations and expanding in powers of π^i.

Indeed, rather than working this through from scratch it is quicker to simply recognize that the fluid case is already contained in the gelatin example worked earlier. To see this, combine Eqs. (14.91) and (14.92), which shows that

$$\mathcal{B} = X^3\left[\frac{9}{2}(1-Y) + Z\right], \tag{14.108}$$

and so the fluid lagrangian (14.102) corresponds to the special instance of (14.93) where

$$F(X,Y,Z) = G\big[\mathcal{B}(X,Y,Z)\big]. \tag{14.109}$$

Consequently, the derivatives F_X, F_Y and F_Z are related to one another, with

$$F_Y = -\frac{9}{2}X^3 G_{\mathcal{B}}(\mathcal{B}), \qquad F_Z = X^3 G_{\mathcal{B}}(\mathcal{B}). \tag{14.110}$$

and

$$F_X = 3X^2\left[\frac{9}{2}(1-Y) + Z\right]G_{\mathcal{B}}(\mathcal{B}), \tag{14.111}$$

where $G_{\mathcal{B}} := \partial G/\partial\mathcal{B}$.

For perfect fluids the quadratic term in powers of π^i that governs the properties of linearized propagation is again given by (14.98). The corresponding sound-propagation speeds for the longitudinal and transverse fields similarly are given by (14.100):

$$c_L^2 = 1 + \left(\frac{2G_{BB}}{G_B}\right)_0 \quad \text{and} \quad c_T^2 = 0, \tag{14.112}$$

which uses

$$\left(\frac{F_{XX}}{F_X}\right)_0 = 2 + 3\left(\frac{G_{BB}}{G_B}\right)_0 \quad \text{and} \quad \left(\frac{F_Y + 3F_Z}{F_X}\right)_0 = -\frac{1}{2} \tag{14.113}$$

once evaluated at $X_0 = Y_0 = Z_0 = 1$ (and so $\mathcal{B}_0 = 1$).

The vanishing of c_T reflects how fluids only support compressional waves. It manifests itself by the absence of a restoring force for $\vec{\pi}_T$ in the quadratic lagrangian, which shows how the absence of restoring forces for some kinds of deformations can make the understanding of fluid motions intrinsically more complicated than for solids.

14.4 Summary

Goldstone bosons are revisited in this chapter, with a view to seeing how their description changes relative to the discussion in Chapter 4 for practical non relativistic many-body applications. Much (but not all) of the earlier description goes through unchanged; in particular, their dominant low-energy behaviour can be robustly derived knowing only general properties of the system's symmetry-breaking pattern $G \to H$.

Three types of applications are briefly described. Of these, the first application describes spin waves – or magnons – in magnetic systems, deriving their properties using only the information that they arise as the Goldstone bosons for the breaking of rotational symmetry in spin space. The properties of spin waves are described separately for ferromagnets and antiferromagnets because these differ significantly due to their differing transformation properties under time-reversal invariance. It is the low-energy magnon spectrum for antiferromagnets that more resembles relativistic dispersion relations, with $E^2 = v^2\mathbf{p}^2$. By contrast, time-reversal breaking implies that the spectrum for ferromagnets is instead $E = k\,\mathbf{p}^2$, making the quantization of these modes more similar to Schrödinger than Klein–Gordon field theory.

The second application examines Goldstone behaviour for broken gauge symmetries, using superconductors as the practical example. Mixing with the gauge boson gives the would-be Goldstone boson a mass through the Anderson–Higgs mechanism (the mother of all Higgs mechanisms). Much can nonetheless be learned by studying the Goldstone mode (*i.e.* by not working in unitary gauge), and it is shown in particular how the symmetry-breaking pattern in this case robustly bakes in many classic low-energy features of superconductors. Having such a model-independent derivation is useful when using the accuracy of these predictions for fundamental tests (such as when using Josephson-junction properties to infer values of the fine-structure constant [400]).

The third application summarizes why phonons arise as Goldstone bosons for the breaking of spacetime symmetries by many-body systems. The main interest in this section is conceptual, partly due to the great generality with which their properties can be determined. Phonons also illustrate (as do magnons for ferromagnets) through concrete examples how the counting of Goldstone modes can be more subtle than for relativistic internal-symmetry applications when it is spacetime symmetries that are broken by the medium (be these rotations, translations, boosts or time-reversal).

Exercises

Exercise 14.1 Use the standard nonlinear realization given in §4.2.2 to derive the two-derivative action appropriate for the Goldstone bosons of the symmetry-breaking pattern $G/H = SU(2)/U(1)$. Show that this agrees with the lagrangian given in (14.8) for antiferromagnets for its predictions for low-energy $2 \to 2$ scattering amongst the Goldstone bosons. (Assume invariance under spatial rotations, so $Z^{ab} \propto \delta^{ab}$.)

Exercise 14.2 Show that for an interaction $H_{\rm int} = -\vec{\mu}_1 \cdot \mathbf{B}(\mathbf{x}_1) - \vec{\mu}_2 \cdot \mathbf{B}(\mathbf{x}_2)$ integrating out the instantaneous electromagnetic field leads to the dipole–dipole interaction

$$H_{dd} = -\vec{\mu}_1 \cdot \left\{ \nabla \times \left[\nabla \times \left(\frac{\vec{\mu}_2}{r} \right) \right] \right\} = -\vec{\mu}_2 \cdot \left\{ \nabla \times \left[\nabla \times \left(\frac{\vec{\mu}_1}{r} \right) \right] \right\}$$

where $\mathbf{r} := \mathbf{x}_1 - \mathbf{x}_2$ and $r = |\mathbf{r}|$. Show that an equivalent way of writing this dipole–dipole interaction is

$$H_{dd} = -\left[\frac{8\pi}{3} \left(\vec{\mu}_1 \cdot \vec{\mu}_2 \right) \delta^3(\mathbf{r}) - \frac{\vec{\mu}_1 \cdot \vec{\mu}_2}{r^3} + \frac{3(\vec{\mu}_1 \cdot \mathbf{r})(\vec{\mu}_2 \cdot \mathbf{r})}{r^5} \right].$$

Use this expression with $\vec{\mu}_1 = \mathfrak{m}_n \Psi^\dagger \sigma \Psi$ and $\vec{\mu}_2 = \mu \, \vec{\rho}$ in Fermi's Golden Rule – *c.f.* Eq. (B.45) – to derive the scattering cross section given in (14.40) for neutron scattering, assuming only the neutron's final momentum and energy are measured.

Exercise 14.3 Assuming a local Landau–Ginzburg description near (but below) a critical point – *i.e.* Eq. (14.68) with $V(\psi^* \psi) = -m^2 \psi^* \psi + g\,(\psi^* \psi)^2$ with $m^2(T) > 0$ for $T < T_c$ – compute the Goldstone-boson lagrangian by integrating out the massive state (or matching) perturbatively in powers of $g \ll 1$.

Exercise 14.4 The London equations for a superconductor can be found by specializing the effective action of §14.2 by assuming it is local in space: *i.e.* $\mathfrak{c}_i(\mathbf{x} - \mathbf{y}) \to \tilde{\mathfrak{c}}_i \, \delta^3(\mathbf{x} - \mathbf{y})$. Write down the lowest-dimension terms in the low-energy superconducting action (or free energy) for ϕ and A_μ assuming the action is local. Use this to derive the field equations and repeat the derivation given in §14.2 of the classic properties of superconductors.

Exercise 14.5 Use the ansatz (14.73) to minimize the local form (14.70) of the Landau–Ginzburg energy functional and derive the coupled ordinary differential equations that must be satisfied by the fields $A_\theta(r)$ and $\chi(r)$ of a vortex solution. What boundary conditions must these fields satisfy as $r \to 0$ in order to ensure they are nonsingular there? Linearize the equations for large r and derive expressions (14.76) for the length scales λ and ξ.

Solve for and plot the profiles $A_\theta(r)$ and $\chi(r)$ numerically. In the special case $\xi \ll \lambda$ show that the magnetic field for radii $r \gg \xi$ is given to a good approximation by the analytic formula [408]

$$B(r) = \frac{\Phi_0}{2\pi\lambda^2}\, K_0(r/\lambda),$$

where the flux quantum is $\Phi_0 = \pi/e$ and $K_0(x)$ is a Hankel function of imaginary argument.

Exercise 14.6 Explicitly solve the Killing equation (14.82) for the Minkowski metric, $g_{\mu\nu} = \eta_{\mu\nu}$, and thereby derive the Poincaré group of transformations $\delta x^\mu = \xi^\mu(x)$. Do the same for the conformal Killing equation in Minkowski space, which modifies the right-hand-side of (14.82) so that $\delta g_{\mu\nu} \propto g_{\mu\nu}$, and thereby derive the 15-parameter conformal group for flat space.

Compute the 10 Killing vectors for de Sitter space in four spacetime dimensions (which is a maximally symmetric spacetime with nonzero curvature). This is most simply done if de Sitter space is regarded as a surface

$$-t^2 + \sum_{i=1}^{4} (x^i)^2 = +L^2$$

for constant L, embedded in five-dimensional Minkowski space.

Exercise 14.7 Compute the leading low-energy interactions for phonons in gelatin – for which the leading description of the dispersion relations is given surrounding Eq. (14.98). Use these to compute the temperature-dependence of the scattering length, l where $1/l := \langle \sigma n \rangle$, for low-energy phonons, assuming only $2 \to 2$ scattering and the phonons to be thermally distributed. Here σ is the leading phonon-phonon scattering cross section and n is the thermal density of phonons.

Exercise 14.8 Repeat the exercise of 14.7 for magnons in a ferromagnet and compute the leading temperature-dependence of the $2 \to 2$ magnon scattering length. Compare your result to [366].

15 Degenerate Systems

The systems considered in this chapter involve fermions that are statistically degenerate, inasmuch as they display a Fermi surface (a short reminder of whose properties appears below). These arise often in practical problems, such as for electrons in materials at room temperature or for nucleons in nuclear matter. What is at face value surprising about such systems is that they are often well-described *even at a quantitative level* in terms of a liquid of weakly interacting fermions. This despite the presence in the microscopic system of strong forces, like Coulomb or nuclear interactions.

This chapter provides the EFT version of the argument as to why the low-energy fermions can be weakly interacting despite the presence of strong forces at a fundamental level. The core idea relies on 'Pauli blocking': that is, the fact that Fermi statistics prevents two particles from sharing the same state. Because of this, statistical degeneracy (the presence of enough particles to ensure that many low-energy states are already occupied) effectively reduces the strength of interactions because reactions cannot proceed if the final states to which they would lead are already occupied by other fermions.

As usual, an advantage of bringing EFT arguments to bear on these issues is the robustness they bring to the conclusions: helping understand why relatively weak interactions for degenerate fermions are generic rather than accidental consequences of special simple-to-analyze situations. A weak-coupling description of the low-energy effective theory is possible given only a few *qualitative* assumptions concerning the spectrum and symmetries relevant at low energies.

Besides showing why the presence of a Fermi surface makes most interactions irrelevant at low energies, the EFT analysis also cleanly identifies possible exceptions to this statement. Later sections in this chapter describe how these exceptions can lead to the instability believed responsible for physical phenomena like superconductivity.

Statistical Degeneracy

First a brief refresher about statistical degeneracy. Consider, to start, a system consisting of $N \gg 1$ non-interacting fermions for whom the single-particle energy levels – states $|n\rangle$ with energies E_n – are labelled by some quantum numbers collectively denoted by n. The label is chosen so that $n_1 < n_2$ implies that $E_{n_1} \leq E_{n_2}$. Because Fermi-Dirac statistics forbid more than one particle from occupying any given state, the lowest-energy multiple-fermion state is obtained by adding particles one at a time to the lowest available unoccupied level.

The Fermi energy, E_F, for this system is defined to be the energy of the state occupied by the last fermion to be added (see Fig. 15.1). It can be regarded as a

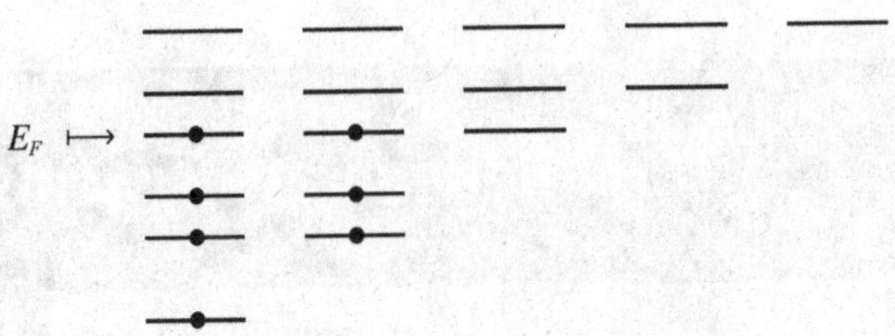

Fig. 15.1 A cartoon of energy levels with each line representing a state, whose energy is portrayed by its vertical position. Dots indicate which levels are populated to produce the ground state of a system of non-interacting fermions. The arrow indicates the Fermi energy.

function of both the level spacing and of the number of particles present, as would be found by eliminating the variable $n_{\max}$ from the two equations

$$\sum_{n=1}^{n_{\max}} 1 = N \quad \text{and} \quad E_F = E_{n_{\max}}. \tag{15.1}$$

For states labelled by continuous momenta, $|\mathbf{p}\rangle$, the Fermi energy is determined by the total particle *density*, N/V with V the system's spatial volume, rather than its total particle number N. To see why, notice that the density of states in momentum space for such particles is given by (see Appendix B.1)

$$\frac{\mathrm{d}N}{\mathrm{d}^3 p} = \frac{V}{(2\pi)^3}\, f(\mathbf{p}), \tag{15.2}$$

where the function $f(\mathbf{p})$ counts the number of states for each momentum, and so is $2s+1$ for free spin-s particles in flat space.

The continuum analog of (15.1) relates E_F to N/V by eliminating the Fermi momentum, p_F, between the equations $E_F = E(p_F)$ and

$$N = \int_0^{p_F} \frac{\mathrm{d}N}{\mathrm{d}^3 p}\, \mathrm{d}^3 p = V \int_0^{p_F} f(\mathbf{p}) \frac{\mathrm{d}^3 p}{(2\pi)^3}. \tag{15.3}$$

The upper limit of integration here is written as if $f(\mathbf{p})$ is a function only of $|\mathbf{p}|$ and not also direction in momentum space, though this is often not true in condensed matter systems due to the breakdown of rotational invariance. When this happens the definition of the Fermi energy is conceptually the same, but calculationally more complicated.

In the simplest instance, where $f(\mathbf{p}) = 2$ (the number of spin states per particle) and $E(\mathbf{p}) = \mathbf{p}^2/2m$ the above formulae give

$$\frac{N}{V} = \frac{p_F^3}{3\pi^2} \quad \text{and} \quad E_F = \frac{p_F^2}{2m} = \frac{1}{2m}\left(\frac{3\pi^2 N}{V}\right)^{2/3}. \tag{15.4}$$

Of course, real electrons and nucleons are not free particles, so why is this definition of E_F useful? Strictly speaking it is not, but it is related (though not equivalent) to the Fermi *level* (as opposed to Fermi *energy*), defined for equilibrium systems as the hypothetical energy for which the probability of finding a fermion becomes precisely one half.

Equivalently, the Fermi level for an equilibrium system is equal to its chemical potential since in thermal equilibrium the probability of finding a fermion at energy E

is given in terms of the system's temperature, T, and chemical potential, μ, by the Fermi-Dirac distribution,

$$\mathfrak{g}(E) = \frac{1}{\exp[(E-\mu)/T]+1}, \tag{15.5}$$

which satisfies $\mathfrak{g}(E) = \frac{1}{2}$ when $E = \mu$. As $T \to 0$ the distribution (15.5) satisfies $\mathfrak{g}(E) \to 1$ if $E < \mu$ and $\mathfrak{g}(E) \to 0$ if $E > \mu$, showing how $\mu \to E_F$ for zero-temperature noninteracting particles. Fermions are statistically degenerate when $\mu \gg T$.

Degenerate fermions add a qualitatively new feature to the story of low-energy effective-theories. Although the interest in effective theories is always in small energies (and small time derivatives), for degenerate systems small energies mean energies near the Fermi level. This generically does not also mean momenta and space derivatives are also small.[1] That is, it is often convenient to define the Fermi level as the zero of energy, since these are the energies that dominate at zero temperature. But then the low-energy limit involves the limit $E \to 0$ with $p \sim p_F$ fixed and not small. This is in contrast with the limit studied to this point in the book, where $E, p \to 0$ together.

Scales Relevant for Conduction Electrons

With later applications to superconductivity in mind, conduction electrons are used as the vehicle for discussing EFTs for statistically degenerate systems. So it is useful to identify what the important energy scales are in this case, since these provide benchmarks relative to which one can judge whether dynamics fits within a low-energy approximation.

As discussed at some length in §12, the basic length scale relevant to atomic energy levels is the Bohr radius: $a_B \sim (\alpha m_e)^{-1}$ (where $m_e \sim 0.5$ MeV is the electron mass and again $\alpha = e^2/4\pi \simeq 1/137$ denotes the fine-structure constant). According to the uncertainty principle, electrons localized to within this size acquire atomic momenta of order

$$\text{momenta:} \quad p_{\rm atom} \sim 1/a_B \sim \alpha\, m_e \quad \text{and kinetic energy:} \quad E_{\rm atom} \sim \alpha^2 m_e, \tag{15.6}$$

making $p_{\rm atom} \sim$ few keV and $E_{\rm atom} \sim 10$ eV, while typical electron speeds are of order $v_{\rm atom} \sim p_{\rm atom}/m_e \sim \alpha \sim 10^{-2}$. The scale $m\alpha^2$ is also of order a typical atomic binding energy because kinetic and potential energies are comparable in the ground state while a_B also characterizes interatomic spacings, at least for closely packed materials like solids.

These scales can be similar to the size of the Fermi level for conducting electrons in a material if atoms are separated by distances of order a_B and if each atom donates a single electron to a particular conduction band and if the dispersion relation for this band is[2] $E(\mathbf{p}) \simeq \mathbf{p}^2/(2m_e)$. Such assumptions imply a density of

[1] Having large momenta associated with small energies is also a generic property of particles in a periodic lattice, due to the periodicity this also implies in momentum space, but these are not the kinematic states of interest in what follows.

[2] None of these assumptions need be true; even the effective mass defined by a conduction band energy, $E(\mathbf{p}) \simeq \mathbf{p}^2/(2m_{\rm eff})$, need not agree with m_e.

$N/V \sim 1/a_B^3 \sim (\alpha m_e)^3$, and as a consequence, equations like (15.4) give a typical Fermi momentum, Fermi level and Fermi velocity of order $p_F \sim \alpha\, m_e \sim$ keV, $E_F \sim \mu \sim \alpha^2 m_e \sim 10$ eV and $v_F \sim p_F/m_e \sim \alpha \sim 10^{-2}$.

In order for thermal energies to compete with $E_F \sim 10$ eV, the temperature would have to be of order $T \sim 10^5$ K; much higher than the temperatures of interest here (room temperature, $T \sim 300$ K, and lower). The system's electrons should therefore be very degenerate, with temperature relevant only for states within order 10 meV or so of the Fermi level.

For comparison, the energy ϵ_{sc}, associated with the superconducting transition in a conventional superconductor may be estimated from the measured temperatures of the superconducting phase transitions, T_c, and are typically of order a few K, making the energy scale $\epsilon_{sc} \sim 10^{-4}$ eV.

The upshot is this: the electronic energies associated with conductivity measurements in general, and superconducting systems in particular, are much lower than the characteristic atomic scales, putting their description well within the domain of EFT methods.

15.1 Fermi Liquids ◇

As a first application of EFT methods to degenerate fermions, following [402, 403], this section uses EFT methods to justify why low-energy states very close to the Fermi level often interact much more weakly than expected based on the strength of the microscopic interactions that are in play. The thrust of the argument is to show that should a fermionic particle (or quasiparticle) ever become sufficiently weakly coupled that its low-energy scaling can be analyzed using perturbative methods, then these methods imply that almost all interactions just become even weaker as one moves to lower and lower energies (closer and closer to the Fermi surface). So (apart from a few marginal interactions – more about which below) the weak-coupling approximation just gets better and better closer and closer to the Fermi level. Systems with degenerate fermions in this kind of weakly coupled regime are called Fermi liquids [401].

The assumption, then, is that there is an energy regime for which the electronic system's low-energy behaviour is well-approximated by weakly interacting quasiparticles with the same quantum numbers as electrons: electric charge $-e$, spin $\frac{1}{2}$, and fermion and lepton number unity. As discussed above, for the interaction energies and temperatures of interest these particles should be statistically degenerate. The rest of this section explores the low-energy interactions of such particles using an effective field theory that contains the most general interactions consistent with these conservation laws, plus parity, time-reversal and charge-conjugation invariance, $SU(2)$ spin rotations, plus whatever discrete symmetries characterize the underlying lattice structure of the atomic nuclei.

15.1.1 EFT Near a Fermi Surface

For notational simplicity it is convenient to choose the zero of energy to lie at the Fermi level, which can be regarded as a surface in momentum space once a dispersion relation $\varepsilon(\mathbf{p})$ for the quasiparticle single-particle states is known. In this language the Fermi surface becomes the surface satisfying $\varepsilon(\mathbf{p}) = 0$, and in what follows all momenta satisfying this condition are denoted by $\{\mathbf{k}\}$.

As seen above, these momenta are typically *not* negligible compared to the inverse inter-atomic spacings: $p_F \sim 1/a_B$. As a result, there is no particular advantage to be gained by working in position space and expanding in powers of spatial derivatives. Consequently, the EFT below is defined directly in momentum space and no appeal is made to locality in position space.

Free Propagation

The assumption of weak interactions states that there is a range of energies for which the effective action can be written $S = S_{\rm free} + S_{\rm int}$, with $S_{\rm free}$ dominating the path integral. The Wilson action can be computed in this energy interval by successively integrating out modes of the quasi-electron, ultimately leaving only those sufficiently close to the Fermi surface, with $S_{\rm free}$ assumed to dominate, at least initially.

The first step is to write down the most general form for $S_{\rm free}$, describing the free propagation of the assumed low-energy quasiparticle. This has the form

$$S_{\rm free} = \int dt\, d^3\mathbf{p}\, \Big[i\psi^\dagger(\mathbf{p})\partial_t\psi(-\mathbf{p}) - \varepsilon(\mathbf{p})\psi^\dagger(\mathbf{p})\psi(-\mathbf{p}) \Big]. \tag{15.7}$$

where $\varepsilon(\mathbf{p})$ gives the quasiparticle dispersion relation, $E = \varepsilon(\mathbf{p})$. This form for $S_{\rm free}$ assumes momentum can be used to label single-particle energy eigenstates (as is true for systems with approximate translation invariance) and uses the freedom to rescale the fields, $\psi(\mathbf{p})$, to normalize the time-derivative term.

The simplest way to integrate out modes of the quasi-electron is to adapt the scaling arguments made in §2.4.2, similar to what was also done for nonrelativistic particles in §11.3.1. The first step is to quantify the notion of proximity to this surface. For momenta near the Fermi surface it is useful to expand the electron dispersion relation about $\mathbf{p} = \mathbf{k}$, writing

$$\mathbf{p} = \mathbf{k} + \mathbf{l}, \tag{15.8}$$

where the vector $\mathbf{k}$ lies on the Fermi surface and $\mathbf{l}$ is perpendicular to it (see Fig. 15.2).

Expanding the single-particle energy in powers of $\mathbf{l}$ in this way gives:

$$\varepsilon(\mathbf{p}) = \mathbf{l}\cdot\mathbf{v}_F + \mathcal{O}(l^2) = \pm\, l\, v_F(\mathbf{k}) + \mathcal{O}(l^2), \tag{15.9}$$

where $l = |\mathbf{l}|$ is the magnitude of $\mathbf{l}$, and

$$\mathbf{v}_F := \nabla_p\, \varepsilon\,\Big|_{FS}, \tag{15.10}$$

with the subscript 'FS' denoting evaluation of the result at the Fermi surface: $\mathbf{p} = \mathbf{k}$. The dot product in Eq. (15.9) is simple to evaluate because the Fermi surface is by definition one along which $\varepsilon(\mathbf{p})$ is constant, so $\mathbf{v}_F$ must be orthogonal to it and point towards increasing energy. Consequently, $\mathbf{v}_F$ is parallel to $\mathbf{l}$ if $\mathbf{p}$ lies above the Fermi surface, and antiparallel to it if $\mathbf{p}$ is below – corresponding to the sign $\pm$ in Eq. (15.9). The $\mathbf{l}$-independent term in Eq. (15.9) vanishes because of the convention that of the Fermi surface is the zero of energy: $\varepsilon(\mathbf{k}) \equiv 0$.

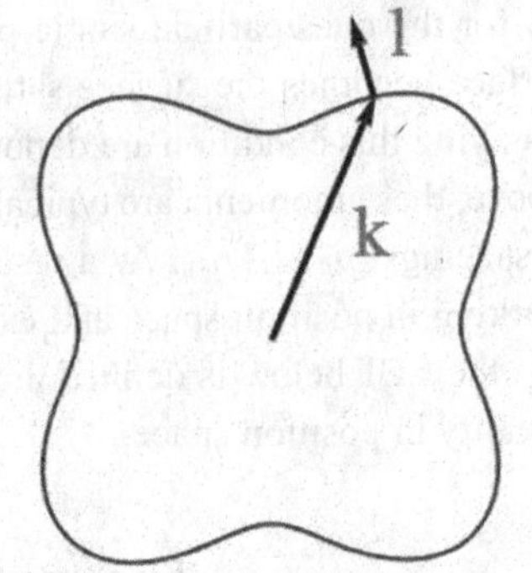

Fig. 15.2 A sketch illustrating the decomposition, $\mathbf{p} = \mathbf{k} + \mathbf{l}$, of a momentum vector into a part, $\mathbf{k}$, on the Fermi surface plus a piece, $\mathbf{l}$, perpendicular to it.

To obtain the near–Fermi-surface Wilson action requires integrating out small frequencies (the shortest times, t) and largest transverse momenta, l. To this end, imagine introducing cutoffs $1/\Lambda_t$ and Λ_l on the t and l integrations appearing in $S_{\rm free}$. Following the logic of §2.4.2, integrating out modes near the cutoff causes them to shrink, but also causes all couplings to evolve so that physical processes do not change. The simplest way to compute these effects is to rescale $t \to t' = t/s_t$ and $l \to l' = s_l l$, so that the cutoffs for the new variables remain unchanged at $1/\Lambda_t$ and Λ_l.

Since the path integral is assumed to be dominated in the energy range of interest by $S_{\rm free}$, these scalings should be done in a way that preserves its form. Inspection of (15.7) using (15.9) and $d^3\mathbf{p} = d^2\mathbf{k}\, dl$ shows this requires $s_t = s_l =: s$, so that the low-energy limit that preserves $S_{\rm free}$ is

$$\mathbf{k} \to \mathbf{k}, \quad l \to s\, l, \quad E \to s\, E \quad \text{and} \quad \psi(\mathbf{p}) \to s^{-1/2}\, \psi(\mathbf{p}) \qquad (15.11)$$

as $s \to 0$.

With these choices the scaling of higher terms in the expansion of $\varepsilon(\mathbf{p})$ in powers of l is determined. A term in $S^{(n)}_{\rm free}$ proportional to l^n in this expansion, scales like $S^{(n)}_{\rm free} \to s^{n-1}\, S^{(n)}_{\rm free}$, and so (as expected) becomes less important as $s \to 0$.

15.1.2 Irrelevance of Fermion Self-Interactions

With scaling properties in hand, it becomes possible to classify how interactions in $S_{\rm int}$ scale as $s \to 0$. This section computes this scaling for arbitrary fermion self-interactions, and shows that the kinematics of scaling near the Fermi surface makes almost all self-interactions scale to zero as $s \to 0$. This shows that these interactions are irrelevant in the weakly coupled fermion theory, and so their effects at low energies are necessarily suppressed by powers of E divided by a UV scale (like E_F). The weak-coupling assumption is self-consistent: if it ever becomes valid in the flow towards low energies, then it becomes ever more valid the lower in energy one goes.

To demonstrate why these conclusions are true, consider first terms in $S_{\rm int}$ that involve four powers of the field ψ that describe two-body electron scattering.

The most general form such an interaction can take consistent with the underlying symmetries is:

$$S^{(4)}_{\rm int} = \int {\rm d}t \prod_{i=1}^{4} {\rm d}^3\mathbf{p}_i \; V_{abcd}(\mathbf{p}_1, \mathbf{p}_2, \mathbf{p}_3, \mathbf{p}_4)\, \delta^3(\mathbf{p}_1 + \mathbf{p}_2 - \mathbf{p}_3 - \mathbf{p}_4) \times \psi^*_d(\mathbf{p}_4)\, \psi^*_c(\mathbf{p}_3)\, \psi_b(\mathbf{p}_2)\, \psi_a(\mathbf{p}_1), \tag{15.12}$$

where there is an implied sum over the indices, $a, b, c, d = \pm\frac{1}{2}$, that label the two components of the electron field in spin space. The coefficients, $V_{abcd}(\mathbf{p}_1, \mathbf{p}_2, \mathbf{p}_3, \mathbf{p}_4)$, must preserve all of the symmetries of the problem, and are generally otherwise arbitrary smooth functions.[3] The only 4-fermion terms omitted in Eq. (15.12) are those involving time derivatives of the electron fields. These are ignored because each additional time derivative suppresses the result by another power of s, making it even more irrelevant than the interactions of Eq. (15.12) turn out to be.

To determine how the interaction of Eq. (15.12) scales as $s \to 0$ as before write ${\rm d}^3\mathbf{p}_i = {\rm d}^2\mathbf{k}_i\, {\rm d}l_i$ and expand the function $V_{abcd}(\mathbf{p}_1, \mathbf{p}_2, \mathbf{p}_3, \mathbf{p}_4)$ about the Fermi surface:

$$V_{abcd}(\mathbf{p}_1, \mathbf{p}_2, \mathbf{p}_3, \mathbf{p}_4) = V_{abcd}(\mathbf{k}_1, \mathbf{k}_2, \mathbf{k}_3, \mathbf{k}_4) + O(l). \tag{15.13}$$

This expansion is useful because each successive term in it scales by a higher power of s (and so is more irrelevant at low energies) than its predecessor. The dominant effect at low energies therefore ignores all of the l-dependent terms.

The determination of the scaling of $S^{(4)}_{\rm int}$ is now a straightforward application of the previously defined scaling rules, apart from one exception. The exception concerns the question as to how the momentum-conserving delta function in Eq. (15.12) should scale. Since each factor of ${\rm d}l_i$ in the integration measure supplies a power of s while the factors of ${\rm d}^2\mathbf{k}$ do not, the issue is whether the delta function can be used to eliminate just ${\rm d}^2\mathbf{k}$ integrals, or whether it is required also to perform one of the ${\rm d}l$ integrals.

As is justified below, generically the delta function can be used to perform only ${\rm d}^2\mathbf{k}$ integrals, in which case it contributes nothing to the scaling of $S^{(4)}_{\rm int}$. There are, however, a few exceptional cases of special kinematics where this is not true. Because the momentum-conserving δ-function in these special cases removes one of the integrations over ${\rm d}l_i$, for them the interaction scales with an additional factor of $1/s$ relative to the generic situation.

Scaling: Generic Kinematics

Consider first the generic case where $\delta^3(\mathbf{p}_1 + \mathbf{p}_2 - \mathbf{p}_3 - \mathbf{p}_4)$ does not scale. Applying the transformations of (15.11) implies that $S^{(4)}_{\rm int}$ scales as

$$S^{(4)}_{\rm int} \to s\, S^{(4)}_{\rm int} \quad \text{(generic kinematics)}, \tag{15.14}$$

with a factor of s^4 coming from the four factors of ${\rm d}l$ and $1/s^3$ coming from ${\rm d}t\,(\psi^*\psi)^2$. For generic kinematics a general two-body interaction is therefore

[3] Although not crucial to the argument being made, smoothness ultimately reflects *cluster decomposition* [404]: the property that probabilities for states widely separated in position space must factorize. The role of cluster decomposition in quantum field theory is outlined in a delightful essay [405], with details in the book [406].

irrelevant. It follows that all other interactions involving more powers of l_i or ∂_t are even more irrelevant.

What happens when additional powers of the fields, $\psi(\mathbf{p})$, are included, to create three- and higher-body interactions? The answer is that these are even more irrelevant than the two-body ones, despite the fact that the fields themselves scale with the *negative* power: $\psi \to s^{-1/2}\psi$. The reason these higher-body interactions nonetheless are irrelevant is because each additional factor of $\psi(\mathbf{p})$ also involves an addition integration measure, $\mathrm{d}^2\mathbf{k}\,\mathrm{d}l$, and the suppression due to the $\mathrm{d}l$ integration overwhelms the enhancement due to the additional power of $\psi(\mathbf{p})$. Furthermore, the symmetries of the theory imply that each factor of ψ is accompanied by a factor of ψ^*, so higher-body interactions are suppressed by at least one additional power of s relative to the generic two-body case just discussed.

The assumption of weakly interacting degenerate fermions is evidently a robust one near the Fermi surface – apart possibly from the cases with special kinematics to be discussed next. *All* of the possible interactions are irrelevant, and so become less and less important at lower energies. Once a system enters into a low-energy regime that is dominated by just this type of quasiparticle, the description in terms of almost free fermions just improves for lower- and lower-energy observables. This is the EFT's way of expressing the suppression of interactions by Pauli blocking that underlies the formalism developed in [401].

Scaling: Exceptional Kinematics

As alluded to above, there are exceptions[4] to the scalings just derived, for which the momentum-conserving delta-function removes one of the $\mathrm{d}l_i$ integrations and so the interaction scales with one fewer power of s. Which is to say

$$S^{(4)}_{\mathrm{int}} \to S^{(4)}_{\mathrm{int}} \qquad \text{(exceptional kinematics)}, \tag{15.15}$$

making these interactions marginal. In a world of irrelevant interactions marginal interactions are king, and so these special cases are of particular interest in the low-energy limit.

Which Kinematics Are Special?

When does the special kinematics with marginal interactions occur? Several common situations are identified in this section. Identifying them requires asking whether or not the momentum-conserving delta function can be used to exclusively perform $\mathrm{d}^2\mathbf{k}$ integrals.

Consider, therefore, the elastic two-body scattering of two electrons mediated by the interaction $S^{(4)}_{\mathrm{int}}$ of (15.12). In a two-body collision the initial momenta, $\mathbf{p}_1$ and $\mathbf{p}_2$, can be regarded as given, and momentum-conservation thought of as three constraints that determine three of the six components of the two final momenta, $\mathbf{p}_3$ and $\mathbf{p}_4$. Energy conservation then provides a fourth constraint, reducing the number of free components of final momenta to two.

[4] The description here is is for $d = 3$ spatial dimensions, but would apply equally well to any $d \geq 2$. $d = 1$ is different because in this case there are no $\mathbf{k}_i$'s and only l_i's, and so the 4-fermi interaction is marginal even for generic kinematics.

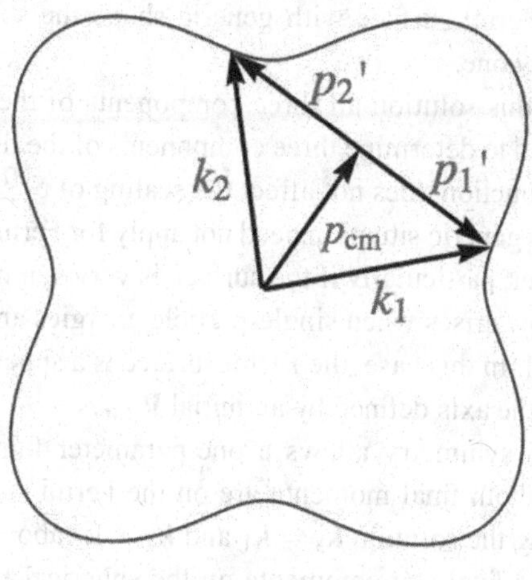

Fig. 15.3 A sketch illustrating the allowed final momenta for 2-body scattering on a Fermi surface.

For momenta near the Fermi surface, the six components of the final momenta are the four components given by $\mathbf{k}_3$ and $\mathbf{k}_4$, plus the two components l_3 and l_4. If all three components of momentum-conservation can be used to perform $d^2\mathbf{k}$ integrals, then the three undetermined parameters in the final state are l_3 and l_4, in addition to one combination of the four $\mathbf{k}_i$.

On the other hand, if it should happen that momentum-conservation determines one combination of l_3 and l_4 then the three unconstrained components of final momentum are the other combination of the l_i, plus *two* components of the final $\mathbf{k}_i$'s.

The strategy to see if it is possible to use the delta-function to perform only $\mathbf{k}_i$ integrals is therefore to solve the conditions

$$\mathbf{k}_1 + \mathbf{k}_2 = \mathbf{k}_3 + \mathbf{k}_4, \tag{15.16}$$

using momenta lying purely on the Fermi surface. If it should happen that there is at most a one-parameter family of solutions to these equations, then all three momentum-conservation conditions can be fulfilled using only the $\mathbf{k}_i$'s. If, however, a two-parameter family of solutions to Eq. (15.16) exists, then one of the l_i's must be constrained by momentum-conservation.

It is simplest to solve Eq. (15.16) graphically. Momentum-conservation is most simply expressed in the CM frame, in which both the initial pair and the final pair of momenta must be equal and opposite. In a general frame $\mathbf{p}_i \equiv \mathbf{P}_{CM} + \mathbf{p}_i'$ with $\mathbf{p}_1' = -\mathbf{p}_2'$ and $\mathbf{p}_3' = -\mathbf{p}_4'$. Pictorially, this means that the set of allowed final-state momenta given an initial pair of momenta, $\mathbf{p}_1$ and $\mathbf{p}_2$, consist of all possible pairs of equal and opposite momenta centred at the point $\mathbf{P}_{CM}$.

If all four momenta are required to lie on the Fermi surface, in addition to satisfying momentum-conservation, then the space of solutions for $\mathbf{k}_3$ and $\mathbf{k}_4$ consists of those pairs of points on the Fermi surface connected by a chord that is bisected by the point $\mathbf{P}_{CM}$ (see Fig. 15.3).

As the figure shows, one solution to this condition always exists: choose $\mathbf{k}_3 = \mathbf{k}_1$ and $\mathbf{k}_4 = \mathbf{k}_2$ (or interchange $1 \leftrightarrow 2$). If l_3 and l_4 are also nonzero then this solution for $\mathbf{p}_3$ and $\mathbf{p}_4$ need not be precisely identical to the initial momenta on the Fermi surface.

For a Fermi surface with generic shape the solution obtained in this way may be the only one.

For this solution all three components of the momentum-conservation condition are used to determine three components of the final $\mathbf{k}_i$, so the momentum-conserving delta function does not affect the scaling of $S^{(4)}_{\text{int}}$ at all. This is the generic situation.

This generic situation need not apply for Fermi surfaces with more specific shapes, however, particularly if the surface is very symmetric. The most extreme example of this type arises when single-particle energies are rotation invariant: $\varepsilon(\mathbf{p}) = \varepsilon(p)$ for $p = |\mathbf{p}|$. In this case, the Fermi surface is a sphere, and so is invariant under rotations about the axis defined by an initial $\mathbf{P}_{CM}$.

This symmetry allows a one-parameter family of solutions to Eq. (15.16), for which both final momenta are on the Fermi surface. The solutions are obtained by rotating the solution $\mathbf{k}_3 = \mathbf{k}_1$ and $\mathbf{k}_4 = \mathbf{k}_2$ about the axis $\mathbf{P}_{CM}$, inscribing a circle of allowed final-state momenta on the spherical Fermi surface. Because only a single parameter remains free of the four potentially independent components of $\mathbf{k}_3$ and $\mathbf{k}_4$, it remains true in this case that all three of the momentum-conservation conditions are saturated using only $\mathrm{d}^2\mathbf{k}_i$ integrations.

In order for momentum-conservation not to constrain three of the components of the final-state $\mathbf{k}_i$'s there must be at least a two-parameter family of solutions to Eq. (15.16). That is, there must locally be a region within momentum space for which an entire patch of the Fermi surface always satisfies the momentum-conservation condition.

BCS Kinematics

The most important example of this type arises whenever the single-particle energies are even or odd functions under momentum reflection:

$$\varepsilon(-\mathbf{p}) = \pm\varepsilon(\mathbf{p}). \tag{15.17}$$

This is not a hypothetical or fine-tuned possibility; for example, it would follow if the underlying system were time-reversal invariant. What matters is that (15.17) ensures that $\varepsilon(\mathbf{k}) = 0$ implies that $\varepsilon(-\mathbf{k}) = 0$. When this is true the entire Fermi surface solves the momentum-conservation condition so long as the initial momenta are chosen to be equal and opposite: $\mathbf{k}_1 + \mathbf{k}_2 = 0$. In this case, for *any* choice for $\mathbf{k}_3$, its opposite $\mathbf{k}_4 = -\mathbf{k}_3$ also lives on the Fermi surface while still satisfying the momentum-conservation condition. The two-parameter family of final $\mathbf{k}_i$'s satisfying momentum-conservation consists of the Fermi-surface itself. Consequently, one of the components of the momentum-conservation condition must constrain the off-surface momenta, l_i, changing the low-energy scaling of this type of two-body interaction in the way found earlier.

Within the theory of superconductivity devised by Bardeen, Cooper and Schrieffer (BCS) [380] it is marginal interactions of this type that play an important role in triggering superconductivity for many materials (as is explored in more detail in §15.2).

Other Special Kinematics

Other instances can also give two-parameter solutions to Eq. (15.16), provided the Fermi surface has particular properties. Two examples of this arise when the Fermi

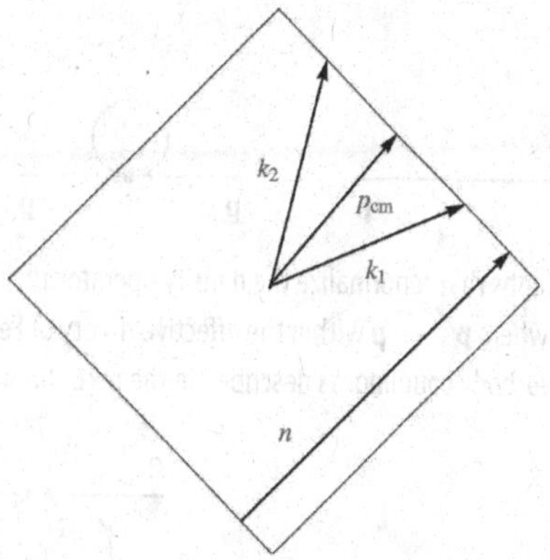

Fig. 15.4 A sketch of a cubic Fermi surface, illustrating two special configurations with marginal scaling. In one the sum $\frac{1}{2}(\mathbf{k}_1 + \mathbf{k}_2)$ lies on a planar part of the Fermi surface. The other special configuration arises when two regions of the Fermi surface (which in general need not be planar) are related by a 'nesting' vector, $\mathbf{n}$.

surface is a cube, such as shown in Fig. 15.4. From this point of view a cubic Fermi surface has two special features: (*i*) it has faces that are planes, and (*ii*) it has opposite faces that are *nested*, inasmuch as one is the translation of the other by a fixed vector, $\mathbf{n}$. Either of these two properties suffices to allow a two-parameter family of solutions to (15.16).

For instance, suppose a region $\mathcal{P}$ of the Fermi surface is planar. Then if the initial momenta are chosen (as illustrated in Fig. 15.4) so that $\mathbf{P}_{CM} = \frac{1}{2}(\mathbf{k}_1 + \mathbf{k}_2)$ lies in this plane, $\mathcal{P}$, then the two-parameter family of allowed final momenta consists of the (2-dimensional) span of all vectors, $\mathbf{k}_3$ and $\mathbf{k}_4$, that lie in $\mathcal{P}$ and sum to $\mathbf{P}_{CM}$.

Nesting works somewhat similarly. The nesting assumption states that two (possibly curved) 2-dimensional regions of the Fermi surface are related by translating by $\mathbf{n}$, so for $\mathbf{p}$ in these regions $\varepsilon(\mathbf{p}) = 0$ implies[5] $\varepsilon(\mathbf{p}+\mathbf{n}) = 0$. In this case, choose initial momenta satisfying $\mathbf{k}_1 + \mathbf{k}_2 = \mathbf{n}$, where $\mathbf{n}$ is the nesting vector relating two regions of the Fermi surface. If $\varepsilon(-\mathbf{p}) = \varepsilon(\mathbf{p})$ then the two-parameter class of solutions to (15.16) is parameterized by an arbitrary $\mathbf{k}_3$ lying in the nested region, since for any such choice $\mathbf{k}_4 = \mathbf{n} - \mathbf{k}_3$ automatically also lies on the Fermi surface since $\varepsilon(\mathbf{k}_4) = \varepsilon(-\mathbf{k}_3) = 0$.

15.1.3 Marginal Interactions

The above sections show that the scalings inferred from the free lagrangian imply that all interactions are irrelevant, apart from a few marginal 4-fermion interactions at a few specific places in momentum space. Fermi liquids describe essentially free particles at very low energies, and this is basically why free-electron (or weakly interacting nucleon) models often work so well.

But the marginal interactions actually do contribute in several specific ways, and this section outlines some of them, using marginal interactions of the BCS type, for which $\mathbf{p}_1 + \mathbf{p}_2 \simeq 0$. Because there are only very specific types of marginal

[5] Repeated shifts do not generate more and more nested regions because lattice symmetries make $\varepsilon(\mathbf{p})$ a periodic function in momentum space, and $2\mathbf{n}$ is typically commensurate with its period.

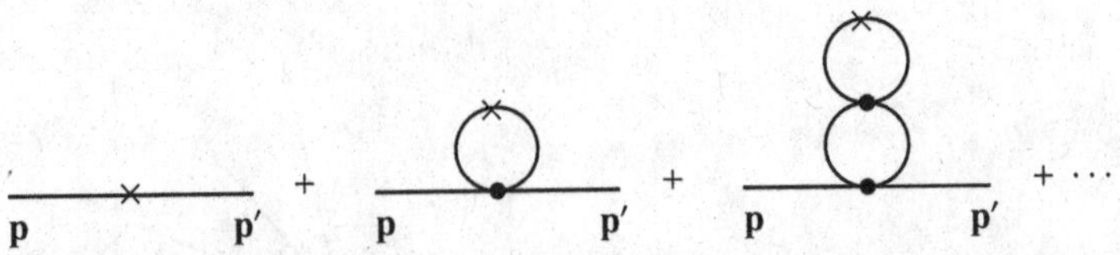

Fig. 15.5 Feynman graphs that renormalize the density operator $\psi^*(\mathbf{p})\psi(\mathbf{p}')$ (represented by the cross) but only in the limit where $\mathbf{p}' \to \mathbf{p}$ within the effective theory of Fermi liquids. The four-point interaction is a marginal two-body coupling, as described in the text.

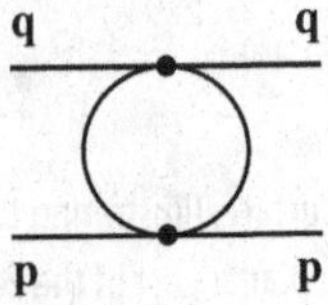

Fig. 15.6 The Feynman graph giving the leading perturbative correction to the marginal two-body interaction strength within the low-energy theory of Fermi liquids.

interactions – *i.e.* 4-fermi interactions with $\mathbf{p}_1 \simeq -\mathbf{p}_2$ and $\mathbf{p}_3 \simeq -\mathbf{p}_4$ – they contribute to observables in very specific ways. For instance, many physical processes involve the density of electrons, $\psi^*(\mathbf{p})\psi(\mathbf{p}')$, and for generic momenta this operator behaves as if the fermion is a free particle inasmuch as it is not renormalized by the marginal interactions.[6]

For position-independent quantities like expectation values, however, the limit $\mathbf{p}' \to \mathbf{p}$ is required, and in this regime the marginal interaction *does* renormalize the matrix elements of the density operator, through graphs of the form given in Fig. 15.5. Furthermore, because the interaction is marginal only when negligible momentum flows from loop to loop, iterations of these corrections can be re-summed.

For later purposes the most interesting corrections due to marginal interactions are the corrections they make to their own scaling. That is, although scaling inferred using the $S_{\rm free}$ implies that these interactions are marginal, the scaling properties themselves receive perturbative corrections which cause the effective couplings to run. Such corrections make the most difference precisely for marginal interactions because they can cause $S^{(4)}_{\rm int}$ to scale nontrivially; potentially tipping it to become marginally relevant or marginally irrelevant rather than exactly marginal.

The graph to compute to determine this is shown in Fig. 15.6, in which the vertices represent the marginal interaction, and this is reflected in the choice of external momenta. In this graph the integration is over modes for which $l > \lambda$, where λ is a floating cutoff that tracks the effects of integrating out successive modes *à la* Wilson. Crossed versions of this graph need not be computed because for these the required interactions are irrelevant except for specific values of the internal momenta. As a result, they do not give a divergent dependence on λ and so do not contribute to $\lambda \mathrm{d}/\mathrm{d}\lambda$ of the result.

[6] This argument, as applied to the electron-phonon vertex for generic momentum transfers $\mathbf{p}' \neq \mathbf{p}$, is at the root of Migdal's theorem [407], which says the electron-phonon vertex is not dressed by low-energy interactions.

Evaluating Fig. 15.6 using the propagator constructed from (15.7) gives the following correction

$$\delta V_{abcd}(\mathbf{p},-\mathbf{p},\mathbf{q},-\mathbf{q}) = 2\mathrm{i}\int \frac{\mathrm{d}^2\mathbf{k}\,\mathrm{d}l\,\mathrm{d}\omega}{(2\pi)^4}\frac{V_{abef}(\mathbf{p},-\mathbf{p},-\mathbf{k},\mathbf{k})\,V_{efcd}(\mathbf{k},-\mathbf{k},\mathbf{q},-\mathbf{q})}{\left[(E-\omega)-v_F(\mathbf{k})\,l+\mathrm{i}\epsilon\right]\left[(E+\omega)-v_F(\mathbf{k})\,l+\mathrm{i}\epsilon\right]},$$
$$= -\int \frac{\mathrm{d}^2\mathbf{k}\,\mathrm{d}l}{(2\pi)^3}\frac{V_{abef}\,V_{efcd}}{\left|v_F(\mathbf{k})\,l-E\right|}, \tag{15.18}$$

in which $E = \varepsilon(\mathbf{p})$ is the energy of the external line and ϵ is the usual infinitesimal parameter used to choose the contour in the complex ω plane. For the internal lines $\varepsilon(\mathbf{k},l) \simeq v_F(\mathbf{k})l$ is used.

For the purposes of seeing how the potential runs, all that matters in (15.18) is that the dl integration diverges logarithmically in the UV. Because the divergence is logarithmic it doesn't really matter much how it is regulated; the formulae below simply cut the divergence off at $l = \lambda$. This suffices for the Wilsonian purpose of identifying the λ-dependence acquired by the potential to compensate for the loss of modes with $l > \lambda$. Neglecting terms $O(1/\lambda)$, the result is:

$$\lambda\frac{\partial}{\partial\lambda}V_{abcd}(\mathbf{p},-\mathbf{p},\mathbf{q},-\mathbf{q}) = \int\frac{\mathrm{d}^2\mathbf{k}}{(2\pi)^3\,v_F(\mathbf{k})}V_{abef}(\mathbf{p},-\mathbf{p},-\mathbf{k},\mathbf{k})\,V_{efcd}(\mathbf{k},-\mathbf{k},\mathbf{q},-\mathbf{q}). \tag{15.19}$$

To get a feel for what this means, approximate the interaction potentials as being ndependent of momenta and take only the spin-singlet combination,

$$V_{abcd}(\mathbf{p},-\mathbf{p},-\mathbf{k},\mathbf{k}) \simeq \frac{V}{2}\left(\delta_{ac}\delta_{bd}-\delta_{ad}\delta_{bc}\right), \tag{15.20}$$

in which case, (15.19) simplifies to

$$\lambda\frac{\partial V}{\partial\lambda} = N_F V^2, \tag{15.21}$$

where the positive quantity

$$N_F := \int\frac{\mathrm{d}^2\mathbf{k}}{(2\pi)^3\,v_F(\mathbf{k})}, \tag{15.22}$$

gives the density of states on the Fermi surface. Eq. (15.21) emphasizes that it is ultimately the dimensionless product $N_F V$ that controls whether the interaction is weak or strong.

Eq. (15.21) integrates to give one of the main results: how $V(\mu)$ evolves with scale:

$$V(\mu) = \frac{V(\mu_0)}{1+N_F V(\mu_0)\,\ln\left(\mu_0/\mu\right)}. \tag{15.23}$$

This result contains a lot of information. For repulsive interactions – for which $V(\mu_0) > 0$ – it implies that $V(\mu) < V(\mu_0)$ for $\mu < \mu_0$, making the interaction weaker at lower energies. It is therefore marginally irrelevant. As energies are lowered, the running logarithmically suppresses the strength of repulsive interactions.

For attractive interactions – for which $V(\mu_0) < 0$ – Eq. (15.23) instead implies that $|V(\mu)| > |V(\mu_0)|$ for $\mu < \mu_0$, making the interaction more important at lower energies; it is marginally relevant.

Growth in the infrared represents an instability of the low-energy theory (similar to what happens in QCD) because at sufficiently low energies the attractive interaction becomes strong enough to leave the perturbative regime, and to qualitatively change what are the important degrees of freedom and how they scale. For the marginal interaction experienced by electron pairs having opposite momenta this is known as the BCS instability [380].

15.2 Superconductivity and Fermion Pairing ♠

The previous discussion indicates a low-energy instability generic to weakly coupled degenerate fermions experiencing an attractive interaction, regardless of how weak this attraction might happen to be at high-energies. The existence of such an instability is very suggestive for the understanding of superconductors, for which pairs of electrons combine into Cooper pairs whose strong interactions ensure a doubly charged order parameter like $\langle\psi(\mathbf{p})\psi(-\mathbf{p})\rangle$ becomes nonzero, thereby spontaneously breaking electromagnetic gauge invariance (with all the consequences described in §14.2). It seems compelling to understand this pairing as due to the BCS instability among low-energy electrons, but how might these electrons come to experience a net attractive interaction in the first place given their underlying Coulomb repulsion?

In conventional superconductors the answer is at first sight surprising: Coulomb repulsion is ultimately swamped by relatively feeble electron-electron interactions caused by phonon exchange [380]. To understand this in the low-energy effective theory requires a detour to see how phonons behave and interact with electrons near the Fermi level.

15.2.1 Phonon Scaling

The phonons in question are the same phonons described in §14.3, though we meet them here in a different guise. As Fig. 15.7 shows, in order to participate significantly in electron interactions near the Fermi surface, the phonons of interest should have momenta, $\mathbf{q}$, comparable to the Fermi momentum p_F. The regime of interest is low energies but not necessarily long wavelengths: potentially far from the extremely long-wavelength limit discussed in §14.3. The rest of this section explores how phonons behave in this large-$\mathbf{q}$ regime.

Free Propagation

Like any Goldstone mode, phonon interactions must taper off at long wavelengths, but how strong can they interact when $\mathbf{q}$ is not so small?

In real solids there are good reasons why phonon interactions are expected to be weak. Phonons in solids describe the vibrations of the atoms or ions making up the lattice itself, since these are the quantities whose ordering is ultimately responsible for breaking the spacetime symmetries. Because phonons are associated with atomic

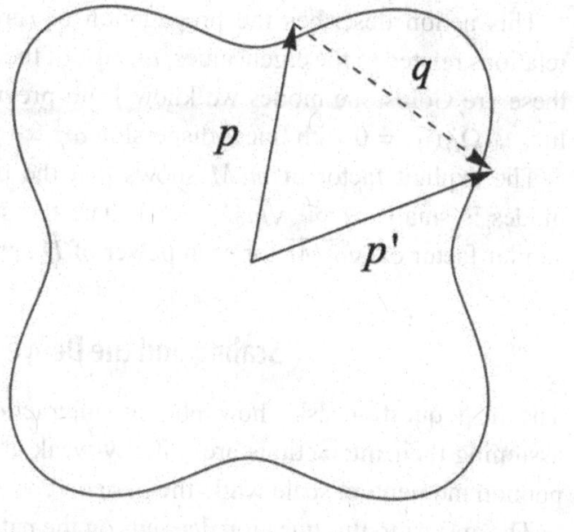

Fig. 15.7 A typical electron-phonon interaction in which emission or absorption of a phonon of momentum $\mathbf{q} = \mathbf{p}' - \mathbf{p}$ causes a transition between two low-energy electrons near the Fermi surface.

motion their properties depend on the mass, M, of the moving atoms, which is enormous compared to the mass, m, of the quasi-electrons whose motion governs electromagnetic response.

For an elemental solid with atomic number A, this mass ratio is $M/m \sim 10^3\,A$. For $A \sim 100$ – as is reasonable for many elements – M/m can be as large as 10^5. Since it costs an infinite amount of energy to start the ions vibrating when $M/m \to \infty$, phonon modes should decouple in this limit. It is therefore important to follow the M-dependence of phonon couplings, particularly if these m/M-suppressed couplings are to be compared in strength to something so strong as the Coulomb force.

To this end, consider first a generic quadratic action describing freely propagating phonons, keeping as usual only the fewest time derivatives:

$$S^{ph}_{\text{free}} = \frac{1}{2}\int \mathrm{d}t\,\mathrm{d}^3\mathbf{q}\Big\{M\,\mathcal{Z}_{ij}(\mathbf{q})\,\dot{u}^i(\mathbf{q})\,\dot{u}^j(-\mathbf{q}) - m\left[\mathcal{Z}^{\frac{1}{2}}\Omega^2\mathcal{Z}^{\frac{1}{2}}\right]_{ij}(\mathbf{q})\,u^i(\mathbf{q})\,u^j(-\mathbf{q})\Big\}, \tag{15.24}$$

where u^i physically represents a measure of local lattice displacement, and the quantities $\mathcal{Z}$ and Ω^2 are matrices of medium-dependent coefficients whose details are not crucial in what follows. There is no point writing this in position space given that the momenta of interest are not so different from the inverse lattice separations.

The factor of M in front of the kinetic term reflects the underlying origin of this term as the kinetic energy of the lattice ions, and a factor of the quasi-electron mass, m, premultiplies the second term for dimensional reasons. m is a natural scale to use since the restoring force between ions arises due to their electronic interactions.

As usual, canonical normalization for the kinetic term is achieved by redefining fields,

$$D_i(\mathbf{q}) \equiv \left[M\,\mathcal{Z}(\mathbf{q})\right]^{\frac{1}{2}}_{ij}\,u_j(\mathbf{q}), \tag{15.25}$$

in terms of which

$$S^{ph}_{\text{free}} = \frac{1}{2}\int \mathrm{d}t\,\mathrm{d}^3\mathbf{q}\left[\dot{\vec{D}}(\mathbf{q})\cdot\dot{\vec{D}}(-\mathbf{q}) - \frac{m}{M}\,\Omega^2_{ij}(\mathbf{q})\,D^i(\mathbf{q})\,D^j(-\mathbf{q})\right]. \tag{15.26}$$

This action describes the propagation of three phonon modes with dispersion relations related to the eigenvalues, $\omega_a(\mathbf{q})$, of the three-by-three matrix $\Omega_{ij}(\mathbf{q})$. Since these are Goldstone modes we know from previous sections that they are gapless: $\lim_{\mathbf{q}\to 0}\Omega_{ij}(\mathbf{q}) = 0$ with linear dispersion $\omega_a^2 \propto c_a^2\,\mathbf{q}^2$ as $\mathbf{q}\to 0$.

The explicit factor of m/M shows that the propagation speed, c_a, for all three modes is small: $c_a \propto \sqrt{m/M} \ll 1$. The transformation (15.25) also generates a similar factor of $\sqrt{m/M}$ for each power of $\vec{D}$ appearing in a phonon interaction.

Scaling and the Debye Frequency

The next question asks how phonon interactions scale in the low-energy limit, assuming their interactions are initially weak at high energies. But how should the phonon momentum scale while the phonon energy scales to zero: $E_{ph} \to s\,E_{ph}$?

The answer to this question depends on the nature of the low-energy physics which is to be studied. For example, when studying the propagation of low-energy sound waves, as in §14.3, the appropriate scaling limit would be $E_{ph} \to s\,E_{ph}$ with $\mathbf{q}\to s\,\mathbf{q}$, since this preserves the phonon dispersion relation at long wavelengths.

For understanding how phonons influence low-energy electron properties, however, a different limit is required. In this case, our interest is in the phonons whose exchange influences electron scattering at low energies. Although the energy transfer available for low-energy electron scattering scales like s, the relevant momentum transfers remain of order p_F, as in Fig. 15.7 which sketches the relevant low-energy kinematics.

The low-energy scaling appropriate to phonons relevant to electron scattering near the Fermi surface is therefore

$$E_{ph}(\mathbf{q}) \to s\,E_{ph}(\mathbf{q}), \qquad \mathbf{q}\to\mathbf{q}. \tag{15.27}$$

With this choice, the phonon kinetic energy, $\partial_t\vec{D}\cdot\partial_t\vec{D}$, is scale-independent provided that the phonon field scales as $\vec{D}\to s^{-1/2}\,\vec{D}$. Notice that this scaling does not also preserve the size of the second term of the free action, Eq. (15.26), because of the choice (15.27) not to scale $\mathbf{q}$. But it is reasonable to treat this term perturbatively given that it comes with the very small factor m/M.

Under these rescalings the second, 'restoring', term in Eq. (15.26) scales to s^{-2} times itself, and so is *relevant*. So even though it starts off suppressed by m/M at the underlying electronic scale, $E_{\rm atom}$, its relevance means that it grows to become $O(1)$ – and so of the same size as the phonon kinetic term – at an energy, ω_D, defined by $(\omega_D/E_{\rm atom})^2 \sim (m/M)$. The resulting scale,

$$\omega_D \sim \left(\frac{m}{M}\right)^{1/2} E_{\rm atom}, \tag{15.28}$$

is called the *Debye energy*. For solids having an atomic number $A\sim 100$, one expects in order of magnitude $\omega_D \sim 10^{-3}\,E_{\rm atom} \sim 10^{-2}$ eV, corresponding to temperatures in the hundreds of K.

Having both terms in (15.26) similar in size at this scale indicates that this is the phonon energy for which phonons of momentum $q\sim p_F$ are *on shell*, in the sense that they satisfy the phonon dispersion relation: $E_{ph}(q) = c\,q$, with $c\propto (m/M)^{1/2}$ the appropriate sound speed.

Scaling below the Debye Scale

At energies *below* the Debye scale, it is the second term of Eq. (15.26) that dominates in the weak-coupling path integral, instead of the kinetic term (which is simply the RG way of saying that for energies $\omega \ll \omega_D$ and $|\mathbf{q}| \sim p_F$ the 4-momentum of an off-shell phonon is dominantly spacelike rather than timelike). As a result, it is this term rather than the kinetic term that controls the size of the fluctuations in $\vec{D}$ below the Debye scale.

The scaling law for $\vec{D}$ in this regime therefore becomes $\vec{D} \to s^{+1/2}\,\vec{D}$, making the total scaling law for phonons

$$\begin{aligned} E_{ph} &\to s\,E_{ph}; && \mathbf{q} \to \mathbf{q}; \\ \vec{D} &\to s^{-1/2}\,\vec{D} && \text{for } E_{ph} > \omega_D, \\ \vec{D} &\to s^{+1/2}\,\vec{D} && \text{for } E_{ph} < \omega_D. \end{aligned} \tag{15.29}$$

These scaling rules determine how interactions for these phonons scale. For instance, a term, $S_{d,f}$, in the phonon action involving d time derivatives and f powers of $\vec{D}$ therefore scales according to

$$\begin{aligned} S_{d,f} &\to s^{-1+d-f/2}\,S_{d,f}, && \text{for } E_{ph} > \omega_D, \\ S_{d,f} &\to s^{-1+d+f/2}\,S_{d,f}, && \text{for } E_{ph} > \omega_D, \end{aligned} \tag{15.30}$$

showing that the dominant terms have no time derivatives ($d = 0$). These dominant self-interactions clearly grow when scaling down from $E \sim E_{\rm atom}$ to $E \sim \omega_D$, but then (because $f > 2$ for any self-interaction term) shrink at energies below this (eventually decoupling as $s \to 0$).

How big do these interactions become at their largest (*i.e.* at the Debye scale)? Keeping in mind that a typical coupling at high energies involving f phonon fields is suppressed by $\kappa_f(E_{\rm atom}) \propto (m/M)^{f/2}$, the scaling rules given in (15.30) imply the maximum strength attained by these couplings is (for $d = 0$):

$$\kappa_f(\omega_D) \sim \left(\frac{E_{\rm atom}}{\omega_D}\right)^{1+f/2} \kappa_f(E_{\rm atom}) \sim \left(\frac{m}{M}\right)^{(f-2)/4}. \tag{15.31}$$

Because $f > 2$ for any interaction term, these interactions remain suppressed by m/M even at their maximum, at the Debye scale, justifying continued treatment of phonons using perturbative methods.

Phonon-Electron Interactions

This is all very nice, but what matters for low-energy electronic properties is how phonon-*electron* interactions scale at low energies. The simplest such interaction, involving the fewest possible electron and phonon fields (and no time derivatives), describes phonon emission and absorption and has the form

$$S_{\rm int}^{e-ph} = \sqrt{\frac{m}{M}} \int {\rm d}t\, {\rm d}^3\mathbf{q}\, {\rm d}^3\mathbf{p}\, {\rm d}^3\mathbf{p}'\; g^i_{ab}(\mathbf{q};\mathbf{p},\mathbf{p}')\, \delta^3(\mathbf{p}-\mathbf{p}'-\mathbf{q}) D_i(\mathbf{q})\, \psi^*_a(\mathbf{p})\psi_b(\mathbf{p}'), \tag{15.32}$$

where the overall factor of $(m/M)^{1/2}$ comes from the canonical redefinition (15.25).

It is straightforward to read off how such interactions scale, by simply applying the scaling transformations of the previous sections. As usual, the dominant contribution at low energies comes from the leading term in the expansion of electron momenta about the Fermi surface,

$$g^i_{ab}(\mathbf{q};\mathbf{p}_1,\mathbf{p}_2) \simeq g^i_{ab}(\mathbf{q};\mathbf{k}_1,\mathbf{k}_2) + O(l). \tag{15.33}$$

Furthermore, for (15.32) the momentum-conserving delta function can always be used to perform the nonscaling $d^3\mathbf{q}$ integration, and so never contributes to overall scaling. The result therefore is

$$\begin{aligned} S^{e-ph}_{\rm int} &\to s^{-1/2}\, S^{e-ph}_{\rm int}, \qquad &&\text{for } E_{ph} > \omega_D, \\ S^{e-ph}_{\rm int} &\to s^{+1/2}\, S^{e-ph}_{\rm int}, \qquad &&\text{for } E_{ph} < \omega_D, \end{aligned} \tag{15.34}$$

which shows relevant scaling for energies above the Debye scale and irrelevant scaling below it. More complicated interactions scale with higher powers of s than these.

The maximum strength attained by this interaction is acquired at $E \sim \omega_D$, and is given by

$$g^i_{ab}(\omega_D) \sim \left(\frac{E_{\rm atom}}{\omega_D}\right)^{1/2} g^i_{ab}(E_{\rm atom}) \sim \left(\frac{m}{M}\right)^{1/4}, \tag{15.35}$$

and so remains suppressed by a positive power of m/M.

The upshot so far is this: phonons complicate the low-energy Fermi-liquid picture of electrons, but only by adding perturbatively small interactions suppressed by positive powers of m/M. Not too surprisingly, the phonons whose couplings matter the most for electron scattering are those whose momenta and energies allow them to be on-shell when interacting with Fermi-surface electrons: $|\mathbf{q}| \sim p_F$ and $E_{ph} \sim \omega_D \sim \sqrt{m/M}\, E_F$.

Because phonon interactions are relevant between $E_{\rm atom}$ and ω_D the powers of m/M suppressing interactions at $E \sim \omega_D$ are systematically lower than those appearing at $E \sim E_{\rm atom}$, but in all cases the power is positive. Phonons therefore do not destroy the overall Fermi-liquid picture in which the low-energy response involves weakly interacting quasi-electrons. These quasi-electrons just end up also coupled to a gas of weakly interacting phonons at low energies.

The validity of this picture can be tested for real materials by comparing the predictions implied for measurable quantities, such as specific heat, thermal and electrical conductivity and so on. For example, a thermal gas of phonons with dispersion relation $E(p) = cp$ has thermal energy density

$$\left(\frac{U}{V}\right)_{\rm ph} = g^{\rm ph}_\star \int \frac{d^3\mathbf{p}}{(2\pi)^3}\, \frac{E(p)}{e^{E(p)/T}-1}, \tag{15.36}$$

where $g^{\rm ph}_\star = 3$ counts the number of internal (spin) states. Differentiating gives a specific heat that agrees well with measurements for many solids when evaluated with the integration cut off at $E_{\rm max} = \omega_D$. In particular, it predicts $dU/dT \propto T^3$ for $T \ll \omega_D$.

Degenerate electrons similarly contribute at low energies an amount

$$\left(\frac{U}{V}\right)_{\rm el} = g_\star^{\rm el} \int \frac{{\rm d}^2{\bf k}{\rm d}l}{(2\pi)^3} \frac{v_F({\bf k})l}{e^{v_F l/T}+1}, \tag{15.37}$$

where now $g_\star^{\rm el} = 2$. Scaling $l \to l/T$ shows this predicts $dU/dT \propto T$ at low energies (and so falls there more slowly than does the phonon contribution). The success of predictions like these (and others) gives an *a posteriori* justification for the Landau Fermi-liquid description of electrons in many materials.

15.2.2 Phonon-Coulomb Competition

This is all still very nice, but does not yet help see how phonon-mediated processes can compete with repulsive Coulomb interactions. To see how this comes about, consider computing phonon exchange to see how this looks as an electron self-interaction.

For instance, single-phonon exchange using the interaction (15.32) in general leads to an interaction that is nonlocal in time, with terms of the form

$$\int {\rm d}t\,{\rm d}t' \prod_{r=1}^{2} {\rm d}^3{\bf q}_r \prod_{j=s}^{4} {\rm d}^3{\bf p}_s\, g^i_{ab}({\bf q}_1;{\bf p}_3,{\bf p}_1)\, g^j_{cd}({\bf q}_2;{\bf p}_4,{\bf p}_2)\, \langle D_i({\bf q}_1,t)D_j({\bf q}_2,t')\rangle$$
$$\times\ \psi^*_a({\bf p}_3,t)\psi_b({\bf p}_1,t)\psi^*_c({\bf p}_4,t')\psi_d({\bf p}_2,t')\,\delta^3({\bf q}_1-{\bf p}_4+{\bf p}_2)\,\delta^3({\bf q}_2-{\bf p}_3+{\bf p}_1), \tag{15.38}$$

where $\langle D_i({\bf q}_1,t)D_j({\bf q}_2,t')\rangle$ denotes the relevant phonon propagator.

Only for very late times, in the regime where the phonons involve energies $E < \omega_D$, can this be regarded as a contribution to a fermion self-interaction like (15.12), since the propagator is then local in time,

$$\langle D_i({\bf q}_1,t)D_j({\bf q}_2,t')\rangle = \mathcal{D}_{ij}({\bf q}_1,{\bf q}_2)\delta(t-t'), \tag{15.39}$$

because the phonon free action is here dominated by the restoring force term $\Omega^2_{ij}D^iD^j$. In this regime $\langle D_i(t)D_j(t)\rangle$ scales like $s\langle D_i(t)D_j(t)\rangle$, and so $\mathcal{D}_{ij}({\bf q}_1,{\bf q}_2)$ in this part of the propagator does not scale.

By contrast, contributions to $\langle D_i(t)D_j(t)\rangle$ have more time structure for shorter separations that receive contributions from the regime $E > \omega_D$, because here the kinetic term in the phonon free action dominates. In this regime $\langle D_i(t)D_j(t)\rangle$ instead has time-dependence consistent with it scaling like $s^{-1}\langle D_i(t)D_j(t)\rangle$.

For the purposes of following factors of m/M, the overall strength of the phonon-induced electron self-interaction (local in time or not) can be written as

$$V_{abcd} \sim \int {\rm d}t'\, g^i_{ab}({\bf q}_1;{\bf p}_3,{\bf p}_1)\, g^j_{cd}({\bf q}_2;{\bf p}_4,{\bf p}_2)\langle D_i({\bf q}_1,t)D_j({\bf q}_2,t')\rangle. \tag{15.40}$$

At the scale ω_D phonon exchange begins to look like a potential interaction, and at energies below this inherits the same scaling from the scaling of the g^i_{ab}'s as was previously inferred directly for instantaneous 4-electron self-interactions. That is, $V_{abcd} \to sV_{abcd}$ for generic momenta, and $V_{abcd} \to s^0 V_{abcd}$ for the special momentum configurations like ${\bf p}_1 = -{\bf p}_2$.

The same distinction between generic momenta and special momentum configurations also changes the scaling of phonon exchange with energies larger than ω_D, for the same reason. When momentum-conserving delta functions are combined in expressions like (15.38) it remains true that for the special configurations one of them removes a dl integral, thereby raising the scaling of the interaction by a factor of $1/s$.

The upshot is this: the strength of the effective phonon-mediated coupling always goes like the square of the scaling of the underlying electron-phonon couplings, g^i_{ab}. Consequently,

$$\begin{aligned} S^{e-ph-e}_{\rm int} &\to s^{-1}\, S^{e-ph-e}_{\rm int} \quad \text{for } E_{ph} > \omega_D \text{ (generic case)} \\ S^{e-ph-e}_{\rm int} &\to s^{+1}\, S^{e-ph-e}_{\rm int} \quad \text{for } E_{ph} < \omega_D \text{ (generic case)} \end{aligned} \tag{15.41}$$

and

$$\begin{aligned} S^{e-ph-e}_{\rm int} &\to s^{-2}\, S^{e-ph-e}_{\rm int} \quad \text{for } E_{ph} > \omega_D \text{ (special config.)} \\ S^{e-ph-e}_{\rm int} &\to s^{0}\, S^{e-ph-e}_{\rm int} \quad \text{for } E_{ph} < \omega_D \text{ (special config.).} \end{aligned} \tag{15.42}$$

What does this mean for the powers of m/M in V_{abcd}? In the generic case at high energies the phonon-electron couplings are $g(E_{\rm atom}) \sim (m/M)^{1/2}$ and so phonon exchange leads to electron interactions that are of order $V(E_{\rm atom}) \sim m/M$. At the Debye scale, the phonon couplings grow to be of order $g(\omega_D) \sim (m/M)^{1/4}$ and the phonon-mediated interaction there is of size $V_{abcd} \sim (m/M)^{1/2}$. This same result is also found from the scaling (15.41) using $V_{abcd} \to s^{-1} V_{abcd}$ with $s \sim \omega_D/E_{\rm atom} \sim (m/M)^{1/2}$. Not surprisingly, in the generic case interactions never become large.

More interesting is the case of the special kinematics for which the initial and final electrons are equal and opposite. It is still true in this case that at high energies the phonon-electron couplings are $g(E_{\rm atom}) \sim (m/M)^{1/2}$ and so phonon-mediated electron interactions are of order $V(E_{\rm atom}) \sim m/M$. In this case, though, the fermion self-interaction gets enhanced relatively to g^2 by the running of (15.42). Whereas phonon self-interactions at the Debye scale are still $g(\omega_D) \sim (m/M)^{1/4}$ in size, the phonon-mediated electron self-interaction at the Debye scale is instead enhanced by an additional power of $1/s$ to be

$$V(\omega_D) \sim [g(\omega_D)]^2 \left(\frac{E_{\rm atom}}{\omega_D}\right) \sim \left(\frac{m}{M}\right)^{1/2} \left(\frac{M}{m}\right)^{1/2} \sim O(1). \tag{15.43}$$

All suppression by powers of m/M cancel, though only for the special BCS kinematics. But for these kinematics the result can compete with other electron-electron interactions.

Just below the Debye scale the dominant implications of phonon exchange for electron scattering can be summarized as an effective attractive two-body interaction, and so the strength of these interactions can be analyzed to lower energies using the purely electron theory discussed in previous sections, using a total interaction of the form

$$V^{tot}_{abcd}(\mathbf{k}, -\mathbf{k}, \mathbf{k}', -\mathbf{k}') = V^{ph}_{abcd}(\mathbf{k}, -\mathbf{k}, \mathbf{k}', -\mathbf{k}') + V^{el}_{abcd}(\mathbf{k}, -\mathbf{k}, \mathbf{k}', -\mathbf{k}'), \tag{15.44}$$

where $V^{el}_{abcd}(\mathbf{k}_1, \mathbf{k}_2, \mathbf{k}_3, \mathbf{k}_4)$ describes all other two-body electron interactions arising from all sources, other than through phonon exchange. This includes the residual screened Coulomb inter-electron interactions and typically represents a repulsive force.

Assuming the total interaction to be weak enough to apply perturbative methods allows the use of the results of §15.1.3 to see how the interaction scales to much lower energies. For simplicity the evolution is here stated assuming all potentials to be momentum-independent constants on the Fermi surface, as in Eq. (15.20). In this case, it is useful to absorb the density of states on the Fermi surface, $\mathcal{N}_F$ – defined by Eq. (15.22) – to define the dimensionless quantities

$$\mu := \mathcal{N}_F V^{el}(E = E_{\rm atom}), \qquad \mu_\star := \mathcal{N}_F V^{el}(E = \omega_D), \tag{15.45}$$

characterizing the electronic interactions at both high energies and ω_D, as well as the phonon-potential strength at ω_D,

$$\nu := -\mathcal{N}_F V^{ph}(E = \omega_D). \tag{15.46}$$

The sign in this expression is chosen so that $\nu > 0$ when phonon-mediated inter-electron forces are attractive. In the end, phonon exchange provides an attractive inter-electron force because it describes how the Coulomb electron-ion attraction slightly distorts the ionic lattice about one electron, thereby providing a small local over-density of positive charge towards which the second electron is attracted.

Assuming V^{el} is repulsive, Eq. (15.23) implies that μ and $\mu_\star$ are related by

$$\mu_\star = \frac{\mu}{1 + \mu \ln(E_{\rm el}/\omega_D)} = \frac{\mu}{1 + \frac{1}{2}\mu \ln(M/m)}. \tag{15.47}$$

The total electron interaction at energies $E \ll \omega_D$ then becomes (again using Eq. (15.23))

$$V^{tot}(E) = \frac{V^{tot}(\omega_D)}{1 - (\nu - \mu_\star)\ln(\omega_D/E)}, \tag{15.48}$$

showing how phonon-mediated attraction competes with the renormalized strength of the other electronic interactions to control the size of the total potential.

Everything rides on whether or not V^{ph} dominates the renomalized value of V^{el} at the Debye scale; *i.e.* whether or not $\nu > \mu_\star$. There are two options:

- If $|V^{el}(\omega_D)| > |V^{ph}(\omega_D)|$ then $\nu < \mu_\star$, and so $V^{tot}(\omega_D)$ starts off repulsive at $E = \omega_D$, and therefore becomes logarithmically weaker at lower energies.
- If, on the other hand, it is V^{ph} that dominates at $E = \omega_D$ – a real possibility since $V^{ph}(\omega_D)$ is not suppressed by powers of m/M – then $\nu > \mu_\star$ and the net force is attractive. As a result, V^{tot} is marginally relevant, becoming more and more attractive at energies lower than the Debye scale.

At low enough energies perturbation theory in a net-attractive interaction eventually fails, invalidating expression (15.48) for the running of V^{tot}. At this point, the quasi-electron pairs experience a strong attractive force, provided their momenta are equal and opposite since it is only for these configurations that instability applies. The ground state can respond dramatically to minimize this newly strong interaction energy, such as by condensing pairs of bound quasi-electrons, giving rise to a superconducting transition.

The scale, E_{sc}, at which the coupling becomes strong is

$$E_{sc} = \omega_D \exp\left[-\left(\frac{1}{\nu - \mu_\star}\right)\right] \sim E_{\rm atom}\left(\frac{m}{M}\right)^{\frac{1}{2}} \exp\left[-\left(\frac{1}{\nu - \mu_\star}\right)\right]. \tag{15.49}$$

Table 15.1 A comparison of some BCS predictions with experiment

Element	Al	Cd	Hg(α)	In	Nb	Pb	Sn	Ta	Tl	V	Zn
$(2\Delta(0)/T_c)_{\rm exp}$	3.3	3.2	4.6	3.6	3.8	4.4	3.5	3.6	3.6	3.4	3.2
T_c (K)	1.14	0.56	4.15	3.40	9.50	7.19	3.72	4.48	2.39	5.38	0.875

This is exponentially small relative to the Debye scale whenever the coupling $v - \mu_\star$ is weak. It is exponentially small because the running of a marginal interaction is logarithmic, and so it takes a long range of energies to make an initially weak coupling into a strong one. This is the understanding of why conventional superconductors have transition temperatures that are so small: of order a few K.

There are two independent ways to access the scale E_{sc} experimentally: the size of the superconducting gap at zero temperature, $2\Delta(0)$, and the value of the transition temperature, T_c. To compare with measurements requires a more detailed calculation than given here, to determine more precisely the numerical prefactor of the exponential for these two observables. The resulting weak-coupling predictions turn out to be given by [408]

$$\Delta(0) \simeq 2\,E_{sc}, \qquad \text{and} \qquad T_c \simeq 1.13\,E_{sc}. \tag{15.50}$$

The theory makes the most accurate predictions for their ratio, $2\Delta(0)/T_c \simeq 3.53$, in which E_{sc} cancels. The success of these predictions can be seen by comparing with the experiment values given in Table 15.1 (taken from [409]).

To the uninitiated, the amazing thing about this comparison is that it works as quantitatively as it does. An intrinsically feeble force is taken to compete with the enormous Coulomb repulsion between electrons, with all other interactions discarded. This sounds at face value like an argument at best trusted to within an order of magnitude, whereas the *quantitative* predictions for the properties of superconductors obtained are in practice quite successful, often working at the 10% level or better.

The great power of this EFT analysis is its ability to understand this accuracy by quantifying the errors being made by neglecting other contributions. The EFT identifies that the accuracy of neglecting all other degrees of freedom besides the BCS-unstable modes works so well because it is at heart a low-energy approximation, due to the irrelevant scaling of all self-interactions for Fermi liquids.

Part of the evidence that phonon-mediated attraction lies at the root of the superconducting story comes from comparing E_{sc} for different materials having different values for the ion mass M, a dependence known as the *isotope effect*. Eq. (15.49) predicts the dominant dependence to arise in two separate ways. First, M enters through the explicit factor $(m/M)^{\frac{1}{2}}$, and second, it also enters implicitly through the M-dependence of the renormalized interaction strength, $\mu_\star$.

This M-dependence is usually characterized by an exponent, with $E_{sc} \propto (m/M)^{\alpha_{sc}}$ and α_{sc} given in the BCS theory by using (15.49) to evaluate

$$\alpha_{sc} := -\frac{M}{E_{sc}}\frac{\partial E_{sc}}{\partial M} = \frac{1}{2}\left[1 - \left(\frac{\mu_\star}{v - \mu_\star}\right)^2\right]. \tag{15.51}$$

Table 15.2 The isotope effect for various superconductors (Numbers taken from reference [409].)

Element	α_{sc}	Element	α_{sc}	Element	α_{sc}
Zn	0.45 ± 0.05	Pb	0.49 ± 0.02	Mo	0.33
Cd	0.32 ± 0.07	Tl	0.61 ± 0.10	Nb_3Sn	0.08 ± 0.02
Sn	0.47 ± 0.02	Ru	0.00 ± 0.05	Mo_3Ir	0.33 ± 0.03
Hg	0.50 ± 0.03	Os	0.15 ± 0.05	Zr	0.00 ± 0.05

For weak-coupling superconductors – those for which $\mu_\star \ll \nu$ – the universal result is $\alpha_{sc} = \frac{1}{2}$. The index systematically decreases as the relative strength of $\mu_\star$ grows in comparison with ν. Some experimental values for α_{sc} are given in Table 15.2 – again taken from [409] – and are seen to be consistent with the prediction that they should lie at or below the value 0.5, with best agreement being for weak-coupling superconductors.

15.3 Quantum Hall Systems ♣

Quantum Hall systems provide a second example where interacting degenerate electrons display a rich and striking phenomenology for which remarkably precise predictions can be made. This section provides a brief sketch of some properties of these systems, together with some steps towards seeing how effective field theories can help understand the robustness of these predictions.

15.3.1 Hall and Ohmic Conductivity

The systems of interest have electronic structure that is very asymmetric and effectively traps low-energy electrons to move in only two dimensions (for an introduction to quantum Hall physics, see [411–414]). When a large magnetic field is applied perpendicular to these two directions and an electric field is applied parallel to them currents flow and transport properties can be measured. Because of the magnetic field the current need not flow parallel to the electric field, with Ohm's law taking the form

$$J_i = \sigma_{ij} E_j, \tag{15.52}$$

where J_i denotes the electromagnetic current density, E_i is the applied electric field and coordinates are chosen so that the electrons move in the $x-y$ plane, so $i, j = x, y$. For rotationally invariant systems – which is assumed henceforth – the conductivities satisfy $\sigma_{yy} = \sigma_{xx}$ and $\sigma_{yx} = -\sigma_{xy}$, making the two observable quantities the *Ohmic* conductivity σ_{xx} and the *Hall* conductivity σ_{xy}. Notice that in two spatial dimensions both J_i and E_i have dimension (mass)^2 in fundamental units, and so σ_{ij} is dimensionless.

The resistivity matrix defined by $E_i = \rho_{ij} J_j$ is also useful, and is given by the inverse of σ_{ij}. Its components are related to those of the conductivity by

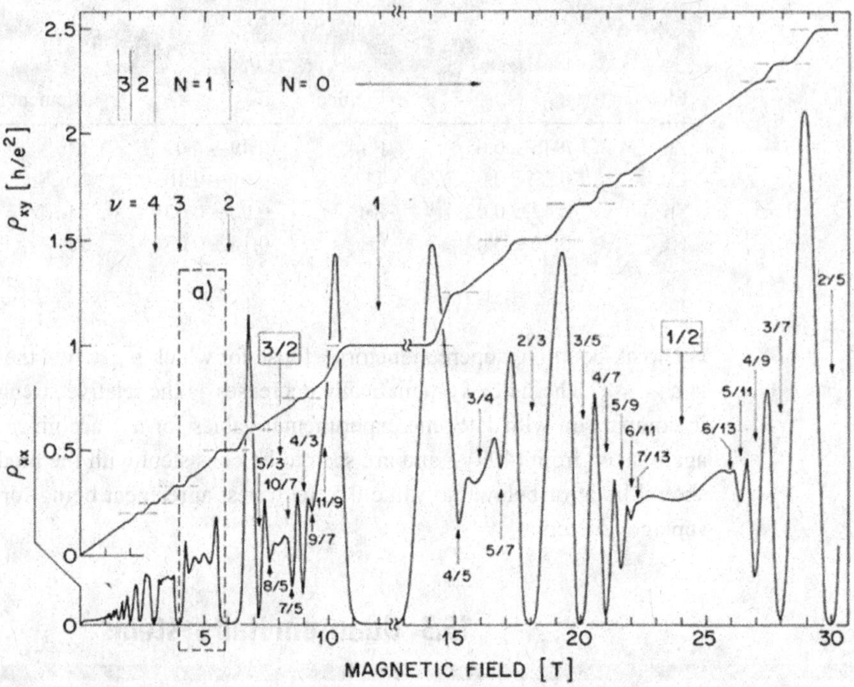

Fig. 15.8 Traces of longitudinal (or Ohmic) resistivity (ρ_{xx}) and Hall resistivity (ρ_{xy}) vs applied magnetic field, with plateaux appearing in the Hall plot. The Ohmic resistivity tends to zero for fields where the Hall resistivity plateaus. Figure taken from [410]

$$\rho_{xx} = \frac{\sigma_{xx}}{\sigma_{xx}^2 + \sigma_{xy}^2} \quad \text{and} \quad \rho_{xy} = -\frac{\sigma_{xy}}{\sigma_{xx}^2 + \sigma_{xy}^2}, \tag{15.53}$$

which can be compactly summarized as $\rho = -1/\sigma$ for the complex variables $\sigma := \sigma_{xy} + i\sigma_{xx}$ and $\rho := \rho_{xy} + i\rho_{xx}$. Since, in general, $\sigma_{xx} \geq 0$ these complex variables can be taken to live in the upper half plane. Notice that when $\sigma_{xy} \neq 0$ the limit $\sigma_{xx} \to 0$ is also the limit $\rho_{xx} \to 0$.

Fig. 15.8 shows what is found when these resistivities are measured (as a function of applied magnetic field) for real quantum Hall systems, and reveals a number of very striking features. In particular, there are regimes of magnetic field for which neither ρ_{xx} nor ρ_{xy} change as B is varied (these are called quantum Hall 'plateaux'). Even more remarkably, ρ_{xx} is consistent with zero on these plateaux while ρ_{xy} is very accurately a ratio of integers q/p when expressed in units of $2\pi/e^2$ (= h/e^2 once Planck's constant is put back in).

If electrons behaved as (electrically charged) billiard balls, simple arguments [413] show that one would expect to find $\sigma_{xy} = \nu(e^2/2\pi)$, where the *filling factor* ν is defined as the number of electrons present divided by the number of flux quanta in the magnetic field:[7]

[7] Notice that the flux quantum here is taken as $\Phi_0 = 2\pi/e$, which is twice as large as used for superconductors in §14.2.1. This difference arises because the charge carriers here are assumed to have charge $\pm e$ (as opposed to charge $\pm 2e$ for paired electrons in a superconductor).

$$\nu := \frac{N_e}{N_\Phi} = \frac{n_e A}{BA/\Phi_0} = \frac{n_e \Phi_0}{B} = \frac{2\pi n_e}{eB}, \tag{15.54}$$

where A is the sample's area. For a Hall plateau (assuming for the minute $\sigma_{xx} = 0$) this implies that $\rho_{xy} = -1/\sigma_{xy}$ is proportional to B. Although the general trend for the Hall resistivity in Fig. (15.8) is indeed linear in B, this proportionality is interrupted by the presence of the plateaux. In terms of conductivities, the plateaux are experimentally characterized by the values

$$\sigma_{xx} = 0 \quad \text{and} \quad \sigma_{xy} = \frac{k\, e^2}{2\pi}, \tag{15.55}$$

where $k = p/q$ is a rational number. The agreement of k with a fraction holds to within the experimental accuracy – which is extremely good, better than a part in 10^9 when k is an integer. The special cases where k is an integer are called 'integer Hall plateaux' and were the first to be discovered [415] while those where k is a non-integer rational number are 'fractional Hall plateaux' [416].

Quantum motion of a free charged particle in a constant magnetic field gives evenly spaced harmonic oscillator energy levels, $E_n = \left(n + \frac{1}{2}\right)(eB/m)$ called Landau levels, and for a degenerate system of free electrons the quantization of Landau-level energies describes well the integer quantum Hall states [417]. What is a puzzle is why this quantization survives so extremely accurately even for strongly interacting electrons and electrons in disordered environments, why it persists unchanged as B ranges through a finite interval, and what decides whether the quantization is an integer or a non-integer fraction. (When $k = p/q$ is a fraction, it also transpires that the denominator q is almost always odd.)[8]

There is a microscopic explanation for these effects that shows what the electrons are doing when these quantum Hall states arise and why their transport properties are quite robust. The integer effect is understood in this way to depend on free electron properties in a magnetic field, coupled to some disorder that acts to localize many electron states (making them not participate in charge transport measurements) [413]. The fractional states are instead understood to arise due to the effects of Coulomb interactions for these electrons [421]. In this picture Coulomb interactions cause dramatic effects because of the enormous degeneracy of states these systems enjoy. Two dimensional electrons in magnetic fields can have Landau levels that are occupied by macroscopically large numbers of electrons (like 10^{11} electrons or more). Such an enormous degeneracy amongst non-interacting states allows the interacting ground state to be very sensitive to electron interactions, leading to strongly correlated ground states.

The remainder of this section explores – following [418–420] – the extent with which this microscopic understanding can be translated into an EFT language, along the lines so successfully done for superconductivity. The use of EFT methods seems appropriate since transport properties (like resistivity or conductivity) near the plateaux take place at very low temperatures and at energies much smaller than generic electronic scales. One can hope for robust properties to arise at low energies

[8] The key word here is 'almost', since plateaux with even denominators also sometimes occur (like $k = 5/2$). But these are the exceptions that prove the rule (and, of course, both types must ultimately be explained).

as simple consequences of the low-energy theory, and not depend on many of the details of the underlying electrons.

15.3.2 Integer Quantum Hall Systems

At very low energies, the effective theory obtained by integrating out all of the higher energy excitations is a function purely of the electromagnetic field, A_μ, used to explore the electromagnetic transport. What is interesting about two spatial dimensions is that this effective theory allows an effective interaction that is *more* important at low energies than is the usual Maxwell action. This effective coupling – called the *Chern-Simons* term [335] – has the form

$$S_{CS} = \frac{k\,e^2}{4\pi}\int_M d^3x\,\epsilon^{\mu\nu\lambda}A_\mu\partial_\nu A_\lambda, \tag{15.56}$$

where $\mathcal{M}$ denotes the 2+1 dimensions of space and time containing the quantum Hall electrons and k is the effective coupling constant (after the factor $e^2/4\pi$ is extracted). Here, $\epsilon^{\mu\nu\lambda}$ is the completely antisymmetric Levi–Civita tensor (see Appendix A.2.1), conventionally chosen so that $\epsilon^{012} = +1$. It is because S_{CS} involves only three powers of fields and derivatives that it can dominate at low energies the Maxwell term (which has four).

The electromagnetic current arising from a system described by Eq. (15.56) is inferred by differentiating with respect to A_μ, giving

$$\langle J^\mu\rangle = \frac{\delta S_{CS}}{\delta A_\mu} = \frac{ke^2}{4\pi}\epsilon^{\mu\nu\lambda}F_{\nu\lambda}, \tag{15.57}$$

which when evaluated as a function of a perturbing electric field $E_a = F_{a0}$ and magnetic field $B = F_{12}$ gives

$$J_a = \frac{ke^2}{2\pi}\,\epsilon_{ab}E_b \quad\text{and}\quad J^0 = \frac{ke^2}{2\pi}\,B. \tag{15.58}$$

Comparing the first of these with Ohm's law – in the form $J_a = \sigma_{ab}E_b$ – implies the conductivities

$$\sigma_{xx} = 0 \quad\text{and}\quad \sigma_{xy} = \frac{ke^2}{2\pi}. \tag{15.59}$$

This shows that k can be interpreted as the Hall conductivity in units of $e^2/2\pi$ and so should be an integer for integer quantum Hall systems and take fractional values in fractional quantum Hall systems. The second equation in (15.58) also gives the number-density of charge carriers because $n_e = |J^0|/e$, and comparing the resulting prediction for the ratio $2\pi n_e/eB$ with (15.54) reproduces the expectation that

$$|k| = \nu, \tag{15.60}$$

also gives the quantum Hall system's proper filling fraction.

The action (15.56) has several novel properties that tell an interesting story about the role that anomalies, topology and boundary physics can play in an effective theory. This story centres on the transformation properties of (15.56) under electromagnetic gauge transformations – *c.f.* Eq. (7.11) – for which $A_\mu \to A_\mu + \partial_\mu\zeta$

with arbitrary $\zeta(x)$. These are *not* a symmetry of the lagrangian density in (15.56), which varies into a total derivative:

$$\epsilon^{\mu\nu\lambda} A_\mu \partial_\nu A_\lambda \to \epsilon^{\mu\nu\lambda} A_\mu \partial_\nu A_\lambda + \partial_\mu\left(\zeta\, \epsilon^{\mu\nu\lambda} \partial_\nu A_\lambda\right), \tag{15.61}$$

a result that uses the antisymmetry of $\epsilon^{\mu\nu\lambda}$ to conclude $\epsilon^{\mu\nu\lambda}\partial_\mu\partial_\nu A_\lambda = 0$. As the next few sections show, this transformation property makes the physics of (15.56) sensitive both to the topology of the Hall sample and to its boundaries.

Quantization of k

Up until this point, the parameter k defined in (15.56) could be any real number. This section now argues that gauge invariance requires that k must be an integer,[9] thereby providing a very robust explanation for the quantization of integer quantum Hall levels. The argument that it is ultimately gauge invariance that is responsible for the precise value of the filling fractions found for integer quantum Hall systems was first made in [422], and is the reason why the quantization of these values is so robust.

Why must k be an integer? The argument is topological, and so to make it it is useful to imagine applying (15.56) in a topologically interesting situation: where the two-dimensional sample is a 2-sphere. For the same reason it is useful also to work at finite temperature (which for the purposes of topology simply means working in Euclidean-signature metric with Euclidean time, τ_E, periodically identified (see Appendix A.2.2): $\tau_E = \tau_E + \beta$ – where $\beta := 1/T$ is the inverse temperature). Notice that it is sufficient that one *could* work in this topology to draw conclusions about k; it is not necessary that every particular quantum Hall application use this topology.

The significance of making the spatial directions be a two-sphere is that this quantizes the value of any homogeneous applied magnetic field, F_{12}. After all, a homogeneous magnetic field on a 2-sphere is the field of a magnetic monopole situated at its centre, and this was shown in another context in §14.1.2 to require

$$\int_{S_2} \mathrm{d}^2 x\, F_{12} = \frac{2\pi n}{e}, \tag{15.62}$$

where e is the smallest unit of free charge and n is an integer.

The significance of making time a Euclidean circle is that it allows a topological class of 'large' $U(1)$ gauge transformations, $g = \exp[ie\zeta(\tau_E)]$, that are single-valued on the circle

$$g(\tau_E + \beta) = g(\tau_E) \quad \text{by having} \quad \zeta(\tau_E + \beta) = \zeta(\tau_E) + \frac{2\pi m}{e}, \tag{15.63}$$

where m is an integer. This can be achieved through the explicit choice $\zeta = (2\pi m/e)(\tau_E/\beta)$, for example. What is important is that the gauge potential also transforms under these large transformations, with

$$A_\tau \to A_\tau + \frac{2\pi m}{e\beta}. \tag{15.64}$$

The quantization condition for k follows from the requirement that (15.56) is invariant under these large gauge transformations of A_τ. To see why, evaluate (15.56)

[9] The changes required for fractional quantum Hall systems are discussed in §15.3.3.

with F_{12} given by a time-independent magnetic monopole configuration and A_τ a constant. With these choices

$$S_{CS} = \frac{ke^2}{4\pi}\int_0^\beta d\tau_E \int_{S_2} d^2x\, \epsilon^{\mu\nu\lambda} A_\mu \partial_\nu A_\lambda = \frac{ke^2\beta}{2\pi} A_\tau \int_{S_2} d^2x\, F_{12} = nke\beta A_\tau, \tag{15.65}$$

which performs several spatial integrations by parts to isolate the A_τ as an overall coefficient, and the last equality uses (15.62) where n is the magnetic monopole quantum. Performing the transformation (15.64) to (15.65) then shows that S_{CS} is not invariant since

$$S_{CS} \to S_{CS} + 2\pi k\, nm. \tag{15.66}$$

What matters is not whether S_{CS} is invariant, but whether $e^{iS_{CS}}$ is, since this is what appears in the path integral,[10] and (15.66) shows that $e^{iS_{CS}}$ remains invariant if k is an integer (as are n and m).

This argument gives a very robust topological reason why both the filling fraction and the Hall conductivity are rigidly quantized at integer values; all that is required is that the lowest-dimension interaction, S_{CS}, dominate at low energies, since the above argument shows that its coefficient k must be quantized on very general grounds whose roots lie in gauge invariance. It also captures the low-energy limit of a more microscopic understanding of the underlying electron system's response to the topology of its environment [423]. Finally, this argument explains why quantities like σ_{xy} and ν are so robust to adiabatic changes to system parameters in integer quantum Hall systems. Because k is quantized, it generically cannot change in a continuous way as other parameters slowly vary.

Surface Currents and Anomaly Matching

Another remarkable consequence of (15.61) can be seen for situations where the quantum Hall sample has a spatial boundary. In the presence of boundaries Eq. (15.61) implies that the action (15.56) cannot be the entire low-energy story. This is because any quantum Hall system that is dominated by S_{CS} at low energies must also contain mobile charge-carrying degrees of freedom localized on its boundaries.

To see why, consider the transformation properties of (15.56) under the transformation $A_\mu \to A_\mu + \partial_\mu \zeta$. If $\mathcal{M}$ has a boundary (as real quantum Hall systems usually do) then S_{CS} is *not* gauge invariant, and instead transforms as

$$S_{CS} \to S_{CS} + \frac{k\,e^2}{4\pi}\int_{\partial\mathcal{M}} d^2x\, \zeta\, \epsilon^{\mu\nu\lambda} n_\mu F_{\nu\lambda}, \tag{15.67}$$

where n_μ is (as usual) the outward-pointing normal on the boundary, $\partial\mathcal{M}$, of the bulk Hall system.

In real systems the failure of gauge invariance implied by (15.67) must be cancelled by a related failure coming from degrees of freedom, ψ, that live exclusively on the boundary. Recall that the boundary in this case is 1+1 dimensional: consisting

[10] Notice that the factor of i remains in the exponential $e^{iS_{CS}}$ even in Euclidean signature because the factor of i cancels between the transformations $dt \to -i d\tau_E$ and $A_t \to iA_\tau$.

of a single spatial dimension plus time. (For concreteness' sake it is useful to imagine the quantum Hall sample being a disc, with the boundary being its circumference.)

Degrees of freedom moving in the single spatial dimension of the boundary can be *chiral*, inasmuch as their motion might be only clockwise or only counterclockwise around the disc. This kind of uni-directional motion, with *e.g.* $\psi = \psi(x - vt)$, is possible when there is only a single spatial direction. What is important is that this kind of chirality for an electrically charged degree of freedom implies a failure of gauge invariance that can be just what is needed to cancel (15.67).

A 2D Dirac fermion is an example of a kind of charge carrier that could be localized on the boundary. In this case, direction of motion is related to the spin of the fermion, as can be seen from the form of the 2D Dirac equation

$$\partial\!\!\!/\psi = \left(\gamma^0\partial_0 + \gamma^1\partial_1\right)\psi = \gamma^0\left(\partial_t + \gamma_3\partial_x\right)\psi = 0, \tag{15.68}$$

where $\gamma^0 = i\sigma_1$ and $\gamma^1 = \sigma_2$ are representations of the 2D gamma matrices in terms of Pauli matrices, and $\gamma_3 := -\gamma^0\gamma^1 = \sigma_3$ is the matrix whose eigenvalues give the spin handedness of the fermion field: $\gamma_3\psi_\pm = \pm\psi_\pm$. The Dirac equation then shows that these fields satisfy the equations

$$\left(\partial_0 \pm \partial_1\right)\psi_\pm = 0 \quad \text{with solutions} \quad \psi_\pm = \psi_\pm(x \mp t), \tag{15.69}$$

which shows how states of definite chirality move in specific directions along the boundary.

For these types of fermions the chiral anomaly implies (see Exercise 15.3) that the currents satisfy

$$\partial_\mu\langle ie\,\overline{\psi}_\pm\gamma^\mu\psi_\pm\rangle = \frac{1}{2}\partial_\mu\langle ie\,\overline{\psi}\gamma^\mu(1 \pm \gamma_3)\psi\rangle = \pm\frac{e^2}{4\pi}\,\epsilon^{ab}F_{ab}. \tag{15.70}$$

This means that their quantum action, Γ_b, transforms anomalously, with [210]

$$\delta\Gamma_b = (N_+ - N_-)\frac{e^2}{4\pi}\int_{\partial M} d^2x\;\zeta\,\epsilon^{ab}F_{ab}. \tag{15.71}$$

This has precisely the form needed to cancel (15.67) provided the integer k agrees with the mismatch between the number of left- and right-moving charge-carrying species on the boundary. A similar construction can also be made using charge-carrying bosons on the boundary. Although the electromagnetic Chern–Simons action requires the presence of such boundary charge-carriers, it is relatively mute about their detailed properties.[11]

This is a practical example of the phenomenon of anomaly matching described in §4.3. The underlying theory in this case involves electrons interacting in four spacetime dimensions with photons and lattice ions and is gauge invariant (including having no anomalies). This must therefore also be true of the low-energy effective description, and so any presence of a term like (15.56) – such as indicated by a nonzero Hall conductivity – necessarily requires the existence of some other boundary degree of freedom to ensure the low-energy description of the whole system remains gauge invariant.

By doing so the effective theory correctly reproduces what happens in more microscopic understandings of integer quantum Hall states, since these also indicate

[11] Indeed, in 1+1 dimensions bosons and fermions can even be equivalent to one another [487–489]; for a simple argument explaining why see [490].

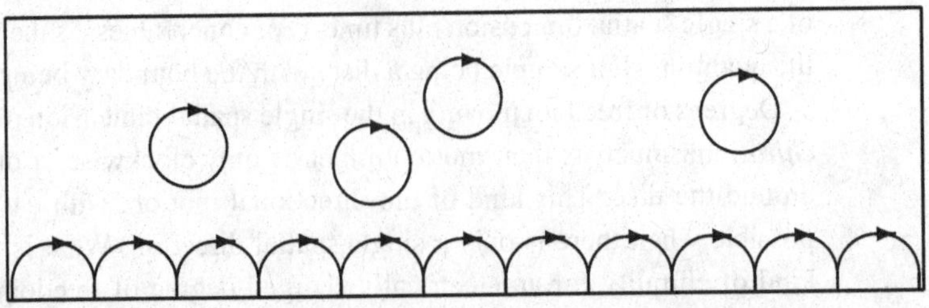

Fig. 15.9 Cartoon of semiclassical Landau motion in a magnetic field, showing how orbits in the interior do not carry charge across a sample's length while surface orbits can if they bounce repeatedly off the sample's edge. Notice that the motion is chiral inasmuch as the circulation goes around the sample in a specific direction. This is a specific mechanism for the origin of surface currents in quantum Hall systems, as are required on general grounds for the low-energy EFT by anomaly matching.

that boundary degrees of freedom play a prominent role transporting charge in quantum Hall systems. Real systems are typically not homogeneous on macroscopic scales and instead often come with the electron fluid dispersed into a collection of quantum Hall domains, along which current percolates through the sample largely moving along domain boundaries. Because of disorder within the samples, electrons experience random potentials that keep many of them from participating in macroscopic charge transport, but this cannot block the mobility of charges moving along the surfaces of the domains. The microscopic origin of these surface currents can already be seen within a semi-classical cartoon of electron motion, for which Landau levels can be regarded as billiard-ball electrons moving in gyromagnetic circles under the influence of the applied magnetic field. This circular motion transports no net charge across the sample, but the same need not be true for electrons near the sample's edge which can move through longer distances by repeatedly reflecting from the surface (as sketched in Fig. 15.9).

15.3.3 Fractional Quantum Hall Systems

So far, the arguments for quantizing k are almost too good: by quantizing k as an integer they seem to preclude the possibility of having fractional quantum Hall states. How do the fractional states arise in the low-energy EFT?

Fractional quantum Hall states change the above arguments by involving new degrees of freedom in the low-energy theory. The required new degrees of freedom turns out to be one or more abelian gauge field, a_μ. This degree of freedom is 'emergent' at low energies inasmuch as it is more indirectly related to the properties of weakly coupled electrons. It also expresses what is called 'topological order' in the sample, inasmuch as the properties of the low-energy field a_μ allow the material to respond to different sample topologies in a way that captures topological features of the underlying electron dynamics (such as how the number of degenerate ground states available depends on sample topology).

The low-energy properties of a gauge field like a_μ should be described by the lowest-dimension interaction terms, and for a_μ in 2+1 dimensions these are again given by the Chern–Simons interaction plus the Maxwell lagrangian plus higher and higher dimensions. Taking the leading terms to be

$$S_a = \int d^3x \left[\frac{k}{4\pi} \epsilon^{\mu\nu\lambda} a_\mu \partial_\nu a_\lambda - \frac{1}{4g^2} f_{\mu\nu} f^{\mu\nu} + \cdots \right], \tag{15.72}$$

where $f_{\mu\nu} := \partial_\mu a_\nu - \partial_\nu a_\mu$ is the abelian field strength for a_μ. This leads to classical equations of motion that predict (see Exercise 15.4) an energy gap

$$E_{\text{gap}} = \frac{kg^2}{\pi}, \tag{15.73}$$

showing that no bulk gapless mode is predicted. At energies much lower than the gap, the Maxwell term may be neglected and the action is simply dominated by the Chern–Simons coupling. Because this coupling is topological it does not affect the system energy (again see Exercise 15.4) or predict new propagating bulk modes.

One must also ask about the invariance of S_a under the new gauge transformations

$$a_\mu \to a_\mu + \partial_\mu \omega, \tag{15.74}$$

and for this the topological character of the Chern–Simons action again plays an informative role. In particular, evaluating the action in a topologically nontrivial background and repeating verbatim the arguments made above for electromagnetism now shows that the coefficient k of the Chern–Simons term in S_a must be an integer.

To see why the introduction of a_μ allows fractional Hall conductivities requires adding the coupling between a_μ and the electromagnetic field. The lowest-dimension action with which this can be done in this case is through a Chern-Simons mixing term. Adopting this coupling – and rescaling a factor of e out of a_μ – replaces (15.56) with

$$S'_{\text{CS}} = \frac{e^2}{2\pi} \int_{\mathcal{M}} d^3x \left[\epsilon^{\mu\nu\lambda} a_\mu \partial_\nu A_\lambda + \frac{k}{2} \epsilon^{\mu\nu\lambda} a_\mu \partial_\nu a_\lambda \right], \tag{15.75}$$

where the first term coupling a_μ to the electromagnetic field A_μ is chosen to ensure invariance under electromagnetic gauge transformations, $\delta A_\mu = \partial_\mu \zeta$.

Performing the gaussian integration over a_μ is equivalent to evaluating at the saddle point, $a_\mu = a^c_\mu$, where

$$f^c_{\mu\nu} = -\frac{1}{k} F_{\mu\nu}, \tag{15.76}$$

obtained by solving $\delta S'_{\text{CS}}/\delta a_\mu = 0$, leading to the following generator of electromagnetic response

$$\Gamma_{\text{CS}} = -\frac{e^2}{4\pi k} \int_{\mathcal{M}} d^3x \, \epsilon^{\mu\nu\lambda} A_\mu \partial_\nu A_\lambda. \tag{15.77}$$

The same arguments that gave (15.59) from (15.56) in the integer quantum Hall case now imply that on a Hall plateau dominated by (15.75) one expects to find

$$\sigma_{xx} = 0 \quad \text{and} \quad \sigma_{xy} = -\frac{e^2}{2\pi k}, \tag{15.78}$$

showing, in particular, a fractional Hall conductivity. For reasons that become clearer below the integer k must be odd for Hall systems built using fermionic charge carriers, but can be even for Hall systems built from bosons. Both options are allowed at this point because the effective theory cannot tell in itself which of these appears in the underlying theory. More complicated fractions $\sigma_{xy} = (p/q)(e^2/2\pi)$ may also arise, and are obtained by adding additional emergent gauge fields – *e.g.* a_μ and b_μ – beyond the single one considered here (see, for example, Exercise 15.5).

One might worry that something has gone wrong in this calculation, since the final result (15.77) has the same form as (15.56) but with the coefficient of $e^2/4\pi$ not quantized to be an integer; a quantization that was argued earlier to be a mandatory result. Strictly speaking, this worry is a legitimate one: the saddle-point configuration (15.76) is inconsistent with the boundary conditions for $F_{\mu\nu}$ and $f_{\mu\nu}$ in cases with nontrivial topology. This emphasizes that it is in general wrong to integrate out a_μ at low energies, despite it not describing propagating bulk degrees of freedom. The topological gauge field a_μ must be included in the low-energy theory since it expresses the topological order of the microscopic theory. That is to say: a_μ brings the news to low energies of how the underlying electrons respond to a sample's topology. However, the conductivity (15.78) is nonetheless the correct result (despite its naive derivation), since the saddle-point integration can be computed for topologically trivial samples for which boundary conditions do not obstruct the use of (15.76).

Edge States

Besides encoding topological information, the Chern–Simons action also carries useful information about boundary behaviour for samples with boundaries. To see what this information is, consider first the action S_a for a_μ without its coupling to electromagnetic fields:

$$S_a = \frac{ke^2}{4\pi} \int_{\mathcal{M}} \mathrm{d}^3x\, \epsilon^{\mu\nu\lambda} a_\mu \partial_\nu a_\lambda, \tag{15.79}$$

where $\mathcal{M}$ denotes the 2+1 dimensional quantum Hall bulk. At the classical level the variation of S_a with respect to δa_μ has the form

$$\delta S_a = \frac{ke^2}{2\pi} \int_{\mathcal{M}} \mathrm{d}^3x\, \epsilon^{\mu\nu\lambda} \delta a_\mu \partial_\nu a_\lambda + \frac{ke^2}{4\pi} \oint_{\partial\mathcal{M}} \mathrm{d}^2x\, \epsilon^{\mu\nu\lambda} a_\mu n_\nu \delta a_\lambda, \tag{15.80}$$

and so requiring $\delta S_a = 0$ for variations in the bulk $\mathcal{M}$ that vanish on the boundary implies that $f_{\mu\nu} = 0$ throughout the bulk (and so, locally, $a_\mu = \partial_\mu \varphi$ for some φ). Similarly, requiring $\delta S_a = 0$ for variations on the boundary $\partial\mathcal{M}$ implies $\epsilon^{\mu\nu\lambda} a_\mu n_\nu \delta a_\lambda = 0$, which says that the components of a_μ and δa_μ that are tangent to the boundary should be parallel to one another. Choosing coordinates $\{t, x, y\}$ throughout the bulk and coordinate $\{t, x\}$ on the boundary, stationarity with respect to variations at the boundary is therefore ensured if

$$a_t + v\, a_x = 0 \quad (\text{on } \partial\mathcal{M}) \tag{15.81}$$

for any value of the parameter v.

The action S_a is only invariant under the gauge transformation $\delta a_\mu = \partial_\mu \omega$ if ω vanishes on $\partial\mathcal{M}$. More generally, one finds

$$\delta S_a = \frac{ke^2}{4\pi} \oint_{\partial\mathcal{M}} \mathrm{d}^2x\, \epsilon^{\mu\nu\lambda} \omega\, n_\mu \partial_\nu a_\lambda. \tag{15.82}$$

Because of this failure of gauge invariance, some of the would-be gauge modes of a_μ become physical, but only on the boundary. As mentioned above, these modes locally have the form $a_\mu = \partial_\mu \varphi$ for a Stueckelberg scalar degree of freedom, φ, whose dynamics is localized on the boundary. The classical equations of motion for the boundary mode φ can be read off from (15.81), which implies that

$$(\partial_t + v\,\partial_x)\varphi = 0, \tag{15.83}$$

whose chiral solutions describe wave motion only in a single direction along the boundary: $\varphi = \varphi(x - vt)$, with speed v.

An action that reproduces this equation of motion for such a field has the form [424]

$$S_\varphi = \frac{ke^2}{4\pi}\int_{\partial M} dt\, dx\left[\partial_t\varphi\,\partial_x\varphi + v\,(\partial_x\varphi)^2\right]. \tag{15.84}$$

These arguments show how the failure of S_a to be gauge invariant under the emergent gauge transformations of the field a_μ again implies the existence of chiral edge modes for fractional quantum Hall states, though in a somewhat more constructive way.

Turning on now the coupling of a_μ to electromagnetic fields – using (15.75) – reveals the role of these edge states in transporting electric charge. In particular, the electromagnetic coupling term in (15.75) does not change the boundary condition (15.81) for a_μ, and so also does not modify the chiral evolution equation (15.83). Varying (15.75) with respect to A_μ gives

$$\delta S'_{\rm CS} = \frac{e^2}{2\pi}\int_M \mathrm{d}^3x\, \epsilon^{\mu\nu\lambda}\partial_\nu a_\lambda\,\delta A_\mu + \frac{e^2}{2\pi}\oint_{\partial M}\mathrm{d}^2x\,\epsilon^{\mu\nu\lambda} n_\nu a_\lambda\,\delta A_\mu, \tag{15.85}$$

which reveals both the bulk and the boundary contributions to the electric current:

$$J^\mu = \frac{e^2}{4\pi}\epsilon^{\mu\nu\lambda} f_{\nu\lambda}\quad \text{(bulk)} \quad\text{and}\quad j^\mu = \frac{e^2}{2\pi}\epsilon^{\mu\nu\lambda} n_\nu a_\lambda \quad \text{(boundary)}, \tag{15.86}$$

and so, in particular, the boundary electromagnetic current is related to φ by

$$j^t = -\frac{e^2}{2\pi}a_x = -\frac{e^2}{2\pi}\partial_x\varphi \quad\text{and}\quad j^x = \frac{e^2}{2\pi}a_0 = \frac{e^2}{2\pi}\partial_t\varphi = -\frac{e^2 v}{2\pi}\partial_x\varphi, \tag{15.87}$$

and the last equality uses the equation of motion (15.83). Notice that the chiral nature of φ evolution guarantees that $j^x = v j^t$.

Bulk Quasiparticles

A final topic in quantum Hall physics explores the implications of the emergent gauge field a_μ for the behaviour of charged quasiparticle excitations. To this end, imagine introducing a first quantized version of a quasiparticle in the quantum Hall fluid that couples to a_μ through an interaction of the form

$$\mathfrak{L}_{\rm int} = a_\mu \mathcal{J}^\mu, \tag{15.88}$$

where for a quasiparticle localized at rest within the bulk we have

$$\mathcal{J}^0(\mathbf{x}) = e\,\delta^2(\mathbf{x}) \quad\text{and}\quad \mathcal{J}^1 = \mathcal{J}^2 = 0, \tag{15.89}$$

and so, in particular, $\partial_\mu \mathcal{J}^\mu = 0$.

Combining this with the Chern–Simons action, (15.79), for a_μ, leads to the classical saddle-point configuration

$$\frac{ke^2}{2\pi} f_{\mu\nu} + \epsilon_{\mu\nu\lambda}\mathcal{J}^\lambda = 0, \tag{15.90}$$

which, using (15.89), implies that $f_{a0} = 0$ and (recalling $\epsilon_{012} = -1$ – *c.f.* Appendix A.2.1)

$$f_{12} = \frac{2\pi}{ke^2}\mathcal{J}^0 = \frac{2\pi}{ek}\,\delta^2(\mathbf{x}). \tag{15.91}$$

This shows that the effect of coupling a quasiparticle to the Chern–Simons gauge field a_μ is to attach a flux line – containing $2\pi/(ek)$ units of flux – to the position of the quasiparticle. This combination of a particle with flux for the Chern–Simons field is sometimes called a 'composite' particle.

What is important is that the flux of a_μ gives the quasiparticle fractional electric charge and fractional statistics. For instance, using the bulk electromagnetic current given in (15.86) shows the quasiparticle's electric charge density is

$$J^0 = \frac{e^2}{2\pi} f_{12} = \frac{e}{k}\,\delta^2(\mathbf{x}) \quad \text{(filling fraction } \nu = 1/k), \tag{15.92}$$

indicating it carries $1/k$ units of charge in the filling fraction $\nu = 1/k$ state. The fractional statistics is similarly obtained using the Aharonov–Bohm effect [425], which states that moving a point charge e_q in a large circle, C, around (but well outside) a flux tube carrying magnetic flux Φ, gives the wave-function a phase

$$\mathcal{P} = \exp\left[ie_q \oint_C \mathrm{d}x^\mu A_\mu\right] = \exp\left[ie_q \Phi(C)\right]. \tag{15.93}$$

For the present case the point source itself carries a_μ charge $e_q = e$ – *c.f.* Eq. (15.89) – and a_μ flux $\Phi = 2\pi/(ek)$ – *c.f.* Eq. (15.91) – so this argument shows that moving the point-charge of one quasiparticle in a large circle about the flux of another quasiparticle brings the state a total phase $\mathcal{P} = \exp[ie(2\pi/ek)] = \exp[2\pi i/k]$. The statistics phase is just half of this because exchanging the position of two particles means only moving them half of the way around the circle. The upshot is the quasiparticle acquires a statistics phase

$$\mathcal{P}_s = \exp[i\pi/k] \quad \text{(filling fraction } \nu = 1/k), \tag{15.94}$$

making them behave like *anyons* having fractional statistics [426]: in this case, the statistics phase of $1/k$ of a fermion.[12]

Both (15.92) and (15.94) agree with microscopic calculations of the charge and statistics of the low-lying excitations about a fractional quantum Hall state with filling fraction $\nu = 1/k$. In the microscopic theory there is a hierarchical framework [427, 428] for understanding the many quantum Hall fractions $\nu = p/q$ in real systems. In this picture the principle Laughlin series of plateaux with $\nu = 1/k$ (and $k = 2n + 1$ odd) are regarded as arising due to the condensation of the

[12] Interchanging two identical particles in a quantum state gives it a phase $|12\rangle = e^{i\theta}|21\rangle$, with bosons and fermions corresponding to the cases $e^{i\theta} = \pm 1$. If this phase is not ± 1 the particle statistics is said to be fractional. Fractional statistics is possible in two spatial dimensions (but not in $d \geq 3$ spatial dimensions).

underlying electrons into a strongly correlated state. The low-energy quasiparticles are then vortices in the resulting quantum Hall fluid and this turns out to give them the fractional charge and statistics computed above. In this language quantum Hall plateaux are plateaux precisely because the effect of adding more magnetic field is to spawn more vortices in the fluid rather than to change its filling fraction. Much like for the Type II superconductors described in §14.2.2, this process continues until eventually there are so many vortices that these vortices themselves condense. The resulting condensed fluid corresponds to a fractional quantum Hall state with filling fraction p/q with $p \neq 1$. Its excitations will also contain vortices with different statistics and charges, which can themselves condense and so on and so on.

Notice that in 2+1 dimensions particles and vortices have the same first-quantized kinematical description since they are both characterized by their two-dimensional centre-of-mass coordinates, plus their charge and their statistics. This suggests the possiblity that the various quantum Hall phases might all resemble one another, with each one being obtainable from the others through a 'duality' transformation that swaps out one sort of particle/vortex charge carrier and replaces it with another (with an associated calculable relationship between their conductivities [420, 429]). The existence of such symmetries – initially proposed for real systems on phenomenological grounds [430, 431] – appears to be borne out theoretically by more and more robust understandings of how duality can arise in 2+1 dimensional conformal field theories [432], and to be supported by some observational evidence [433].

15.4 Summary

This chapter describes two types of EFT descriptions for many-body non-relativistic systems involving degenerate fermions.

The first system studied involves the physics of a Fermi liquid: the phenomenon whereby the mutual interactions of statistically degenerate fermions become less and less important at lower energies. Physically, the interactions become weak at low energies because of 'Pauli blocking': scattering by interactions is reduced because of the presence of fermions blocking access to would be final states. Within the EFT language the same information gets expressed by the fact that generic interactions are irrelevant once the kinematics of proximity to a Fermi surface is included. As a result, interactions scale to zero at low energies and become less and less important. This phenomenon is what underlies the utility of free-fermion models of electronic transport in materials, and of independent nucleon models in nuclei.

The classification of the scaling of low-energy interactions near a Fermi surface also identifies a few exceptions to the generic irrelevance of low-energy interactions. These exceptions are important, and ultimately lead to instabilities at low energies that underly many interesting many-body phenomena like charge- or spin-density waves and superconductivity. The case of the superconducting BCS instability is explored in more detail and in particular the issue of why feeble phonon-mediated interactions can compete successfully with Coulomb interactions is assessed using EFT methods, which provide a simple understanding of why the predictions of these theories can be so robust.

Quantum Hall systems are the second system to be studied using EFT methods. The use of Chern–Simons theories as low-energy approximations to these systems is shown to capture the robustness of many novel features of both integer and fractional quantum Hall materials. The appearance of emergent gauge fields is shown to play an important role, in particular by providing the mechanism whereby the low-energy theory incorporates some novel physical properties, like topological order, in these systems.

Exercises

Exercise 15.1 For continuum normalized momentum eigenstates show that the following matrix element evaluates to

$$\langle N+1|a^\star|N\rangle = \sqrt{1 \pm f},$$

where $f(\mathbf{p})/(2\pi)^3$ is the phase-space density of particles, and the upper (lower) sign corresponds to bosons (fermions).

Consider a scattering process $A(\mathbf{p}) + B(\mathbf{q}) \to A(\tilde{\mathbf{p}}) + B(\tilde{\mathbf{q}})$, mediated by a Hamiltonian $H = H_{\rm free} + H_{\rm int}$ with

$$H_{\rm free} = E_0 + \int \mathrm{d}^3 p \Big[\varepsilon_A(p)\, a_p^\star a_p + \varepsilon_B(p)\, b_p^\star b_p \Big]$$

$$\text{and} \quad H_{\rm int} = \int \mathrm{d}^3 p\, \mathrm{d}^3 q\, \mathrm{d}^3 \tilde{p}\, \mathrm{d}^3 \tilde{q} \Big[\mathfrak{h}(p,q,\tilde{p},\tilde{q})\, a_{\tilde{p}}^\star a_p\, b_{\tilde{q}}^\star b_q + \mathfrak{h}^*(p,q,\tilde{p},\tilde{q}) a_p^\star a_{\tilde{p}}\, b_q^\star b_{\tilde{q}} \Big]\, \delta^3(\mathbf{p}+\mathbf{q}-\tilde{\mathbf{p}}-\tilde{\mathbf{q}}).$$

Suppose this two-body scattering process, $A + B \to A + B$, occurs when N_A particles of type A encounter N_B particles of type B. Compute the rate predicted by Fermi's Golden Rule (see Eq. (B.45)) for the transition for the differential transition to final-state particles within a small momentum-space volume $\mathrm{d}^3\tilde{p}$ and $\mathrm{d}^3\tilde{q}$ of two specific momenta $\tilde{\mathbf{p}}$ and $\tilde{\mathbf{q}}$. Show that this result is proportional to the system volume, $\mathcal{V}$, so it is the transition rate per unit volume that is well-behaved in the $\mathcal{V} \to \infty$ limit. The result for $\mathrm{d}\mathcal{R} := \mathrm{d}\Gamma/\mathcal{V}$ is

$$\mathrm{d}\mathcal{R}[A+B\to A+B] = \frac{1}{(2\pi)^2}\, |\mathfrak{h}(p,q,\tilde{p},\tilde{q})|^2\, \delta^4(p+q-\tilde{p}-\tilde{q}) \times f_A(p) f_B(q)[1 \pm f_A(\tilde{p})][1 \pm f_B(\tilde{q})]\, \mathrm{d}^3 p\, \mathrm{d}^3 q\, \mathrm{d}^3 \tilde{p}\, \mathrm{d}^3 \tilde{q},$$

with the upper (lower) sign applying for bosons (fermions).

As an application of the previous problem's result, take the special case of fermions where $\mathfrak{h} = \mathfrak{h}_0$ is momentum-independent, and the single-particle energies are $\varepsilon_A(p) = \mathbf{p}^2/(2m)$ and $\varepsilon_B(q) = \mathbf{q}^2/(2m)$. Suppose a single A particle is scattered from a collection of N_B B particles who singly occupy all of the energy levels up to the minimum required to contain all the particles. Show that this gives a spherical Fermi sea for B particles whose radius in momentum space, p_F, satisfies

$$n_B := \frac{N_B}{\mathcal{V}} = \int_0^{p_F} \frac{\mathrm{d}^3 p}{(2\pi)^3} = \frac{p_F^3}{6\pi^2},$$

and so the phase-space distribution is a step function,

$$f_B(\mathbf{p}) = \Theta(p_F - |\mathbf{p}|),$$

where $\Theta(x) = 0$ if $x < 0$ and $\Theta(x) = 1$ if $x > 0$, and so $1 - \Theta(x) = \Theta(-x)$.

The rate-per-A-particle, $d\mathcal{R}_A$, is found by dividing the rate-per-unit-volume $d\mathcal{R}$ by the number density $n_A = f_A/(2\pi)^3$. Use this to show that the scattering rate for the incident A particle of momentum $\mathbf{p}$ then is

$$d\mathcal{R}_A[A + B \to A + B] = 2\pi\, |\mathfrak{h}_0|^2\, \delta^4(p + q - \tilde{p} - \tilde{q}) \times \Theta(p_F - |\mathbf{q}|)\, \Theta(|\tilde{\mathbf{q}}| - p_F)\, d^3q\, d^3\tilde{p}\, d^3\tilde{q}.$$

The momenta in this problem must satisfy the four conditions

$$\mathbf{p} + \mathbf{q} = \tilde{\mathbf{p}} + \tilde{\mathbf{q}}, \qquad \mathbf{p}^2 + \mathbf{q}^2 = \tilde{\mathbf{p}}^2 + \tilde{\mathbf{q}}^2$$
$$|\mathbf{q}| < p_F \qquad \text{and} \qquad |\tilde{\mathbf{q}}| > p_F,$$

of which the first two express energy and momentum-conservation and the second two are the constraints of Pauli blocking. The total rate is obtained by integrating over the initial and final momenta, though this integration can be tricky due to the interlacing constraints of conservation and Pauli blocking. Derive the following approximate result in the particular limit $|\mathbf{p}| \ll p_F$,

$$\mathcal{R}_A[A + B \to A + B] \simeq \frac{16\pi^3 m}{15}\, |\mathfrak{h}_0|^2\, p^4,$$

for $p \ll p_F$. Notice this is much smaller than the naive result in the absence of Pauli blocking, for which one expects $\mathcal{R}_A \sim m|\mathfrak{h}_0|^2 p_F^4$.

Exercise 15.2 Evaluate the Feynman graph of Fig. 15.6 using the Feynman rules corresponding to the free action (15.7) and interaction (15.12) and (15.13). Use your result to derive (15.19). Specialize your result to the special case (15.20) and thereby derive the running formula (15.23).

Exercise 15.3 Use dimensional regularization to compute the momentum-space vacuum polarization caused by a Dirac fermion in 1+1 dimensions, and use this to show that the axial current, $J_A^\mu = i\overline{\psi}\gamma_3\gamma^\mu\psi$, satisfies the axial anomaly equation

$$\partial_\mu J_A^\mu = \frac{1}{2\pi}\, \epsilon^{ab} F_{ab},$$

in the presence of a background electromagnetic field. Show also that the vector current, $J_V^\mu = i\overline{\psi}\gamma^\mu\psi$, satisfies $\partial_\mu J_V^\mu = 0$. Use these results to show that the chiral currents $J_\pm^\mu := \frac{1}{2}\left(J_V^\mu \pm J_A^\mu\right)$ satisfy

$$\partial_\mu J_\pm^\mu = \pm\frac{1}{4\pi}\, \epsilon^{ab} F_{ab}.$$

Exercise 15.4 Consider the Cherns–Simons/Maxwell system in 2+1 dimensions for an abelian gauge potential a_μ. Propagation of a_μ waves is governed by the action

$$S_a = \int d^3x \left[\frac{k}{4\pi}\epsilon^{\mu\nu\lambda} a_\mu \partial_\nu a_\lambda - \frac{1}{4g^2} f_{\mu\nu} f^{\mu\nu}\right],$$

where $f_{\mu\nu} := \partial_\mu a_\nu - \partial_\nu a_\mu$. Assuming k is dimensionless what is the dimension of the field a_μ and of the coupling g in powers of energy? Show that the spectrum of waves predicted for this lagrangian exhibits an energy gap of size

$$E_{\text{gap}} = \frac{kg^2}{\pi}.$$

Show that the Chern–Simons action (just the first term of S_a given above) makes no contribution to the system Hamiltonian. One way to do this is to identify where the metric must go in the action to make it generally covariant and then differentiating the result with respect to $g_{\mu\nu}$ to learn its contribution to the stress energy $T^{\mu\nu}$.

Exercise 15.5 Consider the following Chern–Simons coupling between the electromagnetic potential A_μ and two emergent potentials a_μ and b_μ:

$$S = -\int d^3x \left[\frac{1}{2\pi}\epsilon^{\mu\nu\lambda} a_\mu \partial_\nu A_\lambda + \frac{1}{2\pi}\epsilon^{\mu\nu\lambda} b_\mu \partial_\nu a_\lambda + \frac{r}{4\pi}\epsilon^{\mu\nu\lambda} a_\mu \partial_\nu a_\lambda + \frac{s}{4\pi}\epsilon^{\mu\nu\lambda} b_\mu \partial_\nu b_\lambda\right]$$

and show that in topologically trivial configurations it predicts a filling fraction

$$\nu = \frac{1}{r + \frac{1}{s}} = \frac{s}{rs+1}. \tag{15.95}$$

This provides a concrete example of how the low-energy EFT can capture quantum Hall states with filling fraction $\nu = p/q$ with $p \neq 1$.

16 EFTs and Open Systems

A common situation in physics involves measurements that sample only a subset of a system's degrees of freedom. Indeed, having measurements restricted only to a relatively restricted set of variables is typically the rule rather than the exception in physics. A broad class of examples of this type describes the properties of particles moving through almost any type of medium – *e.g.* visible light moving through transparent materials; X-rays irradiating flesh; or neutrinos passing through the Sun or Earth. The focus is often on the properties of the moving particle and less interest in the medium's response to the particle (though not always, as in the example of charged particles moving through the interior of a particle detector). Although the explicit examples given below are usually of this particle-in-a-medium type, the division of systems into a measured subsystem and an unmeasured 'environment' is much more general.

Such problems lend themselves to effective descriptions that describe the properties of the measured subsystem (*e.g.* the moving particle) after coarse-graining over the properties of the unmeasured environment. Although the resulting description shares many features with the Wilsonian effective action used in previous sections, it is not identical and the differences are instructive. This chapter describes the framework within which these types of systems are efficiently described.

The main qualitative difference between these kinds of effective descriptions and the Wilsonian formulation of previous chapters lies in the criteria used to distinguish the part of the system on which measurements are performed (call it system A) from the unmeasured environment (system B). On one hand, in Wilsonian treatments the measured subsystem consists of all low-energy degrees of freedom (or perhaps all low-energy degrees of freedom also carrying a given amount of a conserved quantum number, like charge or baryon number). For more general 'open' systems, on the other hand, the division between the measured system and its environment is not defined in terms of conserved quantities.

For Wilsonian systems the restriction to states defined in terms of conserved things like energy largely ensures that degrees of freedom do not over time cross back and forth[1] between systems A and B. Energy conservation ensures that the time evolution of states restricted to the low-energy sector agrees with the time evolution of the full system. And there is no doubt that this evolution is described by a Hamiltonian, since a Hamiltonian description exists in the full theory. The only question is what the effective Hamiltonian (or effective action) looks like once expressed purely in terms of low-energy fields.

[1] As described in §6, this argument is more subtle when applied to time-dependent backgrounds because of the breaking of energy conservation that such backgrounds allow, and this lies at the root of the requirement, for Wilsonian treatments, that the evolution is adiabatic.

The reasoning is different for open systems, because information can flow between systems A and B and their quantum entanglement can evolve with time. In general, this makes tracking the evolution of system A much more difficult without also tracking how system B evolves, and there is no adiabatic theorem at work ensuring that sector-B evolution is comparatively simple. As a result, the properties of a state at any given time might depend on the entire history of its interactions with B in the past. In particular, time evolution restricted just to system A need not be described in terms of a simple effective Hamiltonian, for instance leading to non-Hamiltonian phenomena such as thermalization or decoherence.[2]

The Good News from this section is this: hierarchies of scale can simplify open systems just as they do in the Wilsonian case, but this simplification might not be usefully expressed in terms of an effective action (or effective Hamiltonian). Open EFTs as described in this section capture some of the ways in which this additional simplicity can occur, and how it can be efficiently exploited.

16.1 Thermal Fluids

A good place to start in this story is (in retrospect) probably also the earliest historical example where EFT methods were used: the thermodynamic description of the macroscopic features of statistical systems of mobile atoms, and its generalization to thermal fluids when these properties vary in space and time. Although the discussion of fluids lies somewhat outside this book's main line of development, it provides an important and relatively familiar example of how dissipative and open systems can be treated in a way that does not pre-assume the existence of a macroscopic Hamiltonian or an action formalism.

Everyday fluids can contain enormous numbers of particles, usually making it prohibitively complicated to describe their dynamics in detail. Things simplify, however, when one takes it slow: asking only how coarse-grained quantities evolve over times $t \gg \tau$, where τ is the typical time between collisions for particles in the fluid.

Conserved quantities play a special role when asking such long-time questions because conservation itself prevents the conserved quantity from changing locally due to collisions of the underlying atoms. In order to change a locally conserved quantity over a macroscopic distance in a fluid requires physically moving something through that distance; an inherently slow process that occurs for times $T \sim L/v$, where L is the distance of interest and v is a typical system speed.

Effective theories always profit by exploiting hierarchies of scale. For fluids (and open systems more generally) a natural hierarchy to exploit is this ratio of time-scales: T/τ.

[2] The absence of Hamiltonian evolution need not also preclude the existence of generalized actions, such as thermodynamic potentials or generating functionals for 1PI (and other) graphs. It also does not preclude there being a Hamiltonian description once sufficient numbers of additional degrees of freedom are 'integrated in' (usually without requiring the extreme case of including back all of the degrees of freedom in sector B).

16.1.1 Statistical Framework♡

The first step is to systematize the intuition that a focus on late times is best organized in terms of conservation laws. This section recaps two ways in which conserved quantities enter into the late-time description of thermal fluids.

The first argument is the one based on detailed balance that identifies the equilibrium distribution functions as being special because they allow reactions to run equally efficiently in both directions, regardless of the details of the form of the underlying microscopic interactions. This identifies these states as being of particular interest for describing situations that remain unchanged over times much longer than the frequency of microscopic collisions. The special role played by conserved quantities in the resulting distribution functions is what identifies the thermodynamic and geometrical variables whose time evolution can be usefully used to characterize the system's late-time coarse-grained behaviour.

The second role played by conservation laws comes from the differential equations that express their local conservation, since these are the equations that govern how the above thermodynamic variables can vary over long distances in space and time.

Equilibrium Distribution Functions

Consider first the constraints on the equilibrium phase-space single-particle distribution functions, $f(\mathbf{k}, \mathbf{x})$, that can be obtained by demanding detailed balance between all possible microscopic scattering processes. It is often the case that the joint distribution functions for multiple particles are products of single-particle distribution functions, and when this is so the rate of change of the single-particle distributions is controlled by the difference between the rates for scattering into and out of the relevant state.

For simplicity, consider a gas containing only one species of particle, with single-particle phase-space probability distribution function $f(\mathbf{k}, \mathbf{x})$ that satisfies

$$\int \frac{\mathrm{d}^3\mathbf{k}}{(2\pi)^3}\, f(\mathbf{k}, \mathbf{x}) = n(\mathbf{x}), \tag{16.1}$$

where $n(\mathbf{x})$ is the particle density. This distribution is initially imagined to be independent of position, though this assumption is re-examined in the next section. The rate of change of this distribution function due to particle collisions is

$$\partial_t f(\mathbf{k}) = \int \mathrm{d}\alpha\, \mathrm{d}\hat{\beta}\, \Big[R_{\rm in}(\alpha \to \beta) - R_{\rm out}(\beta \to \alpha)\Big], \tag{16.2}$$

where $|\beta\rangle$ is some many-particle state that includes a particle with the given particle momentum, $\mathbf{k}$, and $R_{\rm in}$ is the differential rate for producing this state from another state $|\alpha\rangle$ (with $R_{\rm out}$ similarly being the differential rate for scattering out of $|\beta\rangle$ to another state). Eq. (16.2) uses the notation $\mathrm{d}\alpha = \prod_i \mathrm{d}^3\mathbf{p}_i$ for $\alpha = \{\mathbf{p}_1, \cdots, \mathbf{p}_{N_\alpha}\}$ and similarly for the final state writes $\mathrm{d}\hat{\beta} = \prod_j \mathrm{d}^3\mathbf{k}_i$, where $\beta = \{\mathbf{k}, \mathbf{k}_2, \cdots, \mathbf{k}_{N_\beta}\}$. The caret on $\mathrm{d}\hat{\beta}$ indicates that the product runs only over $j \geq 2$ and leaves out the momentum $\mathbf{k}$.

The differential rates $R_{\rm in}$ and $R_{\rm out}$ themselves depend on the distribution functions $f(\mathbf{k})$, with

$$R(\alpha \to \beta) \propto |M(\alpha \to \beta)|^2 \prod_{i\in\alpha} f(\mathbf{p}_i) \prod_{j\in\beta} [1 \pm f(\mathbf{k}_j)], \tag{16.3}$$

which includes a factor of f for each particle in the initial state and a factor of $[1 \pm f]$ for each particle in the final state (with upper sign for bosons and lower sign for fermions). These factors arise from the action of creation/destruction operators on multiply-occupied states:

$$\mathfrak{a}|N\rangle = \sqrt{N}\,|N-1\rangle \quad \text{and} \quad \mathfrak{a}^*|N\rangle = \sqrt{1 \pm N}\,|N+1\rangle, \tag{16.4}$$

with $N = 0, 1$ for fermions and $N = 0, 1, 2, \cdots$ for bosons (see Appendices C.1 and B).

Now comes the main point. The equilibrium state we seek is the one for which the right-hand side of (16.2) vanishes for all momenta and general choices for the matrix-elements $M_{\rm in}(\alpha \to \beta)$ and $M_{\rm out}(\beta \to \alpha)$. Unitarity of the S-matrix implies [406] that the quantity $|M_{\rm in}(\alpha \to \beta)|^2 = |M_{\rm out}(\beta \to \alpha)|^2$ can be extracted as a common factor in (16.2), and once this is done the right-hand side generically vanishes if the f's satisfy the condition

$$\prod_{i\in\alpha} f(\mathbf{p}_i) \prod_{j\in\beta} [1 \pm f(\mathbf{k}_j)] = \prod_{i\in\alpha} [1 \pm f(\mathbf{p}_i)] \prod_{j\in\beta} f(\mathbf{k}_j), \tag{16.5}$$

for all choices of states $|\alpha\rangle$ and $|\beta\rangle$.

This last condition is easiest to solve if regrouped as

$$\prod_{i\in\alpha} \frac{f(\mathbf{p}_i)}{[1 \pm f(\mathbf{p}_i)]} = \prod_{i\in\beta} \frac{f(\mathbf{k}_i)}{[1 \pm f(\mathbf{k}_i)]}, \tag{16.6}$$

since this says that the quantity $\prod_i [f_i/(1 \pm f_i)]$ is conserved during all collisions. Consequently, its logarithm, the quantity $\chi := \sum_i \ln[f_i/(1 \pm f_i)]$, is both conserved and additive (that is, when evaluated for many-particle states it is given by the sum over the value it has for each particle separately: $\chi = \sum_i \chi_i$). A generic solution to (16.6) takes χ to be a linear combination of all of the system's additive conserved charges, such as energy, momentum, electric charge and any others. This gives

$$\chi = \beta_\mu P^\mu + \sum_a \xi_a Q^a = -\beta E + \boldsymbol{\beta} \cdot \mathbf{P} + \sum_a \beta \mu_a Q^a, \tag{16.7}$$

for some coefficients β^μ and ξ_a, where the second equality follows standard convention to write $\xi_a = \beta\mu_a$ where $\beta := \beta^0 = -\beta_0$.

Working in the fluid's rest frame, defined as the frame where the spatial components of β_μ satisfy $\boldsymbol{\beta} = 0$, and solving for f in terms of χ then gives the familiar equilibrium result:

$$f(\mathbf{k}_i) = \frac{1}{\exp[\beta(E_i - \mu_a q_{ai})] \mp 1}, \tag{16.8}$$

where the upper sign is for bosons and the lower sign is for fermions while $E_i = E(\mathbf{k}_i)$ and q_{ai} are the single-particle eigenvalues of the energy and charge Q_a for particle 'i'.

The coefficient $\beta = 1/T$ defines the equilibrium system's temperature, while the μ_a are its chemical potentials. Notice that, strictly speaking, chemical potentials are only defined in this way for conserved charges, which for relativistic systems count the *difference* between particles and antiparticles carrying a given charge.

It is a special feature of the nonrelativistic limit that conserved charges can effectively count particle number, since at low energies it can happen that there are no antiparticles present.[3]

Macroscopic Observables

The above derivation of the statistical distribution functions identifies those properties of the fluid state that remain unchanged by microscopic collisions, and so provide useful macroscopic state labels when tracking the evolution over time frames much longer than typical collision times. This suggests promoting the quantities $\beta(\mathbf{x}, t)$, $\boldsymbol{\beta}(\mathbf{x}, t)$ and $\mu_a(\mathbf{x}, t)$ to functions of position and time when tracking macroscopic fluid evolution. It is often convenient when doing so to define a local fluid 4-velocity, U_μ, satisfying $U^\mu U_\mu = -1$, such that $\beta_\mu =: \beta\, U_\mu$.

To this end, it is useful to coarse-grain the underlying microscopic theory by dividing the original fluid up into mutually exclusive cells, of size ℓ [434]. Since particles can move from cell to cell, each can separately be treated as a grand-canonical ensemble, within which conserved quantities like net charge and energy can fluctuate.

On one hand, the size ℓ must be chosen to be large enough that each cell contains sufficiently many particles that it is reasonably close to the thermodynamic limit (for which fluctuations of thermodynamic quantities can be neglected). This allows the state of the fluid within any given cell to be well-described by thermodynamic averages.

On the other hand, ℓ must be chosen small enough that the various thermodynamic quantities can be regarded as being constant across any given cell. This implies that all thermodynamic relations, such as equations of state and so on, are local in that they do not involve derivatives of thermodynamic quantities. For instance, the equation of state for an ideal gas is $p = nT$, which in local equilibrium can be taken to hold point-wise for all $\mathbf{x}$ and for all times, once coarse-grained over scales much smaller than ℓ.

The existence of a range for ℓ satisfying both criteria is possible precisely because of the existence of a hierarchy of scales; by assumption these tools are only to be used to study fluid evolution over very long times and distances compared with microscopic collision lengths and times.

16.1.2 Evolution through Conservation

The above argument identifies the relevant macroscopic fluid variables, but what governs how these variables evolve in space and time? As motivated above, their macroscopic evolution is governed by local conservation laws [434, 435].

Key conservation laws for these purposes are the conservation of energy and momentum. At a local level these are expressed (courtesy of Noether's theorem) through stress-energy conservation, which states $\partial_\mu T^{\mu\nu} = 0$, for the fluid's stress-energy tensor $T^{\mu\nu} = T^{\nu\mu}$. Current conservation for any internal symmetries,

[3] Chemical potentials are also sometimes used for charges that are only approximately conserved, assuming the symmetry-breaking interactions are weak enough that their equilibration time is much longer than the time scales of interest.

$\partial_\mu J^\mu_a = 0$, is also useful for systems blessed with additional conserved charges (and so possibly also chemical potentials, μ_a).

In order to use conservation laws like these one must first express the currents in terms of the macroscopic variables, β, μ_a and U_μ. Keeping in mind that applications are restricted to long distances and late times this expression is usefully organized into a derivative expansion, with derivative-free terms being dominant.

The most general forms allowed by spacetime symmetries in the absence of derivative terms are

$$T_{\mu\nu} = p\,\eta_{\mu\nu} + (p+\rho)U_\mu U_\nu \quad \text{and} \quad J^\mu_a = n_a\, U^\mu, \tag{16.9}$$

where $p = p(\beta, \mu_a)$, $\rho = \rho(\beta, \mu_a)$ and $n_b(\beta, \mu_a)$ are as-yet unspecified functions of the scalar variables β and μ_a (or, equivalently, $T = 1/\beta$ and μ_a).

To interpret their physical meaning it is useful to define the local fluid velocity by writing $U^\mu =: \{\gamma, \gamma\,\mathbf{v}\}$, where $\gamma := (1-\mathbf{v}^2)^{-1/2}$ is chosen to ensure $U^\mu U_\mu = -1$. Then the components of $T^{\mu\nu}$ and J^μ_a become

$$T^{00} = \frac{\rho + p\,\mathbf{v}^2}{1-\mathbf{v}^2}, \qquad T^{0i} = \frac{(p+\rho)v^i}{1-\mathbf{v}^2}, \qquad T^{ij} = p\,\delta^{ij} + \frac{(p+\rho)v^i v^j}{1-\mathbf{v}^2},$$

$$J^0_a = \frac{n_a}{\sqrt{1-\mathbf{v}^2}} \quad \text{and} \quad J^i_a = \frac{n_a\, v^i}{\sqrt{1-\mathbf{v}^2}}. \tag{16.10}$$

Evaluated in the fluid rest frame, these reveal ρ to be the energy density, p is the pressure and n_a is the charge density for the conserved quantity associated with chemical potential μ_a.

Conservation of these currents – *i.e.* the conditions $\partial_\mu T^{\mu\nu} = \partial_\mu J^\mu_a = 0$ – then provides a set of evolution equations for the macroscopic variables. In particular, the vanishing of $\partial_\mu T^{\mu i} - v^i \partial_\mu T^{\mu 0}$ implies that

$$D_t \mathbf{v} = -\frac{1-\mathbf{v}^2}{p+\rho}\left[\nabla p + \partial_t p\, \mathbf{v}\right], \tag{16.11}$$

where $D_t := \partial_t + \mathbf{v}\cdot\nabla$ is the 'convective' time derivative. Similarly, conservation of J^μ_a gives

$$\partial_\mu(n_a U^\mu) = \partial_t\left[\frac{n_a}{\sqrt{1-\mathbf{v}^2}}\right] + \nabla\cdot\left[\frac{n_a\,\mathbf{v}}{\sqrt{1-\mathbf{v}^2}}\right] = 0 \tag{16.12}$$

and $U_\nu \partial_\mu T^{\mu\nu} = 0$ is equivalent to

$$U^\mu \partial_\mu \rho + (p+\rho)\partial_\mu U^\mu = 0. \tag{16.13}$$

Using (16.12) to eliminate $\partial_\mu U^\mu$ from (16.13) then gives, after a bit of manipulation, the ultimately more useful formula

$$n_a\left[p\, D_t\left(\frac{1}{n_a}\right) + D_t\left(\frac{\rho}{n_a}\right)\right] = 0. \tag{16.14}$$

Navier-Stokes Equations

These expression become very familiar in the nonrelativistic limit, for which $n_a \to n$ often goes over to the total particle number density, and where the overall fluid speed is small, $\mathbf{v}^2 \ll 1$, as is the speed of individual particles. Slow particle speeds ensure $\rho - nm \sim nm\langle\mathbf{v}^2\rangle$, in order of magnitude, since particle kinetic energies are dominated by their rest mass. Similarly, $p \sim nm\langle \mathbf{v}_i^2 \rangle$, since pressure is of order the average internal atomic kinetic energies. As a result, $p/\rho \sim O(\langle \mathbf{v}_i^2 \rangle) \ll 1$ in this limit.

If so, (16.11) and (16.12) are recognizable as old friends. First, (16.11) reduces to the familiar Navier–Stokes equation [435] (with no viscosity – more about which below), which governs the fluid's nonrelativistic momentum flow:

$$\partial_t \mathbf{v} + (\mathbf{v} \cdot \nabla)\mathbf{v} \simeq -\frac{1}{\rho} \nabla p. \tag{16.15}$$

Eq. (16.12) similarly becomes the continuity equation describing conservation of particle number,

$$\partial_t n + \nabla \cdot (n\,\mathbf{v}) \simeq 0. \tag{16.16}$$

Eq. (16.14) also invites emotional involvement when $n_a \to n$. In this case, $\mathfrak{u} := \rho/n$ can be regarded as the energy per particle while $\mathfrak{v} := 1/n$ is the volume per particle, and both are the thermodynamic quantities relevant for a fluid region that follows a fixed number of fluid particles (whose volume typically varies due to the fluid motion). Since (by construction) such a volume has a fixed number of particles, the entropy included in it ($\mathfrak{s} = s/n$, if s is the entropy per unit volume) then satisfies $T\mathrm{d}\mathfrak{s} = \mathrm{d}\mathfrak{u} + p\,\mathrm{d}\mathfrak{v}$, and so

$$T\mathrm{d}\mathfrak{s} = p\,\mathrm{d}\left(\frac{1}{n}\right) + \mathrm{d}\left(\frac{\rho}{n}\right). \tag{16.17}$$

Comparing this with (16.14) then shows

$$D_t \mathfrak{s} := \partial_t \mathfrak{s} + \mathbf{v} \cdot \nabla \mathfrak{s} = 0, \tag{16.18}$$

indicating that conservation of the stress-energy given in (16.9) and (16.10) implies that entropy remains constant for fluid regions that follow particles in the flow.

These arguments show why conservation of the non-derivative part of $T_{\mu\nu}$ describes a perfect fluid [435]. Perfect fluids can also be captured using an action formalism [436–441], along the lines described in §14.3.3. (See also [399] for a discussion of imperfect fluids.)

Although it goes beyond the scope of this book, imperfect fluids can be included in the above treatment by extending Eqs. (16.9) or (16.10) to include terms involving derivatives of the fields, showing there to be no fundamental obstacle to including dissipation along these lines [397, 442]. Robustness of description is a great virtue of this kind of treatment, which bases the long-distance description on the minimal number of physical degrees of freedom and conservation laws (as opposed, say, to adding new fields in search of an action formalism).

Rather than pursuing these issues further for fluids in the hydrodynamic regime, the remainder of this chapter instead focusses on setting up a more general and systematic EFT-like description for open quantum systems.

16.2 Open Systems

This section describes the underlying framework for splitting a quantum system into sectors A and B. To keep applications general (and approximations explicit) this split is first formulated in its most general context [443].

To that end, suppose the quantum system of interest consists of two sectors, A and B, but with measurements only performed in sector A while sector B remains unmeasured. In the spirit of EFT methods the goal is to describe purely within sector A how measurement results evolve for such systems, removing sector B at the outset once and for all in terms of a few appropriately defined effective interactions. In general, such an approach might be prohibitively complicated, but experience with effective theories suggests that it could dramatically simplify in the presence of some sort of hierarchy of scale (more about which below). To keep things concrete in the examples that follow sector A might describe the momentum and/or flavour of a particle moving through a medium, whilst sector B describes all the degrees of freedom of the medium itself.

In practice, it is sometimes useful to consider a slightly more general case where a partial measurement is made on sector B, in addition to the measurements performed in sector A. This can be useful because sometimes it is desirable to focus in sector A on a particle's spin or flavour degrees of freedom (such as for neutrino oscillations) and to lump its momentum in with all of the other environmental degrees of freedom. Yet, clearly, flavour measurements are actually done at specific locations (for instance on Earth) so some implicit momentum information is also included in real measurements (and sometimes neglecting this partial information can lead to misleading conclusions). The slightly more general formulation required to analyze this situation is described in reference [444].

For most later applications it suffices to consider the case where states in the full system's Hilbert space can be written as sums of products of states in sectors A and B, so a basis of states in the full system can be decomposed as

$$|a, b\rangle = |a\rangle \otimes |b\rangle. \tag{16.19}$$

As ever in quantum mechanics, observables are Hermitian operators, and in this case observables involving only sector A can be written[4] $\mathcal{O}_A = O_A \otimes I_B$, with matrix elements $\langle a, b|\mathcal{O}_A|a', b'\rangle = \langle a|O_A|a'\rangle\, \delta_{bb'}$.

16.2.1 Density Matrices ♡

Returning to the problem of describing how measurements evolve when performed only in sector A, it is useful first to recall the distinction between pure and mixed states, since this is important in what follows. Here is a brief review of what these are, and how time-evolution works for both.

Imperfect knowledge means that the exact state of a system cannot be precisely pinned down, and when this is true the system is described by a density matrix, $\rho(t)$. For instance, suppose the system's state is $|\psi\rangle = |I\rangle$ with probability p_I, where $I = \{a, b\}$, provides labels for a complete basis in the full Hilbert space. Then the density matrix describing these probabilities is the operator[5]

$$\hat{\rho} := \sum_I p_I\, |I\rangle\langle I| := \sum_{ab} p_{ab}\, |a, b\rangle\langle a, b|. \tag{16.20}$$

4 Here, and in what follows, I_A (or I_B) denotes the unit operator acting in sector A (or B).
5 Hats are temporarily added here to Schrödinger-picture operators, to distinguish them from the interaction-picture quantities of most interest in later sections.

In this language the requirement that mutually exclusive probabilities sum to unity, $\sum_I p_I = 1$, becomes $\text{Tr}\,\hat{\rho} = 1$.

In the special case where the state $|\psi\rangle$ is exactly known (*i.e.* when the system is in a 'pure' state) it follows that

$$\hat{\rho} = |\psi\rangle\langle\psi| \quad \text{(pure state)}. \tag{16.21}$$

Whenever (16.21) holds for some $|\psi\rangle$ it implies that $\hat{\rho}^2 = \hat{\rho}$. Because the eigenvalues of $\hat{\rho}$ are non-negative probabilities that sum to unity, whenever $\hat{\rho}^2 = \hat{\rho}$ there must exist a state for which (16.21) holds. When $\hat{\rho}^2 \neq \hat{\rho}$ the state is said to be 'mixed' and there is no $|\psi\rangle$ for which (16.21) is true.

As usual in quantum mechanics, observables are hermitian operators, $\hat{O} = \hat{O}^*$, and for a system in a definite state, $|\psi\rangle$, the mean outcome of a series of measurements of O is given by $\langle\hat{O}\rangle = \langle\psi|\hat{O}|\psi\rangle$. For the more general mixed state one must also average over the probability of being in any particular state, so the expectation becomes

$$\langle\hat{O}\rangle = \sum_I p_I \langle I|\hat{O}|I\rangle = \text{Tr}\,(\hat{\rho}\,\hat{O}), \tag{16.22}$$

where $\hat{\rho}$ is the system's density matrix.

Next consider how observables like (16.22) evolve in time. This is described here in the Schrödinger picture, at least to start with, though a transition is made to the (often more practical) interaction picture in later sections. For systems prepared in a specific state, $|\psi(t)\rangle$, time evolution is given in the Schrödinger picture by

$$\mathrm{i}\,\partial_t|\psi(t)\rangle = H|\psi(t)\rangle, \tag{16.23}$$

where H is the system's Hamiltonian.[6] In this picture, operators describing observables are time-independent.

Since observables do not depend on time in the Schrödinger picture, when a system's quantum state is imperfectly known its time evolution is given by knowing how its density matrix, $\hat{\rho}$, evolves with time. Once this is known, it completely specifies how all measurement outcomes – such as (16.22) – change over time.

The time-evolution that $\hat{\rho}$ inherits from (16.23) is called the Liouville equation [445]

$$\mathrm{i}\,\partial_t\hat{\rho} = \left[H\,,\hat{\rho}\,\right], \tag{16.24}$$

and is the starting point for all later discussions about how evolution looks once restricted to sector A. Eq. (16.24) is formally solved by

$$\hat{\rho}(t) = \hat{U}(t,t_0)\,\hat{\rho}_0\,\hat{U}^*(t,t_0), \tag{16.25}$$

where hermiticity of H ensures $\hat{U}(t,t_0) = \exp[-\mathrm{i}H(t-t_0)]$ is unitary, and so satisfies $\hat{U}(t,t_0)\hat{U}^*(t,t_0) = 1$. Furthermore, $\hat{U}(t,t_1)\hat{U}(t_1,t_0) = \hat{U}(t,t_0)$, while

$$\mathrm{i}\,\partial_t\hat{U}(t,t_0) = H\,\hat{U}(t,t_0). \tag{16.26}$$

Regarded as a differential equation for $U(t,t_0)$ this equation should be solved subject to the initial condition $\hat{U}(t_0,t_0) = I$.

6 No hats are used here since H is both the Schrödinger-picture and the Heisenberg-picture Hamiltonian. It need not agree with the interaction-picture Hamiltonian.

Reduced Density Matrix

The next step is to explore how to understand measurements restricted to sector A in a way that refers as much as possible only to the measured sector A. A key tool to this end is the 'reduced' density matrix, defined by tracing the full density matrix over the unmeasured sector B:

$$\hat{\rho}_A := \mathrm{tr}_B\, \hat{\rho} \quad \text{so} \quad \langle a|\, \hat{\rho}_A\, |a'\rangle = \sum_b \langle a, b|\, \hat{\rho}\, |a', b\rangle, \tag{16.27}$$

which defines tr_B as the trace restricted to sector B. The point of this definition is that the matrix $\hat{\rho}_A$ captures the information in the full state relevant to measurements in sector A.

The reduced density matrix, in particular, controls the time-dependence of measurements for any observable, $\hat{O}_A := \hat{O}_A \otimes I_B$, that acts only in sector A. Since, in Schrödinger picture, the burden of time evolution lies with the state, the evolution of the expectation value of a measurement is

$$\begin{aligned} \langle \hat{O}_A \rangle := \mathrm{Tr}\,[\hat{\rho}(t)\, \hat{O}_A] &= \sum_{aa'} \sum_{bb'} \langle a, b|\, \hat{\rho}\, |a', b'\rangle \langle a', b'|\, \hat{O}_A\, |a, b\rangle \\ &= \sum_{aa'} \sum_{b} \langle a, b|\, \hat{\rho}\, |a', b\rangle \langle a'|\, \hat{O}_A\, |a\rangle = \mathrm{tr}_A\, [\hat{\rho}_A(t)\, \hat{O}_A]. \end{aligned} \tag{16.28}$$

Here, the first line inserts a complete set of states and the second line uses $\langle a', b'|\hat{O}_A|a, b\rangle = \langle a'|\hat{O}_A|a\rangle\, \delta_{bb'}$.

16.2.2 Reduced Time Evolution$^\diamond$

The previous section shows how the time-evolution of *any* measurement restricted to A is determined purely in terms of operators acting only in sector A if the evolution of $\hat{\rho}_A(t)$ is known. This puts a premium on understanding how $\hat{\rho}_A(t)$ evolves, and in particular on its sensitivity to any interaction that couples sector A to sector B.

In practice, explicit progress often relies on perturbative methods, and in this section it is the interaction between sectors A and B that is treated perturbatively. Suppose, therefore, that the Hamiltonian for the full system has the form

$$H = H_0 + \hat{H}_{\rm int} = \mathcal{H}_A + \mathcal{H}_B + \hat{H}_{\rm int}, \tag{16.29}$$

where $H_0 = \mathcal{H}_A + \mathcal{H}_B$ describes the separate evolution of sectors A and B,

$$\mathcal{H}_A = H_A \otimes I_B \quad \text{and} \quad \mathcal{H}_B = I_A \otimes H_B, \tag{16.30}$$

and $\hat{H}_{\rm int}$ is the interaction that couples the two sectors together. The goal is to integrate the Liouville equation, (16.24), in powers of $\hat{H}_{\rm int}$.

Interaction Picture

Integration is conveniently done in the interaction picture, defined by performing the unitary transformation $U_0 = \exp[iH_0 t]$ on all states and operators. In particular, the interaction-picture interaction Hamiltonian is denoted V and given by

$$V(t) := e^{iH_0 t}\, \hat{H}_{\rm int}\, e^{-iH_0 t}, \tag{16.31}$$

and the interaction-picture density matrix becomes

$$\rho(t) := e^{iH_0 t}\,\hat{\rho}(t)\,e^{-iH_0 t}, \tag{16.32}$$

which, together with the Liouville equation (16.24) implies that

$$\partial_t\,\rho = -i\big[V(t), \rho\big]. \tag{16.33}$$

This integrates to give the useful integral form

$$\rho(t) = \rho(t_0) - i\int_{t_0}^{t} d\tau\,\big[V(\tau)\,,\rho(\tau)\big], \tag{16.34}$$

showing how an iterative solution can be generated in powers of $V(t)$. Equivalently, the evolution operator $U(t, t_0)$ is defined in the interaction picture by $\rho(t) = U(t, t_0)\,\rho(t_0)\,U^*(t, t_0)$, and so satisfies

$$\partial_t U(t, t_0) = -iV(t)\,U(t, t_0), \tag{16.35}$$

with initial condition $U(t_0, t_0) = 1$. This integrates to give

$$U(t, t_0) = 1 - i\int_{t_0}^{t} d\tau\,V(\tau)\,U(\tau, t_0), \tag{16.36}$$

which when solved iteratively generates the usual perturbative series solution

$$U(t, t_0) = \sum_{n=0}^{\infty}(-i)^n\int_{t_0}^{t} d\tau_1\int_{t_0}^{\tau_1} d\tau_2\cdots\int_{t_0}^{\tau_{n-1}} d\tau_n\;V(\tau_1)\cdots V(\tau_n). \tag{16.37}$$

Evolution of ρ_A

The above formalism for the evolution of $\rho(t)$ can be used to see how the reduced density matrix, $\rho_A(t)$, evolves in time. Direct insertion of (16.34) into the definition (16.27) gives

$$\rho_A(t) = \rho_A(t_0) - i\int_{t_0}^{t} d\tau\;\mathrm{tr}_B\big[V(\tau)\,,\rho(\tau)\big]. \tag{16.38}$$

Alternatively, tracing over B after using (16.37) in $\rho(t) = U(t, t_0)\rho(t_0)U^*(t, t_0)$ gives

$$\rho_A(t) = \rho_A(t_0) - i\int_{t_0}^{t} d\tau\;\mathrm{tr}_B\big[V(\tau)\,,\rho(t_0)\big] + (-i)^2\int_{t_0}^{t} d\tau \times\int_{t_0}^{\tau} d\tau'\;\mathrm{tr}_B\Big[V(\tau)\,,\big[V(\tau')\,,\rho(t_0)\big]\Big] + \cdots. \tag{16.39}$$

For instance, suppose the system starts off initially uncorrelated

$$\rho(t = t_0) = \varrho_A \otimes \varrho_B, \tag{16.40}$$

with $\mathrm{tr}_A\,\varrho_A = \mathrm{tr}_B\,\varrho_B = 1$. Then $\rho_A(t_0) = \varrho_A$ at initial times and in the absence of $V(t)$ (16.39) predicts that ρ_A remains fixed at ϱ_A in the interaction picture. In the Schrödinger picture this means that $\hat{\rho}_A$ evolves only through the action of the Hamiltonian H_A for sector A. If $[H_A, \varrho_A] = 0$, as is usually chosen, then $\hat{\rho}_A$ remains time-independent also in the Schrödinger picture in the absence of the interaction with sector B.

Eq. (16.39) has equally simple implications at linear order in $V(t)$. At this order, the difference between $\rho_A(t)$ and ϱ_A can be neglected (since it is order V). Then (16.39) implies that ρ_A evolves in the same way as it would if it satisfied a Liouville equation,

$$i\partial_t \rho_A = \left[\overline{V}(t), \rho_A\right], \tag{16.41}$$

with an interaction Hamiltonian obtained by averaging[7] $V(t)$ over sector B:

$$\overline{V}(t) := \langle\langle V(t) \rangle\rangle := \operatorname{tr}_B \left[V(t)\, \varrho_B\right]. \tag{16.42}$$

This is the regime appropriate for particles that interact very weakly with their environment, such as discussed in §16.3.2 for neutrinos moving through the interior of the Sun.

Beyond linear order (16.39) begins to differ in its implications from what one would obtain from a Liouville equation like (16.41), regardless of the choice for $\overline{V}(t)$. For instance, Eq. (16.39) predicts that

$$\partial_t\left(\rho_A^2 - \rho_A\right) = -i\left\{\rho_A \operatorname{tr}_B\left[V(t), \rho\right] + \operatorname{tr}_B\left[V(t), \rho\right](\rho_A - I_A)\right\} \tag{16.43}$$

which uses (16.38) in its differential form

$$\partial_t\, \rho_A(t) = -i \operatorname{tr}_B\left[V(t)\,, \rho(t)\right]. \tag{16.44}$$

Whenever $\operatorname{tr}_B[V(t), \rho(t)] = [\overline{V}(t), \rho_A(t)]$ for some effective interaction $\overline{V}(t)$, the right-hand side of Eq. (16.43) is proportional to $[\overline{V}(t), \rho_A^2 - \rho_A]$ and so vanishes for states that initially satisfy $\rho_A^2 = \rho_A$. So in this special case an initially pure state, $\rho_A(t_0) = |\psi\rangle\langle\psi|$, remains pure for all later times, with the pure state evolving through an effective Schrödinger equation of the form $i\partial_t|\psi\rangle = \overline{V}|\psi\rangle$.

In general the right-hand side of (16.43) does not vanish and so the quadratic term of (16.39) evolves initially pure states into mixed states. This shows that, in general, there need not exist an effective interaction $\overline{V}$ acting within sector A whose commutator with ρ_A agrees with $\operatorname{tr}_B[V(t), \rho(t)]$ for all times. An initially pure state that does not remain pure at later times is said to decohere.

These considerations show that understanding the evolution of ρ_A for open systems need not be as simple as just finding some effective Hamiltonian with which to evolve the initial state $|\psi\rangle$ through an effective Schrödinger (or Liouville) equation. Starting at second order in $V(t)$, the time evolution predicted by (16.39) can be complicated and history-dependent; introducing correlations between sectors A and B in the sense that it doesn't preserve the factorized form of Eq. (16.40) to later times. The remainder of this chapter explicitly displays examples of these, and other, non-Wilsonian effects that in general arise as sector A evolves in the presence of sector B.

16.3 Mean Fields and Fluctuations

Because open systems can evolve in ways inconsistent with Hamiltonian evolution – such as when they thermalize, or exhibit decoherence – the basic utility of an

[7] Eq. (16.42) introduces the double-bracket notation, for averaging any operator over sector B.

effective Hamiltonian seems up for grabs. However, experience shows that particles moving through complicated aggregates of atoms are often well-described by motion through a relatively simple effective average environment, characterized by a few collective properties like index of refraction or an effective mass and so on. This section outlines when and why this occurs, arguing that for some systems the effects of an environment can sometimes be captured by an effective hamiltonian, called the mean-field approximation, even for open systems. The initial focus is on what characterizes the effective average description for an arbitrary sector B, as well as how to characterize fluctuations about this average. The question of when an average description is a good approximation, and if so what small parameter controls the size of fluctuations, is then addressed in later sections.

16.3.1 The Mean/Fluctuation Split$^\diamond$

The guiding principle when splitting the time evolution of $\rho_A(t)$ into a mean-field and a fluctuation piece is to demand that time-evolved probabilities can be written as a non-interfering sum of mean-field and fluctuation contributions.

To this end – for systems that are initially uncorrelated, so $\rho(t_0) = \varrho_A \otimes \varrho_B$ as in (16.40) – define the mean evolution operator for sector A as the average of $U(t,t')$ over sector B,

$$\overline{U}(t,t_0) := \langle\!\langle U(t,t_0)\rangle\!\rangle = \operatorname*{tr}_B \left[\varrho_B\, U(t,t_0)\right]. \tag{16.45}$$

Similarly, define the difference

$$\Delta U(t,t_0) := U(t,t_0) - \overline{U}(t,t_0), \tag{16.46}$$

which necessarily satisfies

$$\langle\!\langle \Delta U\rangle\!\rangle = \operatorname*{tr}_B \left[\varrho_B\, \Delta U\right] = 0. \tag{16.47}$$

The point of these definitions is that all expectation values in sector A can be written as the sum of a mean-field piece plus a fluctuation piece (with no interference between these two). That is, for any $\mathcal{O}_A = O_A \otimes I_B$, Eqs. (16.45) and (16.46) imply that

$$\begin{aligned}\langle \mathcal{O}_A\rangle &= \operatorname{Tr}\left[\rho(t)\,\mathcal{O}_A\right] = \operatorname{Tr}\left[U(t,t_0)\,\rho(t_0)U^*(t,t_0)\mathcal{O}_A\right] \\ &= \operatorname*{tr}_A\left[\overline{U}(t,t_0)\,\varrho_A\overline{U}^*(t,t_0)O_A\right] + \operatorname{Tr}\left[\Delta U(t,t_0)\,\rho(t_0)\Delta U^*(t,t_0)\,\mathcal{O}_A\right],\end{aligned} \tag{16.48}$$

where the second line writes $U = \overline{U} + \Delta U$, with all cross terms involving both $\overline{U}(t,t_0)$ and $\Delta U(t,t_0)$ vanishing by virtue of the identity (16.47). Eq. (16.48) is an exact statement, and defines the mean-field and fluctuation parts of any expectation: $\langle \mathcal{O}_A\rangle = \langle \mathcal{O}_A\rangle_{\rm m} + \langle \mathcal{O}_A\rangle_{\rm f}$.

This distinction between mean-field and fluctuation parts of the time evolution can also be made directly for the reduced density matrix itself, with $\rho_A(t) = \rho^{\rm m}_A(t)+\rho^{\rm f}_A(t)$, where

$$\rho^{\rm m}_A(t) := \overline{U}(t,t_0)\,\varrho_A\overline{U}^*(t,t_0) \quad \text{and} \quad \rho^{\rm f}_A(t) := \operatorname*{Tr}_B\left[\Delta U(t,t_0)\,\rho(t_0)\Delta U^*(t,t_0)\right], \tag{16.49}$$

so that

$$\langle O_A \rangle = \mathop{\mathrm{tr}}_A \left[\rho_A^{\mathrm{m}} O_A\right] + \mathop{\mathrm{tr}}_A \left[\rho_A^{\mathrm{f}} O_A\right]. \tag{16.50}$$

Several features about the distinction between mean-field and fluctuation parts of the time evolution bear emphasis. First, this division does not rely on perturbing in powers of V. Second, the mean-field Hamiltonian, V_{m}, is naturally defined as the generator of $\overline{U}$,

$$V_{\mathrm{m}}(t) := \mathrm{i}\, \partial_t \overline{U}\, \overline{U}^{-1}, \tag{16.51}$$

and this differs from the average $\overline{V}$ once one works beyond linear order in V. Explicitly, the leading terms in an expansion of V_{m} and ΔU in powers of V are

$$\Delta U(t, t_0) = -\mathrm{i} \int_{t_0}^{t} \mathrm{d}\tau\ \delta V(\tau) + O(V^2), \tag{16.52}$$

and

$$V_{\mathrm{m}}(t) = \overline{V}(t) - \mathrm{i} \int_{t_0}^{t} \mathrm{d}\tau \langle\!\langle \delta V(t)\ \delta V(\tau) \rangle\!\rangle + O\left(V^3\right), \tag{16.53}$$

with $\delta V(t) := V(t) - \overline{V}(t)$ and $\langle\!\langle (\cdots) \rangle\!\rangle := \mathrm{tr}_B[\varrho_B(\cdots)]$ as before.

Unitarity

The differential evolution of $\rho_A^{\mathrm{m}}(t)$ is given in terms of $V_{\mathrm{m}}(t)$ by differentiating (16.49), leading to a Liouville-like equation,

$$\frac{\partial \rho_A^{\mathrm{m}}}{\partial t} = -i\left[V_{\mathrm{m}}(t)\, \rho_A^{\mathrm{m}}(t) - \rho_A^{\mathrm{m}}(t) V_{\mathrm{m}}^*(t)\right]. \tag{16.54}$$

In this equation $V_{\mathrm{m}}(t)$ is *not* assumed to be hermitian, since the expression (16.53) implies that

$$\begin{aligned} \frac{1}{2}\left(V_{\mathrm{m}} + V_{\mathrm{m}}^*\right) &= \overline{V} - \frac{\mathrm{i}}{2} \int_{t_0}^{t} \mathrm{d}\tau \langle\!\langle [\delta V(t), \delta V(\tau)] \rangle\!\rangle + O\left(V^3\right), \\ -\frac{\mathrm{i}}{2}\left(V_{\mathrm{m}} - V_{\mathrm{m}}^*\right) &= -\frac{1}{2} \int_{t_0}^{t} \mathrm{d}\tau \langle\!\langle \{\delta V(t), \delta V(\tau)\} \rangle\!\rangle + O\left(V^3\right), \end{aligned} \tag{16.55}$$

where as usual $[A, B] = AB - BA$, while $\{A, B\} = AB + BA$. The rate of change of ρ_A^{f} is:

$$\frac{\partial \rho_A^{\mathrm{f}}}{\partial t} = \int_{t_0}^{t} \mathrm{d}\tau\ \mathop{\mathrm{tr}}_B \left[\left(\delta V(t)\, \rho(t_0)\, \delta V(\tau) + \delta V(\tau)\, \rho(t_0)\, \delta V(t)\right)\right] + O\left(V^3\right). \tag{16.56}$$

V_{m} need not be hermitian because $\overline{U}$ need not be unitary, which occurs because probability lost from ρ_A^{m} is gained by ρ_A^{f} (or vice versa). In the examples to follow, where system A describes a beam of particles (*e.g.* photons) moving within a medium B (*e.g.* glass), transfer of probability from ϱ_A^{m} to ϱ_A^{f} might correspond to the diffuse scattering of some particles out of an initially coherent beam, for example. To see the implications of conservation of probability for such a transfer, use the condition $\mathrm{Tr}\ \rho = 1$ for the full system, which implies that the same is true for the reduced density matrix for sector A: $\text{tr}_A\, \rho_A = \text{Tr}\, \rho = 1$. Consequently, the differential expression for conservation of total probability is expressed by

$$0 = \partial_t \operatorname*{tr}_A \rho_A = -\mathrm{i} \operatorname*{tr}_A \left\{\rho_A^{\mathrm{m}}(t)\left[V_{\mathrm{m}}(t) - V_{\mathrm{m}}^*(t)\right]\right\} + \operatorname*{tr}_A (\partial_t\, \rho_A^{\mathrm{f}}). \tag{16.57}$$

Eq. (16.57) is a generalization to open systems of the optical theorem for scattering processes – which as usually formulated relates the imaginary part of the amplitude for forward-scattering to the total scattering cross section. The connection between (16.57) and scattering becomes explicit if system A is restricted to only the initial momentum state for a beam of particles, with all other momentum states for the beam lumped into B together with the environment, since this makes the trace over A appearing on the right-hand side restrict to forward scattering (for more details see §16.3.4).

Recursiveness

Notice that the split between mean-field and fluctuations is recursive, in the following sense. Suppose that sector A is itself divided into independent subsectors, A' and B', with observables only measured if they act in subsector A'. Suppose also that the initial state has no correlations between these two new subsectors: $\rho_A(t_0) = \varrho_A = \varrho_{A'} \otimes \varrho_{B'}$. Then the reduced description of sector A can be further reduced to describe only subsector A', with

$$\rho_{A'}(t) := \operatorname*{tr}_{B'}\left[\rho_A(t)\right] = \operatorname*{tr}_{B'\cup B}\left[\rho(t)\right]. \tag{16.58}$$

The mean-field part of the evolution in sector A' is then equally well defined in terms of either the mean-field evolution in sector A or the full evolution, regardless of whether the trace over sectors B and B' are performed separately or all at once. That is, defining

$$\overline{U}_A := \operatorname*{tr}_B\left[\varrho_B U\right] \quad \text{and} \quad \overline{U}_{A'} := \operatorname*{tr}_{B'}\left[\varrho_{B'}\overline{U}_A\right] = \operatorname*{tr}_{B\cup B'}\left[(\varrho_{B'}\otimes\varrho_B)U\right], \tag{16.59}$$

and

$$\Delta U_A := U - \overline{U}_A \quad \text{and} \quad \Delta U_{A'} \equiv U - \overline{U}_{A'}, \tag{16.60}$$

implies that $\text{tr}_B\left[\varrho_B \Delta U_A\right] = \text{tr}_{B\cup B'}\left[(\varrho_{B'}\otimes\varrho_B)\Delta U_{A'}\right] = 0$. Consequently, the expectation of any observable, $O_{A'} = O_{A'} \otimes I_{B'} \otimes I_B$, restricted to subsector A' is

$$\begin{aligned}
\langle O_{A'}\rangle &:= \text{Tr}\left[\rho(t) O_{A'}\right] \\
&= \operatorname*{tr}_A\left[\overline{U}_A(t,t_0)\varrho_A\overline{U}_A^*(t,t_0)O_{A'}\right] + \text{Tr}\left[\Delta U_A(t,t_0)\rho(t_0)\Delta U_A^*(t,t_0)O_{A'}\right] \qquad (16.61)\\
&= \operatorname*{tr}_{A'}\left[\overline{U}_{A'}(t,t_0)\varrho_{A'}\overline{U}_{A'}^*(t,t_0)O_{A'}\right] + \text{Tr}\left[\Delta U_{A'}(t,t_0)\rho(t_0)\Delta U_{A'}^*(t,t_0)O_{A'}\right],
\end{aligned}$$

and so on.

The effective mean-field hamiltonians, $V_{A\mathrm{m}}(t)$ and $V_{A'\mathrm{m}}(t)$, are then defined, as usual, in terms of $\overline{U}_A(t,t_0)$ and $\overline{U}_{A'}(t,t_0)$, using Eq. (16.51). Using the notation $\langle\!\langle(\cdots)\rangle\!\rangle_{B'} := \text{tr}_{B'}[(\cdots)\varrho_{B'}]$ – and similarly for $\langle\!\langle(\cdots)\rangle\!\rangle_B$ – then leads to the following perturbative expressions

$$\begin{aligned} V_{B'\mathrm{m}}(t) &\simeq \langle\!\langle V_{B\,\mathrm{m}}(t)\rangle\!\rangle_{B'} - \mathrm{i}\int_{t_0}^{t} \mathrm{d}\tau \Big\{ \Big[\langle\!\langle V_{B\,\mathrm{m}}(t) V_{B\,\mathrm{m}}(\tau)\rangle\!\rangle_{B'} - \langle\!\langle V_{B\,\mathrm{m}}(t)\rangle\!\rangle_{B'} \langle\!\langle V_{B\,\mathrm{m}}(\tau)\rangle\!\rangle_{B'} \Big] \\ &\simeq \langle\!\langle V(t)\rangle\!\rangle_{B'\cup B} - \mathrm{i}\int_{t_0}^{t} \mathrm{d}\tau \Big\{ \Big[\langle\!\langle \langle\!\langle V(t) V(\tau)\rangle\!\rangle_{B} - \langle\!\langle V(t)\rangle\!\rangle_{B} \langle\!\langle V(\tau)\rangle\!\rangle_{B} \rangle\!\rangle_{B'} \Big] \\ &\qquad + \Big[\langle\!\langle \langle\!\langle V(t)\rangle\!\rangle_{B} \langle\!\langle V(\tau)\rangle\!\rangle_{B} \rangle\!\rangle_{B'} - \langle\!\langle V(t)\rangle\!\rangle_{B'\cup B} \langle\!\langle V(\tau)\rangle\!\rangle_{B'\cup B} \Big] \Big\}, \\ &\simeq \langle\!\langle V(t)\rangle\!\rangle_{B'\cup B} - \mathrm{i}\int_{t_0}^{t} \mathrm{d}\tau \Big\{ \langle\!\langle V(t) V(\tau)\rangle\!\rangle_{B'\cup B} - \langle\!\langle V(t)\rangle\!\rangle_{B'\cup B} \langle\!\langle V(\tau)\rangle\!\rangle_{B'\cup B} \Big\}, \end{aligned} \tag{16.62}$$

which drop $O(V^3)$ contributions.

The recursive nature of these definitions resembles the recursiveness of the definition of the Wilsonian effective action, and so lends itself to a renormalization-group-like analysis of the effects of a medium on particle propagation in which fluctuations on successively larger distance scales are successively integrated out.

What is missing so far from this discussion of the mean-field limit and fluctuations around it is a quantification of when it is a good approximation to neglect the fluctuations. This is returned to in later sections once some intuition is developed through the exploration of a few practical examples.

The Wilsonian Special Case

Before exploring practical examples of the mean-field/fluctuation split in open systems, it is worth remarking in passing that the simplest example of such a split arises already with the Wilsonian effective theory with which this book started. In a Wilsonian system the observed part of the system (what is called here sector A) is defined relative to the unobserved sector (sector B) in terms of a conserved quantity: energy. Because of this the interaction Hamiltonian V does *not* have any off-diagonal components that link sector A (low-energy states) to sector B (high-energy states). By definition, the Hamiltonian is diagonal in the energy eigenbasis.

Because V vanishes in this special case, it is clear that ΔU also vanishes, although the entire formulation in terms of the interaction picture is also overkill. Because $H = H_A + H_B$ there is no need to perturb in V to find the time evolution, which within the low-energy sector is simply generated by the low-energy part of the Hamiltonian: $U_A = \exp[-\mathrm{i}H_A(t - t_0)]$. This is the simplest way to see how simple Hamiltonian evolution necessarily emerges in the special case that the system is divided up using conserved quantities. (Though it does not in itself directly express H_A in terms of low-energy fields only, which is also required for a useful low-energy limit.)

16.3.2 Neutrinos in Matter

Neutrinos passing through matter provide a simple practical application of the mean-field/fluctuation formalism just described. Neutrino interactions with an environment are particularly simple because the interactions involved are incredibly weak, with typical scattering lengths being longer than a light-year in ordinary matter [446]. Neutrino propagation through matter is also simple because the absence of neutrinos in most types of matter makes the distinction between sectors A and B particularly clean.[8]

Because neutrino interactions are so incredibly weak, first-order perturbation theory usually is all that is required, and as a consequence the fluctuations residing in ΔU are negligible and the mean-field approximation usually suffices for all practical applications [444].

Mean-Field Neutrino Evolution

At first order in the interaction Hamiltonian the mean contribution of the medium to neutrino propagation has the simple form $V_m \simeq \langle\langle V \rangle\rangle$, where V is the neutrino-medium interaction, so the first step in its evaluation is to characterize what is known about V.

As we understand them today, neutrinos are described by three majorana neutrino fields, $\nu_i, i = 1,2,3$, who couple to any medium through the weak interactions. These consist of the charged-current weak interactions, described in §7.1.2, plus the neutral-current weak interactions mediated by Z-boson exchange. The relevant part of these for neutrinos interacting at energies well below the W-boson mass, $M_W \simeq 80$ GeV, is

$$\mathfrak{L}_{\nu\,\rm wk} = \mathrm{i}\,(\bar{\nu}_i \gamma_\mu \gamma_L \nu_j)\, g^a_{ij} J^\mu_a, \tag{16.63}$$

where γ_L and γ_R project onto left- and right-handed spinors, J^μ_a are a set of operators involving the degrees of freedom of the medium (*i.e.* electrons, protons and neutrons) and g^a_{ij} are 3×3 coupling matrices whose coefficients are predicted[9] by the Standard Model.

For neutrino couplings to charged leptons, $\{\ell_m, m = 1,2,3\} = \{e, \mu, \tau\}$, the currents J^μ_a are given by (*c.f.* Eq. (7.60) of §7.3.3):

$$(J^\mu_\pm)_{mn} = \mathrm{i}[\bar{\ell}_m \gamma^\mu (1 \pm \gamma_5)\, \ell_n], \tag{16.64}$$

and the coupling matrices are

$$(g^{mn}_+)_{ij} = \sqrt{2}G_F \left[U_{mj} U^*_{ni} + \delta_{mn}\delta_{ij} \left(-\frac{1}{2} + \sin^2\theta_W \right) \right], \tag{16.65}$$

$$(g^{mn}_-)_{ij} = \sqrt{2}G_F\, \delta_{mn}\delta_{ij} \sin^2\theta_W, \tag{16.66}$$

where $\sin\theta_W$ denotes, as usual, the weak mixing angle and U_{mi} are the charged-current PMNS matrix elements described in §7.1.1 in the presence of nonzero neutrino masses, $m_i \neq 0$ while the remaining terms come from the neutral-current interactions.

Two-neutrino interactions with protons and neutrons arise purely through neutral-current Z-boson exchange with their quark content, leading to currents

$$(J^\mu_\pm)_a = \mathrm{i}\bar{q}_a \gamma^\mu (1 \pm \gamma_5)\, q_a, \tag{16.67}$$

[8] Extreme environments like supernovae are more complicated inasmuch as the extreme densities make interactions more important, and make neutrinos themselves part of the environment.

[9] Strictly speaking, the inclusion of neutrino masses implies generalizing the Standard Model, such as through the inclusion of the SMEFT effective mass term discussed in §9.

with coupling matrices

$$(g^a_+)_{ij} = \sqrt{2}\, G_F\, \delta_{ij} \left(T_{3a} - Q_a \sin^2\theta_w\right),$$
$$(g^a_-)_{ij} = \sqrt{2}\, G_F\, \delta_{ij} \left(-Q_a \sin^2\theta_w\right), \tag{16.68}$$

where T_{3a} and Q_a are the third component of weak isospin and electric charge (in units of e) of the corresponding quark.

For quarks contained within protons and neutrons what matters are the up- and down-type quark contributions, for which $T_{3u} = \frac{1}{2}$, $Q_u = \frac{2}{3}$, $T_{3d} = -\frac{1}{2}$ and $Q_d = -\frac{1}{3}$. The contribution of these quarks to the combination $(g^a_+)_{ij}\,(J^\mu_+)_a + (g^a_-)_{ij}\,(J^\mu_-)_a$ therefore evaluates to

$$\frac{G_F}{\sqrt{2}}\,\delta_{ij}\left[\left(1 - \frac{8}{3}\sin^2\theta_w\right) i(\overline{u}\gamma^\mu u) + \left(-1 + \frac{4}{3}\sin^2\theta_w\right) i(\overline{d}\gamma^\mu d) + i(\overline{u}\gamma^\mu\gamma_5 u) - i(\overline{d}\gamma^\mu\gamma_5 d)\right] = \sqrt{2}\, G_F\, \delta_{ij}\left[j^\mu_{3V} + j^\mu_{3A} - 2\sin^2\theta_w\, j^\mu_{\rm em}\right] \tag{16.69}$$

where

$$j^\mu_{3V} = \frac{i}{2}(\bar{u}\gamma^\mu u - \bar{d}\gamma^\mu d), \quad j^\mu_{3A} = \frac{i}{2}(\bar{u}\gamma^\mu\gamma_5 u - \bar{d}\gamma^\mu\gamma_5 d) \tag{16.70}$$

are Noether currents for vector and axial $SU(2)$ transformations, while

$$j^\mu_{\rm em} = \frac{i}{3}(2\bar{u}\gamma^\mu u - \bar{d}\gamma^\mu d) \tag{16.71}$$

is the contribution of u and d quarks to the electromagnetic current.

For practical applications over scales larger than a fm or so, it is more useful to know these currents in the effective theory built using protons and neutrons than for quarks. As discussed in more detail in §8, the form for these currents can in general be complicated, although it simplifies if the current in question is conserved. Happily enough, both j^μ_{3V} and $j^\mu_{\rm em}$ are conserved (to good approximation, for j^μ_{3V}, and exactly, for electromagnetism) in this way if the Hamiltonian for the medium is dominated by the strong and electromagnetic interactions. In this case, the corresponding effective 'macroscopic' currents can immediately be written in the effective theory involving protons and neutron fields simply by constructing the Noether currents for them in the usual way, starting from the action of these symmetries in the effective theory. This leads to the following expression for the parity-even part of the hadronic weak currents,

$$[(g^a_+)_{ij}\,(J^\mu_+)_a + (g^a_-)_{ij}\,(J^\mu_-)_a]_{\rm even} \simeq \frac{G_F}{\sqrt{2}}\,\delta_{ij}\left[\left(1 - 4\sin^2\theta_w\right) i(\overline{p}\gamma^\mu p) - i(\overline{n}\gamma^\mu n)\right], \tag{16.72}$$

up to terms involving higher derivatives (that turn out not to be required in the applications considered later). The axial terms are not quite so simple because of the breaking of axial $SU(2)$ symmetries by quantum chromodynamics, as is also sketched in §8, but these axial terms are also not required in what follows.

Returning now to neutrinos propagating through a material, consider sector A to comprise the neutrino sector while sector B consists of some sort of ordinary matter built from protons, neutrons and electrons. To first order in the weak interactions the mean-field Hamiltonian acting on the neutrinos is given by

$$V_{\rm m} \simeq \langle\!\langle V \rangle\!\rangle = -\int \mathrm{d}^3x\, g^a_{ij}\left(\mathrm{i}\,\bar{\nu}_i\gamma_\mu\gamma_L\nu_j\right)\langle\!\langle J^\mu_a \rangle\!\rangle, \tag{16.73}$$

showing that it is obtained simply by replacing the interaction current, J^μ_a, with its mean, $\langle\!\langle J^\mu_a \rangle\!\rangle$, evaluated in the medium.

These expectation values can be evaluated relatively simply for materials built from electrons, protons and neutrons dominantly interacting through the strong and electromagnetic interactions. There are two reasons why this evaluation is straightforward. First, as discussed in §8, parity invariance implies that the average of all axial vectors in the medium vanish, meaning that it is only the expectation values of the two currents $j^\mu_{3\nu}$ and $j^\mu_{\rm em}$ that are required.

Second, as also discussed above, the approximate conservation of $SU_\nu(2)$ – and exact conservation of $U_{\rm em}(1)$ – allow use of (16.72) for the vector currents $j^\mu_{3\nu}$ and $j^\mu_{\rm em}$ in terms of protons and neutrons. Including the contributions of electrons, protons and neutrons then gives

$$\langle\!\langle g^a_{ij}\, J^\mu_a \rangle\!\rangle \approx \frac{G_F}{\sqrt{2}}\left\{2U_{je}\,U^*_{ie}\, j^\mu_e(x) - \delta_{ij} j^\mu_n(x) + \delta_{ij}\,(1-4\sin^2\theta_w)\,[j^\mu_p(x) - j^\mu_e(x)]\right\}, \tag{16.74}$$

where $j^\mu_e = \langle\!\langle \mathrm{i}\bar{e}\gamma^\mu e \rangle\!\rangle$, $j^\mu_p = \langle\!\langle \mathrm{i}\bar{p}\gamma^\mu p \rangle\!\rangle$ and $j^\mu_n = \langle\!\langle \mathrm{i}\bar{n}\gamma^\mu n \rangle\!\rangle$ are shorthands for the local mean electron, proton and neutron current densities, respectively.

For nonrelativistic particles the neglect of terms proportional to particle speed v gives a further simplification, since all of the spatial components of the mean currents, j^μ_e, j^μ_n and j^μ_p, vanish as $v \to 0$. This leaves $\langle\!\langle j^\mu_a \rangle\!\rangle \approx \bar{n}_a \delta^\mu_0$ for $a = e, p, n$, as the dominant contribution, where $\bar{n}_a$ is simply the average number density in the medium for each particle type. Furthermore, for systems that are locally electrically neutral over the scales of interest it is also true that $\bar{n}_e(\mathbf{x},t) \simeq \bar{n}_p(\mathbf{x},t)$, allowing the last term of (16.74) to be dropped. (This last term is also suppressed because of the numerical coincidence that the experimental value $\sin^2\theta_w \simeq 0.23$ makes $1 - 4\sin^2\theta_w \simeq 0.08$.)

Matter-Induced Neutrino Mixing

Using the approximations described above allows the leading form of the effective Hamiltonian, $V_{\rm m} \simeq \langle\!\langle V \rangle\!\rangle$, to be expressed in terms of the medium's electron and neutron densities, $\bar{n}_e$ and $\bar{n}_n$. The result is what would have been obtained from an effective interaction lagrangian of the form

$$\mathfrak{L}_{\rm med} \simeq \langle\!\langle \mathfrak{L}_{\nu\,{\rm wk}} \rangle\!\rangle = \frac{G_F}{\sqrt{2}}\,\mathrm{i}(\bar{\nu}_i\gamma_0\gamma_L\nu_j)\left(2U_{je}\,U^*_{ie}\,\bar{n}_e - \delta_{ij}\,\bar{n}_n\right). \tag{16.75}$$

For observational purposes the important term in (16.75) is the charged-current interaction, which distinguishes amongst neutrino flavours. It does so because of the assumption that the medium contains only hadrons and electrons (and no muons or tau leptons, say), since this implies that only electron-type neutrinos experience charged-current interactions with the medium. This matters because the resulting flavour-dependence can mediate neutrino oscillations in matter that differ in character from those that occur in vacuum [448].

To explore the nature of these matter-dependent oscillation parameters it is useful to switch from interaction picture to Heisenberg picture, for which the fields carry the burden of time evolution. In the Heisenberg picture flavour evolution is found by combining (16.75) with the lagrangian $\mathfrak{L}_0$ for free propagation (including neutrino masses, m_i). In an arbitrary flavour basis, the free-propagation lagrangian is

$$\mathfrak{L}_0 = -\bar{\nu}_i \left(\delta_{ij}\partial\!\!\!/ + m_{ij}\gamma_L + m^*_{ij}\gamma_R\right)\nu_j, \tag{16.76}$$

where m is the left-handed neutrino mass matrix in flavour space, which, in general, is a 3×3 complex symmetric matrix.

If $\bar{n}_e$ and $\bar{n}_n$ are independent of spacetime position then the field equation for the neutrino fields coming from $\mathfrak{L}_0 + \mathfrak{L}_{\rm med}$ is satisfied by modes $u_i\, e^{ik\cdot x}$ where the 4-momentum, k^μ, and the flavour-vector, u_i, satisfy the matrix equation

$$\left(i\, k\!\!\!/ + m\,\gamma_L + m^*\gamma_R + i\gamma^0\mu\right) u = 0, \tag{16.77}$$

where

$$\mu = \sqrt{2}\, G_F\,(g_e\,\bar{n}_e + g_n\,\bar{n}_n) \tag{16.78}$$

with $g_n = -\frac{1}{2}\, I$ and $g_e = w\, w^\dagger$ being matrices in flavour space, where w is the vector

$$w = \begin{pmatrix} U_{1e} \\ U_{2e} \\ U_{3e} \end{pmatrix}. \tag{16.79}$$

In particular, $w = (1, 0, 0)^T$ in a weak-interaction basis (which is defined as the basis for which $U_{ia} = \delta_{ia}$, for $i, a = e, \mu, \tau$).

In vacuum, where $\bar{n}_e = \bar{n}_n = 0$ (so $\mu = 0$), Eq. (16.77) implies that $k^\mu k_\mu + m_i^2 = 0$, where m_i^2 is an eigenvalue of the non-negative hermitian matrix $m^\dagger m$, and neutrino propagation eigenstates are the eigenvectors of the matrix $m^\dagger m$. For neutrinos moving through matter the propagation eigenstates typically differ from those found in vacuum because the matrices $m^\dagger m$ and μ in general do not commute. Furthermore, the rotation from vacuum eigenstate to matter eigenstate need not be small if the eigenvalues of $m^\dagger m$ and μ are both small compared to $k := |\mathbf{k}|$, particularly when $m^\dagger m/(2k)$ is similar in size to μ. Exercise 16.2 makes this explicit, showing (for two neutrino species) that the probability of a neutrino of energy $E \gg m$ being produced in one of the weak-interaction eigenstates and then being detected in the other eigenstate after travelling a distance L is

$$P_{e\mu}(E, L) \simeq \sin^2 2\theta_m \sin^2\left(\frac{\delta m^2 L}{4E}\right), \tag{16.80}$$

where $\delta m^2 = m_2^2 - m_1^2 > 0$ is the difference between the neutrino squared-masses in vacuum and the matter mixing angle is given by

$$\sin 2\theta_m = \left(\frac{\delta m^2}{4k\,\mu_0}\right) \sin 2\theta_\nu \tag{16.81}$$

with θ_ν the mixing angle that relates flavour and propagation eigenstates in vacuum and

$$\mu_0 := \left[\left(\frac{G_F \bar{n}_e}{\sqrt{2}}\right)^2 - \left(\frac{G_F \bar{n}_e}{\sqrt{2}}\right)\frac{\delta m^2 \cos 2\theta_\nu}{2k} + \left(\frac{\delta m^2}{4k}\right)^2\right]^{1/2}. \tag{16.82}$$

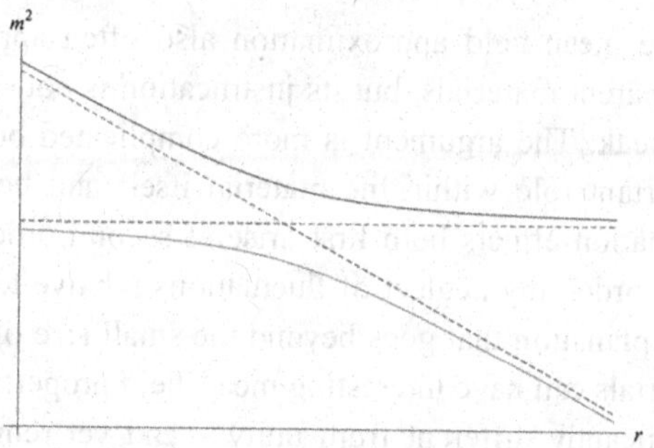

Fig. 16.1 A plot of neutrino mass eigenvalues, m_i^2, (for two species of neutrinos) as a function of radius, r, within the Sun (solid lines) as well as what these masses would be in the absence of vacuum mixing: $\theta_v = 0$ (dashed lines). The plot falls with r for electron neutrinos, since it is proportional to the density $\bar{n}_e$ of electrons within the Sun. Resonance occurs where the dotted lines cross. A neutrino evolving adiabatically through the resonance follows a solid line and so completely converts from one unmixed species to another. Nonadiabatic evolution has a probability P_j of jumping from one branch to the other when passing through the resonance.

MSW Oscillations

Notice that matter oscillations can be maximal (*i.e.* $\sin 2\theta_m = 1$) for some k, regardless of the size of θ_v, and this has important practical implications when neutrinos pass through environments with position-dependent electron densities, $\bar{n}_e = \bar{n}_e(\mathbf{x})$. Position-dependent densities arise, for instance, for neutrinos produced by nuclear reactions deep within the Sun, which encounter a monotonically falling spherically symmetric distribution, $\bar{n}_e(r)$, that eventually vanishes at the solar surface.

In particular, if it should happen that the electron density, $\bar{n}_e(0)$, at the solar centre should satisfy $G_F\bar{n}_e(0)/\sqrt{2} > \delta m^2 \cos 2\theta_v/(4E)$, then the oscillations become maximal when the resonance condition

$$\frac{G_F\bar{n}_e}{\sqrt{2}} = \left(\frac{\delta m^2}{4k}\right) \cos 2\theta_v, \tag{16.83}$$

is eventually satisfied somewhere within the solar interior. If passage through this region takes a sufficiently long time, conversion of neutrino flavour can be extremely efficient even if vacuum oscillations are comparatively small – see Exercise 16.3 and Fig. 16.1 – a phenomenon called the Mikheyev–Smirnov–Wolfenstein (or 'MSW') effect [448]. What is important is that this conversion can be large even though both $G_F\bar{n}_e$ and $\delta m^2/k$ are small compared with k, and so vacuum masses and matter couplings remain treatable as extremely tiny linear perturbations to neutrino energies.

16.3.3 Photons: Mean-Field Evolution♠

Interactions with matter are more complicated for photons than for neutrinos. The previous section argues that neutrinos interact so weakly with the solar environment that working to first order in V suffices for practical applications. Since the effects of fluctuations first arise at second order in V, mean-field methods are justified for neutrinos in the Sun simply because their interactions are so weak [449].

The mean-field approximation also often applies for photons moving through transparent materials, but its justification is not simply because photon interactions are weak. The argument is more complicated both because electric fields play an important role within the material itself and because the leading mean-field and fluctuation effects both first arise at second order in V. Because both arise at the same order any neglect of fluctuations relative to a mean-field description requires an explanation that goes beyond the small size of V. For photons the puzzle is why materials can have interesting mean-field properties – such as an index of refraction significantly different from unity – and yet remain transparent (due to negligible scattering from fluctuations).

Explanations of these properties usually rely on the coherence of photon scattering, which in turn amounts to an expansion in powers of $1/N$ where $N \gg 1$ is the number of scattering sites within what is called a coherence volume. The next two sections explore these issues through the use of some simple examples. The present section starts by recounting the mean-field description, while the next section describes the contributions of fluctuations and why they can sometimes be neglected.

Photons in Polarizable Media

A starting point is the open-system description of photons coupled to a material built from electrically neutral but polarizable atoms. To this end, sector A consists of the Hilbert space of the photons while sector B consists of the states spanned by the medium itself. A second-quantized approach is useful because the materials of interest typically involve many atoms, through which many photons typically pass. Following the developments of this chapter's previous sections, the interactions of main interest are those that couple these two sectors to one another.

The EFT describing light interacting with nonrelativistic neutral particles is discussed in §12.3.1, with $\mathfrak{L} = \mathfrak{L}_0 + \mathfrak{L}_{\rm int}$. Here, $\mathfrak{L}_0$ describes the free propagation of atoms and photons through the usual kinetic terms as well as any internal inter-atom interactions – such as those related to Coulomb forces – that do not involve photon propagation

$$\mathfrak{L}_0 = \frac{1}{2}(\mathbf{E}^2 - \mathbf{B}^2) + i\Psi^* \partial_t \Psi + \frac{1}{2M}\Psi^* \nabla^2 \Psi + \mathfrak{L}_{\rm atom-atom}, \tag{16.84}$$

where $\mathfrak{L}_{\rm atom-atom}$ describes the inter-atomic interactions. The leading photon-atom interactions are given by (12.90) and (12.91), reproduced here for convenience:

$$\mathfrak{L}_{\rm int} = \mathfrak{g}\, \Psi^*\Psi \nabla \cdot \mathbf{E} + \frac{1}{2}\Psi^*\Psi \left(\mathfrak{p}_E\, \mathbf{E}^2 - \mathfrak{p}_B\, \mathbf{B}^2\right) + \cdots . \tag{16.85}$$

This drops terms in Eqs. (12.90) and (12.91) that do not involve both Ψ and electromagnetic fields simultaneously. In fundamental units the effective coupling $\mathfrak{g}$ has dimensions (length)^2 while $\mathfrak{p}_E$ and $\mathfrak{p}_B$ are dimension (length)^3, though for nonrelativistic systems these scales can arise either as small lengths or short times. For many systems the relevant scale for these couplings is set by atomic sizes and energies.

For long-wavelength low-energy photon propagation in neutral media the first term in (16.85) proves to be redundant – in the sense of §2.5 – making the remaining terms of most direct interest. At leading order, these contribute to the mean-field description an amount

$$\langle\langle \mathfrak{L}_{\rm int} \rangle\rangle \simeq \frac{\bar{n}}{2}\left(\mathfrak{p}_E\, \mathbf{E}^2 - \mathfrak{p}_B\, \mathbf{B}^2\right), \tag{16.86}$$

where $\bar{n} = \langle\langle \Psi^* \Psi \rangle\rangle$ is the medium's mean atomic density. This shows how $\mathfrak{p}_E$ and $\mathfrak{p}_B$, respectively, contribute to the medium's bulk dielectric permittivity and permeability as electric and magnetic polarizabilities.

Wave Propagation

To see the implications for photon propagation within the medium consider first a translation-invariant material, for which $\bar{n}$ is independent of $\mathbf{x}$ and t. At face value,[10] using the lagrangian $\mathfrak{L}_0 + \mathfrak{L}_{\rm int}$ and switching to the Heisenberg picture leads to the following evolution equations for the radiation part of the Maxwell field in the presence of the medium,

$$(1 + \bar{n}\, \mathfrak{p}_E) \nabla \cdot \mathbf{E} = 0 \quad \text{and} \quad (1 + \bar{n}\, \mathfrak{p}_B) \nabla \times \mathbf{B} - (1 + \bar{n}\, \mathfrak{p}_E)\, \partial_t \mathbf{E} = 0, \tag{16.87}$$

together with the Bianchi identities

$$\nabla \cdot \mathbf{B} = \nabla \times \mathbf{E} + \partial_t \mathbf{B} = 0. \tag{16.88}$$

Taking the curl of the two vector Maxwell equations, using the identity $\nabla \times (\nabla \times \mathbf{E}) = \nabla(\nabla \cdot \mathbf{E}) - \nabla^2 \mathbf{E}$, gives (in the usual way) the wave equation

$$\partial_t^2 \mathbf{E} - c_m^2 \nabla^2 \mathbf{E} = 0, \tag{16.89}$$

and similarly for $\mathbf{B}$.

Plane-wave mode solutions to these equations, proportional to $e^{ik\cdot x}$, therefore satisfy the dispersion relation

$$\omega(\mathbf{k}) := k^0 = c_m |\mathbf{k}|, \tag{16.90}$$

with wave-propagation phase speed given by

$$c_m = \sqrt{\frac{1 + \bar{n}\, \mathfrak{p}_B}{1 + \bar{n}\, \mathfrak{p}_E}} =: \frac{1}{\mathfrak{n}_m}. \tag{16.91}$$

This last equality (recalling this book's convention that the speed of light in vacuum is $c = 1$) defines the medium's index of refraction, $\mathfrak{n}_m$. Notice $\mathfrak{n}_m = c_m = 1$ for a Lorentz-invariant medium, for which $\mathfrak{p}_E = \mathfrak{p}_B$.

Expression (16.91) simplifies to $c_m \simeq 1 - \frac{1}{2}\, \bar{n}(\mathfrak{p}_E - \mathfrak{p}_B)$ and $\mathfrak{n}_m \simeq 1 + \frac{1}{2}\, \bar{n}(\mathfrak{p}_E - \mathfrak{p}_B)$, if one works to linear order in the interaction with the environment, with perturbation theory requiring $|\bar{n}\mathfrak{p}_E|, |\bar{n}\mathfrak{p}_B| \ll 1$. Extensions of the domain of validity of the above formulae to larger values of $\mathfrak{n}_m - 1$ (as appropriate for many ordinary transparent materials, like glass or water) is explored in §16.3.5.

[10] The 'face value' qualifier is here because the argument to follow treats the interaction with the medium, $\mathfrak{L}_{\rm int}$, on an equal footing with the kinetic terms in $\mathfrak{L}_0$, whereas identifying the mean-field limit simply by averaging $\mathfrak{L}_{\rm int}$ over sector B has to this point been derived only at linear order in the interactions. As it turns out, the arguments to follow have a broader domain of validity than linear-order perturbation theory, a point returned to in §16.3.5 and §16.4.

Reflection and Refraction

Mean quantities need not be translation invariant, and when they are not they generically cause scattering to occur. The simplest example of this arises at the boundary between two media, on each side of which $\mathfrak{n}_m$ differs, as is now explored in more detail. Suppose, for instance, this boundary is a plane with Cartesian coordinates adapted so that it lies at $z = 0$, and denote the index of refraction in the regions on either side by $\mathfrak{n}_\pm$, where $\mathfrak{n}_+$ applies for $z > 0$ while $\mathfrak{n}_-$ applies for $z < 0$.

If $\mathfrak{n}_+ \neq \mathfrak{n}_-$ then translation invariance breaks at $z = 0$, implying momentum need not be conserved there and so incident photons are scattered. For planar surfaces this scattering is particularly simple because unbroken translation invariance in the x and y directions ensures only the component of momentum in the z direction can change at the surface. This suffices to derive the familiar 'equal-angle' law (for reflection) and Snell's law (for refraction).

To see why, suppose a plane wave approaches the surface from the $z > 0$ side with initial momentum $\mathbf{k}_+$, and coordinates are chosen so that $\mathbf{k}_+ = k_x\mathbf{e}_x - k_z\mathbf{e}_z$ lies within the x–z plane. (Here, $\mathbf{e}_i$ denote three mutually orthogonal unit vectors pointing along each coordinate axis.) After encountering the surface, a photon either passes through to the region $z < 0$ (refracts) with momentum $\mathbf{k}'_-$ or reflects back to $z > 0$ with momentum $\mathbf{k}'_+$.

Momentum conservation in the x-y plane ensures these final momenta satisfy $\mathbf{k}'_\pm = k_x\mathbf{e}_x + k'_{z\pm}\mathbf{e}_z$, with components $k'_{z\pm}$ determined by energy conservation: $\omega_+(\mathbf{k}_+) = \omega_\pm(\mathbf{k}'_\pm)$. Using (16.90) and (16.91) for the dispersion relation implies that

$$\frac{|\mathbf{k}_+|}{\mathfrak{n}_+} = \frac{|\mathbf{k}'_\pm|}{\mathfrak{n}_\pm}, \tag{16.92}$$

with the upper sign for reflection and the lower sign for refraction.

For reflection, (16.92) implies that $|\mathbf{k}'_+| = |\mathbf{k}_+|$, which together with momentum conservation in the x-direction requires $k'_{z+} = -k_{z+}$, showing that the normal component of momentum is simply reversed. This becomes the 'equal-angle' law for reflection once expressed in terms of the angle θ between the photon's momentum and the direction of the normal to the surface (which in this example is simply the z axis), because trigonometry gives the initial and final values of θ as

$$\sin\theta' = \frac{|k'_{x+}|}{|\mathbf{k}'_+|} = \frac{|k_x|}{|\mathbf{k}_+|} = \sin\theta \qquad \text{(reflection)}. \tag{16.93}$$

Similarly for refraction, for which conservation of momentum in the x direction and energy conservation (16.92), together imply

$$\sin\theta' = \frac{|k'_{x-}|}{|\mathbf{k}'_-|} = \left(\frac{\mathfrak{n}_+}{\mathfrak{n}_-}\right)\frac{|k_x|}{|\mathbf{k}_+|} = \left(\frac{\mathfrak{n}_+}{\mathfrak{n}_-}\right)\sin\theta \qquad \text{(refraction)}. \tag{16.94}$$

This is recognizable as Snell's law [450], which is often quoted in the special case where $\mathfrak{n}_+ = 1$ (as appropriate when the initial wave is incident from outside any medium). Refraction cannot occur – *i.e.* reflection is total – when $|(\mathfrak{n}_+/\mathfrak{n}_-)\sin\theta| > 1$.

A More Microscopic Picture

For neutral atoms polarizability has its roots in how the atom's electrically charged constituents respond to applied fields. This section sketches how this more microscopic picture of polarizability can also be understood using a mean-field open-system description of photons coupled to these underlying slowly moving charges. Doing so allows an estimate for whether quantities like $\bar{n}\,\mathfrak{p}_E$ can be expected to be large, and shows why they can reasonably be order unity despite all interactions involving factors of the relatively weak electromagnetic coupling e. It is the fact that $\bar{n}\,\mathfrak{p}_E$ need not be small that explains why many materials have an index of refraction satisfying $\mathfrak{n}_m - 1 \simeq \mathcal{O}(1)$.

To this end, the idea is to exploit the recursiveness of EFT methods to drill down so that the theory of the underlying 'full' system is the effective theory of nonrelativistic electrons and nuclei from which the polarizable atoms are made. These are described by an effective lagrangian like the one used in Eqs. (12.37)–(12.40) (plus higher orders), regarded as an expansion in powers of microscopic length scales. For simplicity the relevant nucleus is assumed to be spinless.

Denoting the electron field as Ψ, the nuclear field by Φ, the terms unsuppressed by mass in the Schrödinger effective theory are given by (12.37), reproduced for convenience here:

$$\begin{aligned}\mathfrak{L}_{0,0} &= \mathrm{i}\,\Psi^\dagger D_t\Psi + \mathrm{i}\,\Phi^* D_t\Phi + \frac{1}{2}(\mathbf{E}^2 - \mathbf{B}^2) \\ &= \mathrm{i}\,\Psi^\dagger\partial_t\Psi + \mathrm{i}\,\Phi^*\partial_t\Phi + eA_0\big(Z\,\Phi^*\Phi - \Psi^\dagger\Psi\big) + \frac{1}{2}(\mathbf{E}^2 - \mathbf{B}^2),\end{aligned} \tag{16.95}$$

where (as before) $D_\mu\Psi = (\partial_\mu + \mathrm{i}eA_\mu)\Psi$ and $D_\mu\Phi = (\partial_\mu - \mathrm{i}ZeA_\mu)\Phi$. As above, system A consists of a sector describing a beam of incident photons, say while system B (the environment, or medium) consists of the dynamics of the electrons and nuclei (and any electromagnetic fields that happen to appear there).

Of course, the system described by the lagrangian of (12.37) through (12.40) is broad enough to include a great variety of materials with many kinds of possible electromagnetic response, including conductors, dielectrics, plasmas and so on. Many interactions become unimportant, however, once restricted to sufficiently slowly varying probes, with electrons remaining localized in individual atoms. When this is so, for instance, the material remains locally electrically neutral, and so $Z\langle\!\langle\Phi^\dagger\Phi\rangle\!\rangle \simeq \langle\!\langle\Psi^\dagger\Psi\rangle\!\rangle$, over the scales of interest.

To describe how electromagnetic waves interact with media requires those terms in $\mathfrak{L}$ that involve the field $\mathbf{A}$. The lowest-dimension terms of this sort are given by (12.38) and (12.39), reproduced here as

$$\mathfrak{L}_{1,0} = \frac{1}{2m}\Psi^\dagger\mathbf{D}^2\Psi - \frac{e}{2m}\,c_F\,\mathbf{B}\cdot(\Psi^\dagger\boldsymbol{\sigma}\Psi) + \frac{1}{2M}\Phi^*\mathbf{D}^2\Phi. \tag{16.96}$$

No nuclear magnetic moment here couples Φ to $\mathbf{A}$ because the nucleus is assumed to be spinless. Of these terms it is the interactions involving electrons that dominate at low energies, due to the relatively small size of the electron mass, $m \ll M$. Having a smaller mass allows electrons to move more quickly than do nuclei, given similar applied forces, making their response the dominant effect.

The terms contained in (16.96) governing electronic response to **A** are

$$\mathfrak{L}_{1,0} \supset -\frac{ie}{2m}\,\mathbf{A}\cdot\left[(\nabla\Psi^\dagger)\Psi - \Psi^\dagger\nabla\Psi\right] - \frac{e^2}{2m}\mathbf{A}^2(\Psi^\dagger\Psi) - \frac{e}{2m}\,c_F\,\mathbf{B}\cdot(\Psi^\dagger\sigma\Psi) + \cdots, \tag{16.97}$$

and the polarizabilities are found by computing the mean-field evolution Hamiltonian, $V_{\rm m}$ (or, equivalently, the corresponding lagrangian density), for photons using these interactions, and matching the result to what is obtained using (16.86).

For non-magnetic materials the leading contribution that gives a term proportional to $\mathbf{E}^2$ in the mean-field action arises at second order in the first interaction[11] of (16.97). To compute the polarizabilities requires evaluating the second-order part of (16.53) using the interaction linear in **A** appearing in (16.97).

An Atomic Model

Rather than attempting a full calculation of this type for atomic electrons, it is instructive instead to explore what is found using a relatively simple model for electrons within an atom. To this end, model the atom's electronic states as having two energy levels: an S-wave ground state, $|g\rangle$, plus a P-wave excited state, $|e, \ell_z\rangle$, with $L_3|e,\ell_z\rangle = \ell_z|e,\ell_z\rangle$, where $\ell_z = -1, 0, 1$ labels the excited state's component of angular momentum.

In this model the atomic Hamiltonian for a single atom simplifies to

$$H_i|g\rangle = E_g|g\rangle \qquad \text{and} \qquad H_i|e,\ell_z\rangle = E_e|e,\ell_z\rangle, \tag{16.98}$$

with $\omega_{eg} := E_e - E_g > 0$. For a medium involving N atoms the full Hamiltonian for system B then is

$$H_B = \sum_{i=1}^{N} H_i. \tag{16.99}$$

As discussed above, the interaction V between systems A and B is in principle given by matrix elements within the atoms of the term $\mathbf{A}\cdot\left[(\nabla\Psi^\dagger)\Psi - \Psi^\dagger\nabla\Psi\right]$ from (16.97). As applied to the simple atomic model it is useful to specialize to photons with wavelengths much longer than atomic sizes, for which the dipole approximation of §12.2.2 applies (in particular, see the discussion surrounding Eq. (12.57)). This suggests modelling the photon-atom coupling as

$$V(t) = -\sum_{i=1}^{N} \mathfrak{D}_i(t)\cdot\mathbf{E}(\mathbf{x}_i,t), \tag{16.100}$$

where $\mathbf{E}$ is the electric field evaluated at the atomic position and $\mathfrak{D}(t)$ is the dipole moment operator for the two-level atomic system, given explicitly (in interaction picture) for each atom by

[11] For some systems quadratic terms in **A** can arise at first order in the photon-electron interaction, of the form $\mathfrak{L}_{m1} \simeq -(e^2\bar{n}_e/2m)\mathbf{A}^2$ where $\bar{n}_e$ is the mean local electron density. These describe the appearance of a plasma mass for systems with mobile electrons, rather than the polarizabilities of interest here.

$$\mathfrak{D}(t) = \begin{pmatrix} 0 & \mathbf{d}\, e^{i\omega_{eg}t} \\ \overline{\mathbf{d}}\, e^{-i\omega_{eg}t} & 0 \end{pmatrix}. \tag{16.101}$$

This encodes the requirements of rotational invariance, which implies that $\langle e, \ell_z|\mathfrak{D}|e, \ell_z\rangle = \langle g|\mathfrak{D}|g\rangle = 0$ while $\langle e, \ell_z|\mathfrak{D}|g\rangle = \mathbf{d}(\ell_z)\, e^{i\omega_{eg}t}$ and $\langle g|\mathfrak{D}|e, \ell_z\rangle = \overline{\mathbf{d}}(\ell_z)\, e^{-i\omega_{eg}t}$.

Mean-Field Action

The goal now is to evaluate the second-order expression, (16.53), reproduced here as

$$V_{\rm m}(t) = \overline{V}(t) - i\int_{t_0}^{t} d\tau\, \langle\!\langle \delta V(t)\, \delta V(\tau)\rangle\!\rangle + \mathcal{O}\left(V^3\right), \tag{16.102}$$

for the photon's mean-field Hamiltonian in an environment of these simple two-level atoms. The main interest is in slowly varying electric fields compared with the timescale set by ω_{eg}, and so takes $\mathbf{E}$ to be time-independent.

For sector B assume a time-independent density matrix

$$\varrho_B = \prod_{i=1}^{N} \otimes \begin{pmatrix} p_e I_3/3 & 0 \\ 0 & p_g \end{pmatrix}, \tag{16.103}$$

where I_3 is the 3×3 unit matrix and the two universal numbers p_e and p_g are the same for all atoms. This commutes with the sector B Hamiltonian given in (16.98) and (16.99), and is general enough to include, for example, a thermal state. Clearly, $p_e + p_g = 1$ ensures $\mathrm{tr}_B\, \varrho_B = 1$. Physically, p_e gives the fraction of excited atoms in this environment.

With these assumptions the first-order contribution to $V_{\rm m}$ vanishes, because $\langle\!\langle V\rangle\!\rangle = 0$. The second-order term in (16.102) requires computing the autocorrelation function for $\mathfrak{D}_i$, which is

$$\begin{aligned}\langle\!\langle \mathfrak{D}_i^a(t)\mathfrak{D}_j^b(s)\rangle\!\rangle &= \left[p_g\, \langle g|\mathfrak{D}^a(t)\mathfrak{D}^b(s)|g\rangle + \frac{p_e}{3}\sum_{\ell_z=-1,0,1} \langle e, \ell_z|\mathfrak{D}^a(t)\mathfrak{D}^b(s)|e, \ell_z\rangle\right]\delta_{ij} \\ &= \left[p_g\, e^{-i\omega_{eg}(t-s)} + \frac{p_e}{3}\, e^{i\omega_{eg}(t-s)}\right]\frac{\mathcal{D}}{3}\,\delta^{ab}\,\delta_{ij},\end{aligned} \tag{16.104}$$

where the last line uses the assumed rotational invariance of ϱ_B and defines the non-negative fluctuation coefficient,

$$\mathcal{D} := \sum_{\ell_z=0,\pm1} \overline{\mathbf{d}}(\ell_z)\cdot \mathbf{d}(\ell_z) \geq 0. \tag{16.105}$$

Neglecting the t-dependence of $\mathbf{E}$ allows the time integration in Eq. (16.102) to be performed explicitly to obtain $V_{\rm m}$, giving[12]

[12] An oscillatory term proportional to $e^{-i\omega_{eg}(t-t_0)}$ is dropped when evaluating the integral, assuming E_e has a small negative imaginary part due to the excited state's instability to photon emission through the process $|e\rangle \to |g\rangle + \gamma$ described in §12.2.2.

$$
\begin{aligned}
V_{\rm m}(t) &\simeq -{\rm i}\sum_{ij=1}^{N} E_a\,E_b \int_{t_0}^{t} {\rm d}s \langle\!\langle \mathfrak{D}_i^a(t)\mathfrak{D}_j^b(s)\rangle\!\rangle \\
&= -{\rm i}\sum_{i=1}^{N} \frac{\mathbf{E}^2}{3}\int_{t_0}^{t} {\rm d}s\, \mathcal{D}\left[p_g\, e^{-{\rm i}\omega_{eg}(t-s)} + \frac{p_e}{3}\, e^{{\rm i}\omega_{eg}(t-s)}\right] \\
&= -\frac{\mathcal{D}}{3\,\omega_{eg}}\left(N_g - \frac{N_e}{3}\right)\mathbf{E}^2,
\end{aligned}
\tag{16.106}
$$

where $N_e = p_e N$ and $N_g = p_g N$ are, respectively, the total number of atoms in the excited and ground states. The overall negative sign here indicates that the atom reduces its energy by polarizing in the presence of a local electric field.

In order of magnitude $\omega_{eg} \sim e^2/a_B$ (where a_B is the Bohr radius) is a typical atomic energy scale, which is a UV quantity for the effective description of photon propagation. This makes it reasonable to have it appear in the denominator of the effective coupling strength. Furthermore, on dimensional grounds one also expects $\mathcal{D} \sim \mathbf{d}\bar{\mathbf{d}} \sim (ea_B)^2$ where $-e$ is the electron's charge, leading to the estimate

$$
\frac{\mathcal{D}}{\omega_{eg}} \sim a_B^3. \tag{16.107}
$$

To read off the polarizabilities this should be matched to the internal electrostatic energy density for dielectrics in macroscopic electromagnetic fields, as computed for the Maxwell action coupled to (16.86). For isotropic systems, when the energy density is computed for given external charges specified then this has the familiar positive form [451]

$$
\mathcal{H}_A(\mathbf{D}) := \frac{H_A}{\Omega} = \frac{\mathbf{D}^2}{2\,\varepsilon}, \tag{16.108}
$$

where Ω is the volume of the region of space of interest in the above calculation, $\mathbf{D} = \varepsilon\,\mathbf{E}$ is the macroscopic electric displacement field in the long-wavelength theory, and

$$
\varepsilon = 1 + \bar{n}\,\mathfrak{p}_E \tag{16.109}
$$

is the medium's dielectric permittivity.

By contrast, when the energy density is determined by specifying the electrostatic potentials (or electric fields) it instead is given (for isotropic media) by the Legendre transform, $\widetilde{\mathcal{H}}_A = \mathcal{H}_A - \mathbf{E}\cdot\mathbf{D}$ and so is [451]

$$
\widetilde{\mathcal{H}}_A(\mathbf{E}) := \frac{\widetilde{H}_A}{\Omega} = -\frac{\varepsilon\,\mathbf{E}^2}{2}. \tag{16.110}
$$

Eqs. (16.108) and (16.110) differ in their overall sign because they measure different things. In particular, to fix external electric potentials when changing fields a charge reservoir is required, from which charges flow in or out as required by the condition that potentials remain fixed. It is the latter of these two energies that is relevant when comparing to (16.106) because by taking the external electric fields to be given it implicitly specifies the external potentials rather than charges.

Comparing (16.110) with (16.106) then gives $\mathfrak{p}_B \simeq 0$ (because magnetic response vanishes at the order studied for the two-level model) and

$$\mathfrak{p}_E = \frac{2\mathcal{D}}{3\,\omega_{eg}}\left(p_g - \frac{p_e}{3}\right) = \frac{2\mathcal{D}}{3\,\omega_{eg}}\left(1 - \frac{4p_e}{3}\right), \tag{16.111}$$

where the second equality uses $p_e + p_g = 1$. Using the estimate (16.107) then gives $\mathfrak{p}_E \sim O(a_B^3)$ with $\mathfrak{p}_E$ positive in the absence of a significant population inversion (*i.e.* unless more than $\frac{3}{4}$ of the atoms present happen to be in the metastable excited state).

The predicted index of refraction becomes

$$\mathfrak{n}_m = \sqrt{1 + \bar{n}\,\mathfrak{p}_E} \simeq 1 + \frac{\bar{n}\mathcal{D}}{3\,\omega_{eg}}\left(1 - \frac{4p_e}{3}\right), \tag{16.112}$$

where the approximate equality assumes a dilute system for which $\bar{n}\mathcal{D}/\omega_{eg} \sim \bar{n}\,a_B^3 \ll 1$ (in which case $\mathfrak{n}_m - 1 \ll 1$ as well, such as is true for air at standard temperatures and pressures). It is clear, however, that systems exist for which average inter-particle separations are of order atomic sizes, and so for which $\bar{n} \sim 1/a_B^3$. When this is true (16.112) predicts that $\mathfrak{n}_m - 1$ can easily be of order unity (such as is measured to be true for everyday materials like glass or water, for which $\mathfrak{n}_m \simeq 1.5$ or 1.3, respectively). The reliability of predictions like (16.112) in the regime where $\bar{n}\mathfrak{p}_E$ is not small is a topic returned to in §16.3.5.

16.3.4 Photons: Scattering and Fluctuations♠

So far, the treatment of both neutrino and photon examples of particle propagation within a medium has been done exclusively within the mean-field approximation. Before quantifying what controls this approximation (particularly for photons, where the weakness of the underlying particle-environment couplings is insufficient in itself) it is useful to examine some implications of the fluctuation part of evolution. As a bonus, this examination makes more explicit the connection between the general open-system formalism and standard scattering results.

To this end, consider the situation where sector A describes the single-particle properties (momentum, spin *etc*) of a particle moving through some sort of medium, with sector B describing all of the unmeasured properties of the medium. In this language scattering consists of anything that changes the sector-A momentum. The advantage of framing scattering in terms of the general open-system framework is the robustness with which it identifies precisely what properties of the medium are responsible for particle scattering.

It is not unusual (but also not inevitable) for the mean-field limit of the medium to be translation invariant, at least when coarse-grained on macroscopic scales. If so, the mean-field part of the evolution conserves particle momentum and so cannot contribute to scattering. Scattering then arises (if it does at all) purely because of interactions with the medium's fluctuations about its mean. Scattering from fluctuations is the topic now studied in more detail.

It should be emphasized that the recursive nature of the open-system framework implies that a study of scattering from fluctuations can also capture situations where scattering comes from inhomogeneities in the mean properties. That is, it sometimes happens that mean properties are not translation invariant on some scales, but become so on larger scales.[13] So fluctuations in this section can cover a diverse range of phenomena, starting from fluctuations in particle properties when describing scattering from individual atoms in the medium, and continuing to coherent processes (like refraction and reflection) at larger distances from more complicated position dependence in mean particle properties.

Scattering from Fluctuations

To set up scattering problems using the formalism of open systems, imagine that sector A consists of the single-particle sector corresponding to a particle traversing the medium, and for simplicity assume these to be spanned simply by momentum eigenstates, $|\mathbf{p}\rangle$. Sector B collectively includes all degrees of freedom of the medium. Things are conceptually simplest if the medium does not consist of particles of the type passing through – such as neutrinos or muons (as opposed to electrons, say) passing through ordinary solids or gasses – but this assumption is not strictly necessary.

The initial state is taken to be uncorrelated, $\rho_A(t_0) = \varrho_A \otimes \varrho_B$, with

$$\varrho_B = \sum_b p_B(b)\,|b\rangle\langle b| \quad \text{and} \quad \varrho_A = \sum_a p_A(a)\,|a\rangle\langle a| \tag{16.113}$$

describing the initial state of the environment and the initial particle beam.[14] A fundamental assertion of quantum mechanics states that the probability for measuring the system to be in a range of final states, $A_f := \{|a'\rangle\}$, at time t is then captured by evolving ρ from t_0 to t and computing the expectation value for the hermitian observable $\mathcal{O}_{A_f} = O_{A_f} \otimes I_B$, where I_B is the unit matrix in sector B while

$$O_{A_f} = \sum_{a' \in A_f} |a'\rangle\langle a'| \tag{16.114}$$

is the projection matrix onto the final states of interest in sector A. The validity of this assertion may be seen by explicit evaluation, since

$$\mathrm{Tr}\left[\rho(t)\,\mathcal{O}_{A_f}\right] = \mathop{\mathrm{tr}}_A\left[\rho_A(t)\,O_{A_f}\right] = \sum_{\substack{a\in A_i \\ a'\in A_f}} \sum_{bb'} p_A(a)p_B(b)|\langle a', b'|U(t,t_0)|a,b\rangle|^2. \tag{16.115}$$

With later applications in mind, suppose that $\langle a'|\overline{U}(t,t_0)|a\rangle = 0$ for all $a' \neq a$, so the mean-field contribution does not contribute to scattering (such as is true when $|a\rangle$ are momentum eigenstates and $\overline{U}$ is translation invariant). Then all scattering comes from fluctuations, ΔU, for which the leading expression in powers of V is

$$\Delta U(t,t_0) \simeq -\mathrm{i}\int_{t_0}^{t} \mathrm{d}s\left[V(s) - \langle\!\langle V(s)\rangle\!\rangle_B\right] = -\mathrm{i}\int_{t_0}^{t} \mathrm{d}s\;\delta V(s). \tag{16.116}$$

[13] Clouds furnish a concrete example of this type, for which the water-density profile varies in space – *i.e.* from droplet to droplet – when examined on scales large compared with atomic fluctuations but small compared to droplet sizes. The mean density becomes homogeneous, however, once coarse-grained to scales larger than the droplets, in which case light-scattering by droplets becomes described as scattering from fluctuations.

[14] The discussion starts using discrete states (such as if momenta were normalized within a large but finite-sized box), with the transition to continuum normalization taken later.

The transition probability from states $A_i = \{|a\rangle\}$ to states $A_f = \{|a'\rangle\}$ (where for simplicity A_i and A_f are taken to be disjoint) in this case becomes

$$P_t(A_i \to A_f) := \sum_{\substack{a\in A_i \\ a'\in A_f}} \sum_{bb'} p_{\mathcal{A}}(a) p_{\mathcal{B}}(b) |\langle a', b'|\Delta U(t, t_0)|a, b\rangle|^2. \tag{16.117}$$

In the limit, $t - t_0 \to \infty$, the transition amplitude, $\langle a', b'|\Delta U(t, t_0)|a, b\rangle$, between eigenstates of $H_0 = H_{\mathcal{A}} + H_{\mathcal{B}}$ goes to

$$-\mathrm{i} \int_{-\infty}^{\infty} \mathrm{d}s \, \langle a', b'|\delta V(s)|a, b\rangle = -2\pi\mathrm{i} \, \langle a', b'|\delta V(0)|a, b\rangle \, \delta(E - E'). \tag{16.118}$$

Squaring this last expression and inserting the result into (16.117) to obtain $P_t(A_i \to A_f)$ gives a well-known ill-defined factor $\delta(E = 0)$. For the present purposes, this issue can be side-stepped[15] by instead differentiating to compute the transition rate, $\Gamma(A_i \to A_f) = \partial_t P_t(A_i \to A_f)$. Doing so using the lowest-order expression (16.116) gives

$$\Gamma(A_i \to A_f) = \sum_{\substack{a\in A_i \\ a'\in A_f}} \sum_{bb'} p_{\mathcal{A}}(a) p_{\mathcal{B}}(b) \left[\langle a', b'|\delta V(t)|a, b\rangle \int_{t_0}^{t} \mathrm{d}s \, \langle a', b'|\delta V(s)|a, b\rangle^* + \text{c.c.} \right], \tag{16.119}$$

which in the long-time limit, $t - t_0 \to \infty$, approaches the well-defined result

$$\Gamma(A_i \to A_f) \to 2\pi \sum_{\substack{a\in A_i \\ a'\in A_f}} \sum_{bb'} p_{\mathcal{A}}(a) p_{\mathcal{B}}(b) |\langle a', b'|\delta V(0)|a, b\rangle|^2 \delta(E - E'), \tag{16.120}$$

which is Fermi's golden rule for lowest-order scattering from fluctuations.

Thomson Scattering

To make the above manipulations concrete it is useful to have a simple illustrative application in mind. To this end, consider photons (described by the electromagnetic vector potential $\mathbf{A}$) moving through a medium consisting of massive charged particles (with charge e_q and mass m, represented by a Schrödinger field Ψ). As discussed in §12.1.3, at low energies photon scattering from such particles is described by Thomson scattering, *via* the leading low-energy effective coupling

$$V = \frac{e_q^2}{2m} \int \mathrm{d}^3\mathbf{x} \, \mathbf{A}^2 (\Psi^\dagger \Psi). \tag{16.121}$$

The goal is to evaluate the above expressions in this example to compute the rate with which photons scatter from such a medium.

To leading order, the mean-field Hamiltonian for Thomson scattering therefore is

$$V_{\mathrm{m}} \simeq \langle\!\langle V \rangle\!\rangle = \frac{e_q^2}{2m} \int \mathrm{d}^3\mathbf{x} \, \mathbf{A}^2 \, \bar{n}(\mathbf{x}), \tag{16.122}$$

[15] This little dance is a special case of a more general argument; as is further explored in §16.4.1.

where the mean heavy-particle density, $\bar{n} := \text{tr}_B[\Psi^\dagger\Psi\,\varrho_B]$, is position-independent if the environment is translation-invariant.

Assuming the mean-field limit of the medium is translation invariant – so $\overline{U}$ does not scatter the photons – and switching to continuum normalized momentum eigenstates when using (16.121) in Eq. (16.120) gives the leading contribution to the unpolarized scattering rate as

$$\begin{aligned} \mathrm{d}\Gamma[\gamma(\mathbf{k}) \to \gamma(\mathbf{k}')] &\simeq 2\pi\left(2 \times \frac{e_q^2}{2m}\right)^2 \left[\frac{1}{2\omega_k(2\pi)^3}\right]^2 \frac{1}{2}\sum_{\lambda\tilde{\lambda}} |\tilde{\epsilon}\cdot\epsilon|^2\delta(\omega_k - \omega_{k'}) \\ &\quad \times G_B(\mathbf{k}-\mathbf{k}')\, f_A(\mathbf{k})[1+f_A(\mathbf{k}')]\,\mathrm{d}^3\mathbf{k}\,\mathrm{d}^3\mathbf{k}' \\ &= \frac{1}{8(2\pi)^5}\left(\frac{e_q^2}{m}\right)^2 \sum_{\lambda\tilde{\lambda}} |\tilde{\epsilon}\cdot\epsilon|^2\, G_B(\mathbf{k}-\mathbf{k}')f_A(\mathbf{k})[1+f_A(\mathbf{k}')]\,\mathrm{d}^3\mathbf{k}\,\mathrm{d}\Omega', \end{aligned} \tag{16.123}$$

where in the continuum, $p_A(a) \to f_A(\mathbf{k})$ is the dimensionless phase-space density of initial photons of momentum $\mathbf{k}$, with the Bose-Einstein statistics of photons used when evaluating the matrix element of V. In the final line $\mathrm{d}\Omega'$ is the element of solid angle for the final photon's direction while $\omega_k = |\mathbf{k}|$ and

$$\begin{aligned} G_B(\mathbf{k}-\mathbf{k}') &:= \sum_{bb'} p_B(b) \int \mathrm{d}^3x\,\mathrm{d}^3y\; e^{\mathrm{i}(\mathbf{k}-\mathbf{k}')\cdot(\mathbf{x}-\mathbf{y})}\langle b'|\delta n(\mathbf{x})|b\rangle\langle b|\delta n(\mathbf{y})|b'\rangle \\ &= \int \mathrm{d}^3x\,\mathrm{d}^3y\; e^{\mathrm{i}(\mathbf{k}-\mathbf{k}')\cdot(\mathbf{x}-\mathbf{y})}\langle\!\langle \delta n(\mathbf{x})\delta n(\mathbf{y})\rangle\!\rangle. \end{aligned} \tag{16.124}$$

Here $n(\mathbf{x}) := \Psi^\dagger\Psi(\mathbf{x})$ is the density operator for the field $\Psi(\mathbf{x})$, while $\delta n(\mathbf{x}) = n(\mathbf{x}) - \langle\!\langle n(\mathbf{x})\rangle\!\rangle$ and $\langle\!\langle \delta n(\mathbf{x})\delta n(\mathbf{y})\rangle\!\rangle = \text{tr}_B[\delta n(\mathbf{x})\varrho_B\delta n(\mathbf{y})]$.

For translation-invariant media the correlation function $\langle\!\langle \delta n(\mathbf{x})\delta n(\mathbf{y})\rangle\!\rangle$ is a function only of $\mathbf{s} := \mathbf{x} - \mathbf{y}$, and so is completely independent of $\mathbf{r} := \frac{1}{2}(\mathbf{x}+\mathbf{y})$. Consequently, $G_B(\mathbf{k}-\mathbf{k}') = \int \mathrm{d}^3\mathbf{r}\; \mathcal{G}_B(\mathbf{k}-\mathbf{k}')$ with

$$\mathcal{G}_B(\mathbf{k}-\mathbf{k}') := \int \mathrm{d}^3s\; e^{\mathrm{i}(\mathbf{k}-\mathbf{k}')\cdot\mathbf{s}}\langle\!\langle \delta n(\mathbf{s})\delta n(0)\rangle\!\rangle. \tag{16.125}$$

This shows that $\mathrm{d}\Gamma = \int \mathrm{d}^3\mathbf{r}\;\mathrm{d}\lambda(\mathbf{r})$, where for translation-invariant systems the differential scattering rate per unit volume, $\mathrm{d}\lambda(\mathbf{r})$, is both position-independent and well-defined in the large-volume limit:

$$\begin{aligned} \frac{\mathrm{d}\lambda}{\mathrm{d}\Omega'}[\gamma(\mathbf{k}) \to \gamma(\mathbf{k}')] &\simeq \frac{1}{8}\left(\frac{e_q^2}{m}\right)^2 \frac{1}{(2\pi)^5}\sum_{\lambda\tilde{\lambda}} |\tilde{\epsilon}\cdot\epsilon|^2\, \mathcal{G}_B(\mathbf{k}-\mathbf{k}')f_A(\mathbf{k})[1+f_A(\mathbf{k}')]\,\mathrm{d}^3\mathbf{k} \\ &= f_A(1+f_A')\left(\frac{\mathrm{d}\sigma_{\mathrm{th}}}{\mathrm{d}\Omega'}\right)\mathcal{G}_B(\mathbf{k}-\mathbf{k}')\,\frac{\mathrm{d}^3\mathbf{k}}{(2\pi)^3}, \end{aligned} \tag{16.126}$$

with $\mathrm{d}\sigma_{\mathrm{th}}/\mathrm{d}\Omega'$ the Thomson cross section of (12.34).

In the particular case where B is a fluid of independent nonrelativistic charged fermions, using continuum-normalized momentum eigenstates as a basis shows that $\langle\mathbf{p}'|\delta n(\mathbf{x})|\mathbf{p}\rangle \simeq (2\pi)^{-3}\exp[\mathrm{i}(\mathbf{p}-\mathbf{p}')\cdot\mathbf{x}]$, and so (16.125) reduces to

$$\mathcal{G}(\mathbf{k}-\mathbf{k}') \simeq \int \frac{\mathrm{d}^3p\,\mathrm{d}^3p'}{(2\pi)^3}\, f_B(\mathbf{p})[1-f_B(\mathbf{p}')]\,\delta^3(\mathbf{p}+\mathbf{k}-\mathbf{p}'-\mathbf{k}'). \tag{16.127}$$

For dilute systems (defined by $f_B \ll 1$) this becomes

$$\mathcal{G}(\mathbf{k}-\mathbf{k}') \simeq \int \frac{d^3p}{(2\pi)^3} f_B(\mathbf{p}) = \bar{n}, \tag{16.128}$$

and so using this in (16.126) reproduces the differential scattering rate expected from independent Thomson scattering from a bath of charged particles with total density $\bar{n}$.

Scattering from a dilute thermal distribution corresponds to the special case

$$f_{A,B} = \frac{1}{e^{(E-\mu)/T} \pm 1} \tag{16.129}$$

where μ and T are the chemical potential and temperature (with the upper sign corresponding to Fermi–Dirac statistics and the lower sign to Bose–Einstein statistics). But scattering from more general thermal fluids is also easily captured by this formalism. As shown in Exercise 16.4, scattering from a thermal bath is equally well encoded as a special case of scattering from density fluctuations, where the fluctuations are those of the grand canonical ensemble.

The scattering rate is more usefully expressed in terms of the *extinction coefficient*, $\mathfrak{h}$, defined[16] as the rate for scattering photons into all directions divided by the incident photon flux $\mathcal{F} := f_A(\mathbf{k})\, d^3\mathbf{k}/(2\pi)^3$. For (16.126) this becomes

$$\mathfrak{h} := \frac{1}{\mathcal{F}} \int \left(\frac{d\lambda}{d\Omega'}\right) d\Omega' \simeq \frac{1}{8}\left(\frac{e_q^2}{m}\right)^2 \frac{1}{(2\pi)^2} \sum_{\lambda\tilde{\lambda}} \int d\Omega' \, |\tilde{\epsilon} \cdot \epsilon|^2 \, \mathcal{G}_B(\mathbf{k}-\mathbf{k}')[1 + f_A(\mathbf{k}')]$$
$$\simeq \bar{n}\, \sigma_{\text{th}}, \tag{16.130}$$

where the approximate equality neglects $f_A(\mathbf{k}')$ and specializes to dilute systems with ideal-gas fluctuations.

Macroscopic Fluctuations

The above example shows why particles moving through translation-invariant media are scattered dominantly by the medium's fluctuations. The fluctuations in the medium that matter for this scattering are those of the interaction Hamiltonian, $V(t)$, itself. For interactions that couple to the density of particles in the medium it is density fluctuations that control the scattering rate.

The previous example also shows that when the important fluctuations are thermal and the system is dilute (and so described thermodynamically as an ideal gas, say) then scattering from density fluctuations reproduces the same result as obtained by computing scattering from individual particles in the medium and summing these incoherently to obtain a rate proportional to the mean density of scatterers, $\bar{n}$.

But the power of the open-system derivation is its generality, and nothing requires the fluctuations responsible for scattering to be as simple as thermal. More generically, fluctuations also arise when environments are mixtures of materials of different kinds, or when particles scatter from more complex collective properties. These other cases are often equally well captured by the fluctuation formalism

16 Extinction is sometimes expressed in terms of scattered intensity (or energy) rather than rate for photon scattering, but the difference is immaterial for the elastic scattering considered here.

given above, with the quantities like $\langle\langle \delta n(\mathbf{x})\delta n(\mathbf{y})\rangle\rangle$ (or its analog for other scattering processes) instead computed by averaging over these alternate fluctuations.

As mentioned earlier, clouds provide a concrete example of this type. Indeed, if microscopic thermal fluctuations of the underlying atoms were the whole story one would conclude that clouds must be transparent because they are made of air and water and these are separately transparent. What this simplistic argument misses are the droplets, through which light rays refract and from which they reflect. From a microscopic point of view (smaller than the droplets) this is an example of scattering within the mean-field sector, because the mean properties like density are not translation invariant. But when examined on scales much larger than the droplet sizes the scattering is again from density fluctuations; it is just that it is the drops themselves that furnish the properties of these fluctuations (see Exercise 16.6).

16.3.5 Domain of Validity of Mean-Field Theory

It is now time to return to the question of what controls the validity of mean-field theory: *i.e.* what is the small parameter that controls the size of ΔU evolution relative to $\overline{U}$? In particular, is there a nontrivial regime described using only $\overline{U}$ while neglecting ΔU completely? (One thinks the answer to this should be 'yes', given experience with light in the geometrical optics regime for which rays can undergo multiple refractions and reflections with negligible degradation in intensity.)

For neutrinos the answer appears relatively straightforward, and is rooted in the feebleness of the weak interactions. Since for neutrinos it suffices to work to linear order in V, ΔU plays no role since its first contribution arises at $\mathcal{O}(V^2)$. This reasoning is basically correct, although there is a late-time complication that must be dealt with when discussing oscillations (described in §16.4).

The situation is not quite as simple for photons in dielectrics, for which both $\overline{U}$ and ΔU typically receive their leading contributions at order V^2. If the size of V were the only relevant quantity why would it ever be true that photons could propagate far enough to be refracted or reflected (for which only $\overline{U}$ is relevant) without having also been scattered out of the initial beam (through the scattering in ΔU)?

This section argues that for photons there is a second small parameter in the problem that suppresses ΔU relative to $\overline{U}$ at any fixed order in V. This second parameter is $1/N_{\rm coh}$, where $N_{\rm coh} \gg 1$ is the number of scattering sites within a *coherence volume*, defined as the volume from which a given photon can scatter coherently.

To see what is involved, imagine a photon scattering with amplitude $\mathcal{A}_i, i = 1, \cdots, N$ from N distinct scattering sites, with $N \gg 1$. These sites might be different atoms within a medium, or perhaps different droplets in a more complicated mixture. The total probability for scattering then is

$$P = \left|\sum_{i=1}^{N} \mathcal{A}_i\right|^2 = \sum_{i=1}^{N} |\mathcal{A}_i|^2 + \sum_{i>j=1}^{N} (\mathcal{A}_i^* \mathcal{A}_j + \mathcal{A}_i \mathcal{A}_j^*). \tag{16.131}$$

Denoting the modulus and phase of each scattering amplitude by $\mathcal{A}_i = A\, e^{i\alpha_i}$, with the modulus assumed to be the same for all i, compare the following two extreme situations. First, suppose that the phases, α_i, are also the same for all i. Then the scattering probability would be

$$P \to P_{\rm coh} := A^2 N^2 \tag{16.132}$$

which, for this 'coherent' case, scales quadratically with N.

Alternatively, imagine that the N scattering amplitudes carry different and random phases, $\mathfrak{a}_i$. Then

$$P \to P_{\rm inc} := A^2 \left[N + 2 \sum_{i>j=1}^{N} \cos(\mathfrak{a}_i - \mathfrak{a}_j) \right], \tag{16.133}$$

in which the final sum remains much smaller than N for large N because the phases cancel as often as reinforce one another. In this ('incoherent') case, P instead scales linearly with N, for large N, rather than quadratically. Clearly, $P_{\rm coh} \gg P_{\rm inc}$ when $N \gg 1$. Because scattering amplitudes include phase factors like $e^{i(\mathbf{k}-\mathbf{k}')\cdot\mathbf{x}}$, where $\mathbf{k}$ and $\mathbf{k}'$ are the initial and final photon momenta, coherent scattering usually[17] occurs in the forward direction (for which $\mathbf{k}' = \mathbf{k}$).

The large-N enhancement associated with coherent scattering can be a justification for working nontrivially with $\overline{U}$ while dropping ΔU. It can do so because it is $\overline{U}$ that typically contains the coherent part of a process while ΔU contains only the incoherent part. This distinction can be seen explicitly in the examples examined above. For instance, (16.106) and (16.122) predict $V_{\rm m} \propto N$ and so also predict any order-$V_{\rm m}^2$ contribution in $\overline{U}$ to be proportional to N^2. By contrast, formulae like (16.126) – together with (16.128) – or expression (16.130) are constructed using ΔU at second order in V and are instead proportional to N (through the factor $\bar{n}$).

How large is $N_{\rm coh}$ in specific systems? What counts is the number of scatterers for which the scattering phase remains approximately equal, as in (16.132). Since real materials are usually characterized by a particle density, $\bar{n}$, the number of particles sharing the same scattering phase is normally called the coherence volume, $\Omega_{\rm coh}$, with $N_{\rm coh} = \bar{n}\,\Omega_{\rm coh}$.

Usually, the coherence volume is at minimum of order $\Omega_{\rm coh} \gtrsim \lambda^3$, where λ is the wavelength of the scattering photon. But $\Omega_{\rm coh}$ can be made larger than this, such as by ensuring that the relative phases of the initial photon state remain fixed over many wavelengths (such as by preparing the photons using a laser). For atoms in a material with density $\bar{n} \sim 1/a_{\scriptscriptstyle B}^3$ the typical inter-atomic separation is of order 0.1 nm, so for visible light (with $\lambda \sim 500$ nm) $N_{\rm coh} = \bar{n}\lambda^3 \gtrsim (5000)^3 = 1.25 \times 10^{11}$ can easily be enormous.[18]

16.4 Late Times and Perturbation Theory *

Up to this point, the discussion of open EFTs has been entirely cast in perturbation theory: the influence of the unmeasured sector B on the evolution of observables in sector A is treated in powers of the interaction $V(t)$ that couples them to one another. But there is an important regime where this kind of perturbative reasoning generically breaks down: at very late times. This is potentially dangerous for any

17 Bragg scattering [452] shows that coherence can sometimes also occur for non-forward scattering.

18 By contrast: for MeV neutrinos $\lambda \sim 100$ fm and so $\bar{n}\lambda^3 \ll 1$ for materials right up to near-nuclear densities. This is why coherence is usually not the relevant issue for use of mean-field methods in neutrino physics.

EFT description, given that effective theories are typically aimed at slow processes that act over long distances.

A simple example of this breakdown is already seen in the above examples, in particular in the transition from Eq. (16.117) to (16.120). In that discussion a problem arose because of the divergence of the time integrals appearing in the scattering probability once (16.117) is evaluated in powers of V. But this is really just the tip of the proverbial iceberg; when interacting with an environment, it is ultimately generic that $\exp[-iVt]$ eventually differs significantly from $1 - iVt$, regardless of how small the matrix elements of V might be.

Experience with particles interacting with matter also shows that problems predicting perturbatively at late times are not esoteric or hypothetical. Consider, for example, the example of photons passing through a transparent material like glass or water. In this case, interactions are weak inasmuch as the likelihood of any particular photon interacting with any particular atom is reasonably small. Perturbation theory essentially predicts that to leading approximation nothing happens, so photons dominantly continue on relatively unchanged by the presence of the medium. Although this starts off as a good description, eventually – *e.g.* once one reaches the geometrical optics regime – essentially 100% of photons either refract or reflect and none remain unperturbed in their original trajectories. Evidently, perturbation theory in V must be a bad approximation in this late-time regime.

A similar issue arises with neutrino oscillations. If neutrino interactions really are so weak that going beyond linear order in V is unnecessary, it should never be true that small interaction-induced phase shifts can be reliably exponentiated, as they must to provide the interference inherent in the time-dependence $e^{-i\Delta E t}$ characteristic of oscillations.

The goal of this section is to explore how to make reliable predictions in these kinds of late-time limits. Happily, success making predictions need not require access to any sort of complete non-perturbative understanding (something that is rarely possible in quantum field theory). Instead, robust resummation techniques often exist that allow inference of late-time behaviour without a full solution of the entire theory. This section starts with the simplest examples of this type, and progresses on to more comprehensive techniques.

16.4.1 Late-Time Resummation

To start, it is useful to think through relatively well-understood examples of how nominally failing perturbation theory at late times can be resummed so as to restore the ability to predict. These simple examples are useful since they contain the essence of the core argument that also works in more general settings.

Exponential Decays

Radioactive decay provides perhaps the very simplest example of the apparent breakdown of perturbation theory at late times. In this case, a parent particle spontaneously decays, $\mathcal{P} \to \mathcal{D}_1 + \mathcal{D}_2 + \cdots$ (perhaps through the weak interactions) into a collection of daughter particles.

In the simplest examples this occurs due to the existence of a nonzero matrix element $\langle \mathcal{D}_1, \mathcal{D}_2, \cdots |V|\mathcal{P}\rangle$, and the absence of any conservation laws that forbid the transition $\mathcal{P} \to \mathcal{D}_1, \mathcal{D}_2, \cdots$. Standard expressions like Fermi's golden rule – *c.f.* Eq. (16.120) – then give the decay rate Γ in terms of $|\langle \mathcal{D}_1, \mathcal{D}_2, \cdots |V|\mathcal{P}\rangle|^2$, and are therefore second-order in the weak interaction responsible.

The late-time issue arises once one asks for the survival probability for a given atom as a function of time. This probability is given by the well-known exponential decay law,[19]

$$P(t) := P[\mathcal{P}(t_0) \to \mathcal{P}(t)] = e^{-\Gamma(t-t_0)}, \tag{16.134}$$

which is experimentally verified to be true for times $(t - t_0) \gg \tau = 1/\Gamma$, where τ is the decay's mean lifetime. But how can this exponential be trusted given that Γ is computed only to order V^2? Why is it not compulsory always to write (16.134) as

$$P(t) \simeq 1 - \Gamma(t - t_0) + \cdots, \tag{16.135}$$

with only unreliable predictions once $t - t_0 \gtrsim \tau$? Indeed, the first-principles derivation of the transition probability (as opposed to the rate) actually leads to (16.117), whose failure to converge at large t (at lowest order) corresponds precisely to the linear growth of (16.135).

The argument underlying the validity of the exponential decay law rests on two pillars: (*i*) the decay rate Γ *can* be computed perturbatively and gives a result that is independent of time and (*ii*) for *any* initial time, t_0, there is a finite (and possibly short) window of times t for which $P(t)$ satisfies

$$\frac{dP}{dt} = -\Gamma P, \tag{16.136}$$

which relies only on the likelihood of having a decay in any short time window being independent of the likelihood of there being a decay in any other time windows. (This last property can break down in some specific situations, leading then to non-exponential evolution [453]).

Now comes the main point: even though the time windows used to derive (16.136) have a finite size there is often an unending overlapping set of such windows since Eq. (16.136) itself makes no direct reference to time, and in its derivation t_0 is arbitrary. As a result, the domain of validity of (16.136) – and hence also of its solutions like (16.134) – is actually the union over all t_0 of all overlapping domains on which it can be derived.

At first sight, this seems an odd type of argument. After all, Eq. (16.136) is ultimately derived by differentiating (16.135), and dropping terms higher than quadratic in V. So, normally both equations would be expected to have precisely the same domain of validity. But because (16.135) leads to (16.136) for any choice of t_0, this latter equation actually can remain valid for arbitrarily large times. By contrast, (16.135) itself explicitly depends on t_0 and so is not equally extendable without first passing through (16.136). It is because (16.136) has this extended domain of validity that its solution – the exponential decay law, (16.134) – can be valid over times much longer than $\tau = 1/\Gamma$.

[19] There can also be deviations from exponential decay, under circumstances made clearer below.

The astute reader will notice a similarity between this argument and the renormalization-group argument made when transitioning between Eq. (7.28) and (7.32) in earlier chapters. The basic logic is the same: although differentiation followed by integration seems as if it should not provide new information, it sometimes can if the differential version of a result has a broader domain of validity.

This is a very powerful line of argument, and when it works it allows working to all orders in Γt without having to understand all observables at all orders in V. When applied using the leading-order expression for Γ, use of (16.134) when integrating (16.136) amounts to resumming all orders in Γt while dropping terms that involve extra powers of the interaction V without the corresponding extra powers of t (much like summing the 'leading logs' in a renormalization-group argument).

The remainder of this section sketches how this reasoning also works in a second important class of examples: the propagation of particles through matter, for which interaction phases generated by mean-field evolution $\overline{U}$ are often exponentiated when following the interference of states as they evolve in time (such as for matter-induced neutrino oscillations and the coherent scattering and interference of electromagnetic waves). The following section then frames the argument in an even more general context.

Neutrino Oscillations

As always, the first step in understanding why mean-field interaction phases can be exponentiated consistently lies in recalling why the fluctuation evolution, ΔU, is suppressed relative to $\overline{U}$ in the first place. Previous sections argue that for neutrinos (in environments like the solar interior) this suppression relies only on the weakness of the weak interactions.

To pin this down more precisely, recall that at first order in V mean-field interactions introduce single-particle energy shifts to neutrinos – see *e.g.* the discussion following Eq. (16.77) – of order $V_1 = \langle\langle V \rangle\rangle \sim G_F \bar{n}$, where G_F is Fermi's constant and $\bar{n} = \langle\langle n \rangle\rangle$ is the density of scatterers (in practice, the density of neutrons or electrons).

By contrast, the leading contributions to ΔU are given by the rate of scattering from fluctuations, and taking these to be given by microscopic particle-by-particle fluctuations leads to a depletion of probability from the mean-field sector that is of order $\Gamma \sim \bar{n}\,\sigma$, where σ is the neutrino interaction cross section. An estimate of the cross section at the energies of practical interest is $\sigma \sim G_F^2 mE$, with E the neutrino energy and m the scatterer mass, for scatterers assumed to be approximately at rest.[20] Fluctuations are therefore negligible relative to mean-field evolution to the extent that Γ/V_1 remains small, *i.e.* when

$$G_F mE \simeq 1.2 \times 10^{-8} \left(\frac{m}{\text{GeV}}\right)\left(\frac{E}{\text{MeV}}\right) \ll 1. \tag{16.137}$$

This is clearly satisfied for MeV neutrinos scattering from electrons or nucleons, say.

What of the validity of exponentiating matter-dependent phase-shifts of the form $\delta_m \sim G_F \bar{n}\, t$ when computing interference between the evolution of neutrino eigenstates? Such evolution is required when computing oscillation patterns sensitive to

[20] More precisely, this estimate applies at energies low enough for the Fermi theory of §7.1 to apply (*i.e.* well below 100 GeV).

time-dependent interference terms like $e^{-i(E_1-E_2)t}$ (see, for example, Exercise 16.2), where E_i are the energies of two species of neutrino. Neutrino oscillations require these factors be evaluated with $|E_2 - E_1| \sim \delta m^2/E$, which by assumption is similar to $G_F \bar{n}_e$ for resonant oscillations.

The argument for this very much follows the logic of exponential decays. For single-particle neutrino states the differential version of the evolution is given by the Liouville (or Schrödinger) equation itself, (16.41). This equation is the differential version of the mean-field evolution given in terms of $\overline{U}$ in (16.49), and although $\overline{U}$ can only be computed in powers of V over a finite range, $(t - t_0)$, this can be done for any value of t_0. Consequently, (16.41) has a broader domain of validity, and so its solutions can be trusted even when $t - t_0$ is so large that $\int_{t_0}^{t} ds\, \overline{V}(s)$ is not small and $\overline{U}$ is not arbitrarily close to the unit operator.

Photons

A very similar story also holds for electromagnetic fields within dielectrics, although with different details due to the requirement, discussed above, for coherence when neglecting scatterings from fluctuations.

In more detail, recall that the leading mean-field interaction Hamiltonian arises in this case at second order in the underlying interactions – *c.f.* for example, Eq. (16.106) – and gives single-photon states energy shifts of the form $\delta\omega(\mathbf{k}) \simeq (c_m - 1)k$, where the photon propagation speed within the medium is related to its index of refraction by $c_m = \mathfrak{n}_m^{-1}$.

The validity of keeping only mean-field evolution and dropping ΔU relies on this energy shift being much larger than the extinction rate, which Exercise 16.6 shows is given in the same regime by the Rayleigh scattering expression $\mathfrak{h} \simeq \bar{n}\,\sigma_R \sim \omega^4(\mathfrak{n}_m - 1)^2/\bar{n}$, and so (writing $\omega = 2\pi/\lambda$) requires

$$\frac{\omega^3(1-c_m)}{\bar{n}\,c_m^2} = \frac{(2\pi)^3}{\lambda^3\bar{n}}\left(\frac{1-c_m}{c_m^2}\right) \ll 1. \tag{16.138}$$

Clearly, this allows $1 - c_m$ to be order unity provided that $\lambda^3\bar{n}$ is large;[21] *i.e.* there are a large number of scatterers within the volume set by a cubic wavelength. Neglect of fluctuations is valid when the number of coherent scatterers – $N_{\rm coh}$ defined in the discussion below Eq. (16.133) – is large.

What justifies working beyond linear order in $\delta\omega\, t$ when computing the late-time interference of photons propagating in such a medium? Just like for the previous examples, this is justified because the differential mean-field evolution equation for photon states,

$$i\,\partial_t \rho_A = \big[V_{\rm m}(t)\,,\,\rho_A\big], \tag{16.139}$$

can be derived with a broader domain of validity than can the direct expression for $\overline{U}$ as a function of $V(t)$. And it is this differential equation that ensures photon propagation eigenstates evolve proportional to $e^{-i\omega t}$ with $\omega \simeq c_m k = k/\mathfrak{n}_m$ included in the phase.

[21] Of course, having λ much larger than inter-particle spacings is also part of the limit that allows the medium to be coarse-grained in the first place; it is because the photon interacts with many atoms that its multiple interactions can be well-described in terms of a few medium properties.

16.4.2 Master Equations

The previous sections describe several situations for which late-time behaviour of a reduced state can be computed perturbatively, despite there being generic problems in this regime due to the systematic appearance of powers of the evolution time, $t - t_0$, together with powers of the interaction Hamiltonian, V.

This section closes out the book by exploring more generally how this can happen, in particular extending its treatment to cases where mean-field methods need not play such a central role. This is done through the development of a series of (quantum) *master equations*, defined as evolution equations[22] for the density matrix $\rho_A(t)$, defined by (16.27).

Nakajima–Zwanzig Equation

The first master equation derived is obtained by returning to first principles, starting from the Liouville equation, (16.33), for the full system. The problem with directly tracing the Liouville equation – such as in (16.44) – is that the right-hand side depends on the full density matrix ρ rather than just the reduced matrix ρ_A. Formally, the problem of identifying the evolution of the reduced density matrix, ρ_A, in terms of itself is solved by projecting the Liouville time-evolution of the full density matrix, ρ, onto the subspace A using an appropriate projection operator.

To see how this works, consider the real vector space consisting of hermitian operators acting within the quantum-mechanical Hilbert space. Density matrices live in this vector space and any linear transformation – like time evolution – acting on a density matrix can be regarded as a 'super-operator' acting on this space (as opposed to an 'operator', like ρ, that acts directly on the quantum mechanical Hilbert space) [454].

In particular, when computing the evolution of $\rho_A(t)$ a special role is played by the projection super-operator $\mathcal{P}$, defined to act on a general hermitian operator O by

$$\mathcal{P}(O) := \operatorname*{tr}_B(O) \otimes \varrho_B. \tag{16.140}$$

Here, ϱ_B is a density matrix that characterizes sector B (the environment). In practice, it is chosen equal to the initial B state, $\rho_0 = \rho(t_0) = \varrho_A \otimes \varrho_B$, that arises if the initial state is assumed to be uncorrelated. Because $\operatorname{tr}_B \varrho_B = 1$ this definition defines a projection operator, inasmuch as $\mathcal{P}^2 = \mathcal{P}$. It also satisfies $\mathcal{P}(O_A \otimes \varrho_B) = O_A \otimes \varrho_B$, and so $\mathcal{P}(\rho_0) = \rho_0$ for uncorrelated initial states. More generally, $\mathcal{P}[\rho(t)] = \rho_A(t) \otimes \varrho_B$, where $\rho_A(t) = \operatorname{tr}_B \rho$ is the reduced density matrix whose time-evolution is sought.

Because $\mathcal{P}$ is a projection operator, it follows that $\mathcal{Q} = 1 - \mathcal{P}$ is also a projection operator (*i.e.* also satisfies $\mathcal{Q}^2 = \mathcal{Q}$) and that $\mathcal{P}\mathcal{Q} = \mathcal{Q}\mathcal{P} = 0$. The idea is to compute the evolution of $\mathcal{P}(\rho)$ as a proxy for ϱ_A. The evolution of $\mathcal{P}(\rho)$ is not quite the same as that of ρ because for $\mathcal{P}(\rho) = \rho_A \otimes \varrho_B$ the environment (sector B) does not evolve. However, ρ and $\mathcal{P}(\rho)$ agree with one another (by construction) once sector B is traced out, which suffices to make their predictions agree when restricted to observables that act only in sector A.

[22] More than one type of master equation is derived here because some are only approximate.

In this same language time evolution is also a linear operation, given in the interaction picture by $\partial_t \rho = \mathcal{L}_t(\rho)$, where

$$\mathcal{L}_t(O) := -\mathrm{i}\big[V(t), O\big]. \tag{16.141}$$

The problem of solving for the evolution of $\rho_A(t)$ is then solved by combining $\mathcal{L}_t$ with $\mathcal{P}$ and $\mathcal{Q}$, inasmuch as $[\partial_t \rho_A(t)] \otimes \varrho_B = \mathcal{P}(\partial_t \rho) = \partial_t \mathcal{P}(\rho)$ can be computed using the pair of equations

$$\begin{aligned} \partial_t \mathcal{P}(\rho) &= \mathcal{P}(\partial_t \rho) = \mathcal{P}\mathcal{L}_t(\rho) = \mathcal{P}\mathcal{L}_t\mathcal{P}(\rho) + \mathcal{P}\mathcal{L}_t\mathcal{Q}(\rho) \\ \text{and} \quad \partial_t \mathcal{Q}(\rho) &= \mathcal{Q}(\partial_t \rho) = \mathcal{Q}\mathcal{L}_t(\rho) = \mathcal{Q}\mathcal{L}_t\mathcal{P}(\rho) + \mathcal{Q}\mathcal{L}_t\mathcal{Q}(\rho), \end{aligned} \tag{16.142}$$

where the last equality on each line uses $\mathcal{P} + \mathcal{Q} = 1$. The idea is to use the second of these equations to eliminate $\mathcal{Q}(\rho)$ from the right-hand side of the first equation, thereby obtaining an evolution equation that involves only $\mathcal{P}(\rho)$.

Because the equations are linear, they can be formally solved by successive integrations, as follows. Define $\mathcal{G}(t, s)$ as the solution to $\partial_t \mathcal{G}(t, s) = \mathcal{Q}\mathcal{L}_t\mathcal{G}(t, s)$ with initial condition $\mathcal{G}(t, t) = 1$. Then $\mathcal{G}(t, s)$ is given explicitly by

$$\begin{aligned} \mathcal{G}(t, s) &= 1 + \sum_{n=1}^{\infty} \int_s^t \mathrm{d}s_1 \cdots \int_s^{s_{n-1}} \mathrm{d}s_n \, \mathcal{Q}\mathcal{L}_{s_1} \cdots \mathcal{Q}\mathcal{L}_{s_n} \\ &= 1 + \sum_{n=1}^{\infty} \frac{1}{n!} \int_s^t \mathrm{d}s_1 \cdots \int_s^t \mathrm{d}s_n \, \mathfrak{P}\big[\mathcal{Q}\mathcal{L}_{s_1} \cdots \mathcal{Q}\mathcal{L}_{s_n}\big], \end{aligned} \tag{16.143}$$

where $\mathfrak{P}$ denotes path-ordering (or time-ordering) of the $\mathcal{Q}\mathcal{L}_{s_i}$.

In terms of this the formal solution for $\mathcal{Q}[\rho(t)]$ with initial condition $\mathcal{Q}[\rho(t_0)] = \mathcal{Q}(\rho_0)$ is given by

$$\mathcal{Q}[\rho(t)] = \mathcal{G}(t, t_0)\mathcal{Q}(\rho_0) + \int_{t_0}^t \mathrm{d}s \, \mathcal{G}(t, s)\, \mathcal{Q}\mathcal{L}_s\mathcal{P}[\rho(s)], \tag{16.144}$$

as can be verified by explicit differentiation, using $\partial_t \mathcal{G}(t, s) = \mathcal{Q}\mathcal{L}_t\mathcal{G}(t, s)$. Once this solution is inserted into the first of Eqs. (16.142) one obtains the exact integro-differential equation [455]

$$\partial_t \mathcal{P}[\rho(t)] = \mathcal{P}\mathcal{L}_t\mathcal{P}[\rho(t)] + \mathcal{P}\mathcal{L}_t\mathcal{G}(t, t_0)\mathcal{Q}(\rho_0) + \int_{t_0}^t \mathrm{d}s \, \mathcal{K}(t, s)[\rho(s)], \tag{16.145}$$

which defines the kernel $\mathcal{K}(t, s) = \mathcal{P}\mathcal{L}_t\mathcal{G}(t, s)\mathcal{Q}\mathcal{L}_s\mathcal{P}$. The second term on the right-hand side vanishes for uncorrelated initial conditions, $\rho_0 = \varrho_A \otimes \varrho_B$, since these imply $\mathcal{P}(\rho_0) = \rho_0$ and so $\mathcal{Q}(\rho_0) = 0$.

Eq. (16.145) is called the Nakajima–Zwanzig equation and is an exact consequence of (and so is typically no easier to solve than) the original Liouville equation for $\rho(t)$. Its main virtues are two-fold: first, because a factor of $\mathcal{P}$ stands to the far left of every term, it has the structure $(\cdots) \otimes \varrho_B$ and so only carries content in sector A. Second, because $\rho(t)$ only appears in it through the combination $\mathcal{P}[\rho(t)] = \rho_A(t)$ it really is the equation giving $\partial_t \rho_A$ directly in terms of ρ_A itself.

Eq. (16.145) is made more explicit by writing it out order-by-order in V using the definitions of $\mathcal{L}_t$, $\mathcal{P}$ and $\mathcal{Q}$. It is convenient when doing so to expand V in a basis of operators in product form,

$$V(t) = \sum_n A_n(t) \otimes B_n(t), \tag{16.146}$$

and keep only terms out to second order in V. To this order, it suffices to approximate the kernel by its leading (second-order in V) part, $\mathcal{K} \simeq \mathcal{K}_2 = \mathcal{P}\mathcal{L}_t\mathcal{Q}\mathcal{L}_s\mathcal{P}$.

Choosing an uncorrelated initial condition, $\rho(t_0) = \varrho_A \otimes \varrho_B$, Eq. (16.145) reduces to the following approximate expression

$$\partial_t \rho_A(t) = -i \sum_n \Big[A_n(t), \rho_A(t)\Big] \langle\!\langle B_n(t) \rangle\!\rangle + (-i)^2 \sum_{mn} \int_{t_0}^{t} ds \bigg\{ \Big[A_m(t), A_n(s)\, \rho_A(s)\Big] \times \langle\!\langle \delta B_m(t)\, \delta B_n(s) \rangle\!\rangle - \Big[A_m(t), \rho_A(s)\, A_n(s)\Big] \langle\!\langle \delta B_n(s)\, \delta B_m(t) \rangle\!\rangle \bigg\} + \mathcal{O}(V^3), \tag{16.147}$$

where, as usual, $\langle\!\langle(\cdots)\rangle\!\rangle = \mathrm{tr}_B[(\cdots)\, \varrho_B]$. This is a central result whose consequences are more fully explored in the following sections.

Notice that if $\rho_A(t)$ is re-expressed in terms of its initial value, again dropping all terms beyond V^2, then (16.147) agrees with the B-sector trace of the differential version of Eq. (16.39),

$$\partial_t \rho_A(t) = -i\Big[\overline{V}(t)\,, \rho_A(t_0)\Big] + (-i)^2 \int_{t_0}^{t} d\tau\, \mathop{\mathrm{tr}}_B \Big[V(t)\,, \Big[V(\tau)\,, \rho(t_0)\Big]\Big], \tag{16.148}$$

where $\overline{V} := \langle\!\langle V \rangle\!\rangle$. In what follows it is Eq. (16.147) that is used, and not (16.148), and this is ultimately the source of any late-time resummation that is found.

Returning now to the problem of making late-time predictions, does Eq. (16.147) help? That is, does this equation have a broader domain of validity than does the direct evaluation of $\rho_A(t)$ – such as Eq. (16.39) – from which Eqs. (16.147) and (16.148) can be derived? At face value the answer is 'no', due to the presence in (16.147) of the nonlocal convolution over the region $t_0 \leq s \leq t$. This integration obstructs using the arguments of previous sections to extend the domain of validity of the differential evolution, because of its explicit dependence on the entire evolution history between t_0 and t.

Lindblad Equation

It is perhaps not a surprise that (16.147) does not in itself provide a handle on late-time evolution, given that to this point very little has been used about the properties of ϱ_B describing the initial state of sector B. As is typically the case for EFT methods, simplicity does not come until there is a hierarchy of scales to exploit.

In the present instance a simplifying hierarchy arises if sector B includes the 'fast' degrees of freedom relative to a slower sector A. (Exploiting such hierarchies of time-scales is a venerable tradition that goes back to the Born–Oppenheimer approximation [312].) Such a hierarchy occurs if the correlation functions $\langle\!\langle \delta B_n(t)\, \delta B_m(s) \rangle\!\rangle$ fall off to zero for $t - s$ much larger than a characteristic time-scale, τ_c. Then a useful hierarchy arises if τ_c is much smaller than the times over which the evolution of $\rho_A(t)$ is sought.

In such a circumstance the reduced density matrix $\varrho_A(s)$ can be Taylor expanded about $s = t$, with the logic that it varies more slowly than does the rest of the integrand because of the assumption that $\langle\!\langle \delta B_n(t)\, \delta B_m(s) \rangle\!\rangle$ is sharply peaked about $s = t$. Once this is done $\varrho_A(t)$ (and its derivatives) can be factored out of the integral.

For instance, suppose the sector-B correlation function appearing in (16.147) is approximately local in time,

$$\langle\!\langle \delta B_m(t)\, \delta B_n(s) \rangle\!\rangle \simeq C_{mn}(t)\, \delta(t-s), \tag{16.149}$$

where hermiticity of the B_n's implies that $C^*_{mn} = C_{nm}$. In this case, the integral over s in (16.147) collapses so that $\partial_t\, \rho_A(t)$ is given directly by

$$\partial_t\, \rho_A \simeq -\mathrm{i} \sum_n \Big[A_n, \rho_A\Big] \langle\!\langle B_n \rangle\!\rangle + \frac{1}{2}(-\mathrm{i})^2 \sum_{mn} C_{mn} \Big[A_m A_n\, \rho_A + \rho_A A_m A_n - 2A_n\, \rho_A A_m\Big], \tag{16.150}$$

where the coefficients and operators on the right-hand side all depend on a common time, t. A master equation of this type is called a *Lindblad* – or GKSL (Gorini, Kossakowski, Sudarshan, Lindblad) – equation [458], and is much easier to work with because it is Markovian (in the sense that $\partial_t\, \varrho_A(t)$ depends only on variables at time t and not on the history of evolution prior to this time.

An equation like (16.150) sometimes also allows an inference of late-time behaviour that would normally lie beyond the reach of perturbation theory, much as the exponential decay law (16.134) is ultimately justified by the differential evolution (16.136). An example of where this happens is when the coefficients C_{mn} and operators A_m do not themselves depend explicitly on time, because then (16.150) can have a broader domain of validity than its perturbative derivation would naively seem to allow. The extended domain of validity arises because (16.150) then holds equally well for a sequence of overlapping windows of time, in which case its solutions are valid over the union of these overlapping domains, thereby giving $\rho_A(t)$ at late times. For example, in the system considered below perturbation in V is justified by the small size, $g \ll 1$, of a coupling parameter, and for large t the integration of (16.150) resums all orders in $g^2 t$ while neglecting corrections of order $g^4 t$.

Although the A_n and B_n are hermitian in the above derivation, this is not strictly required to be true for a Lindblad equation, which can be more generally written

$$\partial_t\, \rho_A \simeq -\mathrm{i}\Big[H, \rho_A\Big] + \frac{1}{2}(-\mathrm{i})^2 \sum_{mn} C_{mn} \Big[A^*_m A_n\, \rho_A + \rho_A A^*_m A_n - 2A_n\, \rho_A A^*_m\Big], \tag{16.151}$$

where $H = H^*$ but the same need not be so for the A_n. When the coefficients C_{mn} are hermitian and positive semi-definite,[23] Eq. (16.151) can be shown [458] to be the most general equation of this type that is linear in ρ_A and preserves its hermiticity, positivity and its normalization condition (*i.e.* the condition $\mathrm{tr}_A\, \rho_A = 1$).

Approach to Equilibrium

To make the above discussion more explicit, it is useful to think through an explicit example for which a Lindblad-type equation emerges to describe late-time behaviour.

[23] As the example considered below shows in detail, the derivation of the Lindblad equation from the Nakajima-Zwanzig equation can appear to produce a matrix C_{mn} that is not hermitian and positive semi-definite. This is sometimes dealt with by coarse-graining over any fast oscillations on the right-hand side (often called the 'rotating wave approximation') if such fast oscillations should be present [456]. Strictly speaking, though, such 'sick' values for C_{mn} only arise when applied beyond the domain of validity of the Lindblad derivation, making coarse-graining superfluous (see [457] for further discussion).

To this end, consider the approach to equilibrium of a two-level system (an atom or qubit) coupled to an environment [459] (taken here to be a relativistic real scalar field – meant as a proxy, say, for an electromagnetic field).

The free Hamiltonian is taken to be $H_0 = H_A \otimes I_B + I_A \otimes H_B$, where for the qubit

$$H_A = \frac{\omega}{2}\begin{pmatrix} 1 & 0 \\ 0 & -1 \end{pmatrix}, \tag{16.152}$$

and sector-B has the field Hamiltonian

$$H_B = \frac{1}{2}\int \mathrm{d}^3x \left[(\partial_t\phi)^2 + (\nabla\phi)^2 + m^2\phi^2\right]. \tag{16.153}$$

These are coupled to one another through the interaction-picture interaction

$$V(t) = g(A \otimes B + A^\dagger \otimes B^*) \tag{16.154}$$

with the coupling $g \ll 1$ assumed small enough to justify perturbative methods, while

$$A = \begin{pmatrix} 0 & 0 \\ 1 & 0 \end{pmatrix} e^{-\mathrm{i}\omega t} \quad \text{and} \quad A^\dagger = \begin{pmatrix} 0 & 1 \\ 0 & 0 \end{pmatrix} e^{\mathrm{i}\omega t}, \tag{16.155}$$

and

$$B(t) = \phi[\mathbf{x}_0, t], \tag{16.156}$$

where the field is evaluated at the atom's (static) position $\mathbf{x}(t) = \mathbf{x}_0$. Eq. (16.152) identifies $\omega = E_\uparrow - E_\downarrow$ as the energy difference between the two qubit states, and putting ω into H_0 rather than V (and working with non-degenerate perturbation theory) assumes ω to be much larger than any perturbative field-induced shift in qubit energies.

The field ϕ is imagined prepared in an initial state ϱ_B, chosen to be a thermal state

$$\varrho_B = \frac{1}{Z}\exp[-\beta H_B], \tag{16.157}$$

with temperature $T = 1/\beta$. Here, $Z = \mathrm{tr}_B[\exp(-\beta H_B)]$ using the sector-B Hamiltonian given in (16.153).

As is shown in Exercise 16.7, choosing the field to be in a thermal state implies that its Wightman function

$$W(x,x') := \langle\!\langle \phi(x)\phi(x')\rangle\!\rangle = \mathop{\mathrm{tr}}_B\left[\varrho_B\,\phi(x)\,\phi(x')\right], \tag{16.158}$$

satisfies the KMS condition [460]

$$W(\tau - \mathrm{i}\beta) = W(-\tau), \tag{16.159}$$

where $\tau = t - t'$. For instance, explicit evaluation of $W(x,x')$ for a free massless field in a thermal state at coincident spatial points, $\mathbf{x} = \mathbf{x}'$, gives (see Exercise 16.7)

$$W(\tau) := W(\mathbf{x}_0, t+\tau; \mathbf{x}_0, t) \to -\frac{1}{4\beta^2[\sinh(\pi\tau/\beta) - \mathrm{i}\epsilon]^2} \quad \text{(massless limit)}, \tag{16.160}$$

which satisfies (16.159). Notice the exponential falloff in $W(\tau)$ for $\tau \gg \beta$.

Property (16.159) expresses detailed balance in sector B, and (as shall be seen) ensures that the late-time limit of ϱ_A is also thermal:

$$\varrho_{A\infty} = \begin{bmatrix} \frac{1}{1+e^{\beta\omega}} & 0 \\ 0 & \frac{1}{1+e^{-\beta\omega}} \end{bmatrix} = \begin{bmatrix} e^{-\beta\omega} & 0 \\ 0 & 1 \end{bmatrix} \frac{1}{1+e^{-\beta\omega}}, \tag{16.161}$$

with temperature $T = 1/\beta$. The goal is to reliably calculate the asymptotic relaxation rate towards this state.

Substituting these expressions into the definitions allows the Nakajima–Zwanzig equation (16.147) for the qubit's interaction-picture density matrix to be written (at second order in the coupling g) as

$$\begin{aligned} \frac{\partial \varrho_{\uparrow\uparrow}}{\partial t} &= g^2 \int_{-t}^{t} ds\; W(s)\, e^{-i\omega s} - 4g^2 \int_0^t ds\; \mathrm{Re}[W(s)]\, \cos(\omega s)\, \varrho_{\uparrow\uparrow}(t-s) \\ \frac{\partial \varrho_{\uparrow\downarrow}}{\partial t} &= -4ig^2\, e^{i\omega t} \int_0^t ds\; \mathrm{Re}[W(s)]\, \mathrm{Im}[e^{-i\omega(t-s)} \varrho_{\uparrow\downarrow}(t-s)] \\ &= -2g^2 \int_0^t ds\; \mathrm{Re}[W(s)]\, e^{i\omega s}\, \varrho_{\uparrow\downarrow}(t-s) \\ &\quad + 2g^2\, e^{2i\omega t} \int_0^t ds\; \mathrm{Re}[W(s)]\, e^{-i\omega s}\, \varrho^*_{\uparrow\downarrow}(t-s), \end{aligned} \tag{16.162}$$

showing how the diagonal and off-diagonal components of ϱ_A evolve independent of one another. The derivation of (16.162) uses the relations $\varrho_{\downarrow\downarrow} = 1 - \varrho_{\uparrow\uparrow}$ and $\varrho_{\downarrow\uparrow} = \varrho^*_{\uparrow\downarrow}$ to eliminate $\varrho_{\downarrow\downarrow}$ and $\varrho_{\downarrow\uparrow}$.

Notice also that the initial choice $\varrho_A = |{\downarrow}\rangle\langle{\downarrow}|$, together with a strict interpretation of perturbation theory, would imply dropping all but the first term on the right-hand side of the first of Eqs. (16.162) if one stops at $O(g^2)$. Keeping only this term reproduces the perturbative (small-time) excitation rate of the qubit due to the presence of the field [464]. It is by keeping the other terms that one obtains the information needed to resum perturbation theory at late times, providing at large t information to all orders in $g^2 t$.

Now comes the main point. Late-time behaviour for $t \sim O(1/g^2)$ can be reliably inferred if ϱ_{ab} evolves sufficiently slowly compared with thermal time-scales. In particular, if ϱ_{ab} does not change appreciably over the $O(\beta)$ interval over which $W(t)$ remains nonzero, one can simplify Eqs. (16.162) by expanding $\varrho_{ab}(t-s)$ in powers of s,

$$\varrho_{ab}(t-s) \simeq \varrho_{ab}(t) - s\left(\frac{\partial \varrho_{ab}}{\partial t}\right)_t + \cdots, \tag{16.163}$$

and integrating the result term by term.

For instance, dropping all but the first term in this expansion completely removes the convolution over the qubit's past history, and for $\varrho_{\uparrow\uparrow}$ this leads to an evolution equation of the Markovian form,

$$\frac{\partial \varrho_{\uparrow\uparrow}}{\partial t} \simeq g^2\mathcal{R} - 2g^2 \mathcal{C}\, \varrho_{\uparrow\uparrow}(t), \tag{16.164}$$

with coefficients given by

$$\mathcal{R}(\omega) := \int_{-\infty}^{\infty} d\tau\; W(\tau)\, e^{-i\omega\tau} \to \frac{1}{2\pi} \frac{\omega}{e^{\beta\omega} - 1} \quad \text{(thermal, massless limit)}, \tag{16.165}$$

and

$$C(\omega) := \int_{-\infty}^{\infty} d\tau \, \mathrm{Re}[W(\tau)] \cos(\omega\tau) \to \frac{\omega}{4\pi} \coth\left(\frac{\beta\omega}{2}\right) \quad \text{(thermal, massless limit).} \tag{16.166}$$

Because (16.164) makes no reference to the initial time, its solutions can be trusted for much later times than can straight-up perturbation theory, using the same arguments that justify the exponential decay law (16.134) starting from (16.136). Integrating leads to the prediction

$$\varrho_{\uparrow\uparrow}(t) = \frac{1}{e^{\beta\omega}+1} + \left[\varrho_{\uparrow\uparrow}(0) - \frac{1}{e^{\beta\omega}+1}\right] e^{-t/\xi_T}, \tag{16.167}$$

with

$$\xi_T = \frac{1}{2g^2 C(\omega)} \to \frac{2\pi}{g^2\omega} \tanh\left(\frac{\beta\omega}{2}\right) \quad \text{(thermal, massless limit).} \tag{16.168}$$

For large t this solution can be trusted to all orders in $g^2 t$ but neglects effects that are of order $g^4 t$ and smaller. The solution (16.167) also allows an *ex post facto* quantification of when the Markovian expansion of (16.163) is justified, since it shows that the neglect of subdominant terms requires

$$\xi_T \gg \beta \quad \text{or} \quad g^2 C \ll T, \tag{16.169}$$

a condition that gets better and better the smaller g is. To these must be added the condition

$$g^2 C \ll \omega, \tag{16.170}$$

that is required by the assumption that $V(t)$ is perturbatively small relative to H_A.

A similar story goes through for $\varrho_{\uparrow\downarrow}$ but with an important complication. Although Taylor expanding $\varrho_{\uparrow\downarrow}(t-s)$ also removes the history-dependence of Eq. (16.162), the result *does* refer directly to the initial time because of the last term of the last line, and its explicit dependence on $e^{2i\omega t}$. This dependence appears to obstruct being able to trust the solutions obtained by integrating for very long times, and furthermore oscillates very rapidly compared with the late-time evolution of interest because of condition (16.170).

This problematic rapidly-varying term should be dropped from the evolution equation, however, because it only contributes to the solution by an amount suppressed by $g^2 C/\omega$ (as may be seen, for instance, by direct integration), making it subdominant[24] by virtue of condition (16.170).

The resulting evolution equation then has the Lindblad form, (16.150), which (converting back to the Schrödinger picture) becomes

$$\frac{\partial \varrho_{\uparrow\downarrow}}{\partial t} \simeq -\mathrm{i}(\omega + g^2\Delta)\varrho_{\uparrow\downarrow}(\tau) - g^2 C \, \varrho_{\uparrow\downarrow}(\tau), \tag{16.171}$$

where the coefficients are given by (16.165), (16.166) and

$$\Delta(\omega) := 2\int_0^{\infty} ds \, \mathrm{Re}[W(s)] \sin(\omega s). \tag{16.172}$$

24 In the literature this term is often dropped by appealing to the 'rotating wave approximation' [456], which coarse-grains the evolution equation over times much longer than ω^{-1} but smaller than the relaxation times of interest.

Although the new function Δ diverges[25] in the $s \to 0$ limit, this divergence is renormalized into the atomic frequency: $\omega_R = \omega + g^2\Delta$.

Eq. (16.171) now has the form that can be integrated to late times, so its solution

$$\varrho_{\uparrow\downarrow}(t) = \varrho_{\uparrow\downarrow}(0)\, e^{-t/\xi_D}\, e^{-i\omega_R t}, \tag{16.173}$$

provides a reliable description of late-time behaviour. The relaxation time-scale appearing here is $\xi_D = 2\xi_T$, and so describes slower relaxation than found above for $\varrho_{\uparrow\uparrow}$.

The thermal state (16.161) provides the unique static solution to (16.164) and (16.171) and so is the late-time state to which the above solutions relax. Notice that the late-time relaxation rates ξ_D and ξ_T both differ from the timescale, $\xi_R := 1/(g^2\mathcal{R})$, that describes the early-time perturbative excitation rate [464] out of the ground state caused by the field ϕ.

The lesson is this: hierarchies of scale can provide simplifications even for open systems. For the qubit example straight-up perturbation in g fails at late times, but this failure can be reliably resummed for large t to obtain predictions to all orders in g^2t. This happens because the full Nakajima–Zwanzig evolution is well-described by an approximate Lindblad equation for evolution that is very slow compared with the environment's typical correlation time. Solutions to the resulting Lindblad equation can be trusted at late times if its perturbative derivation works equally well in any small time interval.

16.5 Summary

This final chapter explores how EFT methods work for systems that are open and so *not* Wilsonian, inasmuch as measurements are restricted to a sector that is not isolated (as opposed to the low-energy limit, whose isolation is enforced by conservation laws like energy conservation).

Such systems can exchange information with other, unobserved, degrees of freedom, and, in general, this can lead to qualitatively new phenomena like thermalization or decoherence. Since thermalization and decoherence can evolve pure states into mixed ones, they, in general, need not have a simple description in terms of an effective Hamiltonian.

This section argues that EFT methods can nonetheless apply, in the special case that the ignored degrees of freedom have a characteristic time-scale after which correlations tend to die out. If one's interest is only in very late-time evolution compared with this scale, then the description simplifies making the late-time limit the analog for these systems of the low-energy limit for Wilsonian effective theories.

As the examples in this section illustrate, although, in general, late-time evolution need not be Hamiltonian, it can happen that an effective Hamiltonian is nonetheless possible to construct. Systems that are dominated by average mean-field properties of their environment provide a broad class of situations for which this is true, of which thermal fluids provide perhaps the historically earliest examples of EFT reasoning.

[25] Notice that (surprisingly) the function C does *not* similarly diverge, due to the presence of the $i\epsilon$ factor in the denominator of the Wightman function seen in Eq. (16.160).

In these cases, the breakdown of the mean-field approximation is driven by interactions with fluctuations in the environment, and some effort is made here to quantify when these may be neglected. This is done initially for several practical examples of particles moving through an ambient environment (like neutrinos in the solar interior or photons within transparent media), and then more generally by introducing the quantum master equation and some of its late-time approximations.

Exercises

Problem 16.1 An important role is played by interactions for which the interaction is $V(t) = V_A(t) \otimes V_B(t)$, where fluctuations of $V_B(t)$ in the environment have a short correlation time, τ, for which

$$\langle\!\langle V_B(t)V_B(s)\rangle\!\rangle - \langle\!\langle V_B(t)\rangle\!\rangle\langle\!\langle V_B(s)\rangle\!\rangle \approx \tau\, \sigma_V^2\, \delta(t-s),$$

where $\sigma_V^2 > 0$ is a measure of the instantaneous variance of V in the environment. Working to second order in V, use unitarity to show that interactions with these properties contribute to the growth of $\mathrm{tr}_A(\rho_A^{\rm f})$ with time according to

$$\partial_t \operatorname*{tr}_A(\rho_A^{\rm f}) = \tau\sigma_V^2 \operatorname*{tr}_A\left[\varrho_A(t)V_A^2(t)\right] \geq 0.$$

For a local interaction Hamiltonian, $V(t) = \int \mathrm{d}^3\mathbf{x}\, \mathcal{V}(\mathbf{x},t)$, where $\mathcal{V}(\mathbf{x},t) = \mathcal{V}_A(\mathbf{x},t) \otimes \mathcal{V}_B(\mathbf{x},t)$ with local autocorrelations of $\mathcal{V}_B$ in both space and time

$$\langle\!\langle \delta\mathcal{V}_B(\mathbf{x},t)\, \delta\mathcal{V}_B(\mathbf{x}',t')\rangle\!\rangle = \tau\ell^3\sigma_V^2\, \delta^3(\mathbf{x}-\mathbf{x}')\delta(t-t'),$$

show that at second order in V the growth of $\mathrm{tr}_A(\rho_A^{\rm f})$ with time is extensive, with

$$\partial_t \operatorname*{tr}_A(\rho_A^{\rm f}) = \int \mathrm{d}^3\mathbf{x}\, \tau\ell^3\sigma_V^2 \operatorname*{tr}_A\left[\varrho_A(t)\mathcal{V}_A^2(\mathbf{x},t)\right] \geq 0.$$

Problem 16.2 Consider two species of neutrinos, $\nu_\pm$, for which $U_{-e} = \cos\theta_\nu$ and $U_{+e} = \sin\theta_\nu$, where θ_ν is the vacuum PMNS mixing angle. Show that in the ultra-relativistic regime relevant to neutrino-oscillation experiments, the evolution equation is given by (16.77) and so the energy $k^0 = E$ to linear order in μ satisfies

$$E \simeq k + \frac{m^\dagger m}{2k} + \mu,$$

where μ is defined in (16.78). In the weak interaction basis show that when the symmetric left-handed vacuum neutrino mass matrix m is real it can be written in terms of the mass eigenvalues, $m_\pm^2 = m_0^2 \pm \frac{1}{2}\delta m^2$, and the vacuum mixing angle, θ_ν, by

$$m^\dagger m = m_0^2 \begin{pmatrix} 1 & 0 \\ 0 & 1 \end{pmatrix} + \frac{\delta m^2}{2} \begin{pmatrix} -\cos 2\theta_\nu & \sin 2\theta_\nu \\ \sin 2\theta_\nu & \cos 2\theta_\nu \end{pmatrix}.$$

In the same basis, the matrices appearing in the matter-dependent term is given by

$$g_e = \begin{pmatrix} 1 & 0 \\ 0 & 0 \end{pmatrix}, \quad \text{and} \quad g_n = \begin{pmatrix} -\frac{1}{2} & 0 \\ 0 & -\frac{1}{2} \end{pmatrix}.$$

Show that E is diagonalized by propagation eigenstates $\nu_- = (\cos\theta_m, \sin\theta_m)^T$ and $\nu_+ = (-\sin\theta_m, \cos\theta_m)$, where

$$\sin 2\theta_m = \left(\frac{\delta m^2}{4k\mu_0}\right) \sin 2\theta_\nu$$

with

$$\mu_0 := \left[\left(\frac{G_F\bar{n}_e}{\sqrt{2}}\right)^2 - \left(\frac{G_F\bar{n}_e}{\sqrt{2}}\right)\frac{\delta m^2 \cos 2\theta_\nu}{2k} + \left(\frac{\delta m^2}{4k}\right)^2\right]^{1/2}.$$

Show that the probability of a neutrino of energy $E \gg m$ being produced in one of the weak-interaction eigenstates and then being detected in the other eigenstate after travelling a distance L is

$$P_{e\mu}(E, L) \simeq \sin^2 2\theta_m \, \sin^2\left(\frac{\delta m^2 L}{4E}\right).$$

Prove that the medium-dependent mixing is maximal, $\sin 2\theta_m = 1$, when

$$\frac{G_F\bar{n}_e}{\sqrt{2}} = \left(\frac{\delta m^2}{4k}\right) \cos 2\theta_\nu,$$

in which case the matter oscillations are called 'resonant'.

Problem 16.3 For the two species of neutrinos considered in Exercise 16.2, suppose a neutrino is produced as an electron-type eigenstate at $t = t_0$ by nuclear reactions deep within the Sun, and then passes through the regime of resonant oscillation while escaping the Sun through an exponentially falling electron density, $\bar{n}_e(t) = \bar{n}_e(0)\, e^{-(t-t_0)/h}$, with scale-height h (the neutrino moves at essentially the speed of light). Suppose that $\theta_\nu \ll 1$ and so $\nu_- \simeq \nu_e$ and $\nu_+ \simeq \nu_\mu$ with $m_+^2 > m_-^2$ outside the Sun.

Resonant oscillations occur if $G_F\bar{n}_e(0)/\sqrt{2} \gg \delta m^2/(4k)$. Show that in this case the matter mixing angle starts deep within the Sun with $\sin 2\theta_m(t_0) \ll \sin 2\theta_\nu$ very small and $\cos 2\theta_m(t_0) \simeq -1$. Show also that $\sin 2\theta_m(t) \to \sin 2\theta_\nu$ (and so is also small when θ_ν is small) with $\cos 2\theta_m(t) \to \cos 2\theta_\nu \simeq +1$ at the solar surface (defined as the place where $\bar{n}_e(t) \to 0$).

Use the Landau–Zener formalism describing level crossing in ordinary quantum mechanics to derive the Parke formula for the survival probability that a later measurement (after resonance crossing) at time t also measures the neutrino to be electron-like,

$$P_e(t) \simeq \frac{1}{2} + \left(\frac{1}{2} - P_J\right) \cos 2\theta_m(t_0) \, \cos 2\theta_m(t),$$

once averaged over quickly oscillating factors. Here,

$$P_J := \exp\left[-\frac{\pi}{2}\left(\frac{\sin^2 2\theta_\nu}{\cos 2\theta_\nu}\right)\left(\frac{\delta m^2 h}{2k}\right)\right],$$

is the 'jump' probability for making a non-adiabatic transition when passing through the resonance regime. Use this result to show that the survival probability after passage through the Sun can be written $P_e \simeq \sin^2\theta_v + P_J \cos 2\theta_v$ for resonant oscillations, which can be very small for adiabatic transitions (for which $P_J = 0$).

Problem 16.4 Consider a medium in local equilibrium, divided into cells of volume $\Omega_\ell := \ell^3$ that are small enough that thermodynamic quantities are effectively constant within any one cell, but large enough that fluctuations away from the thermodynamic limit are small. Because particles can move from one cell to another, each such cell can be regarded as being described by a grand canonical ensemble.

Prove that for a grand canonical ensemble the thermal fluctuation in the total number of particles in any one cell is given by

$$\langle\!\langle \delta N \delta N \rangle\!\rangle = T\Omega_\ell \left(\frac{\partial^2 p}{\partial \mu^2}\right)_{T\Omega_\ell} = \frac{\kappa_T \overline{N}^2 T}{\Omega_\ell}$$

where $\overline{N} := \langle\!\langle N \rangle\!\rangle$ is the mean number of particles in the cell, p is the pressure, μ is the chemical potential, T is the temperature and[26]

$$\kappa_T := \frac{1}{\bar{n}\,(\partial p/\partial \bar{n})_T},$$

is the isothermal compressibility, with $\bar{n} := \overline{N}/\Omega_\ell$.

Assuming each cell to be uncorrelated with the others, for distances much larger than ℓ the density-density autocorrelation function for such a fluid becomes approximately local: $\langle\!\langle \delta n(\mathbf{x})\delta n(\mathbf{y})\rangle\!\rangle = \langle\!\langle \delta n(\mathbf{x})\delta n(\mathbf{x})\rangle\!\rangle \Omega_\ell \delta^3(\mathbf{x}-\mathbf{y})$. Use this to show that

$$\langle\!\langle \delta n(\mathbf{x})\delta n(\mathbf{y})\rangle\!\rangle = \kappa_T \bar{n}^2 T\, \delta^3(\mathbf{x}-\mathbf{y}).$$

Evaluate this last expression for an ideal gas (for which the equation of state is $p = \bar{n}T$) and show that in this case it implies that Eq. (16.125) evaluates to

$$\mathcal{G}(\mathbf{k}-\mathbf{k}') = \bar{n},$$

in agreement with (16.128).

Problem 16.5 Unitarity – in the form of Eq. (16.57) – provides another handle on the size of scattering by fluctuations, and directly yields the net loss of mean-field evolution in terms of the imaginary part of the mean-field Hamiltonian. Show that for Thomson scattering in environments for which the correlation times, τ, and correlation distances, ℓ, are small

$$\langle\!\langle \delta n(\mathbf{x},t)\delta n(\mathbf{y},s)\rangle\!\rangle \simeq \tau\,\Omega_\ell \langle\!\langle \delta n(\mathbf{x},t)\delta n(\mathbf{x},t)\rangle\!\rangle \delta^3(\mathbf{x}-\mathbf{y})\delta(t-s)$$

the net transfer of probability from mean-field to 'diffuse' scattering at second order is given by (16.57), in the form

$$\partial_t \operatorname*{tr}_A(\rho_A^{\rm f}) \simeq \frac{e_q^4}{8m^2}\int {\rm d}^3\mathbf{x}\,\tau\,\Omega_\ell \operatorname*{tr}_A\left[\varrho_A \mathbf{A}^4(\mathbf{x},t)\right] \langle\!\langle \delta n(\mathbf{x},t)\delta n(\mathbf{x},t)\rangle\!\rangle$$

26 Recall that fundamental units are used, for which Boltzmann's constant is $k_B = 1$.

again showing how it is the medium's density fluctuations that control this scattering rate. Evaluate this expression for Thomson scattering of individual photons from a dilute medium and verify that it reproduces the extinction coefficient given in (16.130).

Problem 16.6 Consider a dielectric medium in local equilibrium, divided into cells of volume $\Omega_\ell := \ell^3$ that are small enough that thermodynamic quantities are effectively constant within any one cell, but large enough that fluctuations away from the thermodynamic limit are small. Imagine that the dielectric permittivity varies from cell to cell, such as would occur if some cells were droplets full of water while others were full of air.

Imagine the interaction-picture interaction Hamiltonian for photons in such a medium is

$$\delta V(t) = -\frac{1}{2}\int d^3\mathbf{x}\, \delta\varepsilon(\mathbf{x})\, \mathbf{E}^2(\mathbf{x}, t),$$

where $\delta\varepsilon(\mathbf{x})$ is taken to be a gaussian random variable from cell to cell, with translationally invariant mean $\bar{\varepsilon}$ and variance σ_ε^2, where

$$\langle\langle \delta\varepsilon(\mathbf{x})\delta\varepsilon(\mathbf{y})\rangle\rangle = \Omega_\ell \langle\langle \delta\varepsilon(\mathbf{x})\delta\varepsilon(\mathbf{x})\rangle\rangle\, \delta^3(\mathbf{x}-\mathbf{y}) = \sigma_\varepsilon^2\, \delta^3(\mathbf{x}-\mathbf{y}).$$

Compute the scattering rate per unit volume for individual photons – *i.e.* $f_A(\mathbf{k}') \ll 1$ – as a function of ℓ and σ_ε^2, and show that it predicts an extinction coefficient of the Rayleigh-scattering form

$$\mathfrak{h} \simeq \frac{\omega^4}{6\pi}\sigma_\varepsilon^2.$$

For media where fluctuations in ε arise from thermal fluctuations use

$$\delta\varepsilon = \left(\frac{\partial\varepsilon}{\partial n}\right)_T \delta n + \left(\frac{\partial\varepsilon}{\partial T}\right)_n \delta T,$$

to relate σ_ε^2 to $\langle\langle \delta T^2\rangle\rangle$ and $\langle\langle \delta n^2\rangle\rangle$. Compute these fluctuations as functions of thermodynamic variables – and show $\langle\langle \delta T\,\delta n\rangle\rangle = 0$ – and use these to derive Einstein's 1910 formula [461]

$$\mathfrak{h} = \frac{\omega^4}{6\pi}\left[\bar{n}\,T\left(\frac{\partial\bar{n}}{\partial p}\right)_T\left(\frac{\partial\varepsilon}{\partial n}\right)_T^2 + \frac{T^2}{nc_V}\left(\frac{\partial\varepsilon}{\partial T}\right)_n^2\right],$$

where c_V is the heat capacity. This expression shows that photon scattering can become very large near a critical point (a phenomenon called *critical opalescence*), where the isothermal compressibility $(\partial\bar{n}/\partial p)_T$ can grow without bound.

Specialize your result to media for which $(\partial\varepsilon/\partial T)_n \simeq 0$ and for which $\varepsilon - 1$ is proportional to particle density $\bar{n}$ to derive $\bar{n}\,(\partial\varepsilon/\partial\bar{n})_T \simeq \varepsilon - 1 \simeq 2(\mathfrak{n}_m - 1)$, where $\mathfrak{n}_m = \sqrt{\varepsilon}$ is the medium's index of refraction. Hence, derive Rayleigh's original 1881 formula [462]

$$\mathfrak{h} \simeq \frac{2\omega^4}{3\pi\bar{n}}(\mathfrak{n}_m - 1)^2.$$

Show that this agrees with $\mathfrak{h} \simeq \bar{n}\,\sigma_R$, where σ_R is the Rayleigh cross section of (12.95), after using (16.91) in the form $\mathfrak{n}_m \simeq 1 + \frac{1}{2}\bar{n}\,\mathfrak{p}_E$.

Problem 16.7 This exercise works through the transition from Nakajima–Zwanzig equation to Lindblad equation for the simple system defined by Eqs. (16.154) through (16.153). Without using the explicit form of H_B show that when the scalar field is chosen to be in a thermal state its correlation functions satisfy the KMS condition (16.159). Compute the form of $W(\tau)$ explicitly and show that

$$W(\tau) = \frac{m}{4i\pi\beta} \frac{1}{[\sinh(\pi\tau/\beta) - i\epsilon]} K_1\left(\tfrac{\beta m i}{\pi}[\sinh(\pi\tau/\beta) - i\epsilon]\right)$$

where $K_1(z)$ is a Bessel function. Use this to derive the $m \to 0$ limit given in (16.160).

Derive the second-order Nakajima–Zwanzig equation for this system and show that it is given by (16.162). Show that in the late-time limit your result goes over the Markovian form (16.164), doing so keeping a nonzero scalar-field mass, m. Verify that your expressions go over to Eqs. (16.164) through (16.168) in the massless limit.

Problem 16.8 Repeat Exercise 16.7 but this time choose the two-level atom to move along a uniformly accelerated world-line within Minkowski space. Choose the field to be prepared in the standard Minkowski vacuum for a free scalar field. This system describes the response of an accelerating 'Unruh detector' [463, 464].

Prove that all of the results of Exercise 16.7 go through for this system with the replacement $T = 1/\beta \to a/2\pi$, where a is the atom's proper acceleration (for details of this calculation using the formalism described here see [457]).

Adieu

This chapter also brings this book to a close. The book starts in Part I by exploring a variety of low-energy issues, and using these to develop the tools of Wilsonian effective field theory using a particular simple toy model as the vehicle for doing so. The remaining parts of the book apply and extend these principles to a wider class of examples, with the goal of underlining how the same tools are so widely used by so many areas of physics.

Part II starts off this process using examples of EFT reasoning in traditional relativistic systems, such as to low-energy QED, QCD, the Standard Model and General Relativity. The first of these examples is one where the underlying UV theory is completely known and calculable. For such systems EFT methods provide an efficient way to organize and simplify calculations that could have been done in other ways. The second example is a system where the UV theory is known but strongly interacting, making it not known how to explicitly compute many things from first principles. The utility of this example is that it shows that such complications need not obstruct the use of EFT methods, which are indeed very successful in describing low-energy hadron phenomenology. The last two examples are to theories whose UV completion is not known, and so for which EFT methods help identify the kinds of physics that might be expected to arise at low energies for a broad class of completions.

Part III extends the discussion to nonrelativistic systems, which are often the ones of most practical interest in many real-world applications. This part of the book explores how the nonrelativistic expansion in powers of v/c can be regarded as a low-energy limit, and how nonrelativistic Schrödinger effective theories can emerge once antiparticles are integrated out without also doing so for their particle partners. Much of this section is devoted to applications in atomic physics, both using a second-quantized framework and a first-quantized one (for which matching to bulk degrees of freedom can be regarded as the imposition of boundary conditions on the bulk fields).

The book closes out, with Part IV, describing new issues that arise for many-body applications, for which one almost never follows *all* of the low-energy degrees of freedom. For some questions – such as Goldstone boson dynamics or electrons moving near the Fermi surface – this need not matter, and ordinary effective lagrangians can be found that capture aspects of the collective behaviour of the more complicated underlying many-body system. But for other questions the choice to neglect some low-energy modes can be crucial, leading to qualitatively new non-Hamiltonian phenomena like decoherence and thermalization.

As these diverse examples show, the power of low-energy, late-time methods is in the ubiquity of the arguments used. As such, they provide a unifying thread that is deeply woven into the rich tapestry of physics.

Appendix A Conventions and Units

Conventions and units should be one's friend in physics, in that they should both make an analysis more transparent and they should nudge people away from, rather than towards, common mistakes. They are also often acquired without much thought as one grows up.

This section is meant to explain the ones used in this book. The focus here is to list conventional choices for aficionados, with little effort made explaining the field theories involved. A reader seeking more detailed (though still cursory) background information should try their luck with Appendix B (for the quantum mechanics of scattering) and Appendix C (for quantum field theory).

A.1 Fundamental Units

It is common to use specific units adapted to specific problems so that numerical values are not too far from one (such as using the Angstrom – or Rydberg – for atomic electrons, fm for nuclear processes, astronomical units for the solar system or megaparsecs in cosmology). Such choices are mostly not made here, since one of the points of this book is to emphasize the broad utility of EFT methods in many different areas in physics.

Instead, this book uses *fundamental units*, for which the fundamental constants $\hbar$, c and k_B (Planck's constant, the speed of light and Boltzmann's constant) equal 1. For instance, $c = 1$ is ensured by measuring time and distances both in seconds (where a second of distance means a light-second; the distance light travels in a second). Similarly, $\hbar = 1$ if $(\text{energy})^{-1}$ and time are both measured in seconds – where an inverse-second of energy means the amount $\hbar/(1\text{sec}) = 6.58211 \times 10^{-16}$ eV – and so on.

In this book usually the basic unit is taken to be energy, given in eV or multiples thereof. The utility of this choice is that the proton and neutron rest masses in these units are (respectively) 0.938 GeV and 0.940 GeV (which is to say, the energy tied up in the rest mass of a nucleon is just shy of 1 GeV). This is useful because once told that the mass of the earth is $M_\oplus \simeq 3.35 \times 10^{51}$ GeV you also know roughly how many nucleons are in it, since the biggest contributor to an object's mass usually comes from the mass of each nucleon residing in its constituent nuclei.

Fundamental units have the very useful benefit of boiling equations down to relations between physical quantities without cluttering them up with symbols purely to do with units. This is a particularly good virtue when identifying which scales are relevant to any given problem, as is central to the utility of EFT methods. Electromagnetic units are set by using the proton charge e as the unit of charge rather than the Coulomb.

Ordinary units may always be retrieved by putting in any missing factors of $\hbar$, c or k_B as required by dimensional analysis. Useful rules of thumb for this purpose are:

$$1 \text{ fm} \simeq (0.2 \text{ GeV})^{-1} \simeq 3 \times 10^{-24} \text{ sec} \quad \text{and} \quad 1 \text{ K} \simeq 9 \times 10^{-5} \text{ eV}. \tag{A.1}$$

The conversions of other units into powers of eV and to powers of metres are given below.

Length and Time

$1/M_p\ (= G_N/\hbar c)^{\frac{1}{2}}$	=	8.1897×10^{-29}	c^2/eV	=	1.6161×10^{-35}	$mc/\hbar$
$1/m_p$	=	1.0658×10^{-9}	c^2/eV	=	2.1031×10^{-16}	$mc/\hbar$
1 fm	=	5.06773×10^{-9}	$\hbar c/\text{eV}$	=	10^{-15}	m
$1/m_e$	=	1.957×10^{-6}	c^2/eV	=	3.8616×10^{-13}	$mc/\hbar$
$a_0\ (= 1/\alpha m_e)$	=	2.6818×10^{-4}	c^2/eV	=	5.2918×10^{-11}	$mc/\hbar$
1 A	=	5.06773×10^{-4}	$\hbar c/\text{eV}$	=	10^{-10}	m
1 nm	=	5.06773×10^{-3}	$\hbar c/\text{eV}$	=	10^{-9}	m
1 μm	=	5.06773	$\hbar c/\text{eV}$	=	10^{-6}	m
1 cm	=	5.06773×10^{4}	$\hbar c/\text{eV}$	=	0.01	m
1 m	=	5.06773×10^{6}	$\hbar c/\text{eV}$	=	1	m
1 km	=	5.06773×10^{9}	$\hbar c/\text{eV}$	=	10^{3}	m
1 sec	=	1.51927×10^{15}	$\hbar/\text{eV}$	=	2.99792×10^{8}	m/c
1 min	=	9.11562×10^{16}	$\hbar/\text{eV}$	=	1.79875×10^{10}	m/c
1 hr	=	5.46937×10^{18}	$\hbar/\text{eV}$	=	1.07925×10^{12}	m/c
1 day	=	1.31265×10^{20}	$\hbar/\text{eV}$	=	2.59020×10^{13}	m/c
1 yr	=	4.795×10^{22}	$\hbar/\text{eV}$	=	9.461×10^{15}	m/c
1 pc	=	1.564×10^{23}	$\hbar c/\text{eV}$	=	3.08568×10^{16}	m
1 kpc	=	1.564×10^{26}	$\hbar c/\text{eV}$	=	3.08568×10^{19}	m
1 Mpc	=	1.564×10^{29}	$\hbar c/\text{eV}$	=	3.08568×10^{22}	m

Microscopic Energy and Mass

1 eV	=	10^{-9}	GeV	=	5.06773×10^{6}	$\hbar c/m$
1 keV	=	10^{-6}	GeV	=	5.06773×10^{9}	$\hbar c/m$
1 MeV	=	10^{-3}	GeV	=	5.06773×10^{12}	$\hbar c/m$
1 GeV	=	1	GeV	=	5.06773×10^{15}	$\hbar c/m$
αm_e	=	3.7289×10^{-6}	GeV/c^2	=	1.8897×10^{10}	$\hbar/mc$
m_e	=	5.10999×10^{-4}	GeV/c^2	=	2.5896×10^{12}	$\hbar/mc$
	=	9.10939×10^{-28}	g			
m_p	=	0.938272	GeV/c^2	=	4.75491×10^{15}	$\hbar/mc$
	=	1.67262×10^{-24}	g			
	=	1.83615×10^{3}	m_e			

$M_P = (\hbar c/G_N)^{\frac{1}{2}}$	$= 1.22105 \times 10^{19}$	GeV/c^2	$= 6.1879 \times 10^{34}$	$\hbar/mc$
	$= 2.17671 \times 10^{-5}$	g		
	$= 1.30138 \times 10^{19}$	m_p		
$\hat{M}_P = (\hbar c/8\pi G_N)^{\frac{1}{2}}$	$= 2.43564 \times 10^{18}$	GeV/c^2	$= 1.23431 \times 10^{34}$	$\hbar/mc$
	$= 4.34191 \times 10^{-6}$	g		
	$= 2.59588 \times 10^{18}$	m_p		

Ordinary Units Expressed Microscopically

1 g	$= 5.60959 \times 10^{23}$	GeV/c^2	$= 2.84279 \times 10^{39}$	$\hbar/mc$
1 kg	$= 5.60959 \times 10^{26}$	GeV/c^2	$= 2.84279 \times 10^{42}$	$\hbar/mc$
1 Joule = 1 kg m^2/s^2	$= 6.24151 \times 10^{9}$	GeV	$= 3.16303 \times 10^{25}$	$\hbar c/m$
1 erg = 1 g cm^2/s^2	$= 6.24151 \times 10^{2}$	GeV	$= 3.16303 \times 10^{18}$	$\hbar c/m$
$= 10^{-7}$ J				
1 Newton = 1 kg m/s^2	$= 1.23162 \times 10^{-6}$	$\text{GeV}^2/\hbar c$	$= 3.16303 \times 10^{25}$	$\hbar c/\text{m}^2$
	$= 1.23162 \times 10^{12}$	$\text{eV}^2/\hbar c$		
1 dyne = 1 g cm/s^2	$= 1.23162 \times 10^{-11}$	$\text{GeV}^2/\hbar c$	$= 3.16303 \times 10^{20}$	$\hbar c/\text{m}^2$
$= 10^{-5}$ N	$= 1.23162 \times 10^{7}$	$\text{eV}^2/\hbar c$		
1 Watt = 1 J/s	$= 4.10824 \times 10^{-15}$	$\text{GeV}^2/\hbar$	$= 1.05507 \times 10^{17}$	$\hbar c^2/\text{m}^2$
	$= 4.10824 \times 10^{3}$	$\text{eV}^2/\hbar$		
1 Hz = 1/s	$= 6.5821 \times 10^{-25}$	$\text{GeV}/\hbar$	$= 3.3356 \times 10^{-9}$	c/m
1 Kelvin	$= 8.61742 \times 10^{-14}$	GeV/k_B	$= 4.36707 \times 10^{2}$	$\hbar c/\text{m}k_B$
	$= 8.61742 \times 10^{-5}$	eV/k_B	$= 1/11604.4$	eV/k_B

Electromagnetic Units

1 Coulomb	$= 6.24151 \times 10^{18}$	e		
1 Volt = 1 J/C	$= 1$	eV/e	$= 5.06773 \times 10^{6}$	$\hbar c/\text{m}e$
	$= 10^{-9}$	GeV/e		
1 Farad = 1 C/V	$= 6.24151 \times 10^{18}$	e^2/eV	$= 1.23162 \times 10^{12}$	$\text{m}e^2/\hbar c$
1 Ampere = 1 C/s	$= 4.10824 \times 10^{3}$	$\text{eV}e/\hbar$	$= 2.08194 \times 10^{10}$	ec/m
1 Ohm = 1 V/A	$= 2.43413 \times 10^{-4}$	$\hbar/e^2$		
1 Mho = 1/Ohm	$= 4.10824 \times 10^{3}$	$e^2/\hbar$		
1 Weber = 1 V s	$= 1.51927 \times 10^{15}$	$\hbar/e$		
1 Tesla = 1 Weber/m^2	$= 59.1572$	$\text{eV}^2/\hbar e c^2$	$= 1.51927 \times 10^{15}$	$\hbar/e\text{m}^2$
1 Gauss = 10^{-4} Tesla	$= 5.91572 \times 10^{-3}$	$\text{eV}^2/\hbar e c^2$	$= 1.51927 \times 10^{11}$	$\hbar/e\text{m}^2$
$\phi_0 = 2\pi\hbar/e$	$= 6.28319$	$\hbar/e$	$= 4.13567 \times 10^{-15}$	Weber
			$= 1/(2.418 \times 10^{14})$	Weber
$\epsilon_0 = 8.854 \times 10^{-12}$ F/m	$= 10.905$	$e^2/\hbar c$		
$\mu_0 = 4\pi \times 10^{-7}$ N/A^2	$= 0.0917012$	$\hbar/ce^2$	$\epsilon_0\mu_0 = 1/c^2$	
$\alpha = e^2/(4\pi\epsilon_0\hbar c)$	$= 7.2974 \times 10^{-3}$		$1/\alpha = 137.036$	

In these tables m_e denotes the electron mass, m_p is the proton mass and α is the electromagnetic fine-structure constant (evaluated at low energies, $\mu \sim m_e$).

A.2 Conventions

Like religion and politics, conventions are a subject normally avoided in polite company for fear of provoking strong words or fisticuffs. Any practising physicist should usually adopt a set of conventions and stick to them, and (as is the case for many) the ones used here are largely the ones I learned as a student. (Because of this they usually agree with those used in Steven Weinberg's many textbooks.) This section explains my rationale for the main choices made.

A.2.1 Geometrical Conventions

The conventions for vectors are such that Greek indices represent spacetime coordinates in 3+1 dimensions, with $x^\mu = \{x^0, x^1, x^2, x^3\} = \{t, x, y, z\}$ a contravariant vector built from Cartesian coordinates. Spatial indices are denoted by latin letters, such as $x^a = \{x, y, z\}$ or $x^i = \{x, y, z\}$, with letters chosen early or later in the alphabet in a way that distinguishes them from any other indices present (such as those describing internal symmetries, or spacetime spinors, *etc.*).

The Einstein summation convention is used throughout the book, unless explicitly stated otherwise. In this convention any repeated appearance of an index represents a summation of that index over its entire range. So $a^\mu a_\mu = a^0 a_0 + a^1 a_1 + a^2 a_2 + a^3 a_3$ while $a^i a_i = a^1 a_1 + a^2 a_2 + a^3 a_3$, and so on.

Metric Conventions

The spacetime metric is denoted $g_{\mu\nu}(x)$ and defines the invariant line-element by $ds^2 = g_{\mu\nu}(x)\, dx^\mu\, dx^\nu$, that gives the square of the distance ds between two infinitesimally separated points: x^μ and $x^\mu + dx^\mu$. The signature of the metric is $(-+++)$, so the Minkowski metric that describes the flat space of special relativity in Cartesian coordinates is given explicitly by

$$ds^2 = \eta_{\mu\nu}\, dx^\mu\, dx^\nu = -dt^2 + \delta_{ij}\, dx^i\, dx^j = -dt^2 + dx^2 + dy^2 + dz^2. \tag{A.2}$$

This is one of the choices that generates the most heat when discussed, since half the world learns this choice (often called the 'east-coast' or 'mostly plus' or 'right' metric) while the other half adopts the opposite sign for $\eta_{\mu\nu}$ (called the 'west-coast' or 'mostly minus' or 'wrong' metric). Normally, much heat (and not much light) is spent on whether it is more sensible for time intervals or space intervals to be negative. With (A.2) time-like vectors have negative length, while vectors in the three space-like directions have positive length.

A more compelling reason for using the convention (A.2) comes once Wick rotations are made to Euclidean space, such as is often done when discussing thermal systems (for which temperature can often be conveniently regarded as periodicity in imaginary time – see *e.g.* §A.2.2). In this case, $\tau = it$ and so $dt^2 = -d\tau^2$. With the above choice the metric becomes positive definite, as do the lengths of all vectors, like $a^2 := \eta_{\mu\nu}\, a^\mu a^\nu$. With the 'mostly-minus' metric convention all such squares become negative when Euclideanized (and when quantities like a^2 are negative it can be a nightmare finding sign mistakes). As mentioned earlier, your conventions should be your friend, and should nudge you towards making fewer errors rather than more errors.

Since notation is part of language, part of the thinking behind metric conventions is also the practice of the community with which one wishes to communicate. Broadly speaking, most relativists, cosmologists and string theorists use the $(-+++)$ metric used here, while particle physicists are more split, though with a majority of phenomenologists using mostly-minus conventions.

With the above metric choice the action for scalars and gauge bosons (see Appendix B) have negative coefficients, since this is required to have positive kinetic energies. That is

$$\mathfrak{L} = -\frac{1}{2}\,\eta^{\mu\nu}\partial_\mu\phi\,\partial_\nu\phi = \frac{1}{2}\left[(\partial_t\phi)^2 - (\nabla\phi)^2\right] \tag{A.3}$$

while

$$\mathfrak{L} = -\frac{1}{4}\,F_{\mu\nu}F^{\mu\nu} = \frac{1}{2}\left(\mathbf{E}^2 - \mathbf{B}^2\right). \tag{A.4}$$

Curvature Conventions

A natural convention is to define the curvature so that the same sign also applies to the action for the metric in General relativity, which is given – see, for example, (10.1) – by

$$\frac{\mathfrak{L}}{\sqrt{-g}} = -\frac{R}{16\pi G_N}. \tag{A.5}$$

This is ensured if one adopts the curvature conventions $R := g^{\mu\nu}R_{\mu\nu}$ with Ricci tensor defined by $R_{\mu\nu} := R^{\lambda}{}_{\mu\lambda\nu}$ and Riemann curvature tensor given by

$$R^{\mu}{}_{\nu\lambda\rho} = \partial_\rho\Gamma^{\mu}_{\nu\lambda} + \Gamma^{\mu}_{\rho\sigma}\Gamma^{\sigma}_{\nu\lambda} - (\rho \leftrightarrow \lambda). \tag{A.6}$$

Here,

$$\Gamma^{\mu}_{\nu\lambda} = \frac{1}{2}g^{\mu\alpha}\left[\partial_\nu g_{\alpha\lambda} + \partial_\lambda g_{\alpha\nu} - \partial_\alpha g_{\nu\lambda}\right] \tag{A.7}$$

is the Christoffel symbol (of the second kind) built from derivatives of the metric, and its inverse $g^{\mu\nu}$ defined by $g^{\mu\nu}g_{\nu\lambda} = \delta^{\mu}_{\lambda}$.

The above definitions are the same as used in the well-known book [397], and are also almost the same as a very popular choice (often called the 'MTW' – or 'geometrical' – choice, with MTW representing the authors Misner, Thorne and Wheeler, of an influential relativity textbook [396]). 'Almost the same' here means the only difference relative to MTW conventions is the overall sign of the definition (A.6). The motivation for the MTW choice is that it gives a positive curvature for spheres in euclidean space (and negative curvatures to hyperbolae), though at the expense of introducing an unusual gravity-specific sign in the action.

Levi-Civita Conventions

Finally, another useful geometrical tensor (in four spacetime dimensions) is the four-index Levi–Civita completely antisymmetric tensor $\epsilon_{\mu\nu\lambda\rho}$. In flat space this is defined to be completely antisymmetric under the interchange of any two indices (and so to vanish whenever two indices take the same values); to have elements ± 1 when all indices are different. The convention used here takes $\epsilon^{0123} = +1$ and then all other components are dictated by the antisymmetry condition.

On curved space it is worth working with a *vierbein* (or tetrad): defined as a basis of four vector fields, $e^a{}_\mu(x)$, with $a = 0, 1, 2, 3$. The basis is chosen to be orthonormal and complete in the sense that

$$g^{\mu\nu} e^a{}_\mu e^b{}_\nu = \eta^{ab} \quad \text{and} \quad \eta_{ab}\, e^a{}_\mu e^b{}_\nu = g_{\mu\nu}, \tag{A.8}$$

with the Einstein summation convention in full force, and where η^{ab} is the signature-$(-+++)$ Minkowski tensor, and $g^{\mu\nu}$ is the inverse of the spacetime metric $g_{\mu\nu}$, so $g^{\mu\nu} g_{\nu\lambda} = \delta^\mu_\lambda$. Evidently, the matrix $e^a{}_\mu$ is morally the square root of the metric $g_{\mu\nu}$.

The above orthogonality and completeness relations allow the definition of the inverse $e_a{}^\mu$, defined to satisfy $e_a{}^\mu e^b{}_\mu = \delta^b_a$ and $e_a{}^\mu e^a{}_\nu = \delta^\mu_\nu$. Any tensor can then be described by its world-index components, like $T_{\mu\nu}$, or its tangent-frame components, like $T_{ab} = e_a{}^\mu e_b{}^\nu T_{\mu\nu}$, and so on. The basis vectors in the vierbein are not unique, with the freedom to do local Lorentz transformations, $e^a{}_\mu \to \Lambda^a{}_b e^b{}_\mu$, where the position-dependent matrices $\Lambda^a{}_b$ satisfy the Lorentz-group definition: $\eta_{ac}\Lambda^a{}_b\Lambda^c{}_d = \eta_{bd}$. These definitions ensure it is consistent to raise and lower indices with either η_{ab} or $g_{\mu\nu}$ (or their inverses) in arbitrary order, so $T^{ab} = e^a{}_\mu e^b{}_\nu T^{\mu\nu} = \eta^{ac}\eta^{bd} T_{cd}$ and so on.

With these definitions in mind the Minkowski flat-space conventions for the Levi–Civita tensor apply to the tangent-frame components. That is, $\epsilon^{abcd} = +1$ when $a = 0$, $b = 1$, $c = 2$ and $d = 3$, and so $\epsilon_{abcd} = -1$ for the same choices for a, b, c and d. The value for all other choices of indices is then determined by complete antisymmetry under permutations of any pair of indices. With this choice then the world-index versions are defined by

$$\epsilon^{\mu\nu\lambda\rho} := e_a{}^\mu e_b{}^\nu e_c{}^\lambda e_d{}^\rho \epsilon^{abcd}, \tag{A.9}$$

and the also completely antisymmetric

$$\epsilon_{\mu\nu\lambda\rho} := e^a{}_\mu e^b{}_\nu e^c{}_\lambda e^d{}_\rho \epsilon_{abcd} = g_{\mu\alpha} g_{\nu\beta} g_{\lambda\sigma} g_{\rho\zeta} \epsilon^{\alpha\beta\sigma\zeta}, \tag{A.10}$$

which satisfy $\epsilon^{\mu\nu\lambda\rho} = \det[e_a{}^\mu] = \det^{1/2}[-g^{\mu\nu}] = \det^{-1/2}[-g_{\mu\nu}]$ when $\mu = 0$, $\nu = 1$, $\lambda = 2$ and $\rho = 3$. It is conventional to introduce the notation $g := \det[g_{\mu\nu}]$, which is negative given the Lorentzian signature shared by $g_{\mu\nu}$ and η_{ab}. In terms of this $\epsilon_{\mu\nu\lambda\rho} = -\det[e^a{}_\mu] = -\sqrt{-g}$ when $\mu = 0$, $\nu = 1$, $\lambda = 2$ and $\rho = 3$, with all other entries defined by antisymmetry.

As Eq. (A.9) makes clear, the quantity $\epsilon^{\mu\nu\lambda\rho}$ transforms as a rank-4 contravariant tensor under coordinate transformations (or diffeomorphisms) and is invariant under local Lorentz transformations, since the tangent-frame quantity ϵ^{abcd} transforms as a scalar under diffeomorphisms (and is a rank-4 contravariant tensor under local Lorentz transformations). One sometimes encounters in the literature (but never elsewhere in this book) a related tensor, $\varepsilon^{\mu\nu\lambda\rho} := \sqrt{-g}\, \epsilon^{\mu\nu\lambda\rho}$ whose components equal ± 1 when all indices are different (or its covariant version $\varepsilon_{\mu\nu\lambda\rho} := \epsilon_{\mu\nu\lambda\rho}/\sqrt{-g}$), whose nonzero components are also ± 1. Although these quantities have simple components, they transform differently under diffeomorphisms, transforming as a tensor density (of weight $\pm\frac{1}{2}$) due to the additional factor of $\sqrt{-g}$ that is present.

Useful identities use the fact that two Levi–Civitas make a metric, since both are invariant tensors under proper Lorentz transformations (more about which below) while Levi–Civita changes sign under parity and time-reversal (and the metric does not). More precisely,

$$\begin{aligned}
\epsilon_{\mu\nu\lambda\rho}\epsilon^{\alpha\beta\sigma\zeta} &= -\delta^\alpha_\mu\,\delta^\beta_\nu\,\delta^\sigma_\lambda\,\delta^\zeta_\rho \pm \text{(23 other permutations of } \alpha, \beta, \sigma \text{ and } \zeta)\\
\epsilon_{\mu\nu\lambda\rho}\epsilon^{\alpha\beta\sigma\rho} &= -(\delta^\alpha_\mu\,\delta^\beta_\nu\,\delta^\sigma_\lambda + \delta^\beta_\mu\,\delta^\sigma_\nu\,\delta^\alpha_\lambda + \delta^\sigma_\mu\,\delta^\alpha_\nu\,\delta^\beta_\lambda - \delta^\beta_\mu\,\delta^\alpha_\nu\,\delta^\sigma_\lambda - \delta^\sigma_\mu\,\delta^\beta_\nu\,\delta^\alpha_\lambda - \delta^\alpha_\mu\,\delta^\sigma_\nu\,\delta^\beta_\lambda)\\
\epsilon_{\mu\nu\lambda\rho}\epsilon^{\alpha\beta\lambda\rho} &= -2(\delta^\alpha_\mu\,\delta^\beta_\nu - \delta^\beta_\mu\,\delta^\alpha_\nu) \qquad\qquad \text{(A.11)}\\
\epsilon_{\mu\nu\lambda\rho}\epsilon^{\alpha\nu\lambda\rho} &= -3!\;\delta^\alpha_\mu\\
\epsilon_{\mu\nu\lambda\rho}\epsilon^{\mu\nu\lambda\rho} &= -4!\,.
\end{aligned}$$

The right-hand sides of these identities are the most general possible tensors built only out of the Kronecker delta and the metric with the same symmetries as the left-hand side. The numerical coefficients are most easily determined by evaluating both sides using explicit values for the open indices.

A.2.2 Finite Temperature and Euclidean Signature

It is often useful to work with a Euclidean-signature metric, for which all of the eigenvalues of $g_{\mu\nu}$ are positive – also called the $(+ + ++)$ metric. For instance, the metric in rectangular coordinates for 4D flat Euclidean space is

$$\mathrm{d}s^2 = g_{mn}\,\mathrm{d}x^m\,\mathrm{d}x^n = (\mathrm{d}x^1)^2 + (\mathrm{d}x^2)^2 + (\mathrm{d}x^3)^2 + (\mathrm{d}x^4)^2. \qquad \text{(A.12)}$$

One of the great virtues of using a $(- + ++)$ metric in Lorentzian signature is that the positive Euclidean metric is obtained simply by replacing $x^0 = t \to -\mathrm{i}\tau_E$, where $\tau_E = x^4$ is the corresponding Euclidean coordinate. The choice of sign in this transformation ensures that terms in the action transform as

$$\exp\left\{\frac{\mathrm{i}}{2}\int \mathrm{d}t\,\mathrm{d}^3x\,\left[(\partial_t\phi)^2 - (\nabla\phi)^2\right]\right\} \to \exp\left\{-\frac{1}{2}\int \mathrm{d}\tau_E\mathrm{d}^3x\,\left[(\partial_{\tau_E}\phi)^2 + (\nabla\phi)^2\right]\right\} \qquad \text{(A.13)}$$

and so the oscillatory factor $e^{\mathrm{i}S(\phi)}$ in the path integral suppresses large gradients.

Equilibrium calculations at finite temperature provide a concrete situation where Euclidean methods are particularly useful. They are useful because of the resemblance between the thermal density matrix, $\rho \propto e^{-\beta H} = e^{-H/T}$, and the time-evolution operator, $U(t, 0) = \exp[-\mathrm{i}Ht]$. This resemblance makes it look as if a thermal density matrix enters into calculations in the same way as would the time-evolution operator for a shift in imaginary time through a distance $\Delta t = -\mathrm{i}\Delta\tau_E = -\mathrm{i}\,\beta$ and so

$$\exp\left[-\mathrm{i}H\Delta t\right] = \exp\left[-H\Delta\tau_E\right] = \exp\left[-\beta H\right]. \qquad \text{(A.14)}$$

Furthermore, in this language thermal averages, like the partition function

$$Z = \mathrm{Tr}\left[e^{-\beta H}\right] = \sum_N \left\langle N\left|e^{-\beta H}\right|N\right\rangle, \qquad \text{(A.15)}$$

correspond to an evolution of a state $|N\rangle$ through a time interval $-\mathrm{i}\beta$ and then identifying the state obtained with the initial state (and summing). This makes plausible (and can be turned into a proof) that thermal expectation values can be rewritten in terms of field theories in a Euclidean-signature space for which the Euclidean time direction is a circle with circumference β.

Once time becomes a circular direction, boundary conditions must be imposed on the fields in this direction. Standard thermal behaviour is reproduced if integer-spin bosons are chosen to be periodic, $\phi(\tau_E+\beta) = \phi(\tau_E)$, and half-integral spin fermions are chosen to be anti-periodic, $\psi(\tau_E + \beta) = -\psi(\tau_E)$.

A.2.3 Dirac Conventions

For fermions the metric conventions drive related conventional choices for the Dirac matrices, γ^μ, since essentially everyone agrees these should be defined to satisfy the algebra $\{\gamma^\mu, \gamma^\nu\} = 2\eta^{\mu\nu}$ (in Minkowski space). In curved spaces one instead demands

$$\{\gamma^a, \gamma^b\} = 2\eta^{ab}, \tag{A.16}$$

in the tangent frame (defined by the tetrad $e^a{}_\mu$ of the previous section) and then converts to world indices using $\gamma^\mu := e_a{}^\mu \gamma^a$. With these definitions the world-index Dirac matrices satisfy the generally covariant Clifford algebra

$$\{\gamma^\mu, \gamma^\nu\} = 2g^{\mu\nu} \tag{A.17}$$

where $g^{\mu\nu}$ is the inverse metric.

In Minkowski space (or in the tangent frame of a curved space) the use of the $(-+++)$ metric implies $(\gamma^0)^2 = -1$ while $(\gamma^i)^2 = +1$ for $i = x, y, z$. Because this makes γ^0 imaginary (when diagonal) it is useful to define $\beta := \mathrm{i}\gamma^0$ so that $\beta^2 = 1$.

A convenient choice of basis for the Dirac matrices (which diagonalizes $\gamma_5 = -\mathrm{i}\gamma^0\gamma^1\gamma^2\gamma^3$) which satisfies (A.16) is given by

$$\gamma_0 = -\gamma^0 = \begin{pmatrix} 0 & \mathrm{i} \\ \mathrm{i} & 0 \end{pmatrix}, \qquad \gamma_k = \begin{pmatrix} 0 & -\mathrm{i}\sigma_k \\ \mathrm{i}\sigma_k & 0 \end{pmatrix} \tag{A.18}$$

where σ_k are the usual 2×2 Pauli matrices for $k = 1, 2, 3$. In this basis

$$\beta = \begin{pmatrix} 0 & I \\ I & 0 \end{pmatrix} \quad \text{and} \quad \gamma_5 = \begin{pmatrix} I & 0 \\ 0 & -I \end{pmatrix}, \tag{A.19}$$

where I is the 2×2 unit matrix.

The Lorentz generators in this representation are given by $\mathcal{J}_{\mu\nu} = -\frac{\mathrm{i}}{4}[\gamma_\mu, \gamma_\nu]$, and so defining rotations, $\mathcal{J}_k$, and boosts, $\mathcal{K}_k$, by $\mathcal{J}_{0k} = \mathcal{K}_k$ and $\mathcal{J}_{ij} = \epsilon_{ijk}\,\mathcal{J}_k$ allows these generators to be written explicitly as

$$\mathcal{J}_k = \frac{1}{2}\begin{pmatrix} \sigma_k & 0 \\ 0 & \sigma_k \end{pmatrix}, \qquad \mathcal{K}_k = \frac{\mathrm{i}}{2}\begin{pmatrix} -\sigma_k & 0 \\ 0 & \sigma_k \end{pmatrix}. \tag{A.20}$$

Because these are block-diagonal they show that $[\gamma_5, \mathcal{J}_{\mu\nu}] = 0$, and so the 4-dimensional spinor representation is reducible: the two 2-dimensional eigenspaces of γ_5 each furnish separate representations of the Lorentz group. Furthermore, although these two representations agree on their representation of the $\mathcal{J}_k$ (both are spin-half), their representations of the $\mathcal{K}_k$ are conjugates of one another in the precise sense that the Pauli-matrix identity $\sigma_k = -\sigma_2\sigma_k^*\sigma_2$ implies

$$\mathcal{K}_{k\pm} = \sigma_2\, \mathcal{K}^*_{k\mp}\, \sigma_2, \tag{A.21}$$

where the sign in the subscript denotes the eigenvalue of γ_5.

Using this representation for Lorentz boosts allows explicit construction of the spinors $\mathbf{u}(\mathbf{p},\sigma)$ and $\mathbf{v}(\mathbf{p},\sigma)$ appearing in the field expansion of (C.30), reproduced again here:

$$\Psi(x) = \sum_{\sigma=\pm\frac{1}{2}} \int \frac{d^3p}{\sqrt{(2\pi)^3 2E_p}} \left[\mathbf{u}(\mathbf{p},\sigma)\, \mathfrak{c}_{p\sigma}\, e^{ip\cdot x} + \mathbf{v}(\mathbf{p},\sigma)\, \bar{\mathfrak{c}}^*_{p\sigma}\, e^{-ip\cdot x}\right], \tag{A.22}$$

where $E_p = \sqrt{\mathbf{p}^2 + m^2}$ and m is the particle mass.

These spinors satisfy[1] $(\mathrm{i}\,\not{p} + m)\mathbf{u} = (\mathrm{i}\,\not{p} - m)\mathbf{v} = 0$, where as usual the slash denotes contraction with a 4-vector, as in $\not{p} := p_\mu \gamma^\mu$. In the rest frame . . .

$$\mathbf{u}(\mathbf{p},\sigma) = \frac{1}{\sqrt{2}} \begin{pmatrix} \sqrt{E_p+m} - \boldsymbol{\sigma}\cdot\hat{\mathbf{p}}\,\sqrt{E_p-m} & 0 \\ 0 & \sqrt{E_p+m} + \boldsymbol{\sigma}\cdot\hat{\mathbf{p}}\,\sqrt{E_p-m} \end{pmatrix} \begin{pmatrix} \chi(\sigma) \\ \chi(\sigma) \end{pmatrix}, \tag{A.23}$$

where $\hat{\mathbf{p}} = \mathbf{p}/|\mathbf{p}|$. Here, $\chi(\sigma)$ is a 2-component spinor encoding the spin of the particle in its rest frame. If defined as eigenstates of $\mathcal{J}_3$ these become

$$\chi(\sigma = +1/2) = \begin{pmatrix} 1 \\ 0 \end{pmatrix} \quad \text{and} \quad \chi(\sigma = -1/2) = \begin{pmatrix} 0 \\ 1 \end{pmatrix}. \tag{A.24}$$

The spinor $\mathbf{v}$ is found by a similar exercise, or by the action of charge conjugation (see below). A short calculation shows that these spinors satisfy the useful completeness relations

$$\sum_{\sigma=\pm\frac{1}{2}} \mathbf{u}(\mathbf{p},\sigma)\overline{\mathbf{u}}(\mathbf{p},\sigma) = -\mathrm{i}\,\not{p} + m \quad \text{and} \quad \sum_{\sigma=\pm\frac{1}{2}} \mathbf{v}(\mathbf{p},\sigma)\overline{\mathbf{v}}(\mathbf{p},\sigma) = -\mathrm{i}\,\not{p} - m, \tag{A.25}$$

whose right-hand sides reduce in the rest frame to $m(\beta \pm 1)$, projecting onto the appropriate eigenspace of β, as expected.

Weyl and Majorana Spinors

There are two natural ways to reduce the 4-dimensional Dirac spinor to two components in a Lorentz-invariant way. Since left-handed spinors satisfy $\gamma_5\psi_L = \psi_L$ and right-handed spinors satisfy $\gamma_5\psi_R = -\psi_R$ a general Dirac spinor can be written in this basis as

$$\Psi = \begin{pmatrix} \psi_L \\ \psi_R \end{pmatrix}. \tag{A.26}$$

The conditions $\psi_R = 0$ or $\psi_L = 0$ are clearly Lorentz-invariant. Dirac spinors satisfying one of these conditions are called (left- or right-handed) Weyl spinors.

The other Lorentz-invariant way to constrain a Dirac spinor is to demand that ψ_R be the complex conjugate of ψ_L – up to multiplication by σ_2 as in condition (A.21). A spinor satisfying this type of reality condition is called a Majorana spinor,

[1] This follows purely from the consistency of the Poincaré transformation properties of Ψ and $a_{p\sigma}$, but can equivalently be regarded as a consequence of the field equation $(\not{\partial} + m)\Psi = 0$.

$$\Psi_M = \begin{pmatrix} \psi_L \\ -\varepsilon\,\psi_L^* \end{pmatrix} = C\Psi_M^*, \tag{A.27}$$

where $\varepsilon = i\sigma_2 = \begin{pmatrix} 0 & 1 \\ -1 & 0 \end{pmatrix}$ is the real 2×2 antisymmetric matrix. The matrix C is called the charge-conjugation matrix, and is given explicitly in this basis by

$$C = \begin{pmatrix} 0 & \varepsilon \\ -\varepsilon & 0 \end{pmatrix} = -\gamma^2. \tag{A.28}$$

A final convention involves the definition of the Dirac conjugate, which here is given by

$$\overline{\Psi} := \Psi^\dagger \beta = i\Psi^\dagger \gamma^0. \tag{A.29}$$

(Keep in mind the factor of i here when comparing with conventions using the opposite signature for the metric.) When applied to Majorana spinors (A.27) becomes

$$\overline{\Psi}_M := \Psi_M^\dagger \beta = \Psi_M^T C\beta = \Psi_M^T \gamma_5 \epsilon, \tag{A.30}$$

where the superscript 'T' denotes the transpose in spinor space, in the same way that '†' denotes hermitian conjugation in this space. The matrix ϵ defined here is called the time-reversal matrix, given explicitly in this basis by

$$\epsilon = \begin{pmatrix} \varepsilon & 0 \\ 0 & \varepsilon \end{pmatrix}, \tag{A.31}$$

in terms of which $C = \gamma_5 \epsilon \beta$.

The matrices β, ϵ and γ_5 provide a very useful set inasmuch as they characterize a spinor's transformation properties under parity, time-reversal and charge conjugation (more about the definition of these is given in §C.4.3). Chasing through the definitions shows that these symmetries get realized on Dirac spinors as follows:

$$\begin{aligned} \mathcal{P}\,\Psi(x)\,\mathcal{P}^{-1} &= \eta_P\,\beta\,\Psi(x_P) \\ \mathcal{C}\,\Psi(x)\,\mathcal{C}^{-1} &= \eta_c\,C\,\Psi^*(x) \\ \mathcal{T}\,\Psi(x)\,\mathcal{T}^{-1} &= \eta_t\,\epsilon\,\Psi(x_T), \end{aligned} \tag{A.32}$$

where η_P, η_c and η_t are arbitrary phases while $x_P^\mu := P^\mu{}_\nu x^\nu$ and $x_T^\mu := T^\mu{}_\nu x^\nu$ are the parity and time-reversal transforms of the point x^μ (with the matrices $P^\mu{}_\nu$ and $T^\mu{}_\nu$ defined in Eq. (C.64)). In particular, an individual Majorana spinor represents a spin-half particle that is its own antiparticle (in much the same way that a real scalar represents a spin-zero particle that is its own antiparticle).

Spinor Bilinears

Since local lagrangian densities are scalars they are built from combinations of fermion bilinears of the form $\overline{\Psi}_1 M \Psi_2$ for two fields Ψ_1 and Ψ_2. It is useful to expand the arbitrary 4×4 matrix M in terms of a standard basis that transforms covariantly under Lorentz transformations. This basis is conveniently chosen to be the sixteen matrices

$$1\,, \quad \gamma_5\,, \quad \gamma^\mu\,, \quad \gamma_5\gamma^\mu \quad \text{and} \quad \gamma^{\mu\nu} = \frac{1}{2}\left[\gamma^\mu, \gamma^\nu\right]. \tag{A.33}$$

Table A.1 The signs appearing in (A.34) and (A.35) for M one of the basis (A.33) of Dirac matrices

	1	γ^μ	$\gamma^{\mu\nu}$	$\gamma_5\gamma^\mu$	γ_5
ξ	−	−	+	+	−
ζ	+	−	−	−	−
η	+	−	+	−	+
$\xi\zeta$	−	+	−	−	+
λ	+	−	−	+	+
χ	+	+	+	−	−

These satisfy useful symmetry and hermiticity relations together with the parity and time-reversal matrices β and ϵ. In addition to $\gamma_5 = \gamma_5^T = \gamma_5^\dagger$, $\beta = \beta^T = \beta^\dagger$ and $-\epsilon = \epsilon^\dagger = \epsilon^{-1} = \epsilon^T$, one has the identities

$$M^T = \xi(\epsilon M \epsilon)\,, \quad M^\dagger = \zeta(\beta M \beta) \quad M^* = \xi\zeta(\epsilon\beta M \epsilon\beta) \quad \text{and} \quad M = \eta(\gamma_5 M \gamma_5), \tag{A.34}$$

with the signs ξ, ζ and η given for each member of the basis (A.33) in Table A.1.

These identities are useful in that they dictate the reality and symmetry properties of bilinears built from Majorana fermions. That is, if Ψ_1 and Ψ_2 both satisfy condition (A.27), then

$$\overline{\Psi}_1 M \Psi_2 = \lambda(\overline{\Psi}_2 M \Psi_1) \qquad \text{and} \qquad \overline{\Psi}_1 M \Psi_2 = \chi(\overline{\Psi}_1 M \Psi_2)^*, \tag{A.35}$$

with signs λ and χ also given in Table A.1.

Standard-Model Fermions

It is often useful to use real fields when writing down the most general effective couplings, and for spin-half fields this means using Majorana spinors. This section develops the notation for writing a generation of Standard Model fermions in terms of Majorana spinors, following [194].

The types of spin-half 2-component Weyl fermions in a Standard Model generation are

$$\begin{pmatrix} u_L \\ d_L \end{pmatrix}, \quad \begin{pmatrix} \nu_L \\ e_L \end{pmatrix}, \quad u_R, \quad d_R, \quad e_R. \tag{A.36}$$

A 4-component Majorana field for each right-handed particle is then defined by

$$\gamma_R U = u_R\,, \quad \gamma_R D = d_R \quad \text{and} \quad \gamma_R E = e_R. \tag{A.37}$$

The left-handed components of these spinors are simply given by the conjugate fields, so

$$\gamma_L U = u_L^c = \varepsilon\, u_R^*\,, \quad \gamma_L D = d_L^c = \varepsilon\, d_R^* \quad \text{and} \quad \gamma_L E = e_L^c = \varepsilon\, e_R^*, \tag{A.38}$$

with the matrix ε defined below Eq. (A.27).

The same construction applies to the $SU_L(2)$ doublets, starting with the definitions

$$\gamma_L Q = \begin{pmatrix} u_L \\ d_L \end{pmatrix} \quad \text{and} \quad \gamma_L L = \begin{pmatrix} \nu_L \\ e_L \end{pmatrix}, \tag{A.39}$$

and so the right-handed components become

$$\gamma_R Q = \begin{pmatrix} u_R^c \\ d_R^c \end{pmatrix} \quad \text{and} \quad \gamma_R L = \begin{pmatrix} \nu_R^c \\ e_R^c \end{pmatrix}, \tag{A.40}$$

where $d_R^c = -\varepsilon\, d_L^*$ and so on.

A.2.4 Dimensional Regularization

Dimensional regularization is usually the regularization of choice for practical calculations, both because of its comparative simplicity and because it treats symmetries relatively benignly. This section collects some of the useful formulae associated with this regularization used in the main text.

The fundamental formula used in the main text involves a single loop integral of the form

$$J(q) := \int \frac{d^4 p}{(2\pi)^4} \left[\frac{(p^2)^A}{(p^2 + q^2)^B} \right], \tag{A.41}$$

where $p^2 = p_\mu p^\mu$ and q^2 is a Lorentz-invariant function of $q_\mu q^\mu$, and possible low-energy masses. The squares of all 4-momenta are taken in Lorentzian signature, so $J(q)$ is a Lorentz-invariant function of q^μ. At face value, this integral diverges in the ultraviolet for $2A + 4 \geq 2B$, and the goal is to define the integral so as to be able sensibly to evaluate physical quantities before the divergence is ultimately eliminated by absorbing it into the value of a bare parameter when renormalizing.

The denominator of the integrand usually has an implicit $i\varepsilon$ factor that tells how to navigate around any poles in the energy integrations, and the result is the same as what is obtained from Wick rotating the energy to imaginary values, using[2] $p^0 = ip^4$, so $p^\mu p_\mu = -(p^0)^2 + \mathbf{p}^2 = (p^4)^2 + \mathbf{p}^2 \geq 0$, while $d^4 p = i d^4 p_E$, where $d^4 p_E := dp^4 d^3\mathbf{p}$ is the Euclidean integration measure. The rotation occurs because p^4 is integrated through real values rather than imaginary ones. Once this is done, the angular integration over the direction of p^m can be done by inspection, leaving only a divergent one-dimensional integral to be regularized.

The idea of dimensional regularization is to consider the same expression in D dimensions,

$$I_D(q) := \int \frac{d^D p_E}{(2\pi)^D} \left[\frac{p^{2A}}{(p^2 + q^2)^B} \right], \tag{A.42}$$

with the desired answer formally obtained by $J = \lim_{D\to 4} iI_D$. The virtue of introducing D as a variable is that the integral converges in the ultraviolet for $D < 2(B - A)$, with the finite result obtained by explicit integration being

[2] The sign here is chosen by the requirement that the rotation from the real to the imaginary axis avoids the poles at $p^0 = \sqrt{\mathbf{p}^2 + m^2} - i\varepsilon$.

$$I_D(q) = \left[\frac{S_{D-1}}{2(2\pi)^D}\right] \left(q^2\right)^{A-B+D/2} \int_0^\infty dx \left[\frac{x^{A+(D-2)/2}}{(x+1)^B}\right]$$
$$= \left[\frac{\pi^{D/2}}{(2\pi)^D \Gamma\left(\frac{D}{2}\right)}\right] \frac{\Gamma\left(A+\frac{D}{2}\right)\Gamma\left(B-A-\frac{D}{2}\right)}{\Gamma(B)} \left(q^2\right)^{A-B+D/2}, \qquad \text{(A.43)}$$

where S_n is the area of the n-dimensional unit sphere and $\Gamma(z)$ is Euler's Gamma function, defined by analytically continuing the defining relation $\Gamma(z+1) = z\,\Gamma(z)$ to the complex plane.

The last equality here extends the definition of I_D to any complex D except for the poles of $\Gamma\left(A+\frac{D}{2}\right)\Gamma\left(B-A-\frac{D}{2}\right)$, which occur whenever the argument of a Γ-function is a non-positive integer (and so includes the case of real interest where $D = 4$). The regularization is performed by evaluating the result at $D = 4 - 2\epsilon$ for $0 < \epsilon \ll 1$, with the limit $\epsilon \to 0$ taken after renormalization has removed the divergence. The incipient divergence in this limit appears as a pole, arising from asymptotic formulae for the Gamma function like

$$\Gamma[\epsilon] = \frac{1}{\epsilon} - \gamma + \mathcal{O}(\epsilon), \qquad \text{(A.44)}$$

where the Euler–Mascheroni constant, γ, is defined by the limit

$$\gamma := \lim_{n\to\infty}\left[\sum_{k=1}^{n} \frac{1}{k} - \ln n\right] \simeq 0.5772156649015328606065120900824024310 42\ldots. \qquad \text{(A.45)}$$

Poles near negative integers are found by repeatedly using $\Gamma(z+1) = z\,\Gamma(z)$. For instance, choosing $z = -1 + \epsilon$ implies that

$$\Gamma[-1+\epsilon] = \frac{\Gamma[\epsilon]}{-1+\epsilon} \simeq -\frac{1}{\epsilon} + (\gamma - 1) + \mathcal{O}(\epsilon), \qquad \text{(A.46)}$$

and so on.

Renormalization Schemes

The pole in this expression expresses the divergence that the integral possesses when $D = 4$, which (if ultraviolet[3] in origin) is usually absorbed into the renormalization of a bare coupling. This is possible because the full expression for a physical quantity depends on both this bare coupling and the loop integral. Although it is unambiguous to say that the bare coupling cancels the divergent part of a loop integral, there *is* an ambiguity associated with how much of the finite parts of a loop are also subtracted in the same way. A precise statement about how much of the finite part to absorb when cancelling divergences defines what is called the renormalization 'scheme'. There is nothing unique about any scheme, with different choices simply corresponding to different ways for precisely defining the meaning of the coupling in question.

To see how this works in practice, consider the divergences associated with the vacuum polarization of the electromagnetic field. In quantum electrodynamics the

[3] Poles as $D \to 4$ can also arise due to infra-red divergences, and these should *not* be renormalized (as may be seen from the discussion following Eq. (12.19) in the main text).

Fourier transform of the propagator, $\langle T A_\mu(x) A_\nu(y)\rangle$, for non-interacting photons is (in a Lorentz-covariant gauge)

$$\Delta^{(0)}_{\mu\nu}(p) = \frac{1}{p^2 - i\varepsilon}\left[\eta_{\mu\nu} + C_A \frac{p_\mu p_\nu}{p^2}\right], \tag{A.47}$$

where ε is a small positive infinitesimal (*not* the ϵ from $D = 4 - 2\epsilon$) and C_A is a quantity whose precise value depends on the gauge being used (*e.g.* with $C_A = 0$ in Feynman gauge or $C_A = -1$ in Landau gauge). The precise form of C_A is not important since it does not appear in any physical predictions. This propagator has a pole at $p^2 = 0$ that defines the photon's energy-momentum dispersion relation: $\varepsilon_p = p^0 = |\mathbf{p}|$. The freedom to rescale fields, $A_\mu \to \lambda A_\mu$, is used when writing (A.47) to ensure the $\eta_{\mu\nu}$ term has unit residue at this pole (this is an automatic consequence of canonical normalization).

Once interactions are included, the propagator does not remain as simple as in (A.47), but it turns out that Lorentz-covariance and gauge invariance require that its most general form must be

$$\begin{aligned}\Delta_{\mu\nu}(p) &= \frac{1}{(p^2 - i\varepsilon)[1 - \Pi(p^2)]}\left[\eta_{\mu\nu} - \Pi(p^2)\,\frac{p_\mu p_\nu}{p^2}\right] + C_A\,\frac{p_\mu p_\nu}{p^2}\\ &= \frac{\eta_{\mu\nu}}{(p^2 - i\varepsilon)[1 - \Pi(p^2)]} + \widetilde{C}_A\,\frac{p_\mu p_\nu}{p^2},\end{aligned} \tag{A.48}$$

where the function, $\Pi(p^2)$, is known as the vacuum polarization (and the second line defines the quantity $\widetilde{C}_A$, whose gauge-dependent value still does not matter for physical predictions).

Although $\Pi(p^2)$ vanishes for non-interacting photons, it is nonzero once couplings to charged particles are included. If $\Pi(p^2)$ were also to have a pole at $p^2 = 0$, such as if $\Pi(p^2) = A/p^2 + B + \cdots$, then the propagator's pole gets moved to $p^2 = A$ (which, provided A is negative, would imply that the interactions give the photon a nonzero mass). So long as $\Pi(p^2)$ is less singular than this near $p^2 = 0$ the pole in $\Delta_{\mu\nu}$ survives, indicating that no mass gets developed.

In quantum electrodynamics (QED) – the theory of interacting electrons and photons – $\Pi(p^2)$ is obtained by evaluating 1-particle irreducible graphs with two external photon legs (see Fig. 7.4). The absence of reducible photon lines in these graphs precludes them from introducing a pole, and this ensures that loops of virtual electrons do not shift the photon mass. The one-loop vacuum polarization graph with an electron in the loop contributes

$$\begin{aligned}&\Pi(p^2)_{\text{1-loop}}\\ &= -\frac{8e^2}{(4\pi)^{D/2}}\,\Gamma\left(2 - \frac{D}{2}\right)\int_0^1 du\; u(1-u)\left[\frac{m^2 + p^2 u(1-u)}{\mu^2}\right]^{(D-4)/2}\\ &= \frac{e^2}{2\pi^2}\int_0^1 du\, u(1-u)\left\{\frac{1}{(D/2) - 2} + \gamma + \ln\left[\frac{m^2 + p^2 u(1-u)}{4\pi\mu^2}\right] + O(D-4)\right\},\end{aligned} \tag{A.49}$$

once regularized in dimensional regularization. Here, $-e$ is the electron charge and m is the electron mass, and μ is an arbitrary scale introduced by replacing $e^2 \to e^2\,\mu^{4-D}$ so that e remains dimensionless in D spacetime dimensions.

Although this expression has no pole at $p^2 = 0$, neither does it vanish there since

$$\Pi(0)_{1\text{-loop}} = \frac{e^2}{12\pi^2}\left[\frac{1}{(D/2)-2} + \gamma + \ln\left(\frac{m^2}{4\pi\mu^2}\right) + O(D-4)\right]. \qquad \text{(A.50)}$$

Consequently, the propagator's residue at $p^2 = 0$ is no longer unity. To fix this, the field must be rescaled once more – that is to say: 'renormalized' – to ensure unit residue, by taking $A_\mu \to \sqrt{Z_3}\, A_\mu$, after which the propagator rescales to $\Delta^{(0)}_{\mu\nu}(p) \to Z_3\, \Delta^{(0)}_{\mu\nu}(p)$. In perturbation theory writing $Z_3 = 1 + \delta Z$, with $\delta Z \sim O(e^2)$, then shows that the renormalized vacuum polarization becomes

$$\Pi(p^2)_{\text{ren}} = \Pi(p^2)_{1\text{-loop}} - (Z_3 - 1). \qquad \text{(A.51)}$$

Now comes the main point regarding convenient renormalization schemes. The physical renormalization choice (or 'on-shell' scheme) for Z_3 requires $\Pi(0)_{\text{ren}} = 0$ since this guarantees unit residue at the propagator's pole at $p^2 = 0$. This gives

$$Z_3^{\text{phys}} - 1 \simeq \Pi(0)_{1\text{-loop}} = \frac{e^2}{12\pi^2}\left[\frac{1}{(D/2)-2} + \gamma + \ln\left(\frac{m^2}{4\pi\mu^2}\right) + O(D-4)\right], \qquad \text{(A.52)}$$

after which the limit $D \to 4$ can be taken to give

$$\Pi(p^2)^{\text{phys}}_{\text{ren}} = \frac{e^2}{2\pi^2}\int_0^1 du\, u(1-u)\ln\left[1 + \frac{p^2 u(1-u)}{m^2}\right]. \qquad \text{(A.53)}$$

But if the only goal is to subtract off divergences, the minimalist choice – called the *minimal subtraction* or MS scheme – merely subtracts the pole at $D = 4$, so

$$Z_3^{\text{MS}} - 1 \simeq \frac{e^2}{6\pi^2}\left(\frac{1}{D-4}\right), \qquad \text{(A.54)}$$

and so (again taking $D \to 4$)

$$\Pi(p^2)^{\text{MS}}_{\text{ren}} = \frac{e^2}{12\pi^2}\left[\gamma + \ln\left(\frac{m^2}{4\pi\mu^2}\right)\right] + \Pi(p^2)^{\text{phys}}_{\text{ren}}. \qquad \text{(A.55)}$$

A slightly more convenient and equally minimal choice [35–37], called the *modified minimal subtraction* (or $\overline{\text{MS}}$) scheme, subtracts the universal factors γ and $\ln 4\pi$ as well as the divergent pole, leading to

$$Z_3^{\overline{\text{MS}}} - 1 \simeq \frac{e^2}{6\pi^2}\left[\frac{1}{(D/2)-2} + \gamma - \ln(4\pi)\right], \qquad \text{(A.56)}$$

and so

$$\Pi(p^2)^{\overline{\text{MS}}}_{\text{ren}} = \frac{e^2}{12\pi^2}\ln\left(\frac{m^2}{\mu^2}\right) + \Pi(p^2)^{\text{phys}}_{\text{ren}}. \qquad \text{(A.57)}$$

Although these last two renormalization schemes do not use canonically normalized fields, they trade this against the advantage of simplicity for other types of calculations. In particular, they allow more simple direct integration of the renormalization-group evolution of couplings with scale and so simplify the resummation of leading logarithms (such as described in the main text in §7.2.1).

Appendix B Momentum Eigenstates and Scattering

This appendix collects (often only with telegraphic derivation) some useful relations for computing scattering and decay rates. Since some subtleties of continuum normalization for momenta are dealt with by appealing to discrete normalization, this discussion starts with a summary of conventions regarding momentum eigenstates.

B.1 Momentum Eigenstates

There are three different conventions often used: discrete normalization, continuum normalization and relativistic continuum normalization. All three types arise in this book, so this section furnishes a brief reminder of how to convert from one to another. For simplicity this is done here for one spatial dimension, though identical arguments also work in other choices for the number of dimensions.

Discrete normalization corresponds to situations where momentum takes a denumerably infinite set of values, such as occurs if spatial dimensions have finite length, L say, perhaps satisfying periodic boundary conditions so fields satisfy $\psi(x + L) = \psi(x)$ for any x. For momentum eigenstates, $\psi(x) \propto \exp[ipx]$, this condition implies that $p = 2\pi n/L$ for integer n, making p denumerable as required. Normalization and completeness relations for states then take the usual quantum form, such as

$$(p \,|\, q) = \delta_{pq} \quad \text{and} \quad \sum_p |\, p)(p\,| = 1, \tag{B.1}$$

where the sum over p is really a sum over the integer n. Here the 'rounded ket' notation, $|\, p)$, is used to distinguish states normalized this way from non-denumerable situations normalized in the continuum.

Inserting a complete set of position eigenstates and using the wavefunction $\langle x \,|\, p) \propto e^{ipx}$ shows that the orthonormality condition of (B.1) becomes

$$(p \,|\, q) = \int_0^L dx\, (p \,|x\rangle \langle x \,|\, q) = \delta_{pq}\,, \tag{B.2}$$

and so $\langle x \,|\, p) = L^{-1/2}\, e^{ipx}$. This normalization then implies that completeness takes the usual form

$$\langle x \,|\, y\rangle = \sum_p \langle x \,|\, p)(p \,|\, y\rangle = \frac{1}{L}\sum_{n=-\infty}^{\infty} e^{2i\pi n(x-y)/L} = \delta(x-y). \tag{B.3}$$

Continuum Normalization

Continuum-normalized states $|p\rangle$ are obtained from discrete-normalized states in the infinite-volume limit $L \to \infty$. In this limit the spacing, $2\pi/L$, between adjacent levels goes to zero, so the denumerable label p goes over to a continuum one. For L very large but still finite there are $\mathrm{d}N = \mathrm{d}p/(2\pi/L)$ states in a small continuous interval $\mathrm{d}p$, and so the density of states is $\mathrm{d}N/\mathrm{d}p = L/(2\pi)$. Therefore any sum over p goes over to an integral according to the rule

$$\sum_p F(p) = \int \mathrm{d}p\, F(p)\, \frac{\mathrm{d}N}{\mathrm{d}p} = L \int \frac{\mathrm{d}p}{2\pi}\, F(p). \tag{B.4}$$

Using this conversion, for very large L the completeness relation for $|p)$ becomes

$$1 = \sum_p |p)(p| = L \int \frac{\mathrm{d}p}{2\pi}\, |p)(p| =: \int \mathrm{d}p\, |p\rangle\langle p|, \tag{B.5}$$

where the last equality suggests the definition of the continuum-normalized state

$$|p\rangle := \sqrt{\frac{L}{2\pi}}\, |p). \tag{B.6}$$

Multiplying (B.5) through on the right by $|q\rangle$ shows that consistency requires that the continuum state must satisfy the normalization condition

$$\langle p\,|\,q\rangle = \delta(p-q), \tag{B.7}$$

which can also be inferred directly from the definitions using

$$\langle p\,|\,q\rangle = \frac{L}{2\pi}\,(p\,|\,q) = \lim_{L\to\infty} \frac{L}{2\pi}\,\delta_{pq}. \tag{B.8}$$

The right-hand side of this expression is zero if $p \neq q$ and if $p = q$ it goes to infinity as $L \to \infty$. This suggests it is a Dirac delta function, $\delta(p-q)$, up to normalization. To get the normalization notice that the integral over p of (B.8) in this limit is given by

$$\int \mathrm{d}p\, \langle p\,|\,q\rangle = \frac{2\pi}{L} \sum_p \langle p\,|\,q\rangle = \sum_p (p\,|\,q) = 1, \tag{B.9}$$

and so the right-hand side of (B.8) goes to $\delta(p-q)$ as $L \to \infty$, as claimed.

A useful relation when converting between discrete and continuum normalizations is

$$\sum_p |p)(p| = \frac{2\pi}{L} \sum_p \left[\frac{L}{2\pi}\,|p)(p|\right] \to \int \mathrm{d}p\, |p\rangle\langle p|, \tag{B.10}$$

showing that completeness sums are the same, regardless of whether momenta are normalized discretely or in the continuum. Often these kinds of sums arise weighted by quantities like an initial probability distribution for $P(p)$, and when this is so $P(p)$ goes over in the continuum limit to a phase-space distribution, $f(p)$, as follows. If $P(p)$ is the probability of having any one value for p, and varies slowly enough to be regarded as being constant in a short interval $\mathrm{d}p$, then the density of probability, $\mathrm{d}\mathcal{P}(p)$, for finding the particle in $\mathrm{d}p$ is:

$$\mathrm{d}\mathcal{P}(p) = \frac{\mathrm{d}N}{\mathrm{d}p}\, P(p)\, \mathrm{d}p = \frac{L}{2\pi}\, P(p)\, \mathrm{d}p, \tag{B.11}$$

and so the differential probability per-unit-spatial-volume of finding the particle in this momentum region (*i.e.* the phase-space probability density) becomes

$$\frac{f(p)}{2\pi} := \frac{1}{L}\left(\frac{d\mathcal{P}}{dp}\right) = \frac{P(p)}{2\pi}. \tag{B.12}$$

The 2π in the left-hand side's denominator is conventional, and ensures that for a thermal distribution (say) for which the position-space probability density is

$$\mathfrak{p} := \int \frac{dp}{2\pi}\,\frac{1}{e^{E/T} \pm 1}, \tag{B.13}$$

one has $f(p) = (e^{E/T} \pm 1)^{-1}$ with no additional factors of 2π. In the $L \to \infty$ limit one therefore has

$$\sum_p P(p)\,|p)(p\,| \to \int dp\; f(p)\,|p\rangle\langle p\,|. \tag{B.14}$$

For three spatial dimensions identical arguments show that the density of states is $dN/d^3\mathbf{p} = \mathcal{V}/(2\pi)^3$, where $\mathcal{V} := L^3$ is the system's large spatial volume. This means discrete sums go over into 3D integrals according to

$$\sum_{\mathbf{p}} \to \mathcal{V}\int \frac{d^3p}{(2\pi)^3} = \mathcal{V}\int \frac{dp_x dp_y dp_z}{(2\pi)^3}, \tag{B.15}$$

so if $|\mathbf{p}\rangle = [\mathcal{V}/(2\pi)^3]^{1/2}|\mathbf{p})$ then as $\mathcal{V} \to \infty$ the completeness formula (B.10) becomes

$$1 = \sum_{\mathbf{p}} |\mathbf{p})(\mathbf{p}| \to \int d^3p\; |\mathbf{p}\rangle\langle\mathbf{p}|, \tag{B.16}$$

while orthogonality goes over to

$$\langle\mathbf{p}|\mathbf{q}\rangle = \delta^3(\mathbf{p}-\mathbf{q}) = \delta(p_x - q_x)\,\delta(p_y - q_y)\,\delta(p_z - q_z). \tag{B.17}$$

A sum weighted by an initial probability distribution similarly goes over to

$$\sum_{\mathbf{p}} P(\mathbf{p})\,|\mathbf{p})(\mathbf{p}\,| \to \int d^3p\; f(\mathbf{p})\,|\mathbf{p}\rangle\langle\mathbf{p}\,|. \tag{B.18}$$

Covariant Normalization

An additional normalization change is often made for relativistic theories, since for these it can be inconvenient that $|p\rangle$ satisfies a Lorentz non-invariant condition like (B.7) and (B.9). In particular, (B.9) implies $\langle p|q\rangle$ transform inversely to the way the measure dp transforms.

It happens, however, that the combination dp/E_p *is* invariant if $E_p = \sqrt{p^2 + m^2}$ is the energy associated with a given momentum p. This makes it useful to define the covariantly normalized state $|p\,\rangle_r := \sqrt{2E_p}\;|p\rangle$, which satisfies a Lorentz-invariant completeness condition

$$\int \frac{dp}{2E_p}\,|p\,\rangle_r\,{}_r\langle p\,| = \int dp\;|p\rangle\langle p| = 1, \tag{B.19}$$

and orthogonality relation

$$_r\langle p \,|\, q \rangle_r = \sqrt{4E_pE_q}\;\langle p \,|\, q\rangle = 2E_p\,\delta(p-q) = 2E_q\,\delta(p-q). \qquad \text{(B.20)}$$

For three spatial dimensions the relativistic normalization is again defined by $|\mathbf{p}\,\rangle_r := \sqrt{2E(\mathbf{p})}\;|\mathbf{p}\rangle$, which satisfies a Lorentz-invariant completeness condition

$$\int \frac{\mathrm{d}^3p}{2E(\mathbf{p})}\;|\mathbf{p}\,\rangle_r\;{}_r\langle\,\mathbf{p}\,| = \int \mathrm{d}\mathbf{p}\;|\mathbf{p}\rangle\langle\mathbf{p}| = 1, \qquad \text{(B.21)}$$

and orthogonality relation

$$_r\langle\,\mathbf{p}\,|\,\mathbf{q}\,\rangle_r = 2E(\mathbf{p})\,\delta^3(\mathbf{p}-\mathbf{q}) = 2E(\mathbf{q})\,\delta^3(\mathbf{p}-\mathbf{q}). \qquad \text{(B.22)}$$

B.2 Basics of Scattering Theory

Scattering describes interactions for which the particles involved start off as widely separated wave-packets, then approach one another and interact briefly as their wave-packets overlap, and then separate to great distances again. Theoretical simplicity arises because many details are not required, with only the total change in energy and momentum due to the scattering being measured (rather than, say, their detailed trajectories for all times).

In principle, in quantum mechanics a particle moving in an initial (or final) wave-packet cannot be exact momentum (or energy) eigenstates because the uncertainty principle ensures that such eigenstates are not localized in space or time at all. It is nonetheless often possible to approximate the real states using a class of energy eigenstate, since scattering results are often largely insensitive to the details of the wave packets describing the initial states. The idealized energy eigenstates used for this purpose (described below) are called *scattering states*.

The goal is to compute scattering perturbatively in the interaction that dominates when the scattering wave-packets overlap. To this end, suppose the complete Hamiltonian, H, can be written $H = H_0 + V$, where H_0 describes the evolution of the initial and final wave packets before and after the scattering. A key assumption is that the full set of energy eigenvalues for H contains (but need not be identical with) the spectrum of energy eigenstates for H_0. For instance, eigenstates of both H and H_0 can be labelled by their asymptotic momentum in the remote past, or the remote future. States in the spectrum of H but not in H_0 might include bound states whose existence relies on the presence of the interaction V.

Denote the energy eigenstates of H_0 by $|\alpha\rangle$, with α collectively denoting all of the labels required to describe single- and many-particle states and $H_0|\alpha\rangle = E_\alpha|\,\alpha\rangle$. A wave packet of such states can be schematically written

$$|\,\phi_f\rangle := \int \mathrm{d}\alpha\; f(\alpha)|\alpha\rangle, \qquad \text{(B.23)}$$

where $f(\alpha)$ defines a normalizable packet. The label α is treated as continuous because it contains continuum-normalized momentum states (possibly among other labels). Because the spectrum of H_0 lies within the spectrum of H the same labels, α, and energies, E_α, also describe some of the eigenstates of the full system,

$H|\alpha\rangle\rangle = E_\alpha|\alpha\rangle\rangle$ (where – in this Appendix only – a double ket $|(\cdots)\rangle\rangle$ denotes an eigenstate of H).

In the Schrödinger picture the burden of time evolution is carried by the state of the system, and for scattering the time evolution of states prepared in appropriate wave packets, $|\phi_f\rangle$, have essentially the same evolution in the remote past and the remote future when evolved by either H or H_0 (because the particle wave-packets no longer overlap). That is, at very late times there is a state of the full system for which

$$\lim_{t\gg T} e^{-iHt}|\phi_f\rangle\rangle_o = \lim_{t\gg T} e^{-iH_0 t}|\phi_f\rangle, \tag{B.24}$$

and a similar state, $|\phi_f\rangle\rangle_i$, – in general different than $|\phi_f\rangle\rangle_o$ if scattering actually occurs – whose evolution under H agrees with the evolution of a packet $|\phi_f\rangle$ under H_0 in the remote past:

$$\lim_{t\ll -T} e^{-iHt}|\phi_f\rangle\rangle_i = \lim_{t\ll -T} e^{-iH_0 t}|\phi_f\rangle. \tag{B.25}$$

Taking the limiting case of appropriately peaked wave packets, $f(\alpha)$, allows the definition of idealized 'in' and 'out' scattering eigenstates of the full Hamiltonian, $|\alpha\rangle\rangle_{o,i}$, that satisfy

$$\lim_{t\gg T} e^{-iHt}|\alpha\rangle\rangle_o = \lim_{t\gg T} e^{-iH_0 t}|\alpha\rangle \quad \text{and} \quad \lim_{t\ll -T} e^{-iHt}|\alpha\rangle\rangle_i = \lim_{t\ll -T} e^{-iH_0 t}|\alpha\rangle. \tag{B.26}$$

Scattering asks only for transition amplitudes between states that evolve like an eigenstate of H_0 in the remote past to similar states that evolve like eigenstates of H_0 in the remote future. Any such a scattering amplitude can be reconstructed from the matrix of all possible amplitudes between scattering energy eigenstates,

$$S_{\beta\alpha} := {}_o\langle\langle\beta|\alpha\rangle\rangle_i. \tag{B.27}$$

This quantity is called the S-matrix. The scattering operator, S, is defined as that operator whose matrix elements between H_0 eigenstates, $|\alpha\rangle$, reproduce the amplitudes (B.27):

$$\langle\beta|S|\alpha\rangle := S_{\beta\alpha}. \tag{B.28}$$

Formally, S can be computed in terms of the Møller wave operators

$$\Omega(t) := e^{iHt}\, e^{-iH_0 t}, \tag{B.29}$$

because

$$|\alpha\rangle\rangle_o = \lim_{t\gg T}\Omega(t)|\alpha\rangle \quad \text{and} \quad |\alpha\rangle\rangle_i = \lim_{t\ll -T}\Omega(t)|\alpha\rangle. \tag{B.30}$$

The operators $\Omega^\pm = \lim_{t\to\pm\infty}\Omega(t)$ are isometric operators, but strictly speaking are not unitary in the presence of 'bound states' that are contained in the spectrum of H but not in the spectrum of H_0. The S-matrix is then given by

$$S = \lim_{t\to\infty}\lim_{t'\to -\infty}\Omega^*(t)\Omega(t') = (\Omega^+)^*\Omega^-, \tag{B.31}$$

in which the limit $t\to\mp\infty$ must be defined with some care (which is where the appropriately normalized wave packets come in).

B.2.1 Time-Dependent Perturbation Theory

An approximate expression for $\mathcal{S}$ in powers of $V = H - H_0$ can be obtained using standard steps of time-dependent perturbation theory. To this end, notice that $\Omega^*(t)\,\Omega(t') = e^{iH_0 t}\, e^{-iH(t-t')}\, e^{-iH_0 t'}$ satisfies the differential equation

$$i\frac{d}{dt}[\Omega^*(t)\,\Omega(t')] = e^{iH_0 t}(H - H_0)\, e^{-iH(t-t')}\, e^{-iH_0 t'} = V(t)\,\Omega^*(t)\,\Omega(t'), \qquad \text{(B.32)}$$

where this last equality defines the interaction picture operator $V(t) := e^{iH_0 t} V\, e^{-iH_0 t}$. This can be solved iteratively, leading to the following expression for $\mathcal{S} = \lim_{\substack{t\to+\infty \\ t'\to-\infty}} \Omega^*(t)\Omega(t')$:

$$\begin{aligned}\mathcal{S} &= \sum_{n=0}^{\infty} (-i)^n \int_{-\infty}^{\infty} d\tau_1 \int_{-\infty}^{\tau_1} d\tau_2 \cdots \int_{-\infty}^{\tau_{n-1}} d\tau_n\, V(\tau_1)V(\tau_2)\cdots V(\tau_n)\\ &= \sum_{n=0}^{\infty} \frac{(-i)^n}{n!} \int_{-\infty}^{\infty} d^4x_1 \cdots \int_{-\infty}^{\infty} d^4x_n\, T[\mathfrak{H}(x_1)\cdots \mathfrak{H}(x_n)],\end{aligned} \qquad \text{(B.33)}$$

which uses $V(t) = \int d^3x\, \mathfrak{H}(x)$ and introduces the time-ordering operation $T[O_1(t)\, O_2(t')] = \Theta(t - t')O_1(t)O_2(t') + \Theta(t' - t)O_2(t')O_1(t)$ (where Θ denotes a step function).

Using momentum eigenstates, for which $\langle\beta|O(x)|\alpha\rangle = e^{i(p_\alpha - p_\beta)\cdot x}\langle\beta|O(0)|\alpha\rangle$, the S-matrix can therefore be written

$$S_{\beta\alpha} = \delta_{\beta\alpha} - i\mathcal{M}_{\beta\alpha}(2\pi)^4\delta^4(p_\beta - p_\alpha) \qquad \text{(B.34)}$$

with

$$\mathcal{M}_{\beta\alpha} = \langle\beta|\mathfrak{H}_I(0)|\alpha\rangle - \frac{i}{2!}\int d^4x \langle\beta|T\,[\mathfrak{H}_I(x)\mathfrak{H}_I(0)]\,|\alpha\rangle + \cdots. \qquad \text{(B.35)}$$

B.2.2 Transition Rates

The expressions for the S-matrix are proportional to an energy-conserving (and often momentum-conserving) delta function when expressed in terms of energy eigenstates rather than wave packets. This means that the square of S-matrix elements – what should be the transition probabilities – are proportional to $[\delta(E)]^2 = \delta(E)\,\delta(0)$ and so must diverge. Physically, this divergence arises because energy eigenstates are an idealization of the wave packets that scattering really involves. The convenience of using scattering energy eigenstates to compute the S-matrix has as its price the necessity to more carefully sort out the relationship between physical quantities and S-matrix elements.

Going back to the wave-packet description, $|\,\phi_f\rangle\!\rangle_i = \int d\alpha\, f(\alpha)|\alpha\rangle\!\rangle_i$, the probability of finding the system in the final state labeled by β becomes

$$P_f(\beta) = |{}_o\langle\!\langle\beta|\phi_f\rangle\!\rangle_i|^2 = \int d\alpha\, d\alpha'\, f^*(\alpha')f(\alpha)\, {}_o\langle\!\langle\beta|\alpha\rangle\!\rangle_i\, {}_i\langle\!\langle\alpha'|\beta\rangle\!\rangle_o. \qquad \text{(B.36)}$$

In practice, the packet $f(\alpha)$ is peaked about some value $\overline{\alpha}$ with a width about this value that is small compared with experimental resolutions but not so small as to run into trouble from the uncertainty relations. It is also usually true that $S_{\beta\alpha}$

does not have a strong dependence on α in the regime over which $f(\alpha)$ has its support. Concretely, the energy width of a wave packet is usually small compared with the energy dependence of the scattering cross section. (If this were not true the experiment would do a poor job measuring the S-matrix.)

Under these circumstances – and assuming β is distinguishable from all of the α in the support of $f(\alpha)$, so $S_{\beta\alpha} \simeq -i\mathcal{M}_{\beta\alpha}(2\pi)^4\delta^4(p_\beta - p_\alpha)$ – then (B.36) factorizes,

$$P_f(\beta) \approx |\mathcal{M}_{\beta\overline{\alpha}}|^2 \int [d\alpha]\,[d\alpha']\, f^*(\alpha')f(\alpha), \tag{B.37}$$

where $d\alpha\,(2\pi)^4\delta^4(p_\alpha - p_\beta) = [d\alpha]$, and similarly for $[d\alpha']$. The delta functions enforcing energy-momentum conservation are no longer a problem because they are used to perform part of the integration over α and α'.

The Finite-Volume/Finite-Time Trick

What is important about (B.37) is its factorization of reaction probabilities into an interaction part, $|\mathcal{M}_{\beta\overline{\alpha}}|^2$, and a part involving precisely how the initial wave-packet is set up. Given this factorization, it would be useful to identify the interaction part as efficiently as possible without having to set up the wave-packets in detail each time.

A trick for doing so is to directly compute $S_{\beta\alpha}$ with the system imagined to be inside a box with large but finite volume $\mathcal{V}$, and allowing the interactions to last only over a large but finite time interval, T. When this is true, probabilities can be computed as in ordinary quantum mechanics by squaring the relevant transition amplitude, and the result's dependence on $\mathcal{V}$ and T can then be studied to identify what remains physical in the limit $\mathcal{V}, T \to \infty$. Once such a quantity is identified, the temporary theoretical contrivance of the box and interval can be dropped.

Conventions for discrete and continuum normalizations for momentum eigenstates are summarized in §B.1. Following the discussion there, states using discrete normalization for momenta, $(\mathbf{p}|\mathbf{q}) = \delta_{\mathbf{pq}}$, are denoted $|\,\alpha)$, those using nonrelativistic continuum normalization, $\langle\mathbf{p}|\mathbf{q}\rangle = \delta^3(\mathbf{p}-\mathbf{q})$, are denoted $|\,\alpha\rangle$, while those using relativistic normalization, ${}_r\langle\mathbf{p}|\mathbf{q}\rangle_r = 2E_p\,\delta^3(\mathbf{p}-\mathbf{q})$, are denoted $|\,\alpha\rangle_r$.

For a state involving N_α particles in a large but finite box of volume, $\mathcal{V}$, these states have the relative normalization

$$|\,\alpha\rangle = \left[\frac{\mathcal{V}}{(2\pi)^3}\right]^{N_\alpha/2} |\,\alpha] \quad\text{and}\quad |\,\alpha\rangle_r = \left[\frac{2\overline{E}\mathcal{V}}{(2\pi)^3}\right]^{N_\alpha/2} |\,\alpha], \tag{B.38}$$

where $(2\overline{E})^{N_\alpha/2}$ is shorthand for the product $\prod_{i=1}^{N_\alpha}\sqrt{2E_i}$. The corresponding S-matrix elements, $S^{\rm cont}_{\beta\alpha} = \langle\beta|S|\,\alpha\rangle$, $S^{\rm disc}_{\beta\alpha} = (\beta|S|\,\alpha)$ and $S^{\rm rel}_{\beta\alpha} = {}_r\langle\beta|S|\,\alpha\rangle_r$, are therefore related by

$$S^{\rm cont}_{\beta\alpha} = \left[\frac{\mathcal{V}}{(2\pi)^3}\right]^{(N_\alpha+N_\beta)/2} S^{\rm disc}_{\beta\alpha} \quad\text{and}\quad S^{\rm rel}_{\beta\alpha} = \left[\frac{2\overline{E}\mathcal{V}}{(2\pi)^3}\right]^{(N_\alpha+N_\beta)/2} S^{\rm disc}_{\beta\alpha}. \tag{B.39}$$

For translationally invariant theories the S-matrix in each case differs from unity by

$$S_{\beta\alpha} = \delta_{\beta\alpha} - i\mathcal{M}_{\beta\alpha}(2\pi)^4\delta^4_{\mathcal{V}T}(\mathbf{p}_\beta - \mathbf{p}_\alpha), \tag{B.40}$$

where the finite-volume delta functions arise when evaluating in the form,

$$(2\pi)^4 \delta^4_{\mathcal{V}T}(p_\alpha - p_\beta) := \int_{\mathcal{V}T} \mathrm{d}^4 x\, e^{\mathrm{i}(p_\alpha - p_\beta)\cdot x}, \tag{B.41}$$

with the spatial integration over the volume, $\mathcal{V}$, and the time integration runs from $-T/2$ to $+T/2$. As $\mathcal{V}T \to \infty$ the quantity $\delta_{\mathcal{V}T}$ goes to the standard delta-function that enforces energy-momentum conservation.

Now comes the main point. Transition probabilities involve $|S_{\beta\alpha}|^2$ and so necessarily also contain a factor of $\delta^4_{\mathcal{V}T}(0)$. But direct evaluation of (B.41) shows that for finite T and $\mathcal{V}$ this factor is $(2\pi)^4\delta_{\mathcal{V}T}(0) = \mathcal{V}T$. The $\delta_{\mathcal{V}T}(0)$ divergence in $|S_{\beta\alpha}|^2$ is thereby seen to have a physical origin. In a system invariant under translations in space and time (*i.e.* precisely those that generate energy-momentum conserving delta functions), the instantaneous transition probability is the same everywhere and at every instant, making the total rate scale with the volume and time interval over which the initial wave-packets overlap. But this volume and time interval are infinite if the states involved are chosen for convenience to be momentum and energy eigenstates.

This diagnosis suggests a remedy:[1] it is the transition rate per unit volume per unit time that is well-behaved as $\mathcal{V}, T \to \infty$, so this is what should be computed in the continuum limit. Suppose the initial particles have some probability, $P(\alpha)$, to be in a particular region of phase space containing $\Delta\alpha$ states and small enough that $P(\alpha)$ is approximately constant within it. Using discrete normalization, the total transition probability for scattering from $\Delta\alpha$ to a similar small region of final-state particles, $\Delta\beta$, is then simply $\mathcal{P}(\alpha \to \beta) = |S^{\rm disc}_{\beta\alpha}|^2 P(\alpha)\Delta\alpha\Delta\beta$. Since the density of states in momentum space is $\mathcal{V}/(2\pi)^3$, the total number of states in an interval $\mathrm{d}\beta = \prod_{i=1}^{N_\beta} \mathrm{d}^3 p_i$ for an N_β-particle state is $\Delta\beta = [\mathcal{V}/(2\pi)^3]^{N_\beta}\,\mathrm{d}\beta$, and similarly for $\Delta\alpha$. The differential transition rate per-unit-time then becomes

$$\begin{aligned}
\mathrm{d}\Gamma(\alpha \to \beta) &:= \frac{\mathrm{d}\mathcal{P}(\alpha \to \beta)}{T} = \frac{|S^{\rm disc}_{\beta\alpha}|^2}{T} P(\alpha)\,\Delta\alpha\Delta\beta \\
&= \frac{f(\alpha)}{T}\left[|S^{\rm cont}_{\beta\alpha}|^2 \left(\frac{\mathcal{V}}{(2\pi)^3}\right)^{-(N_\alpha+N_\beta)}\right]\left(\frac{\mathcal{V}}{(2\pi)^3}\right)^{(N_\beta+N_\alpha)} \mathrm{d}\alpha\,\mathrm{d}\beta \\
&= \mathcal{V} f(\alpha)\,|\mathcal{M}^{\rm cont}_{\beta\alpha}|^2\,(2\pi)^4\delta^4_{\mathcal{V}T}(p_\beta - p_\alpha)\,\mathrm{d}\alpha\,\mathrm{d}\beta \\
&= \mathcal{V} f(\alpha)\,|\widehat{\mathcal{M}}^{\rm cont}_{\beta\alpha}|^2\,(2\pi)^4\delta^4_{\mathcal{V}T}(p_\beta - p_\alpha)\,\mathrm{d}\hat\alpha\,\mathrm{d}\hat\beta \\
&= \mathcal{V} f(\alpha)\,|\widehat{\mathcal{M}}^{\rm rel}_{\beta\alpha}|^2\,(2\pi)^4\delta^4_{\mathcal{V}T}(p_\beta - p_\alpha)\,\mathrm{d}\hat\alpha_r\,\mathrm{d}\hat\beta_r,
\end{aligned} \tag{B.42}$$

where the replacement $P(\alpha) \to f(\alpha)$ follows the argument leading to (B.14) – where

$$f(\alpha) = \prod_{j\in\alpha} f_j(\alpha), \tag{B.43}$$

is the joint phase-space probability density of initial particles, assumed to be independent of one another. The new notation is

[1] §16.4 provides a more sophisticated version of this same remedy, at least for the growth with T. See also the discussion leading to (16.120).

$$d\hat{\alpha} := \prod_{j\in\alpha}\frac{d^3p_j}{(2\pi)^3},\quad d\hat{\alpha}_r := \prod_{j\in\alpha}\frac{d^3p_j}{(2\pi)^3 2E_j},\quad d\hat{\beta} := \prod_{j\in\beta}\frac{d^3p_j}{(2\pi)^3},\quad d\hat{\beta}_r := \prod_{j\in\beta}\frac{d^3p_j}{(2\pi)^3 2E_j}, \tag{B.44}$$

and $\widehat{\mathcal{M}}_{\beta\alpha} := (2\pi)^{3(N_\alpha+N_\beta)/2}\mathcal{M}_{\beta\alpha}$. $\widehat{\mathcal{M}}_{\beta\alpha}$ is a natural quantity to define because the factor of $(2\pi)^{3/2}$ for each particle cancels the $(2\pi)^{-3/2}$ appearing in the expansion of all fields in terms of creation and annihilation operators – see, for instance, Eqs. (C.28), (C.30), (C.33) or (C.39) – since these appear systematically when evaluating matrix elements like those appearing in (B.35).

Fermi's Golden Rule

The special case where only the leading term of the transition amplitude of (B.35) dominates the rate of (B.42) leads to the very useful formula

$$d\Gamma(\alpha\to\beta) = \mathcal{V} f(\alpha)\,|\langle\beta|\mathfrak{H}_i(0)|\alpha\rangle|^2\,(2\pi)^4\delta^4_{\mathcal{V}T}(p_\beta-p_\alpha)\,d\alpha\,d\beta, \tag{B.45}$$

known as Fermi's Golden Rule.

Decays: $N_\alpha = 1$

Expression (B.42) can be directly used – right out of the box, so to speak – in the special case $N_\alpha = 1$, since in this case $\mathcal{V} f(\alpha)\,d\hat{\alpha} = \mathcal{V} f(\mathbf{p})\,d^3p/(2\pi)^3$ is the probability of the initial particle being in the initial momentum region d^3p, making the limit $\mathcal{V}, T \to \infty$ a trivial one. The resulting expression for the decay rate of a single particle then is

$$d\Gamma(\alpha\to\beta) = \frac{1}{2E_\alpha}|\widehat{\mathcal{M}}^{\rm rel}_{\beta\alpha}|^2(2\pi)^4\delta^4(p_\alpha - p_\beta)\,d\hat{\beta}_r. \tag{B.46}$$

This result is not quite Lorentz-invariant, because of the $1/(2E_\alpha)$ in front. This factor is just what is needed to provide the time-dilation of a fast-moving particle's decay lifetime.

Scattering: $N_\alpha \geq 2$

More generally, for $N_\alpha \geq 2$ the scattering rate per-unit-volume in the large-volume, late-time limit for initially uncorrelated particles is

$$\begin{aligned}\frac{d\Gamma(\alpha\to\beta)}{\mathcal{V}} &= \left[\prod_{k\in\alpha} f_k(\mathbf{p}_k)\,\frac{d^3p_k}{(2\pi)^3}\right]|\widehat{\mathcal{M}}^{\rm cont}_{\beta\alpha}|^2\,(2\pi)^4\delta^4(p_\beta-p_\alpha)\prod_{j\in\beta}\frac{d^3p_j}{(2\pi)^3} \\ &= \left[\prod_{k\in\alpha} f_k(\mathbf{p}_k)\,\frac{d^3p_k}{(2\pi)^3 2E_k}\right]|\widehat{\mathcal{M}}^{\rm rel}_{\beta\alpha}|^2\,(2\pi)^4\delta^4(p_\beta-p_\alpha)\prod_{j\in\beta}\frac{d^3p_j}{(2\pi)^3 2E_j},\end{aligned} \tag{B.47}$$

which (reasonably) is proportional to the phase-space density of each particle in the initial state. The second line makes the Lorentz-transformation properties of this rate manifest.

Using again that $\mathcal{V} f(\alpha)\, \mathrm{d}\hat{\alpha} = \mathcal{V} f(\mathbf{p})\, \mathrm{d}^3p/(2\pi)^3$ is the probability of an initial particle being in the initial momentum region d^3p, expression (B.47) states that the interaction rate per particle, for particle 'B' (for 'beam'), is

$$\mathrm{d}\Gamma_B = (2\pi)^3 \frac{\mathrm{d}\Gamma(\alpha \to \beta)}{\mathcal{V} f_B\, \mathrm{d}^3p_B} \tag{B.48}$$

$$= \frac{1}{2E_B} \left[\prod_{\substack{k\in\alpha \\ k\neq B}} f_k(\mathbf{p}_k) \frac{\mathrm{d}^3p_k}{(2\pi)^3 2E_k} \right] |\widehat{\mathcal{M}}^{\mathrm{rel}}_{\beta\alpha}|^2 (2\pi)^4 \delta^4(p_\beta - p_\alpha) \prod_{j\in\beta} \frac{\mathrm{d}^3p_j}{(2\pi)^3 2E_j}.$$

Two-Body Scattering: $N_\alpha = 2$

In the special case of two-particle scattering, $N_\alpha = 2$, the product over initial particles just involves the 'other' (non-beam) particle, denoted 'T' (for 'target'):

$$\mathrm{d}\Gamma_B = \frac{f_T(\mathbf{p}_T)}{(2\pi)^3} \left[\frac{\mathrm{d}^3p_T}{4E_B E_T} \right] |\widehat{\mathcal{M}}^{\mathrm{rel}}_{\beta\alpha}|^2 (2\pi)^4 \delta^4(p_\beta - p_\alpha) \prod_{j\in\beta} \frac{\mathrm{d}^3p_j}{(2\pi)^3 2E_j}, \tag{B.49}$$

and so depends explicitly on the density of target particles. It is conventional to normalize out the target-dependent factors and define the cross section by $\mathrm{d}\sigma = \mathrm{d}\Gamma_B/F$, where F satisfies two conditions. First, the Lorentz-transformation properties of F are fixed by requiring $\mathrm{d}\sigma$ to be Lorentz invariant. Inspection of (B.49) shows this implies that

$$F = \left(\frac{f_T(\mathbf{p}_T)}{(2\pi)^3} \right) \frac{\mathcal{F}\, \mathrm{d}^3p_T}{4E_T E_B}, \tag{B.50}$$

where $\mathcal{F}$ is a Lorentz-invariant quantity. Second, F should evaluate to the particle flux (as seen by the beam particle), $\mathrm{d}n_T v_{\mathrm{rel}}$, when evaluated in the target-particle rest frame. Here, $\mathrm{d}n_T = f_T\, \mathrm{d}^3p_T/(2\pi)^3$ is the target's ordinary-space particle density and v_{rel} is the (Lorentz-invariant) relative velocity of the initial two particles,

$$v_{\mathrm{rel}} = \sqrt{1 - \frac{m_B^2\, m_T^2}{(p_B \cdot p_T)^2}}, \tag{B.51}$$

where p_B^μ and p_T^μ are the 4-momenta of the two initial particles. The resulting two-body cross section becomes

$$\mathrm{d}\sigma(\alpha \to \beta) = \frac{|\widehat{\mathcal{M}}^{\mathrm{rel}}_{\beta\alpha}|^2}{\mathcal{F}} (2\pi)^4 \delta^4(p_\alpha - p_\beta) \prod_{j\in\beta} \frac{\mathrm{d}^3p_j}{(2\pi)^3 2E_j} \tag{B.52}$$

$$\text{with} \quad \mathcal{F} := (-4p_B \cdot p_T) v_{\mathrm{rel}} = 4\sqrt{(p_B \cdot p_T)^2 - m_B^2\, m_T^2}.$$

Appendix C Quantum Field Theory: A Cartoon

Quantum field theory (QFT) plays a central role in most areas of theoretical physics, but this is not really a deep statement. At one level quantum field theory is merely ordinary quantum mechanics applied to processes that change the total number of particles present, and this makes it particularly useful for relativistic systems since fundamental principles (the consistency of quantum mechanics and special relativity) forbid relativistic interactions from ever leaving the total number of particles unchanged.

But the utility of QFT methods are not restricted to processes that change the total number of particles. It is also useful when framing quantum systems with a fixed number of particles, largely because its main feature – the language of creation and annihilation operators – lends itself to efficiently expressing natural laws in such a way that ensures the validity of a few fundamental principles right from the start. The principles that get baked in in this way include 'unitarity' (which is to say, conservation of probability in quantum evolution) and 'cluster decomposition' (which means the factorization of probabilities for independent events when these events are causally separated from one another in spacetime).

Although this book is not meant as a textbook on quantum field theory, QFT tools are nonetheless often used within these pages. Consequently, this Appendix is offered, both as a quick refresher on some elements of quantum theory of fields, as well as a way to collect together some of the main useful formulae used elsewhere. Although this summary possibly provides a useful reminder for those already with some QFT background, it has insufficient detail to teach the subject to a complete newbie.

C.1 Creation and Annihilation Operators

The goal in quantum field theory is to set up a quantum mechanical framework in which the total number of particles can change. The first step when doing so is to identify the Hilbert space within which quantum operators act.

In elementary single-particle quantum mechanics a basis of states, $|i\rangle$, for the single-particle Hilbert space, $\mathcal{H}_1 = \{|i\rangle\}$, can be chosen consisting of eigenstates of the complete set of commuting observables that label single-particle states. In concrete examples these labels are often chosen to be momentum and any internal quantum numbers, like total spin, s, and its third component, σ: so $|i\rangle = |\mathbf{p}, s, \sigma, \cdots\rangle$.

Ordinary quantum mechanics involving N particles similarly involves a Hilbert space, $\mathcal{H}_N = \{|i_1, i_2, \cdots i_N\rangle\}$, built as products of N copies of the single-particle basis states. For bosons these states are defined to be completely symmetric in the interchange of any two pairs of labels for identical particles, while the states are antisymmetric under this interchange for fermions. For example, for a two-particle state

$$|i_1, i_2\rangle = \pm|i_2, i_1\rangle \tag{C.1}$$

where the upper (lower) sign corresponds to the particles being bosons (fermions).

The Hilbert space for quantum field theory, $\mathcal{H}$, combines the Hilbert spaces $\mathcal{H}_0$, $\mathcal{H}_1$ and $\mathcal{H}_2$ and so on, up to $\mathcal{H}_N$ and beyond, with N arbitrarily large. Here $\mathcal{H}_0 = \{|0\rangle\}$ is the one-dimensional space spanned by the zero-particle state, $|0\rangle$, while $\mathcal{H}_N$ for $N \geq 1$ is defined as above. A Hilbert space constructed in this way is called a Fock space.

When dealing with different kinds of particles it is useful to label states using the *occupation-number representation*. Instead of listing the single-particle labels for all particles in the state, this representation lists the number of *independent* particle labels, together with the number of particles present in the state that carry these labels. For instance, in the occupation-number representation a state containing two particles with momentum $\mathbf{p}$, is denoted $|\mathbf{p}^{(2)}\rangle$ rather than $|\mathbf{p}, \mathbf{p}\rangle$. For general labels '$i$', the occupation-number representation for a five-particle state contining two particles having single-particle quantum number i_1 and three particles with quantum number i_2 is similarly $|i_1^{(2)}, i_2^{(3)}\rangle$ rather than $|i_1, i_1, i_2, i_2, i_2\rangle$.

A very convenient basis of operators acting within the Fock space of quantum field theory is given by *creation* and *annihilation* operators in the following way. The annihilation operator, $\mathfrak{a}_i$, is defined as the operator that removes a particle with quantum number i from a given state. If the state on which $\mathfrak{a}_i$ acts does not contain the particle in question then the operator is defined to give zero. That is,

$$\mathfrak{a}_i|0\rangle = 0\,, \qquad \mathfrak{a}_i|j\rangle = \delta_{ij}|0\rangle\,, \qquad \mathfrak{a}_i|j,k\rangle = \delta_{ij}|k\rangle + (-)^{ij}\delta_{ik}|jl\rangle, \tag{C.2}$$

and so on, where the sign $(-)^{ij}$ is -1 if both particles 'i' and 'j' are fermions and is $+1$ otherwise.

This definition implies that the Hermitian conjugate, $\mathfrak{a}_i^*$, is a *creation operator* for the same particle type, that satisfies

$$\mathfrak{a}_i^*|0\rangle = |i\rangle\,, \qquad \mathfrak{a}_i^*|j\rangle = |i;j\rangle, \tag{C.3}$$

and so on. Together with the normalization convention, $\langle i|j\rangle = \delta_{ij}$, these definitions imply the following properties. For bosons $|i,j\rangle = |j,i\rangle$, and so

$$\left[\mathfrak{a}_i, \mathfrak{a}_j\right] = \left[\mathfrak{a}_i^*, \mathfrak{a}_j^*\right] = 0 \quad \text{and} \quad \left[\mathfrak{a}_i, \mathfrak{a}_j^*\right] = \delta_{ij}. \tag{C.4}$$

For fermions $|i,j\rangle = -|j,i\rangle$, and so

$$\{\mathfrak{a}_i, \mathfrak{a}_j\} = \{\mathfrak{a}_i^*, \mathfrak{a}_j^*\} = 0 \quad \text{and} \quad \{\mathfrak{a}_i, \mathfrak{a}_j^*\} = \delta_{ij}, \tag{C.5}$$

in which $[A,B] = AB - BA$ and $\{A,B\} = AB + BA$, as usual.

In the occupation-number representation the above rules are captured by

$$\mathfrak{a}_i|i_1^{(n_1)}, \cdots, i_r^{(n_r)}\rangle = \sum_{j=1}^{r} s_{ij}\delta_{ii_j}\sqrt{n_j}\,|i_1^{(n_1)}, \cdots, i_j^{(n_j-1)}, \cdots, i_r^{(n_r)}\rangle. \tag{C.6}$$

where $s_{ij} = (-)^{i(n_1 i_1 + \cdots n_{j-1} i_{j-1})}$, and

$$\mathfrak{a}_i^* |i_1^{(n_1)}, \cdots, i_r^{(n_r)}\rangle = |i^{(1)}, i_1^{(n_1)}, \cdots, i^{(n_i+1)}, \cdots, i_r^{(n_r)}\rangle \qquad \text{if } i \neq i_j \text{ for any } j$$
$$= \sqrt{n_{i_j}+1}\; s_{ij}\, |i_1^{(n_1)}, \cdots, i_j^{(n_i+1)}, \cdots, i_r^{(n_r)}\rangle \qquad \text{if } i = i_j. \quad \text{(C.7)}$$

In particular, $\mathfrak{a}_i^* \mathfrak{a}_i$ counts the number of particles with quantum number 'i', because the previous two formulae imply

$$\mathfrak{a}_i^* \mathfrak{a}_i |i_1^{(n_1)}, \cdots, i_r^{(n_r)}\rangle = \left(\sum_{j=1}^{r} \delta_{ii_j} n_j \right) |i_1^{(n_1)}, \cdots, i_r^{(n_r)}\rangle. \quad \text{(C.8)}$$

What is important about the creation and annihilation operators is that they make a very convenient basis, in terms of which *any* operator acting on $\mathcal{H}$ can be expanded:[1]

$$O = A_{0,0} + \sum_i \left[A_{0,1}(i)\, \mathfrak{a}_i + A_{1,0}(i)\, \mathfrak{a}_i^* \right]$$
$$+ \sum_{ij} \left[A_{0,2}(i,j)\, \mathfrak{a}_i \mathfrak{a}_j + A_{1,1}(i,j) \mathfrak{a}_i^* \mathfrak{a}_j + A_{2,0}(i,j)\, \mathfrak{a}_i^* \mathfrak{a}_j^* \right] + \cdots . \quad \text{(C.9)}$$

To see that this is so, it suffices to argue that the coefficient functions $\{A_{0,0}, A_{1,0}(i), A_{0,1}(i), \ldots\}$ can be solved for in terms of the matrix elements: $\langle \psi | O | \phi \rangle$, for all choices for $\langle \psi |$ and $| \phi \rangle$. (This can be shown by induction, starting with $\langle 0 | O | 0 \rangle = A_{0,0}$ (because $\mathfrak{a}_i |0\rangle = \langle 0 | \mathfrak{a}_i^* = 0$) and continuing with $\langle j | O | 0 \rangle = A_{1,0}(j)$, $\langle 0 | O | j \rangle = A_{0,1}(j)$ and so on.)

A system's hamiltonian is an important special case of an operator that can be expanded in this way, and part of the reason creation and annihilation operators are so useful is that this particular expansion is usually an efficient one. For instance, non-interacting particles are ones for which the energy cost of adding N particles is just N times the energy of adding one particle (*i.e.* there is no interaction energy). With (C.8) in mind the hamiltonian for a collection of non-interacting particles is therefore

$$H_{\text{free}} = E_0 + \sum_i \varepsilon_i\, \mathfrak{a}_i^* \mathfrak{a}_i, \quad \text{(C.10)}$$

where the single-particle labels 'i' appearing in the sum are a complete set of mutually commuting labels for single-particle energy eigenstates.

As is easily verified by explicit evaluation, the hamiltonian H_{free} has the occupation-number states as eigenstates, with eigenvalues given by

$$H_{\text{free}} |i_1^{(n_1)}, \cdots, i_r^{(n_r)}\rangle = \left(E_0 + \sum_j n_j \varepsilon_j \right) |i_1^{(n_1)}, \cdots, i_r^{(n_r)}\rangle. \quad \text{(C.11)}$$

This reveals E_0 to be the energy of the zero-particle state (or vacuum), $|0\rangle$, while ε_i is the energy associated with the addition of a single particle having quantum number 'i'. For momentum eigenstates (using standard non-relativistic normalization) this expression for H_{free} becomes

[1] Notice that all instances of $\mathfrak{a}^*$ here stand to the left of all instances of $\mathfrak{a}$ in this expression; something that can be arranged without loss of generality by changing, if needed, the order of operators using the commutation relations (C.4) or (C.5) (a process called 'normal ordering').

$$H_{\rm free} = E_0 + \sum_k \int {\rm d}^3 p\, \varepsilon_k(\mathbf{p})\, \mathfrak{a}^*_{\mathbf{p},k} \mathfrak{a}_{\mathbf{p},k}, \tag{C.12}$$

where k represents all non-momentum single-particle labels (like spin or particle type) and $\varepsilon_k(\mathbf{p})$ is the single-particle dispersion relation (*i.e.* single-particle energy as a function of momentum). In the special case of relativistic systems Lorentz covariance requires that $\mathbf{p}$ and $p^0 = \varepsilon(\mathbf{p})$ must be components of a single 4-momentum vector, p^μ, so $\varepsilon_k(\mathbf{p}) = \sqrt{\mathbf{p}^2 + m_k^2}$ where m_k is the corresponding particle's rest mass.

Interactions have similar representations. For instance, a term in $H_{\rm int}$ describing the emission or absorption of a photon by a charged particle, $f(\mathbf{p}, \sigma) + \gamma(\mathbf{k}, \lambda) \to f(\mathbf{q}, \zeta)$, could be written

$$H_{\rm int} \ni \sum_{\lambda\sigma\zeta} \int {\rm d}^3 p\, {\rm d}^3 q\, {\rm d}^3 k \left[h_{\lambda\sigma\zeta}(\mathbf{p}, \mathbf{q}, \mathbf{k})\, \mathfrak{c}^*_{\mathbf{p}\sigma} \mathfrak{c}_{\mathbf{q}\zeta} \mathfrak{a}_{\mathbf{k}\lambda} + \text{h.c.} \right] \delta^3(\mathbf{p} - \mathbf{q} - \mathbf{k}), \tag{C.13}$$

where 'h.c.' denotes the hermitian conjugate, while $\mathfrak{c}_{\mathbf{p}\sigma}$ denotes the annihilation operator for a charged particle with momentum $\mathbf{p}$ and spin component σ, while $\mathfrak{a}_{\mathbf{k}\lambda}$ is the same for the photon of momentum $\mathbf{k}$ and helicity λ.

The above expression can mediate processes like photon absorption (or, from the 'h.c.' term, emission) because it gives a nonzero matrix element

$$\langle \mathbf{p}, \sigma | H_{\rm int} | \mathbf{q}, \zeta; \mathbf{k}, \lambda \rangle = h_{\lambda\sigma\zeta}(\mathbf{p}, \mathbf{q}, \mathbf{p})\, \delta^3(\mathbf{p} - \mathbf{q} - \mathbf{k}), \tag{C.14}$$

which, when used in expressions like Fermi's golden rule (B.45), can contribute a nonzero transition rate. Eq. (C.14) evaluates the matrix element using (continuum-normalized version of) expressions (C.6) and (C.7) defining the action of creation and annihilation operators.

C.2 Nonrelativistic Free Fields

An important condition satisfied by most physical systems is cluster decomposition: the factorization of probabilities for events that are widely separated in space at a given time.[2] A physical property demanded of H is that this clustering property should be preserved by time evolution.

What is convenient about the creation- and annihilation-operator formalism is that the requirements of cluster decomposition are automatically satisfied if the (possibly complex) coefficient functions in H – like $h_{\lambda\sigma\zeta}(\mathbf{p}, \mathbf{q}, \mathbf{k})$ in (C.13), for example – are sufficiently smooth functions of their momentum arguments (*e.g.* admitting a Taylor expansion in powers of momenta) once the delta-function is extracted that enforces momentum conservation (if this is conserved). These are the momentum-space ways of saying that in position space the hamiltonian is local:

$$H = \int {\rm d}^3 x\, \mathfrak{H}(\mathbf{x}), \tag{C.15}$$

2 This property assumes the system to have started without initial correlations between the particles involved in these events. For relativistic systems the requirement of 'large' spatial separation means events that are sufficiently outside each other's light cones (large *spacelike* separations). For thermal fluids cluster decomposition is included in the condition of local equilibrium.

for some hamiltonian density $\mathfrak{H}(\mathbf{x})$. Locality is related to cluster decomposition because the time-evolution operator, $U(t,t_0) = \exp[-iH(t-t_0)]$, should schematically be a product over positions (and so should preserve factorization of amplitudes for spatially separated events) when H is a sum over positions.

This suggests that interactions should often simplify when expressed in position space, using the Fourier-transforms of the creation and annihilation operators. For spinless particles this leads to defining position-space fields, $\Phi(\mathbf{x})$, of the form

$$\Phi(\mathbf{x}) := \int \frac{d^3p}{(2\pi)^{3/2}}\, \mathfrak{a}_{\mathbf{p}}\, e^{i\mathbf{p}\cdot\mathbf{x}}. \tag{C.16}$$

Such a position-space quantum field can similarly be defined for the destruction operator of each separate type of particle. This last equation is written in Schrödinger representation (for which operators do not evolve in time), but it is often more usefully written in the Interaction representation (for which the operators evolve in time using the free field equations). In interaction picture (C.16) becomes

$$\Phi(\mathbf{x},t) := \int \frac{d^3p}{(2\pi)^{3/2}}\, \mathfrak{a}_{\mathbf{p}}\, e^{-iE_pt+i\mathbf{p}\cdot\mathbf{x}}, \tag{C.17}$$

where E_p is the single-particle energy.

The point of expressing H in terms of $\Phi(\mathbf{x})$ and $\Phi^*(\mathbf{x})$ rather than $\mathfrak{a}_{\mathbf{p}\sigma}$ and $\mathfrak{a}^*_{\mathbf{p}\sigma}$ is that the condition of cluster-decomposition – *i.e.* smoothness of coefficients like $h_{\sigma\lambda\zeta}(\mathbf{p},\mathbf{q},\mathbf{k})$ in (C.13) – is ensured by the locality requirement, (C.15), where $\mathfrak{H}(\mathbf{x})$ is built from sums of local monomials of $\Phi(\mathbf{x})$, $\Phi^*(\mathbf{x})$ and their derivatives, all evaluated at the same spatial point.

For example, rather than using (C.12) for a system of non-interacting spinless particles with dispersion relation $\varepsilon(\mathbf{p}) = \mathbf{p}^2/2m$, its Hamiltonian could equally well be written

$$H = \frac{1}{2m}\int d^3x\, \nabla\Phi^*(\mathbf{x})\cdot\nabla\Phi(\mathbf{x}). \tag{C.18}$$

Because this hamiltonian is invariant under rephasing, $\Phi(\mathbf{x}) \to e^{i\omega}\Phi(\mathbf{x})$, Noether's theorem implies that it necessarily commutes with total particle number, measured by the operator

$$N = \int d^3x\, \Phi^*(\mathbf{x})\,\Phi(\mathbf{x}) = \int d^3p\, \mathfrak{a}^*_{\mathbf{p}}\mathfrak{a}_{\mathbf{p}}. \tag{C.19}$$

Sometimes momentum is not a good single-particle label for energy eigenstates, such as for particles that interact with an external potential, $V(\mathbf{x})$. For example, a system of such particles interacting only with the potential (which do not mutually interact with one another) can be written

$$H = \int d^3x \left[\frac{1}{2m}\nabla\Phi^*(\mathbf{x})\cdot\nabla\Phi(\mathbf{x}) + V(\mathbf{x})\,\Phi^*(\mathbf{x})\Phi(\mathbf{x})\right]. \tag{C.20}$$

As is easy to verify the hamiltonian of Eq. (C.20) can be written in the diagonal form (C.11) by generalizing (C.16) to

$$\Phi(\mathbf{x}) := \sum_i \mathfrak{a}_i\,\varphi_i(\mathbf{x}), \tag{C.21}$$

where the mode-function $\varphi_i(\mathbf{x})$ is defined as an eigenstate satisfying the single-particle Schrödinger equation,

$$\left[-\frac{1}{2m}\nabla^2 + V(\mathbf{x})\right]\varphi_i(\mathbf{x}) = \varepsilon_i\varphi_i(\mathbf{x}). \tag{C.22}$$

Using (C.21) and (C.22) in (C.20) then puts it into diagonal form

$$H = E_0 + \sum_i \mathfrak{a}_i^* \mathfrak{a}_i \, \varepsilon_i. \tag{C.23}$$

An example of a system for which interactions are *not* local, and so do not cluster, is the case where particles interact through long-range forces, such as in

$$H = \int d^3x \left[\frac{1}{2m}\nabla\Phi^*(\mathbf{x})\cdot\nabla\Phi(\mathbf{x}) + \int d^3y\, U(\mathbf{x}-\mathbf{y})\,\Phi^*(\mathbf{x})\,\Phi(\mathbf{x})\,\Phi^*(\mathbf{y})\,\Phi(\mathbf{y})\right]. \tag{C.24}$$

Examples like this are not normally regarded as counter-examples of the requirement that physical systems cluster, since the long-range interaction $U(\mathbf{x}-\mathbf{y})$ usually arises once a massless (or very light) degree of freedom is integrated out. A famous example is the generation of the Coulomb potential once electromagnetic interactions are integrated out. In all such cases, however, the Wilson action for the relevant effective theory (from which massless or light particles are *not* integrated out) is local, and so does cluster.

C.2.1 Nonrelativistic Fields with Spin

Position-space fields can also incorporate spin, by requiring them to transform under rotations, such as by using a spinor field, a vector field or some other finite-dimensional representation of the rotation group. Generalizing (C.16) to a field Ψ_a transforming in one such a representation leads to

$$\Psi_a(\mathbf{x}) := \sum_{\sigma=-s}^{s}\int \frac{d^3p}{(2\pi)^{3/2}}\, u_a(\sigma)\,\mathfrak{a}_{\mathbf{p}\sigma}\, e^{i\mathbf{p}\cdot\mathbf{x}}, \tag{C.25}$$

where the sum is over the 3rd-component-of-spin quantum number, σ, of the particle state (assumed to have spin s). Here the 'polarization tensor' $u_a(\sigma)$ is the Clebsch–Gordan coefficient required to ensure consistency between the assumed transformation properties under rotations of the field and of the particle states.

For spin-half particles $s = \frac{1}{2}$ and Ψ_a (with $a = 1, 2$) can be a two-component spinor, which under rotations with infinitesimal parameter ω^k transform as $\delta\Psi = \frac{i}{2}\,\omega^k\sigma_k\,\Psi$ (in addition to its action on $\mathbf{x}$), where σ_k are the usual 2×2 Pauli matrices. In this case, the spin sum runs over $\sigma = \pm\frac{1}{2}$, with

$$u(+1/2) = \begin{pmatrix} 1 \\ 0 \end{pmatrix} \quad \text{and} \quad u(-1/2) = \begin{pmatrix} 0 \\ 1 \end{pmatrix}. \tag{C.26}$$

A spin-one particle has $s = 1$ and can similarly be represented by a vector field, V_k, with $k = x, y, z$. In this case, rotations act as $\delta V_k = \epsilon_{klm}\omega^l V^m$ (in addition to their action on $\mathbf{x}$), and $\sigma = 0, \pm 1$. In this case, $u_k(\sigma)$ becomes the polarization vector appropriate for each of the three choices for σ. For instance, for a particle with momentum parallel to the z axis, $\mathbf{p} = \mathbf{e}_z$, one finds $\mathbf{u}(0) = \mathbf{e}_z$ and $\mathbf{u}(\pm 1) = \frac{1}{\sqrt{2}}(\mathbf{e}_x \pm i\mathbf{e}_y)$, so

$$\mathbf{V}(\mathbf{x}) = \sum_{\sigma=-1}^{1} \int \frac{d^3p}{(2\pi)^{3/2}}\, \mathbf{u}(\sigma)\, \mathfrak{a}_{\mathbf{p}\sigma}\, e^{i\mathbf{p}\cdot\mathbf{x}}. \tag{C.27}$$

C.3 Relativistic Free Fields

A similar story goes through for position-space fields in relativistic theories, but with two important differences. The simplest difference simply recognizes that the fields must transform in finite-dimensional representations of the Lorentz group rather than just rotations. The more subtle difference is that all position-space fields must come with both destruction and creation parts, in a way that is elaborated below.

It is this second condition that underlies many of the profound consequences – like the existence of antiparticles, the spin-statistics theorem, the CPT theorem and crossing symmetry – of combining quantum mechanics with special relativity. Although it goes beyond this summary to derive these consequences in detail, both types of differences are illustrated in the low-spin examples below.

C.3.1 Relativistic Spin-0 Fields

Scalar fields can be used for spin-zero particles[3] and in this case the expansion in terms of creation and annihilation operators generalizes the nonrelativistic result (C.17) to

$$\phi(x) = \int \frac{d^3p}{\sqrt{(2\pi)^3 2E_p}} \left[\mathfrak{a}_{\mathbf{p}}\, e^{ip\cdot x} + \bar{\mathfrak{a}}^*_{\mathbf{p}}\, e^{-ip\cdot x}\right], \tag{C.28}$$

where $p \cdot x$ is short for $p_\mu x^\mu = -E_p t + \mathbf{p}\cdot\mathbf{x}$, where $E_p = \sqrt{\mathbf{p}^2 + m^2}$ is the relativistic particle energy and m its rest mass. This expression normalizes momentum eigenstates in the same way as does (C.17) – *i.e.* using the nonrelativistic normalization $\langle \mathbf{p}|\mathbf{q}\rangle = \delta^3(\mathbf{p}-\mathbf{q})$, where $|\mathbf{p}\rangle = \mathfrak{a}^*_{\mathbf{p}}|0\rangle$ – and so the factor of $\sqrt{E_p}$ in the denominator is precisely what is required to make the left-hand side transform as a Lorentz scalar. (If $\mathfrak{a}^*_{\mathbf{p}}|0\rangle = |\mathbf{p}\rangle_r$ were instead normalized covariantly, as in (B.20), then the measure appearing in (C.28) would be the Lorentz-invariant combination d^3p/E_p, as expected.)

It is the term involving $\bar{\mathfrak{a}}^*_{\mathbf{p}}$ that is the new 'creation part' of the field alluded to above. Here $\bar{\mathfrak{a}}_{\mathbf{p}}$ is the destruction operator for the *antiparticle* for the particle destroyed by $\mathfrak{a}_{\mathbf{p}}$, whose properties are dictated by the requirement that the commutator $[\phi(x), \phi(y)]$ vanish for all spacelike separations, $(x-y)^2 > 0$. If this commutator would not vanish, then neither would the same commutator built using

[3] The choice of field representation is not unique for any given spin, with the general condition known since the 1960s [14, 51, 465, 466]. Different choices of representation typically do not define physically different theories. For instance, a 4-vector field – instead of a scalar field – can represent a spinless particle, but the 4-vector in this case is simply the gradient of the scalar: $V_\mu = \partial_\mu \phi$.

the hamiltonian density, $[\mathfrak{H}(x), \mathfrak{H}(y)]$, and if this does not vanish then the time-orderings of H appearing in the S-matrix (see §B.2) become problematic given that different observers can disagree on the time-ordering of x^0 and y^0 for spacelike-separated points.

In particular, this condition requires that the antiparticle have precisely the same mass as does the particle, and it must carry precisely the opposite charge for any symmetry that multiplies the field $\phi(x)$ by a phase: $\phi(x) \to e^{i\omega}\phi(x)$. A particle can be its own antiparticle if $\mathfrak{a}_\mathbf{p}$ is used instead of $\bar{\mathfrak{a}}_\mathbf{p}$ in (C.28). But $\phi(x)$ is then real, and so particle and antiparticle can only be the same in that they carry no additive conserved charges. (The photon is an example of a particle of this type that is the same as its antiparticle.)

Cluster decomposition is ensured if local interactions are built as before from powers of fields and their derivatives at a single point, and for relativistic systems this is more conveniently done using the action, $S = \int d^4x\, \mathfrak{L}$, than with the Hamiltonian, $H = \int d^3x\, \mathfrak{H}$, because the action is Lorentz invariant while the energy is not. For free charged fields this action is quadratic and the fields can always be defined in such a way that the action becomes

$$S^{\rm free}_{\rm spin\,0} = -\int d^4x \left[\partial_\mu\phi^*\,\partial^\mu\phi + m^2\phi^*\phi\right], \tag{C.29}$$

where m is the rest mass appearing in the dispersion relation: $\varepsilon(\mathbf{p}) = E_p = \sqrt{\mathbf{p}^2+m^2}$. The fields in this action – and in (C.28) – are chosen to be 'canonical' inasmuch as the canonical equal-time commutation relation, $[\Pi(\mathbf{x},t), \phi(\mathbf{y},t)] = -i\delta^3(\mathbf{x}-\mathbf{y})$, agrees with the creation/annihilation algebra $[a_\mathbf{p}, a^*_\mathbf{q}] = \delta^3(\mathbf{p}-\mathbf{q})$ when $\Pi = \delta S/\delta(\partial_t\phi)$ is the canonical momentum. (The numerical factors in (C.28) are also chosen to ensure that canonical commutation relations agree with the $a_\mathbf{p}, a^*_\mathbf{p}$ algebra.) Unlike in nonrelativistic systems, relativity makes it necessary to quantize spinless particles using bose statistics – a consequence of the spin-statistics theorem: all integer-spin particles must be bosons and all half-odd spin particles must be fermions.

Interactions are similarly built using non-quadratic (but local) terms in ϕ and its derivatives. It is the requirement that all interactions be built from $\phi(x) \propto \mathfrak{a} + \bar{\mathfrak{a}}^*$ that implies that relativistic interactions *never* preserve particle number, in contrast with the interactions written above for the nonrelativistic case.

C.3.2 Relativistic Spin-1/2 Fields

Spin-half particles are represented using Lorentz-spinor fields, $\psi_a(x)$, which are taken to be distinct from their antiparticles and so represented by 4-component Dirac spinors ($a = 1, \cdots, 4$; see the discussion above Eq. (A.27) (or Appendix §A.2.3) for the distinction between these and Majorana or Weyl spinors).

$$\psi_a(x) = \sum_{\sigma=\pm\frac{1}{2}} \int \frac{d^3p}{\sqrt{(2\pi)^3 2E_p}} \left[u_a(\mathbf{p},\sigma)\,\mathfrak{c}_{\mathbf{p}\sigma}\,e^{ip\cdot x} + v_a(\mathbf{p},\sigma)\,\bar{\mathfrak{c}}^*_{\mathbf{p}\sigma}\,e^{-ip\cdot x}\right], \tag{C.30}$$

where the destruction operator is labelled $\mathfrak{c}$ to distinguish it from the spinless destruction operator described above. For spin-half particles consistency *requires* these operators to anti-commute,

$$\{\mathfrak{c}^*_{\mathbf{p}\sigma}, \mathfrak{c}_{\mathbf{q}\zeta}\} = \delta^3(\mathbf{p}-\mathbf{q})\,\delta_{\sigma\zeta}, \tag{C.31}$$

to ensure that the hamiltonian density, $\mathfrak{H}(x)$, can continue to commute with itself when evaluated at spacelike-separated points in spacetime. $u(\mathbf{p},\sigma)$ and $v(\mathbf{p},\sigma)$ are spinors that are chosen to ensure that both sides of (C.30) transform the same way under Lorentz transformations. This implies they satisfy the Dirac conditions $(\mathrm{i}\,\not{p} + m)u(\mathbf{p},\sigma) = (\mathrm{i}\,\not{q} - m)v(\mathbf{p},\sigma) = 0$. Here, the 'slash' notation denotes $\not{p} = p_\mu\gamma^\mu$, where the Dirac conventions used for the gamma matrices γ^μ are outlined in §A.2.3. Notice that in the particle rest frame these conditions become the projections $\mathrm{i}\gamma^0 u = +u$ for particles and $\mathrm{i}\gamma^0 v = -v$ for antiparticles, and in any other frame they are found by applying the appropriate Lorentz boost to these conditions.

The action that captures these conditions for free fields is

$$S^{\text{free}}_{\text{spin }1/2} = -\int \mathrm{d}^4x\,\overline{\psi}(\not{\partial} + m)\psi, \tag{C.32}$$

where again fields are chosen to be canonically normalized, and m is the particle rest mass that enters its dispersion relation $\varepsilon(\mathbf{p}) = E_p = \sqrt{\mathbf{p}^2 + m^2}$.

C.3.3 Relativistic Spin-1 Fields

For spin one and higher the field content needed to treat massive and massless states differs. This is because a massive spin-s state contains $2s+1$ spin components, $\sigma = -s, -s+1, \cdots, s-1, s$, while a minimal massless spin-s state usually contains only two helicities: $\lambda = \pm s$. Although these two options have the same number of states for $s = 0$ and $s = \frac{1}{2}$, they differ from one another for $s \geq 1$.

Massive Spin-1 Fields

Consider first the massive case. The smallest fields that can be used for massive spin-one particles are vector fields, $V_\mu(x)$, and consistency of the 4-vector Lorentz-transformation rule with the transformations of creation and annihilation operators for massive spin-one particles implies

$$V_\mu(x) = \sum_{\lambda=-1}^{1}\int \frac{\mathrm{d}^3p}{\sqrt{(2\pi)^3 2E_p}}\left[\varepsilon_\mu(\mathbf{p},\lambda)\,\mathfrak{a}_{\mathbf{p}\lambda}\,e^{ip\cdot x} + \varepsilon^*_\mu(\mathbf{p},\lambda)\,\bar{\mathfrak{a}}^*_{\mathbf{p}\lambda}\,e^{-ip\cdot x}\right], \tag{C.33}$$

where, as before, $E_p = \sqrt{\mathbf{p}^2 + m^2}$ with m the particle's rest mass. This form is also only consistent with $[\mathfrak{H}(x), \mathfrak{H}(y)] = 0$ for spacelike separations if the particles are bosons, so $[\mathfrak{a}_{\mathbf{p}}, \mathfrak{a}^*_{\mathbf{q}}] = [\bar{\mathfrak{a}}_{\mathbf{p}}, \bar{\mathfrak{a}}^*_{\mathbf{q}}] = \delta^3(\mathbf{p}-\mathbf{q})$.

The polarization vector $\varepsilon_\mu(\mathbf{p},\lambda)$ satisfies $p^\mu\varepsilon_\mu = 0$, and so for momentum pointing up the z axis, the polarization vector's spatial part can be chosen to be

$$\varepsilon^\mu(\lambda = \pm 1) = \frac{1}{\sqrt{2}}\begin{pmatrix} 0 \\ \mathbf{e}_x \pm \mathrm{i}\,\mathbf{e}_y \end{pmatrix}, \quad \varepsilon^\mu(\lambda = 0) = \begin{pmatrix} p \\ E_p\,\mathbf{e}_z \end{pmatrix} \text{ when } p^\mu = \begin{pmatrix} E_p \\ p\,\mathbf{e}_z \end{pmatrix}. \tag{C.34}$$

One rationale for the condition $p^\mu\varepsilon_\mu(\mathbf{p},\lambda) = 0$ is that if $V_\mu = \partial_\mu\phi$ were a gradient, it would actually represent a spin-0 particle, and this option must be projected out (as the condition $\partial^\mu V_\mu = 0$ indeed does).

A free lagrangian for this type of particle is

$$S^{\text{free}}_{\text{spin 1}} = -\int d^4x \left[\frac{1}{2} F^*_{\mu\nu} F^{\mu\nu} + m^2 V^\mu V^*_\mu\right], \tag{C.35}$$

where $F_{\mu\nu} := \partial_\mu V_\nu - \partial_\nu V_\mu$. Notice that the field equation obtained from this by varying V^*_μ is

$$\partial_\mu F^{\mu\nu} - m^2 V^\nu = (\Box - m^2) V^\nu - \partial^\nu(\partial_\mu V^\mu) = 0, \tag{C.36}$$

which, when acted on again by ∂_ν implies $m^2 \partial_\mu V^\mu = 0$. When $m \neq 0$ these field equations both project out the spinless part (*i.e.* ensure $\partial \cdot V = 0$) and – once this is used in (C.36) – ensure $p^\mu p_\mu + m^2 = -E_p^2 + \mathbf{p}^2 + m^2 = 0$, thereby identifying m as the particle's rest mass.

Massless Spin-1 Fields

Next consider the massless case. It happens that the absence of the longitudinal mode, $\lambda = 0$, precludes also using a 4-vector field like $V_\mu(x)$ to represent a massless spin-one particle. In this case, the smallest finite-dimension representation of the Lorentz group that can be used to represent the two helicity states of a massless spin-one field turns out to be an antisymmetric tensor, $F_{\mu\nu} = -F_{\nu\mu}$. The two separate helicities are represented by the self-dual and anti-self-dual parts, $F^\pm_{\mu\nu} = F_{\mu\nu} \pm \mathrm{i}\,\widetilde{F}_{\mu\nu}$, where the 'dual' field strength is defined by

$$\widetilde{F}_{\mu\nu} = \frac{1}{2}\,\epsilon_{\mu\nu\lambda\rho}\, F^{\lambda\rho}, \tag{C.37}$$

for $\epsilon_{\mu\nu\lambda\rho}$ the completely antisymmetric Levi-Civita tensor of Eq. (A.10).

The mode functions that ensure the consistency of the transformation rule for $F_{\mu\nu}$ and for a massless spin-one particle turn out to imply [51, 54]

$$F_{\mu\nu} = \partial_\mu A_\nu - \partial_\nu A_\mu, \tag{C.38}$$

with

$$A_\mu(x) = \sum_{\lambda=\pm1} \int \frac{d^3p}{\sqrt{(2\pi)^3 2E_p}} \left[\, \varepsilon_\mu(\mathbf{p},\lambda)\, \mathfrak{a}_{\mathbf{p}}\, e^{ip\cdot x} + \varepsilon^*_\mu(\mathbf{p},\lambda)\, \mathfrak{a}^*_{\mathbf{p}}\, e^{-ip\cdot x}\right], \tag{C.39}$$

where $E_p = |\mathbf{p}|$ is the relativistic particle energy for massless particles (and, with photons in mind, the antiparticle is chosen to be identical to the particle). This looks a lot like (C.33), but with the important omission of the $\lambda = 0$ polarization.

At this point, the astute reader asks: 'How can (C.38) and (C.39) be consistent with the earlier statement that $F_{\mu\nu}$ is the smallest field whose Lorentz-transformation properties are consistent with representing a massless spin-one particle? Why not simply use the 4-vector A_μ instead?' From this point of view what is important about Eqs. (C.38) and (C.39) is this: performing a Lorentz transformation on the creation and annihilation operators to accomplish a Lorentz transformation that takes $p^\mu \to \Lambda^\mu{}_\nu p^\nu$ indeed implies the field $F_{\mu\nu}$ transforms into $\Lambda_\mu{}^\lambda \Lambda_\nu{}^\rho F_{\lambda\rho}$, as should a rank-two tensor.

But the field A_μ does *not* transform as a 4-vector, because of the omission of the $\lambda = 0$ mode. It instead satisfies

$$A_\mu \to \Lambda^\nu{}_\mu A_\nu + \partial_\mu \Omega \tag{C.40}$$

for some scalar function Ω. Although the first term on the right-hand side corresponds to the transformation of a covariant 4-vector, the second term does not. Lorentz-invariant actions for massless spin-one particles can only be built using $A_\mu(x)$ instead of $F_{\mu\nu}(x)$ if they are also chosen to be *gauge invariant*, that is, invariant under the shift $A_\mu \to A_\mu + \partial_\mu \Omega$, for general Ω. It is the freedom to do just such a transformation that allows the $\lambda = 0$ spin-state to be removed, as required for a massless spin-one state.

C.3.4 Massless Spin-2 Fields

A similar story holds for massless spin-two particles for which there are only two spin states, with helicities $\lambda = \pm 2$. A field that can represent this kind of particle is a tensor $C_{\mu\nu\lambda\rho}$ with the same symmetries as the Riemann tensor: $C_{\mu\nu\lambda\rho} = C_{\lambda\rho\mu\nu} = -C_{\nu\mu\lambda\rho} = -C_{\mu\nu\rho\lambda}$, and in addition a trace-free condition, $\eta^{\lambda\rho} C_{\lambda\mu\rho\nu} = 0$.

Requiring the transformation properties of this field to be consistent with what is obtained once expanded in terms of creation and annihilation operators for a massless spin-two field implies [51] $C_{\mu\nu\lambda\rho}$ is obtained as two derivatives of a field

$$h_{\mu\nu}(x) = \sum_{\lambda=\pm 2} \int \frac{d^3p}{\sqrt{(2\pi)^3 2E_p}} \left[\varepsilon_{\mu\nu}(\mathbf{p}, \lambda)\, \mathfrak{a}_\mathbf{p}\, e^{ip\cdot x} + \varepsilon^*_{\mu\nu}(\mathbf{p}, \lambda)\, \mathfrak{a}^*_\mathbf{p}\, e^{-ip\cdot x} \right], \tag{C.41}$$

where $E_p = |\mathbf{p}|$ is the relativistic particle energy for massless particles and (as for photons) antiparticle is identified with the particle.

When a Lorentz transformation is performed on the particle creation and annihilation operators such that their 4-momentum transforms as $p^\mu \to \Lambda^\mu{}_\nu p^\nu$, then $h_{\mu\nu}$ does not transform as a tensor. It only does so up to a gauge transformation of the form

$$h_{\mu\nu} \to \partial_\mu \Omega_\nu + \partial_\nu \Omega_\mu, \tag{C.42}$$

for some field Ω_μ. This gauge symmetry must be preserved if interactions are to be built directly using $h_{\mu\nu}$ rather than $C_{\mu\nu\lambda\rho}$.

Precisely this kind of structure is obtained when the field equations of General Relativity are expanded about a flat background spacetime. In this case, the metric is written $g_{\mu\nu} = \eta_{\mu\nu} + 2\kappa\, h_{\mu\nu}$ with $\kappa^2 = 8\pi G_N$ related to Newton's constant, and under coordinate transformations $\delta x^\mu = \xi^\mu(x)$ the fluctuation field $h_{\mu\nu}$ transforms as $\delta h_{\mu\nu} = \partial_\mu \xi_\nu + \partial_\nu \xi_\mu$, as above. The field $C_{\mu\nu\lambda\rho}$ is then the linearized Weyl tensor, defined as the completely trace-free part of the Riemann tensor built from $g_{\mu\nu}$.

C.4 Global Symmetries

Symmetries play an important role in quantum mechanics, just as they do in classical mechanics. Symmetries are special because their existence allows exact statements to be made about transition probabilities and about energy eigenstates.

In quantum mechanics it is a theorem [41] that transformations that do not change any transition probabilities can always be represented in terms of unitary operators.[4] Being represented by a unitary operator means that the action of the symmetry on any state can be written as

$$|\,\psi\rangle \to |\,\tilde{\psi}\rangle = U\,|\,\psi\rangle, \tag{C.43}$$

where $U^*U = UU^* = I$, with I being the identity operator in the Hilbert space. Probabilities remain unchanged, because $\langle\tilde{\psi}_1|\,\tilde{\psi}_2\rangle \to \langle\psi_1|U^*U|\,\psi_2\rangle = \langle\psi_1|\,\psi_2\rangle$. Matrix elements of operators, A, remain unchanged under such transformations because $A \to \tilde{A} = UAU^*$, and so its matrix elements become: $\langle\tilde{\psi}_1|\tilde{A}|\,\tilde{\psi}_2\rangle = \langle\psi_1|U^*UAU^*U|\psi_2\rangle = \langle\psi_1|A|\,\psi_2\rangle$.

The complete set of transformations, $\{g_a\}$, that preserve all matrix elements in this way forms a group, where group multiplication, g_1g_2, consists of convolution (*i.e.* successive performance of the two individual transformations). Furthermore, there can be (but need not be) an independent operator $U(g)$ for each such transformation. These operators form a unitary representation inasmuch as $U(g_1)U(g_2) = U(g_1g_2)$.

A *symmetry* is defined to be any transformation of this type that does not change the system's hamiltonian:

$$\widetilde{H} = UHU^* = H, \tag{C.44}$$

and so $[U, H] = 0$. These form a subgroup of the group of matrix-element-preserving transformations. Two classic consequences follow immediately from Eq. (C.44).

- *Conservation:* Writing each unitary symmetry as $U = \exp[iQ]$ defines a hermitian charge, Q, whose quantum numbers for any state are conserved in time. This follows because $[U, H] = 0$ implies $[Q, H] = 0$ and so – because the time-evolution operator is $U(t, t_0) = \exp[-iH(t - t_0)]$ – then $[Q, U(t, t_0)] = 0$ as well. Therefore, if $Q|\,\psi(t = t_0)\rangle = q|\,\psi(t_0)\rangle$ then (in Schrödinger picture)

$$Q\,|\,\psi(t)\rangle = QU(t,t_0)|\,\psi(t_0)\rangle = U(t,t_0)Q\,|\,\psi(t_0)\rangle = q\,|\,\psi(t)\rangle. \tag{C.45}$$

  This expresses conservation in the usual sense that once a state is prepared to have a particular value of Q then it has this same value for all later times.
- *Spectrum degeneracy:* It also follows from (C.44) that if two energy eigenstates are related by a symmetry then they must have the same energy. That is, because $[U, H] = 0$ if $|\,\psi_2\rangle = U|\,\psi_2\rangle$ and $H|\,\psi_i\rangle = E_i\,|\,\psi_i\rangle$, then

$$E_2|\,\psi_2\rangle = H|\,\psi_2\rangle = HU|\,\psi_1\rangle = UH|\,\psi_1\rangle = E_1(U|\,\psi_1\rangle) = E_1|\,\psi_2\rangle, \tag{C.46}$$

  and because $|\,\psi_2\rangle \neq 0$ this means $E_1 = E_2$. Physically, this says that if a transformation is a symmetry it should not affect energies; *e.g.* in a rotationally invariant world a ruler has the same total energy regardless of whether it is laid along the x, y or z axes.

For relativistic systems, it is usually true that symmetries commute with momentum as well as energy, $[U, P^\mu] = 0$. If so, then because $E_i^2(\mathbf{p}) = \mathbf{p}^2 + m_i^2$ it follows that any two states related by U must have the same rest mass.

[4] Or anti-unitary operators, as is the case for time-reversal (more about which in §C.4.3).

All of these statements go over for quantum field theories as well, although for some conclusions (like spectral degeneracy) it matters whether the system's ground state is invariant under the symmetry or not (that is, whether the symmetry is spontaneously broken).

Just as for ordinary quantum mechanics the action of symmetries (besides time-reversal) on states and operators is given by a unitary operator U, with $|\tilde{\psi}\rangle = U|\psi\rangle$ and $\widetilde{O} = UOU^*$. In particular, for creation and annihilation operators $\tilde{\mathfrak{a}}_{\mathbf{p}} = U\mathfrak{a}_{\mathbf{p}}U^*$ and $\tilde{\mathfrak{a}}^*_{\mathbf{p}} = U\mathfrak{a}^*_{\mathbf{p}}U^*$, and so field operators transform as $\tilde{\phi} = U\phi U^*$ and so on.

If $U|0\rangle = |0\rangle$ (which multiplying through by U^* shows also implies $U^*|0\rangle = |0\rangle$) then these transformations amongst creation operators also imply that the corresponding particle states – call them $|\psi(\mathbf{p})\rangle = \mathfrak{a}^*_{\mathbf{p}}|0\rangle$ and $|\tilde{\psi}(\mathbf{p})\rangle = \tilde{\mathfrak{a}}^*_{\mathbf{p}}|0\rangle$ – are related by the action of U, since

$$|\tilde{\psi}(\mathbf{p})\rangle = \tilde{\mathfrak{a}}^*_{\mathbf{p}}|0\rangle = U\mathfrak{a}^*_{\mathbf{p}}U^*|0\rangle = U\mathfrak{a}^*_{\mathbf{p}}|0\rangle = U|\psi(\mathbf{p})\rangle. \tag{C.47}$$

When this is true (C.46) implies these particles share the same energy. The above argument shows why this implication also generally fails when $U|0\rangle \neq |0\rangle$ (*i.e.* when the symmetry is spontaneously broken). In this case, particles need not align into degenerate multiples for a spontaneously broken symmetry, and the symmetry instead partially acts to shift the vacuum itself (which by assumption is not invariant).[5]

When a symmetry is not spontaneously broken then it can be linearly realized on the fields themselves, as described in the main text in §4.2.1, with

$$\phi^i \to \tilde{\phi}^i := U(g)\,\phi^i U^*(g) = \phi^j\,\mathcal{G}_j{}^i. \tag{C.48}$$

Applying two transformations in succession and using $U(g_1g_2) = U(g_2)U(g_2)$ then shows that the matrices $\mathcal{G}_i{}^j$ satisfy $\mathcal{G}_i{}^j(g_1g_2) = \mathcal{G}_i{}^k(g_1)\,\mathcal{G}_k{}^j(g_2)$.

It is not always true that symmetries can be realized linearly in this linear way, with the general case being a nonlinear realization

$$\phi^i \to \tilde{\phi}^i = U(g)\,\phi^i U^*(g) = \xi^i(\phi, g), \tag{C.49}$$

where $\xi^i(\phi, g)$ is potentially a nonlinear function of the ϕ^i. This is the case to which one is led when a symmetry is spontaneously broken, as described in §4.2.2 (with more details given in Appendix C.6).

C.4.1 Lie Algebra Summary

Many of the symmetries of practical interest are enumerated using continuous parameters (like translations, rotations, chiral symmetries and isospin or gauge transformations), making them Lie groups from the mathematical point of view. This section steps back from the main line of development to summarize a few facts about these groups, together with their related Lie algebras.

Of particular interest in physical applications are often explicit representations of Lie groups and algebras in terms of matrices (that are often important in specific physical applications). For the present purposes representations are simply examples

[5] For translation-invariant ground states the very definitions of Q and U become delicate for spontaneously broken symmetries, at least for field theories in the infinite-volume limit – *c.f.* Eq. (C.82).

of matrices or operators in a physical problem whose matrix multiplication rules furnish examples of the underlying group multiplication rule. Eq. (C.48) provides an example of this, with $U(g_1g_2) = U(g_1)U(g_2)$ providing a unitary representation of the group in the quantum Hilbert space, while the $N \times N$ matrices $\mathcal{G}(g_1g_2) = \mathcal{G}(g_1)\mathcal{G}(g_2)$ provide a finite-dimensional representation of the group on the space of N fields ϕ^i. Although (C.48) leaves ambiguous where the fields ϕ^i and $\tilde{\phi}^i$ are evaluated, for simplicity in what follows they are taken to be evaluated at the same spacetime point (making the symmetry an 'internal' symmetry, as opposed to a 'spacetime' symmetry – see §C.4.2).

For Lie groups the abstract group elements and their explicit realizations are labelled by continuous real parameters, ω^a, with $a = 1, \cdots, N_g$, and both $U(\omega)$ and $\mathcal{G}(\omega)$ are infinitely differentiable. (For instance, for rotations in three dimensions the ω_a could correspond to the angles of rotation about three orthogonal axes.)

Continuous symmetries are often efficiently characterized by their generators, t_a, defined by examining transformations arbitrarily close to the identity element: $g(\omega) = I + \mathrm{i}\omega^a t_a + O(\omega^2)$. It can be shown that the parameters ω^a can be defined in such a way that any group element that is continuously deformable to the identity element can be written as an exponential of the generators: $g(\omega) = \exp[\mathrm{i}\omega^a t_a]$. The statement that a group is closed under multiplication implies that these generators satisfy a set of commutation relations of the form

$$[t_a, t_b] = \mathrm{i}c^d{}_{ab}\, t_d. \tag{C.50}$$

The span of all linear combinations of such generators is called the Lie algebra associated with the Lie group. The coefficients $c^d{}_{ab} = -c^d{}_{ba}$ appearing here are called 'structure constants', whose form encodes the multiplication law that defines the underlying group.

Explicit representations of the Lie group also provide representations for the corresponding Lie algebra. For infinitesimal transformations, $g = 1 + \mathrm{i}\omega^a t_a$, the unitary operator in the quantum Hilbert space becomes $U(g) = I + \mathrm{i}\omega^a T_a$ and the representation matrices for the fields become $\mathcal{G} = I + \mathrm{i}\omega^a \mathcal{T}_a$. These act on the fields so that $\delta\phi^i = \tilde{\phi}^i - \phi^i$ has the form

$$\delta\phi^i(x) = \mathrm{i}\omega^a [T_a\,, \phi^i(x)] = \mathrm{i}\omega^a (\mathcal{T}_a)_j{}^i\, \phi^j(x). \tag{C.51}$$

Because the operators U and the matrices $\mathcal{G}$ satisfy the same group multiplication rule as do the group elements g, the operators T_a and the matrices $\mathcal{T}_a$ satisfy the same commutation relations as do the generators t_a: that is, $[T_a, T_b] = \mathrm{i}c^d{}_{ab}T_d$ and $[\mathcal{T}_a, \mathcal{T}_b] = \mathrm{i}c^d{}_{ab}\mathcal{T}_d$, with the same structure constants as in (C.50).

Conjugate and Adjoint Representations

Any explicit representation of a Lie algebra, $\{\mathcal{T}_a\}$ say, satisfying

$$[\mathcal{T}_a, \mathcal{T}_b] = \mathrm{i}c^d{}_{ab}\, \mathcal{T}_d, \tag{C.52}$$

can be used to define two other representations. The first of these is found by taking the transpose of (C.52), which shows that the operators $\mathcal{S}_a = -\mathcal{T}_a^T$ (where the superscript 'T' denotes taking the transpose) also satisfy (C.52). For unitary representations (those for which the matrices $\mathcal{G}$ are unitary) the $\mathcal{T}_a$ are Hermitian matrices and so $\mathcal{S}_a = -\mathcal{T}_a^T = -\mathcal{T}_a^*$ are also related by complex conjugation (not Hermitian conjugation) to the $\mathcal{T}_a$'s.

A second related representation can be built from the $c^d{}_{ab}$'s themselves (and so is more an intrinsic property of the group than of its specific representation in terms of the $\mathcal{T}_a$. To see why, notice that the associative property of matrix multiplication ensures that the quantity $[A,[B,C]] + [B,[C,A]] + [C,[A,B]]$ identically vanishes for any three matrices A, B and C. Applying this Jacobi identity to three generators of the Lie algebra then implies that

$$\begin{aligned} 0 &= [\mathcal{T}_a,[\mathcal{T}_b,\mathcal{T}_c]] + [\mathcal{T}_b,[\mathcal{T}_c,\mathcal{T}_a]] + [\mathcal{T}_c,[\mathcal{T}_a,\mathcal{T}_b]] \\ &= -\left(c^d{}_{bc}c^e{}_{ad} + c^d{}_{ca}c^e{}_{bd} + c^d{}_{ab}c^e{}_{cd}\right)\mathcal{T}_e. \end{aligned} \tag{C.53}$$

The bracket on the right-hand side of this equation therefore vanishes for any set of structure constants. One way to read this identity is to say that the matrices $\mathcal{A}_a$ with components $(\mathcal{A}_a)^b{}_c := \mathrm{i}c^b{}_{ac}$ satisfy the commutation relation $[\mathcal{A}_a, \mathcal{A}_b] = \mathrm{i}c^d{}_{ab}\mathcal{A}_d$, with precisely the same structure constants as in (C.52), and so therefore furnish another representation – called the *adjoint representation* – of the same Lie algebra.

Finite-Dimensional Unitary Representations

In physical situations continuous symmetry groups often arise as explicit finite-dimensional unitary matrices, such as for the 3×3 orthogonal matrices – *i.e.* $O(3)$ transformations – describing rotations in space, or more generally the internal $N \times N$ unitary matrices – *i.e.* $U(N)$ transformations – amongst N complex fields, ψ^i. This turns out to mean that a special role is often played in physics by *compact groups*, for which the parameter space of the group is a compact set.

Compact groups play a special role because it is a theorem that only compact groups have finite-dimensional, unitary and faithful matrix representations.[6] (The Lorentz group, for instance, is *not* compact although its subgroup of spatial rotations is. Consequently, as found explicitly in §A.2.3 say, although rotations can be represented using finite-dimensional unitary transformations, any finite-dimensional representations of boosts cannot be unitary.)

This section summarizes some useful properties satisfied by representations built from finite-dimensional unitary matrices, for which $\mathcal{G}^\dagger = \mathcal{G}^{-1}$ and so $\mathcal{T}_a^\dagger = \mathcal{T}_a$ if $\mathcal{G}(\omega) = \exp[\mathrm{i}\omega^a\mathcal{T}_a]$. Because the generators are finite-dimensional and hermitian, the quantity

$$\gamma_{ab} = \mathrm{Tr}\,(\mathcal{T}_a\mathcal{T}_b), \tag{C.54}$$

is both symmetric and positive definite (and so can be regarded as a metric, called the group's *Killing metric*). Linear combinations of the generators can always be chosen to ensure that

$$\gamma_{ab} = \delta_{ab} \tag{C.55}$$

[6] A representation is faithful if there is a one-to-one correspondence between the group elements and the matrices which represent them. If the group of interest is *defined* by a finite-dimensional and unitary representation, this representation is by definition faithful.

(such as is true for the standard Pauli-matrix representation of $SU(2)$ – *i.e.* 2×2 unitary matrices with unit determinant – for which $\mathcal{T}_a = \frac{1}{2}\sigma_a$ for $a = 1, 2, 3$). This convention for the generators of compact groups is usually assumed to have been made throughout this book.

The metric γ_{ab} can be used to build a completely covariant version of the structure constants:

$$c_{bca} := c^d{}_{bc}\,\gamma_{da}. \tag{C.56}$$

Its definition automatically implies that $c_{abd} = -c_{bad}$, but when the generators are chosen so that $\gamma_{ab} = \delta_{ab}$ it turns out that c_{abd} is completely antisymmetric under the interchange of *any* two of its indices.

Real Unitary Representations

There is also no loss of generality in assuming representation matrices to be real: $g = g^*$, because any complex representation can always be decomposed into its real and imaginary parts. Although this can always be done, the resulting representation need *not* be irreducible. For reducible representations there is a basis in which all group elements can be written in a block-diagonal form:

$$\mathcal{G} = \begin{pmatrix} \mathcal{G}_{(1)} & & \\ & \ddots & \\ & & \mathcal{G}_{(n)} \end{pmatrix}. \tag{C.57}$$

The unitarity and reality of the group elements, $\mathcal{G}$, then imply the matrices $\mathcal{T}_a$ are antisymmetric and imaginary:

$$\mathcal{T}_a = \mathcal{T}_a^\dagger = -\mathcal{T}_a^* = -\mathcal{T}_a^T. \tag{C.58}$$

Subgroups and Subalgebras

When describing a symmetry-breaking pattern where G breaks to $H \subset G$ it is convenient to choose a basis of generators, T_a, for G that includes the generators, t_i, of H as a subset. To this end, decompose the generators T_a, $a = 1, \cdots, N_G$ into the subset t_i, $i = 1, \cdots, N_H$ and X_α, $\alpha = N_H + 1, \cdots, N_G$, so the X_α's constitute a basis of generators not included in the unbroken subalgebra. Here $N_G = \dim G$ is the number of linearly independent generators of the Lie algebra of G and $N_H = \dim H$ is its counterpart for H. Since H is itself a group, its closure under multiplication – *i.e.* the statement that $h_1, h_2 \in H$ implies $h_1 h_2 \in H$ – ensures that

$$t_i\, t_j - t_j\, t_i = i\, c_{ijk}\, t_k, \tag{C.59}$$

with no X_α's on the right-hand-side, or (schematically) $c_{ij\alpha} = 0$.

The X_α are not contained in the algebra of H and need not themselves generate a group. Instead, they are said to generate the space, G/H, of *cosets*. A coset is an equivalence class defined to contain all of the elements of G that are related by the multiplication by an element of H. In the applications of §C.6, the X_α's represent those generators of the symmetry group, G, that are spontaneously broken, and (in relativistic applications for internal symmetries) a Goldstone mode is expected for each independent choice of α.

When the generators of G are chosen to ensure the complete antisymmetry of the c_{abd}'s then the group property of H (summarized above as $c_{ija} = 0$) also implies that $c_{iaj} = 0$. This says

$$[t_i, X_\alpha] = i\, c_{i\alpha\beta} X_\beta, \tag{C.60}$$

with no t_j's on the right-hand side. Equivalently, this states that the X_α's form a (possibly reducible) representation of H. Once exponentiated into a statement about group multiplication, the condition $[t, X] \sim X$ implies that for any $h \in H$

$$h X_\alpha h^{-1} = L^\beta{}_\alpha X_\beta. \tag{C.61}$$

where the coefficients, $L^\beta{}_\alpha(h)$, form a representation of H.

By contrast, the commutator $[X_\alpha, X_\beta]$ need not have a particularly simple form, and can be proportional to both X_γ's and t_i's. (The special case of a coset G/H for which $[X_\alpha, X_\beta]$ does not contain any X_γ's is called a *symmetric* space.)

C.4.2 Internal vs Spacetime Symmetries

Notice that the above discussion distinguishes unitary transformations (those that preserve matrix elements) from symmetries (those unitary transformations that commute with the hamiltonian). This notion of symmetry is adequate for internal symmetries – *i.e.* those that do not act on spatial position or time, so $U\phi(x)U^* = \tilde\phi(x)$ with both sides evaluated at the same position.

A broader definition is needed for spacetime symmetries, for which the transformations act both on the fields and the spacetime point: $U\phi(x)U^* = \tilde\phi(\tilde x)$, with $\tilde x^\mu \neq x^\mu$. Lorentz transformations are simple examples where this matters, since for these H generally is not invariant, since it is part of a 4-vector: $UP_\mu U^* = \Lambda^\nu{}_\mu P_\nu$, with $P^0 = H$. In this case a symmetry is defined instead by the invariance of the action, $S = \int dt\, L$, rather than of H. For scattering problems transformations that are symmetries in this sense also commute with the S-matrix.

The Coleman–Mandula theorem provides an important constraint on the kinds of continuous spacetime symmetries that can be present within interacting relativistic quantum field theories. The Coleman–Mandula theorem [315] states that the most general possible non-Grassman[7] transformations that commute with a (nontrivial – *i.e.* $S \neq I$) S-matrix are:

$$U = \exp\left[\frac{i}{2}\omega^{\mu\nu} J_{\mu\nu} + i a^\mu P_\mu + i\omega^a Q_a\right] \tag{C.62}$$

with generators P_μ, $J_{\mu\nu} = -J_{\nu\mu}$, and Q_a.

Ten of these are no surprise in a relativistic theory: the six generators $J_{\mu\nu}$ satisfy the commutation relations appropriate to the Lorentz group and the four P_μ generate spacetime translations and so mutually commute (and fill out the usual 4-momentum operator). The commutation relations between $J_{\mu\nu}$ and P_μ fill out the algebra of

[7] This is the assumption that supersymmetric theories violate; see [467] for the generalization to this case.

the Poincaré group of Lorentz transformations and translations, making these the defining symmetries of special relativity.[8]

The power of the Coleman–Mandula theorem is what it says about the remaining generators: the Q_a's. These must be internal symmetries, and although they can fail to commute quietly amongst themselves, $[Q_a, Q_b] = ic^d{}_{ab}T_d$, the theorem states that they must always commute with the spacetime symmetries (*i.e.* the Poincaré generators): $[P_\mu, Q_a] = 0$ and $[J_{\mu\nu}, Q_a] = 0$. The theorem is proven by assuming it to be false, and then showing that the additional conservation laws for the spacetime symmetries are so strong that they generically force the scattering matrix to be trivial: $S = I$.

C.4.3 Discrete Symmetries

Discrete symmetries (those that are not described by continuous parameters) are also important for physics. Some of these can be internal symmetries, such as an example like $\phi(x) \to -\phi(x)$, which defines a discrete Z_2 symmetry in field space. Such symmetries constrain the kinds of interactions that can arise (forbidding, in the Z_2 example, terms involving odd powers of fields). Their representations can also be used to classify states (in the Z_2 case states can be chosen to either change sign or be invariant under the group's action).

Spacetime discrete symmetries are also important. These are defined to be those Poincaré transformations that cannot be continuously deformed to the unit element. There are two such discrete transformations within the Lorentz group. To see why, recall that the general Lorentz transformation is defined by the condition

$$\Lambda^\mu{}_\nu \Lambda^\rho{}_\lambda \eta_{\mu\rho} = \eta_{\nu\lambda}. \tag{C.63}$$

This condition throws up two obstructions to being able to deform $\Lambda^\mu{}_\nu$ to the identity transformation. One of these arises because (C.63) implies that the determinant of the matrix $\Lambda^\mu{}_\nu$ must be ± 1, but only those with determinant $+1$ can be continuously connected to the identity matrix. Similarly, (C.63) requires $|\Lambda^0{}_0| \geq 1$ and so any matrix with $\Lambda^0{}_0 < -1$ also cannot be continuously related to the identity matrix.

A general solution $\Lambda^\mu{}_\nu$ to (C.63) can be written as a combination of a matrix continuously connected to the identity, $\Lambda^\mu{}_\nu = (e^\omega)^\mu{}_\nu$ (called a 'proper' Lorentz transformation) times a product of one or both of the two specific matrices

$$P^\mu{}_\nu := \begin{pmatrix} 1 & & & \\ & -1 & & \\ & & -1 & \\ & & & -1 \end{pmatrix} \quad \text{and} \quad T^\mu{}_\nu := \begin{pmatrix} -1 & & & \\ & 1 & & \\ & & 1 & \\ & & & 1 \end{pmatrix}, \tag{C.64}$$

where P (parity) acts to reflect all spatial coordinates while T (time-reversal) flips the sign of time.

These matrices show why time reversal is the lone symmetry that cannot be represented by a unitary operator. Denoting by $U(\Lambda, a)$ the representation of the

[8] For theories involving only massless particles this symmetry group is sometimes a bit larger; comprising the conformal group that also includes rescalings, $x^\mu \to sx^\mu$, and conformal boosts.

Lorentz transformation $\Lambda^\mu{}_\nu$ and spacetime translation a^μ, the group multiplication law for the Poincaré group implies

$$U(\Lambda_1, a_1)U(\Lambda_2, a_2) = U(\Lambda_1\Lambda_2, \Lambda_1 a_2 + a_1). \tag{C.65}$$

Denoting the operator that represents time-reversal by $\mathcal{T} = U(T, 0)$, this implies that

$$\mathcal{T}U(\Lambda, a)\mathcal{T}^{-1} = U(T\Lambda T^{-1}, Ta), \tag{C.66}$$

and so for $\Lambda = I$ and infinitesimal a^μ this says $\mathcal{T}(\mathrm{i}P_\mu)\mathcal{T}^{-1} = T^\nu{}_\mu \mathrm{i}P_\nu$ and so $\mathcal{T}\mathrm{i}H\mathcal{T}^{-1} = -\mathrm{i}H$ and $\mathcal{T}\mathrm{i}\mathbf{P}\mathcal{T}^{-1} = \mathrm{i}\mathbf{P}$.

Now comes the main point. If $\mathcal{T}$ were unitary then it would satisfy $\mathcal{T}H\mathcal{T}^{-1} = -H$, which is inconsistent with H being bounded from below (as it typically is for stable systems). But if it is antiunitary then $\mathcal{T}\mathrm{i}H\mathcal{T}^{-1} = -\mathrm{i}\mathcal{T}H\mathcal{T}^{-1}$, allowing $\mathcal{T}$ to commute with H. For antiunitary $\mathcal{T}$ it then follows that $\mathcal{T}\mathbf{P}\mathcal{T}^{-1} = -\mathbf{P}$.

A third important discrete symmetry interchanges particles with antiparticles (with momenta and spins held fixed). This acts on creation and destruction operators by $C\,\mathfrak{a}_i\,C^{-1} = \eta_c\,\bar{\mathfrak{a}}_i$, an operation called charge-conjugation. Here η_c is a phase that can differ for different particle types. The action of charge conjugation on fields is found by applying the definition to the expansion of fields in terms of creation and annihilation operators. For example, for a scalar field this leads to

$$\begin{aligned} C\,\phi(x)\,C^{-1} &= \int \frac{\mathrm{d}^3p}{\sqrt{(2\pi)^3 2E_p}} \Big[C\,\mathfrak{a}_\mathbf{p}\,C^{-1} e^{ip\cdot x} + C\,\bar{\mathfrak{a}}^*_\mathbf{p}\,C^{-1} e^{-ip\cdot x}\Big] \\ &= \eta_c \int \frac{\mathrm{d}^3p}{\sqrt{(2\pi)^3 2E_p}} \Big[\bar{\mathfrak{a}}_\mathbf{p}\,e^{ip\cdot x} + \mathfrak{a}^*_\mathbf{p}\,e^{-ip\cdot x}\Big] = \eta_c\phi^*(x), \end{aligned} \tag{C.67}$$

and so acts as complex conjugation.

The three discrete symmetries, C, P and T, are individually symmetries of electromagnetism, gravity and the strong interactions, but all three are separately broken by the weak interactions. The action of each of these three symmetries on various familiar physical quantities is summarized in Table C.1. (For A_μ the signs given in this table include the phases – like η_c in (C.67) – appearing in the parity, time-reversal and charge-conjugation transformations of the electromagnetic field.)

For relativisitic systems it turns out to be a theorem that any real and local action, $S = \int \mathrm{d}^4x\,\mathfrak{L}$, turns out to be always invariant under the combined combination of all three symmetries: CPT (not surprisingly, a result called the CPT theorem). There is a

Table C.1 The transformation properties of common quantities under parity (P), time-reversal (T) and charge-conjugation (C)

Quantity		P	T	C	Quantity		P	T	C
Position	$\mathbf{x}$	−	+	+	Momentum	$\mathbf{p}$	−	−	+
Spin	$\mathbf{s}$	+	−	+	Helicity	$\mathbf{p}\cdot\mathbf{s}$	−	+	+
Current	$\mathbf{j}$	−	−	−	Charge density	j^0	+	+	−
Vector potential	$\mathbf{A}$	−	−	−	Scalar potential	A^0	+	+	−
Electric field	$\mathbf{E}$	−	+	−	Magnetic field	$\mathbf{B}$	+	−	−

simple reason for this. The action of C complex conjugates all the fields in $\mathfrak{L}$ and the anti-unitary nature of time-reversal complex conjugates all couplings in $\mathfrak{L}$, so their combined effect takes $\mathfrak{L} \to \mathfrak{L}^*$. But the lagrangian is hermitian so this has no effect. Time reversal and parity also together act to reverse the sign of all components of any 4-vector, $(PT)^\mu{}_\nu V^\nu = -V^\mu$. But this also has no effect because there are always an even number of such 4-vectors because this must be true if $\mathfrak{L}$ is to be a Lorentz scalar.

C.5 Gauge Interactions

Consider next a collection of quantum fields, ϕ^i, that transform under linearly realized infinitesimal internal symmetries of the form $\phi^i(x) \to \tilde{\phi}^i(x)$ with

$$\delta\phi^i(x) := \tilde{\phi}^i(x) - \phi^i(x) = \mathrm{i}\omega^a(\mathcal{T}_a)^i{}_j\,\phi^j(x). \tag{C.68}$$

This is a global (or rigid) symmetry when the transformation parameter ω^a is independent of spacetime position.[9] But an important role is also played by local symmetries, for which $\omega^a = \omega^a(x)$ is a function of spacetime [468].

Since global symmetries are special cases of local ones, it is more difficult to make a theory invariant under a local symmetry than for a global one. To see this explicitly, consider an action for a collection of fields, ϕ^i, where $\mathfrak{L} = \mathfrak{L}(\phi, \partial\phi)$ is a function of both the fields and their first derivatives. The variation of $\mathfrak{L}$ under (C.68) is

$$\begin{aligned}\delta\mathfrak{L} &= \left(\frac{\partial\mathfrak{L}}{\partial\phi^i}\right)\mathrm{i}\omega^a(\mathcal{T}_a)^i{}_j\,\phi^j + \left(\frac{\partial\mathfrak{L}}{\partial(\partial_\mu\phi^i)}\right)\partial_\mu\left[\mathrm{i}\omega^a(\mathcal{T}_a)^i{}_j\,\phi^j\right] \\ &= \left(\frac{\partial\mathfrak{L}}{\partial\phi^i}\mathrm{i}\omega^a(\mathcal{T}_a)^i{}_j\,\phi^j + \frac{\partial\mathfrak{L}}{\partial(\partial_\mu\phi^i)}\mathrm{i}\omega^a(\mathcal{T}_a)^i{}_j\,\partial_\mu\phi^j\right) + \left(\frac{\partial\mathfrak{L}}{\partial(\partial_\mu\phi^i)}\right)\mathrm{i}\partial_\mu\omega^a(\mathcal{T}_a)^i{}_j\,\phi^j.\end{aligned} \tag{C.69}$$

The first two terms on the right-hand side vanish whenever $\mathcal{L}$ is invariant under a global symmetry like (C.68), but with the parameter ω^a spacetime independent.

Eq. (C.69) shows that even if a lagrangian is arranged to be invariant under global transformations, it need not be invariant under local ones, transforming in a universal way

$$\delta\mathfrak{L} = \left(\frac{\partial\mathfrak{L}}{\partial(\partial_\mu\phi^i)}\right)\mathrm{i}\partial_\mu\omega^a(\mathcal{T}_a)^i{}_j\,\phi^j = j^\mu_a\,\partial_\mu\omega^a, \tag{C.70}$$

where $j^\mu_a = [\partial\mathfrak{L}/\partial(\partial_\mu\phi^i)]\mathrm{i}(\mathcal{T}_a)^i{}_j\phi^j$ is the Noether current for the global symmetry, as defined by Eq. (4.7) in §4.1.1. The universal form of this transformation suggests a way to build a locally invariant lagrangian. Juxtaposing the fact that (C.70) involves $\partial_\mu\omega^a$ and that massless spin-one particles can only be represented by a field A_μ if it transforms as $\delta A_\mu = \partial_\mu\omega$, as in Eq. (C.40), suggests that a locally invariant lagrangian might be constructed by adding a new term, $\mathfrak{L}_j \doteq -j^\mu_a A^a_\mu$, in whose variation the transformation $\delta A^a_\mu = \partial_\mu\omega^a$ would cancel (C.70).

Adding $\mathfrak{L}_j$ need not be the whole story, because, in general, the current j^μ_a also transforms under the transformation (C.68). This transformation can also be inferred universally since the symmetry generators, T_a, themselves are obtained by

[9] The attentive reader will notice that the matrix $\mathcal{T}_a$ used in (C.68) is the transpose of the one used in (C.51). This is done so that the signs found in this section agree with those widely used in the literature.

integrating j_a^0 over all space. Since the generators satisfy $[T_a, T_b] = \mathrm{i}c^d{}_{ab}\, T_d$, the currents must transform (up to terms that vanish when integrated over space) as

$$\delta j_a^\mu = \mathrm{i}\omega^b \left[T_b\,, j_a^\mu\right] = c^d{}_{ab}\, \omega^b\, j_d^\mu\,. \tag{C.71}$$

This suggests modifying the transformation rule for A_μ^a to also transform in the adjoint representation:

$$\delta A_\mu^a = \partial_\mu \omega^a + c^a{}_{bc}\, \omega^b\, A_\mu^c\,. \tag{C.72}$$

so that $\delta(j_a^\mu A_\mu^a) = j_a^\mu \partial_\mu \omega^a$.

To see whether this works, start with a more general lagrangian density $\mathfrak{L} = \mathfrak{L}(\phi, \partial\phi, A, \partial A)$ and ask whether it can be invariant under the transformations (C.68) and (C.72). The variation of $\mathfrak{L}$ then is

$$\begin{aligned}\delta\mathfrak{L} = &\left(\frac{\partial \mathfrak{L}}{\partial \phi^i}\right) \mathrm{i}\omega^a (\mathcal{T}_a)^i{}_j\, \phi^j + \left(\frac{\partial \mathfrak{L}}{\partial(\partial_\mu \phi^i)}\right) \partial_\mu \left[\mathrm{i}\omega^a (\mathcal{T}_a)^i{}_j\, \phi^j\right] \\ &+ \left(\frac{\partial \mathfrak{L}}{\partial A_\mu^a}\right) \left(\partial_\mu \omega^a + c^a{}_{bc}\, \omega^b\, A_\mu^c\right) + \left(\frac{\partial \mathfrak{L}}{\partial(\partial_\nu A_\mu^a)}\right) \partial_\nu \left(\partial_\mu \omega^a + c^a{}_{bc}\, \omega^b\, A_\mu^c\right),\end{aligned} \tag{C.73}$$

and for this to vanish for arbitrary local functions $\omega^a(x)$ the coefficients in it of ω^a, $\partial_\mu \omega^a$ and $\partial_\mu \partial_\nu \omega^a$ must separately vanish. The coefficient of $\partial_\mu \partial_\nu \omega^a$ vanishes if

$$\frac{\partial \mathfrak{L}}{\partial(\partial_\mu A_\nu^a)} = -\frac{\partial \mathfrak{L}}{\partial(\partial_\nu A_\mu^a)}, \tag{C.74}$$

which means that A_μ^a appears differentiated in $\mathfrak{L}$ only through the antisymmetric combination $f_{\mu\nu}^a := \partial_\mu A_\nu^a - \partial_\nu A_\mu^a$.

Changing independent variable from $\partial_\mu A_\nu^a$ to $f_{\mu\nu}^a$, the coefficient of $\partial_\mu \omega^a$ vanishes when

$$\left(\frac{\partial \mathfrak{L}}{\partial(\partial_\mu \phi^i)}\right) \mathrm{i}(\mathcal{T}_a)^i{}_j\, \phi^j + \frac{\partial \mathfrak{L}}{\partial A_\mu^a} + 2\left(\frac{\partial \mathfrak{L}}{\partial f_{\mu\nu}^b}\right) c^b{}_{ac}\, A_\nu^c = 0. \tag{C.75}$$

To extract the implications of this condition, consider first terms in $\mathfrak{L}$ that do not depend on ϕ^i or its derivative at all. In this case, (C.75) states that $f_{\mu\nu}^a$ and A_μ^a can only appear together in $\mathfrak{L}$, though the one combination

$$F_{\mu\nu}^a = f_{\mu\nu}^a + c^a{}_{bc} A_\mu^b A_\nu^c = \partial_\mu A_\nu^a - \partial_\nu A_\mu^a + c^a{}_{bc} A_\mu^b A_\nu^c. \tag{C.76}$$

What is special about this quantity is that the dependence on $\partial_\mu \omega^a$ cancels when it is transformed using (C.72), leaving $\delta F_{\mu\nu}^a = c^a{}_{bc} \omega^b F_{\mu\nu}^c$.

Re-introducing a dependence on ϕ^i and trading $\partial\mathfrak{L}/\partial A_\mu^a$ for $(\partial\mathfrak{L}/\partial A_\mu^a)_F$ (with the subscript indicating the derivative is taken at fixed $F_{\mu\nu}^a$ instead of fixed $\partial_\mu A_\nu^a$ or $f_{\mu\nu}^a$), condition (C.75) becomes

$$\left(\frac{\partial \mathfrak{L}}{\partial(\partial_\mu \phi^i)}\right) \mathrm{i}(\mathcal{T}_a)^i{}_j\, \phi^j + \left(\frac{\partial \mathfrak{L}}{\partial A_\mu^a}\right)_F = 0, \tag{C.77}$$

which implies $\partial_\mu \phi^i$ must always appear together with A_μ^a through the covariant-derivative combination

$$(D_\mu \phi)^i := \partial_\mu \phi^i - \mathrm{i}(\mathcal{T}_a)^i{}_j A_\mu^a \phi^j. \tag{C.78}$$

As is easily verified, $\partial_\mu \omega^a$ also cancels in the transformation rule $\delta(D_\mu\phi)^i = i\omega^a(\mathcal{T}_a)^i{}_j(D_\mu\phi)^j$. The lesson from the $\partial_\mu\omega^a$ term is that A^a_μ only appears in $\mathfrak{L}$ as part of $(D_\mu\phi)^i$ or $F^a_{\mu\nu}$, and never on its own. Notice that the covariant derivative and field strength are related by the following easily proven identity

$$[D_\mu, D_\nu]\phi = -i(\mathcal{T}_a\phi)\, F^a_{\mu\nu}. \tag{C.79}$$

Finally, setting the coefficient of undifferentiated ω^a to zero gives

$$\left(\frac{\partial\mathfrak{L}}{\partial\phi^i}\right) i(\mathcal{T}_a)^i{}_j\,\phi^j + \left(\frac{\partial\mathfrak{L}}{\partial(D_\mu\phi^i)}\right) i(\mathcal{T}_a)^i{}_j\,(D_\mu\phi)^j + \left(\frac{\partial\mathfrak{L}}{\partial F^b_{\mu\nu}}\right) c^b{}_{ac}\, F^c_{\mu\nu} = 0, \tag{C.80}$$

which simply states that $\mathfrak{L}$ should be built from ϕ^i, $(D_\mu\phi)^i$ and $F^a_{\mu\nu}$ in a way that is invariant under *global* transformations (for which ω^a is a constant).

Here is the point: a global internal symmetry can be promoted to a local internal symmetry by introducing a new massless spin-one particle for each symmetry generator, and then building the lagrangian out of undifferentiated fields like ϕ^i, covariant derivatives like $(D_\mu\phi)^i$ and covariant field strengths like $F^a_{\mu\nu}$.

C.5.1 Higgs Mechanism

Historically, when promoting global to local symmetries the need for massless spin-one particles was seen as a handicap. Although it worked splendidly for the massless photon in quantum electrodynamics, the prospects for applications elsewhere seemed limited (the phenomenon of confinement prevented the discovery of massless gluons until later).

The modern understanding wherein all fundamental spin-one particles, massless or not, are gauge bosons had to await the discovery (by Brout and Englert [55], Guralnik, Hagan and Kibble [56, 57], Higgs [58, 59] and others, building on earlier work by Anderson [60] for nonrelativistic systems) of what is widely called the *Higgs* mechanism. This mechanism shows why the spin-one particles can be massive, provided they are associated with local symmetries that are spontaneously broken.

It is fundamental that systems with spontaneously broken symmetries do not have unique ground states, because by assumption the action of $U(g)$ on one ground state gives a different state, $|\tilde{0}\rangle := U|0\rangle \neq |0\rangle$. But because a symmetry satisfies $UH = HU$ the state $|\tilde{0}\rangle$ has precisely the same energy as does $|0\rangle$, making it a second 'ground' state.

Related to this, the operators $U(g) = \exp[iQ]$, and their generators Q, are less useful when dealing with spontaneously broken symmetries in field theories, particularly in situations where the spatial directions are infinitely large and the ground state is translation invariant [43]. This is because in a field theory Noether's theorem ensures that

$$Q = \int d^3x\; j^0(x), \tag{C.81}$$

arises as the integral over a local current density. It follows then that the state $Q|0\rangle$ is not normalizable (and so does not lie within the Fock space built apon $|0\rangle$), because

$$||Q|0\rangle||^2 = \langle 0|QQ|0\rangle = \int d^3x\; \langle 0|Qj^0(x)|0\rangle = \int d^3x\; \langle 0\,|Qj^0(0)|0\rangle, \tag{C.82}$$

where the first equality uses $Q^* = Q$, the second equality uses (C.81) and the last equality uses the representation of translation symmetries to write $j^0(x) = e^{iP\cdot x} j^0(0) e^{-iP\cdot x}$, together with the translation invariance of Q and the ground state: $[Q, P_\mu] = 0$ and $P_\mu|0\rangle = 0$. The final result diverges like the volume of space because the integrand does not depend at all on x^μ.

Because of this it is preferable to have a more useful proxy for spontaneous symmetry-breaking than the evaluation of $U(g)|0\rangle$ or $Q|0\rangle$. The existence of a nonzero *order parameter* fills this role, by providing a simpler-to-use criterion for the non-invariance of the vacuum. For example, imagine two fields that are related by a symmetry, such as if[10]

$$\psi(x) = \mathrm{i}[Q, \phi(x)]. \tag{C.83}$$

Then $\psi(x)$ is an order parameter for the symmetry generated by Q if its vacuum expectation value (*vev*) is nonzero: $v := \langle 0|\,\psi(x)|0\rangle \neq 0$. A nonzero *vev* is a proxy for spontaneous symmetry-breaking because an unbroken symmetry implies $Q|0\rangle = 0$ – and its conjugate $\langle 0|Q = 0$ – and both of these arise in the right-hand side of (C.83) once its vacuum expectation-value is taken. Since unbroken symmetry implies that $v = 0$, it follows that nonzero v implies that the symmetry must be broken.

In order not to break any spacetime symmetries the order-parameter field must be a Lorentz-scalar and independent of x^μ. To see how this works at weak couplings, where semiclassical reasoning is valid, consider then a collection of scalar fields, $\phi^i, i = 1, \cdots, N$, which without loss of generality can be chosen to be real. Suppose the particles represented by these fields couple to a collection of spin-one particles represented by A_μ^a, with local symmetry group $\delta\phi^i = \mathrm{i}\omega^a (\mathcal{T}_a)^i{}_j \phi^j$ and $\delta A_\mu^a = \partial_\mu \omega^a + c^a{}_{bc}\omega^b A_\mu^c$. A lagrangian density for these particles involving only up to two derivatives is

$$\mathfrak{L} = -V(\phi) - \frac{1}{2} Z_{ij} (D_\mu\phi)^i (D^\mu\phi)^j - \frac{1}{4g^2}\gamma_{ab} F_{\mu\nu}^a F^{b\mu\nu}, \tag{C.84}$$

where Z_{ij} are a collection of numerical coefficients (that can be set to δ_{ij} by appropriately redefining the fields), γ_{ab} is the Killing metric of Eq. (C.54), the covariant derivative is $D_\mu\phi = \partial_\mu\phi - \mathrm{i}\mathcal{T}_a A_\mu^a \phi$ and $F_{\mu\nu}^a$ is as defined in (C.76) with $c^a{}_{bc}$ the structure constants associated with the generators $\mathcal{T}_a$.

For the present purposes the important feature is to have a potential energy, $V(\phi)$, whose minimum occurs for $\phi^i \neq 0$. This is easily arranged following the example of the toy model of §1.1. For instance, for symmetries that preserve the quantity $\phi^T\phi = \Sigma_i(\phi^i)^2$, the potential

$$V(\phi) = \frac{\lambda}{4}\left(\phi^T\phi - v^2\right)^2, \tag{C.85}$$

does the job. For positive real parameters λ and v^2 this potential is strictly non-negative and vanishes for the minimizing surface $\phi^T\phi = v^2$. This does not pick a unique solution for ϕ^i because it contains all configurations related by the

[10] Notice that commutators like $\delta\phi(x) = \mathrm{i}[Q, \phi(x)]$ are usually well-defined even if the action of Q on $|0\rangle$ is not. This is because the equal-time commutators of fields are usually local, such as the canonical commutation relations $\Pi(\mathbf{x}, t), \phi(\mathbf{y}, t)] = -\mathrm{i}\delta^3(\mathbf{x} - \mathbf{y})$, for which the delta-function removes the otherwise diverging spatial integration.

symmetries that preserve $\phi^T\phi$. Each of these provides an equally good vacuum, and all are equivalent to the extent that they are related by symmetries.

For concreteness' sake choose the vacuum to be the one with $\langle\phi^1\rangle = v$ and all others zero, and expand all quantum fields about this semi-classical vacuum expectation value: $\phi^1 = v + \hat{\phi}^1$ and $\phi^i = \hat{\phi}^i$ for $i \neq 1$. The leading correction to the classical limit keeps only terms quadratic in the $\hat{\phi}^i$ and A^a_μ.

The revealing terms in this expansion are those arising within the scalar kinetic term,

$$\mathfrak{L}_{\rm sk} = -\frac{1}{2} Z_{ij} D^\mu \phi^i D_\mu \phi^j = -\frac{1}{2} Z_{ij} v^2 (\mathcal{T}_a)^i{}_1 (\mathcal{T}_b)^j{}_1 A^a_\mu A^{b\mu} + i Z_{ij} v (\mathcal{T}_a)^i{}_1 A^a_\mu \partial^\mu \hat{\phi}^j + \cdots . \tag{C.86}$$

The second term on the right-hand side is unusual inasmuch as it mixes scalar and vector degrees of freedom. The good news is that it is always possible to perform a gauge transformation to completely remove this term (a choice called 'unitary gauge'). The gauge transformation required to reach this choice absorbs one scalar degree of freedom into A^a_μ for each independent symmetry generator that is broken by the vacuum. It is the addition of these new states that provides the missing longitudinal spin states required to promote the two spin-states of a massless spin-one particle to the three spin states of a massive one.

Once this is removed, the first term on the right-hand side is revealed as a spin-one mass term – compare with (C.35). Canonically normalizing fields (which sets $Z_{ij} = \delta_{ij}$) and computing the particle energies at zero momentum gives the spin-one mass matrix

$$\mu^2_{ab} = V^T \mathcal{T}_a \mathcal{T}_b V, \tag{C.87}$$

where $V^i = \langle\phi^i\rangle$ denotes the field-vector containing the field vacuum expectation values. The spin-one particles indeed acquire a mass when their associated gauge symmetry becomes spontaneously broken.

C.5.2 General Relativity

A short summary of the basics of General Relativity (GR) is also appropriate here, since gravitational interactions arise at several points within the main text. Although a proper discussion goes well beyond the scope of this book, this section suffices to collect some of the main formulae.

There is a strong analogy between GR and gauge theories of massless nonabelian spin-one particles, like QCD. Both involve massless states (though the gluons of QCD, unlike the graviton of GR, are prevented from escaping to infinity as massless states due to the growth of the strong force with distance). Both involve non-abelian local symmetries: for QCD these are the local $SU_c(3)$ colour transformations of the Standard Model, while for GR these are a combination of local diffeomorphisms (and local Lorentz transformations, when coupled to fields with spin). Both also involve nonlinear self-interactions wherein the force carriers themselves carry charges (that is, gluons carry colour and gravitons carry energy and momentum). This makes them unlike abelian massless spin-one particles like photons, which do not carry electric charge.

The basic field for GR is the spacetime metric itself: $g_{\mu\nu}(x)$. The local symmetries in this case correspond to local diffeomorphisms of the type $x^\mu \to x^\mu + \xi^\mu(x)$, under which $g_{\mu\nu}$ transforms linearly, like a covariant rank-two tensor,

$$\delta g_{\mu\nu} = \mathcal{L}_\xi g_{\mu\nu} := \xi^\lambda \partial_\lambda g_{\mu\nu} + \partial_\mu \xi^\lambda g_{\lambda\nu} + \partial_\nu \xi^\lambda g_{\mu\lambda}, \tag{C.88}$$

where the right-hand side defines the *Lie derivative* $\mathcal{L}_\xi$ of the metric. Other fields, such as scalar or vector fields, similarly transform under diffeomorphisms as their index content suggests

$$\delta\phi = \mathcal{L}_\xi \phi := \xi^\lambda \partial_\lambda \phi \quad \text{and} \quad \delta V_\mu = \mathcal{L}_\xi V_\mu = \xi^\lambda \partial_\lambda V_\mu + \partial_\mu \xi^\lambda V_\lambda, \tag{C.89}$$

and so on.

Covariant Derivatives and Curvatures

Just like for local gauge invariance, local lagrangian densities that are invariant under these transformations can be built by starting with a lagrangian that is invariant under a global symmetry (in this case, the spacetime symmetry of Poincaré invariance) with two provisos: all ordinary derivatives get promoted to covariant derivatives, $\partial_\mu \to D_\mu$, and the gauge field itself (in this case $g_{\mu\nu}$) appears through a covariant field strength – in this case, the Riemann tensor (see below) – and its (covariant) derivatives.

The covariant derivatives appropriate for scalars and vectors transforming as in (C.89) are

$$D_\mu\phi := \partial_\mu\phi\,, \quad D_\mu V_\nu := \partial_\mu V_\nu - \Gamma^\lambda_{\mu\nu} V_\lambda \quad \text{and} \quad D_\mu V^\nu := \partial_\mu V^\nu + \Gamma^\nu_{\mu\lambda} V^\lambda, \tag{C.90}$$

where the Christoffel symbol is defined by (A.7), reproduced for convenience here:

$$\Gamma^\mu_{\nu\lambda} = \frac{1}{2} g^{\mu\alpha} \left[\partial_\nu g_{\alpha\lambda} + \partial_\lambda g_{\alpha\nu} - \partial_\alpha g_{\nu\lambda}\right]. \tag{C.91}$$

Here, $g^{\mu\nu}$ denotes the inverse metric, defined by the condition $g^{\mu\nu} g_{\nu\lambda} = \delta^\mu_\lambda$. With the above definitions the metric is covariantly constant:

$$D_\mu g_{\nu\lambda} = 0 = D_\mu g^{\nu\lambda}. \tag{C.92}$$

Notice that the definitions (C.90) ensure that covariant derivatives satisfy the usual product rule for derivatives: *e.g.*

$$\partial_\mu(V_\lambda W^\lambda) = D_\mu(V_\lambda W^\lambda) = (D_\mu V_\lambda) W^\lambda + V_\lambda (D_\mu W^\lambda). \tag{C.93}$$

Notice also that antisymmetrized ordinary derivatives are already covariant, inasmuch as

$$D_\mu V_\nu - D_\nu V_\mu = \partial_\mu V_\nu - \partial_\nu V_\mu, \tag{C.94}$$

so (for example) the relation between electromagnetic field strength and vector potential does not change in the presence of a gravitational field. It is this observation about how Christoffel symbols cancel in antisymmetric tensors that underlies the study of differential forms and exterior derivatives: covariant quantities that can be defined without making reference to a metric.

The covariant field strength containing derivatives of $g_{\mu\nu}$ appropriate for diffeomorphisms is the Riemann tensor, $R^\mu{}_{\nu\lambda\rho}$, as defined by (A.6), again reproduced here:

$$R^\mu{}_{\nu\lambda\rho} = \partial_\rho \Gamma^\mu_{\nu\lambda} + \Gamma^\mu_{\rho\sigma}\Gamma^\sigma_{\nu\lambda} - (\rho \leftrightarrow \lambda). \tag{C.95}$$

This definition implies that the covariant version of this tensor, $R_{\mu\nu\lambda\rho} = g_{\mu\sigma}R^\sigma{}_{\nu\lambda\rho}$, has the important symmetry properties $R_{\mu\nu\lambda\rho} = R_{\lambda\rho\mu\nu} = -R_{\nu\mu\lambda\rho} = -R_{\mu\nu\rho\lambda}$ as well as satisfying the 'Bianchi' identities

$$R_{\mu\nu\lambda\rho} + R_{\mu\lambda\rho\nu} + R_{\mu\rho\nu\lambda} = 0, \tag{C.96}$$

and

$$D_\sigma R_{\mu\nu\lambda\rho} + D_\lambda R_{\mu\nu\rho\sigma} + D_\rho R_{\mu\nu\sigma\lambda} = 0. \tag{C.97}$$

Finally, the Riemann tensor is related to the commutator of two covariant derivatives; a straightforward use of the definitions implies the gravitational analog of (C.79),

$$[D_\mu, D_\nu]V^\lambda = R^\lambda{}_{\nu\rho\mu}V^\rho. \tag{C.98}$$

Generally Covariant Actions

A local action arises as an integral over a lagrangian density, $S = \int d^4x\, \mathfrak{L}$. The lagrangian density cannot be a scalar under diffeomorphisms, however, because $\mathfrak{L}$ must transform in such a way as to cancel the transformation of the measure d^4x. This is accomplished if $\mathfrak{L} = \sqrt{-g}\, L$, where $g = \det(g_{\mu\nu}) < 0$ is the determinant of the metric and L is a scalar under diffeomorphisms (*i.e.* transforms as a scalar field).

The appearance of $\sqrt{-g}$ in the lagrangian density makes the following identity very useful:

$$\partial_\mu(\sqrt{-g}\, V^\mu) = D_\mu(\sqrt{-g}\, V^\mu) = \sqrt{-g}\, D_\mu V^\mu \tag{C.99}$$

for any 4-vector V^μ. This shows that integrals of the form $\int d^4x\, \sqrt{-g}\, D_\mu V^\mu$ are total divergences and so depend only on boundary information.

Because the Riemann tensor already involves two derivatives of the metric, it should appear linearly in the kinetic term for the metric. Because of the symmetries there are two types of tensors that can be built by taking traces of the Riemann tensor. The first is the Ricci tensor, $R_{\mu\nu} := R^\lambda{}_{\mu\lambda\nu} = R_{\nu\mu}$, and the second is the Ricci scalar $R = g^{\mu\nu}R_{\mu\nu}$. The Einstein–Hilbert lagrangian for gravity coupled to matter is then given by

$$\mathfrak{L} = \sqrt{-g}\left[-\frac{1}{2\kappa^2}R + L_m(\phi, A_\mu)\right], \tag{C.100}$$

where $\kappa^2 = 8\pi G_N$ and L_m denotes the generally covariant action for matter fields, given (for example) for a charged scalar field and electromagnetism by

$$L_m = -V(\phi^*\phi) - g^{\mu\nu}D_\mu\phi^*\, D_\nu\phi - \frac{1}{4}g^{\mu\nu}g^{\lambda\rho}F_{\mu\lambda}F_{\nu\rho}, \tag{C.101}$$

where $D_\mu\phi = (\partial_\mu - iqA_\mu)\phi$ and $F_{\mu\nu} = \partial_\mu A_\nu - \partial_\nu A_\mu$.

The Einstein equations obtained from varying this action with respect to $g_{\mu\nu}$ are

$$R^{\mu\nu} - \frac{1}{2} R g^{\mu\nu} + \kappa^2 T^{\mu\nu} = 0, \tag{C.102}$$

where

$$T^{\mu\nu} := \frac{2}{\sqrt{-g}} \left(\frac{\delta S_m}{\delta g_{\mu\nu}} \right), \tag{C.103}$$

and so on.

C.5.3 Spacetime Symmetries Reloaded

Once the metric is recognized as being a dynamical field it is worth revisiting the idea of a spacetime symmetry. Recall that in the bulk of this book spacetime symmetries are regarded as those transformations

$$\delta x^\mu = \xi^\mu(x) \tag{C.104}$$

that leave the Minkowski metric invariant: $\delta(\eta_{\mu\nu}\, dx^\mu\, dx^\nu) = 0$, or

$$\delta\eta_{\mu\nu} = \mathcal{L}_\xi \eta_{\mu\nu} := \xi^\lambda \partial_\lambda \eta_{\mu\nu} + \partial_\mu \xi^\lambda \eta_{\lambda\nu} + \partial_\nu \xi^\lambda \eta_{\mu\lambda} = 0 \tag{C.105}$$

(compare with Eq. (C.88) which defines the transformation for a general metric). The general solution to this condition led to the Poincaré group: $\xi^\mu = a^\mu + \omega^\mu{}_\nu x^\nu$ where a^μ and $\omega_{\mu\nu} = -\omega_{\nu\mu}$ (with $\omega_{\mu\nu} = \eta_{\mu\lambda}\omega^\lambda{}_\nu$) are constant parameters representing translations and Lorentz transformations.

Generally covariant theories provide a new context for these transformations, because for these the action is invariant under a much broader set of transformations: general diffeomorphisms corresponding to (C.104) and (C.88) for general $g_{\mu\nu}(x)$ and $\xi^\mu(x)$. Within this new context $g_{\mu\nu} = \eta_{\mu\nu}$ is a specific solution to the field equations and so can be regarded as being the analog of a field expectation value: $\langle g_{\mu\nu}(x)\rangle = \eta_{\mu\nu}$ in much the same way that the field ϕ acquires a nonzero expectation value $\langle\phi(x)\rangle = v$ in the ground state of the toy model of §1.1.

From this point of view Eq. (C.105) simply identifies that subset of symmetry transformations that leaves the metric's expectation value unchanged – that is, are not spontaneously broken by $\langle g_{\mu\nu}\rangle = \eta_{\mu\nu}$. More generally, the diffeomorphisms that leave a generic metric unchanged are called *isometries* and must satisfy $\delta g_{\mu\nu} = \mathcal{L}_\xi g_{\mu\nu} = 0$, and so

$$\xi^\lambda \partial_\lambda g_{\mu\nu} + \partial_\mu \xi^\lambda g_{\lambda\nu} + \partial_\nu \xi^\lambda g_{\mu\lambda} = D_\mu \xi_\nu + D_\nu \xi_\mu = 0, \tag{C.106}$$

where the first equality follows from the definition of the covariant derivative and uses the definition $\xi_\mu := g_{\mu\lambda}\xi^\lambda$. Any solution ξ^μ to (C.106) is called a Killing vector field, and such fields need not exist for arbitrary metrics. From this point of view (C.105) states that Poincaré transformations are the isometries of Minkowski spacetime.

Conserved Currents

These observations provide another way to identify (and count) conserved currents, at least for gauge symmetries whose transformation parameters are spacetime-independent.

Consider first (for simplicity) an abelian internal local symmetry that acts only on some matter fields through a transformation rule, $\delta\phi^i = \omega(x) f^i(\phi)$, and on the gauge potential $\delta A_\mu = \partial_\mu \omega$. The matter action, $S_m[\phi, A_\mu]$, for the fields ϕ^i must be invariant under the gauge symmetry, and this makes it depend on A_μ through the covariant derivative $D_\mu \phi^i$. Invariance means S_m satisfies

$$\delta S_m = \int \mathrm{d}^4 x \left[\frac{\delta S_m}{\delta \phi^i(x)} \, \omega(x) f^i + \frac{\delta S_m}{\delta A_\mu(x)} \partial_\mu \omega(x) \right] = 0 \tag{C.107}$$

for any field configurations $\phi^i(x)$ and $A_\mu(x)$ and for any symmetry parameter $\omega(x)$. If this is specialized to a solution to the ϕ^i field equation, $\delta S_m / \delta \phi^i = 0$, then the first term vanishes leaving the result[11]

$$0 = \int \mathrm{d}^4 x \; J^\mu \partial_\mu \omega = - \int \mathrm{d}^4 x \; \omega \, (\partial_\mu J^\mu), \tag{C.108}$$

where the second equality performs an integration by parts (and discards the surface term), and defines the current

$$J^\mu := \frac{\delta S_m}{\delta A_\mu(x)}. \tag{C.109}$$

Since (C.108) must vanish for *any* $\omega(x)$ it must be true that J^μ as defined in (C.109) is conserved, in the sense that the ϕ^i equations of motion imply that $\partial_\mu J^\mu = 0$. It is easy to verify in simple examples that this definition of the current agrees with the Noether-current derivation for internal symmetries given in §4.1.1.

The same logic also goes through for spacetime symmetries in generally covariant systems, and provides a more systematic way to count currents. In this case, it is the metric, $g_{\mu\nu}$, that plays the role of the gauge potential, but otherwise the argument goes through identically. Consider then a matter action $S_m[\phi^i, g_{\mu\nu}]$ that is generally covariant in the sense that it is unchanged by some transformation $\delta\phi^i = \mathcal{L}_\xi \phi^i$ and $\delta g_{\mu\nu} = \mathcal{L}_\xi g_{\mu\nu}$:

$$\delta S_m = \int \mathrm{d}^4 x \left[\frac{\delta S_m}{\delta \phi^i(x)} \; \mathcal{L}_\xi \phi^i + \frac{\delta S_m}{\delta g_{\mu\nu}(x)} \; \mathcal{L}_\xi g_{\mu\nu} \right] = 0. \tag{C.110}$$

Specializing to configurations satisfying the ϕ^i equations of motion and using the definition of $\mathcal{L}_\xi g_{\mu\nu}$ given in the first equality of (C.106) then allows (after integration by parts) Eq. (C.110) to be rewritten as

$$0 = \int \mathrm{d}^4 x \; \sqrt{-g} \; T^{\mu\nu} D_\mu \xi_\nu = - \int \mathrm{d}^4 x \; \sqrt{-g} \; \xi_\nu \, D_\mu T^{\mu\nu}, \tag{C.111}$$

and so (because ξ_ν is arbitrary) the stress-energy tensor, $T^{\mu\nu}$, defined by (C.103), must be covariantly conserved, in the sense that

$$D_\mu T^{\mu\nu} = 0. \tag{C.112}$$

[11] The gauge field does not in general satisfy $\delta S_m / \delta A_\mu = 0$ even for classical fields because S_m only consists of the matter action and does not include, for example, the Maxwell action $-\frac{1}{4} F_{\mu\nu} F^{\mu\nu}$.

This definition of the conserved stress-energy has the enormous advantage that it is what appears in the Einstein equations, (C.102), and so is precisely what gravity couples to.

For any specific metric that has an isometry, in the sense that the second equality of (C.106) is satisfied for some ξ_μ, a standard conserved current can be defined for each isometry,

$$j^\mu = T^{\mu\nu}\xi_\nu, \tag{C.113}$$

that satisfies $D_\mu j^\mu = 0$ – as can be seen using Eqs. (C.112) and $D_\mu \xi_\nu + D_\nu \xi_\mu = 0$ (*i.e.* Eq. (C.106)). In the special case of flat Minkowski space these are the conserved currents for Poincaré invariance, but (C.113) shows that they are all really built from one basic quantity: the stress-energy tensor, $T^{\mu\nu}$.

C.6 Nonlinear Realizations

The nonlinear realization used to implement spontaneously broken symmetries in an effective theory can be less intuitive than is the linear realization used for unbroken symmetries. But it is worth understanding given the widespread appearance of Goldstone bosons throughout physics.

This appendix derives the 'standard' nonlinear realization for the general case of an internal symmetry group G spontaneously broken down to a subgroup $H \subset G$. Following steps initially taken by [12, 13] and using the notation of [107] this is done by generalizing the arguments used for the abelian broken symmetry presented in the toy model of the main text. Since half the art of constructing nonlinear realizations involves choosing variables that transform conveniently, the first steps in this construction motivate the choices to be made by describing a simple non-abelian version of the toy model.

C.6.1 A Nonabelian Toy Model

To set up the standard transformation law, consider N real scalar fields, $\phi^i, i = 1, \ldots, N$, arranged for convenience into an N-component column vector, Φ. There is no loss of generality in using real fields, since any complex fields can be decomposed into real and imaginary parts.

The non-abelian toy model is defined by the lagrangian density

$$\mathfrak{L} = -\frac{1}{2}\,\partial_\mu \Phi^T \partial^\mu \Phi - V(\Phi), \tag{C.114}$$

where the superscript 'T' denotes the transpose, and where $V(\Phi)$ is a potential whose detailed form is not important in what follows. The lagrangian's kinetic term is manifestly invariant under the $O(N)$ group ($N \times N$ orthogonal matrices, $O^T O = 1$) of global rotations: $\Phi \to O\Phi$, where the O's are independent of spacetime position, $\partial_\mu O = 0$. Because the fields are chosen to be real, all generators of these symmetries are simultaneously hermitian, imaginary and antisymmetric: $T_a^\dagger = T_a = -T_a^T = -T_a^*$.

In general, the potential $V(\Phi)$ need not also be $O(N)$-invariant, but it is assumed to preserve a subgroup $G \subset O(N)$, in the sense that

$$V(g\,\Phi) = V(\Phi) \qquad \text{for all } g \in G \text{ and all } \Phi. \tag{C.115}$$

The potential V is assumed to satisfy two properties. First, its parameters are assumed to be chosen to allow a weak-coupling semiclassical treatment of the model's predictions. Second, it is assumed to be minimized at field values $\langle\Phi\rangle \neq 0$, for which the symmetry group G is generically spontaneously broken to a subgroup $H \subset G$ defined by: $h\langle\Phi\rangle = \langle\Phi\rangle$, for all $h \in H$. It is convenient to choose generators, t_i, of H as part of the basis of the Lie algebra of G, writing $\{T_a\} = \{t_i, X_\alpha\}$ where T_a are a basis of generators of the algebra of G while X_α are the broken generators of the coset G/H (for more about the nomenclature see §C.4.1).

A Choice of Variables

The idea is to identify the Goldstone and non-Goldstone degrees of freedom in this model and to identify how these each realize the model's symmetries. Within a semiclassical framework this involves sorting the fields $\Phi = \{\phi^i\}$ into a set of Goldstone modes, $\Xi = \{\xi^a\}$, plus an orthogonal set of remaining physical fields, $\mathcal{X} = \{\chi^n\}$.

As usual, Goldstone modes are obtained by performing symmetry transformations on the ground state, and for infinitesimal transformations this corresponds to the directions $X_\alpha\langle\Phi\rangle$ in field space. That is, the components of Φ in this direction, $\langle\Phi\rangle^T X_\alpha \Phi$, are the ones that create and destroy Goldstone particles. It is straightforward to verify that the G-invariance of the potential ensures the masslessness of these modes in the semiclassical approximation. This gives precisely one Goldstone mode for each generator of G/H.

Experience with the abelian symmetries of the toy model of §1.1 suggests that the variables $\langle\Phi\rangle^T X_\alpha \Phi$ need not be the most efficient for making Goldstone properties manifest, however. In particular, the low-energy decoupling of Goldstone modes are most manifest if the freedom to redefine fields is used to arrange that they do not appear at all in the scalar potential. This is most easily arranged by writing

$$\Phi = U(\xi)\,\mathcal{X}, \tag{C.116}$$

where

$$U(\xi) = \exp[i\xi^\alpha(x) X_\alpha], \tag{C.117}$$

is a spacetime-dependent symmetry transformation in the direction of the broken generators, X_α. Since $U(\xi)$ is an element of G, this definition ensures the ξ^α drop out of the scalar potential because G-invariance requires that the potential must satisfy $V(U\mathcal{X}) = V(\mathcal{X})$. Consequently, all terms in $\mathfrak{L}$ involving the Goldstone bosons, ξ^α, vanish when $\partial_\mu \xi^\alpha = 0$, and Eq. (C.116) is the change of variables that makes manifest low-energy properties of the Goldstone bosons.

In order for Eq. (C.116) not to over-count the N original fields in Φ the variables $\mathcal{X}$ must satisfy a constraint that keeps them orthogonal (in field space) to the Goldstone directions, such as:

$$\langle\Phi\rangle^T X_\alpha \mathcal{X} = 0, \qquad \text{for all } X_\alpha \tag{C.118}$$

everywhere in spacetime. As a reality check, notice that this constraint ensures the vanishing of the cross terms, proportional to $\partial_\mu \xi^\alpha \partial^\mu \hat{\chi}^n$, in the quadratic part of the expansion of the kinetic terms about the ground state configuration: $\mathcal{X} = \langle \Phi \rangle + \hat{\mathcal{X}}$. (Proving this uses the identity $\langle \Phi \rangle^\tau X_\alpha \langle \Phi \rangle = 0$, that is a consequence of the antisymmetry of the X_α's.) It can be shown [107] that it is always possible to change to variables satisfying (C.118) from any originally smooth configuration for Φ.

C.6.2 The Nonlinear Realization

The next step asks how the variables ξ^α and χ^n transform under the group G given the simple linear representation of G carried by Φ,

$$\Phi \to \widetilde{\Phi} := g\,\Phi \qquad \text{where} \qquad g = \exp[i\omega^a T_a] \in G. \qquad (C.119)$$

This leads to the standard transformation rules widely used when studying Goldstone boson properties.

The transformation rule implied for the new variables, $\xi^\alpha \to \tilde{\xi}^\alpha$ and $\chi^n \to \tilde{\chi}^n$, is found by writing $\Phi = U(\xi)\mathcal{X}$ and $\widetilde{\Phi} = U(\tilde{\xi})\widetilde{\mathcal{X}}$ in (C.119), and so

$$g\,U(\xi)\mathcal{X} = U(\tilde{\xi})\widetilde{\mathcal{X}}, \qquad (C.120)$$

for any $g \in G$.

The standard nonlinear transformation law therefore becomes:

$$\xi^\alpha \to \tilde{\xi}^\alpha(\xi, g) \qquad \text{and} \qquad \chi^n \to \tilde{\chi}^n(\xi, g, \chi), \qquad (C.121)$$

where

$$g\,e^{i\xi^\alpha X_\alpha} = e^{i\tilde{\xi}^\alpha X_\alpha}\,e^{iu^i t_i} \qquad \text{and} \qquad \widetilde{\mathcal{X}} = e^{iu^i t_i}\,\mathcal{X}. \qquad (C.122)$$

The first of Eqs. (C.122) should be read as defining the nonlinear functions $\tilde{\xi}^\alpha(\xi, g)$ and $u^i(\xi, g)$. One first finds the element, $g\,e^{i\xi \cdot X} \in G$, and then defines the functions $\tilde{\xi}^\alpha$ and u^i by decomposing this matrix into the product of a factor, $e^{i\tilde{\xi} \cdot X}$, lying in G/H times an element, $e^{iu \cdot t}$, in H. The second of Eqs. (C.122) then defines the transformation rule for the non-Goldstone fields, χ^n.

These transformation laws are generically nonlinear in the Goldstone fields, ξ^α. They nonetheless realize the symmetry group G in that $\tilde{\xi}(\theta, g_1 g_2) = \tilde{\xi}(\tilde{\xi}(\xi, g_2), g_1)$, as can be verified using the definitions of Eqs. (C.122) or by noticing that this property is inherited from the original linear representation of G on Φ.

The transformations (C.121) and (C.122) remain linear in the special case where $g = h$ lies in the unbroken sector H. In this case, the solution for u^i and $\tilde{\xi}^\alpha$ are easily seen to be: $e^{iu \cdot t} = h$ and $\widetilde{U} = hUh^{-1}$ since in this case $hU = \widetilde{U}e^{iu \cdot t}$, as required. Both χ^n and ξ^α therefore transform *linearly* under the unbroken symmetry transformations of H, with:

$$\xi^\alpha X_\alpha \to \tilde{\xi}^\alpha X_\alpha = h(\xi^\alpha X_\alpha)h^{-1} = \xi^\alpha L^\beta{}_\alpha X_\beta,$$
$$\mathcal{X} \to \widetilde{\mathcal{X}} = h\mathcal{X}, \qquad (C.123)$$

where the last equality in the first line uses (C.61).

It is harder to be equally explicit for general $g \in G/H$, but closed forms are possible for infinitesimal transformations, $g = 1 + i\omega^\alpha X_\alpha + \cdots$, if one works with a basis of generators that satisfy (C.55). In this case, writing $\gamma = 1 + \mathrm{i}u^i(\xi,\omega)t_i + \cdots$, and $U(\tilde{\xi}) = U(\xi)[1 + \mathrm{i}\Delta^a(\xi,\omega)X_a + \cdots]$ and using (C.122) implies that $u^i(\xi,\omega)$ and $\Delta^a(\xi,\omega)$ are given (at linear order in ω^a) explicitly by:

$$\Delta_a = \mathrm{Tr}\left[X_a e^{-\mathrm{i}\xi\cdot X}(\omega\cdot X)e^{\mathrm{i}\xi\cdot X}\right] \simeq \omega_a - c_{a\beta\gamma}\omega^\beta\xi^\gamma + O(\omega\,\xi^2), \tag{C.124}$$

and

$$u_i = \mathrm{Tr}\left[t_i e^{-\mathrm{i}\xi\cdot X}(\omega\cdot X)e^{\mathrm{i}\xi\cdot X}\right] \approx -c_{i\alpha\beta}\omega^\alpha\xi^\beta + O(\omega\,\xi^2). \tag{C.125}$$

These expressions liberally use the conventional choices $\mathrm{Tr}(X_\alpha X_\beta) = \delta_{\alpha\beta}$, $\mathrm{Tr}(t_i t_j) = \delta_{ij}$ and $\mathrm{Tr}(t_i X_\alpha) = 0$ for the basis of generators of the Lie algbra of G.

In particular, the transformation rules for the ξ^a under broken symmetries implied by (C.124) are

$$\delta\xi^a = \omega^a - c^a{}_{\beta\gamma}\,\omega^\beta\xi^\gamma + O(\omega\,\xi^2). \tag{C.126}$$

This transformation rule is both inhomogeneous (*i.e.* includes a shift) and acts nonlinearly on the fields ξ^a. Inhomogeneous transformations are characteristic of Goldstone bosons because shifts show that a symmetry necessarily changes the vacuum (it changes because the *vev* of the Goldstone boson field – *i.e.* the relevant order parameter – changes). It is the shift component of the symmetry that precludes ξ^a from appearing undifferentiated in the lagrangian and so enforces the low-energy decoupling of Goldstone states. The nonlinearity allows low-energy interactions to arise involving two derivatives; fewer than are possible in the abelian case studied in the toy model of §1.1.

C.6.3 Invariant Lagrangians

The transformation rules allow the construction of G-invariant Lagrangians built directly using the ξ^a and χ^n fields. The main complication arises from the construction of the kinetic terms, since the nonlinearity of the transformation rules for the fields makes them more like local than global transformations due to the spacetime-dependence of the fields.

Connections and Vielbeins

The toy model provides insight into how to construct G-invariant lagrangians. The kinetic term of the toy model is proportional to $\partial_\mu\Phi^T\partial^\mu\Phi$ and so is manifestly G invariant. This must remain so after performing the change of variables to ξ^a and $\mathcal{X}^n$, and it is instructive to see how this comes about.

To this end, notice that the replacement $\Phi = U(\xi)\mathcal{X}$ implies that $\partial_\mu\Phi = U(\partial_\mu\mathcal{X} + U^{-1}\partial_\mu U\mathcal{X})$. This suggests defining the combination

$$\mathcal{D}_\mu\mathcal{X} = \partial_\mu\mathcal{X} + U^{-1}\partial_\mu U\mathcal{X}, \tag{C.127}$$

as a covariant derivative for $\mathcal{X}$. Applying the transformations (C.121) and (C.122) to this shows that it transforms covariantly: $\mathcal{D}_\mu\mathcal{X} \to \mathrm{h}\,\mathcal{D}_\mu\mathcal{X}$, where $\mathrm{h} := e^{\mathrm{i}u\cdot t}$. It does so because $U^{-1}\partial_\mu U$ transforms like a gauge potential:

$$U^{-1}\partial_\mu U \to \tilde{U}^{-1}\partial_\mu\tilde{U} = \mathrm{h}\,(U^{-1}\partial_\mu U)\,\mathrm{h}^{-1} - (\partial_\mu\mathrm{h})\,\mathrm{h}^{-1}. \tag{C.128}$$

More information emerges if $U^{-1}\partial_\mu U$ is separated into a piece proportional to X_a plus one proportional to t_i since the inhomogeneous term, $(\partial_\mu \mathrm{h})\mathrm{h}^{-1}$, is purely proportional to t_i. Defining $\mathcal{A}^i_\mu$ and e^a_μ by

$$U^{-1}\partial_\mu U = -\mathrm{i}\mathcal{A}^i_\mu t_i + \mathrm{i}e^a_\mu X_a, \tag{C.129}$$

Eq. (C.128) implies that each of $\mathcal{A}^i_\mu(\xi)$ and $e^a_\mu(\xi)$ have separate transformation rules,

$$\mathcal{A}^i_\mu(\xi)t_i \to \mathcal{A}^i_\mu(\tilde{\xi})t_i = \mathrm{h}\,[\mathcal{A}^i_\mu(\xi)t_i]\,\mathrm{h}^{-1} - \mathrm{i}\partial_\mu \mathrm{h}\,\mathrm{h}^{-1},$$
$$\text{and}\quad e^a_\mu(\xi)X_a \to e^a_\mu(\tilde{\xi})X_a = \mathrm{h}\,[e^a_\mu(\xi)X_a]\,\mathrm{h}^{-1}. \tag{C.130}$$

The quantity $\mathcal{A}^i_\mu$ therefore transforms as if it were a gauge potential for local H transformations. To see this more explicitly, for infinitesimal $g \simeq 1 + \mathrm{i}\omega^a X_a$ and $\mathrm{h}(\xi, g) \simeq 1 + \mathrm{i}u^i(\xi,\omega)\,t_i$ the above definitions give (compare with[12] Eq. (C.72))

$$\delta\mathcal{A}^i_\mu(\xi) = \partial_\mu u^i(\xi,\omega) - c^i{}_{jk}u^j(\xi,\omega)\mathcal{A}^k_\mu(\xi), \tag{C.131}$$

for structure constants $c^i{}_{jk}$ purely within the Lie algebra of H.

Similarly, $e^a_\mu(\xi)$ transforms covariantly under the transformations, with

$$\delta e^a_\mu(\xi) = -c^a{}_{i\beta}u^i(\xi,\omega)\, e^\beta_\mu(\xi). \tag{C.132}$$

In this last expression, the structure constants define representation matrices, $(\mathcal{T}_i)^a{}_\beta = \mathrm{i}c^a{}_{i\beta}$.

More explicit formulae for $\mathcal{A}^i_\mu$ and e^a_μ can be found by first extracting the overall factor of $\partial_\mu\xi^a$ – so that $\mathcal{A}^i_\mu = \mathcal{A}^i_a(\xi)\,\partial_\mu\xi^a$ and $e^a_\mu = e^a{}_\beta(\xi)\,\partial_\mu\xi^\beta$. Then the useful identity[13]

$$e^{-\mathrm{i}A}e^{\mathrm{i}(A+B)} = 1 + \mathrm{i}\int_0^1 \mathrm{d}s\, e^{-\mathrm{i}sA}B\,e^{\mathrm{i}s(A+B)} = 1 + \mathrm{i}\int_0^1 \mathrm{d}s\, e^{-\mathrm{i}sA}B\,e^{\mathrm{i}sA} + O(B^2) \tag{C.133}$$

for square matrices A and B leads to the following expressions

$$\mathcal{A}^i_a(\xi) = -\int_0^1 ds\ \mathrm{Tr}\left[t^i e^{-\mathrm{i}s\,\xi\cdot X}X_a e^{\mathrm{i}s\,\xi\cdot X}\right] \simeq \frac{1}{2}c^i{}_{a\beta}\xi^\beta + O(\xi^2), \tag{C.134}$$

and

$$e^a{}_\beta(\xi) = \int_0^1 ds\ \mathrm{Tr}\left[X^a e^{-\mathrm{i}s\,\xi\cdot X}X_\beta e^{\mathrm{i}s\,\xi\cdot X}\right] \simeq \delta^a{}_\beta - \frac{1}{2}c^a{}_{\beta\gamma}\xi^\gamma + O(\xi^2), \tag{C.135}$$

where the approximate equalities expand in powers of ξ^a.

In the same way that $\mathcal{A}^i_a$ is used to build G-covariant derivatives like $\mathcal{D}_\mu X$, the n-bein $e^a{}_\beta$ can also be used to build G-invariant self-interactions for the ξ^a. To see how, notice that the covariant quantity, $e^a_\mu = e^a{}_\beta\,\partial_\mu\xi^\beta$, transforms very simply under G: $e_\mu \cdot X \to \mathrm{h}(e_\mu\cdot X)\ \mathrm{h}^{-1}$. Its covariant derivative is constructed from $\mathcal{A}^i_\mu t_i$:

$$(\mathcal{D}_\mu e_\nu)^a = \partial_\mu e^a_\nu + c^a{}_{i\beta}\mathcal{A}^i_\mu\, e^\beta_\nu, \tag{C.136}$$

[12] The sign mismatch between these equations is to do with representing the group using generators that are the transpose of those used for matter in (C.72) (regarding which, see also footnote[9] after Eq. (C.68)).

[13] This identity is derived by setting up and solving a first-order differential equation for $U(s) := e^{-\mathrm{i}sA}e^{\mathrm{i}s(A+B)}$.

which transforms in the same way as does e^α_μ: $\delta(\mathcal{D}_\mu e_\nu)^\alpha = -c^\alpha{}_{i\beta} u^i (\mathcal{D}_\mu e_\nu)^\beta$.

The most general G-invariant lagrangian then is $\mathfrak{L}(e_\mu, \mathcal{D}_\mu e_\nu, \ldots)$, where the ellipses denote terms involving higher covariant derivatives and the lagrangian is constrained to be globally H invariant:

$$\mathfrak{L}(he_\mu h^{-1}, h\mathcal{D}_\mu e_\nu h^{-1}, \ldots) \equiv \mathfrak{L}(e_\mu, \mathcal{D}_\mu e_\nu, \ldots). \tag{C.137}$$

Whenever $\mathfrak{L}$ satisfies (C.137) for constant h, the definitions of $e^\alpha{}_\beta$ and $\mathcal{A}^i_\alpha$ ensure it is also *automatically* invariant under global G transformations of the form of Eqs. (C.121) and (C.122).

For a Poincaré invariant system, the term involving the fewest derivatives found in this way is

$$\mathfrak{L}_{GB} = -\frac{1}{2} f_{\alpha\beta}\, \eta^{\mu\nu}\; e^\alpha_\mu\, e^\beta_\nu = -\frac{1}{2}\, g_{\alpha\beta}(\xi)\, \partial^\mu \xi^\alpha \partial_\mu \xi^\beta, \tag{C.138}$$

where the second equality defines the target-space metric $g_{\alpha\beta} := f_{\gamma\delta}\, e^\gamma{}_\alpha e^\delta{}_\beta$. Here, global H-invariance requires that the constant positive-definite matrix $f_{\alpha\beta}$ must satisfy

$$f_{\lambda\beta} c^\lambda{}_{i\alpha} + f_{\alpha\lambda} c^\lambda{}_{i\beta} = 0. \tag{C.139}$$

Eq. (C.139) can be solved fairly generally. To see how, recall the discussion around (C.61), where it is pointed out that the matrices X_α fill out a linear representation of the unbroken subgroup H with representation matrices given by $(\mathcal{T}_i)^\alpha{}_\beta = c^\alpha{}_{i\beta}$. In terms of these matrices (C.139) states that the commutators, $[\mathcal{T}_i, f]$, vanish in this representation, for all of the generators, $\mathcal{T}_i$, in the Lie algebra of H. If this representation of H is irreducible then, by Schur's lemma, this implies $f_{\alpha\beta}$ must be proportional to the unit matrix, with positive coefficient: $f_{\alpha\beta} = F^2 \delta_{\alpha\beta}$. Otherwise, if this representation can be reduced into n irreducible blocks, then $f_{\alpha\beta}$ need only be block-diagonal, with each diagonal element proportional to a unit matrix:

$$f_{\alpha\beta} = \begin{pmatrix} F_1^2 \delta_{\alpha_1\beta_1} & & \\ & \ddots & \\ & & F_n^2 \delta_{\alpha_n\beta_n} \end{pmatrix}, \tag{C.140}$$

for n independent positive constants, F_n^2.

A similar construction gives the action for the $\mathcal{X}$ fields (and for any other fields that happen to be present at low energies). Because the symmetry H is unbroken, these fields all transform linearly under H: $\mathcal{X} \to h\mathcal{X}$, where the constant matrices $\{h\}$ form a (possibly reducible) representation of H.

In this case, the general coupling of these fields to the Goldstone bosons again starts with an arbitrary, globally H-invariant lagrangian: $\mathfrak{L}(\mathcal{X}, \partial_\mu \mathcal{X}, \ldots) = \mathfrak{L}(h\mathcal{X}, h\partial_\mu \mathcal{X}, \ldots)$, for constant $h \in H$. This lagrangian is automatically promoted to become G-invariant by appropriately coupling the Goldstone bosons.

The promotion to G invariance proceeds by assigning to χ the nonlinear G-transformation rule: $\mathcal{X} \to \mathtt{h}\mathcal{X}$, where $\mathtt{h} = \mathtt{h}(\xi, g) = e^{iu\cdot t} \in H$ is the field-dependent H matrix which is defined by the nonlinear realization, Eq. (C.122). An arbitrary globally H-invariant $\mathcal{X}$-lagrangian then becomes G invariant if all derivatives are replaced by the ξ-dependent covariant derivative: $\partial_\mu \mathcal{X} \to \mathcal{D}_\mu \mathcal{X} = \partial_\mu \mathcal{X} - i\mathcal{A}^i t_i \mathcal{X}$, since this ensures $\mathcal{D}_\mu \mathcal{X} \to \mathtt{h}\mathcal{D}_\mu \mathcal{X}$ and so transforms covariantly under G transformations.

Combining all of the above constructions, a general G-invariant lagrangian has the form $\mathfrak{L}(e_\mu, \mathcal{X}, \mathcal{D}_\mu e_\nu, \mathcal{D}_\mu \mathcal{X}, \dots)$, provided only that $\mathfrak{L}$ is constrained to be invariant under global H transformations:

$$\mathfrak{L}(h e_\mu h^{-1}, h\mathcal{X}, h\mathcal{D}_\mu e_\nu h^{-1}, h\,\mathcal{D}_\mu \mathcal{X}, \dots) \equiv \mathfrak{L}(e_\mu, \mathcal{X}, \mathcal{D}_\mu e_\nu, \mathcal{D}_\mu \mathcal{X}, \dots). \quad \text{(C.141)}$$

In summary, the general statement for nonlinear realizations is this: when a global internal symmetry group G is broken to a subgroup H then the low-energy action is found by constructing the most general H-invariant local lagrangian built from the low-energy field content. This lagrangian is then 'for free' promoted to be G-invariant by coupling the Goldstone bosons in the way dictated by replacing ordinary derivative by covariant derivatives, $\partial_\mu \mathcal{X} \to \mathcal{D}_\mu \mathcal{X}$ and $\partial_\mu e^\alpha_\nu \to \mathcal{D}_\mu e^\alpha_\nu$.

Uniqueness

Although the above construction defines a G-invariant local lagrangian for the fields ξ^a and X, is this the most general way such an action can be built? This section closes with a proof of uniqueness for the construction.

To prove uniqueness assume the existence of a general lagrangian density of the form, $\mathfrak{L}(\xi, \partial_\mu \xi, \chi, \partial_\mu \chi)$, involving the fields ξ^a, χ^n and their derivatives. (The extension to lagrangians depending on second and higher derivatives is straightforward.) It is actually more convenient to trade the dependence of $\mathfrak{L}$ on $\partial_\mu \xi$ for a dependence on the combinations $e^\alpha_\mu = e^\alpha{}_\beta(\xi)\, \partial_\mu \xi^\beta$ and $\mathcal{A}^i_\mu = \mathcal{A}^i_\alpha(\xi)\, \partial_\mu \xi^\alpha$. There is no loss of generality in doing so, since any function of ξ and $\partial_\mu \xi$ can always be written as a function of ξ, e^α_μ and $\mathcal{A}^i_\mu$. This equivalence is most easily seen in terms of the matrix variable $U(\xi) = e^{i\xi \cdot X}$ since any function of ξ and $\partial_\mu \xi$ can equally well be written as a function of U and $\partial_\mu U$, or equivalently as a function of U and $U^{-1}\partial_\mu U$. But expression (C.129) shows that an arbitrary function of $U^{-1}\partial_\mu U$ is equivalent to a general function of e^α_μ and $\mathcal{A}^i_\mu$.

The condition that a general function, $\mathfrak{L}(\xi^a, e^\alpha_\mu, \mathcal{A}^i_\mu, \chi, \partial_\mu \chi)$, is invariant with respect to G transformations then is

$$\delta \mathfrak{L} = \frac{\partial \mathfrak{L}}{\partial \xi^a}\, \delta \xi^a + \frac{\partial \mathfrak{L}}{\partial e^\alpha_\mu}\, \delta e^\alpha_\mu + \frac{\partial \mathfrak{L}}{\partial \mathcal{A}^i_\mu}\, \delta \mathcal{A}^i_\alpha + \frac{\partial \mathfrak{L}}{\partial \chi^n}\, \delta \chi^n + \frac{\partial \mathfrak{L}}{\partial(\partial_\mu \chi^n)}\, \delta \partial_\mu \chi^n = 0. \quad \text{(C.142)}$$

To see what this means, first specialize to the special case where the symmetry transformation lies in H: $g = e^{i\eta \cdot t} \in H$ by using in Eq. (C.142) the transformations:

$$\delta \xi^\alpha = -c^\alpha{}_{i\beta}\eta^i \xi^\beta, \qquad \delta e^\alpha_\mu = -c^\alpha{}_{i\beta}\eta^i e^\beta_\mu, \qquad \delta \mathcal{A}^i_\mu = -c^i{}_{jk}\eta^i \mathcal{A}^k_\mu,$$
$$\text{and} \qquad \delta \chi^n = i\eta^i (t_i \chi)^n, \qquad \delta \partial_\mu \chi^n = i\eta^i (t_i \partial_\mu \chi)^n. \quad \text{(C.143)}$$

Requiring $\delta \mathfrak{L} = 0$ for all possible transformation parameters, η^i, then implies that

$$\frac{\partial \mathfrak{L}}{\partial \xi^\alpha}\, c^\alpha{}_{i\beta}\xi^\beta + \frac{\partial \mathfrak{L}}{\partial e^\alpha_\mu}\, c^\alpha{}_{i\beta} e^\beta_\mu + \frac{\partial \mathfrak{L}}{\partial \mathcal{A}^j_\mu}\, c^j{}_{ik}\mathcal{A}^k_\mu - \frac{\partial \mathfrak{L}}{\partial \chi^n}\, i(t_i \chi)^n - \frac{\partial \mathfrak{L}}{\partial(\partial_\mu \chi^n)}\, i(t_i \partial_\mu \chi)^n = 0, \quad \text{(C.144)}$$

which simply states that $\mathfrak{L}$ must be an H-invariant function of its arguments for global linear H transformations.

Next consider transformations that are not in H, $g = e^{i\omega\cdot X} \in G/H$, and instead evaluate Eq. (C.142) using the transformations

$$\delta\xi^\alpha = \xi^\alpha{}_\beta\omega^\beta, \qquad \delta e^\alpha_\mu = -c^\alpha{}_{i\beta}u^i e^\beta_\mu, \qquad \delta\mathcal{A}^i_\mu = \partial_\mu u^i - c^i{}_{jk}u^i\mathcal{A}^k_\mu,$$
$$\text{and} \qquad \delta\chi^n = \mathrm{i}u^i(t_i\chi)^n \qquad \delta\partial_\mu\chi^n = \mathrm{i}u^i(t_i\partial_\mu\chi)^n, \tag{C.145}$$

where $\xi^\alpha = \xi^\alpha{}_\beta(\xi)\,\omega^\beta$ and $u^i = u^i_\alpha(\xi)\,\omega^\alpha$ are the nonlinear functions of ξ defined by Eq. (C.122), or (C.125) and (C.126). Using these in Eq. (C.142), and simplifying the resulting expression using Eq. (C.144), leads to the remaining condition for G invariance:

$$\frac{\partial\mathfrak{L}}{\partial\xi^\alpha}\left(\xi^\alpha{}_\beta + c^\alpha{}_{i\gamma}u^i_\beta\xi^\gamma\right) + \frac{\partial\mathfrak{L}}{\partial\mathcal{A}^j_\mu}\,\partial_\mu u^i_\beta + \frac{\partial\mathfrak{L}}{\partial(\partial_\mu\chi^n)}\,\mathrm{i}\partial_\mu u^i_\beta(t_i\chi)^n = 0. \tag{C.146}$$

To see what this means, first specialize to $\xi^\alpha = 0$, in which case $\partial_\mu u^i_\beta = \partial_\alpha u^i_\beta\,\partial_\mu\xi^\alpha$ vanishes. Then since Eq. (C.126) implies that $\xi^\alpha{}_\beta(\xi = 0) = \delta^\alpha_\beta$, it follows that

$$\left.\frac{\partial\mathfrak{L}}{\partial\xi^\alpha}\right|_{\xi=0} = 0. \tag{C.147}$$

But since the group transformation law for ξ^α is inhomogeneous, it is always possible to perform a symmetry transformation to ensure that $\xi^\alpha = 0$ for *any* point within G/H, and so Eq. (C.147) also implies the more general result

$$\frac{\partial\mathfrak{L}}{\partial\xi^\alpha} \equiv 0 \quad \text{for all } \xi^\alpha \in G/H. \tag{C.148}$$

The rest of the information contained in Eq. (C.146) is extracted by simplifying using $\partial\mathfrak{L}/\partial\xi^\alpha = 0$. This leads to

$$\left(\frac{\partial\mathfrak{L}}{\partial\mathcal{A}^j_\mu} + \frac{\partial\mathfrak{L}}{\partial(\partial_\mu\chi^n)}\,\mathrm{i}(t_i\chi)^n\right)\partial_\mu u^i_\beta = 0, \tag{C.149}$$

which states that the two variables, $\mathcal{A}^i_\mu$ and $\partial_\mu\chi^n$, can only appear in $\mathfrak{L}$ through the one combination: $(\mathcal{D}_\mu\chi)^n \equiv \partial_\mu\chi^n - \mathrm{i}\mathcal{A}^i_\mu(t_i\chi)^n$. That is, χ can appear differentiated in $\mathfrak{L}$ only through the covariant derivative, $\mathcal{D}_\mu\chi$.

We see from these arguments that the G-invariance of $\mathfrak{L}$ is equivalent to the statement that $\mathfrak{L}$ must be an H-invariant function constructed from the covariantly transforming variables e^α_μ, χ and $\mathcal{D}_\mu\chi$. If higher derivatives of ξ had been considered, then the vanishing of the terms in $\delta\mathfrak{L}$ that are proportional to more than one derivative of u^i would similarly imply that derivatives of e^α_μ must also only appear through its covariant derivative, $(\mathcal{D}_\mu e_\nu)^\alpha$, defined by Eq. (C.136).

Since these are the constructions for invariant lagrangians used in earlier sections, this earlier construction must be unique.

C.7 LSZ Reduction and Bound-State Energies

In §12.2.4 of the main text conclusions are drawn about the size of particular contributions to bound-state energies for positronium. These conclusions are drawn using Feynman rules for a correlation function $\langle T\,[\Psi^*_{i_1}(x_1)\,\Phi^*_{i_2}(x_2)\,\Psi_{i_3}(x_3)\,\Phi_{i_4}(x_4)]\rangle$, and this section of the appendix aims to fill in some of the missing steps that relate the correlation function to bound-state energies.

The main observation is this: the Fourier transform of a time-ordered vacuum correlation function of a field operator has poles at the positions of the energies of states that can be created from the vacuum by the operator in question. To see how this works, consider the vacuum time-ordered correlation function for any local field operator $O(x)$:

$$iG(\omega,\mathbf{q}) := \frac{1}{(2\pi)^3}\int d^4x\, \langle\Omega|T[O(0)\,O^*(x)]|\Omega\rangle\, e^{-i\omega x^0+i\mathbf{q}\cdot\mathbf{x}}. \tag{C.150}$$

Imagine evaluating this by inserting a complete set of momentum eigenstates:

$$\begin{aligned} iG(\omega,\mathbf{q}) &= \int d^4x\, e^{iq\cdot x}\sum_N\int\frac{d^3\mathbf{k}}{(2\pi)^3}\Big[\Theta(x^0)\,\langle\Omega|O(0)|N(\mathbf{k})\rangle\langle N(\mathbf{k})|O^*(x)|\Omega\rangle \\ &\qquad + \Theta(-x^0)\,\langle\Omega|O^*(x)|N(\mathbf{k})\rangle\langle N(\mathbf{k})|O(0)|\Omega\rangle\Big] \\ &= \int d^4x\, e^{iq\cdot x}\sum_N\int\frac{d^3\mathbf{k}}{(2\pi)^3}\,\Theta(x^0)\,\langle\Omega|O(0)|N(\mathbf{k})\rangle\langle N(\mathbf{k})|O^*(x)|\Omega\rangle, \end{aligned} \tag{C.151}$$

where N contains all other labels besides momentum, and $\Theta(u) = \{0$ if $u < 0$ and 1 if $u > 0\}$ is the usual Heaviside step function. The last equality assumes that $O(x)$ carries a conserved charge and the quantum numbers are such that it is $\langle N(\mathbf{k})|O^*(x)|\Omega\rangle$ that is nonzero (and so $\langle\Omega|O^*(x)|N(\mathbf{k})\rangle$ vanishes). Spacetime translation invariance implies $\langle N(\mathbf{k})|O^*(x)|\Omega\rangle = \langle N(\mathbf{k})|O^*(0)|\Omega\rangle\, e^{-ik_N\cdot x}$ and so

$$\begin{aligned} iG(\omega,\mathbf{q}) &= \sum_N\int\frac{d^3\mathbf{k}}{(2\pi)^3}\,|\langle\Omega|O(0)|N(\mathbf{k})\rangle|^2\int d^4x\,\Theta(x^0)\,e^{i(q-k_N)\cdot x} \\ &= i\sum_N\frac{|\langle\Omega|O(0)|N(\mathbf{q})\rangle|^2}{E_N(\mathbf{q})-\omega+i\epsilon}, \end{aligned} \tag{C.152}$$

where $q^\mu = (\omega,\mathbf{q})$ and $k_N^\mu = [E_N(\mathbf{k}),\mathbf{k}]$, and the Fourier representation of $\Theta(u)$ is used:

$$\Theta(u) = \int\frac{dw}{2\pi}\left(\frac{i}{w+i\epsilon}\right)e^{-iwu}. \tag{C.153}$$

For the present purposes what is important about (C.152) is the pole it reveals at $\omega = E_N(\mathbf{q}) + i\epsilon$. This argument as applied to many-field correlation functions is related to the 'Lehmann-Symanzik-Zimmermann' (LSZ) reduction formula [323], which further argues that the residue at these poles gives S-matrix elements for transitions amongst states corresponding to the fields involved in the correlation function.

Notice that nothing in the above derivation assumes $O(x)$ is a particular 'elementary' field for particle type N; *any* operator for which $\langle\Omega|O(0)|N(\mathbf{k})\rangle$ is nonzero – usually called an 'interpolating field' for N – will do. In particular, for applications to two-body bound states it is usually convenient to focus on interpolating fields that are bilinears of the 'fundamental' fields:

$$O(x_1,x_2) = \Psi(x_1)\,\Phi(x_2), \tag{C.154}$$

and choose equal times, $x_1^0 = x_2^0 =: x^0$, so that $O(x_1, x_2) = O(x^0, \mathbf{x}\ \mathbf{X})$, where $\mathbf{X}$ is the centre-of-mass coordinate for $\mathbf{x}_1$ and $\mathbf{x}_2$ while $\mathbf{x} = \mathbf{x}_1 - \mathbf{x}_2$ is the relative separation. In this case, the starting point for the above argument would be the correlation function

$$iG(\omega, \mathbf{q}) := \frac{1}{(2\pi)^6} \int dx^0 d^3\mathbf{x}\, d^3\mathbf{X}\, \langle \Omega | T[O(0, 0; 0)\, O^*(x^0, \mathbf{x}; \mathbf{X})] | \Omega \rangle\, e^{-i\omega x^0 + i\mathbf{q}\cdot\mathbf{x}}, \tag{C.155}$$

rather than (C.150), where the integration over $\mathbf{X}$ has the effect of projecting onto zero centre-of-mass momentum.

In practice, the above correlation function is computed by perturbing about an approximate solution for the bound state, in which case it is useful to write the near-pole behaviour as

$$G(\omega, \mathbf{q}) \simeq \frac{|R(\omega, \mathbf{q})|^2}{\omega - E_N} \tag{C.156}$$

where $R(\omega, \mathbf{q}) = \langle \Omega | O(0) | N(\mathbf{q}) \rangle$ has no pole at $\omega = \varepsilon_N$, and expand $E_N = \varepsilon_N + \delta E_N$ and $R(\omega, \mathbf{q}) = \mathcal{R}(\mathbf{q}) + \delta R(\omega, \mathbf{q})$. Then

$$\begin{aligned} G(\omega, \mathbf{q}) &\simeq \frac{|\mathcal{R}(\mathbf{q}) + \delta R(\omega, \mathbf{q})|^2}{\omega - \varepsilon_N - \delta E_N} \\ &\simeq \frac{|\mathcal{R}(\mathbf{q})|^2}{\omega - \varepsilon_N} + \frac{\mathcal{R}^*(\mathbf{q})\, \delta R(\omega, \mathbf{q}) + \mathcal{R}(\mathbf{q})\, \delta R^*(\omega, \mathbf{q})}{\omega - \varepsilon_N} \\ &\quad + \frac{\mathcal{R}^*(\mathbf{q})}{\omega - \varepsilon_N}\, \delta E_N\, \frac{\mathcal{R}(\mathbf{q})}{\omega - \varepsilon_N} + \cdots, \end{aligned} \tag{C.157}$$

which shows that the leading corrections to δE_N can be read off by amputating the two external bound-state propagators – that is to say, by multiplying by a factor of $(\omega - \varepsilon_N)/\mathcal{R}$ and its complex conjugate – and evaluating the result at $\omega = \varepsilon_N$. In the nonrelativistic applications of Chapter 12 this amounts to evaluating the amputated graph and taking the expectation value of the result using the zeroeth-order (Schrödinger–Coulomb) wavefunction.

Appendix D Further Reading

This book touches only briefly on each of the applications of effective field theories (EFTs) throughout physics in order to emphasize the great generality of EFT techniques. But this also means that many readers are likely to be dissatisfied with the level of detail used to describe each application. This section aims to help with this by providing some further reading for those interested in quenching a more fundamental thirst for knowledge in each of the areas touched.

The bibliography given here is not meant to be an exhaustive survey of the literature, about parts of which I am sure I am relatively poorly informed (and I apologize in advance for any gems I may have missed). Instead, I list references that I have found useful myself, and include review articles to which the reader should go for more detailed referencing in each area.

Many of these papers (at least those published since the development of the World-Wide Web in the early 1990s) are available for free online. In particular, references like

[arXiv:hep-ph/9708416] or [arXiv:1704.02751]

are shorthands (respectively) for the links:

https://arxiv.org/abs/hep-ph/9708416
or https://arxiv.org/abs/1704.02751.

D.1 Quantum Field Theory

The main prerequisite for reading this book is an understanding of quantum field theory (QFT). At face value, QFT is only a convenient formalism for handing many-particle quantum mechanics, including in particular processes like emission and absorption, that change the number of particles. But it is also the ubiquitous language of physics, since it makes it simple to bake in basic properties like unitarity and cluster decomposition from the get-go when trying to guess a system's dynamics. QFT is particularly useful for relativistic applications because it is a basic fact of relativistic quantum mechanics that *all* interactions involve components that change the number of particles (due to the inevitable presence of antiparticles).

For this reason the appendices are largely devoted to providing a very brief summary of the basic facts of quantum field theory. Inevitably, an interested reader will want more, and here are a few suggestions for further reading (organized roughly by topics).

There are very many good textbooks on quantum field theory, not all of which can be named here. Some useful textbooks on general-purpose relativistic quantum field theory that I have used are:

- 'An Introduction to Quantum Field Theory', by M. Peskin and D. Schroeder, Westview, 1995. An excellent and readable book on field theory (though with an unfortunate choice of metric conventions). The first four chapters of the book give all required background preparation for this book, and the remaining chapters give the tools needed to take the material presented in this book to the next level (renormalization, higher order effects and a more solid theoretical foundation).
- 'The Quantum Theory of Fields, I–III', by S. Weinberg, Cambridge Press, 2000. An original and encyclopedic presentation of quantum field theory from one of the masters who helped systematize much of it. In particular, Volume I addresses many of the foundational arguments that underpin quantum field theory, while volumes II and III are more dedicated to applications. Most of the field theoretical arguments alluded to in this book are laid out in detail here. You will learn something new every time you read it, probably for the rest of your life. But it is likely not as good for novices as is an introductory text like Peskin and Schroeder's book.
- 'Quantum Field Theory in a Nutshell', by A. Zee, Princeton Press, 2010. This book fits into a special niche in that its emphasis is more on concepts and less on calculational tools. Not a bad place for a learner to start, but probably also not enough in itself for someone seeking a practical hands-on calculational ability.
- 'Quantum Field Theory', by L. Ryder, Cambridge Press, 1996 (2nd ed). An older and somewhat more introductory text on field theory, providing more than enough background material to understand this book.
- 'Quantum Field Theory', by L. Brown, Cambridge Press, 1994. A clear introduction to quantum field theory with an interestingly novel choice of topics that gives an extremely solid underpinning (though does not cover nonabelian gauge theories).
- 'Quantum Field Theory', by G. Sterman, Cambridge Press, 1993. A clear and systematic exposition of modern field theoretic techniques which includes a number of topics (like infrared divergences and factorization) not covered in other texts.
- 'Advanced Topics in Quantum Field Theory', by M. Shifman, Cambridge University Press 2010, gives an authoritative discussion of QFT with an emphasis on non-perturbativee methods. This book covers many topics often not encountered in QFT textbooks.
- For detailed (but advanced and somewhat more mathematical) discussions of C, P, and T symmetries, the spin-statistics theorem and related topics, try 'PCT, Spin and Statistics, and All That', by A. Wightman and R. Streater, Princeton University Press, 2000.

Other books that are more aimed at particle physics are also listed below in the section devoted to the Standard Model. Many of these books (and those mentioned later), particularly Weinberg's, advocate an effective field theory point of view, though this is usually not their main focus.

D.2 EFT Framework

In this book Part I develops the main EFT formalism used throughout the rest of the text. The main logic explored throughout the entire EFT program is largely laid out in the paper entitled:

- 'Phenomenological Lagrangians' [*Physica A* **96** (1979) 327] by Steven Weinberg [2]. This paper has held up remarkably well to the passage of time and remains worth reading as a statement of purpose for those taking up the subject anew.

A book with similar goals to the one you are reading, whose scope directly aims at EFT methods (but with a complementary choice of topics), is:

- 'Effective Field Theories', by Alexei Petrov and Andrew Blechman, World Scientific, 2015. This book aims more directly at high-energy physics and nonrelativistic applications like NRQED and effective theories of gravity than the one you are now reading.

Quantum Actions

The framework of generating functionals goes back into the mists of time in the mid-twentieth century when quantum field theory was relatively young. The specific use of the 1PI quantum action (often in the old days also called an 'effective action' though this term is now normally reserved for the Wilson action) came in the mid-1960s, where it was introduced within perturbation theory as the formal sum over 1PI graphs [5]. The non-perturbative definition used here came a bit later in [15].

In retrospect, much of the formalism of field theory used today was systematized in the 1960s. A comprehensive one-stop-shopping source for much of these developments is:

- 'The Quantum Theory of Fields, vol I', by Steven Weinberg, Cambridge Press, 2000. This book (already mentioned above) authoritatively lays out the foundations of quantum field theory, straight from the proverbial horse's mouth. Unlike most books on quantum field theory, this book ('Vol-I', for short) does not start off assuming quantum field theory is the right subject to study. The goal instead is to study what it means for quantum mechanics to be consistent with special relativity (Poincaré-invariance), and Vol-I systematically makes the case that this *is* quantum field theory.
- 'Aspects of Symmetry', by Sidney Coleman, Cambridge Press, 2010. This is a collection of lectures given by Sidney Coleman over the years at the summer school in Erice in Sicily. All of these are well-known as masterful expositions of different topics in field theory, and include a very clear explanation of the generators $W[J]$ and $\Gamma[\varphi]$ of connected and 1PI correlation functions.
- 'What Is Quantum Field Theory, and What Did We Think It Is?', also by Steven Weinberg [a contribution to the proceedings of the conference *Conceptual foundations of quantum field theory*, Boston 1996, pp. 241–251 [405]), also available online at hep-th/9702027]. This is less of a 'shut up and calculate' description of quantum field theory, and more of a retrospective view of what quantum field

theory is and how this has changed over the years. In particular, it provides a chatty and easy to read summary of the modern picture, wherein quantum field theory is what emerges when you combine special relativity, quantum mechanics and 'cluster decomposition' (the principle that probabilities for independent events widely separated in space must factorize).

This last article explicitly enunciates the basic modern point of view: quantum field theory in itself has very little content, in that the most general field theory consistent with the analyticity properties of scattering amplitudes is the same as the most general physics that is consistent with these properties (subject to a few motherhood principles like conservation of probability (*i.e.* unitarity) and cluster decomposition). Although largely taken for granted now, this was controversial in the 1960s when it was felt that quantum field theory could not describe the strong interactions. This led to a program that based itself only on the analytic properties of the *S*-matrix, a summary of which can be found in the review

- 'Regge Poles and S-matrix Theory', by Steven Frautschi, New York: W. A. Benjamin, Inc., 1963,

and which has echoes in more recent lines of research [469].

The formalism of coarse-graining short distances and the related renormalization group also has a long history. It starts off with the study of renormalization and scaling in particle physics (and QED in particular) in the early 1950s [470, 471]. A big improvement in generality came with more explicit formulations of how to split low- and high-energy degrees of freedom (coarse-graining) in the 1960s and early 70s, starting within condensed-matter physics [472–474] and moving from there back to particle physics [139, 140]. Extensive reviews of these developments can be found in these sources:

- 'The Renormalization group and the epsilon expansion', by Ken Wilson [*Physics Reports* **12** (1974) 75] provides an excellent contemporary survey of these techniques by an inventor.
- 'Field Theory, the Renormalization Group and Critical Phenomena', by Daniel Amit, World Scientific 1984. This book aims more at condensed matter applications of renormalization methods, and later editions (with Victor Martin-Mayor) also include discussions of strong-coupling, lattice models and numerical methods.
- 'Statistical Field Theory', by G. Parisi, Addison-Wesley 1988. This book gives a high-level and influential discussion that is accessible to people with both particle and condensed matter backgrounds.
- 'Why the Renormalization Group Is a Good Thing', published in *Asymptotic Realms Of Physics,* 1–19 Cambridge 1981 [40]. This is a contribution to the proceedings of the festschrift for Francis Low by Steven Weinberg, clearly summarizing some of the history and ideas. (A bonus is the statement found in this article of the Three Laws of Theoretical Physics.)

Later progress in formulating and using coarse-grained techniques starts with Polchinski's formulation of the exact renormalization group [25], and continuing with later refinements [26, 27, 475]. In the meantime parallel developments separately begin to apply EFT ideas to more and more areas of physics (aspects of which are largely the subject of this book and so are described below).

Much of the discussion in this chapter is based on the presentation given in some review lectures on effective theories

- 'Introduction to Effective Field Theory', by C. P. Burgess [*Ann.l Rev. Nucl. Part. Sci.* **57** (2007) 329, arXiv:hep-th/0701053]. These are my own lectures and so, not surprisingly, they overlap in their layout with what is found in this book, including the development of the 1LPI action and the use of the toy model as a useful vehicle for illustrating more general features.
- 'Five Lectures on Effective Field Theory', by David B. Kaplan (arXiv:nucl-th/0510023), which is a very clear survey that sets up the framework quite broadly and then narrows in to applications more focussed on nuclear and nonrelativistic physics. In particular, the discussion of scaling given here is largely as presented in these lectures.

Power Counting

Chapter 3 deals with power-counting with effective lagrangians, using dimensional analysis to estimate the dependence of generic Feynman graphs in terms of the scales appearing in the couplings of the effective theory. The arguments made parallel the power counting arguments used when deciding the superficial degree of divergence of Feynman graphs, such as when proving the renormalizability of a field theory (like QED) [128, 476]. The dimensional analysis likely comes across as cavalier inasmuch as the relevant graphs really give multidimensional integrals and one might worry whether their behaviour is well-captured by naive one-dimensional estimates. As usual, Weinberg's textbook (Vol-I *ibid*) is an invaluable – though fairly compact – resource for these arguments.

The justifications for these arguments ultimately rely on Weinberg's theorem [477], which underpins the proofs of renormalizability, and clarify why naive arguments properly capture the multidimensional complications. An authoritative summary of the issues, with historical commentary, can be found in Vol I of Weinberg's 'Quantum Field Theory' trilogy, cited above. A more recent (though also not that recent) and exhaustive treatment can also be found in

- 'Renormalization', by John Collins, Cambridge Press, 1984. This book provides a very detailed treatment of renormalization in the post-dimensional regularization age.

The spirit of power counting from an EFT framework is already in the 'Phenomenological Lagrangians' paper cited above [2], though I follow in this book the notation and presentation outlined in my own review [24].

The 'method of regions' is a very useful technique for identifying how different scales can enter a calculation when using dimensional regularization. This is described in some detail in the book

- 'Introduction to Soft-Collinear Effective Theory', by Thomas Becher, Alessandro Broggio and Andrea Ferroglia, Springer, 2015. More generally, this book is a useful handbook for techniques that arise when using dimensional regularization within an EFT analysis.

Symmetries

For some reason physicists tend to pick up much of their group theory on the streets. Two very useful introductions to the theory of Lie groups for physicists, and a very useful reference with extensive tables, are

- 'Lie Algebras in Particle Physics', by H. Georgi, Perseus, 1999 (2nd ed). This is a very thorough treatment of group theory for particle physics, essential for those who find Appendix C.4.1 either too telegraphic or incomplete.
- 'Semi-Simple Lie Algebras and Their Representations', by R. N. Cahn, Benjamin-Cummings, 1984. This book picks up where the previous suggestion leaves off, presenting more of the properties of groups and their representations.
- 'Group Theory for Unified Model Building', R. Slansky, *Phys. Rep.* **79** (1981) 1–128. This provides a very useful summary of the properties of the representations of Lie groups, including detailed tables showing how representations decompose in terms of representations of subgroups.

The development of the theory of nonlinear realizations starts with the nonlinear sigma model for pion physics [11], which was then generalized to general groups in the standard form used today in [12, 13].

An in-depth discussion of nonlinear realizations and their historical development, as well as a systematic derivation of anomalies both from the point of view of triangle diagrams and of path integral measure, including also the Wess–Zumino consistency relations [478, 479] and their solution using descent equations [480], is given in

- 'The Quantum Theory of Fields, vol II', by Steven Weinberg, Cambridge Press, 2000. This, the second volume of the Quantum Theory of Fields trilogy ('Vol-II' for short), picks up where Vol-I leaves off, touching on most of the higher topics of quantum field theory. Besides detailed derivations this book has many historical commentaries from one of the central participants of the time.

A summary of consistency conditions and the descent equations that emphasizes more geometrical methods is given in the review by Bruno Zumino in 'Relativity, Groups and Topology II', edited by B. S. de Witt and R. Stora, Elsevier, Amsterdam, 1984.

Time-Dependent Backgrounds

The power of making low-energy arguments and the relative simplicity of the low-energy limit in quantum mechanics has been known for a long time, going back to the Born–Oppenheimer approximation [312]. Effective theories were first systematically developed for field theories with applications to particle scattering in mind [48] (though a parallel line of development was also underway in condensed matter physics [401]). Because time-dependent backgrounds often do not arise in these applications, work to develop a formalism for describing classical time evolution within a Wilsonian effective theory was historically not a priority (so far as I can see).

The beginnings of precision cosmology with the measurement of primordial fluctuations [481] provided a big incentive for having controlled low-energy approximations in time-dependent environments, and most developments trace back to this. (See the discussion below concerning Chapter 10 for more references to the cosmology literature.) More recently, the prospect of measuring gravitational waves also stimulated a Wilsonian reformulation [359] for calculations of time-dependent classical motion, such as those describing the radiation of inspiraling, nonrelativistic, gravitating objects like black holes or neutron stars.

The discussion in Chapter 6 follows the logic of my own reviews [24] and [482]. This differs somewhat from much of the cosmology literature, for which EFTs often zero in more specifically to the study of fluctuations about a cosmological background along the lines developed in [104]. In particular, the toy model discussion follows [89], which was itself stimulated by related work on cosmological fluctuations [483].

The discussion of well-posedness of the initial-value problem is only now starting to sink in to the EFT community, largely driven by the desire to describe and test modifications to general relativity in the strong-field regime revealed by gravitational wave observations. References [98] and [99] quoted in the main text provide good summaries of these issues both for gravitational and fluid physics.

D.3 Relativistic Applications

Part II begins a discussion of relativistic applications. The Fermi theory of the weak interactions is the poster child for how effective theories arise in nature, and so is discussed in a variety of EFT reviews such as:

- 'Weak Interactions and Modern Particle Theory', by H. Georgi, Benjamin Cummings, 1984. A very physical discussion of much of the standard model and some of the techniques used to compute with it, with EFT methods squarely in mind. The treatment of the weak interactions includes various loop corrections to the Fermi lagrangian and survives well despite its age.

The notation and description of the weak interactions and QED used in this section is partly taken from my own book

- 'The Standard Model: A Modern Primer', by Guy Moore and me, Cambridge Press, 2007, post–Higgs-discovery revision 2013 uses the Standard Model as a vehicle for learning quantum field theory, mostly at the level of tree graphs, but includes sections on QED, infrared effects, hadrons and chiral perturbation theory. Modern EFT methods are also included, and used to organize the treatment of Beyond the Standard Model (BSM) physics.

There are many classic texts on Quantum Electrodynamics, though usually of a vintage that predates the widespread adoption of EFT reasoning. More modern discussions can be found in some of the books listed above on quantum field theory.

A discussion of the $U_A(1)$ problem and the role of anomalies in resolving it is given by [484], as well as in the review 'Uses of Instantons', by Coleman in Aspects of Symmetry [17].

Chiral Perturbation Theory

Chiral perturbation theory was the place where EFT methods, and the low-energy treatment of Goldstone bosons, were first systematized in the particle-physics literature. A discussion appears in Weinberg (Quantum Theory of Fields Vol II *ibid*) with historical notes. A very comprehensive and instructive book on the subject is

- 'Dynamics of the Standard Model', by J. Donoghue, E. Golowich and B. Holstein, Cambridge Press, 1992. This contains an advanced discussion of the standard model, with particular emphasis on bound states in QCD, chiral symmetry and radiative corrections.

Many good review articles also exist on chiral perturbation theory, a selection of which is listed in reference [485].

Standard Model

There are many books on the Standard Model, which is a well-developed subject. In addition to the ones listed above, two other noteworthy examples are

- 'Quarks and Leptons: An Introductory Course in Particle Physics', by F. Halzen and A. Martin, Wiley, 1984. This book is an elementary introduction aimed at developing the computational tools and getting people calculating, with a minimum of formal baggage. An excellent introduction to utilitarian field theory.
- 'Quantum Field Theory and the Standard Model', by Matthew Schwartz, Cambridge Press 2014, is a more recent and modern treatment of quantum field theory as applied to the Standard Model, that also draws heavily on EFT methods to organize calculations.
- 'The Standard Model and Beyond', by Paul Langacker, CRC Press 2010, is a masterful summary of particle physics by one of the field's masters.

Supersymmetry is a well-developed topic in its own right and there are a number of books that review its various aspects. Among those aimed at possible implications for particle phenomenology are

- 'The Quantum Theory of Fields, vol III', by S. Weinberg, Cambridge Press, 2000. This presentation has the nice feature that it builds directly from the tools built in vols I and II of this sequence. This is particularly nice for supersymmetry, since many of the other presentations of supersymmetry use completely different notation for spinors used in supersymmetry compared to spinors used elsewhere in physics. Weinberg's vols I–III are uniform in their treatment of spinors in all aspects of their use.
- 'Supersymmetry in Particle Physics', by Ian Aitchison, Cambridge Press, 2007, is a more recent and modern treatment of supersymmetry as aimed at particle physics applications.
- 'Theory and Phenomenology of Sparticles: An Account of Four-Dimensional N = 1 Supersymmetry in High Energy Physics', by Manuel Drees, Rohini

Godbole and Probir Roy, World Scientific Press, 2005 provides a phenomenologist's eye view of N = 1 supersymmetry in 4 dimensions.

The above are complemented by discussions of supersymmetry that emphasize the more formal strong-coupling, gravity and string theory connections.

- 'Modern Supersymmetry, Dynamics and Duality', by John Terning, Oxford Press, 2006, is a treatment of supersymmetry as aimed at applications to dualities and many of the modern issues associated with supersymmetric systems.
- 'Supersymmetry and String Theory, Beyond the Standard Model', by Michael Dine, Cambridge Press, 2007, is a treatment of supersymmetry as aimed at more fundamental applications, such as to string theory.
- 'Supergravity', by A. van Proeyen and D. Z. Freedman, Cambridge Press, 2012, is a modern treatment of supergravity by those that invented much of it, covering many topics not usually treated.
- 'Introduction to Supersymmetry and Supergravity', by Peter West, World Scientific, 1990. This is a treatment that includes a discussion of superspace methods and extended supergravity, topics often skirted over quickly in textbooks (but not here).
- 'Supersymmetry and Supergravity', by Julius Wess and Jon Bagger, Princeton Press, 1992. This is one of the standard textbooks by some of the authors who helped define the subject.

General Relativity and Cosmology

More and more, physicists in all fields are expected to be knowledgeable about gravitational physics, and the geometrical techniques used in its study. Some useful texts for these purposes are:

- 'Gravitation and Cosmology: Principles and Applications of the General Theory of Relativity', S. Weinberg, Wiley, 1972. An oldie but a goodie: a very physical introduction to general relativity and its applications in astrophysics, the solar system and cosmology.
- 'Gravitation', C. W. Misner, K. S. Thorne and J. A. Wheeler, Freeman, 1973. The classic book with the quirky style, which sets the standard for its comprehensive and modern treatment of geometrical techniques.
- 'General Relativity', R. M. Wald, University of Chicago Press, 1984. A modern update of the two previous classics, containing more of the modern mathematical techniques.
- 'Spacetime and Geometry: An Introduction to General Relativity', S. Carroll, Cambridge Press, 2019. A re-release of a modern and very readable book that is a good place to start.

The last few decades have seen cosmology turn from a very speculative to a data-rich subject. Much of the evidence that the Standard Model is incomplete comes from the unified picture of cosmology that this data has spurred, making a good knowledge of this area also mandatory for many areas of physics. Some of the books I have learned from myself are

- 'Principles of Physical Cosmology', P. J. E. Peebles, Princeton Press, 1993. This is a classic written by one of the inventors of the modern picture of physical cosmology.
- 'Physical Foundations of Cosmology', V. Mukhanov, Cambridge Press, 2005. This is a comprehensive and very clear description of the theory of fluctuations and structure formation, by one of its inventors.
- 'Modern Cosmology', S. Dodelson, Elsevier, 2003. This is a modern textbook on cosmology including the discussion of fluctuations.
- 'Cosmology', S. Weinberg, Oxford, 2008. This is a modern treatment that definitively updates Weinberg's earlier book on Gravitation and Cosmology.
- 'Introduction to Cosmology', B. Ryden, Pearson, 2002. This introductory book is aimed at undergraduates, and so does not presuppose as much background. Yet it is also thorough and detailed, so a good place to start for beginners.

The presentation of the effective theory of gravity and cosmology used in this book follows some of my own review articles on EFTs in cosmology, mentioned above [482]. String theory provides a concrete example of what might take place at the highest energies, and this is partly what makes it interesting. Some references are

- 'Superstring Theory, vols I and II', by M. Green, J. Schwarz and E. Witten, Cambridge Press, 1987. Strings and superstrings described by the masters. This book has provided the first exposure to the field for many novice learners.
- 'String Theory, vols I and II', by Joe Polchinski, Cambridge Press, 1998. Strings and branes (including the origins of the word 'brane' for membranes [87]) most often arise in supergravity and string theory, and this book is the classic textbook by the discoverer of D-branes [88].
- 'String Theory and M-Theory', by K. Becker, M. Becker and J. Schwarz, Cambridge Press, 2007. A modern update on the subject.
- 'String Theory and Particle Physics', by L. E. Ibáñez and A. M. Uranga, Cambridge Press, 2012. This is a recent and high-level introduction to the phenomenological aspects of string theory and branes.
- 'A First Course in String Theory', by B. Zweibach, Cambridge Press, 2009. This is an undergraduate level introduction to string theory, aimed at those who do not already have an exposure to quantum field theory.

D.4 Nonrelativistic Applications

The last two parts of the book are aimed at nonrelativistic applications of EFT methods, both for collections of just a few slowly moving particles and for full-blown many body systems. Given here are a selection of references that I have found useful myself when learning these areas. As the level of detail below (compared with above) indicates, these topics bring me further from my own area of expertise, making it inevitable that I have missed many other gems (and my apologies if this is so).

HQET, NRQED and All That

There are a variety of review articles for heavy-particle EFTs, some of which are included among the reviews of EFT methods quoted earlier. Some others are

- 'Review of selected topics in HQET', A. I. Vainshtein (hep-ph/9512419).
- 'Heavy-quark effective theory', M. Neubert, in 20th Johns Hopkins Workshop on Current Problems in Particle Theory (hep-ph/9610385).
- 'Heavy-Quark and Soft-collinear Effective Field Theory', by C. W. Bauer and M. Neubert, in PDG review K. A. Olive et al., *Chin. Phys.* **C38** (2014) 090001 (http://pdg.lbl.gov).
- 'An Introduction to the heavy quark effective theory', F. Hussain and G. Thompson, (hep-ph/9502241).
- 'Precision study of positronium: Testing bound state QED theory', S. G. Karshenboim, *Int. J. Mod. Phys.* **A19** (2004) 3879 (hep-ph/0310099).
- 'An Introduction to NRQED', G. Paz, *Mod. Phys. Lett.* **A30** (2015) 1550128 (arXiv:1503.07216 (hep-ph)).

First-Quantized Methods

There are fewer surveys of first-quantized methods within an effective field theory context. Discussions of collective coordinates can be found in books on solitons, such as in Coleman's 'Aspects of Symmetry' mentioned earlier, or (for example)

- 'Magnetic Monopoles', by Ja. Schnir, Springer-Verlag, 2005.
- 'Advanced Topics in Quantum Field Theory', by M. Shifman, mentioned above, discusses many aspects of solitons, including collective coordinates for monopoles, vortices and instantons.

For reviews on the quantum mechanics of the inverse-square potential and the phenomenon of 'fall to the centre', see, for example

- B. Holstein, 'Anomalies for Pedestrians', *Am. J. Phys.* **61** (1993) 142;
- A. M. Essin and D. J. Griffiths, 'Quantum Mechanics of the $1/x^2$ Potential', *Am. J. Phys.* **74** (2006) 109.
- M. W. Frank, D. J. Land and R. M. Spector, 'Singular Potentials', *Rev. Mod. Phys.* **43** (1971) 36.

Atomic Physics

A number of books on atomic methods have proven useful over the years, including

- 'Quantum Mechanics of One- and Two-Electron Atoms', by H. A. Bethe and E. E. Salpeter, Springer-Verlag 1957; Plenum Publishing, 1977;
- 'Rydberg Atoms', by T. F. Gallagher, Cambridge Press, 1994;
- 'Many-Body Atomic Physics', ed. by J. J. Boyle and M. S. Pindzola, Cambridge Press, 1998.
- 'Atom-Photon Interactions', C. Cohen-Tannoudji, J. Dupont-Roc and G. Grynberg, Wiley Press, 2004.
- 'Bose-Einstein Condensation in Dilute Gases', C. J. Pethick and H. Smith, Cambridge Press, 2001.

Goldstone Bosons in Nonrelativistic Systems

The different types of Goldstone counting for spacetime symmetries in nonrelativistic systems is another one of those things that has long been in the air (since at least the early 1980s, when I was a graduate student), and because these issues arise more commonly for condensed matter systems were appreciated there much earlier. See, for example

- 'Concepts in Solids: Lectures on the Theory of Solids', by Philip W. Anderson, World Scientific, Singapore, 1997.

For particle physicists perhaps the most familiar examples where these issues arise are spin waves in ferromagnets and antiferromagnets [107] (see also Chapter 14.1), whose unusual properties eventually became systematized in [105] and [108].

Condensed Matter Surveys

Condensed matter physics is a vast area of research, for which there are a number of good textbook treatments, including:

- 'Principles of Condensed Matter Physics', by Paul Chaikin and Tom Lubensky, Cambridge Press, 1995. This both gives a thorough treatment of condensed matter physics, with an emphasis on its 'soft' side, and is very accessible to those of the unwashed who are not professional condensed matter physicists (myself included). This book includes many instances of topological defects and domain walls of various types.
- 'Solid State Physics', N. W. Ashcroft and N. D. Mermin, Harcourt, 1976. This is a classic undergraduate textbook on condensed matter physics.
- 'Introduction to Solid State Physics, 8th Edition', C. Kittel, Wiley, 2004. This is the other classic undergraduate textbook.
- 'Introduction to Superconductivity', M. Tinkham, McGraw-Hill Press, 1975. This is a book aimed more explicitly at the phenomenon of superconductivity.
- 'Quantum Field Theory and Condensed Matter', R. Shankar, Cambridge Press, 2017. This is a more modern treatment of many of the ideas handled in the older texts.
- 'Quantum Field Theory of Many Body Systems', X. G. Wen, Oxford Press, 2004. Another modern treatment of quantum field theory for a modern condensed matter audience.

Degenerate Systems

The treatment of degenerate systems described here follows the wonderful reviews,

- 'Effective field theory and the Fermi surface'. J. Polchinski, In the proceedings of the TASI school *Recent directions in particle theory* (hep-th/9210046) [402].
- 'Renormalization group approach to interacting fermions', R. Shankar, *Rev. Mod. Phys.* **66** (1994) 129 (cond-mat/9307009) [403].

A classic treatment of Fermi liquids that predates EFT methods is

- 'Quantum Theory of Many-Particle Systems', A. L. Fetter and J. D. Walecka, McGraw-Hill, 1971 (Dover, 2002).
- 'Methods of Quantum Field Theory in Statistical Physics', by A.A. Abrikosov, L.P. Gorkov and I.P. Dzyaloshinski, Dover Press 1963. A venerable classic applying QFT methods to condensed matter problems.
- 'Theory if Interacting Fermi Systems', by P. Nozières, Addison-Wesley 1964 and 1997, a nicely written and more detailed treatment about degenerate and interacting systems of fermions.

There are a number of good reviews about Quantum Hall systems, both from the point of view of EFT methods as well as from a more fundamental point of view. Some textbook descriptions are in

- 'Field Theories of Condensed Matter Physics', E. Fradkin, Cambridge Press, 2013.
- 'The Quantum Hall Effect', R. E. Prange and S. M. Girvin, Springer-Verlag, 1987.
- 'Perspectives in Quantum Hall Effects: Novel Quantum Liquids in Low-Dimensional Semiconductor Structures', S. Das Sarma and A. Pinczuk, John Wiley & Sons, 2004.

Some lecture notes that I have found very useful are

- 'The Quantum Hall Effect: Novel Excitations and Broken Symmetries', S. M. Girvin, Lectures delivered at *École d'Éte des Houches*, July 1998 (arXiv:cond-mat/9907002 (cond-mat.mes-hall)).
- 'Quantum Hall Fluids', A. Zee, (cond-mat/9501022).
- 'Topological Orders and Edge Excitations in FQH States', X. G. Wen (cond-mat/9506066).
- 'Introduction to the Physics of the Quantum Hall Regime', A.H. MacDonald (cond-mat/9410047).
- 'Lectures on the Quantum Hall Effect', D. Tong, (arXiv:1606.06687 (hep-th)).
- 'Three Lectures on Topological Phases of Matter', E. Witten. lectures given at the PITP school 2015. Published in *Riv. Nuovo Cim.* **39** (2016) 313.

Some other useful references are cited in the main text.

Open Systems

The treatment of fluids goes back to the nineteenth century, and is the birthplace of many EFT methods. Very useful textbooks are

- 'An Introduction to Fluid Dynamics' G. K. Batchelor, Cambridge Press, 1967.
- 'Fluid mechanics', L. D. Landau and E. M. Lifshitz, *A Course of Theoretical Physics* (2nd revised ed.) Vol 6, Pergamon Press, 1987.

A textbook study of how electromagnetic fields interact with media is [451]:

- 'Electrodynamics of Continuous Media', L. D. Landau, and E. M. Lifshitz, in *A Course of Theoretical Physics* Vol 8, Pergamon Press, 1960.

The theory of open systems is also a well-studied field, and some textbook treatments are given by

- 'The Theory of Open Quantum Systems', H. P. Breuer and F. Petruccione, Oxford Press, 2002. This is a book from which I have learned some of the open-system techniques described in this book. It is very user-friendly to those not in the area.

Two other good textbooks for this area are

- 'An Open Systems Approach to Quantum Optics', H. Carmichael, Springer-Verlag, 1991.
- Quantum Dynamical Semigroups and Applications', R. Alicki and K. Lendi, Springer, 2007.

References

[1] For an early demonstration for renormalizable theories see:
Appelquist, T. and Carazzone, J. 1975.
Infrared Singularities and Massive Fields.
Physical Review **D11** (1974) 2856.

[2] Weinberg, S. 1979.
Phenomenological Lagrangians.
Physica A **96** (1979) 327.

[3] Nambu, Y. 1960.
Quasiparticles and Gauge Invariance in the Theory of Superconductivity.
Physical Review **117** (1960) 648.

[4] Goldstone, J. 1961.
Field Theories with Superconductor Solutions.
Nuovo Cimento **19** (1961) 154.

[5] Goldstone, J., Salam, Abdus and Weinberg, S. 1962.
Broken Symmetries.
Physical Review **127** (1962) 965.

[6] Heisenberg, W. 1938.
Zur Theorie der explosionsartigen Schauer in der kosmischen Strahlung. II
Zeitschrift für Physik **113** 61.

[7] Sakata, S., Umezawa, H. and Kamefuchi, S. 1952.
On the Green-Functions of the Quantum Electrodynamics.
Progress in Theoretical Physics **7** 327.

[8] Borchers, H. 1960.
Über die mannigfaltigkeit der interpolierenden felder zu einer kausalen S-matrix.
Nuovo Cimento **15** (1960) 784.

[9] Chisholm, J. S. R. 1961.
Change of Variables in Quantum Field Theories.
Nuclear Physics **26** (1961) 469.

[10] Kamefuchi, S., O'Raifeartaigh, L. and Salam, A. 1961.
Change of Variables and Equivalence Theorems in Quantum Field Theories.
Nuclear Physics **28** (1961) 529.

[11] Weinberg, S. 1968.
Nonlinear Realizations of Chiral Symmetry.
Physical Review **166** (1968) 1568.

[12] Coleman, S. R., Wess J., and Zumino, B. 1969.
Structure of Phenomenological Lagrangians, 1.
Physical Review **177** (1969) 2239.

[13] Callan, C. G., Coleman, S. R., Wess J., and Zumino, B. 1969.
Structure of Phenomenological Lagrangians, 2.
Physical Review **177** (1969) 2247.

[14] Weinberg, S. 1969.
Feynman Rules for Any Spin III.
Physical Review **181** (1969) 1893.

[15] Jona-Lasinio, G. 1964.
Relativistic Field Theories with Symmetry Breaking Solutions.
Nuovo Cimento **34** (1964) 1790.

[16] Symanzik, K. 1970.
Renormalizable Models with Simple Symmetry Breaking 1.
Symmetry Breaking by a Source Term.
Communications in Mathematical Physics **16** (1970) 48.

[17] Coleman, S. 2010.
Aspects of Symmetry: Selected Erice Lectures.
Cambridge University Press, 2010.

[18] Wick, G. C. 1950.
The Evaluation of the Collision Matrix.
Physical Review **80** (1950) 268.

[19] Kaplan, D. B. 2005.
Five Lectures on Effective Field Theory.
(arXiv:nucl-th/0510023).

[20] Politzer, H. D. 1980.
Power Corrections at Short Distances.
Nuclear Physics **B172** (1980) 349.

[21] Iliopoulos, J., Itzykson, C. and Martin, A. 1975.
Functional Methods and Perturbation Theory.
Reviews of Modern Physics **47** (1975) 165.

[22] Taylor, J. C. 1971.
Ward Identities and Charge Renormalization of the Yang-Mills Field.
Nuclear Physics **B33** (1971) 436.

[23] Slavnov, A. A. 1972.
Ward Identities in Gauge Theories.
Theoretical and Mathematical Physics **10** (1972) 99
(*Teoreticheskaya i Matematischeskaya Fizika* **10** (1972) 153).

[24] Burgess, C. P. 2004.
Quantum Gravity in Everyday Life: General Relativity as an Effective Field Theory.
Living Reviews of Relativity **7** (2004) 5 (arXiv:gr-qc/0311082).

[25] Polchinski, J. 1984.
Renormalization and Effective Lagrangians.
Nuclear Physics **B231** (1984) 269.

[26] Wetterich, C. 1993.
Exact Evolution Equation for the Effective Potential.
Physics Letters **B301** (1993) 90(arXiv:1710.05815).

[27] Morris, T. 1993.
Exact Renormalization Group and Approximate Solutions.

International Journal of Modern Physics **A9** (1993) 2411 (arXiv:hep-ph/9308265).

[28] Burgess, C. P. and London, D. 1993.
Uses and Abuses of Effective Lagrangians.
Physical Review **D48** (1993) 4337 (arXiv:hep-ph/9203216).

[29] Beneke, M. and Smirnov, V. A. 1998.
Asymptotic Expansion of Feynman Integrals Near Threshold.
Nuclear Physics **B522** (1998) 321344 (hep-ph/9711391).
Smirnov, V. A. 1999.
Problems of the Strategy of Regions.
Physics Letters **B465** (1999) 226234 (hep-ph/9907471).

[30] Becher, T., Broggio, A. and Ferroglia, A. 2014.
Introduction to Soft-Collinear Effective Theory.
Lecture Notes in Physics **896** (2015) 1 (arXiv:1410.1892 (hep-ph)).

[31] Weinberg, S. 1980.
Effective Gauge Theories.
Physics Letters **91B** (1980) 51.

[32] Gasser, J. and Leutwyler, H. 1984.
Chiral Perturbation Theory to One Loop.
Annals of Physics **158** (1984) 142.

[33] Bollini, C. and Giambiagi, J. J. 1972.
Dimensional Renormalization:
The Number of Dimensions as a Regularizing Parameter.
Nuovo Cimento **B12** (1972) 20.

[34] 't Hooft, G. and Veltman, M. 1972.
Regularization and Renormalization of Gauge Fields.
Nuclear Physics **B44** (1972) 189.

[35] 't Hooft, G. 1973.
Dimensional Regularization and the Renormalization Group.
Nuclear Physics **B61** (1973) 455.

[36] 't Hooft, G. 1973.
An Algorithm for the Poles at Dimension Four
in the Dimensional Regularization Procedure.
Nuclear Physics **B62** (1973) 444.

[37] Weinberg, S. 1973.
New Approach to the Renormalization Group.
Physical Review **D8** (1973) 3497.

[38] Landau, L. D. 1959.
On Analytic Properties of Vertex Parts in Quantum Field Theory.
Nuclear Physics **13** 181.

[39] 't Hooft, G. and Veltman, M. J. G. 1974.
Diagrammar.
NATO Science Series **B4** (1974) 177.

[40] Weinberg, S. 1981.
Why the Renormalization Group Is a Good Thing.
Cambridge, Proceedings Asymptotic Realms of Physics (1981) 1–19.

[41] Wigner, E. P. 1931.
Gruppentheorie und ihre Anwendung auf die Quanten-mechanik der Atomspektren.
Braunschweig, 1931. (English translation, Academic Press Inc, New York 1959).

[42] Noether, E. 1918.
Invariante Variationsprobleme.
Nachrichten von der Gesellschaft der Wissenschaften, Gottingen, Mathematisch-Physikalische Klasse **2** (1918) 98.
(English translation: in *Transport Theory and Statistical Physics* **1** 186.
See also (arXiv:physics/0503066).)

[43] Fabri, E. and Picasso, L. E. 1966.
Quantum Field Theory and Approximate Symmetries.
Physical Review Letters **16** (1966) 408.

[44] Weinberg, S. 1972.
Approximate Symmetries and PseudoGoldstone Bosons.
Physical Review Letters **29** (1972)1698.

[45] Gell-Mann, M. 1964.
The Symmetry Group of Vector and Axial Vector Currents.
Physics Physique Fizika **1** (1964) 63.

[46] Weisberger, W. 1965.
Renormalization of the Weak Axial-Vector Coupling Constant.
Physical Review Letters **14** (1965) 1047.

[47] Adler, S. 1965.
Calculation of the Axial-Vector Coupling Constant Renormalization in Beta Decay.
Physical Review Letters **14** (1965) 1051.

[48] Weinberg, S. 1966.
Pion Scattering Lengths.
Physical Review Letters **17** (1966) 616.

[49] Weinberg, S. 1966.
Dynamical Approach to Current Algebra.
Physical Review Letters **18** (1966) 188.

[50] Cartan, E. 1904.
Sur la structure des groupes infinis de transformation.
Annales scientifiques de lécole normal supërieur 3e srie, tome 21, (1904) p. 153.

[51] Weinberg, S. 1964.
Feynman Rules for Any Spin 2: Massless Particles.
Physical Review **134** (1964) B882.

[52] Weinberg, S. 1964.
Photons and Gravitons in S Matrix Theory: Derivation of Charge Conservation and Equality of Gravitational and Inertial Mass.
Physical Review **135** (1964) B1049.

[53] Weinberg, S. 1965.
Infrared Photons and Gravitons.
Physical Review **140** (1965) B516.

[54] Weinberg, S. and Witten, E. 1980.
Limits on Massless Particles.
Physics Letters **96B** (1980) 59.

[55] Brout, R. and Englert, F. 1964.
Broken Symmetry and the Mass of Gauge Vector Mesons.
Physical Review Letters **13** (1964) 321.

[56] Guralnik, G. S., Hagen, C. R. and Kibble, T. W. B. 1964.
Global Conservation Laws and Massless Particles.
Physical Review Letters **13** (1964) 585.

[57] Kibble, T. W. B. 1967.
Symmetry Breaking in NonAbelian Gauge Theories.
Physical Review **155** (1967) 1554.

[58] Higgs, P. W. 1964.
Broken Symmetries, Massless Particles and Gauge Fields
Physics Letters **12** (1964) 132.

[59] Higgs, P. W. 1964.
Broken Symmetries and the Masses of Gauge Bosons.
Physical Review Letters **13** (1964) 508.

[60] Anderson, P. W. 1963.
Plasmons, Gauge Invariance, and Mass.
Physical Review **130** (1963) 439.

[61] Weinberg, S. 1967.
A Model of Leptons.
Physical Review Letters **19** (1967) 1264.

[62] Olive, K. A. *et al.* (Particle Data Group). 2014.
Chinese Physics **C38**, 090001 (2014).

[63] Stueckelberg, E. 1938.
Die Wechselwirkungskräfte in der Elektrodynamik und in der Feldtheorie der Kräfte.
Helvetica Physica Acta **11** (1938) 225.

[64] Weinberg, S. 1971.
Physical Processes in a Convergent Theory of the
Weak and Electromagnetic Interactions.
Physical Review Letters **27** (1971) 1688.

[65] Weinberg, S. 1973.
General Theory of Broken Local Symmetries.
Physical Review **D7** (1973) 1068.

[66] Fujikawa, K., Lee, B. W. and Sanda, A. 1972.
Generalized Renormalizable Gauge Formulation of
Spontaneously Broken Gauge Theories.
Physical Review **D6** (1972) 2923.

[67] Lee, B. W., Quigg, C. and Thacker, H. B. 1977.
Weak Interactions at Very High-Energies: The Role of the Higgs Boson Mass.
Physical Review **D16** (1977) 1519.

[68] Froissart, M. 1961.
Asymptotic Behavior and Subtractions in the Mandelstam Representation.
Physical Review **123** 1053.

[69] Cornwall, J. M., Levin, D. N. and Tiktopoulos, G. 1974.
Derivation of Gauge Invariance from High-Energy Unitarity Bounds on the S Matrix.
Physical Review **D10** 1145; Erratum: (*Phys. Rev.* **D11** (1975) 972).

[70] Bell, J. S. and Jackiw, R. 1969.
A PCAC Puzzle: $\pi^0 \to \gamma\gamma$ in the σ Model.
Nuovo Cimento **A60** (1969) 47.

[71] Adler, S. L. 1969.
Axial Vector Vertex in Spinor Electrodynamics.
Physical Review **177** (1969) 2426.

[72] Bardeen, W. A. 1969.
Anomalous Ward Identities in Spinor Field Theories.
Physical Review **184** (1969) 1848.

[73] Adler, S. L. and Bardeen, W. A. 1969.
Absence of Higher Order Corrections in the Anomalous Axial Vector Divergence Equation.
Physical Review **182** (1969) 1517.

[74] Coleman, S. R. and Grossman, B. 1982.
't Hooft's Consistency Condition as a Consequence of Analyticity and Unitarity.
Nuclear Physics **B203** (1982) 205.

[75] Capper, D. M. and Duff, M. J. 1974.
Trace Anomalies in Dimensional Regularization.
Nuovo Cimento **A23** (1974) 173.

[76] Fujikawa, K. 1979.
Path Integral Measure for Gauge Invariant Fermion Theories,.
Physical Review Letters **42** (1979) 1195.

[77] Green, M. B. and Schwarz, J. H. 1984.
Anomaly Cancellation in Supersymmetric D = 10
Gauge Theory and Superstring Theory.
Physics Letters **149B** (1984) 117.

[78] Wess, J. and Zumino, B. 1971.
Consequences of Anomalous Ward Identities.
Physics Letters **B37** (1971) 95.

[79] Weisskopf, V. F. 1939.
On the Self-Energy and the Electromagnetic Field of the Electron.
Physical Review **56** (1039) 72.

[80] 't Hooft, G. 1980.
Naturalness, Chiral Symmetry, and Spontaneous Chiral Symmetry Breaking.
In the proceedings of 'Recent Developments in Gauge Theories' NATO Advanced Study Institute, Cargese, *NATO Science Series* **B59** (1980) 135.

[81] Finkelstein, R. J. 1947.
The γ-Instability of Mesons.
Physical Review **72** (1947) 415.

[82] Fukuda, H. and Miyamoto, Y. 1949.
On the γ-Decay of Neutral Meson.
Progress of Theoretical Physics **4** (1949) 347.

[83] Steinberger, J. 1949.
On the Use of Subtraction Fields and the Lifetimes of Some Types of Meson Decay.
Physical Review **76** (1949) 1180.

[84] Schwinger, J. S. 1951.
On Gauge Invariance and Vacuum Polarization.
Physical Review **82** (1951) 664.

[85] Sutherland, D. G. 1967.
Current Algebra and Some Nonstrong Mesonic Decays.
Nuclear Physics **B2** (1967) 433.

[86] Veltman, M. 1967.
Theoretical Aspects of High Energy Neutrino Interactions.
Proceedings of the Royal Society of London **A301** (1967) 107.

[87] Duff, M. J., Inami, T., Pope, C. N., Sezgin E. and Stelle, K. S. 1988.
Semiclassical Quantization of the Supermembrane.
Nuclear Physics **B297** (1988) 515.

[88] Polchinski, J. 1995.
Dirichlet Branes and Ramond-Ramond charges.
Physical Review Letters **75** (1995) 4724 (arXiv:hep-th/9510017).

[89] Burgess, C. P., Horbatsch, M. W. and Patil, S. P. 2013.
Inflating in a Trough: Single-Field Effective Theory from Multiple-Field Curved Valleys.
Journal of High Energy Physics **1301** (2013) 133 (arXiv:1209.5701 (hep-th)).

[90] Ostrogradsky, M. 1850.
Mémoires sur les équations différentielles, relatives au problème des isopéimètres.
Mémoires de l'Acădémie impériale des sciences de St. Pétersbourg **VI 4** (1850) 385.

[91] Woodard, R. P. 2015.
Ostrogradsky's Theorem on Hamiltonian Instability.
Scholarpedia **10** (2015) 32243 (arXiv:1506.02210 (hep-th)).

[92] Motohashi, H. and Suyama, T. 2015.
Third Order Equations of Motion and the Ostrogradsky Instability.
Physical Review **D91** (2015) 085009 (arXiv:1411.3721 (physics.class-ph)).

[93] Burgess, C. P. and Williams, M. 2014.
Who You Gonna Call? Runaway Ghosts, Higher Derivatives and Time-Dependence in EFTs.
Journal of High Energy Physics **1408** (2014) 074 (arXiv:1404.2236 (gr-qc)).

[94] Horndeski, G. W. 1974.
Second-Order Scalar-Tensor Field Equations in a Four-Dimensional Space.
International Journal of Theoretical Physics **10** (1974) 363.

[95] Nicolis, A., Rattazzi, R. and Trincherini, E. 2009.
The Galileon as a Local Modification of Gravity.
Physical Review **D79** (2009) 064036 (arXiv:0811.2197 (hep-th)).

[96] Deffayet, C., Gao, X., Steer, D. A. and Zahariade, G. 2011.
From k-Essence to Generalised Galileons.
Physical Review **D84** (2011) 064039 (arXiv:1103.3260 (hep-th)).

[97] Solomon, A. R. and Trodden, M. 2018.
Higher-Derivative Operators and Effective Field Theory for General Scalar-Tensor Theories.
Journal of Cosmology and Astroparticle Physics **1802** (2018) 031 (arXiv:1709.09695 (hep-th)).

[98] Sarbach, O. and Tiglio, M. 2012.
Continuum and Discrete Initial-Boundary-Value Problems and Einsteins Field Equations.
Living Reviews of Relativity **15** (2012) 100 (arXiv:1203.6443 (gr-qc)).

[99] Kreiss, H.-O. and Lorenz, J. 1989.
Initial-Boundary Value Problems and the Navier-Stokes Equations.
Pure and Applied Mathematics 136, Academic Press, San Diego, 1989.

[100] Hadamard, J. 1902.
Sur les problèmes aux dérivées partielles et leur signification physique.
Princeton University Bulletin **13** (1902) 4952.

[101] Papallo, G. and Reall, H. S. 2017.
On the Local Well-Posedness of Lovelock and Horndeski Theories.
Physical Review **D96** (2017) 044019 (arXiv:1705.04370 (gr-qc)).

[102] Tikhonov, A. N. and Arsenin, V. Y. 1977.
Solutions of Ill-Posed Problems.
Winston Press, New York, 1977.

[103] Allwright, G. and Lehner, L. 2018.
Towards the Nonlinear Regime in Extensions to GR: Assessing Possible Options.
(arXiv:1808.07897 (gr-qc)).

[104] Cheung, C., Creminelli, P., Fitzpatrick, A. L., Kaplan J. and Senatore, L. 2008.
The Effective Field Theory of Inflation.
Journal of High Energy Physics **0803** (2008) 014 (arXiv:0709.0293 (hep-th)).

[105] Nicolis, A. and Piazza, F. 2011.
Spontaneous Symmetry Probing.
Journal of High Energy Physics **1206** (2012) 025 (arXiv:1112.5174 (hep-th)).

[106] Volkov, D. V. 1973.
Phenomenological Lagrangians.
Soviet Journal of Nuclear Physics **4** (1973) 1;
(*Fizika Elementarnykh Chastits i Atomnogo Yadra* **4** (1973) 3).

[107] Burgess, C. P. 2000.
Goldstone and Pseudo-Goldstone Bosons in Nuclear, Particle and Condensed Matter Physics.
Physics Reports **330** (2000) 193 (hep-th/9808176).

[108] Watanabe, H. and Murayama, H. 2012.
Unified Description of Nambu-Goldstone Bosons without Lorentz Invariance.
Physical Review Letters **108** (2012) 251602 (arXiv:1203.0609 (hep-th)).

[109] Glashow, S. L. 1961.
Partial Symmetries of Weak Interactions.
Nuclear Physics **22** (1961) 579.

[110] Salam, A. 1968.
Weak and Electromagnetic Interactions.
Conference Proceedings **C 680519** (1968) 367.

[111] Glashow, S. L., Iliopoulos, J. and Maiani, L. 1970.
Weak Interactions with Lepton-Hadron Symmetry.
Physical Review **D2** (1970) 1285.

[112] Cabibbo, N. 1963.
Unitary Symmetry and Leptonic Decays.
Physical Review Letters **10** (1963) 531.

[113] Kobayashi, M. and Maskawa, T. 1973.
CP Violation in the Renormalizable Theory of Weak Interaction.
Progress in Theoretical Physics **49** (1973) 652.

[114] Pontecorvo, B. 1957.
Inverse Beta Processes and Nonconservation of Lepton Charge.
Zhurnal Éksperimental'noĭ i Teoreticheskoĭ Fiziki **34** 247
(*Soviet Physics JETP* **7** (1958) 172).

[115] Maki, Z., Nakagawa, M. and Sakata S. 1962.
Remarks on the Unified Model of Elementary Particles.
Progress of Theoretical Physics **28** 870.

[116] Fermi, E. 1933.
Tentativo di una teoria dei raggi β.
La Ricerca Scientifica **2** (12). *Il Nuovo Cimento* **11** 1.

[117] Fermi, E. 1934.
Versuch einer Theorie der beta-Strahlen. I.
Zeitschrift für Physik **88** 161. For an English translation see Wilson, F. L. 1968.
Fermi's Theory of Beta Decay.
American Journal of Physics **36** 1150.

[118] Feynman, R. P. and Gell-Mann, M. 1958.
Theory of the Fermi Interaction.
Physical Review **109** 193.

[119] Sudarshan, E. C. and Marshak, R. E. 1958.
Chirality Invariance and the Universal Fermi Interaction.
Physical Review **109** 1860.

[120] Dirac, P. A. M. 1927.
The Quantum Theory of the Emission and Absorption of Radiation.
Proceedings of the Royal Society of London **A 114** 243.

[121] Fermi, E. 1932.
Quantum Theory of Radiation.
Reviews of Modern Physics **4** 87.

[122] Dirac, P. A. M. 1928.
The Quantum Theory of the Electron.
Proceedings of the Royal Society A: **117** 610.

[123] Tomonaga, S. 1946.
On a Relativistically Invariant Formulation of the Quantum Theory of Wave Fields.
Progress of Theoretical Physics **1** 27.

[124] Schwinger, J. 1948.
On Quantum-Electrodynamics and the Magnetic Moment of the Electron.
Physical Review **73** 416.

[125] Feynman, R. P. 1949.
SpaceTime Approach to Quantum Electrodynamics.
Physical Review **76** 769.
The Theory of Positrons.
Physical Review **76** 749.

[126] Dyson, F. 1949.
The Radiation Theories of Tomonaga, Schwinger, and Feynman.
Physical Review **75** 486.

[127] Furry, W. H. 1937.
A Symmetry Theorem in Positron Theory.
Physical Review **51** 125.

[128] Dyson, F. J. 1949.
The S-matrix in Quantum Electrodynamics.
Physical Review **75** (1949) 1736.

[129] Rutherford, E. 1911.
The Scattering of α and β rays by Matter and the Structure of the Atom.
Philosophical Magazine **6** 21.

[130] Euler, H. and Kockel, B. 1935.
Über die Streuung von Licht an Licht nach der Diracschen Theorie.
Die Naturwissenschaften **23** 246.

[131] Heisenberg, W. and Euler, H. 1936.
Folgerungen aus der Diracschen Theorie des Positrons.
Zeitschrift für Physik **98** 714.

[132] Karplus, R. and Neuman, M. 1951.
The Scattering of Light by Light.
Physical Review **83** (1951) 776.

[133] Low, F. E. 1954.
Scattering of Light of Very Low Frequency by Systems of Spin 1/2.
Physical Review **96** (1954) 1428.

[134] Gell-Mann, M. and Goldberger, M. L. 1954.
Scattering of Low-Energy Photons by Particles of Spin 1/2.
Physical Review **96** (1954) 1433.

[135] Bloch, F. and Nordsieck, A. 1937.
Note on the Radiation Field of the Electron.
Physical Review **52** (1937) 54.

[136] Yennie, D. R., Frautschi, S. C. and Suura, H. 1955.
The Infrared Divergence Phenomena and High-Energy Processes.
Annals of Physics (NY) **13** (1955) 379.

[137] Kinoshita, T. 1962.
Mass Singularities of Feynman Amplitudes.
Journal of Mathematical Physics **3** (1962) 650.

[138] Lee, T.-D. and Nauenberg, M. 1964.
Degenerate Systems and Mass Singularities.
Physical Review **D133** (1964) B1549,

[139] Callan, C. G. 1970.
Broken Scale Invariance in Scalar Field Theory.
Physical Review **D2** (1970) 1541.

[140] Symanzik, K. 1970.
Small Distance Behaviour in Field Theory and Power Counting.
Communications in Mathematical Physics **18** (1970) 227.

[141] Gell-Mann, M. 1961.
The Reaction $\gamma + \gamma \rightarrow nu + \bar{v}$.
Physical Review Letters **6** (1961) 70.

[142] Dicus, D. A. and Repko, W. W. 1993.
Photon Neutrino Scattering.
Physical Review **D48** (1993) 5106 (arXiv:hep-ph/9305284).

[143] Dicus, D. A and Repko, W. W. 1997.
Photon – Neutrino Interactions.
Physical Review Letters **79** (1997) 569 (arXiv:hep-ph/9703210).

[144] Aghababaie, Y. and Burgess, C. P. 2000.
Two Neutrino Five Photon Scattering at Low-Energies.
Physical Review **D 63** (2001) 113006 (arXiv:hep-ph/0006165).

[145] Yang, C. N. 1950.
Selection Rules for the Dematerialization of a Particle into Two Photons.
Physical Review **77** (1950) 242.

[146] Bowick, M. J. and Travesset, A. 2000.
The Statistical Mechanics of Membranes.
Physics Reports **344** (2001) 255 (cond-mat/0002038 (cond-mat.soft)).

[147] Aghababaie, Y. and Burgess, C. P. 2003.
Effective Actions, Boundaries and Precision Calculations of Casimir Energies.
Physical Review **D70** (2004) 085003 (hep-th/0304066).

[148] Casimir, H. B. G. 1948.
On the Attraction between Two Perfectly Conducting Plates.
Proceedings of the Koninklijke Nederlandse Akademie van Wetenschappen **51** 793.

[149] Jaffe, R. 2005.
Casimir Effect and the Quantum Vacuum.
Physical Review **D72** (2005) 021301 (arXiv:hep-th/0503158).

[150] Sparnaay, M. J. 1957.
Attractive Forces between Flat Plates.
Nature **180** (1957) 334.

[151] Lamoreaux, S. K. 1997.
Demonstration of the Casimir Force in the 0.6 to 6 μm Range.
Physical Review Letters **78** 5.

[152] Mohideen, U. and Roy, A. 1998.
Precision Measurement of the Casimir Force from 0.1 to 0.9 μm.
Physical Review Letters **81** 4549. (arXiv:physics/9805038).

[153] Brown, L. S. and Maclay, G. J. 1969.
Vacuum Stress between Conducting Plates: An Image Solution.
Physical Review **184** (1969) 1272.

[154] Ravndal, F. and Thomassen, J. B. 2001.
Radiative Corrections to the Casimir Energy and Effective Field Theory.
Physical Review **D63** (2001) 113007 (hep-th/0101131).

[155] Bordag, M., Wieczorek, E. and Robaschik, D. 1984.
Radiation Corrections to the Casimir Effect (in Russian).
Soviet Journal of Nuclear Physics **39** (1984) 663 (*Yadernaya Fizika* **39** (1984) 1053).

[156] Bordag, M., Wieczorek, E. and Robaschik, D. 1985.
Quantum Field Theoretic Treatment of the Casimir Effect.
Annals of Physics **165** (1985) 192.

[157] Gell-Mann, M. 1964.
A Schematic Model of Baryons and Mesons.
Physics Letters **8** (1964) 214.

[158] Zweig, G. 1964.
An SU(3) Model for Strong Interaction Symmetry and Its Breaking. Version 1. (preprint CERN-TH-401).
An SU(3) Model for Strong Interaction Symmetry and Its Breaking. Version 2. in *Developments in the Quark Theory of Hadrons, Volume 1*, ed. by D. Lichtenberg and S. Rosen, pp. 22–101 (preprint CERN-TH-412).

[159] Bjorken, B. J. and Glashow, S. L. 1964.
Elementary Particles and SU(4).
Physics Letters **11** 255.

[160] Feynman, R. P. 1969.
Very High-Energy Collisions of Hadrons.
Physical Review Letters **23** (1969) 1415.

[161] Greenberg, O. W. 1964.
Spin and Unitary Spin Independence in a Paraquark Model of Baryons and Mesons.
Physical Review Letters **13** (1964) 598.

[162] Han, M. Y. and Nambu, Y. 1965.
Three-Triplet Model with Double SU(3) Symmetry.
Physical Review **139** (1965) B1006.

[163] Struminsky, B. V. 1965.
Magnetic Moments of Baryons in the Quark Model.
JINR-Preprint P-1939 Dubna, Russia.

[164] Fritzsch, H., Gell-Mann, M. and Leutwyler, H. 1973.
Advantages of the Color Octet Gluon Picture.
Physics Letters **47B** (1973) 365.

[165] Vanyashin, V. S. and Terent'ev, M. V. 1965.
The Vacuum Polarization of a Charged Vector Field.
Journal of Experimental and Theoretical Physics **21** (1965) 375.

[166] Khriplovich, I. B. 1970.
Green's Functions in Theories with Non-Abelian Gauge Group.
Soviet Journal of Nuclear Physics **10** (1970) 235.

[167] 't Hooft, G. 1972.
Unpublished talk at the Marseille conference on *Renormalization of YangMills Fields and Applications to Particle Physics.*

[168] Gross, D. J. and Wilczek, F. 1973.
Ultraviolet Behavior of Non-Abelian Gauge Theories.
Physical Review Letters **30** 1343.

[169] Politzer, H. D. 1973.
Reliable Perturbative Results for Strong Interactions.
Physical Review Letters **30** 1346.

[170] Schael, S. *et al.* (ALEPH and DELPHI and L3 and OPAL and SLD Collaborations and LEP Electroweak Working Group and SLD Electroweak Group and SLD Heavy Flavour Group). 2005.
Precision Electroweak Measurements on the Z Resonance.
Physics Reports **427** (2006) 257 (hep-ex/0509008).

[171] Gross, D. J. and Wilczek, F. 1973.
Asymptotically Free Gauge Theories I.
Physical Review **D8** (1973) 3633.

[172] Weinberg, S. 1973.
Nonabelian Gauge Theories of the Strong Interactions.
Physical Review Letters **31** (1973) 494.

[173] Glashow, S. L., Jackiw, R. and Shei, S.-S. 1969.
Electromagnetic Decays of Pseudoscalar Mesons.
Physical Review **187** (1969) 1916.

[174] t Hooft, G. 1976.
Symmetry Breaking through Bell-Jackiw Anomalies.
Physical Review Letters **37** (1976) 8.

[175] Jackiw, R. and Rebbi, C. 1976.
Physical Review Letters **37** (1976) 172.

[176] Callan, C. G., Dashen, R. F. and Gross, D. J. 1976.
The Structure of the Gauge Theory Vacuum.
Physics Letters **63B** (1976) 334.

[177] Vafa, C. and Witten, E. 1984.
Parity Conservation in QCD.
Physical Review Letters **53** (1984) 535.

[178] Heisenberg, W. 1932.
Über den Bau der Atomkerne.
Zeitschrift für Physik **77** 1.

[179] Wigner, E. 1937.
On the Consequences of the Symmetry of the Nuclear Hamiltonian on the Spectroscopy of Nuclei.
Physical Review **51** 106.

[180] Adler, S. L. 1965.
Calculation of the Axial Vector Coupling Constant Renormalization in Beta Decay.
Physical Review Letters **14** (1965) 1051;

[181] Weisberger, W. I. 1965.
Renormalization of the Weak Axial Vector Coupling Constant.
Physical Review Letters **14** (1965) 1047.

[182] Gell-Mann, M., Oakes, R. J. and Renner, B. 1968.
Behavior of Current Divergences under SU3?SU3.
Physical Review **175** (1968) 2195.

[183] Bardeen, W. A., Bijnens, J. and Gérard, J.-M. 1989.
Hadronic Matrix Elements and the $\pi^+ - \pi^0$ Mass Difference.
Physical Review Letters **62** (1989) 1343.

[184] Donoghue, J. F. and Perez, A. F. 1997.
The Electromagnetic Mass Differences of Pions and Kaons.
Physical Review **D55** (1997) 7075 (hep-ph/9611331).

[185] Ruderman, M. and Finkelstein, R. 1949.
Note on the Decay of the π-Meson *Physical Review* **76** (1949) 1458.

[186] Goldberger, M. L. and Treiman, S. 1958.
Decay of the Pi Meson.
Physical Review **110** (1958) 1178.

[187] Jenkins E. E. and Manohar, A. V. 1990.
Baryon Chiral Perturbation Theory Using a Heavy Fermion Lagrangian.
Physics Letters **B255** (1991) 558.

[188] Goldberger, M. L. and Treiman, S. 1958.
Form Factors in β Decay and μ Capture.
Physical Review **111** (1958) 354.

[189] Donoghue, J. F., Golowich, E. and Holstein, B. R. 1992.
Dynamics of the Standard Model.
Cambridge Monographs in Particle Physics, Nuclear Physics and Cosmology **2** (1992) 1 (2nd edition: *Cambridge Monographs in Particle Physics, Nuclear Physics and Cosmology* **35** (2014)).

[190] Glashow, S. and Weinberg, S. 1968.
Breaking Chiral Symmetry.
Physical Review Letters **20** (1968) 224.

[191] Gell-Mann, M. 1961.
Caltech Synchrotron Laboratory Report CTSL-20 (1961), reproduced in the book *The Eightfold Way*, by M. Gall=Mann and Y. Ne'eman (Benjamin, New York, 1964).

[192] Okubo, S. 1962.
Note on Unitary Symmetry in Strong Interactions.
Progress in Theoretical Physics **27** (1962) 949.

[193] Dashen, R. 1969.
Chiral SU(3) × SU(3) As a Symmetry of the Strong Interactions.
Physical Review **183** (1969) 1245.

[194] Burgess, C. P. and Moore, G. D.
The Standard Model: A Modern Primer
(Cambridge Press, 2007; post-Higgs update 2013).

[195] Weinberg, S. 1979.
Baryon and Lepton Nonconserving Processes.
Physical Review Letters **43** (1979) 1566.

[196] Perl, M. *et al.* 1975.
Evidence for Anomalous Lepton Production in $e^+e^?$ Annihilation.
Physical Review Letters **35** (1975) 1489.

[197] Zel' dovich, Ia. B. 1952.
Doklady Akademii Nauk SSSR **86** (1952) 505 (1952).

[198] Davis, R., Harmer, D. S. and Hoffman, K. C. 1968.
Search for Neutrinos from the Sun.
Physical Review Letters **20** (1968) 1205.

[199] Fukudae, Y. *et al.* (Super-Kamiokande Collaboration) 1998.
Evidence for Oscillation of Atmospheric Neutrinos.
Physical Review Letters **81** (1998) 1562 (arXiv:hep-ex/9807003).

[200] Ahmad, Q. R. *et al.* (SNO Collaboration) 2001.
Measurement of the Rate of $\nu_e + d \to p + p + e^-$ Interactions
Produced by ^{8}B Solar Neutrinos at the Sudbury Neutrino Observatory.
Physical Review Letters **87** (1968) 071301 (arXiv:nucl-ex/0106015).

[201] Abe, Y. *et al.* (Double Chooz collaboration) 2012.
Indication for the Disappearance of Reactor Electron
Antineutrinos in the Double Chooz Experiment.
Physical Review Letters **108** (2012) 131801 (arXiv:1112.6353).

[202] An, F. P. *et al.* (Daya Bay Collaboration) 2012.
Observation of Electron-Antineutrino Disappearance at Daya Bay.
Physical Review Letters **108** (2012) 171803 (arXiv:1203.1669).

[203] Pontecorvo, B. 1957.
Mesonium and Anti-Mesonium.
Zhurnal Éksperimental'noĭ i Teoreticheskoĭ Fiziki **33** (1957) 549
(*Soviet Physics JETP* **6** (1957) 429431).

[204] Pontecorvo, B. 1968.
Neutrino Experiments and the Problem of Conservation of Leptonic Charge.
Zhurnal Éksperimental'noĭ i Teoreticheskoĭ Fiziki **53** (1968) 1717
(*Soviet Physics JETP* **26** (1968) 984).

[205] Chivukula, R. S. and Georgi, H. 1987.
Composite Technicolor Standard Model.
Physics Letters **B188** (1987) 99.

[206] DAmbrosio, G., Giudice, G. F., Isidori, G. and Strumia, A. 2002.
Minimal Flavour Violation: An Effective Field Theory Approach.
Nuclear Physics **B645** (2002) 155 (arXiv:hep-ph/0207036).

[207] Georgi, H. and Glashow, S. L. 1972.
Gauge Theories without Anomalies.
Physical Review **D6** (1972) 429.

[208] Bouchiat, C., Iliopoulos, J. and Meyer, P. 1972.
An Anomaly Free Version of Weinberg's Model.
Physics Letters **38B** (1972) 519.

[209] Gross, D. J. and Jackiw, R. 1972.
Effect of Anomalies on Quasirenormalizable Theories.
Physical Review **D6** (1972) 477.

[210] Alvarez-Gaume, L. and Witten, E.
Gravitational Anomalies.
Nuclear Physics **B234** (1984) 269.

[211] Minahan, J. A., Ramond, P. and Warner, R. C. 1990.
A Comment on Anomaly Cancellation in the Standard Model.
Physical Review **D41** (1990) 715.

[212] Weinberg, S. 1980.
Varieties of Baryon and Lepton Nonconservation.
Physical Review **D22** (1980) 1694.

[213] Minkowski, P. 1977.
$\mu \to e\gamma$ at a Rate of One Out of a Billion Muon Decays?
Physics Letters **67B** (1977) 421.

[214] Gell-Mann, M., Ramond, P. and Slansky, R. 1979.
in *Supergravity*, ed. by D. Freedman and P. Van Nieuwenhuizen,
North Holland, Amsterdam (1979), p. 315.

[215] Wilczek, F. and Zee, A. 1979.
Operator Analysis of Nucleon Decay.
Physical Review Letters **43** (1979) 1571.

[216] Abbott, L. F. and Wise, M. B. 1980.
The Effective Hamiltonian for Nucleon Decay.
Physical Review **D22** (1980) 2208.

[217] Buchmuller W. and Wyler D. 1986.
Effective Lagrangian Analysis of New Interactions and Flavor Conservation.
Nuclear Physics **B268** (1986) 621.

[218] Grzadkowski B., Iskrzynski M., Misiak M. and Rosiek J. 2010.
Dimension-Six Terms in the Standard Model Lagrangian.
Journal of High Energy Physics **1010** (2010) 085 (arXiv:1008.4884 (hep-ph)).

[219] Georgi, H. and Glashow, S. L. 1974.
Unity of All Elementary Particle Forces.
Physical Review Letters **32** (1974) 438.

[220] Georgi, H., Quinn, H. R. and Weinberg, S. 1974.
Hierarchy of Interactions in Unified Gauge Theories.
Physical Review Letters **33** (1974) 451.

[221] Dimopoulos, S., Raby, S. and Wilczek, F. 1981.
Supersymmetry and the Scale of Unification.
Physical Review **D24** (1981) 1681.

[222] Ibanez, L. E. and Ross, G. G. 1981.
Low-Energy Predictions in Supersymmetric Grand Unified Theories.
Physics Letters **B105** (1981) 439.

[223] Dimopoulos, S. and Georgi, H. 1981.
Softly Broken Supersymmetry and SU(5).
Nuclear Physics **B193** (1981) 150.

[224] Peccei, R. D. and Quinn, H. R. 1977.
CP Conservation in the Presence of Pseudoparticles.
Physical Review Letters **38** (1977) 1440.

[225] Weinberg, S. 1977.
A New Light Boson? *Physical Review Letters* **40** (1978) 223.

[226] Wilczek, F. 1977.
Problem of Strong *P* and *T* Invariance in the Presence of Instantons.
Physical Review Letters **40** (1978) 279.

[227] Weinberg, S. 1975.
Implications of Dynamical Symmetry Breaking.
Physical Review **D13** (1976) 974 Addendum: (Phys. Rev. **D19** (1979) 1277).

[228] Susskind, L. 1978.
Dynamics of Spontaneous Symmetry Breaking in the Weinberg-Salam Theory.
Physical Review **D20** (1979) 2619.

[229] Gervais, J.-L. and Sakita, B. 1971.
Field Theory Interpretation of Supergauges in Dual Models.
Nuclear Physics **B34** 632.

[230] Volkov, D. V. and Akulov, V. P. 1972.
Possible Universal Neutrino Interaction.
Pisma Zh.Eksp.Teor.Fiz. **16** (1972) 621;
(JETP Letters **16** (1972) 438)
Is the Neutrino a Goldstone Particle?
Physics Letters **46B** (1973) 109;
Teor.Mat.Fiz. **18** (1974) 39.

[231] Wess, J. and Zumino, B. 1974.
Supergauge Transformations in Four Dimensions.
Nuclear Physics **B70** 39.

[232] Farrar, G. R. and Fayet, P. 1978.
Phenomenology of the Production, Decay, and Detection of New Hadronic States Associated with Supersymmetry.
Physics Letters **76B** (1978) 575.

[233] Fayet, P. 1977.
Spontaneously Broken Supersymmetric Theories of Weak, Electromagnetic and Strong Interactions.
Physics Letters **69B** (1977) 489.

[234] Witten, E. 1981.
Dynamical Breaking of Supersymmetry.
Nuclear Physics **B188** (1981) 513.

[235] Alvarez-Gaume, L., Polchinski, J. and Wise, M. B. 1983.
Minimal Low-Energy Supergravity.
Nuclear Physics **B221** (1983) 495.

[236] Arkani-Hamed, N., Dimopoulos, S. and Dvali, G. 1998.
The Hierarchy Problem and New Dimensions at a Millimeter.
Physics Letters **B429** (1998) 263 (arXiv:hep-ph/9803315);
Arkani-Hamed, N., Dimopoulos, S. and Dvali, G. 1999.
Phenomenology, Astrophysics and Cosmology of Theories with

Submillimeter Dimensions and TeV Scale Quantum Gravity.
Physical Review **D59** (1999) 086004 (arXiv:hep-ph/9807344).

[237] Randall, L. and Sundrum, R. 1999.
Large Mass Hierarchy from a Small Extra Dimension.
Physical Review Letters **83** (1999) 33703373 (arXiv:hep-ph/9905221).
Randall, L. and Sundrum, R. 1999.
An Alternative to Compactification.
Physical Review Letters **83** (1999) 4690 (arXiv:hep-th/9906064).

[238] Riess, A. G. *et al.* 1998.
Observational Evidence from Supernovae for an
Accelerating Universe and a Cosmological Constant.
Astronomical Journal **116** (1998) 100938 (arXiv:astro-ph/9805201).

[239] Perlmutter, S. *et al.* 1999.
Measurements of Omega and Lambda from 42 High Redshift Supernovae.
Astrophysical Journal **517** (1999) 565 (arXiv:astro-ph/9812133).

[240] Weinberg, S. 1989.
The Cosmological Constant Problem.
Reviews of Modern Physics **61** (1989) 1.

[241] Einstein, E. 1915.
Erklärung der Perihelbewegung des Merkur aus der allgemeinen Relativitätstheorie.
Königlich Preußische Akademie der Wissenschaften (Berlin). Sitzungsberichte, 10301085.
Einstein, E. 1915.
Die Feldgleichungen der Gravitation.
ibid., 844847.

[242] Einstein, E. 1916.
Die Grundlage der allgemeinen Relativitätstheorie.
Annalen der Physik **49** 769822.

[243] Feynman, R. P. 1963.
Quantum Theory of Gravitation.
Acta Physica Polonica **24** (1963) 697.

[244] DeWitt, B. S. 1967.
Quantum Theory of Gravity. 1. The Canonical Theory.
Physical Review **160** (1967) 1113.
DeWitt, B. S. 1967.
Quantum Theory of Gravity. 2. The Manifestly Covariant Theory.
Physical Review **162** (1967) 1195.
DeWitt, B. S. 1967.
Quantum Theory of Gravity. 3. Applications of the Covariant Theory.
Physical Review **162** (1967) 1239.

[245] Hilbert, D. 1915.
Die Grundlagen der Physik.
Königliche Gesellschaft der Wissenschaften zu Göttingen Mathematisch-physikalische Klasse. Nachrichten, 395407.

[246] Lovelock, D. 1971.
The Einstein Tensor and Its Generalizations.

Journal of Mathematical Physics **12** 498501.
Lovelock, D. 1972.
The Four-Dimensionality of Space and the Einstein Tensor.
Journal of Mathematical Physics **13** 874876.

[247] Donoghue, J. F. 1994.
Leading Quantum Correction to the Newtonian Potential.
Physical Review Letters **72** (1994) 2996 (gr-qc/9310024).
Donoghue, J. F. 1994.
General Relativity as an Effective Field Theory: The Leading Quantum Corrections.
Physical Review **D50** (1994) 3874 (gr-qc/9405057).

[248] Gross, D. J. and Sloan, J. H. 1986.
The Quartic Effective Action for the Heterotic String.
Nuclear Physics **B291** (1987) 41.

[249] Simon, J. Z. 1991.
The Stability of Flat Space, Semiclassical Gravity, and Higher Derivatives.
Physical Review **D43** (1991) 3308.

[250] Burgess, C. P., Holman, R., Tasinato, G. and Williams, M. 2014.
EFT Beyond the Horizon: Stochastic Inflation and How Primordial Quantum Fluctuations Go Classical.
Journal of High Energy Physics **03** 090 (arXiv:1408.5002 (hep-th)).

[251] Burgess, C.P., Holman, R. and Tasinato, G. 2015.
Open EFTs, IR effects & late-time resummations: systematic corrections in stochastic inflation.
Journal of High Energy Physics **01** 153 (arXiv:1512.00169 (gr-qc)).

[252] Will, C. M. 2014.
The Confrontation between General Relativity and Experiment.
Living Reviews of Relativity **17** (2014) 4 (arXiv:1403.7377 (gr-qc)).

[253] Berti, E. *et al.* 2015.
Testing General Relativity with Present and Future Astrophysical Observations.
Classical and Quantum Gravity **32** (2015) 243001 (arXiv:1501.07274 (gr-qc)).

[254] Hawking, S. W. 1974.
Black Hole Explosions?
Nature **248** (5443) 3031.
Hawking, S. W. 1974.
Particle Creation by Black Holes.
Communications in Mathematical Physics **43** (1975) 199
Erratum: (*Communications in Mathematical Physics* **46** (1976) 206).

[255] See, for example:
Peebles, P. J. E. 1980.
The Large-Scale Structure of the Universe.
(Princeton Press, 1980).
Peebles, P. J. E. 1993.
Principles of Physical Cosmology.
(Princeton Press, 1993).

Mukhanov, V. F. 2005.
Physical Foundations of cosmology.
(Cambridge Press, 2005).

[256] Jeans, J. H. 1902.
The Stability of a Spherical Nebula.
Philosophical Transactions of the Royal Society **A 199** 153.

[257] Mukhanov, V. F. and Chibisov, G. V. 1981.
Quantum Fluctuations and a Nonsingular Universe.
Pisma Zh.Eksp.Teor.Fiz. **33** (1981) 549;
(JETP Letters **33** (1981) 532).
Guth, A. H. and Pi, S. Y. 1982.
Fluctuations in the New Inflationary Universe.
Physical Review Letters **49** (1982) 1110.
Starobinsky, A. A. 1982.
Dynamics of Phase Transition in the New Inflationary
Universe Scenario and Generation of Perturbations.
Physics Letters **117B** (1982) 175.
Hawking, S. W. 1982.
The Development of Irregularities in a Single Bubble Inflationary Universe.
Physics Letters **115B** (1982) 295.
Linde, A. D. 1982.
Scalar Field Fluctuations in Expanding Universe
and the New Inflationary Universe Scenario.
Physics Letters **116B** (1982) 335.
Bardeen, J. M., Steinhardt, P. J. and Turner, M. S. 1983.
Spontaneous Creation of Almost Scale – Free Density
Perturbations in an Inflationary Universe.
Physical Review **D28** (1983) 679.

[258] Aghanim, N. *et al.* (Planck Collaboration) 2018.
Planck 2018 Results. VI. Cosmological parameters.
(arXiv:1807.06209 (astro-ph.CO)).

[259] Guth, A. H. 1981.
The Inflationary Universe: A Possible Solution to the Horizon and Flatness Problems.
Physical Review **D23** (1981) 347–356.
Linde, A. D. 1982.
A New Inflationary Universe Scenario: A Possible Solution of the Horizon, Flatness, Homogeneity, Isotropy and Primordial Monopole Problems.
Physics Letters **108B** (1982), 389–393.
Albrecht, A. and Steinhardt, P. J. 1982.
Cosmology for Grand Unified Theories with Radiatively Induced Symmetry Breaking.
Physical Review Letters **48** (1982) 1220–1223.
Linde, A. D. 1983.
Chaotic Inflation.
Physics Letters **129B** (1983) 177.

[260] Burgess, C. P., Lee H. M. and Trott, M. 2009.
Power-Counting and the Validity of the Classical Approximation during Inflation.
Journal of High Energy Physics **0909** (2009) 103 (arXiv:0902.4465 (hep-ph)).
Adshead, P., Burgess, C. P., Holman R. and Shandera, S. 2018.
Power-Counting during Single-Field Slow-Roll Inflation.
Journal of Cosmology and Astroparticle Physics **1802** (2018) 016 (arXiv:1708.07443 (hep-th)).

[261] Liddle, A. R. and Lyth, D. H. 1992.
COBE, Gravitational Waves, Inflation and Extended Inflation.
Physics Letters **291B** (1992) 391 (astro-ph/9208007).

[262] Akrami, Y. *et al.* (Planck Collaboration). 2018.
Planck 2018 Results. X. Constraints on Inflation.
(arXiv:1807.06211 (astro-ph.CO)).

[263] See for example: Oriti, D. 2009.
Approaches to Quantum Gravity: Toward a New Understanding of Space, Time and Matter (Cambridge Press, 2009).

[264] Nambu, Y. 1970.
Quark model and the Factorization of the Veneziano Amplitude.
In *Symmetries and Quark Models: Proceedings of the International Conference* World Scientific 1969 (pp. 258–267).
Nielsen, H. B. 1969.
An almost physical interpretation of the dual N point function.
Nordita preprint (1969) unpublished.

[265] Susskind, L. 1969.
Harmonic Oscillator Analogy for the Veneziano Amplitude.
Physical Review Letters **23** 545547.
Susskind, L. 1970.
Structure of Hadrons Implied by Duality.
Physical Review **D1** 11821186.

[266] Ramond, P. 1971.
Dual Theory for Free Fermions.
Physical Review **D3** 2415.
Neveu, A. and Schwarz, J. 1971.
Tachyon-Free Dual Model with a Positive-Intercept Trajectory.
Physics Letters **34B** 517518.
Gliozzi, F., Scherk J. and Olive, D. I. 1977.
Supersymmetry, Supergravity Theories and the Dual Spinor Model.
Nuclear Physics **B122** (1977) 253.

[267] Duff, M., Howe, P., Inami, T. and Stelle, K. 1987.
Superstrings in D = 10 from Supermembranes in D = 11.
Nuclear Physics **B191** 7074.
Witten, E. 1995.
String Theory Dynamics in Various Dimensions.
Nuclear Physics **B443** (1995) 85 (hep-th/9503124).

Horava, P. and Witten, E. 1995.
Heterotic and Type I String Dynamics from Eleven Dimensions.
Nuclear Physics B **460** 506524 (arXiv:hep-th/9510209).
Duff, M. 1996.
M-theory (the Theory Formerly Known As Strings).
International Journal of Modern Physics A **11** 652341 (arXiv:hep-th/9608117).

[268] Green, M. B. and Schwarz, J. H. 1982.
Supersymmetrical String Theories.
Physics Letters **109B** 444448.
Green, M. B. and Schwarz, J. H. 1984.
Anomaly Cancellations in Supersymmetric D = 10 Gauge Theory and Superstring Theory.
Physics Letters **149B** 117122.
Gross, D. J., Harvey, J. A., Martinec, E. J. and Rohm, R. 1985.
Heterotic String Theory. 1. The Free Heterotic String.
Nuclear Physics **B256** (1985) 253.

[269] Gross, D. J., Harvey, J. A., Martinec, E. J. and Rohm, R. 1986.
Heterotic String Theory. 2. The Interacting Heterotic String.
Nuclear Physics **B267** (1986) 75.

[270] Witten, E. 1984.
Some Properties of O(32) Superstrings.
Physics Letters **149B** (1984) 351.

[271] Witten, E. 1986.
New Issues in Manifolds of SU(3) Holonomy.
Nuclear Physics **B268** (1986) 79.
Burgess, C. P., Font, A. and Quevedo, F. 1986.
Low-Energy Effective Action for the Superstring.
Nuclear Physics **B272** (1986) 661.
Burgess, C. P., Escoda, C. and Quevedo, F. 2006.
Nonrenormalization of Flux Superpotentials in String Theory.
Journal of High Energy Physics **0606** (2006) 044 (hep-th/0510213).

[272] Veneziano, G. 1968.
Construction of a Crossing-Symmetric,
Regge-Behaved Amplitude for Linearly Rising Trajectories.
Nuovo Cimento **A57** 1907.

[273] Virasoro, M. 1969.
Alternative Constructions of Crossing-Symmetric Amplitudes with Regge Behavior.
Physical Review **177** 23092311.
Shapiro, J. A. 1970.
Electrostatic Analogue for the Virasoro Model.
Physics Letters **33B** 361362.

[274] Gross, D. J. and Witten, E. 1986.
Superstring Modifications of Einstein's Equations.
Nuclear Physics **B277** (1986) 1.

[275] Candelas, P., Horowitz, G. T., Strominger, A. and Witten, E. 1985.
Vacuum Configurations for Superstrings.
Nuclear Physics **B258** (1985) 46.
Giddings, S. B., Kachru, Ŝ. and Polchinski, J. 2002.
Hierarchies from Fluxes in String Compactifications.
Physical Review **D66** (2002) 106006 (hep-th/0105097).

[276] Kaluza, T. 1921.
Zum Unitätsproblem in der Physik.
Sitzungsberichte der Königlich Preussischen Akademie der Wissenschaften Berlin (Math. Phys.) 966972.
Klein, O. 1926.
Quantentheorie und funfdimensionale Relativittstheorie.
Zeitschrift für Physik **A37** 895906.
Klein, O. 1926.
The Atomicity of Electricity as a Quantum Theory Law.
Nature **118** 516.
Witten, E. 1981.
Search for a Realistic Kaluza–Klein Theory.
Nuclear Physics **B186** (1981) 412.

[277] Cremmer, E. and Scherk, J. 1976.
Spontaneous Compactification of Space in an Einstein Yang–Mills Higgs Model.
Nuclear Physics **B108** (1976) 409.
Freund P. G. O. and Rubin M. A. 1980.
Dynamics of Dimensional Reduction.
Physics Letters **97B** (1980) 233.
Candelas P. and Weinberg S. 1984.
Calculation of Gauge Couplings and Compact Circumferences from Self-Consistent Dimensional Reduction.
Nuclear Physics **B237** (1984) 397.
Salam A. and Sezgin E. 1984.
Chiral Compactification on Minkowski $\times S^2$ of $N = 2$ Einstein-Maxwell Supergravity in Six-Dimensions.
Physics Letters **147B** (1984) 47.

[278] Christensen, S. M. and Duff, M. J. 1979.
New Gravitational Index Theorems and Supertheorems.
Nuclear Physics **B154** (1979) 301.
Hoover D. and Burgess, C. P. 2005.
Ultraviolet Sensitivity in Higher Dimensions.
Journal of High Energy Physics **0601** (2006) 058 (hep-th/0507293).

[279] Burgess, C. P. and Hoover, D. 2005.
UV Sensitivity in Supersymmetric Large Extra Dimensions: The Ricci-Flat Case.
Nuclear Physics **B772** (2007) 175 (hep-th/0504004).

[280] Kaloper, N., March-Russell, J., Starkman, G. D. and Trodden, M. 2000.
Compact Hyperbolic Extra Dimensions: Branes, Kaluza–Klein Modes and

Cosmology.
Physical Review Letters **85** (2000) 928 (hep-ph/0002001).

[281] Einstein, A. 1916.
Näherungsweise Integration der Feldgleichungen der Gravitation.
Sitzungsberichte der Königlich Preussischen Akademie der Wissenschaften Berlin
part 1: 688–696.
Einstein, A. 1918.
Über Gravitationswellen.
Sitzungsberichte der Königlich Preussischen Akademie der Wissenschaften Berlin
part 1: 154–167.

[282] Lichnerowicz, A. 1961.
Propagateurs et commutateurs en relativité générale.
Publications Mathématiques de l'Institute des Hautes Études Scientifiques **10** (1961) 293–344.

[283] Kachru, S., Kallosh, R., Linde, A. D. and Trivedi, S. P. 2003.
De Sitter Vacua in String Theory.
Physical Review **D68** (2003) 046005 (hep-th/0301240).
Balasubramanian, V., Berglund, P., Conlon, J. P. and Quevedo, F. 2005.
Systematics of Moduli Stabilisation in Calabi-Yau Flux Compactifications.
Journal of High Energy Physics **0503** (2005) 007 (hep-th/0502058).

[284] Antoniadis, I., Arkani-Hamed, N., Dimopoulos S. and Dvali, G. R. 1998.
New Dimensions at a Millimeter to a Fermi and Superstrings at a TeV.
Physics Letters **436B** (1998) 257 (hep-ph/9804398).

[285] Arkani-Hamed, N., Dimopoulos S. and Dvali, G. R. 1998.
The Hierarchy Problem and New Dimensions at a Millimeter.
Physics Letters **429B** (1998) 263 (hep-ph/9803315).

[286] Kapner, D. J. *et al.* (Eot-Wash collaboration). 2007.
Tests of the Gravitational Inverse-Square Law below the Dark-Energy Length Scale.
Physical Review Letters **98** (2007) 021101 (hep-ph/0611184).

[287] Hawking, S. W. 1976.
Breakdown of Predictability in Gravitational Collapse.
Physical Review **D14** (1976) 2460.

[288] De Witt, B. S. 1965.
Dynamical Theory of Groups and Fields.
In *Relativity, Groups and Topology,* ed. by B. S. De Witt and C. De Witt, (New York, Gordon and Breach, 1965).

[289] Gilkey, P. B. 1975.
The Spectral Geometry of a Riemannian Manifold.
Journal of Differential Geometry **10** 601.

[290] Barvinsky, A. O. and Vilkovisky, G. A. 1985.
The Generalized Schwinger-DeWitt Technique in Gauge Theories and Quantum Gravity.
Physics Reports **119** 1.

[291] Vassilevich, D. V. 2003.
Heat Kernel Expansion: Users Manual.
Physics Reports **388** 279 [hep-th/0306138].

[292] Foldy, L. L. and Wouthuysen, S. A. 1950.
On the Dirac Theory of Spin 1/2 Particles and Its Non-Relativistic Limit.
Physical Review **78** 2936.
Foldy, L. L. 1952.
The Electromagnetic Properties of the Dirac Particles.
Physical Review **87** 688693.

[293] Pauli, W. 1927.
Zur Quantenmechanik des magnetischen Elektrons.
Zeitschrift für Physik **43** 601–623.

[294] Caswell, W. E. and Lepage, G. P. 1986.
Effective Lagrangians for Bound State Problems in
QED, QCD, and Other Field Theories.
Physics Letters **167B** (1986) 437.

[295] Labelle, P. 1996.
Effective Field Theories for QED Bound States: Extending
Nonrelativistic QED to Study Retardation Effects.
Physical Review **D58** (1998) 093013 (hep-ph/9608491).

[296] Luke, M. E. and Manohar, A. V. 1996.
Bound States and Power Counting in Effective Field Theories.
Physical Review **D55** (1997) 4129 (hep-ph/9610534).

[297] Grinstein, B. and Rothstein, I. Z. 1997.
Effective Field Theory and Matching in Nonrelativistic Gauge Theories.
Physical Review **D57** (1998) 78 (hep-ph/9703298).

[298] Pineda, A. and Soto, J. 1998.
Effective Field Theory for Ultrasoft Momenta in NRQCD and NRQED.
Nuclear Physics Proceedings Supplement **64** (1998) 428 (hep-ph/9707481).

[299] Bauer, C. W., Fleming, S., Pirjol, D. and Stewart, I. W. 2001.
An Effective Field Theory for Collinear and Soft Gluons: Heavy to Light Decays.
Physical Review **D63** (2001) 114020 (arXiv:hep-ph/0011336).
Bauer, C. W., Pirjol, D. and Stewart, I. W. 2002.
Soft-Collinear Factorization in Effective Field Theory.
Physical Review **D65** (2002) 054022 (arXiv:hep-ph/0109045).
Bauer, C. W., Pirjol, D. and Stewart, I. W. 2002.
Power Counting in the Soft-Collinear Effective Theory.
Physical Review **D66** (2002) 054005 (arXiv:hep-ph/0205289).
Beneke, M., Chapovsky, A. P., Diehl, M. and Feldmann, T. 2002.
Soft Collinear Effective Theory and Heavy to Light Currents beyond Leading Power.
Nuclear Physics **B643** (2002) 431 (hep-ph/0206152).

[300] Goldberger, W. D. and Rothstein, I. Z. 2004.
An Effective Field Theory of Gravity for Extended Objects.
Physical Review **D73** (2006) 104029 (hep-th/0409156).

[301] Bethe, H. A. and Salpeter, E. E. 1957.
Quantum Mechanics of One- and Two-Electron Atoms.
Springer Verlag doi:10.1007/978-3-662-12869-5.

[302] Sudakov, V. V. 1956.
Vertex Parts at Very High-Energies in Quantum Electrodynamics.
Zhurnal Éksperimental'noĭ i Teoreticheskoĭ Fiziki **30** (1956) 87–95
(*Soviet Physics JETP* **3** (1956) 65–71).

[303] Bauer, C. W., Fleming, S. and Luke, M. 2000.
Summing Sudakov Logarithms in $B \to X_s \gamma$ in Effective Field Theory.
Physical Review **D63** (2000) 014006 (arXiv:hep-ph/0005275).

[304] Luke M. E. and Manohar, A. V. 1992.
Reparametrization Invariance Constraints on Heavy Particle Effective Field Theories.
Physics Letters **B286** (1992) 348 (hep-ph/9205228).

[305] Manohar, A. V. 1997.
The HQET / NRQCD Lagrangian to Order α/m^3.
Physical Review **D56** (1997) 230 (hep-ph/9701294).

[306] Heinonen, J., Hill, R. J. and Solon, M. P. 2012.
Lorentz Invariance in Heavy Particle Effective Theories.
Physical Review **D86** (2012) 094020 (arXiv:1208.0601 (hep-ph)).

[307] Labelle P. and Zebarjad, S. M. 1996.
Derivation of the Lamb Shift Using an Effective Field Theory.
Canadian Journal of Physics **77** (1999) 267 (hep-ph/9611313).

[308] Hill, R. J., Lee, G., Paz, G. and Solon, M. P. 2012.
NRQED Lagrangian at Order $1/M^4$.
Physical Review **D87** (2013) 053017 (arXiv:1212.4508 (hep-ph)).

[309] Rosenbluth, M. N. 1950.
High Energy Elastic Scattering of Electrons on Protons.
Physical Review **79** (1950) 615.

[310] Hofstadter, R. 1956.
Electron Scattering and Nuclear Structure.
Reviews of Modern Physics **28** (1956) 214.

[311] Klein, O. and Nishina, Y. 1929.
Über die Streuung von Strahlung durch freie Elektronen nach der neuen relativistischen Quantendynamik von Dirac.
Zeitschrift fur Physik **52** (1929) 853.

[312] Born, M. and Oppenheimer, J. R. 1927.
Zur Quantentheorie der Molekeln.
Annalen der Physik **389** (20) 457484.

[313] Kinoshita, T. and Nio, M. 1995.
Radiative Corrections to the Muonium Hyperfine Structure: The $\alpha^2(Z\alpha)$ Correction.
Physical Review **D53** (1996) 4909 (hep-ph/9512327).

[314] Politzer, H. D. and Wise, M. B. 1988.
Effective Field Theory Approach to Processes Involving Both Light and

Heavy Fields.
Physics Letters **B208** (1988) 504.
Eichten, E. and Hill, B. R. 1990.
An Effective Field Theory for the Calculation of Matrix Elements Involving Heavy Quarks.
Physics Letters **B234** (1990) 511.
Georgi, H. 1990.
An Effective Field Theory for Heavy Quarks at Low-energies.
Physics Letters **B240** (1990) 447.
Falk, A. F., Georgi, H., Grinstein, B. and Wise, M. B. 1990.
Heavy Meson Form-factors From QCD.
Nuclear Physics **B343** (1990) 1.

[315] Coleman, S., Mandula, J. 1967.
All Possible Symmetries of the S Matrix.
Physical Review **159** 1251.

[316] Isgur, N. and Wise, M. B. 1989.
Weak Decays of Heavy Mesons in the Static Quark Approximation.
Physics Letters **B232** (1989) 113.

[317] Voloshin, M. B. and Shifman, M. A. 1987.
On Annihilation of Mesons Built from Heavy and Light Quark and anti-B0 ↔ B0 Oscillations.
Soviet Journal of Nuclear Physics **45** (1987) 292 (*Yadernaya Fizika* **45** (1987) 463).
Bauer, C. and Manohar, A. V. 1997.
Renormalization Group Scaling of the $1/m^2$ HQET Lagrangian.
Physical Review **D57** (1998) 337 (hep-ph/9708306).

[318] Labelle, P., Zebarjad, Z. M. and Burgess, C. P. 1997.
NRQED and Next-to-Leading Hyperfine Splitting in Positronium.
Physical Review **D56** (1997) 8053 (hep-ph/9706449).

[319] Pineda, A. and Soto, J. 1998.
Potential NRQED: The Positronium Case.
Physical Review **D59** (1999) 016005 (hep-ph/9805424).

[320] Fierz, M. 1937.
Zur Fermischen Theorie des β-Zerfalls.
Zeitschrift fur Physik **104** (1937) 553.

[321] Abalmasov, V. A. 1998.
Comment on Nonrelativistic QED and Next-to-Leading Hyperfine Splitting in Positronium.
Physical Review **D58** (1998) 128701.

[322] Pineda, A. and Soto, J. 1997.
Effective Field Theory for Ultrasoft Momenta in NRQCD and NRQED.
Nuclear Physics Proceedings Supplement **64** (1998) 428 (hep-ph/9707481);
Matching at one loop for the four quark operators in NRQCD.
Physical Review **D58** (1998) 114011 (hep-ph/9802365).

[323] Lehmann, H., Symanzik, K. and Zimmerman, W. 1955.
Zur Formulierung quantisierter Feldtheorien.
Nuovo Cimento **1** (1955) 205.

[324] Wheeler, J. A. 1946.
Polyelectrons.
Annals of the New York Academy of Sciences **46** (1946) 221.
Harris, I. and Brown, L. M. 1957.
Radiative Corrections to Pair Annihilation.
Physical Review **105** (1957) 1656.
Czarnecki, A., Melnikov, K. and Yelkhovsky, A. 1999.
α^2 Corrections to Parapositronium Decay.
Physical Review Letters **83** (1999) 1135 (hep-ph/9904478).
Physical Review **A61** (2000) 052502 (hep-ph/9910488).

[325] Ore, A. and Powell, J. L. 1949.
Three Photon Annihilation of an Electron Positron Pair.
Physical Review **75** (1949) 1696.
Adkins, G. S. 1996.
Analytic Evaluation of the Orthopositronium-to-Three-Photon Decay Amplitudes to One-Loop Order.
Physical Review Letters **76** (1996) 4903 (hep-ph/0506213).
Hill, R. J. and Lepage, G. P. 2000.
Order ($\alpha^2\Gamma$, $\alpha^3\Gamma$) Binding Effects in Orthopositronium Decay.
Physical Review **D62** (2000) 111301(R) (hep-ph/0003277).
Kniehl, B. and Penin, A. A. 2000.
Order $\alpha^3 \ln(1/\alpha)$ Corrections to Positronium Decays.
Physical Review Letters **85** (2000) 1210; Erratum **85** 3065 (hep-ph/0004267).
K. Melnikov and A. Yelkhovsky, O($\alpha^3 \ln \alpha$) Corrections to Positronium Decay Rates.
Physical Review **D62** (2000) 116003 (hep-ph/0008099).

[326] Karplus, R. and Klein, A. 1952.
Electrodynamics Displacement of Atomic Energy Levels 3.
The Hyperfine Structure of Positronium.
Physical Review **87** (1952) 848.
Gupta, S. N., Repko, W. W. and Suchyta, C. G. 1989.
Muonium and Positronium Potentials.
Physical Review **D40** (1989) 4100.
Pachucki, K. and Karshenboim, S. G. 1998.
Complete Result for Positronium Energy Levels at Order $m\alpha^6$.
Physical Review Letters **80** (1998) 2101 (hep-ph/9709387).

[327] Lepage, G. P. and Thacker, B. A. 1988.
Effective Lagrangians for Simulating Heavy Quark Systems.
Nuclear Physics Proceedings Supplement **4** (1988) 199.
Thacker, B. A. and Lepage, G. P. 1991.
Heavy Quark Bound States in Lattice QCD.
Physical Review **D43** (1991) 196.
Bodwin, G. T., Braaten, E. and Lepage, G. P. 1995.
Rigorous QCD Analysis of Inclusive Annihilation and Production of Heavy Quarkonium.
Physical Review **D51** (1995) 1125
Erratum: (Physical Review **D55** (1997) 5853) (hep-ph/9407339).

[328] Zel'dovich, Ya. B. 1957.
Parity Nonconservation in the First Order in the Weak-Interaction Constant in Electron Scattering and Other Effects.
Zhurnal Éksperimental'noĭ i Teoreticheskoĭ Fiziki **33** 1531 (*Soviet Physics JETP* **6** (1957) 1184).

[329] Borie, E. 2012.
Lamb Shift in Light Muonic Atoms: Revisited.
Annals of Physics (NY) **327** (2012) 733.

[330] Pineda, A. and Soto, J. 1998.
The Lamb Shift in Dimensional Regularization.
Physics Letters **B420** (1998) 391 (hep-ph/9711292).

[331] Coleman, S. R. 1977.
Classical Lumps and Their Quantum Descendents.
Subnuclear Series **13** (1977) 297.
(reproduced in *Aspects of Symmetry: Selected Erice Lectures* ref. [17] above).

[332] Salam, A. and Strathdee, J. 1969.
Nonlinear Realizations 1: The Role of Goldstone Bosons.
Physical Review **184** (1969) 1750.

[333] Nielsen, H. B. and Olesen, P. 1973.
Vortex-Line Models for Dual Strings.
Nuclear Physics **B61** (1973) 45–61.

[334] Hughes, J. and Polchinski, J. 1986.
Partially Broken Global Supersymmetry and the Superstring.
Nuclear Physics **278** (1986) 147.

[335] Chern, S. S. and Simons, J. 1974.
Characteristic Forms and Geometric Invariants.
Annals of Mathematics **99** (1974) 4869.

[336] Schwarz, A. 1979.
The Partition Function of a Degenerate Functional.
Communications in Mathematical Physics **67** (1979) 1.
Witten, E. 1988.
Topological Quantum Field Theory.
Communications in Mathematical Physics **117** (1988) 353.
Atiyah, M. 1988.
Topological Quantum Field Theories.
Publications Mathématiques de l'IHÉS **68** (1988) 175.

[337] Deser, S. and Zumino, B. 1976.
A Complete Action for the Spinning String.
Physics Letters **B65** (1976) 369.

[338] Brink, L., Di Vecchia, P. and Howe, P. S. 1976.
A Locally Supersymmetric and Reparametrization Invariant Action for the Spinning String.
Physics Letters **B65** (1976) 471.

[339] Polyakov, A. M. 1981.
Quantum Geometry of the Bosonic String.
Physics Letters **B103** (1981) 207.

[340] Grassmann, H. 1844.
Die Lineale Ausdehnungslehre Ein neuer Zweig der Mathematik.
(Verlag, Leipzig, 1844)

[341] Lw, R., Weimer, H., Nipper, J. and Balewski, J. B., Butscher, B., Büchler, H.-P. and Pfau, T. 2012.
An Experimental and Theoretical Guide to Strongly Interacting Rydberg Gases.
Journal of Physics B: Atomic, Molecular and Optical Physics **45** (2012) 113001.
Gallagher, T. F. 1994.
Rydberg Atoms.
(Cambridge Press, 1994)

[342] Burgess, C. P., Hayman, P., Williams, M. and Zalavari, L. 2017.
Point-Particle Effective Field Theory I: Classical Renormalization and the Inverse-Square Potential.
Journal of High Energy Physics **1704** (2017) 106 (arXiv:1612.07313 (hep-ph)).

[343] de Boer, J., Verlinde, E. P. and Verlinde, H. L. 2000.
On the Holographic Renormalization Group.
Journal of High Energy Physics **0008** (2000) 003 (hep-th/9912012).

[344] Maldacena, J. M. 1997.
The Large N Limit of Superconformal Field Theories and Supergravity.
International Journal of Theoretical Physics **38** (1999) 1113 (*Advances in Theoretical and Mathematical Physics* **2** (1998) 231) (hep-th/9711200).
Witten, E. 1998.
Anti-de Sitter Space and Holography.
Advances in Theoretical and Mathematical Physics **2** (1998) 253 (hep-th/9802150).
Gubser, S. S., Klebanov, I. R. and Polyakov, A. M. 1998.
Gauge Theory Correlators from Noncritical String Theory.
Physics Letters **B428** (1998) 105 (hep-th/9802109).

[345] Preston, M. A. and Bhaduri, R. K. 1982.
Structure of the Nucleus.
(Addison-Wesley, Reading Massachusetts, 1975; 2nd printing 1982).

[346] Kaplan, D. B., Savage, M. J. and Wise, M. B. 1998.
Two Nucleon Systems from Effective Field Theory.
Nuclear Physics **B534** (1998) 329 (nucl-th/9802075).
E. Braaten and Hammer, H.-W. 2006.
Universality in Few-Body Systems with Large Scattering Length.
Physics Reports **428** (2006) 259 (cond-mat/0410417).

[347] Feshbach, H. 1958.
Unified Theory of Nuclear Reactions.
Annals of Physics **5** (1958) 357.

Fano, U. 1961.
Effects of Configuration Interaction on Intensities and Phase Shifts.
Physical Review **124** (1961) 1866.

[348] Jackiw, R. 1991.
Delta Function Potentials in Two-Dimensional and
Three-Dimensional Quantum Mechanics.
In Jackiw, R.: *Diverse Topics in Theoretical and Mathematical Physics*
35–53 (1991).

[349] Weyl, H. 1910.
Über gewöhnliche Differentialgleichungen mit Singularitäten
und de zeugehörigen Entwicklungen willkürlicher Funktionen.
Mathematische Annalen **68** (1910) 220.
von Neumann, J. 1929.
Allgemeine Eigenwertheorie Hermitescher Funktionaloperatoren.
Mathematische Annalen **102** (1929) 49–131.
Stone, M. H. 1932.
On One-Parameter Unitary Groups in Hilbert Space.
Annals of Mathematics **33** (1932) 643–648.
Berezin, F. A. and Faddeev, L. D. 1961.
A Remark on Schrodingers Equation with a Singular Potential.
Proceedings of the Soviet Academy of Sciences **2** (1961) 372
(*Doklady Akademii Nauk (Ser. Fiz.)* **137** (1961) 1011).

[350] Plestid, R., Burgess, C. P. and O'Dell, D. H. J. 2018.
Fall to the Centre in Atom Traps and Point-Particle EFT for Absorptive Systems.
Journal of High Energy Physics **1808** (2018) 059
(arXiv:1804.10324 (hep-ph)).

[351] Efimov, V. 1970.
Energy Levels Arising from Resonant Two-Body Forces in a Three-Body System.
Physics Letters **B33** (1970) 563–564.
Braaten, E. and Hammer, H. W. 2007.
Efimov Physics in Cold Atoms.
Annals of Physics **322** (2007) 120 (cond-mat/0612123).

[352] Burgess, C. P., Plestid, R. and Rummel, M.
Effective Field Theory of Black Hole Echoes.
Journal of High Energy Physics **1809** (2018) 113 (arXiv:1808.00847 (gr-qc)).

[353] Erickson, G. W. 1977.
Energy Levels of One electron Atoms.
Journal of Physical Chemistry Reference Data **6** (1977) 831.
Friar, J. L. 1979.
Nuclear Finite Size Effects in Light Muonic Atoms.
Annals of Physics **122** (1979) 151.
Friar, J. L. and Sick, I. 2005.
Muonic Hydrogen and the Third Zemach Moment.
Physical Review **A72** (2005) 040502 (nucl-th/0508025).

[354] Backenstoss, G. 1970.
Pionic Atoms.
Annual Reviews of Nuclear and Particle Science **20** (1970) 467.

[355] Deser, S., Goldberger, M. L., Baumann, K. and Thirring, W. E. 1954.
Energy Level Displacements in Pi-Mesonic Atoms.
Physical Review **96** (1954) 774.

[356] Burgess, C. P., Hayman, P., Rummel, M. and Zalavari, L. 2017.
Point-Particle Effective Field Theory III: Relativistic Fermions and the Dirac Equation.
Journal of High Energy Physics **1709** (2017) 007 (arXiv:1706.01063 (hep-ph)).

[357] Kaplan, D. B., Lee, J. W., Son, D. T. and Stephanov, M. A. 2009.
Conformality Lost.
Physical Review **D80** (2009) 125005 (arXiv:0905.4752 (hep-th)).

[358] Arnold, V. I., Kozlov, V. V. and Neishtadt, A. I. 1988.
Dynamical Systems III: Mathematical Aspects of Classical and Celestial Mechanics.
Encyclopedia of Mathematical Science **3** (Springer, Berlin, 1988).
Lochack, P. and Meunier, C. 1988.
Multiphase Averaging for Classical Systems.
(Springer, New York, 1988.)
Shapere, A. D. and Wilczek, F. 1989.
Geometric Phases in Physics.
Advanced Series in Mathematical Physics **5** (1989) 1.
Berry, M. V. and Robins, J. M. 1993.
Classical Electromagnetic Forces of Reaction: An Exactly Solvable Model.
Proceedings of the Royal Academy of Sciences **A44** (1993) 631.

[359] Goldberger, W. D. and Rothstein, I. Z. 2004.
An Effective Field Theory of Gravity for Extended Objects.
Physical Review **D73** (2006) 104029 (hep-th/0409156);
Dissipative Effects in the Worldline Approach to Black Hole Dynamics.
Physical Review **D73** (2006) 104030 (hep-th/0511133).
Porto, R. A., Ross, A. and Rothstein, I. Z. 2012.
Spin Induced Multipole Moments for the Gravitational Wave Amplitude from Binary Inspirals to 2.5 Post-Newtonian Order.
Journal of Cosmology and Astroparticle Physics **1209** (2012) 028 (arXiv:1203.2962 (gr-qc));

[360] Bethe, H. 1935.
Theory of Disintegration of Nuclei by Neutrons.
Physical Review **47** (1935) 747.

[361] Greene, B. R., Shapere, A. D., Vafa, C. and Yau, S. T. 1990.
Stringy Cosmic Strings and Noncompact Calabi-Yau Manifolds.
Nuclear Physics **B337** (1990) 1.

[362] Bayntun, A., Burgess, C. P. and van Nierop, L. 2010.
Codimension-2 Brane-Bulk Matching: Examples from Six and Ten

Dimensions.
New Journal of Physics **12** (2010) 075015 (arXiv:0912.3039 (hep-th)).

[363] Polchinski, J. 1998.
String Theory. Vol. 1: An Introduction to the Bosonic String.
String Theory. Vol. 2: Superstring Theory and beyond.
(Cambridge Press, 1998).

[364] Phillips, T. G. and Rosenberg, H. M. 1966.
Spin Waves in Ferromagnets.
Reports on Progress in Physics **29** (1966) 285

[365] Heisenberg, W. G. 1928.
Zur Theorie des Ferromagnetismus.
Zeitschrift für Physik **49** (1928) 619.
Holstein, T. and Primakoff, H. 1940.
Field Dependence of the Intrinsic Domain Magnetization of a Ferromagnet.
Physical Review **58** (1940) 1098.

[366] Dyson, F. J. 1956.
General Theory of Spin-Wave Interactions.
Physical Review **102** (1956) 1217.

[367] Néel, L. 1932.
Influence des fluctuations du champ moléculaire sur les propriétés magnetiques des corps.
Annales de Physique **18** (1932) 5.
Bitter, F. 1937.
A Generalization of the Theory of Ferromagnetism.
Physical Review **54** (1937) 79.
van Vleck, J. H. 1941.
On the Theory of Antiferromagnetism.
Journal of Chemical Physics **9** (1941) 85.

[368] Brouwer, L. E. J. 1912.
Uber Abbildung von Mannigfaltigkeiten.
Mathematische Annalen **71** (1912) 97.
Milnor, J. 1978.
Analytic Proofs of the 'Hairy Ball Theorem' and the Brouwer Fixed Point Theorem.
The American Mathematical Monthly **85** (1978) 521.

[369] Dirac, P. 1931.
Quantised Singularities in the Electromagnetic Field.
Proceedings of the Royal Society (London) **A133** (1931) 60.

[370] Landau, L. D. and Lifshitz, E. M. 1935.
Theory of the Dispersion of Magnetic Permeability in Ferromagnetic Bodies.
Physikalische Zeitschrift der Sowjetunion **8** (1935) 153.
Herring, C. and Kittel, C. 1951.
On the Theory of Spin Waves in Ferromagnetic Media.
Physical Review **81** (1951) 869.

[371] Brockhouse, B. N. 1957.
Scattering of Neutrons by Spin Waves in Magnetite.
Physical Review **106** (1957) 859.

[372] See *e.g.*: Zaliznyak, I. and Lee, S. 2005.
Magnetic Neutron Scattering.
In *Modern Techniques for Characterizing Magnetic Materials*, pp. 3–64 (Springer, 2005).

[373] Wightman, A. S. 1956.
Quantum Field Theory in Terms of Vacuum Expectation Values.
Physical Review **101** (1956) 860.

[374] Bloch, F. 1930.
Zur Theorie des Ferromagnetismus.
Zeitschrift für Physik **61** (1930) 206.

[375] Argyle, B. E., Charap, S. H. and Pugh, E. W. 1963.
Deviations from $T^{3/2}$ Law for Magnetization of Ferrometals: Ni, Fe, and Fe +3% Si.
Physical Review **132** (1963) 2051.

[376] Onnes, H. K. 1911.
The Resistance of Pure Mercury at Helium Temperatures.
Communications from the Laboratory of Physics at the University of Leiden **12** (1911) 120.

[377] Meissner, W. and Ochsenfeld, R. 1933.
Ein neuer Effekt bei Eintritt der Supraleitfähigkeit.
Naturwissenschaften **21** (1933) 787.

[378] London, F. 1948.
On the Problem of the Molecular Theory of Superconductivity.
Physical Review **74** (1948) 562.

[379] Deaver, B. and Fairbank, W. 1961.
Experimental Evidence for Quantized Flux in Superconducting Cylinders.
Physical Review Letters **7** (1961) 43.
Doll, R. and Näbauer, M. 1961.
Experimental Proof of Magnetic Flux Quantization in a Superconducting Ring.
Physical Review Letters **7** (1961): 51.

[380] Bardeen, J., Cooper, L. N. and Schrieffer, J. R. 1957.
Theory of Superconductivity.
Physical Review **108** (1957) 1175.

[381] Bednorz, J. G. and Müller, K. A. 1986.
Possible High TC Superconductivity in the Ba-La-Cu-O System.
Zeitschrift für Physik **B64** (1986) 189.

[382] Weinberg, S. 1986.
Superconductivity for Particular Theorists.
Progress in Theoretical Physics (Supplement) **86** (1986) 43.

[383] London, F. and London, H. 1935.
The Electromagnetic Equations of the Supraconductor.
Proceedings of the Royal Society **A149** (1935) (866) 71.

[384] Josephson, B. D. 1962.
Possible New Effects in Superconductive Tunnelling.
Physics Letters **1** (1962) 251.

[385] Anderson, P. W. and Rowell, J. M. 1963.
Probable Observation of the Josephson Tunnel Effect.
Physical Review Letters **10** (1963) 230.

[386] Shapiro, S. 1963.
Josephson Currents in Superconducting Tunneling:
the Effects of Microwaves and Other Observations.
Physical Review Letters **11** (1963) 80.

[387] Ginzburg, V. L. and Landau, L. D. 1950.
Zhurnal Eksperimentalnoi i Teoreticheskoi Fiziki **20** (1950) 1064.
Abrikosov, A. A. 1957.
Zhurnal Eksperimentalnoi i Teoreticheskoi Fiziki **32** (1957) 1442.
Gor'kov, L. P. 1959.
Zhurnal Eksperimentalnoi i Teoreticheskoi Fiziki **36** (1959) 1364.

[388] Rjabinin, J. N. and Schubnikow, L. W. 1935.
Magnetic Properties and Critical Currents of Superconducting Alloys.
Physikalische Zeitschrift der Sowjetunion **7** (1935), no.1, pp. 122125.
Magnetic Properties and Critical Currents of Supra-conducting Alloys.
Nature **135** (1935) 581.

[389] Abrikosov, A. A. 1957.
The Magnetic Properties of Superconducting Alloys.
Journal of Physics and Chemistry of Solids **2** (1957) 199.

[390] Leutwyler, H. 1996.
Phonons as Goldstone Bosons.
Helvetica Physica Acta **70** (1997) 275 (hep-ph/9609466).

[391] DeWitt, B. 1984.
in *Relativity, Groups and Topology II* (proceedings of the Les Houches School, ed. C. DeWitt and B. DeWitt) (Elsevier 1984).

[392] Girvin, S. M. and Yang K. 2019.
Modern Condensed Matter Physics.
(Cambridge Press, 2019).

[393] Brown, J. D. 1993.
Action Functionals for Relativistic Perfect Fluids.
Classical and Quantum Gravity **10** (1993) 1579 (gr-qc/9304026).

[394] Endlich, S., Nicolis, A., Rattazzi, R. and Wang, J. 2010.
The Quantum Mechanics of Perfect Fluids.
Journal of High Energy Physics **1104** (2011) 102 (arXiv:1011.6396 (hep-th)).

[395] Dubovsky, S., Hui, L., Nicolis, A. and Son, D. T. 2011.
Effective Field Theory for Hydrodynamics:
Thermodynamics, and the Derivative Expansion.
Physical Review **D85** (2012) 085029 (arXiv:1107.0731 (hep-th)).

[396] Misner, C. W., Thorne, K. S. and Wheeler, J. A. 1973.
Gravitation.
(Freeman – Princeton Press, 1973).

[397] Weinberg, S. 1972.
Gravitation and Cosmology: Principles and
Applications of the General Theory of Relativity.
(John Wiley and Sons, 1972).

[398] Wald, R. M. 1984.
General Relativity.
(University of Chicago Press, 1984).

[399] Crossley, M., Glorioso, P. and Liu, H. 2015.
Effective Field Theory of Dissipative Fluids.
Journal of High Energy Physics **1709** (2017) 095 (arXiv:1511.03646
(hep-th)).

[400] Mohr, P. J., Newell, D. B. and Taylor, B. N. 2014.
CODATA Recommended Values of the Fundamental Physical Constants:
2014.
Reviews of Modern Physics **88** (2016) 035009 (arXiv:1507.07956
(physics.atom-ph)).

[401] Landau, L. D. 1956.
The Theory of a Fermi Liquid.
Zhurnal Éksperimental'noǐ i Teoreticheskoǐ Fiziki **30** (1956) 1058
(*Soviet Physics JETP* **3** (1957) 920).

[402] Polchinski, J. 1992.
Effective Field Theory and the Fermi Surface.
In proceedings of the TASI school *Recent Directions in Particle Theory*
(hep-th/9210046).

[403] Shankar, R. 1993.
Renormalization Group Approach to Interacting Fermions.
Reviews of Modern Physics **66** (1994) 129 (cond-mat/9307009).

[404] Wichmann, E. H. and Crichton, J. H. 1963.
Cluster Decomposition Properties of the S Matrix.
Physical Review **132** (1963) 2788.

[405] Weinberg, S. 1997.
What is Quantum Field Theory, and What Did We Think It Is?
In *Conceptual Foundations of Quantum Field Theory*, 241–251,
Boston 1996 (hep-th/9702027).

[406] Weinberg, S. 2000.
The Quantum Theory of Fields, vol I.
(Cambridge Press, 2000).

[407] Migdal, A. B. 1958.
Interaction between Electrons and Lattice Vibrations in a Normal Metal.
Zhurnal Éksperimental'noǐ i Teoreticheskoǐ Fiziki **34** 1438
(*Soviet Physics JETP* **7** (1958) 996).

[408] Tinkham, M. 1975.
Introduction to Superconductivity.
(McGraw-Hill Press, 1975).

[409] Kittel, C. 2004.
Introduction to Solid State Physics, 8th Edition.
(Wiley, 2004).

[410] Willett, R., Eisenstein, J. P., Störmer, H. L., Tsui, D. C., Gossard, A. C. and English J. H. 1987.
Observation of an Even-Denominator Quantum Number in the Fractional Quantum Hall Effect.
Physical Review Letters **59** (1987) 1776.

[411] MacDonald, A.H. 1994.
Introduction to the Physics of the Quantum Hall Regime.
(cond-mat/9410047).

[412] Das Sarma, S. and Pinczuk, A. 1996.
Perspectives in Quantum Hall Effects: Novel Quantum Liquids in Low-Dimensional Semiconductor Structures.
(John Wiley & Sons, 2004).

[413] Girvin, S. M. 1999.
The Quantum Hall Effect: Novel Excitations and Broken Symmetries.
Lectures delivered at *École d'Éte Les Houches*, July 1998
(arXiv:cond-mat/9907002 (cond-mat.mes-hall)).

[414] MacDonald, A.H. 2010.
Anomalous Hall Effect.
Reviews of Modern Physics **82** (2010) 1539.

[415] v. Klitzing, K., Dorda, G. and Pepper, M. 1980.
New Method for High-Accuracy Determination of
the Fine-Structure Constant Based on Quantized Hall Resistance.
Physical Review Letters **45** (1980) 494.

[416] Tsui, D. C., Störmer, H. L. and Gossard, A. C. 1982.
Two-Dimensional Magnetotransport in the Extreme Quantum Limit.
Physical Review Letters **48** (1982) 1559.

[417] Ando, T., Matsumoto, Y. and Uemura, Y. 1975.
Theory of Hall Effect in a Two-Dimensional Electron System.
Journals of the Physical Society of Japan **39** (1975) 279.

[418] Tong, D. 2016.
Lectures on the Quantum Hall Effect (arXiv:1606.06687 (hep-th)).

[419] Fradkin, E. 2013.
Field Theories of Condensed Matter Physics.
(Cambridge Press, 2013).

[420] Burgess, C. P. and Dolan, B. P. 2001.
Particle Vortex Duality and the Modular Group:
Applications to the Quantum Hall Effect and other 2-D Systems.
Physical Review **B63** (2001) 155309 (hep-th/0010246).

[421] Laughlin, R. B. 1983.
Anomalous Quantum Hall Effect: An Incompressible
Quantum Fluid with Fractionally Charged Excitations.
Physical Review Letters **50** (1983) 1395.

[422] Laughlin, R. B. 1981.
Quantized Hall Conductivity in Two Dimensions.
Physical Review **B23** (1981) 5632.

[423] Thouless, D. J., Kohomoto, M., Nightingale, M. P. and den Nijs, M. 1982.
Quantized Hall Conductance in a Two-Dimensional Periodic Potential.
Physical Review Letters **49** (1982) 405.

[424] Floreanini, R. and Jackiw, R. 1987.
Self-Dual Fields As Charge-Sensity Solitons.
Physical Review Letters **59** (1987) 1873.

[425] Aharonov, Y. and Bohm, D. 1959.
Significance of Electromagnetic Potentials in Quantum Theory.
Physical Review **115** (1959) 485.

[426] Leinaas, J. M. and Myrheim, J. 1977.
On the Theory of Identical Particles.
Il Nuovo Cimento **B37** (19977) 1.
Wilczek, F. 1982.
Quantum Mechanics of Fractional-Spin Particles.
Physical Review Letters **49** (1982) 957.

[427] Zhang, S. C., Hansson, T. and Kivelson, S. 1989.
Effective-Field-Theory Model for the Fractional Quantum Hall Effect.
Physical Review Letters **62** (1989) 82.

[428] Jain, J. K. 1989.
Composite-Fermion Approach for the Fractional Quantum Hall Effect.
Physical Review Letters **63** (1989) 199.

[429] Lee, D. H. and Fisher, M. P. A. 1989.
Anyon Superconductivity and the Fractional Quantum Hall Effect.
Physical Review Letters **63** (1989) 903.

[430] Kivelson, S., Lee, D. H. and Zhang, S. C. 1992.
Global Phase Diagram in the Quantum Hall Effect.
Physical Review **B46** (1992) 2223.

[431] Lutken, C. A. and Ross, G. G. 1992.
Duality in the Quantum Hall System.
Physical Review **B45** (1992) 11837.

[432] Witten, E. 2003.
SL(2,Z) Sction on Three-Dimensional Conformal Field Theories with Abelian Symmetry.
In Shifman, M. (ed.) *et al. From Fields to Strings, vol. 2*, pp. 1173–1200 (hep-th/0307041).
Seiberg, N., Senthil, T., Wang, C. and Witten, E. 2016.
A Duality Web in 2+1 Dimensions and Condensed Matter Physics.
Annals of Physics **374** (2016) 395 (arXiv:1606.01989 (hep-th)).
Karch, A., Tong, D. and Turner, C. 2019.
A Web of 2d Dualities: $\mathbf{Z}_2$ Gauge Fields and Arf Invariants (arXiv:1902.05550 (hep-th)).

[433] Bayntun, A., Burgess, C. P., Dolan, B. P. and Lee, S. S.
AdS/QHE: Towards a Holographic Description of Quantum Hall Experiments.
New Journal of Physics **13** (2011) 035012 (arXiv:1008.1917 (hep-th)).

Lutken, C. A. and Ross, G. G.
Experimental probes of Emergent Symmetries in the Quantum Hall System.
Nuclear Physics **B850** (2011) 321 (arXiv:1008.5257 (cond-mat.str-el)).

[434] See for example, Kreuzer, H. R. 1984.
Non-Equilibrium Thermodynamics and Its Statistical Foundations,
(Monographs on the Physics and Chemistry of Materials, Oxford Press, 1984.)

[435] For classic textbook treatments see, for example:
Batchelor, G. K. 1967.
An Introduction to Fluid Dynamics.
(Cambridge Press, 1967).
Landau, L. D. and Lifshitz, E. M. 1987.
Fluid Mechanics.
A Course of Theoretical Physics (2nd revised ed.) Vol 6,
(Pergamon Press, 1987).

[436] Herglotz, G. 1911.
Über die Mechanik des deformierbaren Körpers vom Standpunkte der Relativitätstheorie
Annalen der Physik **341** (1911) 493.

[437] Taub, A. H. 1954.
General Relativistic Variational Principle for Perfect Fluids.
Physical Review **94** (1954) 1468.

[438] Salmon, R. 1988.
Hamilton's Principle and the Vorticity Laws for a Relativistic Perfect Fluid.
Geophysical and Astrophysical Fluid Dynamics **43** (1988) 167.

[439] Jackiw, R., Nair, V. P., Pi, S. Y. and Polychronakos, A. P. 2004.
Perfect Fluid Theory and Its Extensions.
Journal of Physics **A 37** (2004) R327 (arXiv:hep-ph/0407101).

[440] Andersson, N. and Comer, G. 2007.
Relativistic Fluid Dynamics: Physics for Many Different Scales.
Living Reviews of Relativity **10** (2007) 1 (arXiv:gr-qc/0605010).

[441] Nicolis, A. 2011.
Low-Energy Effective Field Theory for Finite-Temperature Relativistic Superfluids (arXiv:1108.2513 (hep-th)).

[442] For a classic discussions of dissipation in relativistic fluids see:
Tolman, R. C. 1934.
Relativity, Thermodynamics, and Cosmology.
(Oxford: Clarendon Press, 1934. Reissued Dover, New York, 1984).
For a more recent view see, for example:
Arnold, P., Romatschke, P. and van der Schee, W. 2014.
Absence of a Local Rest Frame in Far From Equilibrium Quantum Matter.
Journal of High Energy Physics **1410** (2014) 110 (arXiv:1408.2518 (hep-th)).

[443] For textbook discussions see:
Carmichael, H. 1991.
An Open Systems Approach to Quantum Optics.
(Springer Verlag, 1991).
Breuer, H.-P. and Petruccione, F. 2002.
The Theory of Open Quantum Systems.

(Oxford Press, 2002).
Alicki, R. and Lendi, K. 2007.
Quantum Dynamical Semigroups and Applications.
(Springer, 2007).

[444] See, for example, Burgess, C. P. and Michaud, D. 1996.
Neutrino Propagation in a Fluctuating Sun.
Annals of Physics **256** (1997) 1 (hep-ph/9606295).

[445] Liouville, J. 1838.
Observations Sur un Mémoire de M. Libri, relatif à la Théorie de la Chaleur
Journal de Mathématiques **3** (1838) 349.

[446] This is textbook material:
Bahcall, J. N. 1989.
Neutrino Astrophysics.
(Cambridge Press, 1989).
Kayser, B., Gibrat-Debu, F. and Perrier, F. 1989.
Physics of Massive Neutrinos. World Sciientific Lecture Notes in Physics **25** (1989) 1.
Raffelt, G. G. 1996).
Stars as Laboratories for Fundamental Physics:
The Astrophysics of Neutrinos, Axions, and Other Weakly Interacting Particles.
(Chicago Press, 1996).
Fukugita, M. and Yanagida, T. 2003.
Physics of Neutrinos and Applications to Astrophysics.
(Springer, 2003).
Mohapatra, R. N. and Pal, P. B. 2004.
Massive Neutrinos in Physics and Astrophysics. Second edition.
World Scientific Lecture Notes in Physics **72** (2004) 1.
Giunti, C. and Kim, C. W. 2007.
Fundamentals of Neutrino Physics and Astrophysics.
(Oxford Press, 2007).
Zuber, K. 2012.
Neutrino Physics.
(CRC Press, 2012).

[447] For a recent discussion see: Janka, H. T. 2017.
Neutrino Emission from Supernovae (arXiv:1702.08713 (astro-ph.HE)).

[448] Wolfenstein, L. 1978.
Neutrino Oscillations in Matter.
Physical Review **D17** (1978) 2369.
Mikheyev, S. P. and Smirnov, A. Yu. 1985.
Resonance Enhancement of Oscillations in Matter and Solar Neutrino Spectroscopy.
Soviet Journal of Nuclear Physics **42** (1985) 913.

[449] Burgess, C. P., Maltoni, M., Rashba, T. I.,
Semikoz, V. B., Tortola, M. A. and Valle, J. W. F. 2003.
Cornering Solar Radiative Zone Fluctuations with KamLAND and SNO Salt.

Journal of Cosmology and Astroparticle Physics **0401** (2004) 007 (hep-ph/0310366).
Semikoz, V. B., Burgess, C. P., Dzhalilov, N. Z., Rashba, T. I. and Valle, J. W. F. 2004.
MHD Origin of Density Fluctuations Deep within the Sun and Their Influence on Neutrino Oscillation Parameters in LMA MSW Scenario.
Yadernaya Fizika **67** (2004) 1172;
(*Physics of Atoms and Nuclei* **67** (2004) 1147).

[450] Ibn Sahl. 984.
On Burning Mirrors and Lenses (Baghdad).
For a textbook treatment see:
Born, M. and Wolf, E. 1959.
Principles of Optics.
(Cambridge 1959, latest reprint 2002).

[451] Landau, L. D. and Lifshitz, E. M. 1960.
Electrodynamics of Continuous Media.
in *A Course of Theoretical Physics* Vol 8, (Pergamon Press, 1960).

[452] Bragg, W. H. and Bragg, W. L. 1913.
The Reflexion of X-rays by Crystals.
Proceedings of the Royal Society of London **A88** (1913) 428.

[453] Misra, B. and Sudarshan, E. C. G. 1977.
The Zeno's paradox in quantum theory.
Journal of Mathematical Physics **18** 756.

[454] Nielsen, M. A. and Chuang, I. L. 2001.
Quantum Computation and Quantum Information.
(Cambridge Press, 2001).

[455] Nakajima, S. 1958.
On Quantum Theory of Transport Phenomena.
Progress in Theoretical Physics **20** (1958) 948.
Zwanzig, R. 1960.
Ensemble Method in the Theory of Irreversibility.
Journal of Chemical Physics **33** (1960) 1338.

[456] Davies, E. 1974.
Markovian Master Equations.
Communications in Mathematical Physics **39** (1974) 91.
Davies, E. 1976.
Markovian Master Equations II.
Mathematische Annalen **219** (1976) 147.
Dumcke, R. and Spohn, H. 1979.
Proper Form of the Generator in the Weak Coupling Limit.
Zeitschrift für Physik **B34** (1979) 419.

[457] Kaplanek, G. and Burgess, C. P. 2019.
Hot Accelerated Qubits: Decoherence, Thermalization, Secular Growth and Reliable Late-time Predictions.
Journal of High Energy Physics **03** (2020) 008 (arXiv:1912.12951 (hep-th)).
Kaplanek, G. and Burgess, C. P. 2019.
Hot Cosmic Qubits: Late-Time de Sitter Evolution and Critical Slowing

Down.
Journal of High Energy Physics **02** (2020) 053 (arXiv:1912.12955 (hep-th)).

[458] Kossakowski, A. 1972.
On Quantum Statistical Mechanics of Non-Hamiltonian Systems.
Reports on Mathematical Physics **3** (1972) 247.
Lindblad, G. 1976.
On the Generators of Quantum Dynamical Semigroups.
Communications in Mathematical Physics **48** (1976) 119.
Gorini, V. , Kossakowski, A. and Sudarshan, E. C. G. 1976.
Completely Positive Semigroups of N-Level Systems.
Journal of Mathematical Physics **17** (1976) 821.

[459] Belavin, A. A., Zel'dovich, B. Ya., Perelomov, A. M. and Popov, V. S. 1969.
Relaxation of Quantum Systems with Equidistant Spectra.
Zhurnal Éksperimental'noĭ i Teoreticheskoĭ Fiziki **56** 264
(*Soviet Physics JETP* **29** (1969)145).
Boyanovsky, D. and de Vega, H. J. 2003.
Dynamical Renormalization Group Approach to Relaxation in Quantum Field Theory.
Annals of Physics **307** (2003) 335 (hep-ph/0302055).

[460] Kubo, R. 1957.
Statistical-Mechanical Theory of Irreversible Processes I.
General Theory and Simple Applications to Magnetic and Conduction Problems.
Journal of the Physical Society of Japan **12** (1957) 570.
Martin, P. C., Schwinger, J. 1959.
Theory of Many-Particle Systems I.
Physical Review **115** (1959) 1342.
Haag, R., Winnink, M. and Hugenholtz, N. M. 1967.
On the Equilibrium States in Quantum Statistical Mechanics.
Communications in Mathematical Physics **5** (1967) 215.

[461] Einstein, A. 1910.
Theorie der Opaleszenz von homogenen Flüssigkeiten und
Flüssigkeitsgemischen in der Nähe des kritischen Zustandes.
(The Theory of the Opalescence of Homogeneous Fluids
and Liquid Mixtures near the Critical State.)
Annalen der Physik **33** (1910) 1275.

[462] Rayleigh, Lord. 1881.
On the Electromagnetic Theory of Light.
The London, Edinburgh, and Dublin Philosophical Magazine and Journal of Science **12** (1881) 81.
Rayleigh, Lord. 1899.
On the Transmission of Light through an Atmosphere Containing
Small Particles in Suspension, and on the Origin of the Blue of the Sky.
The London, Edinburgh, and Dublin Philosophical Magazine and Journal of Science **47** (1899) 375.

[463] Fulling, S. A. 1973.
Nonuniqueness of Canonical Field Quantization in Riemannian Space-Time.

Physical Review **D7** (1973) 2850.
Unruh, W. G. 1976.
Notes on Black-Hole Evaporation.
Physical Review **D14** (1976) 870.
DeWitt, B. S. 1979.
Quantum Gravity: The New Synthesis.
in *General Relativity, An Einstein Centenary Survey*, edited by S. W. Hawking and W. Israel (Cambrdige Press, 1979).

[464] Sciama, D. W., Candelas, P. and Deutsch, D. 1981.
Quantum Field Theory, Horizons and Thermodynamics.
Advances in Physics **30** (1981) 327.

[465] Weinberg, S. 1964.
Feynman Rules for Any Spin.
Physical Review **133** (1964) B1318.

[466] Scadron, M. 1968.
Covariant Propagators and Vertex Functions for Any Spin.
Physical Review **165** (1968) 1640.

[467] Haag, R., Sohnius, M. and Łopuszański, J. T. 1975.
All Possible Generators of Supersymmetries of the S-Matrix.
Nuclear Physics **B88** 257.

[468] Yang, C. N. and Mills, R. 1954.
Conservation of Isotopic Spin and Isotopic Gauge Invariance.
Physical Review **96** (1954) 1.

[469] Arkani-Hamed, N. Cachazo, F. and Kaplan, J. JHEP **1009** (2010) 016 doi:10.1007/JHEP09(2010)016 (arXiv:0808.1446 (hep-th)).

[470] Stueckelberg, E. C. G. and Petermann, A. 1953.
La renormalisation des constants dans la théorie de quanta.
Helvetica Physica Acta **26** (1953) 499.

[471] Gell-Mann, M. and Low, F. E. 1954.
Quantum Electrodynamics at Small Distances.
Physical Review **95** (1954) 1300.

[472] Kadanoff, L. P. 1966.
Scaling Laws for Ising Models near T_c.
Physics **2** (1966) 263.

[473] Wilson, K. G. 1971.
Renormalization Group and Critical Phenomena 1.
Renormalization Group and the Kadanoff Scaling Picture.
Physical Review **B4** (1971) 3174.
Renormalization Group and Critical Phenomena 2.
Phase Space Cell Analysis of Critical Behavior.
Physical Review **B4** (1971) 3184.

[474] Wilson, K. G. and Fisher, M. E. 1972.
Critical Exponents in 3.99 Dimensions.
Physical Review Letters **28** (1972) 240.

[475] Reuter, M. and Wetterich, C. 1994.
Effective Average Action for Gauge Theories and Exact Evolution Equations.
Nuclear Physics **B417** (1994) 181.

[476] Salam, A. 1951.
Overlapping Divergences and the S-Matrix.
Physical Review **82** (1951) 217.

[477] Weinberg, S. 1959.
High-Energy Behavior in Quantum Field Theory.
Physical Review **118** (1959) 838.

[478] Wess, J. and Zumino, B. 1971.
Consequences of Anomalous Ward Identities.
Physics Letters **37B** (1971) 95.

[479] Bardeen, W. A. and Zumino, B. 1984.
Consistent and Covariant Anomalies in Gauge and Gravitational Theories.
Nuclear Physics **B244** (1984) 421.

[480] Manes, J., Stora, R. and Zumino, B. 1985.
Algebraic Study of Chiral Anomalies.
Communications in Mathematical Physics **102** (1985) 157.

[481] Smoot, G. F. *et al.* (COBE Collaboration). 1992.
Structure in the COBE Differential Microwave Radiometer First Year Maps.
Astrophysical Journal **396** (1992) L1.

[482] Burgess, C. P. 2017.
Intro to Effective Field Theories and Inflation.
In the proceedings of the Les Houches Summer School, 'Effective Field Theory in Particle Physics and Cosmology' (arXiv:1711.10592 (hep-th)).

[483] Achucarro, A. Gong, J. O., Hardeman, S., Palma, G. A. and Patil, S. P. 2010.
Mass Hierarchies and Non-Decoupling in Multi-Scalar Field Dynamics.
Physical Review **D84** (2011) 043502 (arXiv:1005.3848 (hep-th)).

[484] 't Hooft, G. 1986.
How Instantons Solve the U(1) Problem.
Physics Reports **142** (1986) 357.

[485] Gasser, J. and Leutwyler, H. 1982.
Quark Masses.
Physics Reports **87** (1982) 77.
Meissner, U. G. 1993.
Recent Developments in Chiral Perturbation Theory.
Reports on Progress in Physics **56** (1993) 903 (hep-ph/9302247).
Leutwyler, H. 1994.
In the proceedings of Hadron Physics 94 (hep-ph/9406283)
Kaplan, D. 1995.
In the proceedings of the 7th Summer School in Nuclear Physics Symmetries, Seattle, WA, 1995 (nucl-th/9506035)
Georgi, H. 1995.
Effective Field Theory.
Annual Review of Nuclear and Particle Science **43** (1995) 205
Pich, A. 1998.
Effective Field Theories.

In the proceedings of the Les Houches Summer School in Theoretical Physics: Probing the Standard Model of Particle Interactions (hep-ph/9806303);
Rothstein, I. Z. 2003.
TASI Lectures on Effective Field Theories (hep-ph/0308266);
Manohar, A. 2017.
Introduction to Effective Field Theories.
In the proceedings of the 2017 Les Houches Summer School on Effective Field Theories (arXiv:1804.05863).

[486] C. Burgess, 2013. The Cosmological Constant Problem: Why Its Hard to Get Dark Energy from Micro-physics. In the proceedings of the Les Houches School on Post-Planck Cosmology 2013 (arXiv:1309.4133 (hep-th)).

[487] Luther, A. and Peschel, I. 1974.
Single-Particle States, Kohn Anomaly, and Pairing Fluctuations in One Dimension.
Physical Review **B9** 2911.

[488] Coleman, S. 1975.
Quantum sine-Gordon Equation as the Massive Thirring model.
Physical Review **D11** 2088.

[489] Mandelstam, S. 1975.
Soliton Operators for the Quantized sine-Gordon Equation.
Physical Review **D11** 3026.

[490] Burgess, C. P. and Quevedo, F. 1993.
Bosonization as duality.
Nuclear Physics **B421** 373 (arXiv:hep-th/9401105 (hep-th)).

Index